PRINCIPIA MATHEMATICA

PRINCIPIA MATHEMATICA

TO ∗56

BY

ALFRED NORTH WHITEHEAD

AND

BERTRAND RUSSELL, F.R.S.

PUBLISHED BY THE PRESS SYNDICATE OF THE UNIVERSITY OF CAMBRIDGE
The Pitt Building, Trumpington Street, Cambridge, United Kingdom

CAMBRIDGE UNIVERSITY PRESS
The Edinburgh Building, Cambridge CB2 2RU, UK www.cup.cam.ac.uk
40 West 20th Street, New York, NY 10011–4211, USA www.cup.org
10 Stamford Road, Oakleigh, Melbourne 3166, Australia
Ruiz de Alarcón 13, 28014 Madrid, Spain

First published 1910
Second edition 1927
Reprinted 1950, 1957, 1960
Paperback edition to *56 1962
Reprinted 1964, 1967, 1970, 1973, 1976, 1978,
1980, 1987, 1990, 1993, 1995
Reprinted in the Cambridge Mathematical Library 1997, 1999

A catalogue record for this book is available from the British Library

ISBN 0 521 62606 4 paperback

Transferred to digital printing 2002

PREFACE

THE mathematical treatment of the principles of mathematics, which is the subject of the present work, has arisen from the conjunction of two different studies, both in the main very modern. On the one hand we have the work of analysts and geometers, in the way of formulating and systematising their axioms, and the work of Cantor and others on such matters as the theory of aggregates. On the other hand we have symbolic logic, which, after a necessary period of growth, has now, thanks to Peano and his followers, acquired the technical adaptability and the logical comprehensiveness that are essential to a mathematical instrument for dealing with what have hitherto been the beginnings of mathematics. From the combination of these two studies two results emerge, namely (1) that what were formerly taken, tacitly or explicitly, as axioms, are either unnecessary or demonstrable; (2) that the same methods by which supposed axioms are demonstrated will give valuable results in regions, such as infinite number, which had formerly been regarded as inaccessible to human knowledge. Hence the scope of mathematics is enlarged both by the addition of new subjects and by a backward extension into provinces hitherto abandoned to philosophy.

The present work was originally intended by us to be comprised in a second volume of *The Principles of Mathematics.* With that object in view, the writing of it was begun in 1900. But as we advanced, it became increasingly evident that the subject is a very much larger one than we had supposed; moreover on many fundamental questions which had been left obscure and doubtful in the former work, we have now arrived at what we believe to be satisfactory solutions. It therefore became necessary to make our book independent of *The Principles of Mathematics.* We have, however, avoided both controversy and general philosophy, and made our statements dogmatic in form. The justification for this is that the chief reason in favour of any theory on the principles of mathematics must always be inductive, *i.e.* it must lie in the fact that the theory in question enables us to deduce ordinary mathematics. In mathematics, the greatest degree of self-evidence is usually not to be found quite at the beginning, but at some later point; hence the early deductions, until they reach this point, give reasons rather for believing the premisses because true consequences follow from them, than for believing the consequences because they follow from the premisses.

In constructing a deductive system such as that contained in the present work, there are two opposite tasks which have to be concurrently performed. On the one hand, we have to analyse existing mathematics, with a view to discovering what premisses are employed, whether these premisses are mutually consistent, and whether they are capable of reduction to more fundamental premisses. On the other hand, when we have decided upon our premisses, we have to build up again as much as may seem necessary of the data previously analysed, and as many other consequences of our premisses as are of sufficient general interest to deserve statement. The preliminary labour of analysis does not appear in the final presentation, which merely sets forth the outcome of the analysis in certain undefined ideas and undemonstrated propositions. It is not claimed that the analysis could not have been carried farther: we have no reason to suppose that it is impossible to find simpler ideas and axioms by means of which those with which we start could be defined and demonstrated. All that is affirmed is that the ideas and axioms with which we start are sufficient, not that they are necessary.

In making deductions from our premisses, we have considered it essential to carry them up to the point where we have proved as much as is true in whatever would ordinarily be taken for granted. But we have not thought it desirable to limit ourselves too strictly to this task. It is customary to consider only particular cases, even when, with our apparatus, it is just as easy to deal with the general case. For example, cardinal arithmetic is usually conceived in connection with *finite* numbers, but its general laws hold equally for infinite numbers, and are most easily proved without any mention of the distinction between finite and infinite. Again, many of the properties commonly associated with series hold of arrangements which are not strictly serial, but have only some of the distinguishing properties of serial arrangements. In such cases, it is a defect in logical style to prove for a particular class of arrangements what might just as well have been proved more generally. An analogous process of generalization is involved, to a greater or less degree, in all our work. We have sought always the most general reasonably simple hypothesis from which any given conclusion could be reached. For this reason, especially in the later parts of the book, the importance of a proposition usually lies in its hypothesis. The conclusion will often be something which, in a certain class of cases, is familiar, but the hypothesis will, whenever possible, be wide enough to admit many cases besides those in which the conclusion is familiar.

We have found it necessary to give very full proofs, because otherwise it is scarcely possible to see what hypotheses are really required, or whether our results follow from our explicit premisses. (It must be remembered that we are not affirming merely that such and such propositions are true, but also that the axioms stated by us are sufficient to prove them.) At the same time, though full proofs are necessary for the avoidance of errors, and for convincing those who may feel doubtful as to our correctness, yet the proofs of propositions may usually be omitted by a reader who is not specially interested in that part of the subject concerned, and who feels no doubt of our substantial accuracy on the matter in hand. The reader who is specially interested in some particular portion of the book will probably find it sufficient, as regards earlier portions, to read the summaries of previous parts, sections, and numbers, since these give explanations of the ideas involved and statements of the principal propositions proved. The proofs in Part I, Section A, however, are necessary, since in the course of them the manner of stating proofs is explained. The proofs of the earliest propositions are given without the omission of any step, but as the work proceeds the proofs are gradually compressed, retaining however sufficient detail to enable the reader by the help of the references to reconstruct proofs in which no step is omitted.

The order adopted is to some extent optional. For example, we have treated cardinal arithmetic and relation-arithmetic before series, but we might have treated series first. To a great extent, however, the order is determined by logical necessities.

A very large part of the labour involved in writing the present work has been expended on the contradictions and paradoxes which have infected logic and the theory of aggregates. We have examined a great number of hypotheses for dealing with these contradictions; many such hypotheses have been advanced by others, and about as many have been invented by ourselves. Sometimes it has cost us several months' work to convince ourselves that a hypothesis was untenable. In the course of such a prolonged study, we have been led, as was to be expected, to modify our views from time to time; but it gradually became evident to us that some form of the doctrine of types must be adopted if the contradictions were to be avoided. The particular form of the doctrine of types advocated in the present work is not logically indispensable, and there are various other forms equally compatible with the truth of our deductions. We have particularized, both because the form of the doctrine which we advocate appears to us the most probable, and because it was necessary to give at least one perfectly definite theory which avoids the contradictions. But hardly anything in our book would be changed by the adoption of a different form of the doctrine of types. In fact, we may go farther, and say that, supposing some other way of avoiding the contradictions to exist, not very much of our book, except what explicitly deals with types, is dependent upon the adoption of the doctrine of types in any form, so soon as it has been shown (as we claim that we have shown) that it is *possible* to construct a mathematical logic which does not lead to contradictions. It should be observed that the whole effect of the doctrine of types is negative: it forbids certain inferences which would otherwise be valid, but does not permit any which would otherwise be invalid. Hence we may reasonably expect that the inferences which the doctrine of types permits would remain valid even if the doctrine should be found to be invalid.

Our logical system is wholly contained in the numbered propositions, which are independent of the Introduction and the Summaries. The Introduction and the Summaries are wholly explanatory, and form no part of the chain of deductions. The explanation of the hierarchy of types in the Introduction differs slightly from that given in ∗12 of the body of the work. The latter explanation is stricter and is that which is assumed throughout the rest of the book.

The symbolic form of the work has been forced upon us by necessity: without its help we should have been unable to perform the requisite reasoning. It has been developed as the result of actual practice, and is not an excrescence introduced for the mere purpose of exposition. The general method which guides our handling of logical symbols is due to Peano. His great merit consists not so much in his definite logical discoveries nor in the details of his notations (excellent as both are), as in the fact that he first showed how symbolic logic was to be freed from its undue obsession with the forms of ordinary algebra, and thereby made it a suitable instrument for research. Guided by our study of his methods, we have used great freedom in constructing, or reconstructing, a symbolism which shall be adequate to deal with all parts of the subject. No symbol has been introduced except on the ground of its practical utility for the immediate purposes of our reasoning.

A certain number of forward references will be found in the notes and explanations. Although we have taken every reasonable precaution to secure the accuracy of these forward references, we cannot of course guarantee their accuracy with the same confidence as is possible in the case of backward references.

Detailed acknowledgments of obligations to previous writers have not very often been possible, as we have had to transform whatever we have borrowed, in order to adapt it to our system and our notation. Our chief obligations will be obvious to every reader who is familiar with the literature of the subject. In the matter of notation, we have as far as possible followed Peano, supplementing his notation, when necessary, by that of Frege or by that of Schröder. A great deal of the symbolism, however, has had to be new, not so much through dissatisfaction with the symbolism of others, as through the fact that we deal with ideas not previously symbolised. In all questions of logical analysis, our chief debt is to Frege. Where we differ from him, it is largely because the contradictions showed that he, in common with all other logicians ancient and modern, had allowed some error to creep into his premisses; but apart from the contradictions, it would have been almost impossible to detect this error. In Arithmetic and the theory of series, our whole work is based on that of Georg Cantor. In Geometry we have had continually before us the writings of v. Staudt, Pasch, Peano, Pieri, and Veblen.

A. N. W.

B. R.

CAMBRIDGE,
November, 1910.

CONTENTS

PAGE

PREFACE v

ALPHABETICAL LIST OF PROPOSITIONS REFERRED TO BY NAMES xii

INTRODUCTION TO THE SECOND EDITION xiii

INTRODUCTION 1

CHAPTER I. PRELIMINARY EXPLANATIONS OF IDEAS AND NOTATIONS . 4

CHAPTER II. THE THEORY OF LOGICAL TYPES 37

CHAPTER III. INCOMPLETE SYMBOLS 66

PART I. MATHEMATICAL LOGIC

Summary of Part I 87

SECTION A. THE THEORY OF DEDUCTION 90

*1. Primitive Ideas and Propositions 91
*2. Immediate Consequences of the Primitive Propositions . 98
*3. The Logical Product of two Propositions 109
*4. Equivalence and Formal Rules 115
*5. Miscellaneous Propositions 123

SECTION B. THEORY OF APPARENT VARIABLES 127

*9. Extension of the Theory of Deduction from Lower to Higher Types of Propositions 127
*10. Theory of Propositions containing one Apparent Variable . 138
*11. Theory of two Apparent Variables 151
*12. The Hierarchy of Types and the Axiom of Reducibility . 161
*13. Identity 168
*14. Descriptions 173

SECTION C. CLASSES AND RELATIONS 187

*20. General Theory of Classes 187
*21. General Theory of Relations 200
*22. Calculus of Classes 205
*23. Calculus of Relations 213
*24. The Universal Class, the Null Class, and the Existence of Classes 216
*25. The Universal Relation, the Null Relation, and the Existence of Relations 228

PAGE

SECTION D. LOGIC OF RELATIONS 231
*30. Descriptive Functions 232
*31. Converses of Relations 238
*32. Referents and Relata of a given Term with respect to a given Relation 242
*33. Domains, Converse Domains, and Fields of Relations . . 247
*34. The Relative Product of two Relations 256
*35. Relations with Limited Domains and Converse Domains . 265
*36. Relations with Limited Fields 277
*37. Plural Descriptive Functions 279
*38. Relations and Classes derived from a Double Descriptive Function 296
Note to Section D 299

SECTION E. PRODUCTS AND SUMS OF CLASSES 302
*40. Products and Sums of Classes of Classes 304
*41. The Product and Sum of a Class of Relations 315
*42. Miscellaneous Propositions 320
*43. The Relations of a Relative Product to its Factors . . 324

PART II. PROLEGOMENA TO CARDINAL ARITHMETIC
Summary of Part II, Section A 328

SECTION A UNIT CLASSES AND COUPLES 329
*50. Identity and Diversity as Relations 331
*51. Unit Classes 338
*52. The Cardinal Number 1 345
*53. Miscellaneous Propositions involving Unit Classes . . . 350
*54. Cardinal Couples 357
*55. Ordinal Couples 364
*56. The Ordinal Number 2_r 375

APPENDIX A
*8. The Theory of Deduction for Propositions containing Apparent Variables 385

APPENDIX C
Truth-Functions and others 401

LIST OF DEFINITIONS 409

NOTE

All cross-references in the text, including references to definitions and propositions, relate to the Second Edition (1927) *and may not necessarily be found in this abridged edition.*

ALPHABETICAL LIST OF PROPOSITIONS REFERRED TO BY NAMES

Name	Number	
Abs	***2·01.**	$\vdash : p \supset \sim p . \supset . \sim p$
Add	***1·3.**	$\vdash : q . \supset . p \vee q$
Ass	***3·35.**	$\vdash : p . p \supset q . \supset . q$
Assoc	***1·5.**	$\vdash : p \vee (q \vee r) . \supset . q \vee (p \vee r)$
Comm	***2·04.**	$\vdash :. p . \supset . q \supset r : \supset : q . \supset . p \supset r$
Comp	***3·43.**	$\vdash :. p \supset q . p \supset r . \supset : p . \supset . q . r$
Exp	***3·3.**	$\vdash :. p . q . \supset . r : \supset : p . \supset . q \supset r$
Fact	***3·45.**	$\vdash :. p \supset q . \supset : p . r . \supset . q . r$
Id	***2·08.**	$\vdash . p \supset p$
Imp	***3·31.**	$\vdash :. p . \supset . q \supset r : \supset : p . q . \supset . r$
Perm	***1·4.**	$\vdash : p \vee q . \supset . q \vee p$
Simp	***2·02.**	$\vdash : q . \supset . p \supset q$
"	***3·26.**	$\vdash : p . q . \supset . p$
"	***3·27.**	$\vdash : p . q . \supset . q$
Sum	***1·6.**	$\vdash :. q \supset r . \supset : p \vee q . \supset . p \vee r$
Syll	***2·05.**	$\vdash :. q \supset r . \supset : p \supset q . \supset . p \supset r$
"	***2·06.**	$\vdash :. p \supset q . \supset : q \supset r . \supset . p \supset r$
"	***3·33.**	$\vdash : p \supset q . q \supset r . \supset . p \supset r$
"	***3·34.**	$\vdash : q \supset r . p \supset q . \supset . p \supset r$
Taut	***1·2.**	$\vdash : p \vee p . \supset . p$
Transp	***2·03.**	$\vdash : p \supset \sim q . \supset . q \supset \sim p$
"	***2·15.**	$\vdash : \sim p \supset q . \supset . \sim q \supset p$
"	***2·16.**	$\vdash : p \supset q . \supset . \sim q \supset \sim p$
"	***2·17.**	$\vdash : \sim q \supset \sim p . \supset . p \supset q$
"	***3·37.**	$\vdash :. p . q . \supset . r : \supset : p . \sim r . \supset . \sim q$
"	***4·1.**	$\vdash : p \supset q . \equiv . \sim q \supset \sim p$
"	***4·11.**	$\vdash : p \equiv q . \equiv . \sim p \equiv \sim q$

INTRODUCTION TO THE SECOND EDITION*

In preparing this new edition of *Principia Mathematica*, the authors have thought it best to leave the text unchanged, except as regards misprints and minor errors†, even where they were aware of possible improvements. The chief reason for this decision is that any alteration of the propositions would have entailed alteration of the references, which would have meant a very great labour. It seemed preferable, therefore, to state in an introduction the main improvements which appear desirable. Some of these are scarcely open to question; others are, as yet, a matter of opinion.

The most definite improvement resulting from work in mathematical logic during the past fourteen years is the substitution, in Part I, Section A, of the one indefinable "p and q are incompatible" (or, alternatively, "p and q are both false") for the two indefinables "not-p" and "p or q." This is due to Dr H. M. Sheffer‡. Consequentially, M. Jean Nicod§ showed that one primitive proposition could replace the five primitive propositions ∗1·2·3·4·5·6.

From this there follows a great simplification in the building up of molecular propositions and matrices; ∗9 is replaced by a new chapter, ∗8, given in Appendix A to this Volume.

Another point about which there can be no doubt is that there is no need of the distinction between real and apparent variables, nor of the primitive idea "assertion of a propositional function." On all occasions where, in *Principia Mathematica*, we have an asserted proposition of the form "$\vdash . fx$" or "$\vdash . fp$," this is to be taken as meaning "$\vdash . (x) . fx$" or "$\vdash . (p) . fp$." Consequently the primitive proposition ∗1·11 is no longer required. All that is necessary, in order to adapt the propositions as printed to this change, is the convention that, when the scope of an apparent variable is the whole of the asserted proposition in which it occurs, this fact will not be explicitly indicated unless "some" is involved instead of "all." That is to say, "$\vdash . \phi x$" is to mean "$\vdash . (x) . \phi x$"; but in "$\vdash . (\exists x) . \phi x$" it is still necessary to indicate explicitly the fact that "some" x (not "all" x's) is involved.

It is possible to indicate more clearly than was done formerly what are the novelties introduced in Part I, Section B as compared with Section A.

* In this introduction, as well as in the Appendices, the authors are under great obligations to Mr F. P. Ramsey of King's College, Cambridge, who has read the whole in MS. and contributed valuable criticisms and suggestions.

† In regard to these we are indebted to many readers, but especially to Drs Behmann and Boscovitch, of Göttingen.

‡ *Trans. Amer. Math. Soc.* Vol. XIV. pp. 481—488.

§ "A reduction in the number of the primitive propositions of logic," *Proc. Camb. Phil. Soc.* Vol. XIX.

They are three in number, two being essential logical novelties, and the third merely notational.

(1) For the "p" of Section A, we substitute "ϕx," so that in place of "$\vdash . (p) . fp$" we have "$\vdash . (\phi, x) . f(\phi x)$." Also, if we have "$\vdash . f(p, q, r, \ldots)$," we may substitute ϕx, ϕy, ϕz, ... for p, q, r, ... or ϕx, ϕy for p, q, and ψz, ... for r, ..., and so on. We thus obtain a number of new general propositions different from those of Section A.

(2) We introduce in Section B the new primitive idea "$(\exists x) . \phi x$," *i.e.* existence-propositions, which do not occur in Section A. In virtue of the abolition of the real variable, general propositions of the form "$(p) . fp$" do occur in Section A, but "$(\exists p) . fp$" does not occur.

(3) By means of definitions, we introduce in Section B general propositions which are molecular constituents of other propositions; thus "$(x) . \phi x . \vee . p$" is to mean "$(x) . \phi x \vee p$."

It is these three novelties which distinguish Section B from Section A.

One point in regard to which improvement is obviously desirable is the axiom of reducibility (∗12·1·11). This axiom has a purely pragmatic justification: it leads to the desired results, and to no others. But clearly it is not the sort of axiom with which we can rest content. On this subject, however, it cannot be said that a satisfactory solution is as yet obtainable. Dr Leon Chwistek* took the heroic course of dispensing with the axiom without adopting any substitute; from his work, it is clear that this course compels us to sacrifice a great deal of ordinary mathematics. There is another course, recommended by Wittgenstein† for philosophical reasons. This is to assume that functions of propositions are always truth-functions, and that a function can only occur in a proposition through its values. There are difficulties in the way of this view, but perhaps they are not insurmountable‡. It involves the consequence that all functions of functions are extensional. It requires us to maintain that "A believes p" is not a function of p. How this is possible is shown in *Tractatus Logico-Philosophicus* (*loc. cit.* and pp. 19—21). We are not prepared to assert that this theory is certainly right, but it has seemed worth while to work out its consequences in the following pages. It appears that everything in Vol. I remains true (though often new proofs are required); the theory of inductive cardinals and ordinals survives; but it seems that the theory of infinite Dedekindian and well-ordered series largely collapses, so that irrationals, and real numbers generally, can no longer be adequately dealt with. Also Cantor's proof that $2^n > n$ breaks down unless n is finite. Perhaps some further axiom, less objectionable than the axiom of reducibility, might give these results, but we have not succeeded in finding such an axiom.

* In his "Theory of Constructive Types." See references at the end of this Introduction.

† *Tractatus Logico-Philosophicus*, ∗5·54 ff.

‡ See Appendix C.

It should be stated that a new and very powerful method in mathematical logic has been invented by Dr H. M. Sheffer. This method, however, would demand a complete re-writing of *Principia Mathematica*. We recommend this task to Dr Sheffer, since what has so far been published by him is scarcely sufficient to enable others to undertake the necessary reconstruction.

We now proceed to the detailed development of the above general sketch.

I. ATOMIC AND MOLECULAR PROPOSITIONS

Our system begins with "atomic propositions." We accept these as a datum, because the problems which arise concerning them belong to the philosophical part of logic, and are not amenable (at any rate at present) to mathematical treatment.

Atomic propositions may be defined negatively as propositions containing no parts that are propositions, and not containing the notions "all" or "some." Thus "this is red," "this is earlier than that," are atomic propositions.

Atomic propositions may also be defined positively—and this is the better course—as propositions of the following sorts:

$R_1(x)$, meaning "x has the predicate R_1";

$R_2(x, y)$ [or xR_2y], meaning "x has the relation R_2 (in intension) to y";

$R_3(x, y, z)$, meaning "x, y, z have the triadic relation R_3 (in intension)";

$R_4(x, y, z, w)$, meaning "x, y, z, w have the tetradic relation R_4 (in intension)";

and so on *ad infinitum*, or at any rate as long as possible. Logic does not know whether there are in fact n-adic relations (in intension); this is an empirical question. We know as an empirical fact that there are at least dyadic relations (in intension), because without them series would be impossible. But logic is not interested in this fact; it is concerned solely with the *hypothesis* of there being propositions of such-and-such a form. In certain cases, this hypothesis is itself of the form in question, or contains a part which is of the form in question; in these cases, the fact that the hypothesis can be framed proves that it is true. But even when a hypothesis occurs in logic, the fact that it can be framed does not itself belong to logic.

Given all true atomic propositions, together with the fact that they are all, every other true proposition can theoretically be deduced by logical methods. That is to say, the apparatus of crude fact required in proofs can all be condensed into the true atomic propositions together with the fact that every true atomic proposition is one of the following: (here the list should follow). If used, this method would presumably involve an infinite enumeration, since it seems natural to suppose that the number of true atomic propositions is infinite, though this should not be regarded as certain. In practice, generality is not obtained by the method of complete enumeration, because this method requires more knowledge than we possess.

We must now advance to molecular propositions. Let p, q, r, s, t denote, to begin with, atomic propositions. We introduce the primitive idea

$$p \mid q,$$

which may be read "p is incompatible with q,"* and is to be true whenever either or both are false. Thus it may also be read "p is false or q is false"; or again, "p implies not-q." But as we are going to define disjunction, implication, and negation in terms of $p \mid q$, these ways of reading $p \mid q$ are better avoided to begin with. The symbol "$p \mid q$" is pronounced: "p stroke q." We now put

$$\sim p \,.=.\, p \mid p \quad \text{Df},$$

$$p \supset q \,.=.\, p \mid \sim q \quad \text{Df},$$

$$p \vee q \,.=.\, \sim p \mid \sim q \quad \text{Df},$$

$$p \,.\, q \,.=.\, \sim(p \mid q) \quad \text{Df}.$$

Thus all the usual truth-functions can be constructed by means of the stroke. Note that by the above,

$$p \supset q \,.=.\, p \mid (q \mid q) \quad \text{Df}.$$

We find that

$$p \,.\supset.\, q \,.\, r \,.\equiv.\, p \mid (q \mid r).$$

Thus $p \supset q$ is a degenerate case of a function of *three* propositions.

We can construct new propositions indefinitely by means of the stroke; for example, $(p \mid q) \mid r$, $p \mid (q \mid r)$, $(p \mid q) \mid (r \mid s)$, and so on. Note that the stroke obeys the permutative law $(p \mid q) \equiv (q \mid p)$ but not the associative law $(p \mid q) \mid r \equiv p \mid (q \mid r)$. (These of course are results to be proved later.) Note also that, when we construct a new proposition by means of the stroke, we cannot know its truth or falsehood unless either (*a*) we know the truth or falsehood of some of its constituents, or (*b*) at least one of its constituents occurs several times in a suitable manner. The case (*a*) interests logic as giving rise to the *rule of inference*, viz.

Given p and $p \mid (q \mid r)$, we can infer r.

This or some variant must be taken as a primitive proposition. For the moment, we are applying it only when p, q, r are atomic propositions, but we shall extend it later. We shall consider (*b*) in a moment.

In constructing new propositions by means of the stroke, we assume that the stroke can have on either side of it any proposition so constructed, and need not have an atomic proposition on either side. Thus given three atomic propositions p, q, r, we can form, first, $p \mid q$ and $q \mid r$, and thence $(p \mid q) \mid r$ and $p \mid (q \mid r)$. Given four, p, q, r, s, we can form

$$\{(p \mid q) \mid r\} \mid s, \quad (p \mid q) \mid (r \mid s), \quad p \mid \{q \mid (r \mid s)\}$$

and of course others by permuting p, q, r, s. The above three are substantially

* For what follows, see Nicod, "A reduction in the number of the primitive propositions of logic," *Proc. Camb. Phil. Soc.* Vol. XIX. pp. 32—41.

different propositions. We have in fact

$$\{(p \mid q) \mid r\} \mid s \,.\, \equiv \,:.\, \sim p \vee \sim q \,.\, r : \vee : \sim s,$$

$$(p \mid q) \mid (r \mid s) \,.\, \equiv : p \,.\, q \,.\, \vee \,.\, r \,.\, s,$$

$$p \mid \{q \mid (r \mid s)\} \,.\, \equiv \,:.\, \sim p : \vee : q \,.\, \sim r \vee \sim s.$$

All the propositions obtained by this method follow from one rule: in "$p \mid q$," substitute, for p or q or both, propositions already constructed by means of the stroke. This rule generates a definite assemblage of new propositions out of the original assemblage of atomic propositions. All the propositions so generated (excluding the original atomic propositions) will be called "molecular propositions." Thus molecular propositions are all of the form $p \mid q$, but the p and q may now themselves be molecular propositions. If p is $p_1 \mid p_2$, p_1 and p_2 may be molecular; suppose $p_1 = p_{11} \mid p_{12}$. p_{11} may be of the form $p_{111} \mid p_{112}$, and so on; but after a finite number of steps of this kind, we are to arrive at atomic constituents. In a proposition $p \mid q$, the stroke between p and q is called the "principal" stroke; if $p = p_1 \mid p_2$, the stroke between p_1 and p_2 is a secondary stroke; so is the stroke between q_1 and q_2 if $q = q_1 \mid q_2$. If $p_1 = p_{11} \mid p_{12}$, the stroke between p_{11} and p_{12} is a tertiary stroke, and so on.

Atomic and molecular propositions together are "elementary propositions." Thus elementary propositions are atomic propositions together with all that can be generated from them by means of the stroke applied any finite number of times. This is a definite assemblage of propositions. We shall now, until further notice, use the letters p, q, r, s, t to denote elementary propositions, not necessarily atomic propositions. The rule of inference stated above is to hold still; *i.e.*

If p, q, r are elementary propositions, given p and $p \mid (q \mid r)$, we can infer r.

This is a primitive proposition.

We can now take up the point (*b*) mentioned above. When a molecular proposition contains repetitions of a constituent proposition in a suitable manner, it can be known to be true without our having to know the truth or falsehood of any constituent. The simplest instance is

$$p \mid (p \mid p),$$

which is always true. It means "p is incompatible with the incompatibility of p with itself," which is obvious. Again, take "$p \,.\, q \,.\, \supset \,.\, p$." This is

$$\{(p \mid q) \mid (p \mid q)\} \mid (p \mid p).$$

Again, take "$\sim p \,.\, \supset \,.\, \sim p \vee \sim q$." This is

$$(p \mid p) \mid \{(p \mid q) \mid (p \mid q)\}.$$

Again, "$p \,.\, \supset \,.\, p \vee q$" is

$$p \mid [\{(p \mid p) \mid (q \mid q)\} \mid \{(p \mid p) \mid (q \mid q)\}].$$

All these are true however p and q may be chosen. It is the fact that we can build up invariable truths of this sort that makes molecular propositions important to logic. Logic is helpless with atomic propositions, because their truth or falsehood can only be known empirically. But the truth of molecular propositions of suitable form can be known universally without empirical evidence.

The laws of logic, so far as elementary propositions are concerned, are all assertions to the effect that, whatever elementary propositions $p, q, r, \ldots$ may be, a certain function

$$F(p, q, r, \ldots),$$

whose values are molecular propositions, built up by means of the stroke, is always true. The proposition "$F(p)$ is true, whatever elementary proposition p may be" is denoted by

$$(p) \,.\, F(p).$$

Similarly the proposition "$F(p, q, r, \ldots)$ is true, whatever elementary propositions $p, q, r, \ldots$ may be" is denoted by

$$(p, q, r, \ldots) \,.\, F(p, q, r, \ldots).$$

When such a proposition is *asserted*, we shall omit the "$(p, q, r, \ldots)$" at the beginning. Thus

$$\text{"}\vdash .\, F(p, q, r, \ldots)\text{"}$$

denotes the assertion (as opposed to the hypothesis) that $F(p, q, r, \ldots)$ is true whatever elementary propositions $p, q, r, \ldots$ may be.

(The distinction between real and apparent variables, which occurs in Frege and in *Principia Mathematica*, is unnecessary. Whatever appears as a real variable in *Principia Mathematica* is to be taken as an apparent variable whose scope is the whole of the asserted proposition in which it occurs.)

The rule of inference, in the form given above, is never required within logic, but only when logic is applied. Within logic, the rule required is different. In the logic of propositions, which is what concerns us at present, the rule used is:

Given, whatever elementary propositions p, q, r may be, both "$\vdash .\, F(p, q, r, \ldots)$" and "$\vdash .\, F(p, q, r, \ldots) \mid \{G(p, q, r, \ldots) \mid H(p, q, r, \ldots)\}$," we can infer "$\vdash .\, H(p, q, r, \ldots)$."

Other forms of the rule of inference will meet us later. For the present, the above is the form we shall use.

Nicod has shown that the logic of propositions (∗1—∗5) can be deduced, by the help of the rule of inference, from two primitive propositions

$$\vdash .\, p \mid (p \mid p)$$

and

$$\vdash : p \supset q \,.\, \supset .\, s \mid q \supset p \mid s.$$

The first of these may be interpreted as "p is incompatible with not-p," or as "p or not-p," or as "not (p and not-p)," or as "p implies p." The second may be interpreted as

$$p \supset q \,.\, \supset : q \supset \sim s \,.\, \supset .\, p \supset \sim s,$$

which is a form of the principle of the syllogism. Written wholly in terms of the stroke, the principle becomes

$$\{p \mid (q \mid q)\} \mid [\{(s \mid q) \mid ((p \mid s) \mid (p \mid s))\} \mid \{(s \mid q) \mid ((p \mid s) \mid (p \mid s))\}].$$

Nicod has shown further that these two principles may be replaced by one. Written wholly in terms of the stroke, this one principle is

$$\{p \mid (q \mid r)\} \mid [\{t \mid (t \mid t)\} \mid \{(s \mid q) \mid ((p \mid s) \mid (p \mid s))\}].$$

It will be seen that, written in this form, the principle is less complex than the second of the above principles written wholly in terms of the stroke. When interpreted into the language of implication, Nicod's one principle becomes

$$p \,.\supset.\, q \,.\, r :\supset.\, t \supset t \,.\, s \mid q \supset p \mid s.$$

In this form, it looks more complex than

$$p \supset q \,.\supset.\, s \mid q \supset p \mid s,$$

but in itself it is less complex.

From the above primitive proposition, together with the rule of inference, everything that logic can ascertain about elementary propositions can be proved, provided we add one other primitive proposition, viz. that, given a proposition $(p, q, r, \ldots) \,.\, F(p, q, r, \ldots)$, we may substitute for $p, q, r, \ldots$ functions of the form

$$f_1(p, q, r, \ldots), \quad f_2(p, q, r, \ldots), \quad f_3(p, q, r, \ldots)$$

and assert

$$(p, q, r, \ldots) \,.\, F\{f_1(p, q, r, \ldots), f_2(p, q, r, \ldots), f_3(p, q, r, \ldots), \ldots\},$$

where $f_1, f_2, f_3, \ldots$ are functions constructed by means of the stroke. Since the former assertion applied to all elementary propositions, while the latter applies only to some, it is obvious that the former implies the latter.

A more general form of this principle will concern us later.

II. ELEMENTARY FUNCTIONS OF INDIVIDUALS

1. *Definition of "individual"*

We saw that atomic propositions are of one of the series of forms:

$$R_1(x), \ R_2(x, y), \ R_3(x, y, z), \ R_4(x, y, z, w), \ \ldots .$$

Here $R_1, R_2, R_3, R_4, \ldots$ are each characteristic of the special form in which they are found: that is to say, R_n cannot occur in an atomic proposition $R_m(x_1, x_2, \ldots x_m)$ unless $n = m$, and then can only occur as R_m occurs, not as $x_1, x_2, \ldots x_m$ occur. On the other hand, any term which can occur as the x's occur in $R_n(x_1, x_2, \ldots x_n)$ can also occur as one of the x's in $R_m(x_1, x_2, \ldots x_m)$ even if m is not equal to n. Terms which can occur in any form of atomic proposition are called "individuals" or "particulars"; terms which occur as the R's occur are called "universals."

We might state our definition compendiously as follows: An "individual" is anything that can be the subject of an atomic proposition.

Given an atomic proposition $R_n(x_1, x_2, \ldots x_n)$, we shall call any of the x's a "constituent" of the proposition, and R_n a "component" of the proposition*. We shall say the same as regards any molecular proposition in which $R_n(x_1, x_2, \ldots x_n)$ occurs. Given an elementary proposition $p \mid q$, where p and q may be atomic or molecular, we shall call p and q "parts" of $p \mid q$; and any parts of p or q will in turn be called parts of $p \mid q$, and so on until we reach the atomic parts of $p \mid q$. Thus to say that a proposition r "occurs in" $p \mid q$ and to say that r is a "part" of $p \mid q$ will be synonymous.

2. *Definition of an elementary function of an individual*

Given any elementary proposition which contains a part of which an individual a is a constituent, other propositions can be obtained by replacing a by other individuals in succession. We thus obtain a certain assemblage of elementary propositions. We may call the original proposition ϕa, and then the propositional function obtained by putting a variable x in the place of a will be called ϕx. Thus ϕx is a function of which the argument is x and the values are elementary propositions. The essential use of "ϕx" is that it collects together a certain set of propositions, namely all those that are its values with different arguments.

We have already had various special functions of propositions. If p is a part of some molecular proposition, we may consider the set of propositions resulting from the substitution of other propositions for p. If we call the original molecular proposition fp, the result of substituting q is called fq.

When an individual or a proposition occurs twice in a proposition, three functions can be obtained, by varying only one, or only another, or both, of the occurrences. For example, $p \mid p$ is a value of any one of the three functions $p \mid q$, $q \mid p$, $q \mid q$, where q is the argument. Similar considerations apply when an argument occurs more than twice. Thus $p \mid (p \mid p)$ is a value of $q \mid (r \mid s)$, or $q \mid (r \mid q)$, or $q \mid (q \mid r)$, or $q \mid (r \mid r)$, or $q \mid (q \mid q)$. When we assert a proposition "$\vdash . (p) . Fp$," the p is to be varied whenever it occurs. We may similarly assert a proposition of the form "$(x) . \phi x$," meaning "all propositions of the assemblage indicated by ϕx are true"; here also, every occurrence of x is to be varied.

3. "*Always true*" and "*sometimes true*"

Given any function, it may happen that all its values are true; again, it may happen that at least one of its values is true. The proposition that all the values of a function $\phi(x, y, z, \ldots)$ are true is expressed by the symbol

$$\text{"}(x, y, z, \ldots) . \phi(x, y, z, \ldots)\text{"}$$

unless we wish to assert it, in which case the assertion is written

$$\text{"}\vdash . \phi(x, y, z, \ldots).\text{"}$$

* This terminology is taken from Wittgenstein.

We have already had assertions of this kind where the variables were elementary propositions. We want now to consider the case where the variables are individuals and the function is elementary, *i.e.* all its values are elementary propositions. We no longer wish to confine ourselves to the case in which it is *asserted* that all the values of $\phi(x, y, z, \ldots)$ are true; we desire to be able to make the proposition

$$(x, y, z, \ldots) \,.\, \phi(x, y, z, \ldots)$$

a part of a stroke function. For the present, however, we will ignore this desideratum, which will occupy us in Section III of this Introduction.

In addition to the proposition that a function ϕx is "always true" (*i.e.* $(x) \,.\, \phi x$), we need also the proposition that ϕx is "sometimes true," *i.e.* is true for at least one value of x. This we denote by

$$\text{“}(\exists x) \,.\, \phi x.\text{”}$$

Similarly the proposition that $\phi(x, y, z, \ldots)$ is "sometimes true" is denoted by

$$\text{“}(\exists x, y, z, \ldots) \,.\, \phi(x, y, z, \ldots).\text{”}$$

We need, in addition to $(x, y, z, \ldots) \,.\, \phi(x, y, z, \ldots)$ and $(\exists x, y, z, \ldots) \,.\, \phi(x, y, z, \ldots)$, various other propositions of an analogous kind. Consider first a function of two variables. We can form

$$(\exists x) : (y) \,.\, \phi(x, y), \; (x) : (\exists y) \,.\, \phi(x, y), \; (\exists y) : (x) \,.\, \phi(x, y), \; (y) : (\exists x) \,.\, \phi(x, y).$$

These are substantially different propositions, of which no two are always equivalent. It would seem natural, in forming these propositions, to regard the function $\phi(x, y)$ as formed in two stages. Given $\phi(a, b)$, where a and b are constants, we can first form a function $\phi(a, y)$, containing the one variable y; we can then form

$$(y) \,.\, \phi(a, y) \text{ and } (\exists y) \,.\, \phi(a, y).$$

We can now vary a, obtaining again a function of one variable, and leading to the four propositions

$$(x) : (y) \,.\, \phi(x, y), \; (\exists x) : (y) \,.\, \phi(x, y), \; (x) : (\exists y) \,.\, \phi(x, y), \; (\exists x) : (\exists y) \,.\, \phi(x, y).$$

On the other hand, we might have gone from $\phi(a, b)$ to $\phi(x, b)$, thence to $(x) \,.\, \phi(x, b)$ and $(\exists x) \,.\, \phi(x, b)$, and thence to

$$(y) : (x) \,.\, \phi(x, y), \; (\exists y) : (x) \,.\, \phi(x, y), \; (y) : (\exists x) \,.\, \phi(x, y), \; (\exists y) : (\exists x) \,.\, \phi(x, y).$$

All of these will be called "general propositions"; thus eight general propositions can be derived from the function $\phi(x, y)$. We have

$$(x) : (y) \,.\, \phi(x, y) : \equiv : (y) : (x) \,.\, \phi(x, y),$$

$$(\exists x) : (\exists y) \,.\, \phi(x, y) : \equiv : (\exists y) : (\exists x) \,.\, \phi(x, y).$$

But there are no other equivalences that always hold. For example, the distinction between "$(x) : (\exists y) \,.\, \phi(x, y)$" and "$(\exists y) : (x) \,.\, \phi(x, y)$" is the same as the distinction in analysis between "For every ϵ, however small, there is a δ such that..." and "There is a δ such that, for every ϵ, however small, ..."

Although it might seem easier, in view of the above considerations, to regard every function of several variables as obtained by successive steps, each involving only a function of one variable, yet there are powerful considerations on the other side. There are two grounds in favour of the step-by-step method; first, that only functions of *one* variable need be taken as a primitive idea; secondly, that such definitions as the above seem to require *either* that we should first vary x, keeping y constant, *or* that we should first vary y, keeping x constant. The former seems to be involved when "(y)" or "$(\exists y)$" appears to the left of "(x)" or "$(\exists x)$," the latter in the converse case. The grounds against the step-by-step method are that it interferes with the method of matrices, which brings order into the successive generation of types of propositions and functions demanded by the theory of types, and that it requires us, from the start, to deal with such propositions as $(y) . \phi(a, y)$, which are not elementary. Take, for example, the proposition "$\vdash : q . \supset . p \vee q$." This will be

$$\vdash :. (p) :. (q) : q . \supset . p \vee q,$$

or

$$\vdash :. (q) :. (p) : q . \supset . p \vee q,$$

and will thus involve all values of either

$(q) : q . \supset . p \vee q$ considered as a function of p,

or $(p) : q . \supset . p \vee q$ considered as a function of q.

This makes it impossible to start our logic with elementary propositions, as we wish to do. It is useless to enlarge the definition of elementary propositions, since that only increases the values of q or p in the above functions. Hence it seems necessary to start with an elementary function

$$\phi(x_1, x_2, x_3, \ldots x_n),$$

before which we write, for each x_r, either "(x_r)" or "$(\exists x_r)$," the variables in this process being taken in any order we like. Here $\phi(x_1, x_2, x_3, \ldots x_n)$ is called the "matrix," and what comes before it is called the "prefix." Thus in

$$(\exists x) : (y) . \phi(x, y)$$

"$\phi(x, y)$" is the matrix and "$(\exists x) : (y)$" is the prefix. It thus appears that a matrix containing n variables gives rise to $n!\,2^n$ propositions by taking its variables in all possible orders and distinguishing "(x_r)" and "$(\exists x_r)$" in each case. (Some of these, however, are equivalent.) The process of obtaining such propositions from a matrix will be called "generalization," whether we take "all values" or "some value," and the propositions which result will be called "general propositions."

We shall later have occasion to consider matrices containing variables that are not individuals; we may therefore say:

A "matrix" is a function of any number of variables (which may or may not be individuals), which has elementary propositions as its values, and is used for the purpose of generalization.

A "general proposition" is one derived from a matrix by generalization. We shall add one further definition at this stage:

A "first-order proposition" is one derived by generalization from a matrix in which all the variables are individuals.

4. *Methods of proving general propositions*

There are two fundamental methods of proving general propositions, one for universal propositions, the other for such as assert existence. The method of proving universal propositions is as follows. Given a proposition

$$\text{“}\vdash . F(p, q, r, \ldots),\text{”}$$

where F is built up by the stroke, and $p, q, r, \ldots$ are elementary, we may replace them by elementary functions of individuals in any way we like, putting

$$p = f_1(x_1, x_2, \ldots x_n),$$
$$q = f_2(x_1, x_2, \ldots x_n),$$

and so on, and then assert the result for all values of $x_1, x_2, \ldots x_n$. What we thus assert is less than the original assertion, since $p, q, r, \ldots$ could originally take all values that are elementary propositions, whereas now they can only take such as are values of $f_1, f_2, f_3, \ldots$. (Any two or more of $f_1, f_2, f_3, \ldots$ may be identical.)

For proving existence-theorems we have two primitive propositions, namely

***8·1.** $\vdash . (\exists x, y) . \phi a \mid (\phi x \mid \phi y)$ and

***8·11.** $\vdash . (\exists x) . \phi x \mid (\phi a \mid \phi b)$

Applying the definitions to be given shortly, these assert respectively

$$\phi a . \supset . (\exists x) . \phi x$$

and

$$(x) . \phi x . \supset . \phi a . \phi b.$$

These two primitive propositions are to be assumed, not only for one variable, but for any number. Thus we assume

$$\phi(a_1, a_2, \ldots a_n) . \supset . (\exists x_1, x_2, \ldots x_n) . \phi(x_1, x_2, \ldots x_n),$$
$$(x_1, x_2, \ldots x_n) . \phi(x_1, x_2, \ldots x_n) . \supset . \phi(a_1, a_2, \ldots a_n) . \phi(b_1, b_2, \ldots b_n).$$

The proposition $(x) . \phi x . \supset . \phi a . \phi b$, in this form, does not look suitable for proving existence-theorems. But it may be written

$$(\exists x) . \sim \phi x . \vee . \phi a . \phi b$$

or

$$\sim \phi a \vee \sim \phi b . \supset . (\exists x) . \sim \phi x,$$

in which form it is identical with *9·11, writing ϕ for $\sim \phi$. Thus our two primitive propositions are the same as *9·1 and *9·11.

For purposes of inference, we still assume that from $(x) . \phi x$ and $(x) . \phi x \supset \psi x$ we can infer $(x) . \psi x$, and from p and $p \supset q$ we can infer q, even when the functions or propositions involved are not elementary.

Existence-theorems are very often obtained from the above primitive propositions in the following manner. Suppose we know a proposition

$$\vdash . f(x, x).$$

Since $\phi x . \supset . (\exists y) . \phi y$, we can infer

$$\vdash . (\exists y) . f(x, y),$$

i.e. $\vdash : (x) : (\exists y) . f(x, y).$

Similarly $\vdash : (y) : (\exists x) . f(x, y).$

Again, since $\phi (x, y) . \supset . (\exists z, w) . \phi (z, w)$, we can infer

$$\vdash . (\exists x, y) . f(x, y)$$

and $\vdash . (\exists y, x) . f(x, y).$

We may illustrate the proofs both of universal and of existence propositions by a simple example. We have

$$\vdash . (p) . p \supset p.$$

Hence, substituting ϕx for p,

$$\vdash . (x) . \phi x \supset \phi x.$$

Hence, as in the case of $f(x, x)$ above,

$$\vdash : (x) : (\exists y) . \phi x \supset \phi y,$$

$$\vdash : (y) : (\exists x) . \phi x \supset \phi y,$$

$$\vdash . (\exists x, y) . \phi x \supset \phi y.$$

Apart from special axioms asserting existence-theorems (such as the axiom of reducibility, the multiplicative axiom, and the axiom of infinity), the above two primitive propositions give the sole method of proving existence-theorems in logic. They are, in fact, always derived from general propositions of the form $(x) . f(x, x)$ or $(x) . f(x. x, x)$ or etc., by substituting other variables for some of the occurrences of x.

III. GENERAL PROPOSITIONS OF LIMITED SCOPE

In virtue of a primitive proposition, given $(x) . \phi x$ and $(x) . \phi x \supset \psi x$, we can infer $(x) . \psi x$. So far, however, we have introduced no notation which would enable us to state the corresponding *implication* (as opposed to *inference*). Again, $(\exists x) . \phi x$ and $(x, y) . \phi x \supset \psi y$ enable us to infer $(y) . \psi y$; here again, we wish to be able to state the corresponding implication. So far, we have only defined occurrences of general propositions as complete asserted propositions. Theoretically, this is their only use, and there is no need to define any other. But practically, it is highly convenient to be able to treat them as parts of stroke-functions. This is entirely a matter of definition. By introducing suitable definitions, first-order propositions can be shown to satisfy all the propositions of ∗1—∗5. Hence in using the propositions of ∗1—∗5, it will no longer be necessary to assume that $p, q, r, \ldots$ are elementary.

The fundamental definitions are given below.

When a general proposition occurs as part of another, it is said to have limited scope. If it contains an apparent variable x, the scope of x is said to be limited to the general proposition in question. Thus in $p \mid \{(x) \,.\, \phi x\}$, the scope of x is limited to $(x) \,.\, \phi x$, whereas in $(x) \,.\, p \mid \phi x$ the scope of x extends to the whole proposition. Scope is indicated by dots.

The new chapter ∗8 (given in Appendix A) should replace ∗9 in *Principia Mathematica.* Its general procedure will, however, be explained now.

The occurrence of a general proposition as part of a stroke-function is defined by means of the following definitions:

$$\{(x) \,.\, \phi x\} \mid q \,.=.\, (\exists x) \,.\, \phi x \mid q \quad \text{Df},$$

$$\{(\exists x) \,.\, \phi x\} \mid q \,.=.\, (x) \,.\, \phi x \mid q \quad \text{Df},$$

$$p \mid \{(y) \,.\, \psi y\} \,.=.\, (\exists y) \,.\, p \mid \psi y \quad \text{Df},$$

$$p \mid \{(\exists y) \,.\, \psi y\} \,.=.\, (y) \,.\, p \mid \psi y \quad \text{Df}.$$

These define, in the first place, only what is meant by the stroke when it occurs between two propositions of which one is elementary while the other is of the first order. When the stroke occurs between two propositions which are both of the first order, we shall adopt the convention that the one on the left is to be eliminated first, treating the one on the right as if it were elementary; then the one on the right is to be eliminated, in each case, in accordance with the above definitions. Thus

$$\{(x) \,.\, \phi x\} \mid \{(y) \,.\, \psi y\} \,.=:\, (\exists x) : \phi x \mid \{(y) \,.\, \psi y\} :$$

$$=: (\exists x) : (\exists y) \,.\, \phi x \mid \psi y,$$

$$\{(x) \,.\, \phi x\} \mid \{(\exists y) \,.\, \psi y\} \,.=:\, (\exists x) : \phi x \mid \{(\exists y) \,.\, \psi y\} :$$

$$=: (\exists x) : (y) \,.\, \phi x \mid \psi y,$$

$$\{(\exists x) \,.\, \phi x\} \mid \{(y) \,.\, \psi y\} \,.=:\, (x) : (\exists y) \,.\, \phi x \mid \psi y.$$

The rule about the order of elimination is only required for the sake of definiteness, since the two orders give equivalent results. For example, in the last of the above instances, if we had eliminated y first we should have obtained

$$(\exists y) : (x) \,.\, \phi x \mid \psi y,$$

which requires either $(x) \,.\, \sim\phi x$ or $(\exists y) \,.\, \sim\psi y$, and is then true.

And
$$(x) : (\exists y) \,.\, \phi x \mid \psi y$$

is true in the same circumstances. This possibility of changing the order of the variables in the prefix is only due to the way in which they occur, *i.e.* to the fact that x only occurs on one side of the stroke and y only on the other. The order of the variables in the prefix is indifferent whenever the occurrences of one variable are all on one side of a certain stroke, while those of the other are all on the other side of it. We do not have in general

$$(\exists x) : (y) \,.\, \chi(x, y) : \equiv : (y) : (\exists x) \,.\, \chi(x, y);$$

here the right-hand side is more often true than the left-hand side. But we do have

$$(\exists x) : (y) \,.\, \phi x \mid \psi y : \equiv : (y) : (\exists x) \,.\, \phi x \mid \psi y.$$

The possibility of altering the order of the variables in the prefix when they are separated by a stroke is a primitive proposition. In general it is convenient to put on the left the variables of which "all" are involved, and on the right those of which "some" are involved, after the elimination has been finished, always assuming that the variables occur in a way to which our primitive proposition is applicable.

It is not necessary for the above primitive proposition that the stroke separating x and y should be the principal stroke, *e.g.*

$$\begin{aligned} p \mid [\{(\exists x) \,.\, \phi x\} \mid \{(y) \,.\, \psi y\}] \,.\, = \,.\, p \mid [(x) : (\exists y) \,.\, \phi x \mid \psi y] \,. \\ = : (\exists x) : (y) \,.\, p \mid (\phi x \mid \psi y) : \\ \equiv : (y) : (\exists x) \,.\, p \mid (\phi x \mid \psi y). \end{aligned}$$

All that is necessary is that there should be *some* stroke which separates x from y. When this is not the case, the order cannot in general be changed. Take *e.g.* the matrix

$$\phi x \vee \psi y \,.\, \sim \phi x \vee \sim \psi y.$$

This may be written $\quad (\phi x \supset \psi y) \mid (\psi y \supset \phi x)$

or $\quad \{\phi x \mid (\psi y \mid \psi y)\} \mid \{\psi y \mid (\phi x \mid \phi x)\}.$

Here there is no stroke which separates all the occurrences of x from all those of y, and in fact the two propositions

$$(y) : (\exists x) \,.\, \phi x \vee \psi y \,.\, \sim \phi x \vee \sim \psi y$$

and $\quad (\exists x) : (y) \,.\, \phi x \vee \psi y \,.\, \sim \phi x \vee \sim \psi y$

are not equivalent except for special values of ϕ and ψ.

By means of the above definitions, we are able to derive all propositions, of whatever order, from a matrix of elementary propositions combined by means of the stroke. Given any such matrix, containing a part p, we may replace p by ϕx or $\phi(x, y)$ or etc., and proceed to add the prefix (x) or $(\exists x)$ or (x, y) or $(x) : (\exists y)$ or $(y) : (\exists x)$ or etc. If p and q both occur, we may replace p by ϕx and q by ψy, or we may replace both by ϕx, or one by ϕx and another by some stroke-function of ϕx.

In the case of a proposition such as

$$p \mid \{(x) : (\exists y) \,.\, \psi(x, y)\}$$

we must treat it as a case of $p \mid \{(x) \,.\, \phi x\}$, and first eliminate x. Thus

$$p \mid \{(x) : (\exists y) \,.\, \psi(x, y)\} \,.\, = : (\exists x) : (y) \,.\, p \mid \psi(x, y).$$

That is to say, the definitions of $\{(x) \,.\, \phi x)\} \mid q$ etc. are to be applicable unchanged when ϕx is not an elementary function.

The definitions of $\sim p$, $p \vee q$, $p \,.\, q$, $p \supset q$ are to be taken over unchanged. Thus

$$\begin{aligned}
\sim\{(x) \,.\, \phi x\} \,.\, = : \{(x) \,.\, \phi x\} \mid \{(x) \,.\, \phi x\} &: \\
= : (\exists x) : \phi x \mid \{(x) \,.\, \phi x\} &: \\
= : (\exists x) : (\exists y) \,.\, (\phi x \mid \phi y)&, \\
\sim\{(\exists x) \,.\, \phi x\} \,.\, = : (x) : (y) \,.\, (\phi x \mid \phi y)&, \\
p \,.\, \supset . (x) \,.\, \phi x : = : p \mid [\{(x) \,.\, \phi x\} \mid \{(x) \,.\, \phi x\}] &: \\
= : p \mid \{(\exists x) : (\exists y) \,.\, (\phi x \mid \phi y)\} &: \\
= : (x) : (y) \,.\, p \mid (\phi x \mid \phi y)&, \\
(x) \,.\, \phi x \,.\, \supset . p : = : \{(x) \,.\, \phi x\} \mid (p \mid p) &: \\
= : (\exists x) \,.\, \phi x \mid (p \mid p) : = : (\exists x) \,.\, \phi x \supset p&, \\
(x) \,.\, \phi x \,.\, \vee . p : = : [\sim\{(x) \,.\, \phi x\}] \mid \sim p &: \\
= : \{(\exists x) : (\exists y) \,.\, (\phi x \mid \phi y)\} \mid (p \mid p) &: \\
= : (x) \,.\, \{(\exists y) \,.\, (\phi x \mid \phi y)\} \mid (p \mid p) &: \\
= : (x) : (y) \,.\, (\phi x \mid \phi y) \mid (p \mid p)&, \\
p \,.\, \vee . (x) \,.\, \phi x : = : (x) : (y) \,.\, (p \mid p) \mid (\phi x \mid \phi y)&.
\end{aligned}$$

It will be seen that in the above two variables appear where only one might have been expected. We shall find, before long, that the two variables can be reduced to one; *i.e.* we shall have

$$(\exists x) : (\exists y) \,.\, \phi x \mid \phi y : \equiv . (\exists x) \,.\, \phi x \mid \phi x,$$

$$(x) : (y) \,.\, \phi x \mid \phi y : \equiv . (x) \,.\, \phi x \mid \phi x.$$

These lead to

$$\sim\{(x) \,.\, \phi x\} \,.\, \equiv . (\exists x) \,.\, \sim \phi x,$$

$$\sim\{(\exists x) \,.\, \phi x\} \,.\, \equiv . (x) \,.\, \sim \phi x.$$

But we cannot prove these propositions at our present stage; nor, if we could, would they be of much use to us, since we do not yet know that, when two general propositions are equivalent, either may be substituted for the other as part of a stroke-proposition without changing the truth-value.

For the present, therefore, suppose we have a stroke-function in which p occurs several times, say $p \mid (p \mid p)$, and we wish to replace p by $(x) \,.\, \phi x$, we shall have to write the second occurrence of p "$(y) \,.\, \phi y$," and the third "$(z) \,.\, \phi z$." Thus the resulting proposition will contain as many separate variables as there are occurrences of p.

The primitive propositions required, which have been already mentioned, are four in number. They are as follows:

(1) $\vdash . (\exists x, y) \,.\, \phi a \mid (\phi x \mid \phi y)$, *i.e.* $\vdash : \phi a \,.\, \supset . (\exists x) \,.\, \phi x$.

(2) $\vdash . (\exists x) \,.\, \phi x \mid (\phi a \mid \phi b)$, *i.e.* $\vdash : (x) \,.\, \phi x \,.\, \supset . \phi a \,.\, \phi b$.

(3) The extended rule of inference, *i.e.* from $(x) \,.\, \phi x$ and $(x) \,.\, \phi x \supset \psi x$ we can infer $(x) \,.\, \psi x$, even when ϕ and ψ are not elementary.

(4) If all the occurrences of x are separated from all the occurrences of y by a certain stroke, the order of x and y can be changed in the prefix; *i.e.*

For $(\exists x):(y).\phi x \mid \psi y$ we can substitute $(y):(\exists x).\phi x \mid \psi y$, and *vice versa*, even when this is only a part of the whole asserted proposition.

The above primitive propositions are to be assumed, not only for one variable, but for any number.

By means of the above primitive propositions it can be proved that all the propositions of ∗1—∗5 apply equally when one or more of the propositions $p, q, r, \ldots$ involved are not elementary. For this purpose, we make use of the work of Nicod, who proved that the primitive propositions of ∗1 can all be deduced from

$$\vdash . p \supset p$$

and
$$\vdash . p \supset q . \supset . s \mid q \supset p \mid s$$

together with the rule of inference: "Given p and $p \mid (q \mid r)$, we can infer r."

Thus all we have to do is to show that the above propositions remain true when p, q, s, or some of them, are not elementary. This is done in ∗8 in Appendix A.

IV. FUNCTIONS AS VARIABLES

The essential use of a variable is to pick out a certain assemblage of elementary propositions, and enable us to assert that all members of this assemblage are true, or that at least one member is true. We have already used functions of individuals, by substituting ϕx for p in the propositions of ∗1—∗5, and by the primitive propositions of ∗8. But hitherto we have always supposed that the function is kept constant while the individual is varied, and we have not considered cases where we have "$\exists\phi$," or where the scope of "ϕ" is less than the whole asserted proposition. It is necessary now to consider such cases.

Suppose a is a constant. Then "ϕa" will denote, for the various values of ϕ, all the various elementary propositions of which a is a constituent. This is a different assemblage of elementary propositions from any that can be obtained by variation of individuals; consequently it gives rise to new general propositions. The values of the function are still elementary propositions, just as when the argument is an individual; but they are a new assemblage of elementary propositions, different from previous assemblages.

As we shall have occasion later to consider functions whose values are not elementary propositions, we will distinguish those that have elementary propositions for their values by a note of exclamation between the letter denoting the function and the letter denoting the argument. Thus "$\phi ! x$" is a function of two variables, x and $\phi ! \hat{z}$. It is a matrix, since it contains no apparent variable and has elementary propositions for its values. We shall henceforth write "$\phi ! x$" where we have hitherto written ϕx.

If we replace x by a constant a, we can form such propositions as

$$(\phi) . \phi ! a, (\exists\phi) . \phi ! a.$$

These are not elementary propositions, and are therefore not of the form $\phi ! a$. The assertion of such propositions is derived from matrices by the method of ✱8. The primitive propositions of ✱8 are to apply when the variables, or some of them, are elementary functions as well as when they are all individuals.

*A function can only appear in a matrix through its values**. To obtain a matrix, proceed, as before, by writing $\phi ! x$, $\psi ! y$, $\chi ! z$, ... in place of $p, q, r, \ldots$ in some molecular proposition built up by means of the stroke. We can then apply the rules of ✱8 to $\phi, \psi, \chi, \ldots$ as well as to $x, y, z, \ldots$. The difference between a function of an individual and a function of an elementary function of individuals is that, in the former, the passage from one value to another is effected by making the same statement about a different individual, while in the latter it is effected by making a different statement about the same individual. Thus the passage from "Socrates is mortal" to "Plato is mortal" is a passage from $f ! x$ to $f ! y$, but the passage from "Socrates is mortal" to "Socrates is wise" is a passage from $\phi ! a$ to $\psi ! a$. Functional variation is involved in such a proposition as: "Napoleon had all the characteristics of a great general."

Taking the collection of elementary propositions, every matrix has values all of which belong to this collection. Every general proposition results from some matrix by generalization†. Every matrix intrinsically determines a certain classification of elementary propositions, which in turn determines the scope of the generalization of that matrix. Thus "x loves Socrates" picks out a certain collection of propositions, generalized in "$(x) . x$ loves Socrates" and "$(\exists x) . x$ loves Socrates." But "$\phi !$ Socrates" picks out those, among elementary propositions, which mention Socrates. The generalizations "$(\phi) . \phi !$ Socrates" and "$(\exists \phi) . \phi !$ Socrates" involve a class of elementary propositions which cannot be obtained from an individual-variable. But any value of "$\phi !$Socrates" is an ordinary elementary proposition; the novelty introduced by the variable ϕ is a novelty of classification, not of material classified. On the other hand, $(x) . x$ loves Socrates, $(\phi) . \phi !$ Socrates, etc. are new propositions, not contained among elementary propositions. It is the business of ✱8 to show that these propositions obey the same rules as elementary propositions. The method of proof makes it irrelevant what the variables are, so long as all the functions concerned have values which are elementary propositions. The variables may themselves be elementary propositions, as they are in ✱1—✱5.

A variable function which has values that are not elementary propositions starts a new set. But variables of this sort seem unnecessary. Every elementary proposition is a value of $\phi ! \hat{x}$; therefore

$$(p) . fp . \equiv . (\phi, x) . f(\phi ! x) : (\exists p) . fp . \equiv . (\exists \phi, x) . f(\phi ! x).$$

* This assumption is fundamental in the following theory. It has its difficulties, but for the moment we ignore them. It takes the place (not quite adequately) of the axiom of reducibility. It is discussed in Appendix C.

† In a proposition of logic, all the variables in the matrix must be generalized. In other general propositions, such as "all men are mortal," some of the variables in the matrix are replaced by constants.

Hence all second-order propositions in which the variable is an elementary proposition can be derived from elementary matrices. The question of other second-order propositions will be dealt with in the next section. A function of two variables, say $\phi(x, y)$, picks out a certain class of classes of propositions. We shall have the class $\phi(a, y)$, for given a and variable y; then the class of all classes $\phi(a, y)$ as a varies. Whether we are to regard our function as giving classes $\phi(a, y)$ or $\phi(x, b)$ depends upon the order of generalization adopted. Thus "$(\exists x):(y)$" involves $\phi(a, y)$, but "$(y):(\exists x)$" involves $\phi(x, b)$.

Consider now the matrix $\phi ! x$, as a function of two variables. If we first vary x, keeping ϕ fixed (which seems the more natural order), we form a class of propositions $\phi ! x, \phi ! y, \phi ! z, \ldots$ which differ solely by the substitution of one individual for another. Having made one such class, we make another, and so on, until we have done so in all possible ways. But now suppose we vary ϕ first, keeping x fixed and equal to a. We then first form the class of all propositions of the form $\phi ! a$, *i.e.* all elementary propositions of which a is a constituent; we next form the class $\phi ! b$; and so on. The set of propositions which are values of $\phi ! a$ is a set not obtainable by variation of individuals, *i.e.* not of the form fx [for constant f and variable x]. This is what makes ϕ a new sort of variable, different from x. This also is why generalization of the form $(\phi) . F!(\phi ! \hat{z}, x)$ gives a function not of the form $f ! x$ [for constant f]. Observe also that whereas a is a constituent of $f ! a$, f is not; thus the matrix $\phi ! x$ has the peculiarity that, when a value is assigned to x, this value is a constituent of the result, but when a value is assigned to ϕ, this value is absorbed in the resulting proposition, and completely disappears. We may define a function $\phi ! \hat{x}$ as that kind of similarity between propositions which exists when one results from the other by the substitution of one individual for another.

We have seen that there are matrices containing, as variables, functions of individuals. We may denote any such matrix by

$$f!(\phi ! \hat{z}, \psi ! \hat{z}, \chi ! \hat{z}, \ldots x, y, z, \ldots).$$

Since a function can only occur through its values, $\phi ! \hat{z}$ (*e.g.*) can only occur in the above matrix through the occurrence of $\phi ! x, \phi ! y, \phi ! z, \ldots$ or of $\phi ! a$, $\phi ! b, \phi ! c, \ldots$, where a, b, c are constants. Constants do not occur in logic, that is to say, the a, b, c which we have been supposing constant are to be regarded as obtained by an extra-logical assignment of values to variables. They may therefore be absorbed into the $x, y, z, \ldots$. Now x, y, z themselves will only occur in logic as arguments to variable functions. Hence any matrix which contains the variables $\phi ! \hat{z}, \psi ! \hat{z}, \chi ! \hat{z}, x, y, z$ and no others, if it is of the sort that can occur explicitly in logic, will result from substituting $\phi ! x, \phi ! y, \phi ! z$, $\psi ! x, \psi ! y, \psi ! z, \chi ! x, \chi ! y, \chi ! z$, or some of them, for elementary propositions in some stroke-function.

It is necessary here to explain what is meant when we speak of a "matrix that can occur explicitly in logic," or, as we may call it, a "logical matrix." A logical matrix is one that contains no constants. Thus $p \mid q$ is a logical matrix; so is $\phi ! x$, where ϕ and x are both variable. Taking any elementary proposition, we shall obtain a logical matrix if we replace all its components and constituents by variables. Other matrices result from logical matrices by assigning values to some of their variables. There are, however, various ways of analysing a proposition, and therefore various logical matrices can be derived from a given proposition. Thus a proposition which is a value of $p \mid q$ will also be a value of $(\phi ! x) \mid (\psi ! y)$ and of $\chi ! (x, y)$. Different forms are required for different purposes; but all the forms of matrices required explicitly in logic are logical matrices as above defined. This is merely an illustration of the fact that logic aims always at complete generality. The test of a logical matrix is that it can be expressed without introducing any symbols other than those of logic, *e.g.* we must not require the symbol "Socrates." Consider the expression

$$f ! (\phi ! \hat{z}, \psi ! \hat{z}, \chi ! \hat{z}, \ldots x, y, z).$$

When a value is assigned to f, this represents a matrix containing the variables $\phi, \psi, \chi, \ldots x, y, z, \ldots$. But while f remains unassigned, it is a matrix of a new sort, containing the new variable f. We call f a "second-order function," because it takes functions among its arguments. When a value is assigned, not only to f, but also to $\phi, \psi, \chi, \ldots x, y, z, \ldots$, we obtain an elementary proposition; but when a value is assigned to f alone, we obtain a matrix containing as variables only first-order functions and individuals. This is analogous to what happens when we consider the matrix $\phi ! x$. If we give values to both ϕ and x, we obtain an elementary proposition; but if we give a value to ϕ alone, we obtain a matrix containing only an individual as variable.

There is no logical matrix of the form $f ! (\phi ! \hat{z})$. The only matrices in which $\phi ! \hat{z}$ is the only argument are those containing $\phi ! a, \phi ! b, \phi ! c, \ldots$, where $a, b, c, \ldots$ are constants; but these are not logical matrices, being derived from the logical matrix $\phi ! x$. Since ϕ can only appear through its values, it must appear, in a logical matrix, with one or more variable arguments. The simplest logical functions of ϕ alone are $(x) . \phi ! x$ and $(\exists x) . \phi ! x$, but these are not matrices. A logical matrix

$$f ! (\phi ! \hat{z}, x_1, x_2, \ldots x_n)$$

is always derived from a stroke-function

$$F(p_1, p_2, p_3, \ldots p_n)$$

by substituting $\phi ! x_1, \phi ! x_2, \ldots \phi ! x_n$ for $p_1, p_2, \ldots p_n$. This is the sole method of constructing such matrices. (We may however have $x_r = x_s$ for some values of r and s.)

Second-order functions have two connected properties which first-order functions do not have. The first of these is that, when a value is assigned to

f, the result may be a logical matrix; the second is that certain constant values of f can be assigned without going outside logic.

To take the first point first: $f ! (\phi ! \hat{z}, x)$, for example, is a matrix containing three variables, f, ϕ, and x. The following logical matrices (among an infinite number) result from the above by assigning a value to f: $\phi ! x$, $(\phi ! x) | (\phi ! x)$, $\phi ! x \supset \phi ! x$, etc. Similarly $\phi ! x \supset \phi ! y$, which is a logical matrix, results from assigning a value to f in $f ! (\phi ! \hat{z}, x, y)$. In all these cases, the constant value assigned to f is one which can be expressed in logical symbols alone (which was the second property of f). This is not the case with $\phi ! x$: in order to assign a value to ϕ, we must introduce what we may call "empirical constants," such as "Socrates" and "mortality" and "being Greek." The functions of x that can be formed without going outside logic must involve a function as a generalized variable; they are (in the simplest case) such as $(\phi) . \phi ! x$ and $(\exists \phi) . \phi ! x$.

To some extent, however, the above peculiarity of functions of the second and higher orders is arbitrary. We might have adopted in logic the symbols

$$R_1 (x), R_2 (x, y), R_3 (x, y, z), \ldots,$$

where R_1 represents a variable predicate, R_2 a variable dyadic relation (in intension), and so on. Each of the symbols $R_1 (x)$, $R_2 (x, y)$, $R_3 (x, y, z)$, ... is a logical matrix, so that, if we used them, we should have logical matrices not containing variable functions. It is perhaps worth while to remind ourselves of the meaning of "$\phi ! a$," where a is a constant. The meaning is as follows. Take any finite number of propositions of the various forms $R_1 (x)$, $R_2 (x, y)$, ... and combine them by means of the stroke in any way desired, allowing any one of them to be repeated any finite number of times. If at least one of them has a as a constituent, *i.e.* is of the form

$$R_n (a, b_1, b_2, \ldots b_{n-1}),$$

then the molecular proposition we have constructed is of the form $\phi ! a$, *i.e.* is a value of "$\phi ! a$" with a suitable ϕ. This of course also holds of the proposition $R_n (a, b_1, b_2, \ldots b_{n-1})$ itself. It is clear that the logic of propositions, and still more of general propositions concerning a given argument, would be intolerably complicated if we abstained from the use of variable functions; but it can hardly be said that it would be impossible. As for the question of matrices, we could form a matrix $f ! (R_1, x)$, of which $R_1 (x)$ would be a value. That is to say, the properties of second-order matrices which we have been discussing would also belong to matrices containing variable universals. They cannot belong to matrices containing only variable individuals.

By assigning $\phi ! \hat{z}$ and x in $f ! (\phi ! \hat{z}, x)$, while leaving f variable, we obtain an assemblage of elementary propositions not to be obtained by means of variables representing individuals and first-order functions. This is why the new variable f is useful.

We can proceed in like manner to matrices

$$F!\{f!(\hat{\phi}!\hat{z},\hat{x}), g!(\hat{\phi}!\hat{z},\hat{x}), \ldots \psi!\hat{z}, \chi!\hat{z}, \ldots x, y, \ldots\}$$

and so on indefinitely. These merely represent new ways of grouping elementary propositions, leading to new kinds of generality.

V. FUNCTIONS OTHER THAN MATRICES

When a matrix contains several variables, functions of some of them can be obtained by turning the others into apparent variables. Functions obtained in this way are not matrices, and their values are not elementary propositions. The simplest examples are

$$(y) \,.\, \phi!(x,y) \text{ and } (\exists y) \,.\, \phi!(x,y).$$

When we have a general proposition $(\phi) \,.\, F\{\phi!\hat{z}, x, y, \ldots\}$, the only values ϕ can take are matrices, so that functions containing apparent variables are not included. We can, if we like, introduce a new variable, to denote not only functions such as $\phi!\hat{x}$, but also such as

$$(y) \,.\, \phi!(\hat{x},y),\ (y,z) \,.\, \phi!(\hat{x},y,z),\ \ldots\ (\exists y) \,.\, \phi!(\hat{x},y),\ \ldots;$$

in a word, all such functions of one variable as can be derived by generalization from matrices containing only individual-variables. Let us denote any such function by $\phi_1 x$, or $\psi_1 x$, or $\chi_1 x$, or etc. Here the suffix 1 is intended to indicate that the values of the functions may be first-order propositions, resulting from generalization in respect of individuals. In virtue of ∗8, no harm can come from including such functions along with matrices as values of single variables.

Theoretically, it is unnecessary to introduce such variables as ϕ_1, because they can be replaced by an infinite conjunction or disjunction. Thus *e.g.*

$$(\phi_1) \,.\, \phi_1 x \,.\equiv:\, (\phi) \,.\, \phi!x : (\phi, y) \,.\, \phi!(x,y) : (\phi) : (\exists y) \,.\, \phi!(x,y) : \text{etc.},$$

$$(\exists\phi_1) \,.\, \phi_1 x \,.\equiv:\, (\exists\phi) \,.\, \phi!x :\vee: (\exists\phi) : (y) \,.\, \phi!(x,y) :\vee: (\exists\phi, y) \,.\, \phi!(x,y) :\vee: \text{etc.},$$

and generally, given any matrix $f!(\phi!\hat{z}, x)$, we shall have the following process for interpreting $(\phi_1) \,.\, f!(\phi_1\hat{z}, x)$ and $(\exists\phi_1) \,.\, f!(\phi_1\hat{z}, x)$. Put

$$(\phi^1) \,.\, f!(\phi^1\hat{z}, x) \,.=:\, (\phi) \,.\, f!\{(y) \,.\, \phi!(\hat{z},y), x\} : (\phi) \,.\, f!\{(\exists y) \,.\, \phi!(\hat{z},y), x\},$$

where $f!\{(y) \,.\, \phi!(\hat{z},y), x\}$ is constructed as follows: wherever, in $f!\{\phi!\hat{z}, x\}$, a value of ϕ, say $\phi!a$, occurs, substitute $(y) \,.\, \phi!(a,y)$, and develop by the definitions at the beginning of ∗8. $f!\{(\exists y) \,.\, \phi!(\hat{z},y), x\}$ is similarly constructed. Similarly put

$$(\phi^2) \,.\, f!(\phi^2!\hat{z}, x) \,.=:\, (\phi) \,.\, f!\{(y,w) \,.\, \phi!(\hat{z},y,w), x\} :$$
$$(\phi) \,.\, f!\{(y) : (\exists w) \,.\, \phi!(\hat{z},y,w), x\} : \text{etc.},$$

where "etc." covers the prefixes $(\exists y) : (w) \,.,\ (\exists y, w) \,.,\ (w) : (\exists y)$. We define $\phi^3, \phi^4, \ldots$ similarly. Then

$$(\phi_1) \,.\, f!(\phi_1\hat{z}, x) \,.=:\, (\phi^1) \,.\, f!(\phi^1\hat{z}, x) : (\phi^2) \,.\, f!(\phi^2\hat{z}, x) : \text{etc.}$$

This process depends upon the fact that $f!(\phi!\hat{z}, x)$, for each value of ϕ and x, is a proposition constructed out of elementary propositions by the stroke, and that ∗8 enables us to replace any of these by a proposition which is not elementary. $(\exists\phi_1) \,.\, f!(\phi_1\hat{z}, x)$ is defined by an exactly analogous *disjunction.*

It is obvious that, in practice, an infinite conjunction or disjunction such as the above cannot be manipulated without assumptions *ad hoc.* We can work out results for any segment of the infinite conjunction or disjunction, and we can "see" that these results hold throughout. But we cannot prove this, because mathematical induction is not applicable. We therefore adopt certain primitive propositions, which assert only that what we can prove in each case holds generally. By means of these it becomes possible to manipulate such variables as ϕ_1.

In like manner we can introduce $f_1(\phi_1\hat{z}, \hat{x})$, where any number of individuals and functions $\psi_1, \chi_1, \ldots$ may appear as apparent variables.

No essential difficulty arises in this process so long as the apparent variables involved in a function are not of higher order than the argument to the function. For example, $x \,\epsilon\, \mathrm{D}\text{‘}R$, which is $(\exists y) \,.\, xRy$, may be treated without danger as if it were of the form $\phi!x$. In virtue of ∗8, $\phi_1 x$ may be substituted for $\phi!x$ without interfering with the truth of any logical proposition which $\phi!x$ is a part. Similarly whatever logical proposition holds concerning $f!(\phi_1\hat{z}, x)$ will hold concerning $f_1(\phi_1\hat{z}, x)$.

But when the apparent variable is of higher order than the argument, a new situation arises. The simplest cases are

$$(\phi) \,.\, f!(\phi!\hat{z}, x), \quad (\exists\phi) \,.\, f!(\phi!\hat{z}, x).$$

These are functions of x, but are obviously not included among the values for $\phi!x$ (where ϕ is the argument). If we adopt a new variable ϕ_2 which is to include functions in which $\phi!\hat{z}$ can be an apparent variable, we shall obtain other new functions

$$(\phi_2) \,.\, f!(\phi_2\hat{z}, x), \quad (\exists\phi_2) \,.\, f!(\phi_2\hat{z}, x),$$

which are again not among values for $\phi_2 x$ (where ϕ_2 is the argument), because the totality of values of $\phi_2\hat{z}$, which is now involved, is different from the totality of values of $\phi!\hat{z}$, which was formerly involved. However much we may enlarge the meaning of ϕ, a function of x in which ϕ occurs as apparent variable has a correspondingly enlarged meaning, so that, however ϕ may be defined,

$$(\phi) \,.\, f!(\phi\hat{z}, x) \text{ and } (\exists\phi) \,.\, f!(\phi\hat{z}, x)$$

can never be values for ϕx. To attempt to make them so is like attempting to catch one's own shadow. It is impossible to obtain one variable which embraces among its values all possible functions of individuals.

We denote by $\phi_2 x$ a function of x in which ϕ_1 is an apparent variable, but there is no variable of higher order. Similarly $\phi_3 x$ will contain ϕ_2 as apparent variable, and so on.

The essence of the matter is that a variable may travel through any well-defined totality of values, provided these values are all such that any one can replace any other significantly in any context. In constructing $\phi_1 x$, the only totality involved is that of individuals, which is already presupposed. But when we allow ϕ to be an apparent variable in a function of x, we enlarge the totality of functions of x, however ϕ may have been defined. It is therefore always necessary to specify what sort of ϕ is involved, whenever ϕ appears as an apparent variable.

The other condition, that of significance, is fully provided for by the definitions of ∗8, together with the principle that a function can only occur through its values. In virtue of the principle, a function of a function is a stroke-function of values of the function. And in virtue of the definitions in ∗8, a value of any function can significantly replace any proposition in a stroke-function, because propositions containing any number of apparent variables can always be substituted for elementary propositions and for each other in any stroke-function. What is necessary for significance is that every complete asserted proposition should be derived from a matrix by generalization, and that, in the matrix, the substitution of constant values for the variables should always result, ultimately, in a stroke-function of atomic propositions. We say "ultimately," because, when such variables as $\phi_2 \hat{z}$ are admitted, the substitution of a value for ϕ_2 may yield a proposition still containing apparent variables, and in this proposition the apparent variables must be replaced by constants before we arrive at a stroke-function of atomic propositions. We may introduce variables requiring several such stages, but the end must always be the same: a stroke-function of atomic propositions.

It seems, however, though it might be difficult to prove formally, that the functions ϕ_1, f_1 introduce no propositions that cannot be expressed without them. Let us take first a very simple illustration. Consider the proposition

$$(\exists \phi_1) . \phi_1 x . \phi_1 a, \text{ which we will call } f(x, a).$$

Since ϕ_1 includes all possible values of $\phi\,!$ and also a great many other values in its range, $f(x, a)$ might seem to make a smaller assertion than would be made by

$$(\exists \phi) . \phi\,! x . \phi\,! a, \text{ which we will call } f_0(x, a).$$

But in fact $f(x, a) . \supset . f_0(x, a)$. This may be seen as follows: $\phi_1 x$ has one of the various sets of forms:

$$(y) . \phi\,!(x, y), (y, z) . \phi\,!(x, y, z), \ldots,$$

$$(\exists y) . \phi\,!(x, y), (\exists y, z) . \phi\,!(x, y, z), \ldots,$$

$$(y) : (\exists z) . \phi\,!(x, y, z), (\exists y) : (z) . \phi\,!(x, y, z), \ldots.$$

Suppose first that $\phi_1 x . = . (y) . \phi\,!(x, y)$. Then

$$\begin{aligned} \phi_1 x . \phi_1 a . \equiv &: (y) . \phi\,!(x, y) : (y) . \phi\,!(a, y) : \\ \supset &: \phi\,!(x, b) . \phi\,!(a, b) : \\ \supset &: (\exists \phi) . \phi\,! x . \phi\,! a. \end{aligned}$$

Next suppose $\phi_1 x . = . (\exists y) . \phi ! (x, y)$. Then

$$\phi_1 x . \phi_1 a . \equiv : (\exists y) . \phi ! (x, y) : (\exists z) . \phi ! (a, z) :$$
$$\supset : (\exists y, z) : \phi ! (x, y) \vee \phi ! (x, z) . \phi ! (a, y) \vee \phi ! (a, z) :$$
$$\supset : (\exists \phi) . \phi ! x . \phi ! a,$$

because $\phi ! (x, y) \vee \phi ! (x, z)$ is of the form $\phi ! x$, when y and z are fixed. It is obvious that this method of proof applies to the other cases mentioned above. Hence

$$(\exists \phi_1) . \phi_1 x . \phi_1 a . \equiv . (\exists \phi) . \phi ! x . \phi ! a.$$

We can satisfy ourselves that the same result holds in the general form

$$(\exists \phi_1) . f ! (\phi_1 \hat{z}, x) . \equiv . (\exists \phi) . f ! (\phi ! \hat{z}, x)$$

by a similar argument. We know that $f ! (\phi ! \hat{z}, x)$ is derived from some stroke-function

$$F(p, q, r, \ldots)$$

by substituting $\phi ! x$, $\phi ! a$, $\phi ! b$, ... (where a, b, ... are constants) for some of the propositions p, q, r, ... and $g_1 ! x$, $g_2 ! x$, $g_3 ! x$, ... (where g_1, g_2, g_3, ... are constants) for others of p, q, r, ..., while replacing any remaining propositions p, q, r, ... by constant propositions. Take a typical case; suppose

$$f ! (\phi ! \hat{z}, x) . = . (\phi ! a) \mid \{(\phi ! x) \mid (\phi ! b)\}.$$

We then have to prove

$$\phi_1 a \mid (\phi_1 x \mid \phi_1 b) . \supset . (\exists \phi) . \phi ! a \mid (\phi ! x \mid \phi ! b),$$

where $\phi_1 x$ may have any of the forms enumerated above.

Suppose first that $\phi_1 x . = . (y) . \phi ! (x, y)$. Then

$$\phi_1 a \mid (\phi_1 x \mid \phi_1 b) . = : (\exists y) : (z, w) . \phi ! (a, y) \mid \{\phi ! (x, z) \mid \phi ! (b, w)\} :$$
$$\supset : (\exists y) . \phi ! (a, y) \mid \{\phi ! (x, y) \mid \phi ! (b, y)\} :$$
$$\supset : (\exists \phi) . \phi ! a \mid (\phi ! x \mid \phi ! b)$$

because, for a given y, $\phi ! (x, y)$ is of the form $\phi ! x$.

Suppose next that $\phi_1 x . = . (\exists y) . \phi ! (x, y)$. Then

$$\phi_1 a \mid (\phi_1 x \mid \phi_1 b) . = : (y) : (\exists z, w) . \phi ! (a, y) \mid \{\phi ! (x, z) \mid \phi ! (b, w)\} :$$
$$\supset : (\exists \psi) . \psi ! a \mid (\psi ! x \mid \psi ! b),$$

putting $\psi ! x . = . \phi ! (x, z) \vee \phi ! (x, w)$. Similarly the other cases can be dealt with. Hence the result follows.

Consider next the correlative proposition

$$(\phi_1) . f ! (\phi_1 \hat{z}, x) . \equiv . (\phi) . f ! (\phi ! \hat{z}, x).$$

Here it is the converse implication that needs proving, *i.e.*

$$(\phi) . f ! (\phi ! \hat{z}, x) . \supset . (\phi_1) . f ! (\phi_1 \hat{z}, x).$$

This follows from the previous case by transposition. It can also be seen independently as follows. Suppose, as before, that

$$f ! (\phi_1 \hat{z}, x) . = . (\phi_1 a) \mid (\phi_1 x \mid \phi_1 b),$$

and put first $\quad \phi_1 x . = . (y) . \phi ! (x, y).$

Then $\quad (\phi_1 a) \mid (\phi_1 x \mid \phi_1 b) . = : (\exists y) : (z, w) . \phi ! (a, y) \mid \{\phi ! (x, z) \mid \phi ! (b, w)\}.$

Thus we require that, given

$$(\psi) . (\psi ! a) \mid (\psi ! x \mid \psi ! b),$$

we should have $(\exists y) : (z, w) . \phi ! (a, y) \mid \{\phi ! (x, z) \mid \phi ! (b, w)\}$.

Now

$$(\psi) . \psi ! a \mid (\psi ! x \mid \psi ! b) . \supset :. \phi ! (a, z) . \supset . \phi ! (x, z) . \phi ! (b, z) :$$
$$\phi ! (a, w) . \supset . \phi ! (x, w) . \phi ! (b, w) :.$$
$$\supset :. \phi ! (a, z) . \phi ! (a, w) . \supset . \phi ! (x, z) . \phi ! (b, w) :.$$
$$\supset :. \phi ! (a, w) . \supset : \phi ! (a, z) . \supset . \phi ! (x, z) . \phi ! (b, w) \quad (1)$$

Also $$\sim \phi ! (a, w) . \supset : \phi ! (a, w) . \supset . \phi ! (x, z) . \phi ! (b, w) \quad (2)$$

$$(1) . (2) . \supset :. (\psi) . \psi ! a \mid (\psi ! x \mid \psi ! b) : \supset :. (\exists y) : \phi ! (a, y) . \supset . \phi ! (x, z) . \phi ! (b, w)$$

which was to be proved.

Put next $$\phi_1 x . = . (\exists y) . \phi ! (x, y).$$

Then $(\phi_1 a) \mid (\phi_1 x \mid \phi_1 b) . = : (y) : (\exists z, w) . \phi ! (a, y) \mid \{\phi ! (x, z) \mid \phi ! (b, w)\}$.

In this case we merely put $z = w = y$ and the result follows.

The method will be the same in any other case. Hence generally:

$$(\phi_1) . f ! (\phi_1 \hat{z}, x) . \equiv . (\phi) . f ! (\phi ! \hat{z}, x).$$

Although the above arguments do not amount to formal proofs, they suffice to make it clear that, in fact, any general propositions about $\phi ! \hat{z}$ are also true about $\phi_1 \hat{z}$. This gives us, so far as such functions are concerned, all that could have been got from the axiom of reducibility.

Since the proof can only be conducted in each separate case, it is necessary to introduce a primitive proposition stating that the result holds always. This primitive proposition is

$$\vdash : (\phi) . f ! (\phi ! \hat{z}, x) . \supset . f ! (\phi_1 \hat{z}, x) \quad \text{Pp.}$$

As an illustration: suppose we have proved some property of all classes defined by functions of the form $\phi ! \hat{z}$, the above primitive proposition enables us to substitute the class $\text{D}'R$, where R is the relation defined by $\phi ! (\hat{x}, \hat{y})$, or by $(\exists z) . \phi ! (\hat{x}, \hat{y}, z)$, or etc. Wherever a class or relation is defined by a function containing no apparent variables except individuals, the above primitive proposition enables us to treat it as if it were defined by a matrix.

We have now to consider functions of the form $\phi_2 x$, where

$$\phi_2 x . = . (\phi) . f ! (\phi ! \hat{z}, x) \text{ or } \phi_2 x . = . (\exists \phi) . f ! (\phi ! \hat{z}, x).$$

We want to discover whether, or under what circumstances, we have

$$(\phi) . g ! (\phi ! \hat{z}, x) . \supset . g ! (\phi_2 \hat{z}, x). \quad \text{(A)}$$

Let us begin with an important particular case. Put

$$g ! (\phi ! \hat{z}, x) . = . \phi ! a \supset \phi ! x.$$

Then $(\phi) . g ! (\phi ! \hat{z}, x) . = . x = a$, according to *13·1.

We want to prove

$$(\phi) . \phi ! a \supset \phi ! x . \supset . \phi_2 a \supset \phi_2 x,$$

i.e. $$(\phi) . \phi ! a \supset \phi ! x . \supset : (\phi) . f ! (\phi ! \hat{z}, a) . \supset . (\phi) . f ! (\phi ! \hat{z}, x) : (\exists \phi) . f ! (\phi ! \hat{z}, a) . \supset . (\exists \phi) . f ! (\phi ! \hat{z}, x).$$

Now $f ! (\phi ! \hat{z}, x)$ must be derived from some stroke-function

$$F(p, q, r, \ldots)$$

by substituting for some of $p, q, r, \ldots$ the values $\phi ! x, \phi ! b, \phi ! c, \ldots$ where $b, c, \ldots$ are constants. As soon as ϕ is assigned, this is of the form $\psi ! x$. Hence

$$(\phi) . \phi ! a \supset \phi ! x . \supset : (\phi) : f ! (\phi ! \hat{z}, a) . \supset . f ! (\phi ! \hat{z}, x) :$$
$$\supset : (\phi) . f ! (\phi ! \hat{z}, a) . \supset . (\phi) . f ! (\phi ! \hat{z}, x) :$$
$$(\exists \phi) . f ! (\phi ! \hat{z}, a) . \supset . (\exists \phi) . f ! (\phi ! \hat{z}, x).$$

Thus generally $(\phi) . \phi ! a \supset \phi ! x . \supset . (\phi_2) . \phi_2 a \supset \phi_2 x$ without the need of any axiom of reducibility.

It must not, however, be assumed that (A) is always true. The procedure is as follows: $f ! (\phi ! \hat{z}, x)$ results from some stroke-function

$$F(p, q, r, \ldots)$$

by substituting for some of $p, q, r, \ldots$ the values $\phi ! x, \phi ! a, \phi ! b, \ldots$ ($a, b, \ldots$ being constants). We assume that, *e.g.*

$$\phi_2 x . = . (\phi) . f ! (\phi ! \hat{z}, x).$$

Thus $$\phi_2 x . = . (\phi) . F(\phi ! x, \phi ! a, \phi ! b, \ldots). \qquad \text{(B)}$$

What we want to discover is whether

$$(\phi) . g ! (\phi ! \hat{z}, x) . \supset . g ! (\phi_2 \hat{z}, x).$$

Now $g ! (\phi ! \hat{z}, x)$ will be derived from a stroke-function

$$G(p, q, r, \ldots)$$

by substituting $\phi ! x, \phi ! a', \phi ! b', \ldots$ for some of $p, q, r, \ldots$. To obtain $g ! (\phi_2 \hat{z}, x)$, we have to put $\phi_2 x, \phi_2 a', \phi_2 b', \ldots$ in $G(p, q, r, \ldots)$, instead of $\phi ! x, \phi ! a', \phi ! b', \ldots$. We shall thus obtain a new matrix.

If $(\phi) . g ! (\phi ! \hat{z}, x)$ is known to be true because $G(p, q, r, \ldots)$ is always true, then $g ! (\phi_2 \hat{z}, x)$ is true in virtue of ✻8, because it is obtained from $G(p, q, r, \ldots)$ by substituting for some of $p, q, r, \ldots$ the propositions $\phi_2 x$, $\phi_2 a', \phi_2 b', \ldots$ which contain apparent variables. Thus in this case an inference is warranted.

We have thus the following important proposition:

Whenever $(\phi) . g ! (\phi ! \hat{z}, x)$ is known to be true because $g ! (\phi ! \hat{z}, x)$ is always a value of a stroke-function

$$G(p, q, r, \ldots),$$

which is true for all values of $p, q, r, \ldots$, then $g ! (\phi_2 \hat{z}, x)$ is also true, and so (of course) is $(\phi_2) . g ! (\phi_2 \hat{z}, x)$.

This, however, does not cover the case where $(\phi) . g ! (\phi ! \hat{z}, x)$ is not a truth of logic, but a hypothesis, which may be true for some values of x and false for others. When this is the case, the inference to $g ! (\phi_2 \hat{z}, x)$ is sometimes legitimate and sometimes not; the various cases must be investigated separately. We shall have an important illustration of the failure of the inference in connection with mathematical induction.

VI. CLASSES

The theory of classes is at once simplified in one direction and complicated in another by the assumption that functions only occur through their values and by the abandonment of the axiom of reducibility.

According to our present theory, all functions of functions are extensional, *i.e.*

$$\phi x \equiv_x \psi x . \supset . f(\phi \hat{z}) \equiv f(\psi \hat{z}).$$

This is obvious, since ϕ can only occur in $f(\phi \hat{z})$ by the substitution of values of ϕ for $p, q, r, \ldots$ in a stroke-function, and, if $\phi x \equiv \psi x$, the substitution of ϕx for p in a stroke-function gives the same truth-value to the truth-function as the substitution of ψx. Consequently there is no longer any reason to distinguish between functions and classes, for we have, in virtue of the above,

$$\phi x \equiv_x \psi x . \supset . \phi \hat{x} = \psi \hat{x}.$$

We shall continue to use the notation $\hat{x}(\phi x)$, which is often more convenient than $\phi \hat{x}$; but there will no longer be any difference between the meanings of the two symbols. Thus classes, as distinct from functions, lose even that shadowy being which they retain in ∗20. The same, of course, applies to relations in extension. This, so far, is a simplification.

On the other hand, we now have to distinguish classes of different orders composed of members of the same order. Taking classes of individuals as the simplest case, $\hat{x}(\phi ! x)$ must be distinguished from $\hat{x}(\phi_2 x)$ and so on. In virtue of the proposition at the end of the last section, the general logical properties of classes will be the same for classes of all orders. Thus *e.g.*

$$\alpha \subset \beta . \beta \subset \gamma . \supset . \alpha \subset \gamma$$

will hold whatever may be the orders of α, β, γ respectively. In other kinds of cases, however, trouble arises. Take, as a first instance, $p‘\kappa$ and $s‘\kappa$. We have

$$x \in p‘\kappa . \equiv : \alpha \in \kappa . \supset_\alpha . x \in \alpha.$$

Thus $p‘\kappa$ is a class of higher order than any of the members of κ. Hence the hypothesis $(\alpha) . f\alpha$ may not imply $f(p‘\kappa)$, if α is of the order of the members of κ. There is a kind of proof invented by Zermelo, of which the simplest example is his second proof of the Schröder-Bernstein theorem (given in ∗73). This kind of proof consists in defining a certain class of classes κ, and then showing that $p‘\kappa \in \kappa$. On the face of it, "$p‘\kappa \in \kappa$" is impossible, since $p‘\kappa$ is not of the same order as members of κ. This, however, is not all that is to be said. A class of classes κ is always defined by some function of the form

$$(x_1, x_2, \ldots) : (\exists y_1, y_2, \ldots) . F(x_1 \epsilon \alpha, x_2 \epsilon \alpha, \ldots y_1 \epsilon \alpha, y_2 \epsilon \alpha, \ldots),$$

where F is a stroke-function, and "$\alpha \epsilon \kappa$" means that the above function is true. It may well happen that the above function is true when $p‘\kappa$ is substituted for α, and the result is interpreted by *8. Does this justify us in asserting $p‘\kappa \epsilon \kappa$?

Let us take an illustration which is important in connection with mathematical induction. Put

$$\kappa = \hat{\alpha}(\breve{R}“\alpha \subset \alpha . a \epsilon \alpha).$$

Then $$\breve{R}“p‘\kappa \subset p‘\kappa . a \epsilon p‘\kappa \quad \text{(see *40·81)}$$

so that, in a sense, $p‘\kappa \epsilon \kappa$. That is to say, if we substitute $p‘\kappa$ for α in the defining function of κ, and apply *8, we obtain a true proposition. By the definition of *90,

$$\overleftarrow{R_*}‘a = p‘\kappa.$$

Thus $\overleftarrow{R_*}‘a$ is a second-order class. Consequently, if we have a hypothesis $(\alpha) . f\alpha$, where α is a first-order class, we cannot assume

$$(\alpha) . f\alpha . \supset . f(\overleftarrow{R_*}‘a). \qquad \text{(A)}$$

By the proposition at the end of the previous section, if $(\alpha) . f\alpha$ is deduced by logic from a universally-true stroke-function of elementary propositions, $f(\overleftarrow{R_*}‘a)$ will also be true. Thus we may substitute $\overleftarrow{R_*}‘a$ for α in any asserted proposition "$\vdash . f\alpha$" which occurs in *Principia Mathematica*. But when $(\alpha) . f\alpha$ is a hypothesis, not a universal truth, the implication (A) is not, *prima facie*, necessarily true.

For example, if $\kappa = \hat{\alpha}(\breve{R}“\alpha \subset \alpha . a \epsilon \alpha)$, we have

$$\alpha \epsilon \kappa . \supset : \alpha \cap \beta \epsilon \kappa . \equiv . \breve{R}“(\alpha \cap \beta) \subset \beta . a \epsilon \beta.$$

Hence $$\alpha \epsilon \kappa . \breve{R}“(\alpha \cap \beta) \subset \beta . a \epsilon \beta . \supset . p‘\kappa \subset \beta \qquad (1)$$

In many of the propositions of *90, as hitherto proved, we substitute $p‘\kappa$ for α, whence we obtain

$$\breve{R}“(\beta \cap p‘\kappa) \subset \beta . a \epsilon \beta . \supset . p‘\kappa \subset \beta \qquad (2)$$

i.e.

$$z \epsilon \beta . aR_* z . \supset_{z,w} . w \epsilon \beta : a \epsilon \beta . aR_* x : \supset . x \epsilon \beta$$

or $$aR_* x . \supset :. z \epsilon \beta . aR_* z . \supset_{z,w} . w \epsilon \beta : a \epsilon \beta : \supset . x \epsilon \beta.$$

This is a more powerful form of induction than that used in the definition of $aR_* x$. But the proof is not valid, because we have no right to substitute $p‘\kappa$ for α in passing from (1) to (2). Therefore the proofs which use this form of induction have to be reconstructed.

It will be found that the form to which we can reduce most of the fallacious inferences that seem plausible is the following:

Given "$\vdash . (x) . f(x, x)$" we can infer "$\vdash : (x) : (\exists y) . f(x, y)$." Thus given "$\vdash . (\alpha) . f(\alpha, \alpha)$" we can infer "$\vdash : (\alpha) : (\exists \beta) . f(\alpha, \beta)$." But this depends upon the possibility of $\alpha = \beta$. If, now, α is of one order and β of another, we do not know that $\alpha = \beta$ is possible. Thus suppose we have

$$\alpha \epsilon \kappa . \supset_\alpha . g\alpha$$

and we wish to infer $g\beta$, where β is a class of higher order satisfying $\beta \epsilon \kappa$. The proposition

$$(\beta) :. \alpha \epsilon \kappa . \supset_\alpha . g\alpha : \supset : \beta \epsilon \kappa . \supset . g\beta$$

becomes, when developed by $*8$,

$$(\beta) :: (\exists \alpha) :. \alpha \epsilon \kappa . \supset . g\alpha : \supset : \beta \epsilon \kappa . \supset . g\beta.$$

This is only valid if $\alpha = \beta$ is possible. Hence the inference is fallacious if β is of higher order than α.

Let us apply these considerations to Zermelo's proof of the Schröder-Bernstein theorem, given in $*73\cdot8$ ff. We have a class of classes

$$\kappa = \hat{\alpha}(\alpha \subset \mathrm{D}'R . \beta - \text{Ɑ}'R \subset \alpha . \breve{R}``\alpha \subset \alpha)$$

and we prove $p'\kappa \epsilon \kappa$ ($*73\cdot81$), which is admissible in the limited sense explained above. We then add the hypothesis

$$x \sim\epsilon (\beta - \text{Ɑ}'R) \cup \breve{R}``p'\kappa$$

and proceed to prove $p'\kappa - \iota'x \epsilon \kappa$ (in the fourth line of the proof of $*73\cdot82$). This also is admissible in the limited sense. But in the next line of the same proof we make a use of it which is not admissible, arguing from $p'\kappa - \iota'x \epsilon \kappa$ to $p'\kappa \subset p'\kappa - \iota'x$, because

$$\alpha \epsilon \kappa . \supset_\alpha . p'\kappa \subset \alpha.$$

The inference from

$$\alpha \epsilon \kappa . \supset_\alpha . p'\kappa \subset \alpha \quad \text{to} \quad p'\kappa - \iota'x \epsilon \kappa . \supset . p'\kappa \subset p'\kappa - \iota'x$$

is only valid if $p'\kappa - \iota'x$ is a class of the same order as the members of κ. For, when $\alpha \epsilon \kappa . \supset_\alpha . p'\kappa \subset \alpha$ is written out it becomes

$$(\alpha) ::: (\exists \beta) ::. (x) :: \alpha \epsilon \kappa . \supset :. \beta \epsilon \kappa . \supset . x \epsilon \beta : \supset . x \epsilon \alpha.$$

This is deduced from

$$\alpha \epsilon \kappa . \supset :. \alpha \epsilon \kappa . \supset . x \epsilon \alpha : \supset . x \epsilon \alpha$$

by the principle that $f(\alpha, \alpha)$ implies $(\exists \beta) . f(\alpha, \beta)$. But here the β must be of the same order as the α, while in our case α and β are not of the same order, if $\alpha = p'\kappa - \iota'x$ and β is an ordinary member of κ. At this point, therefore, where we infer $p'\kappa \subset p'\kappa - \iota'x$, the proof breaks down.

It is easy, however, to remedy this defect in the proof. All we need is

$$x \sim\epsilon (\beta - \text{Ɑ}'R) \cup \breve{R}``p'\kappa . \supset . x \sim\epsilon p'\kappa$$

or, conversely,

$$x \epsilon p'\kappa . \supset . x \epsilon (\beta - \text{Ɑ}'R) \cup \breve{R}``p'\kappa.$$

Now

$$x \epsilon p'\kappa . \supset :. \alpha \epsilon \kappa . \supset_\alpha : \alpha - \iota'x \sim \epsilon \kappa :$$
$$\supset_\alpha : \sim(\beta - \text{G}'R \subset \alpha - \iota'x) . \vee . \sim\{\breve{R}''(\alpha - \iota'x) \subset \alpha - \iota'x\} :$$
$$\supset_\alpha : x \epsilon \beta - \text{G}'R . \vee . x \epsilon \breve{R}''(\alpha - \iota'x)$$
$$\supset :. x \epsilon \beta - \text{G}'R : \vee : \alpha \epsilon \kappa . \supset_\alpha . x \epsilon \breve{R}''\alpha.$$

Hence, by $*72{\cdot}341$,

$$x \epsilon p'\kappa . \supset . x \epsilon (\beta - \text{G}'R) \cup \breve{R}''p'\kappa$$

which gives the required result.

We assume that $\alpha - \iota'x$ is of no higher order than α; this can be secured by taking α to be of at least the second order, since $\iota'x$, and therefore $-\iota'x$, is of the second order. We may always assume our classes raised to a given order, but not raised indefinitely.

Thus the Schröder-Bernstein theorem survives.

Another difficulty arises in regard to sub-classes. We put

$$\text{Cl}'\alpha = \hat{\beta}(\beta \subset \alpha) \quad \text{Df.}$$

Now "$\beta \subset \alpha$" is significant when β is of higher order than α, provided its members are of the same type as those of α. But when we have

$$\beta \subset \alpha . \supset_\beta . f\beta,$$

the β must be of some definite type. As a rule, we shall be able to show that a proposition of this sort holds whatever the type of β, if we can show that it holds when β is of the same type as α. Consequently no difficulty arises until we come to Cantor's proposition $2^n > n$, which results from the proposition

$$\sim\{(\text{Cl}'\alpha) \text{ sm } \alpha\}$$

which is proved in $*102$. The proof is as follows:

$$R \epsilon 1 \rightarrow 1 . \text{D}'R = \alpha . \text{G}'R \subset \text{Cl}'\alpha . \xi = \hat{x}\{x \epsilon \alpha - \breve{R}'x\} . \supset :$$
$$y \epsilon \alpha . y \epsilon \breve{R}'y . \supset_y . y \sim \epsilon \xi : y \epsilon \alpha . y \sim \epsilon \breve{R}'y . \supset_y . y \epsilon \xi : \supset : y \epsilon \alpha . \supset_y . \xi \neq \breve{R}'y :$$
$$\supset : \xi \sim \epsilon \text{G}'R.$$

As this proposition is crucial, we shall enter into it somewhat minutely.

Let $\alpha = \hat{x}(\text{A} \,!\, x)$, and let

$$xR\{\hat{z}(\phi \,!\, z)\} . = . f \,!\, (\phi \,!\, \hat{z}, x).$$

Then by our data,

$$\text{A} \,!\, x . \supset . (\exists\phi) . f \,!\, (\phi \,!\, \hat{z}, x),$$
$$f \,!\, (\phi \,!\, \hat{z}, x) . \supset . \text{A} \,!\, x . \phi \,!\, y \supset_y \text{A} \,!\, y,$$
$$f \,!\, (\phi \,!\, \hat{z}, x) . f \,!\, (\phi \,!\, \hat{z}, y) . \supset . x = y,$$
$$f \,!\, (\phi \,!\, \hat{z}, x) . f \,!\, (\psi \,!\, \hat{z}, x) . \supset . \phi \,!\, y \equiv_y \psi \,!\, y.$$

With these data,

$$x \epsilon \alpha - \breve{R}'x . \equiv : \text{A} \,!\, x : f \,!\, (\phi \,!\, \hat{z}, x) . \supset_\phi . \sim\phi \,!\, x.$$

Thus $\qquad \xi = \hat{x}\{(\phi) : \text{A} \,!\, x : f \,!\, (\phi \,!\, \hat{z}, x) . \supset . \sim\phi \,!\, x\}.$

Thus ξ is defined by a function in which ϕ appears as apparent variable. If we enlarge the initial range of ϕ, we shall enlarge the range of values involved in the definition of ξ. There is therefore no way of escaping from the result that ξ is of higher order than the sub-classes of α contemplated in the definition of Cl'α. Consequently the proof of $2^n > n$ collapses when the axiom of reducibility is not assumed. We shall find, however, that the proposition remains true when n is finite.

With regard to relations, exactly similar questions arise as with regard to classes. A relation is no longer to be distinguished from a function of two variables, and we have

$$\phi(\hat{x}, \hat{y}) = \psi(\hat{x}, \hat{y}) \,.\, \equiv : \phi(x, y) \,.\, \equiv_{x,y} \,.\, \psi(x, y).$$

The difficulties as regards $\dot{p}$'λ and Rl'P are less important than those concerning p'κ and Cl'α, because $\dot{p}$'λ and Rl'P are less used. But a very serious difficulty occurs as regards similarity. We have

$$\alpha \operatorname{sm} \beta \,.\, \equiv \,.\, (\exists R) \,.\, R \in 1 \rightarrow 1 \,.\, \alpha = \text{D}\text{'}R \,.\, \beta = \text{Œ}\text{'}R.$$

Here R must be confined within some type; but whatever type we choose, there may be a correlator of higher type by which α and β can be correlated. Thus we can never prove $\sim(\alpha \operatorname{sm} \beta)$, except in such special cases as when either α or β is finite. This difficulty was illustrated by Cantor's theorem $2^n > n$, which we have just examined. Almost all our propositions are concerned in proving that two classes *are* similar, and these can all be interpreted so as to remain valid. But the few propositions which are concerned with proving that two classes are *not* similar collapse, except where one at least of the two is finite.

VII. MATHEMATICAL INDUCTION

All the propositions on mathematical induction in Part II, Section E and Part III, Section C remain valid, when suitably interpreted. But the proofs of many of them become fallacious when the axiom of reducibility is not assumed, and in some cases new proofs can only be obtained with considerable labour. The difficulty becomes at once apparent on observing the definition of "xR_*y" in $*90$. Omitting the factor "$x \in C\text{'}R$," which is irrelevant for our purposes, the definition of "xR_*y" may be written

$$zRw \,.\, \supset_{z,w} \,.\, \phi\,!\,z \supset \phi\,!\,w : \supset_\phi \,.\, \phi\,!\,x \supset \phi\,!\,y, \qquad \text{(A)}$$

i.e. "y has every elementary hereditary property possessed by x." We may, instead of elementary properties, take any other order of properties; as we shall see later, it is advantageous to take third-order properties when R is one-many or many-one, and fifth-order properties in other cases. But for preliminary purposes it makes no difference what order of properties we take, and therefore for the sake of definiteness we take elementary properties to begin with. The difficulty is that, if ϕ_2 is any second-order property, we cannot deduce from (A)

$$zRw \,.\, \supset_{z,w} \,.\, \phi_2 z \supset \phi_2 w : \supset \,.\, \phi_2 x \supset \phi_2 y. \qquad \text{(B)}$$

Suppose, for example, that $\phi_2 z . = . (\phi) . f!(\phi ! \hat{z}, z)$; then from (A) we can deduce

$$zRw . \supset_{z,w} . f!(\phi ! \hat{z}, z) \supset_\phi f!(\phi ! \hat{z}, w) : \supset : f!(\phi ! \hat{z}, x) . \supset_\phi . f!(\phi ! \hat{z}, y) :$$
$$\supset : \phi_2 x . \supset . \phi_2 y. \qquad \text{(C)}$$

But in general our hypothesis here is not implied by the hypothesis of (B). If we put $\phi_2 z . = . (\exists \phi) . f!(\phi ! \hat{z}, z)$, we get exactly analogous results.

Hence in order to apply mathematical induction to a second-order property, it is not sufficient that it should be itself hereditary, but it must be composed of hereditary elementary properties. That is to say, if the property in question is $\phi_2 z$, where $\phi_2 z$ is either

$$(\phi) . f!(\phi ! \hat{z}, z) \text{ or } (\exists \phi) . f!(\phi ! \hat{z}, z),$$

it is not enough to have

$$zRw . \supset_{z,w} . \phi_2 z \supset \phi_2 w,$$

but we must have, for each elementary ϕ,

$$zRw . \supset_{z,w} . f!(\phi ! \hat{z}, z) \supset f!(\phi ! \hat{z}, w).$$

One inconvenient consequence is that, *primâ facie*, an inductive property must not be of the form

$$xR_* z . \phi ! z$$

or

$$S \epsilon \text{Potid}\text{‘}R . \phi ! S$$

or

$$\alpha \epsilon \text{NC induct} . \phi ! \alpha.$$

This is inconvenient, because often such properties are hereditary when ϕ alone is not, *i.e.* we may have

$$xR_* z . \phi ! z . zRw . \supset_{z,w} . xR_* w . \phi ! w$$

when we do not have

$$\phi ! z . zRw . \supset_{z,w} . \phi ! w,$$

and similarly in the other cases.

These considerations make it necessary to re-examine all inductive proofs. In some cases they are still valid, in others they are easily rectified; in still others, the rectification is laborious, but it is always possible. The method of rectification is explained in Appendix B to this volume.

There is, however, so far as we can discover, no way by which our present primitive propositions can be made adequate to Dedekindian and well-ordered relations. The practical uses of Dedekindian relations depend upon ∗211·63—·692, which lead to ∗214·3—·34, showing that the series of segments of a series is Dedekindian. It is upon this that the theory of real numbers rests, real numbers being defined as segments of the series of rationals. This subject is dealt with in ∗310. If we were to regard as doubtful the proposition that the series of real numbers is Dedekindian, analysis would collapse.

The proofs of this proposition in *Principia Mathematica* depend upon the axiom of reducibility, since they depend upon ∗211·64, which asserts

$$\lambda \subset D\text{‘}P_\epsilon . \supset . s\text{‘}\lambda \epsilon D\text{‘}P_\epsilon.$$

For reasons explained above, if α is of the order of members of λ, $(\alpha) \, . \, f\alpha$ may not imply $f(s{}^{\backprime}\lambda)$, because $s{}^{\backprime}\lambda$ is a class of higher order than the members of λ. Thus although we have

$$D{}^{\backprime}P_\epsilon = \hat{\alpha}\,\{(\exists\beta) \, . \, \alpha = P{}^{\backprime\backprime}\beta\},$$
$$s{}^{\backprime}\lambda = P{}^{\backprime\backprime}s{}^{\backprime}\breve{P}_\epsilon{}^{\backprime\backprime}\lambda,$$

yet we cannot infer $s{}^{\backprime}\lambda \, \epsilon \, D{}^{\backprime}P_\epsilon$ except when $s{}^{\backprime}\lambda$ or $s{}^{\backprime}\breve{P}_\epsilon{}^{\backprime\backprime}\lambda$ is, for some special reason, of the same order as the members of λ. This will be the case when λ is finite, but not necessarily otherwise. Hence the theory of irrationals will require reconstruction.

Exactly similar difficulties arise in regard to well-ordered series. The theory of well-ordered series rests on the definition $*250{\cdot}01$:

$$\text{Bord} = \hat{P}\,(\text{Cl ex}{}^{\backprime}C{}^{\backprime}P \subset \text{Ɑ}{}^{\backprime}\text{min}_P) \quad \text{Df},$$

whence $\quad P \, \epsilon \, \text{Bord} \, . \equiv : \alpha \subset C{}^{\backprime}P \, . \, \exists \, ! \, \alpha \, . \supset_\alpha . \, \exists \, ! \, \alpha - \breve{P}{}^{\backprime\backprime}\alpha.$

In making deductions, we constantly substitute for α some constructed class of higher order than $C{}^{\backprime}P$. For instance, in $*250{\cdot}122$ we substitute for α the class $C{}^{\backprime}P \cap p{}^{\backprime}\overleftarrow{P}{}^{\backprime\backprime}(\alpha \cap C{}^{\backprime}P)$, which is in general of higher order than α. If this substitution is illegitimate, we cannot prove that a class contained in $C{}^{\backprime}P$ and having successors must have an immediate successor, without which the theory of well-ordered series becomes impossible. This particular difficulty might be overcome, but it is obvious that many important propositions must collapse.

It might be possible to sacrifice infinite well-ordered series to logical rigour, but the theory of real numbers is an integral part of ordinary mathematics, and can hardly be the object of a reasonable doubt. We are therefore justified in supposing that some logical axiom which is true will justify it. The axiom required may be more restricted than the axiom of reducibility, but, if so, it remains to be discovered.

The following are among the contributions to mathematical logic since the publication of the first edition of *Principia Mathematica*.

D. Hilbert. Axiomatisches Denken, *Mathematische Annalen*, Vol. 78. Die logischen Grundlagen der Mathematik, *ib.* Vol. 88. Neue Begründung der Mathematik, *Abhandlungen aus dem mathematischen Seminar der Hamburgischen Universität*, 1922.

P. Bernays. Ueber Hilbert's Gedanken zur Grundlegung der Arithmetik, *Jahresbericht der deutschen Mathematiker-Vereinigung*, Vol. 31.

H. Behmann. Beiträge zur Algebra der Logik. *Mathematische Annalen*, Vol. 86.

L. Chwistek. Ueber die Antinomien der Prinzipien der Mathematik, *Mathematische Zeitschrift*, Vol. 14. The Theory of Constructive Types. *Annales de la Société Mathématique de Pologne*, 1923. (Dr Chwistek has kindly allowed us to read in MS. a longer work with the same title.)

H. Weyl. *Das Kontinuum*, Veit, 1918. Ueber die neue Grundlagenkrise der Mathematik, *Mathematische Zeitschrift*, Vol. 10. Randbemerkungen zu Hauptproblemen der Mathematik, *Mathematische Zeitschrift*, Vol. 20.

L. E. J. Brouwer. Begründung der Mengenlehre unabhängig vom logischen Satz des ausgeschlossenen Dritten. *Verhandelingen d. K. Akademie v. Wetenschappen*, Amsterdam, 1918, 1919. Intuitionistische Mengenlehre, *Jahresbericht der deutschen Mathematiker-Vereinigung*, Vol. 28.

A. Tajtelbaum-Tarski. Sur le terme primitif de la logistique, *Fundamenta Mathematicae*, Tom. IV. Sur les "truth-functions" au sens de MM. Russell et Whitehead, *ib.* Tom. V. Sur quelques théorèmes qui équivalent à l'axiome du choix, *ib.*

F. Bernstein. Die Mengenlehre Georg Cantor's und der Finitismus, *Jahresbericht der deutschen Mathematiker-Vereinigung*, Vol. 28.

J. König. *Neue Grundlagen der Logik, Arithmetik und Mengenlehre*, Veit, 1914.

C. I. Lewis. *A Survey of Symbolic Logic*, University of California, 1918.

H. M. Sheffer. Total determinations of deductive systems with special reference to the Algebra of Logic. *Bulletin of the American Mathematical Society*, Vol. XVI. *Trans. Amer. Math. Soc.* Vol. XIV. pp. 481—488. *The general theory of notational relativity*, Cambridge, Mass. 1921.

J. G. P. Nicod. A reduction in the number of the primitive propositions of logic. *Proc. Camb. Phil. Soc.* Vol. XIX.

L. Wittgenstein. *Tractatus Logico-Philosophicus*, Kegan Paul, 1922.

M. Schönwinkel. Ueber die Bausteine der mathematischen Logik, *Math. Annalen*, Vol. 92.

INTRODUCTION

The mathematical logic which occupies Part I of the present work has been constructed under the guidance of three different purposes. In the first place, it aims at effecting the greatest possible analysis of the ideas with which it deals and of the processes by which it conducts demonstrations, and at diminishing to the utmost the number of the undefined ideas and undemonstrated propositions (called respectively *primitive* ideas and *primitive* propositions) from which it starts. In the second place, it is framed with a view to the perfectly precise expression, in its symbols, of mathematical propositions: to secure such expression, and to secure it in the simplest and most convenient notation possible, is the chief motive in the choice of topics. In the third place, the system is specially framed to solve the paradoxes which, in recent years, have troubled students of symbolic logic and the theory of aggregates; it is believed that the theory of types, as set forth in what follows, leads both to the avoidance of contradictions, and to the detection of the precise fallacy which has given rise to them.

Of the above three purposes, the first and third often compel us to adopt methods, definitions, and notations which are more complicated or more difficult than they would be if we had the second object alone in view. This applies especially to the theory of descriptive expressions (∗14 and ∗30) and to the theory of classes and relations (∗20 and ∗21). On these two points, and to a lesser degree on others, it has been found necessary to make some sacrifice of lucidity to correctness. The sacrifice is, however, in the main only temporary: in each case, the notation ultimately adopted, though its real meaning is very complicated, has an apparently simple meaning which, except at certain crucial points, can without danger be substituted in thought for the real meaning. It is therefore convenient, in a preliminary explanation of the notation, to treat these apparently simple meanings as primitive ideas, *i.e.* as ideas introduced without definition. When the notation has grown more or less familiar, it is easier to follow the more complicated explanations which we believe to be more correct. In the body of the work, where it is necessary to adhere rigidly to the strict logical order, the easier order of development could not be adopted; it is therefore given in the Introduction. The explanations given in Chapter I of the Introduction are such as place lucidity before correctness; the full explanations are partly supplied in succeeding Chapters of the Introduction, partly given in the body of the work.

The use of a symbolism, other than that of words, in all parts of the book which aim at embodying strictly accurate demonstrative reasoning, has been forced on us by the consistent pursuit of the above three purposes. The reasons for this extension of symbolism beyond the familiar regions of number and allied ideas are many:

(1) The ideas here employed are more abstract than those familiarly considered in language. Accordingly there are no words which are used mainly in the exact consistent senses which are required here. Any use of words would require unnatural limitations to their ordinary meanings, which would be in fact more difficult to remember consistently than are the definitions of entirely new symbols.

(2) The grammatical structure of language is adapted to a wide variety of usages. Thus it possesses no unique simplicity in representing the few simple, though highly abstract, processes and ideas arising in the deductive trains of reasoning employed here. In fact the very abstract simplicity of the ideas of this work defeats language. Language can represent complex ideas more easily. The proposition "a whale is big" represents language at its best, giving terse expression to a complicated fact; while the true analysis of "one is a number" leads, in language, to an intolerable prolixity. Accordingly terseness is gained by using a symbolism especially designed to represent the ideas and processes of deduction which occur in this work.

(3) The adaptation of the rules of the symbolism to the processes of deduction aids the intuition in regions too abstract for the imagination readily to present to the mind the true relation between the ideas employed. For various collocations of symbols become familiar as representing important collocations of ideas; and in turn the possible relations—according to the rules of the symbolism—between these collocations of symbols become familiar, and these further collocations represent still more complicated relations between the abstract ideas. And thus the mind is finally led to construct trains of reasoning in regions of thought in which the imagination would be entirely unable to sustain itself without symbolic help. Ordinary language yields no such help. Its grammatical structure does not represent uniquely the relations between the ideas involved. Thus, "a whale is big" and "one is a number" both look alike, so that the eye gives no help to the imagination.

(4) The terseness of the symbolism enables a whole proposition to be represented to the eyesight as one whole, or at most in two or three parts divided where the natural breaks, represented in the symbolism, occur. This is a humble property, but is in fact very important in connection with the advantages enumerated under the heading (3).

(5) The attainment of the first-mentioned object of this work, namely the complete enumeration of all the ideas and steps in reasoning employed

in mathematics, necessitates both terseness and the presentation of each proposition with the maximum of formality in a form as characteristic of itself as possible.

Further light on the methods and symbolism of this book is thrown by a slight consideration of the limits to their useful employment:

(α) Most mathematical investigation is concerned not with the analysis of the complete process of reasoning, but with the presentation of such an abstract of the proof as is sufficient to convince a properly instructed mind. For such investigations the detailed presentation of the steps in reasoning is of course unnecessary, provided that the detail is carried far enough to guard against error. In this connection it may be remembered that the investigations of Weierstrass and others of the same school have shown that, even in the common topics of mathematical thought, much more detail is necessary than previous generations of mathematicians had anticipated.

(β) In proportion as the imagination works easily in any region of thought, symbolism (except for the express purpose of analysis) becomes only necessary as a convenient shorthand writing to register results obtained without its help. It is a subsidiary object of this work to show that, with the aid of symbolism, deductive reasoning can be extended to regions of thought not usually supposed amenable to mathematical treatment. And until the ideas of such branches of knowledge have become more familiar, the detailed type of reasoning, which is also required for the analysis of the steps, is appropriate to the investigation of the general truths concerning these subjects.

CHAPTER I

PRELIMINARY EXPLANATIONS OF IDEAS AND NOTATIONS

THE notation adopted in the present work is based upon that of Peano, and the following explanations are to some extent modelled on those which he prefixes to his *Formulario Mathematico.* His use of dots as brackets is adopted, and so are many of his symbols.

Variables. The idea of a variable, as it occurs in the present work, is more general than that which is explicitly used in ordinary mathematics. In ordinary mathematics, a variable generally stands for an undetermined number or quantity. In mathematical logic, any symbol whose meaning is not determinate is called a *variable*, and the various determinations of which its meaning is susceptible are called the *values* of the variable. The values may be any set of entities, propositions, functions, classes or relations, according to circumstances. If a statement is made about "Mr A and Mr B," "Mr A" and "Mr B" are variables whose values are confined to men. A variable may either have a conventionally-assigned range of values, or may (in the absence of any indication of the range of values) have as the range of its values all determinations which render the statement in which it occurs significant. Thus when a text-book of logic asserts that "A is A," without any indication as to what A may be, what is meant is that *any* statement of the form "A is A" is true. We may call a variable *restricted* when its values are confined to some only of those of which it is capable; otherwise, we shall call it *unrestricted.* Thus when an unrestricted variable occurs, it represents any object such that the statement concerned can be made significantly (*i.e.* either truly or falsely) concerning that object. For the purposes of logic, the unrestricted variable is more convenient than the restricted variable, and we shall always employ it. We shall find that the unrestricted variable is still subject to limitations imposed by the manner of its occurrence, *i.e.* things which can be said significantly concerning a proposition cannot be said significantly concerning a class or a relation, and so on. But the limitations to which the unrestricted variable is subject do not need to be explicitly indicated, since they are the limits of significance of the statement in which the variable occurs, and are therefore intrinsically determined by this statement. This will be more fully explained later*.

To sum up, the three salient facts connected with the use of the variable are: (1) that a variable is ambiguous in its denotation and accordingly undefined; (2) that a variable preserves a recognizable identity in various occurrences throughout the same context, so that many variables can occur together in the

* Cf. Chapter II of the Introduction.

same context each with its separate identity; and (3) that either the range of possible determinations of two variables may be the same, so that a possible determination of one variable is also a possible determination of the other, or the ranges of two variables may be different, so that, if a possible determination of one variable is given to the other, the resulting complete phrase is meaningless instead of becoming a complete unambiguous proposition (true or false) as would be the case if all variables in it had been given any *suitable* determinations.

The uses of various letters. Variables will be denoted by single letters, and so will certain constants; but a letter which has once been assigned to a constant by a definition must not afterwards be used to denote a variable. The small letters of the ordinary alphabet will all be used for variables, except p and s after ∗40, in which constant meanings are assigned to these two letters. The following capital letters will receive constant meanings: B, C, D, E, F, I and J. Among small Greek letters, we shall give constant meanings to ϵ, ι and (at a later stage) to η, θ and ω. Certain Greek capitals will from time to time be introduced for constants, but Greek capitals will not be used for variables. Of the remaining letters, p, q, r will be called *propositional letters*, and will stand for variable propositions (except that, from ∗40 onwards, p must not be used for a variable); f, g, ϕ, ψ, χ, θ and (until ∗33) F will be called *functional letters*, and will be used for variable functions.

The small Greek letters not already mentioned will be used for variables whose values are classes, and will be referred to simply as *Greek letters*. Ordinary capital letters not already mentioned will be used for variables whose values are relations, and will be referred to simply as *capital letters*. Ordinary small letters other than p, q, r, s, f, g will be used for variables whose values are not known to be functions, classes, or relations; these letters will be referred to simply as *small Latin letters*.

After the early part of the work, variable propositions and variable functions will hardly ever occur. We shall then have three main kinds of variables: variable classes, denoted by small Greek letters; variable relations, denoted by capitals; and variables not given as necessarily classes or relations, which will be denoted by small Latin letters.

In addition to this usage of small Greek letters for variable classes, capital letters for variable relations, small Latin letters for variables of type wholly undetermined by the context (these arise from the possibility of "systematic ambiguity," explained later in the explanations of the theory of types), the reader need only remember that all letters represent variables, unless they have been defined as constants in some previous place in the book. In general the structure of the context determines the scope of the variables contained in it; but the special indication of the nature of the variables employed, as here proposed, saves considerable labour of thought.

The fundamental functions of propositions. An aggregation of propositions, considered as wholes not necessarily unambiguously determined, into a single proposition more complex than its constituents, is a function *with propositions as arguments.* The general idea of such an aggregation of propositions, or of variables representing propositions, will not be employed in this work. But there are four special cases which are of fundamental importance, since all the aggregations of subordinate propositions into one complex proposition which occur in the sequel are formed out of them step by step.

They are (1) the Contradictory Function, (2) the Logical Sum, or Disjunctive Function, (3) the Logical Product, or Conjunctive Function, (4) the Implicative Function. These functions in the sense in which they are required in this work are not all independent; and if two of them are taken as primitive undefined ideas, the other two can be defined in terms of them. It is to some extent—though not entirely—arbitrary as to which functions are taken as primitive. Simplicity of primitive ideas and symmetry of treatment seem to be gained by taking the first two functions as primitive ideas.

The Contradictory Function with argument p, where p is any proposition, is the proposition which is the contradictory of p, that is, the proposition asserting that p is not true. This is denoted by $\sim p$. Thus $\sim p$ is the contradictory function with p as argument and means the negation of the proposition p. It will also be referred to as the proposition not-p. Thus $\sim p$ means not-p, which means the negation of p.

The Logical Sum is a propositional function with two arguments p and q, and is the proposition asserting p or q disjunctively, that is, asserting that at least one of the two p and q is true. This is denoted by $p \vee q$. Thus $p \vee q$ is the logical sum with p and q as arguments. It is also called the logical sum of p and q. Accordingly $p \vee q$ means that at least p or q is true, not excluding the case in which both are true.

The Logical Product is a propositional function with two arguments p and q, and is the proposition asserting p and q conjunctively, that is, asserting that both p and q are true. This is denoted by $p \,.\, q$, or—in order to make the dots act as brackets in a way to be explained immediately—by $p : q$, or by $p :. q$, or by $p :: q$. Thus $p \,.\, q$ is the logical product with p and q as arguments. It is also called the logical product of p and q. Accordingly $p \,.\, q$ means that both p and q are true. It is easily seen that this function can be defined in terms of the two preceding functions. For when p and q are both true it must be false that either $\sim p$ or $\sim q$ is true. Hence in this book $p \,.\, q$ is merely a shortened form of symbolism for

$$\sim(\sim p \vee \sim q).$$

If any further idea attaches to the proposition "both p and q are true," it is not required here.

The Implicative Function is a propositional function with two arguments p and q, and is the proposition that either not-p or q is true, that is, it is the proposition $\sim p \vee q$. Thus if p is true, $\sim p$ is false, and accordingly the only alternative left by the proposition $\sim p \vee q$ is that q is true. In other words if p and $\sim p \vee q$ are both true, then q is true. In this sense the proposition $\sim p \vee q$ will be quoted as stating that p implies q. The idea contained in this propositional function is so important that it requires a symbolism which with direct simplicity represents the proposition as connecting p and q without the intervention of $\sim p$. But "implies" as used here expresses nothing else than the connection between p and q also expressed by the disjunction "not-p or q." The symbol employed for "p implies q," *i.e.* for "$\sim p \vee q$," is "$p \supset q$." This symbol may also be read "if p, then q." The association of implication with the use of an apparent variable produces an extension called "formal implication." This is explained later: it is an idea derivative from "implication" as here defined. When it is necessary explicitly to discriminate "implication" from "formal implication," it is called "material implication." Thus "material implication" is simply "implication" as here defined. The process of inference, which in common usage is often confused with implication, is explained immediately.

These four functions of propositions are the fundamental constant (*i.e.* definite) propositional functions with *propositions as arguments*, and all other constant propositional functions with propositions as arguments, so far as they are required in the present work, are formed out of them by successive steps. No *variable* propositional functions of this kind occur in this work.

Equivalence. The simplest example of the formation of a more complex function of propositions by the use of these four fundamental forms is furnished by "equivalence." Two propositions p and q are said to be "equivalent" when p implies q and q implies p. This relation between p and q is denoted by "$p \equiv q$." Thus "$p \equiv q$" stands for "$(p \supset q) . (q \supset p)$." It is easily seen that two propositions are equivalent when, and only when, they are both true or are both false. Equivalence rises in the scale of importance when we come to "formal implication" and thus to "formal equivalence." It must not be supposed that two propositions which are equivalent are in any sense identical or even remotely concerned with the same topic. Thus "Newton was a man" and "the sun is hot" are equivalent as being both true, and "Newton was not a man" and "the sun is cold" are equivalent as being both false. But here we have anticipated deductions which follow later from our formal reasoning. Equivalence in its origin is merely mutual implication as stated above.

Truth-values. The "truth-value" of a proposition is *truth* if it is true, and *falsehood* if it is false*. It will be observed that the truth-values of

* This phrase is due to Frege.

$p \vee q$, $p . q$, $p \supset q$, $\sim p$, $p \equiv q$ depend only upon those of p and q, namely the truth-value of "$p \vee q$" is truth if the truth-value of either p or q is truth, and is falsehood otherwise; that of "$p . q$" is truth if that of both p and q is truth, and is falsehood otherwise; that of "$p \supset q$" is truth if either that of p is falsehood or that of q is truth; that of "$\sim p$" is the opposite of that of p; and that of "$p \equiv q$" is truth if p and q have the same truth-value, and is falsehood otherwise. Now the only ways in which propositions will occur in the present work are ways derived from the above by combinations and repetitions. Hence it is easy to see (though it cannot be formally proved except in each particular case) that if a proposition p occurs in any proposition $f(p)$ which we shall ever have occasion to deal with, the truth-value of $f(p)$ will depend, not upon the particular proposition p, but only upon its truth-value; *i.e.* if $p \equiv q$, we shall have $f(p) \equiv f(q)$. Thus whenever two propositions are known to be equivalent, either may be substituted for the other in any formula with which we shall have occasion to deal.

We may call a function $f(p)$ a "truth-function" when its argument p is a proposition, and the truth-value of $f(p)$ depends only upon the truth-value of p. Such functions are by no means the only common functions of propositions. For example, "A believes p" is a function of p which will vary its truth-value for different arguments having the same truth-value: A may believe one true proposition without believing another, and may believe one false proposition without believing another. Such functions are not excluded from our consideration, and are included in the scope of any general propositions we may make about functions; but the particular functions of propositions which we shall have occasion to construct or to consider explicitly are all truth-functions. This fact is closely connected with a characteristic of mathematics, namely, that mathematics is always concerned with extensions rather than intensions. The connection, if not now obvious, will become more so when we have considered the theory of classes and relations.

Assertion-sign. The sign "$\vdash$," called the "assertion-sign," means that what follows is asserted. It is required for distinguishing a complete proposition, which we assert, from any subordinate propositions contained in it but not asserted. In ordinary written language a sentence contained between full stops denotes an asserted proposition, and if it is false the book is in error. The sign "$\vdash$" prefixed to a proposition serves this same purpose in our symbolism. For example, if "$\vdash (p \supset p)$" occurs, it is to be taken as a complete assertion convicting the authors of error unless the proposition "$p \supset p$" is true (as it is). Also a proposition stated in symbols without this sign "$\vdash$" prefixed is not asserted, and is merely put forward for consideration, or as a subordinate part of an asserted proposition.

Inference. The process of inference is as follows: a proposition "p" is asserted, and a proposition "p implies q" is asserted, and then as a sequel the proposition "q" is asserted. The trust in inference is the belief that if the two former assertions are not in error, the final assertion is not in error. Accordingly whenever, in symbols, where p and q have of course special determinations,

$$\text{"}\vdash p\text{" and "}\vdash (p \supset q)\text{"}$$

have occurred, then "$\vdash q$" will occur if it is desired to put it on record. The process of the inference cannot be reduced to symbols. Its sole record is the occurrence of "$\vdash q$." It is of course convenient, even at the risk of repetition, to write "$\vdash p$" and "$\vdash (p \supset q)$" in close juxtaposition before proceeding to "$\vdash q$" as the result of an inference. When this is to be done, for the sake of drawing attention to the inference which is being made, we shall write instead

$$\text{"}\vdash p \supset \vdash q\text{,"}$$

which is to be considered as a mere abbreviation of the threefold statement

$$\text{"}\vdash p\text{" and "}\vdash (p \supset q)\text{" and "}\vdash q\text{."}$$

Thus "$\vdash p \supset \vdash q$" may be read "$p$, therefore q," being in fact the same abbreviation, essentially, as this is; for "p, therefore q" does not explicitly state, what is part of its meaning, that p implies q. An inference is the dropping of a true premiss; it is the dissolution of an implication.

The use of dots. Dots on the line of the symbols have two uses, one to bracket off propositions, the other to indicate the logical product of two propositions. Dots immediately preceded or followed by "$\vee$" or "$\supset$" or "$\equiv$" or "$\vdash$," or by "(x)," "(x, y)," "(x, y, z)"... or "$(\exists x)$," "$(\exists x, y)$," "$(\exists x, y, z)$"... or "$[(\iota x)(\phi x)]$" or "$[R\text{'}y]$" or analogous expressions, serve to bracket off a proposition; dots occurring otherwise serve to mark a logical product. The general principle is that a larger number of dots indicates an outside bracket, a smaller number indicates an inside bracket. The exact rule as to the scope of the bracket indicated by dots is arrived at by dividing the occurrences of dots into three groups which we will name I, II, and III. Group I consists of dots adjoining a sign of implication ($\supset$) or of equivalence ($\equiv$) or of disjunction ($\vee$) or of equality by definition ($=$ Df). Group II consists of dots following brackets indicative of an apparent variable, such as (x) or (x, y) or $(\exists x)$ or $(\exists x, y)$ or $[(\iota x)(\phi x)]$ or analogous expressions*. Group III consists of dots which stand between propositions in order to indicate a logical product. Group I is of greater force than Group II, and Group II than Group III. The scope of the bracket indicated by any collection of dots extends backwards or forwards beyond any *smaller* number of dots, or any *equal* number from a group of less force, until we reach either the end of the asserted proposition or a *greater* number of dots or an *equal* number belonging to a group of equal or superior force. Dots indicating a logical product have a scope which works both backwards and forwards; other dots only work away from the

* The meaning of these expressions will be explained later, and examples of the use of dots in connection with them will be given on pp. 16, 17.

adjacent sign of disjunction, implication, or equivalence, or forward from the adjacent symbol of one of the other kinds enumerated in Group II.

Some examples will serve to illustrate the use of dots.

"$p \vee q . \supset . q \vee p$" means the proposition "'p or q' implies 'q or p.'" When we *assert* this proposition, instead of merely considering it, we write

"$\vdash : p \vee q . \supset . q \vee p$,"

where the two dots after the assertion-sign show that what is asserted is the whole of what follows the assertion-sign, since there are not as many as two dots anywhere else. If we had written "$p : \vee : q . \supset . q \vee p$," that would mean the proposition "either p is true, or q implies 'q or p.'" If we wished to assert this, we should have to put three dots after the assertion-sign. If we had written "$p \vee q . \supset . q : \vee : p$," that would mean the proposition "either 'p or q' implies q, or p is true." The forms "$p . \vee . q . \supset . q \vee p$" and "$p \vee q . \supset . q . \vee . p$" have no meaning.

"$p \supset q . \supset : q \supset r . \supset . p \supset r$" will mean "if p implies q, then if q implies r, p implies r." If we wish to assert this (which is true) we write

"$\vdash :. p \supset q . \supset : q \supset r . \supset . p \supset r$."

Again "$p \supset q . \supset . q \supset r : \supset . p \supset r$" will mean "if '$p$ implies q' implies 'q implies r,' then p implies r." This is in general untrue. (Observe that "$p \supset q$" is sometimes most conveniently read as "p implies q," and sometimes as "if p, then q.") "$p \supset q . q \supset r . \supset . p \supset r$" will mean "if p implies q, and q implies r, then p implies r." In this formula, the first dot indicates a logical product; hence the scope of the second dot extends backwards to the beginning of the proposition. "$p \supset q : q \supset r . \supset . p \supset r$" will mean "$p$ implies q; and if q implies r, then p implies r." (This is not true in general.) Here the two dots indicate a logical product; since two dots do not occur anywhere else, the scope of these two dots extends backwards to the beginning of the proposition, and forwards to the end.

"$p \vee q . \supset :. p . \vee . q \supset r : \supset . p \vee r$" will mean "if either p or q is true, then if either p or 'q implies r' is true, it follows that either p or r is true." If this is to be asserted, we must put four dots after the assertion-sign, thus:

"$\vdash :: p \vee q . \supset :. p . \vee . q \supset r : \supset . p \vee r$."

(This proposition is proved in the body of the work; it is ∗2·75.) If we wish to assert (what is equivalent to the above) the proposition: "if either p or q is true, and either p or 'q implies r' is true, then either p or r is true," we write

"$\vdash :. p \vee q : p . \vee . q \supset r : \supset . p \vee r$."

Here the first pair of dots indicates a logical product, while the second pair does not. Thus the scope of the second pair of dots passes over the first pair, and back until we reach the three dots after the assertion-sign.

Other uses of dots follow the same principles, and will be explained as they are introduced. In reading a proposition, the dots should be noticed first, as they show its structure. In a proposition containing several signs of implication or equivalence, the one with the greatest number of dots before or after it is the *principal* one: everything that goes before this one is stated by the proposition to imply or be equivalent to everything that comes after it.

Definitions. A definition is a declaration that a certain newly-introduced symbol or combination of symbols is to mean the same as a certain other combination of symbols of which the meaning is already known. Or, if the defining combination of symbols is one which only acquires meaning when combined in a suitable manner with other symbols*, what is meant is that any combination of symbols in which the newly-defined symbol or combination of symbols occurs is to have that meaning (if any) which results from substituting the defining combination of symbols for the newly-defined symbol or combination of symbols wherever the latter occurs. We will give the names of *definiendum* and *definiens* respectively to what is defined and to that which it is defined as meaning. We express a definition by putting the *definiendum* to the left and the *definiens* to the right, with the sign "=" between, and the letters "Df" to the right of the *definiens*. It is to be understood that the sign "=" and the letters "Df" are to be regarded as together forming one symbol. The sign "=" without the letters "Df" will have a different meaning, to be explained shortly.

An example of a definition is

$$p \supset q . = . \sim p \vee q \quad \text{Df.}$$

It is to be observed that a definition is, strictly speaking, no part of the subject in which it occurs. For a definition is concerned wholly with the symbols, not with what they symbolise. Moreover it is not true or false, being the expression of a volition, not of a proposition. (For this reason, definitions are not preceded by the assertion-sign.) Theoretically, it is unnecessary ever to give a definition: we might always use the *definiens* instead, and thus wholly dispense with the *definiendum*. Thus although we employ definitions and do not define "definition," yet "definition" does not appear among our primitive ideas, because the definitions are no part of our subject, but are, strictly speaking, mere typographical conveniences. Practically, of course, if we introduced no definitions, our formulae would very soon become so lengthy as to be unmanageable; but theoretically, all definitions are superfluous.

In spite of the fact that definitions are theoretically superfluous, it is nevertheless true that they often convey more important information than is contained in the propositions in which they are used. This arises from two causes. First, a definition usually implies that the *definiens* is worthy of careful consideration. Hence the collection of definitions embodies our choice

* This case will be fully considered in Chapter III of the Introduction. It need not further concern us at present.

of subjects and our judgment as to what is most important. Secondly, when what is defined is (as often occurs) something already familiar, such as cardinal or ordinal numbers, the definition contains an analysis of a common idea, and may therefore express a notable advance. Cantor's definition of the continuum illustrates this: his definition amounts to the statement that what he is defining is the object which has the properties commonly associated with the word "continuum," though what precisely constitutes these properties had not before been known. In such cases, a definition is a "making definite": it gives definiteness to an idea which had previously been more or less vague.

For these reasons, it will be found, in what follows, that the definitions are what is most important, and what most deserves the reader's prolonged attention.

Some important remarks must be made respecting the variables occurring in the *definiens* and the *definiendum*. But these will be deferred till the notion of an "apparent variable" has been introduced, when the subject can be considered as a whole.

Summary of preceding statements. There are, in the above, three primitive ideas which are not "defined" but only descriptively explained. Their primitiveness is only relative to our exposition of logical connection and is not absolute; though of course such an exposition gains in importance according to the simplicity of its primitive ideas. These ideas are symbolised by "$\sim p$" and "$p \vee q$," and by "$\vdash$" prefixed to a proposition.

Three definitions have been introduced:

$$p \,.\, q \,. = . \sim(\sim p \vee \sim q) \quad \text{Df},$$
$$p \supset q \,. = . \sim p \vee q \quad \text{Df},$$
$$p \equiv q \,. = . \, p \supset q \,.\, q \supset p \quad \text{Df}.$$

Primitive propositions. Some propositions must be assumed without proof, since all inference proceeds from propositions previously asserted. These, as far as they concern the functions of propositions mentioned above, will be found stated in ∗1, where the formal and continuous exposition of the subject commences. Such propositions will be called "primitive propositions." These, like the primitive ideas, are to some extent a matter of arbitrary choice; though, as in the previous case, a logical system grows in importance according as the primitive propositions are few and simple. It will be found that owing to the weakness of the imagination in dealing with simple abstract ideas no very great stress can be laid upon their obviousness. They are obvious to the instructed mind, but then so are many propositions which cannot be quite true, as being disproved by their contradictory consequences. The proof of a logical system is its adequacy and its coherence. That is: (1) the system must embrace among its deductions all those propositions which we believe to be true and capable of deduction from logical premisses alone, though possibly they may require some slight limitation in the form of an increased stringency of enunciation; and (2) the system must lead to no contradictions, namely in pursuing our inferences we must never be led to assert both p and not-p, *i.e.* both "$\vdash . p$" and "$\vdash . \sim p$" cannot legitimately appear.

The following are the primitive propositions employed in the calculus of propositions. The letters "Pp" stand for "primitive proposition."

(1) Anything implied by a true premiss is true Pp.

This is the rule which justifies inference.

(2) $\vdash : p \vee p . \supset . p$ Pp,

i.e. if p or p is true, then p is true.

(3) $\vdash : q . \supset . p \vee q$ Pp,

i.e. if q is true, then p or q is true.

(4) $\vdash : p \vee q . \supset . q \vee p$ Pp,

i.e. if p or q is true, then q or p is true.

(5) $\vdash : p \vee (q \vee r) . \supset . q \vee (p \vee r)$ Pp,

i.e. if either p is true or "q or r" is true, then either q is true or "p or r" is true.

(6) $\vdash :. q \supset r . \supset : p \vee q . \supset . p \vee r$ Pp,

i.e. if q implies r, then "p or q" implies "p or r."

(7) Besides the above primitive propositions, we require a primitive proposition called "the axiom of identification of real variables." When we have separately asserted two different functions of x, where x is undetermined, it is often important to know whether we can identify the x in one assertion with the x in the other. This will be the case—so our axiom allows us to infer—if both assertions present x as the argument to some one function, that is to say, if ϕx is a constituent in both assertions (whatever propositional function ϕ may be), or, more generally, if $\phi (x, y, z, \ldots)$ is a constituent in one assertion, and $\phi (x, u, v, \ldots)$ is a constituent in the other. This axiom introduces notions which have not yet been explained; for a fuller account, see the remarks accompanying ✱3·03, ✱1·7, ✱1·71, and ✱1·72 (which is the statement of this axiom) in the body of the work, as well as the explanation of propositional functions and ambiguous assertion to be given shortly.

Some simple propositions. In addition to the primitive propositions we have already mentioned, the following are among the most important of the elementary properties of propositions appearing among the deductions.

The law of excluded middle:

$$\vdash . p \vee \sim p.$$

This is ✱2·11 below. We shall indicate in brackets the numbers given to the following propositions in the body of the work.

The law of contradiction (✱3·24):

$$\vdash . \sim (p . \sim p).$$

The law of double negation (∗4·13):

$$\vdash . p \equiv \sim(\sim p).$$

The principle of *transposition*, *i.e.* "if p implies q, then not-q implies not-p," and vice versa: this principle has various forms, namely

$$(∗4·1)\quad \vdash : p \supset q . \equiv . \sim q \supset \sim p,$$

$$(∗4·11)\quad \vdash : p \equiv q . \equiv . \sim p \equiv \sim q,$$

$$(∗4·14)\quad \vdash :. p . q . \supset . r : \equiv : p . \sim r . \supset . \sim q,$$

as well as others which are variants of these.

The law of tautology, in the two forms:

$$(∗4·24)\quad \vdash : p . \equiv . p . p,$$

$$(∗4·25)\quad \vdash : p . \equiv . p \vee p,$$

i.e. "p is true" is equivalent to "p is true and p is true," as well as to "p is true or p is true." From a formal point of view, it is through the law of tautology and its consequences that the algebra of logic is chiefly distinguished from ordinary algebra.

The law of absorption:

$$(∗4·71)\quad \vdash :. p \supset q . \equiv : p . \equiv . p . q,$$

i.e. "p implies q" is equivalent to "p is equivalent to $p . q$." This is called the law of absorption because it shows that the factor q in the product is absorbed by the factor p, if p implies q. This principle enables us to replace an implication ($p \supset q$) by an equivalence ($p . \equiv . p . q$) whenever it is convenient to do so.

An analogous and very important principle is the following:

$$(∗4·73)\quad \vdash :. q . \supset : p . \equiv . p . q.$$

Logical addition and multiplication of propositions obey the associative and commutative laws, and the distributive law in two forms, namely

$$(∗4·4)\quad \vdash :. p . q \vee r . \equiv : p . q . \vee . p . r,$$

$$(∗4·41)\quad \vdash :. p . \vee . q . r : \equiv : p \vee q . p \vee r.$$

The second of these distinguishes the relations of logical addition and multiplication from those of arithmetical addition and multiplication.

Propositional functions. Let ϕx be a statement containing a variable x and such that it becomes a proposition when x is given any fixed determined meaning. Then ϕx is called a "propositional function"; it is not a proposition, since owing to the ambiguity of x it really makes no assertion at all. Thus "x is hurt" really makes no assertion at all, till we have settled who x is. Yet owing to the individuality retained by the ambiguous variable x, it is an ambiguous example from the collection of propositions arrived at by giving all possible determinations to x in "x is hurt" which yield a proposition, true or false. Also if "x is hurt" and "y is hurt" occur *in the same context*, where y is another variable, then according to the determinations given to x and y, they can be settled to be (possibly) the same proposition or (possibly) different propositions. But apart from some determination given to x and y, they retain in that context their ambiguous differentiation. Thus "x is hurt" is an ambiguous "value" of a propositional function. When we wish to speak of the propositional function corresponding to "x is hurt," we shall write "$\hat{x}$ is hurt." Thus "$\hat{x}$ is hurt" is the propositional function and "x is hurt" is an ambiguous value of that function. Accordingly though "x is hurt" and "y is hurt" *occurring in the same context* can be distinguished, "$\hat{x}$ is hurt" and "$\hat{y}$ is hurt" convey no distinction of meaning at all. More generally, ϕx is an ambiguous value of the propositional function $\phi\hat{x}$, and when a definite signification a is substituted for x, ϕa is an unambiguous value of $\phi\hat{x}$.

Propositional functions are the fundamental kind from which the more usual kinds of function, such as "$\sin x$" or "$\log x$" or "the father of x," are derived. These derivative functions are considered later, and are called "descriptive functions." The functions of propositions considered above are a particular case of propositional functions.

The range of values and total variation. Thus corresponding to any propositional function $\phi\hat{x}$, there is a range, or collection, of values, consisting of all the propositions (true or false) which can be obtained by giving every possible determination to x in ϕx. A value of x for which ϕx is true will be said to "satisfy" $\phi\hat{x}$. Now in respect to the truth or falsehood of propositions of this range three important cases must be noted and symbolised. These cases are given by three propositions of which one at least must be true. Either (1) all propositions of the range are true, or (2) some propositions of the range are true, or (3) no proposition of the range is true. The statement (1) is symbolised by "$(x) . \phi x$," and (2) is symbolised by "$(\exists x) . \phi x$." No definition is given of these two symbols, which accordingly embody two new primitive ideas in our system. The symbol "$(x) . \phi x$" may be read "ϕx always," or "ϕx is always true," or "ϕx is true for all possible values of x." The symbol "$(\exists x) . \phi x$" may be read "there exists an x for which ϕx is true," or "there exists an x satisfying $\phi\hat{x}$," and thus conforms to the natural form of the expression of thought.

Proposition (3) can be expressed in terms of the fundamental ideas now on hand. In order to do this, note that "$\sim \phi x$" stands for the contradictory of ϕx. Accordingly $\sim \phi\hat{x}$ is another propositional function such that each value of $\phi\hat{x}$ contradicts a value of $\sim \phi\hat{x}$, and vice versa. Hence "$(x) . \sim \phi x$" symbolises the proposition that every value of $\phi\hat{x}$ is untrue. This is number (3) as stated above.

It is an obvious error, though one easy to commit, to assume that cases (1) and (3) are each other's contradictories. The symbolism exposes this fallacy at once, for (1) is $(x) . \phi x$, and (3) is $(x) . \sim \phi x$, while the contradictory of (1) is $\sim \{(x) . \phi x\}$. For the sake of brevity of symbolism a definition is made, namely

$$\sim (x) . \phi x . = . \sim \{(x) . \phi x\} \quad \text{Df.}$$

Definitions of which the object is to gain some trivial advantage in brevity by a slight adjustment of symbols will be said to be of "merely symbolic import," in contradistinction to those definitions which invite consideration of an important idea.

The proposition $(x) . \phi x$ is called the "total variation" of the function $\phi \hat{x}$.

For reasons which will be explained in Chapter II, we do not take negation as a primitive idea when propositions of the forms $(x) . \phi x$ and $(\exists x) . \phi x$ are concerned, but we *define* the negation of $(x) . \phi x$, *i.e.* of "ϕx is always true," as being "ϕx is sometimes false," *i.e.* "$(\exists x) . \sim \phi x$," and similarly we *define* the negation of $(\exists x) . \phi x$ as being $(x) . \sim \phi x$. Thus we put

$$\sim \{(x) . \phi x\} . = . (\exists x) . \sim \phi x \quad \text{Df},$$

$$\sim \{(\exists x) . \phi x\} . = . (x) . \sim \phi x \quad \text{Df}.$$

In like manner we define a disjunction in which one of the propositions is of the form "$(x) . \phi x$" or "$(\exists x) . \phi x$" in terms of a disjunction of propositions not of this form, putting

$$(x) . \phi x . \vee . p : = . (x) . \phi x \vee p \quad \text{Df},$$

i.e. "either ϕx is always true, or p is true" is to mean "'ϕx or p' is always true," with similar definitions in other cases. This subject is resumed in Chapter II, and in *9 in the body of the work.

Apparent variables. The symbol "$(x) . \phi x$" denotes one definite proposition, and there is no distinction in meaning between "$(x) . \phi x$" and "$(y) . \phi y$" when they occur in the same context. Thus the "x" in "$(x) . \phi x$" is not an ambiguous constituent of any expression in which "$(x) . \phi x$" occurs; and such an expression does not cease to convey a determinate meaning by reason of the ambiguity of the x in the "ϕx." The symbol "$(x) . \phi x$" has some analogy to the symbol

$$\text{"}\int_a^b \phi(x)\, dx\text{"}$$

for definite integration, since in neither case is the expression a function of x.

The range of x in "$(x) . \phi x$" or "$(\exists x) . \phi x$" extends over the complete field of the values of x for which "ϕx" has meaning, and accordingly the meaning of "$(x) . \phi x$" or "$(\exists x) . \phi x$" involves the supposition that such a field is determinate. The x which occurs in "$(x) . \phi x$" or "$(\exists x) . \phi x$" is called (following Peano) an "apparent variable." It follows from the meaning of "$(\exists x) . \phi x$" that the x in this expression is also an apparent variable. A proposition in which x occurs as an apparent variable is not a function of x. Thus *e.g.* "$(x) . x = x$" will mean "everything is equal to itself." This is an absolute constant, not a function of a variable x. This is why the x is called an *apparent* variable in such cases.

Besides the "*range*" of x in "$(x) . \phi x$" or "$(\exists x) . \phi x$," which is the field of the values that x may have, we shall speak of the "*scope*" of x, meaning the function of which all values or some value are being affirmed. If we are asserting all values (or some value) of "ϕx," "ϕx" is the scope of x; if we are asserting all values (or some value) of "$\phi x \supset p$," "$\phi x \supset p$" is the scope of x; if we are asserting all values (or some value) of "$\phi x \supset \psi x$," "$\phi x \supset \psi x$" will be the scope of x, and so on. The scope of x is indicated by the number of dots after the "(x)" or "$(\exists x)$"; that is to say, the scope extends forwards until we reach an equal number of dots not indicating a logical product, or a greater number indicating a logical product, or the end of the asserted proposition in which the "(x)" or "$(\exists x)$" occurs, whichever of these happens first*. Thus *e.g.*

$$\text{"}(x) : \phi x . \supset . \psi x\text{"}$$

will mean "ϕx always implies ψx," but

$$\text{"}(x) . \phi x . \supset . \psi x\text{"}$$

will mean "if ϕx is always true, then ψx is true for the argument x."

Note that in the proposition

$$(x) . \phi x . \supset . \psi x$$

the two x's have no connection with each other. Since only one dot follows the x in brackets, the scope of the first x is limited to the "ϕx" immediately following the x in brackets. It usually conduces to clearness to write

$$(x) . \phi x . \supset . \psi y$$

rather than $(x) . \phi x . \supset . \psi x$,

since the use of different letters emphasises the absence of connection between the two variables; but there is no logical necessity to use different letters, and it is *sometimes* convenient to use the same letter.

Ambiguous assertion and the real variable. Any value "ϕx" of the function $\phi \hat{x}$ can be asserted. Such an assertion of an ambiguous member of the values of $\phi \hat{x}$ is symbolised by

$$\text{"}\vdash . \phi x.\text{"}$$

Ambiguous assertion of this kind is a primitive idea, which cannot be defined in terms of the assertion of propositions. This primitive idea is the one which embodies the use of the variable. Apart from ambiguous assertion, the consideration of "ϕx," which is an ambiguous member of the values of $\phi \hat{x}$, would be of little consequence. When we are considering or asserting "ϕx," the variable x is called a "real variable." Take, for example, the law of excluded middle in the form which it has in traditional formal logic:

"a is either b or not b."

Here a and b are real variables: as they vary, different propositions are expressed, though all of them are true. While a and b are undetermined, as in the above enunciation, no one definite proposition is asserted, but what is asserted is *any* value of the propositional function in question. This can only

* This agrees with the rules for the occurrences of dots of the type of Group II as explained above, pp. 9 and 10.

be legitimately asserted if, whatever value may be chosen, that value is true, *i.e.* if all the values are true. Thus the above form of the law of excluded middle is equivalent to

"$(a, b) . a$ is either b or not b,"

i.e. to "it is always true that a is either b or not b." But these two, though equivalent, are not identical, and we shall find it necessary to keep them distinguished.

When we assert something containing a real variable, as in *e.g.*

$$\text{"}\vdash . x = x,\text{"}$$

we are asserting *any* value of a propositional function. When we assert something containing an apparent variable, as in

$$\text{"}\vdash . (x) . x = x\text{"}$$

or

$$\text{"}\vdash . (\exists x) . x = x,\text{"}$$

we are asserting, in the first case *all* values, in the second case *some* value (undetermined), of the propositional function in question. It is plain that we can only legitimately assert "*any* value" if *all* values are true; for otherwise, since the value of the variable remains to be determined, it might be so determined as to give a false proposition. Thus in the above instance, since we have

$$\vdash . x = x$$

we may infer

$$\vdash . (x) . x = x.$$

And generally, given an assertion containing a real variable x, we may transform the real variable into an apparent one by placing the x in brackets at the beginning, followed by as many dots as there are after the assertion-sign.

When we assert something containing a real variable, we cannot strictly be said to be asserting a *proposition*, for we only obtain a definite proposition by assigning a value to the variable, and then our assertion only applies to one definite case, so that it has not at all the same force as before. When what we assert contains a real variable, we are asserting a wholly undetermined one of all the propositions that result from giving various values to the variable. It will be convenient to speak of such assertions as *asserting a propositional function*. The ordinary formulae of mathematics contain such assertions; for example

$$\text{"}\sin^2 x + \cos^2 x = 1\text{"}$$

does not assert this or that particular case of the formula, nor does it assert that the formula holds for *all* possible values of x, though it is equivalent to this latter assertion; it simply asserts that the formula holds, leaving x wholly undetermined; and it is able to do this legitimately, because, however x may be determined, a true proposition results.

Although an assertion containing a real variable does not, in strictness, assert a proposition, yet it will be spoken of as asserting a proposition except when the nature of the ambiguous assertion involved is under discussion.

Definition and real variables. When the *definiens* contains one or more real variables, the *definiendum* must also contain them. For in this case we have a function of the real variables, and the *definiendum* must have the same meaning as the *definiens* for all values of these variables, which requires that the symbol which is the *definiendum* should contain the letters representing the real variables. This rule is not always observed by mathematicians, and its infringement has sometimes caused important confusions of thought, notably in geometry and the philosophy of space.

In the definitions given above of "$p \,.\, q$" and "$p \supset q$" and "$p \equiv q$," p and q are real variables, and therefore appear on both sides of the definition. In the definition of "$\sim\{(x) \,.\, \phi x\}$" only the function considered, namely $\phi\hat{z}$, is a real variable; thus so far as concerns the rule in question, x need not appear on the left. But when a real variable is a function, it is necessary to indicate how the argument is to be supplied, and therefore there are objections to omitting an apparent variable where (as in the case before us) this is the argument to the function which is the real variable. This appears more plainly if, instead of a general function $\phi\hat{x}$, we take some particular function, say "$\hat{x} = a$," and consider the definition of $\sim\{(x) \,.\, x = a\}$. Our definition gives

$$\sim\{(x) \,.\, x = a\} \,.\!=.\, (\exists x) \,.\, \sim(x = a) \quad \text{Df.}$$

But if we had adopted a notation in which the ambiguous value "$x = a$," containing the apparent variable x, did not occur in the *definiendum*, we should have had to construct a notation employing the function itself, namely "$\hat{x} = a$." This does not involve an apparent variable, but would be clumsy in practice. In fact we have found it convenient and possible—except in the explanatory portions—to keep the explicit use of symbols of the type "$\phi\hat{x}$," either as constants [*e.g.* $\hat{x} = a$] or as real variables, almost entirely out of this work.

Propositions connecting real and apparent variables. The most important propositions connecting real and apparent variables are the following:

(1) "When a propositional function can be asserted, so can the proposition that all values of the function are true." More briefly, if less exactly, "what holds of any, however chosen, holds of all." This translates itself into the rule that when a real variable occurs in an assertion, we may turn it into an apparent variable by putting the letter representing it in brackets immediately after the assertion-sign.

(2) "What holds of all, holds of any," *i.e.*

$$\vdash : (x) \,.\, \phi x \,.\supset.\, \phi y.$$

This states "if ϕx is always true, then ϕy is true."

(3) "If ϕy is true, then ϕx is sometimes true," *i.e.*

$$\vdash : \phi y \,.\supset.\, (\exists x) \,.\, \phi x.$$

An asserted proposition of the form "$(\exists x) . \phi x$" expresses an "existence-theorem," namely "there exists an x for which ϕx is true." The above proposition gives what is in practice the only way of proving existence-theorems: we always have to find some particular y for which ϕy holds, and thence to infer "$(\exists x) . \phi x$." If we were to assume what is called the multiplicative axiom, or the equivalent axiom enunciated by Zermelo, that would, in an important class of cases, give an existence-theorem where no particular instance of its truth can be found.

In virtue of "$\vdash : (x) . \phi x . \supset . \phi y$" and "$\vdash : \phi y . \supset . (\exists x) . \phi x$," we have "$\vdash : (x) . \phi x . \supset . (\exists x) . \phi x$," *i.e.* "what is always true is sometimes true." This would not be the case if nothing existed; thus our assumptions contain the assumption that there is something. This is involved in the principle that what holds of all, holds of any; for this would not be true if there were no "any."

(4) "If ϕx is always true, and ψx is always true, then '$\phi x . \psi x$' is always true," *i.e.*

$$\vdash :. (x) . \phi x : (x) . \psi x : \supset . (x) . \phi x . \psi x.$$

(This requires that ϕ and ψ should be functions which take arguments of the same *type*. We shall explain this requirement at a later stage.) The converse also holds; *i.e.* we have

$$\vdash :. (x) . \phi x . \psi x . \supset : (x) . \phi x : (x) . \psi x.$$

It is to some extent optional which of the propositions connecting real and apparent variables are taken as primitive propositions. The primitive propositions assumed, on this subject, in the body of the work (*9), are the following:

(1) $\vdash : \phi x . \supset . (\exists z) . \phi z.$

(2) $\vdash : \phi x \vee \phi y . \supset . (\exists z) . \phi z,$

i.e. if either ϕx is true, or ϕy is true, then $(\exists z) . \phi z$ is true. (On the necessity for this primitive proposition, see remarks on *9·11 in the body of the work.)

(3) If we can assert ϕy, where y is a real variable, then we can assert $(x) . \phi x$; *i.e.* what holds of any, however chosen, holds of all.

Formal implication and formal equivalence. When an implication, say $\phi x . \supset . \psi x$, is said to hold always, *i.e.* when $(x) : \phi x . \supset . \psi x$, we shall say that ϕx *formally implies* ψx; and propositions of the form "$(x) : \phi x . \supset . \psi x$" will be said to state *formal implications*. In the usual instances of implication, such as "'Socrates is a man' implies 'Socrates is mortal,'" we have a proposition of the form "$\phi x . \supset . \psi x$" in a case in which "$(x) : \phi x . \supset . \psi x$" is true. In such a case, we feel the implication as a particular case of a formal implication. Thus it has come about that implications which are not particular cases of formal implications have not been regarded as implications at all. There is also a practical ground for the neglect of such implications, for, speaking generally, they can only be *known* when it is already known either that their hypothesis is false or that their conclusion is true; and in neither of these cases do they serve to make us know the conclusion, since in the first case the conclusion need not be true, and in the second it is known already. Thus such implications do not serve the purpose for which implications are chiefly useful, namely that of making us know, by deduction, conclusions of which we were previously ignorant. *Formal* implications, on the contrary, do serve this purpose, owing to the psychological fact that we often know "$(x):\phi x.\supset.\psi x$" and ϕy, in cases where ψy (which follows from these premisses) cannot easily be known directly.

These reasons, though they do not warrant the complete neglect of implications that are not instances of formal implications, are reasons which make formal implication very important. A formal implication states that, for all possible values of x, if the hypothesis ϕx is true, the conclusion ψx is true. Since "$\phi x.\supset.\psi x$" will always be true when ϕx is false, it is only the values of x that make ϕx true that are *important* in a formal implication; what is effectively stated is that, for all these values, ψx is true. Thus propositions of the form "all α is β," "no α is β" state formal implications, since the first (as appears by what has just been said) states

$$(x): x \text{ is an } \alpha.\supset.x \text{ is a } \beta,$$

while the second states

$$(x): x \text{ is an } \alpha.\supset.x \text{ is not a } \beta.$$

And any formal implication "$(x):\phi x.\supset.\psi x$" may be interpreted as: "All values of x which satisfy* ϕx satisfy ψx," while the formal implication "$(x):\phi x.\supset.\sim\psi x$" may be interpreted as: "No values of x which satisfy ϕx satisfy ψx."

We have similarly for "some α is β" the formula

$$(\exists x).x \text{ is an } \alpha.x \text{ is a } \beta,$$

and for "some α is not β" the formula

$$(\exists x).x \text{ is an } \alpha.x \text{ is not a } \beta.$$

Two functions ϕx, ψx are called *formally equivalent* when each always implies the other, *i.e.* when

$$(x):\phi x.\equiv.\psi x,$$

and a proposition of this form is called a *formal equivalence*. In virtue of what was said about truth-values, if ϕx and ψx are formally equivalent, either may replace the other in any truth-function. Hence for all the purposes of mathematics or of the present work, $\phi\hat{z}$ may replace $\psi\hat{z}$ or vice versa in any proposition with which we shall be concerned. Now to say that ϕx and ψx are formally equivalent is the same thing as to say that $\phi\hat{z}$ and $\psi\hat{z}$ have the same *extension*, *i.e.* that any value of x which satisfies either satisfies the other.

* A value of x is said to *satisfy* ϕx or $\phi\hat{x}$ when ϕx is true for that value of x.

Thus whenever a constant function occurs in our work, the truth-value of the proposition in which it occurs depends only upon the extension of the function. A proposition containing a function $\phi\hat{z}$ and having this property (*i.e.* that its truth-value depends only upon the extension of $\phi\hat{z}$) will be called an *extensional* function of $\phi\hat{z}$. Thus the functions of functions with which we shall be specially concerned will all be extensional functions of functions.

What has just been said explains the connection (noted above) between the fact that the functions of propositions with which mathematics is specially concerned are all truth-functions and the fact that mathematics is concerned with extensions rather than intensions.

Convenient abbreviation. The following definitions give alternative and often more convenient notations:

$$\phi x . \supset_x . \psi x : = : (x) : \phi x . \supset . \psi x \quad \text{Df},$$

$$\phi x . \equiv_x . \psi x : = : (x) : \phi x . \equiv . \psi x \quad \text{Df}.$$

This notation "$\phi x . \supset_x . \psi x$" is due to Peano, who, however, has no notation for the general idea "$(x) . \phi x$." It may be noticed as an exercise in the use of dots as brackets that we might have written

$$\phi x \supset_x \psi x . = . (x) . \phi x \supset \psi x \quad \text{Df},$$

$$\phi x \equiv_x \psi x . = . (x) . \phi x \equiv \psi x \quad \text{Df}.$$

In practice however, when $\phi\hat{x}$ and $\psi\hat{x}$ are special functions, it is not possible to employ fewer dots than in the first form, and often more are required.

The following definitions give abbreviated notations for functions of two or more variables:

$$(x, y) . \phi(x, y) . = : (x) : (y) . \phi(x, y) \quad \text{Df},$$

and so on for any number of variables;

$$\phi(x, y) . \supset_{x,y} . \psi(x, y) : = : (x, y) : \phi(x, y) . \supset . \psi(x, y) \quad \text{Df},$$

and so on for any number of variables.

Identity. The propositional function "x is identical with y" is expressed by

$$x = y.$$

This will be defined (cf. ∗13·01), but, owing to certain difficult points involved in the definition, we shall here omit it (cf. Chapter II). We have, of course,

$$\vdash . x = x \quad \text{(the law of identity)},$$

$$\vdash : x = y . \equiv . y = x,$$

$$\vdash : x = y . y = z . \supset . x = z.$$

The first of these expresses the *reflexive* property of identity: a relation is called *reflexive* when it holds between a term and itself, either universally, or whenever it holds between that term and some term. The second of the above propositions expresses that identity is a *symmetrical* relation: a relation is called *symmetrical* if, whenever it holds between x and y, it also holds between y and x. The third proposition expresses that identity is a *transitive* relation: a relation is called *transitive* if, whenever it holds between x and y and between y and z, it holds also between x and z.

We shall find that no new definition of the sign of equality is required in mathematics: all mathematical equations in which the sign of equality is used in the ordinary way express some identity, and thus use the sign of equality in the above sense.

If x and y are identical, either can replace the other in any proposition without altering the truth-value of the proposition; thus we have

$$\vdash : x = y \,.\supset .\, \phi x \equiv \phi y.$$

This is a fundamental property of identity, from which the remaining properties mostly follow.

It might be thought that identity would not have much importance, since it can only hold between x and y if x and y are different symbols for the same object. This view, however, does not apply to what we shall call "descriptive phrases," *i.e.* "the so-and-so." It is in regard to such phrases that identity is important, as we shall shortly explain. A proposition such as "Scott was the author of Waverley" expresses an identity in which there is a descriptive phrase (namely "the author of Waverley"); this illustrates how, in such cases, the assertion of identity may be important. It is essentially the same case when the newspapers say "the identity of the criminal has not transpired." In such a case, the criminal is known by a descriptive phrase, namely "the man who did the deed," and we wish to find an x of whom it is true that "x=the man who did the deed." When such an x has been found, the identity of the criminal has transpired.

Classes and relations. A *class* (which is the same as a *manifold* or *aggregate*) is all the objects satisfying some propositional function. If α is the class composed of the objects satisfying $\phi\hat{x}$, we shall say that α is the class *determined* by $\phi\hat{x}$. Every propositional function thus determines a class, though if the propositional function is one which is always false, the class will be *null*, *i.e.* will have no members. The class determined by the function $\phi\hat{x}$ will be represented by $\hat{z}(\phi z)$*. Thus for example if ϕx is an equation, $\hat{z}(\phi z)$ will be the class of its roots; if ϕx is "x has two legs and no feathers," $\hat{z}(\phi z)$ will be the class of men; if ϕx is "$0 < x < 1$," $\hat{z}(\phi z)$ will be the class of proper fractions, and so on.

It is obvious that the same class of objects will have many determining functions. When it is not necessary to specify a determining function of a class, the class may be conveniently represented by a single Greek letter. Thus Greek letters, other than those to which some constant meaning is assigned, will be exclusively used for classes.

* Any other letter may be used instead of z.

There are two kinds of difficulties which arise in formal logic; one kind arises in connection with classes and relations and the other in connection with descriptive functions. The point of the difficulty for classes and relations, so far as it concerns classes, is that a class cannot be an object suitable as an argument to any of its determining functions. If α represents a class and $\phi \hat{x}$ one of its determining functions [so that $\alpha = \hat{z}(\phi z)$], it is not sufficient that $\phi \alpha$ be a false proposition, it must be nonsense. Thus a certain classification of what appear to be objects into things of essentially different types seems to be rendered necessary. This whole question is discussed in Chapter II, on the theory of types, and the formal treatment in the systematic exposition, which forms the main body of this work, is guided by this discussion. The part of the systematic exposition which is specially concerned with the theory of classes is *20, and in this Introduction it is discussed in Chapter III. It is sufficient to note here that, in the complete treatment of *20, we have avoided the decision as to whether a class of things has in any sense an existence as one object. A decision of this question in either way is indifferent to our logic, though perhaps, if we had regarded some solution which held classes and relations to be in some real sense objects as both true and likely to be universally received, we might have simplified one or two definitions and a few preliminary propositions. Our symbols, such as "$\hat{x}(\phi x)$" and α and others, which represent classes and relations, are merely defined in their use, just as ∇^2, standing for

$$\frac{\partial^2}{\partial x^2} + \frac{\partial^2}{\partial y^2} + \frac{\partial^2}{\partial z^2},$$

has no meaning apart from a suitable function of x, y, z on which to operate. The result of our definitions is that the way in which we use classes corresponds in general to their use in ordinary thought and speech; and whatever may be the ultimate interpretation of the one is also the interpretation of the other. Thus in fact our classification of types in Chapter II really performs the single, though essential, service of justifying us in refraining from entering on trains of reasoning which lead to contradictory conclusions. The justification is that what seem to be propositions are really nonsense.

The definitions which occur in the theory of classes, by which the idea of a class (at least in use) is based on the other ideas assumed as primitive, cannot be understood without a fuller discussion than can be given now (cf. Chapter II of this Introduction and also *20). Accordingly, in this preliminary survey, we proceed to state the more important simple propositions which result from those definitions, leaving the reader to employ in his mind the ordinary unanalysed idea of a class of things. Our symbols in their usage conform to the ordinary usage of this idea in language. It is to be noticed that in the systematic exposition our treatment of classes and relations requires no new primitive ideas and only two new primitive propositions, namely the two forms of the "Axiom of Reducibility" (cf. next Chapter) for one and two variables respectively.

The propositional function "x is a member of the class α" will be expressed, following Peano, by the notation

$$x \epsilon \alpha.$$

Here ϵ is chosen as the initial of the word ἐστί. "$x \epsilon \alpha$" may be read "x is an α." Thus "$x \epsilon$ man" will mean "x is a man," and so on. For typographical convenience we shall put

$$x \sim\epsilon \alpha . = . \sim(x \epsilon \alpha) \quad \text{Df},$$

$$x, y \epsilon \alpha . = . x \epsilon \alpha . y \epsilon \alpha \quad \text{Df}.$$

For "class" we shall write "Cls"; thus "$\alpha \epsilon \text{Cls}$" means "$\alpha$ is a class."

We have

$$\vdash : x \epsilon \hat{z}(\phi z) . \equiv . \phi x,$$

i.e. "'x is a member of the class determined by $\phi\hat{z}$' is equivalent to 'x satisfies $\phi\hat{z}$,' or to 'ϕx is true.'"

A class is wholly determinate when its membership is known, that is, there cannot be two different classes having the same membership. Thus if ϕx, ψx are formally equivalent functions, they determine the same class; for in that case, if x is a member of the class determined by $\phi\hat{x}$, and therefore satisfies ϕx, it also satisfies ψx, and is therefore a member of the class determined by $\psi\hat{x}$. Thus we have

$$\vdash :. \hat{z}(\phi z) = \hat{z}(\psi z) . \equiv : \phi x . \equiv_x . \psi x.$$

The following propositions are obvious and important:

$$\vdash :. \alpha = \hat{z}(\phi z) . \equiv : x \epsilon \alpha . \equiv_x . \phi x,$$

i.e. α is identical with the class determined by $\phi\hat{z}$ when, and only when, "x is an α" is formally equivalent to ϕx;

$$\vdash :. \alpha = \beta . \equiv : x \epsilon \alpha . \equiv_x . x \epsilon \beta,$$

i.e. two classes α and β are identical when, and only when, they have the same membership;

$$\vdash . \hat{x}(x \epsilon \alpha) = \alpha,$$

i.e. the class whose determining function is "x is an α" is α, in other words, α is the class of objects which are members of α;

$$\vdash . \hat{z}(\phi z) \epsilon \text{Cls},$$

i.e. the class determined by the function $\phi\hat{z}$ is a class.

It will be seen that, according to the above, any function of one variable can be replaced by an equivalent function of the form "$x \epsilon \alpha$." Hence any extensional function of functions which holds when its argument is a function of the form "$\hat{z} \epsilon \alpha$," whatever possible value α may have, will hold also when its argument is any function $\phi\hat{z}$. Thus variation of classes can replace variation of functions of one variable in all the propositions of the sort with which we are concerned.

In an exactly analogous manner we introduce dual or dyadic relations, *i.e.* relations between two terms. Such relations will be called simply "relations"; relations between more than two terms will be distinguished as *multiple* relations, or (when the number of their terms is specified) as triple, quadruple,... relations, or as triadic, tetradic,... relations. Such relations will not concern us until we come to Geometry. For the present, the only relations we are concerned with are *dual* relations.

Relations, like classes, are to be taken in *extension*, *i.e.* if R and S are relations which hold between the same pairs of terms, R and S are to be identical. We may regard a relation, in the sense in which it is required for our purposes, as a class of couples; *i.e.* the couple (x, y) is to be one of the class of couples constituting the relation R if x has the relation R to y*. This view of relations as classes of couples will not, however, be introduced into our symbolic treatment, and is only mentioned in order to show that it is possible so to understand the meaning of the word *relation* that a relation shall be determined by its extension.

Any function $\phi(x, y)$ determines a relation R between x and y. If we regard a relation as a class of couples, the relation determined by $\phi(x, y)$ is the class of couples (x, y) for which $\phi(x, y)$ is true. The relation determined by the function $\phi(x, y)$ will be denoted by

$$\hat{x}\hat{y}\phi(x, y).$$

We shall use a capital letter for a relation when it is not necessary to specify the determining function. Thus whenever a capital letter occurs, it is to be understood that it stands for a relation.

The propositional function "x has the relation R to y" will be expressed by the notation

$$xRy.$$

This notation is designed to keep as near as possible to common language, which, when it has to express a relation, generally mentions it between its terms, as in "x loves y," "x equals y," "x is greater than y," and so on. For "relation" we shall write "Rel"; thus "$R \epsilon \text{Rel}$" means "R is a relation."

Owing to our taking relations in extension, we shall have

$$\vdash :. \hat{x}\hat{y}\phi(x, y) = \hat{x}\hat{y}\psi(x, y) . \equiv : \phi(x, y) . \equiv_{x,y} . \psi(x, y),$$

i.e. two functions of two variables determine the same relation when, and only when, the two functions are formally equivalent.

We have $\vdash . z\{\hat{x}\hat{y}\phi(x, y)\}w . \equiv . \phi(z, w),$

* Such a couple has a *sense*, *i.e.* the couple (x, y) is different from the couple (y, x), unless $x = y$. We shall call it a "couple with sense," to distinguish it from the class consisting of x and y. It may also be called an *ordered* couple.

i.e. "z has to w the relation determined by the function $\phi(x, y)$" is equivalent to $\phi(z, w)$;

$$\vdash :. R = \hat{x}\hat{y}\phi(x, y) . \equiv : xRy . \equiv_{x,y} . \phi(x, y),$$
$$\vdash :. R = S . \equiv : xRy . \equiv_{x,y} . xSy,$$
$$\vdash . \hat{x}\hat{y}(xRy) = R,$$
$$\vdash . \{\hat{x}\hat{y}\phi(x, y)\} \epsilon \text{Rel}.$$

These propositions are analogous to those previously given for classes. It results from them that any function of two variables is formally equivalent to some function of the form xRy; hence, in extensional functions of two variables, variation of relations can replace variation of functions of two variables.

Both classes and relations have properties analogous to most of those of propositions that result from negation and the logical sum. The *logical product* of two classes α and β is their common part, *i.e.* the class of terms which are members of both. This is represented by $\alpha \cap \beta$. Thus we put

$$\alpha \cap \beta = \hat{x}(x \epsilon \alpha . x \epsilon \beta) \quad \text{Df.}$$

This gives us $\vdash : x \epsilon \alpha \cap \beta . \equiv . x \epsilon \alpha . x \epsilon \beta,$

i.e. "x is a member of the logical product of α and β" is equivalent to the logical product of "x is a member of α" and "x is a member of β."

Similarly the *logical sum* of two classes α and β is the class of terms which are members of either; we denote it by $\alpha \cup \beta$. The definition is

$$\alpha \cup \beta = \hat{x}(x \epsilon \alpha . \vee . x \epsilon \beta) \quad \text{Df,}$$

and the connection with the logical sum of propositions is given by

$$\vdash :. x \epsilon \alpha \cup \beta . \equiv : x \epsilon \alpha . \vee . x \epsilon \beta.$$

The *negation* of a class α consists of those terms x for which "$x \epsilon \alpha$" can be *significantly and truly* denied. We shall find that there are terms of other types for which "$x \epsilon \alpha$" is neither true nor false, but nonsense. These terms are not members of the negation of α.

Thus the *negation* of a class α is the class of terms of suitable type which are not members of it, *i.e.* the class $\hat{x}(x \sim\epsilon \alpha)$. We call this class "$-\alpha$" (read "not-$\alpha$"); thus the definition is

$$-\alpha = \hat{x}(x \sim\epsilon \alpha) \quad \text{Df,}$$

and the connection with the negation of propositions is given by

$$\vdash : x \epsilon - \alpha . \equiv . x \sim\epsilon \alpha.$$

In place of implication we have the relation of *inclusion*. A class α is said to be included or contained in a class β if all members of α are members of β, *i.e.* if $x \epsilon \alpha . \supset_x . x \epsilon \beta$. We write "$\alpha \subset \beta$" for "$\alpha$ is contained in β." Thus we put

$$\alpha \subset \beta . = : x \epsilon \alpha . \supset_x . x \epsilon \beta \quad \text{Df.}$$

Most of the formulae concerning $p \,.\, q$, $p \vee q$, $\sim p$, $p \supset q$ remain true if we substitute $\alpha \cap \beta$, $\alpha \cup \beta$, $-\alpha$, $\alpha \subset \beta$. In place of equivalence, we substitute identity; for "$p \equiv q$" was defined as "$p \supset q \,.\, q \supset p$," but "$\alpha \subset \beta \,.\, \beta \subset \alpha$" gives "$x \in \alpha \,.\equiv_x.\, x \in \beta$," whence $\alpha = \beta$.

The following are some propositions concerning classes which are analogues of propositions previously given concerning propositions:

$$\vdash \,.\, \alpha \cap \beta = -(-\alpha \cup -\beta),$$

i.e. the common part of α and β is the negation of "not-α or not-β";

$$\vdash \,.\, x \in (\alpha \cup -\alpha),$$

i.e. "x is a member of α or not-α";

$$\vdash \,.\, x \sim\in (\alpha \cap -\alpha),$$

i.e. "x is not a member of both α and not-α";

$$\vdash \,.\, \alpha = -(-\alpha),$$

$$\vdash : \alpha \subset \beta \,.\equiv.\, -\beta \subset -\alpha,$$

$$\vdash : \alpha = \beta \,.\equiv.\, -\alpha = -\beta,$$

$$\vdash : \alpha = \alpha \cap \alpha,$$

$$\vdash : \alpha = \alpha \cup \alpha.$$

The two last are the two forms of the law of tautology.

The law of absorption holds in the form

$$\vdash : \alpha \subset \beta \,.\equiv.\, \alpha = \alpha \cap \beta.$$

Thus for example "all Cretans are liars" is equivalent to "Cretans are identical with lying Cretans."

Just as we have $\quad \vdash : p \supset q \,.\, q \supset r \,.\supset.\, p \supset r,$

so we have $\quad \vdash : \alpha \subset \beta \,.\, \beta \subset \gamma \,.\supset.\, \alpha \subset \gamma.$

This expresses the ordinary syllogism in Barbara (with the premisses interchanged); for "$\alpha \subset \beta$" means the same as "all α's are β's," so that the above proposition states: "If all α's are β's, and all β's are γ's, then all α's are γ's." (It should be observed that syllogisms are traditionally expressed with "therefore," as if they asserted both premisses and conclusion. This is, of course, merely a slipshod way of speaking, since what is really asserted is only the connection of premisses with conclusion.)

The syllogism in Barbara when the minor premiss has an individual subject is

$$\vdash : x \in \beta \,.\, \beta \subset \gamma \,.\supset.\, x \in \gamma,$$

e.g. "if Socrates is a man, and all men are mortals, then Socrates is a mortal." This, as was pointed out by Peano, is not a particular case of "$\alpha \subset \beta \,.\, \beta \subset \gamma \,.\supset.\, \alpha \subset \gamma$," since "$x \in \beta$" is not a particular case of "$\alpha \subset \beta$." This point is important, since traditional logic is here mistaken. The nature and magnitude of its mistake will become clearer at a later stage.

For relations, we have precisely analogous definitions and propositions. We put

$$R \mathbin{\dot{\cap}} S = \hat{x}\hat{y}\,(xRy \,.\, xSy) \quad \text{Df},$$

which leads to

$$\vdash : x\,(R \mathbin{\dot{\cap}} S)\,y \,.\equiv .\, xRy \,.\, xSy.$$

Similarly

$$R \mathbin{\dot{\cup}} S = \hat{x}\hat{y}\,(xRy \,.\vee.\, xSy) \quad \text{Df},$$

$$\dot{-}\, R = \hat{x}\hat{y}\,\{\sim(xRy)\} \quad \text{Df},$$

$$R \mathbin{\dot{\subset}} S \,.=:\, xRy \,.\supset_{x,y}.\, xSy \quad \text{Df}.$$

Generally, when we require analogous but different symbols for relations and for classes, we shall choose for relations the symbol obtained by adding a dot, in some convenient position, to the corresponding symbol for classes. (The dot must not be put on the line, since that would cause confusion with the use of dots as brackets.) But such symbols require and receive a special definition in each case.

A class is said to *exist* when it has at least one member: "α exists" is denoted by "$\exists\,!\,\alpha$." Thus we put

$$\exists\,!\,\alpha \,.=.\, (\exists x) \,.\, x \in \alpha \quad \text{Df}.$$

The class which has no members is called the "null-class," and is denoted by "Λ." Any propositional function which is always false determines the null-class. One such function is known to us already, namely "x is not identical with x," which we denote by "$x \neq x$." Thus we may use this function for defining Λ, and put

$$\Lambda = \hat{x}\,(x \neq x) \quad \text{Df}.$$

The class determined by a function which is always true is called the *universal class*, and is represented by V; thus

$$\mathrm{V} = \hat{x}\,(x = x) \quad \text{Df}.$$

Thus Λ is the negation of V. We have

$$\vdash .\, (x) \,.\, x \in \mathrm{V},$$

i.e. "'x is a member of V' is always true"; and

$$\vdash .\, (x) \,.\, x \sim\in \Lambda,$$

i.e. "'x is a member of Λ' is always false." Also

$$\vdash : \alpha = \Lambda \,.\equiv .\, \sim\exists\,!\,\alpha,$$

i.e. "α is the null-class" is equivalent to "α does not exist."

For relations we use similar notations. We put

$$\dot{\exists}\,!\,R \,.=.\, (\exists x, y) \,.\, xRy,$$

i.e. "$\dot{\exists}\,!\,R$" means that there is at least one couple x, y between which the relation R holds. $\dot{\Lambda}$ will be the relation which never holds, and $\dot{\mathrm{V}}$ the relation which always holds. $\dot{\mathrm{V}}$ is practically never required; $\dot{\Lambda}$ will be the relation $\hat{x}\hat{y}\,(x \neq x \,.\, y \neq y)$. We have

$$\vdash .\, (x, y) \,.\, \sim(x \,\dot{\Lambda}\, y),$$

and

$$\vdash : R = \dot{\Lambda} \,.\equiv .\, \sim\dot{\exists}\,!\,R.$$

There are no classes which contain objects of more than one type. Accordingly there is a universal class and a null-class proper to each type of object. But these symbols need not be distinguished, since it will be found that there is no possibility of confusion. Similar remarks apply to relations.

Descriptions. By a "description" we mean a phrase of the form "*the* so-and-so" or of some equivalent form. For the present, we confine our attention to *the* in the singular. We shall use this word strictly, so as to imply uniqueness; *e.g.* we should not say "A is *the* son of B" if B had other sons besides A. Thus a description of the form "the so-and-so" will only have an application in the event of there being one so-and-so and no more. Hence a description requires some propositional function $\phi\hat{x}$ which is satisfied by one value of x and by no other values; then "the x which satisfies $\phi\hat{x}$" is a description which definitely describes a certain object, though we may not know what object it describes. For example, if y is a man, "x is the father of y" must be true for one, and only one, value of x. Hence "the father of y" is a description of a certain man, though we may not know *what* man it describes. A phrase containing "the" always presupposes some initial propositional function not containing "the"; thus instead of "x is the father of y" we ought to take as our initial function "x begot y"; then "the father of y" means the one value of x which satisfies this propositional function.

If $\phi\hat{x}$ is a propositional function, the symbol "$(\iota x)(\phi x)$" is used in our symbolism in such a way that it can always be read as "the x which satisfies $\phi\hat{x}$." But we do not define "$(\iota x)(\phi x)$" as standing for "the x which satisfies $\phi\hat{x}$," thus treating this last phrase as embodying a primitive idea. Every use of "$(\iota x)(\phi x)$," where it apparently occurs as a constituent of a proposition in the place of an object, is defined in terms of the primitive ideas already on hand. An example of this definition in use is given by the proposition "$E\,!\,(\iota x)(\phi x)$" which is considered immediately. The whole subject is treated more fully in Chapter III.

The symbol should be compared and contrasted with "$\hat{x}(\phi x)$" which in use can always be read as "the x's which satisfy $\phi\hat{x}$." Both symbols are incomplete symbols defined only in use, and as such are discussed in Chapter III. The symbol "$\hat{x}(\phi x)$" always has an application, namely to the class determined by ϕx; but "$(\iota x)(\phi x)$" only has an application when $\phi\hat{x}$ is only satisfied by one value of x, neither more nor less. It should also be observed that the meaning given to the symbol by the definition, given immediately below, of $E\,!\,(\iota x)(\phi x)$ does not presuppose that we know the meaning of "one." This is also characteristic of the definition of any other use of $(\iota x)(\phi x)$.

We now proceed to define "$E\,!\,(\iota x)(\phi x)$" so that it can be read "the x satisfying ϕx exists." (It will be observed that this is a different meaning of existence from that which we express by "$\exists$.") Its definition is

$$E\,!\,(\iota x)(\phi x) \,.=:\, (\exists c) : \phi x \,.\equiv_x .\, x = c \quad \text{Df},$$

i.e. "the x satisfying $\phi\hat{x}$ exists" is to mean "there is an object c such that ϕx is true when x is c but not otherwise."

The following are equivalent forms:

$$\vdash :. E!(\imath x)(\phi x) . \equiv : (\exists c) : \phi c : \phi x . \supset_x . x = c,$$

$$\vdash :. E!(\imath x)(\phi x) . \equiv : (\exists c) . \phi c : \phi x . \phi y . \supset_{x,y} . x = y,$$

$$\vdash :. E!(\imath x)(\phi x) . \equiv : (\exists c) : \phi c : x \neq c . \supset_x . \sim \phi x.$$

The last of these states that "the x satisfying $\phi\hat{x}$ exists" is equivalent to "there is an object c satisfying $\phi\hat{x}$, and every object other than c does not satisfy $\phi\hat{x}$."

The kind of existence just defined covers a great many cases. Thus for example "the most perfect Being exists" will mean:

$$(\exists c) : x \text{ is most perfect} . \equiv_x . x = c,$$

which, taking the last of the above equivalences, is equivalent to

$$(\exists c) : c \text{ is most perfect} : x \neq c . \supset_x . x \text{ is not most perfect.}$$

A proposition such as "Apollo exists" is really of the same logical form, although it does not explicitly contain the word *the*. For "Apollo" means really "the object having such-and-such properties," say "the object having the properties enumerated in the Classical Dictionary*." If these properties make up the propositional function ϕx, then "Apollo" means "$(\imath x)(\phi x)$," and "Apollo exists" means "$E!(\imath x)(\phi x)$." To take another illustration, "the author of Waverley" means "the man who (or rather, the object which) wrote Waverley." Thus "Scott is the author of Waverley" is

$$\text{Scott} = (\imath x)(x \text{ wrote Waverley}).$$

Here (as we observed before) the importance of *identity* in connection with descriptions plainly appears.

The notation "$(\imath x)(\phi x)$," which is long and inconvenient, is seldom used, being chiefly required to lead up to another notation, namely "$R\text{`}y$," meaning "the object having the relation R to y." That is, we put

$$R\text{`}y = (\imath x)(xRy) \quad \text{Df.}$$

The inverted comma may be read "of." Thus "$R\text{`}y$" is read "the $R$ of $y$." Thus if $R$ is the relation of father to son, "$R\text{`}y$" means "the father of y"; if R is the relation of son to father, "$R\text{`}y$" means "the son of $y$," which will only "exist" if $y$ has one son and no more. $R\text{`}y$ is a function of y, but not a propositional function; we shall call it a *descriptive* function. All the ordinary functions of mathematics are of this kind, as will appear more fully in the sequel. Thus in our notation, "sin y" would be written "sin $\text{`}y$," and "sin" would stand for the relation which sin $\text{`}y$ has to y. Instead of a variable descriptive function fy, we put $R\text{`}y$, where the variable relation R takes the

* The same principle applies to many uses of the proper names of existent objects, *e.g.* to all uses of proper names for objects known to the speaker only by report, and not by personal acquaintance.

place of the variable function f. A descriptive function will in general exist while y belongs to a certain domain, but not outside that domain; thus if we are dealing with positive rationals, $\sqrt{y}$ will be significant if y is a perfect square, but not otherwise; if we are dealing with real numbers, and agree that "$\sqrt{y}$" is to mean the *positive* square root (or, is to mean the negative square root), $\sqrt{y}$ will be significant provided y is positive, but not otherwise; and so on. Thus every descriptive function has what we may call a "domain of definition" or a "domain of existence," which may be thus defined: If the function in question is $R\text{‘}y$, its domain of definition or of existence will be the class of those arguments y for which we have $E!\,R\text{‘}y$, *i.e.* for which $E!\,(\imath x)(xRy)$, *i.e.* for which there is one x, and no more, having the relation R to y.

If R is any relation, we will speak of $R\text{‘}y$ as the "associated descriptive function." A great many of the constant relations which we shall have occasion to introduce are only or chiefly important on account of their associated descriptive functions. In such cases, it is easier (though less correct) to begin by assigning the meaning of the descriptive function, and to deduce the meaning of the relation from that of the descriptive function. This will be done in the following explanations of notation.

Various descriptive functions of relations. If R is any relation, the *converse* of R is the relation which holds between y and x whenever R holds between x and y. Thus *greater* is the converse of *less*, *before* of *after*, *cause* of *effect* *husband* of *wife*, etc. The converse of R is written* $\text{Cnv}\text{‘}R$ or $\breve{R}$. The definition is

$$\breve{R} = \hat{x}\hat{y}\,(yRx) \quad \text{Df,}$$

$$\text{Cnv}\text{‘}R = \breve{R} \qquad \text{Df.}$$

The second of these is not a formally correct definition, since we ought to define "Cnv" and deduce the meaning of $\text{Cnv}\text{‘}R$. But it is not worth while to adopt this plan in our present introductory account, which aims at simplicity rather than formal correctness.

A relation is called *symmetrical* if $R = \breve{R}$, *i.e.* if it holds between y and x whenever it holds between x and y (and therefore vice versa). Identity, diversity, agreement or disagreement in any respect, are symmetrical relations. A relation is called *asymmetrical* when it is incompatible with its converse, *i.e.* when $R \dot{\cap} \breve{R} = \dot{\Lambda}$, or, what is equivalent,

$$xRy \,.\, \supset_{x,y} .\, \sim(yRx).$$

Before and after, greater and less, ancestor and descendant, are asymmetrical, as are all other relations of the sort that lead to *series*. But there are many asymmetrical relations which do not lead to series, for instance, that of

* The second of these notations is taken from Schröder's *Algebra und Logik der Relative*.

wife's brother*. A relation may be neither symmetrical nor asymmetrical; for example, this holds of the relation of inclusion between classes: $\alpha \subset \beta$ and $\beta \subset \alpha$ will both be true if $\alpha = \beta$, but otherwise only one of them, at most, will be true. The relation *brother* is neither symmetrical nor asymmetrical, for if x is the brother of y, y may be either the brother or the sister of x.

In the propositional function xRy, we call x the *referent* and y the *relatum*. The class $\hat{x}(xRy)$, consisting of all the x's which have the relation R to y, is called the class of referents of y with respect to R; the class $\hat{y}(xRy)$, consisting of all the y's to which x has the relation R, is called the class of relata of x with respect to R. These two classes are denoted respectively by $\overrightarrow{R}\text{‘}y$ and $\overleftarrow{R}\text{‘}x$. Thus

$$\overrightarrow{R}\text{‘}y = \hat{x}(xRy) \quad \text{Df},$$

$$\overleftarrow{R}\text{‘}x = \hat{y}(xRy) \quad \text{Df}.$$

The arrow runs towards y in the first case, to show that we are concerned with things having the relation R *to* y; it runs away from x in the second case, to show that the relation R goes *from* x to the members of $\overleftarrow{R}\text{‘}x$. It runs in fact *from* a referent and *towards* a relatum.

The notations $\overrightarrow{R}\text{‘}y$, $\overleftarrow{R}\text{‘}x$ are very important, and are used constantly. If R is the relation of parent to child, $\overrightarrow{R}\text{‘}y$ = the parents of y, $\overleftarrow{R}\text{‘}x$ = the children of x. We have

$$\vdash : x \in \overrightarrow{R}\text{‘}y \,.\equiv .\, xRy$$

and

$$\vdash : y \in \overleftarrow{R}\text{‘}x \,.\equiv .\, xRy.$$

These equivalences are often embodied in common language. For example, we say indiscriminately "x is an inhabitant of London" or "x inhabits London." If we put "R" for "inhabits," "x inhabits London" is "xR London," while "x is an inhabitant of London" is "$x \in \overrightarrow{R}\text{‘}$ London."

Instead of $\overrightarrow{R}$ and $\overleftarrow{R}$ we sometimes use sg‘R, gs‘R, where "sg" stands for "sagitta," and "gs" is "sg" backwards. Thus we put

$$\text{sg‘}R = \overrightarrow{R} \quad \text{Df},$$

$$\text{gs‘}R = \overleftarrow{R} \quad \text{Df}.$$

These notations are sometimes more convenient than an arrow when the relation concerned is represented by a combination of letters, instead of a single letter such as R. Thus *e.g.* we should write sg‘$(R \dot{\cap} S)$, rather than put an arrow over the whole length of $(R \dot{\cap} S)$.

The class of all terms that have the relation R to something or other is called the *domain* of R. Thus if R is the relation of parent and child, the

* This relation is not strictly asymmetrical, but is so except when the wife's brother is also the sister's husband. In the Greek Church the relation is strictly asymmetrical.

domain of R will be the class of parents. We represent the domain of R by "$\text{D}{}^{\backprime}R$." Thus we put

$$\text{D}{}^{\backprime}R = \hat{x}\{(\exists y) . xRy\} \quad \text{Df.}$$

Similarly the class of all terms to which something or other has the relation R is called the *converse domain* of R; it is the same as the domain of the converse of R. The converse domain of R is represented by "$\text{G}{}^{\backprime}R$"; thus

$$\text{G}{}^{\backprime}R = \hat{y}\{(\exists x) . xRy\} \quad \text{Df.}$$

The sum of the domain and the converse domain is called the *field*, and is represented by $C{}^{\backprime}R$: thus

$$C{}^{\backprime}R = \text{D}{}^{\backprime}R \cup \text{G}{}^{\backprime}R \quad \text{Df.}$$

The *field* is chiefly important in connection with series. If R is the ordering relation of a series, $C{}^{\backprime}R$ will be the class of terms of the series, $\text{D}{}^{\backprime}R$ will be all the terms except the last (if any), and $\text{G}{}^{\backprime}R$ will be all the terms except the first (if any). The first term, if it exists, is the only member of $\text{D}{}^{\backprime}R \cap - \text{G}{}^{\backprime}R$, since it is the only term which is a predecessor but not a follower. Similarly the last term (if any) is the only member of $\text{G}{}^{\backprime}R \cap - \text{D}{}^{\backprime}R$. The condition that a series should have no end is $\text{G}{}^{\backprime}R \subset \text{D}{}^{\backprime}R$, *i.e.* "every follower is a predecessor"; the condition for no beginning is $\text{D}{}^{\backprime}R \subset \text{G}{}^{\backprime}R$. These conditions are equivalent respectively to $\text{D}{}^{\backprime}R = C{}^{\backprime}R$ and $\text{G}{}^{\backprime}R = C{}^{\backprime}R$.

The *relative product* of two relations R and S is the relation which holds between x and z when there is an intermediate term y such that x has the relation R to y and y has the relation S to z. The relative product of R and S is represented by $R \mid S$; thus we put

$$R \mid S = \hat{x}\hat{z}\{(\exists y) . xRy . ySz\} \quad \text{Df,}$$

whence $\vdash : x(R \mid S)z . \equiv . (\exists y) . xRy . ySz.$

Thus "paternal aunt" is the relative product of *sister* and *father*; "paternal grandmother" is the relative product of *mother* and *father*; "maternal grandfather" is the relative product of *father* and *mother*. The relative product is not commutative, but it obeys the associative law, *i.e.*

$$\vdash . (P \mid Q) \mid R = P \mid (Q \mid R).$$

It also obeys the distributive law with regard to the logical addition of relations, *i.e.* we have

$$\vdash . P \mid (Q \mathbin{\dot\cup} R) = (P \mid Q) \mathbin{\dot\cup} (P \mid R),$$

$$\vdash . (Q \mathbin{\dot\cup} R) \mid P = (Q \mid P) \mathbin{\dot\cup} (R \mid P).$$

But with regard to the logical *product*, we have only

$$\vdash . P \mid (Q \mathbin{\dot\cap} R) \mathbin{\dot\subset} (P \mid Q) \mathbin{\dot\cap} (P \mid R),$$

$$\vdash . (Q \mathbin{\dot\cap} R) \mid P \mathbin{\dot\subset} (Q \mid P) \mathbin{\dot\cap} (Q \mid R).$$

The relative product does not obey the law of tautology, *i.e.* we do not have in general $R \mid R = R$. We put

$$R^2 = R \mid R \quad \text{Df.}$$

Thus paternal grandfather = (father)^2,
maternal grandmother = (mother)^2.

A relation is called *transitive* when $R^2 \subset R$, *i.e.* when, if xRy and yRz, we always have xRz, *i.e.* when

$$xRy \,.\, yRz \,.\, \supset_{x,y,z} \,.\, xRz.$$

Relations which generate series are always transitive; thus *e.g.*

$$x > y \,.\, y > z \,.\, \supset_{x,y,z} \,.\, x > z.$$

If P is a relation which generates a series, P may conveniently be read "precedes"; thus "$xPy \,.\, yPz \,.\, \supset_{x,y,z} \,.\, xPz$" becomes "if x precedes y and y precedes z, then x always precedes z." The class of relations which generate series are partially characterized by the fact that they are transitive and asymmetrical, and never relate a term to itself.

If P is a relation which generates a series, and if we have not merely $P^2 \subset P$, but $P^2 = P$, then P generates a series which is *compact* (*überall dicht*), *i.e.* such that there are terms between any two. For in this case we have

$$xPz \,.\, \supset \,.\, (\exists y) \,.\, xPy \,.\, yPz,$$

i.e. if x precedes z, there is a term y such that x precedes y and y precedes z, *i.e.* there is a term between x and z. Thus among relations which generate series, those which generate compact series are those for which $P^2 = P$.

Many relations which do not generate series are transitive, for example, identity, or the relation of inclusion between classes. Such cases arise when the relations are not asymmetrical. Relations which are transitive and symmetrical are an important class: they may be regarded as consisting in the possession of some common property.

Plural descriptive functions. The class of terms x which have the relation R to some member of a class α is denoted by $R``\alpha$ or $R_\epsilon`\alpha$. The definition is

$$R``\alpha = \hat{x}\{(\exists y) \,.\, y \in \alpha \,.\, xRy\} \quad \text{Df.}$$

Thus for example let R be the relation of *inhabiting*, and α the class of towns; then $R``\alpha$ = inhabitants of towns. Let $R$ be the relation "less than" among rationals, and $\alpha$ the class of those rationals which are of the form $1 - 2^{-n}$, for integral values of $n$; then $R``\alpha$ will be all rationals less than some member of α, *i.e.* all rationals less than 1. If P is the generating relation of a series, and α is any class of members of the series, $P``\alpha$ will be predecessors of $\alpha$'s, *i.e.* the segment defined by $\alpha$. If $P$ is a relation such that $P`y$ always exists when $y \in \alpha$, $P``\alpha$ will be the class of all terms of the form $P`y$ for values of y which are members of α; *i.e.*

$$P``\alpha = \hat{x}\{(\exists y) \,.\, y \in \alpha \,.\, x = P`y\}.$$

Thus a member of the class "fathers of great men" will be the father of y, where y is some great man. In other cases, this will not hold; for instance, let P be the relation of a number to any number of which it is a factor: then

$P``$ (even numbers) = factors of even numbers, but this class is not composed of terms of the form "*the* factor of x," where x is an even number, because numbers do not have only one factor apiece.

Unit classes. The class whose only member is x might be thought to be identical with x, but Peano and Frege have shown that this is not the case. (The reasons why this is not the case will be explained in a preliminary way in Chapter II of the Introduction.) We denote by "$\iota`x$" the class whose only member is x: thus

$$\iota`x = \hat{y}\,(y = x) \quad \text{Df},$$

i.e. "$\iota`x$" means "the class of objects which are identical with x."

The class consisting of x and y will be $\iota`x \cup \iota`y$; the class got by adding x to a class α will be $\alpha \cup \iota`x$; the class got by taking away $x$ from a class $\alpha$ will be $\alpha - \iota`x$. (We write $\alpha - \beta$ as an abbreviation for $\alpha \cap -\beta$.)

It will be observed that unit classes have been defined without reference to the number 1; in fact, we use unit classes to define the number 1. This number is defined as the class of unit classes, *i.e.*

$$1 = \hat{\alpha}\,\{(\exists x)\,.\,\alpha = \iota`x\} \quad \text{Df}.$$

This leads to

$$\vdash :.\, \alpha \in 1 \,.\equiv :\, (\exists x) : y \in \alpha \,.\equiv_y .\, y = x.$$

From this it appears further that

$$\vdash : \alpha \in 1 \,.\equiv .\, \text{E}\,!\,(\imath x)\,(x \in \alpha),$$

whence

$$\vdash : \hat{z}\,(\phi z) \in 1 \,.\equiv .\, \text{E}\,!\,(\imath x)\,(\phi x),$$

i.e. "$\hat{z}\,(\phi z)$ is a unit class" is equivalent to "the x satisfying $\phi\hat{x}$ exists."

If $\alpha \in 1$, $\breve{\iota}`\alpha$ is the only member of $\alpha$, for the only member of $\alpha$ is the only term to which $\alpha$ has the relation $\breve{\iota}$. Thus "$\breve{\iota}`\alpha$" takes the place of "$(\imath x)\,(\phi x)$," if α stands for $\hat{z}\,(\phi z)$. In practice, "$\breve{\iota}`\alpha$" is a more convenient notation than "$(\imath x)\,(\phi x)$," and is generally used instead of "$(\imath x)\,(\phi x)$."

The above account has explained most of the logical notation employed in the present work. In the applications to various parts of mathematics, other definitions are introduced; but the objects defined by these later definitions belong, for the most part, rather to mathematics than to logic. The reader who has mastered the symbols explained above will find that any later formulae can be deciphered by the help of comparatively few additional definitions.

CHAPTER II

THE THEORY OF LOGICAL TYPES

THE theory of logical types, to be explained in the present Chapter, recommended itself to us in the first instance by its ability to solve certain contradictions, of which the one best known to mathematicians is Burali-Forti's concerning the greatest ordinal. But the theory in question is not wholly dependent upon this indirect recommendation: it has also a certain consonance with common sense which makes it inherently credible. In what follows, we shall therefore first set forth the theory on its own account, and then apply it to the solution of the contradictions.

I. *The Vicious-Circle Principle.*

An analysis of the paradoxes to be avoided shows that they all result from a certain kind of vicious circle*. The vicious circles in question arise from supposing that a collection of objects may contain members which can only be defined by means of the collection as a whole. Thus, for example, the collection of *propositions* will be supposed to contain a proposition stating that "all propositions are either true or false." It would seem, however, that such a statement could not be legitimate unless "all propositions" referred to some already definite collection, which it cannot do if new propositions are created by statements about "all propositions." We shall, therefore, have to say that statements about "all propositions" are meaningless. More generally, given any set of objects such that, if we suppose the set to have a total, it will contain members which presuppose this total, then such a set cannot have a total. By saying that a set has "no total," we mean, primarily, that no significant statement can be made about "all its members." Propositions, as the above illustration shows, must be a set having no total. The same is true, as we shall shortly see, of propositional functions, even when these are restricted to such as can significantly have as argument a given object a. In such cases, it is necessary to break up our set into smaller sets, each of which is capable of a total. This is what the theory of types aims at effecting.

The principle which enables us to avoid illegitimate totalities may be stated as follows: "Whatever involves *all* of a collection must not be one of the collection"; or, conversely: "If, provided a certain collection had a total, it would have members only definable in terms of that total, then the said collection has no total." We shall call this the "vicious-circle principle," because it enables us to avoid the vicious circles involved in the assumption of illegitimate totalities. Arguments which are condemned by the vicious-circle

* See the last section of the present Chapter. Cf. also H. Poincaré, "Les mathématiques et la logique," *Revue de Métaphysique et de Morale*, Mai 1906, p. 307.

principle will be called "vicious-circle fallacies." Such arguments, in certain circumstances, may lead to contradictions, but it often happens that the conclusions to which they lead are in fact true, though the arguments are fallacious. Take, for example, the law of excluded middle, in the form "all propositions are true or false." If from this law we argue that, because the law of excluded middle is a proposition, therefore the law of excluded middle is true or false, we incur a vicious-circle fallacy. "All propositions" must be in some way limited before it becomes a legitimate totality, and any limitation which makes it legitimate must make any statement about the totality fall outside the totality. Similarly, the imaginary sceptic, who asserts that he knows nothing, and is refuted by being asked if he knows that he knows nothing, has asserted nonsense, and has been fallaciously refuted by an argument which involves a vicious-circle fallacy. In order that the sceptic's assertion may become significant, it is necessary to place some limitation upon the things of which he is asserting his ignorance, because the things of which it is possible to be ignorant form an illegitimate totality. But as soon as a suitable limitation has been placed by him upon the collection of propositions of which he is asserting his ignorance, the proposition that he is ignorant of every member of this collection must not itself be one of the collection. Hence any significant scepticism is not open to the above form of refutation.

The paradoxes of symbolic logic concern various sorts of objects: propositions, classes, cardinal and ordinal numbers, etc. All these sorts of objects, as we shall show, represent illegitimate totalities, and are therefore capable of giving rise to vicious-circle fallacies. But by means of the theory (to be explained in Chapter III) which reduces statements that are verbally concerned with classes and relations to statements that are concerned with propositional functions, the paradoxes are reduced to such as are concerned with propositions and propositional functions. The paradoxes that concern propositions are only indirectly relevant to mathematics, while those that more nearly concern the mathematician are all concerned with *propositional functions*. We shall therefore proceed at once to the consideration of propositional functions.

II. *The Nature of Propositional Functions.*

By a "propositional function" we mean something which contains a variable x, and expresses a *proposition* as soon as a value is assigned to x. That is to say, it differs from a proposition solely by the fact that it is ambiguous: it contains a variable of which the value is unassigned. It agrees with the ordinary functions of mathematics in the fact of containing an unassigned variable; where it differs is in the fact that the values of the function are propositions. Thus *e.g.* "x is a man" or "$\sin x = 1$" is a propositional function. We shall find that it is possible to incur a vicious-circle fallacy at the very outset, by admitting as possible arguments to a propositional function terms which presuppose the function. This form of the fallacy is very instructive, and its avoidance leads, as we shall see, to the hierarchy of types.

The question as to the nature of a function* is by no means an easy one. It would seem, however, that the essential characteristic of a function is *ambiguity*. Take, for example, the law of identity in the form "A is A," which is the form in which it is usually enunciated. It is plain that, regarded psychologically, we have here a single judgment. But what are we to say of the object of the judgment? We are not judging that Socrates is Socrates, nor that Plato is Plato, nor any other of the definite judgments that are instances of the law of identity. Yet each of these judgments is, in a sense, within the scope of our judgment. We are in fact judging an ambiguous instance of the propositional function "A is A." We appear to have a single thought which does not have a definite object, but has as its object an undetermined one of the values of the function "A is A." It is this kind of ambiguity that constitutes the essence of a function. When we speak of "ϕx," where x is not specified, we mean one value of the function, but not a definite one. We may express this by saying that "ϕx" *ambiguously denotes* ϕa, ϕb, ϕc, etc., where ϕa, ϕb, ϕc, etc., are the various values of "ϕx."

When we say that "ϕx" ambiguously denotes ϕa, ϕb, ϕc, etc., we mean that "ϕx" means one of the objects ϕa, ϕb, ϕc, etc., though not a definite one, but an undetermined one. It follows that "ϕx" only has a well-defined meaning (well-defined, that is to say, except in so far as it is of its essence to be ambiguous) if the objects ϕa, ϕb, ϕc, etc., are well-defined. That is to say, a function is not a well-defined function unless all its values are already well-defined. It follows from this that no function can have among its values anything which presupposes the function, for if it had, we could not regard the objects ambiguously denoted by the function as definite until the function was definite, while conversely, as we have just seen, the function cannot be definite until its values are definite. This is a particular case, but perhaps the most fundamental case, of the vicious-circle principle. A function is what ambiguously denotes some one of a certain totality, namely the values of the function; hence this totality cannot contain any members which involve the function, since, if it did, it would contain members involving the totality, which, by the vicious-circle principle, no totality can do.

It will be seen that, according to the above account, the values of a function are presupposed by the function, not vice versa. It is sufficiently obvious, in any particular case, that a value of a function does not presuppose the function. Thus for example the proposition "Socrates is human" can be perfectly apprehended without regarding it as a value of the function "x is human." It is true that, conversely, a function can be apprehended without

* When the word "function" is used in the sequel, "propositional function" is always meant. Other functions will not be in question in the present Chapter.

its being necessary to apprehend its values severally and individually. If this were not the case, no function could be apprehended at all, since the number of values (true and false) of a function is necessarily infinite and there are necessarily possible arguments with which we are unacquainted. What is necessary is not that the values should be given individually and extensionally, but that the totality of the values should be given intensionally, so that, concerning any assigned object, it is at least theoretically determinate whether or not the said object is a value of the function.

It is necessary practically to distinguish the function itself from an undetermined value of the function. We may regard the function itself as that which ambiguously denotes, while an undetermined value of the function is that which is ambiguously denoted. If the undetermined value is written "ϕx," we will write the function itself "$\phi \hat{x}$." (Any other letter may be used in place of x.) Thus we should say "ϕx is a proposition," but "$\phi \hat{x}$ is a propositional function." When we say "ϕx is a proposition," we mean to state something which is true for every possible value of x, though we do not decide what value x is to have. We are making an ambiguous statement about any value of the function. But when we say "$\phi \hat{x}$ is a function," we are not making an ambiguous statement. It would be more correct to say that we are making a statement about an ambiguity, taking the view that a function is an ambiguity. The function itself, $\phi \hat{x}$, is the single thing which ambiguously denotes its many values; while ϕx, where x is not specified, is one of the denoted objects, with the ambiguity belonging to the manner of denoting.

We have seen that, in accordance with the vicious-circle principle, the values of a function cannot contain terms only definable in terms of the function. Now given a function $\phi \hat{x}$, the values for the function* are all propositions of the form ϕx. It follows that there must be no propositions, of the form ϕx, in which x has a value which involves $\phi \hat{x}$. (If this were the case, the values of the function would not all be determinate until the function was determinate, whereas we found that the function is not determinate unless its values are previously determinate.) Hence there must be no such thing as the value for $\phi \hat{x}$ with the argument $\phi \hat{x}$, or with any argument which involves $\phi \hat{x}$. That is to say, the symbol "$\phi (\phi \hat{x})$" must not express a proposition, as "ϕa" does if ϕa is a value for $\phi \hat{x}$. In fact "$\phi (\phi \hat{x})$" must be a symbol which does not express anything: we may therefore say that it is not significant. Thus given any function $\phi \hat{x}$, there are arguments with which the function has no value, as well as arguments with which it has a value. We will call the arguments with which $\phi \hat{x}$ has a value "possible values of x." We will say that $\phi \hat{x}$ is "significant with the argument x" when $\phi \hat{x}$ has a value with the argument x.

* We shall speak in this Chapter of "values *for* $\phi \hat{x}$" and of "values *of* ϕx," meaning in each case the same thing, namely ϕa, ϕb, ϕc, etc. The distinction of phraseology serves to avoid ambiguity where several variables are concerned, especially when one of them is a function.

When it is said that *e.g.* "$\phi(\phi\hat{z})$" is meaningless, and therefore neither true nor false, it is necessary to avoid a misunderstanding. If "$\phi(\phi\hat{z})$" were interpreted as meaning "the value for $\phi\hat{z}$ with the argument $\phi\hat{z}$ is true," that would be not meaningless, but false. It is false for the same reason for which "the King of France is bald" is false, namely because there is no such thing as "the value for $\phi\hat{z}$ with the argument $\phi\hat{z}$." But when, with some argument a, we assert ϕa, we are not meaning to assert "the value for $\phi\hat{x}$ with the argument a is true"; we are meaning to assert the actual proposition which *is* the value for $\phi\hat{x}$ with the argument a. Thus for example if $\phi\hat{x}$ is "$\hat{x}$ is a man," ϕ (Socrates) will be "Socrates is a man," *not* "the value for the function '$\hat{x}$ is a man,' with the argument Socrates, is true." Thus in accordance with our principle that "$\phi(\phi\hat{z})$" is meaningless, we cannot legitimately deny "the function '$\hat{x}$ is a man' is a man," because this is nonsense, but we can legitimately deny "the value for the function '$\hat{x}$ is a man' with the argument '$\hat{x}$ is a man' is true," not on the ground that the value in question is false, but on the ground that there is no such value for the function.

We will denote by the symbol "$(x).\phi x$" the proposition "ϕx always*," *i.e.* the proposition which asserts *all* the values for $\phi\hat{x}$. This proposition involves the function $\phi\hat{x}$, not merely an ambiguous value of the function. The assertion of ϕx, where x is unspecified, is a different assertion from the one which asserts all values for $\phi\hat{x}$, for the former is an ambiguous assertion, whereas the latter is in no sense ambiguous. It will be observed that "$(x).\phi x$" does not assert "ϕx with all values of x," because, as we have seen, there must be values of x with which "ϕx" is meaningless. What is asserted by "$(x).\phi x$" is all propositions which are values for $\phi\hat{x}$; hence it is only with such values of x as make "ϕx" significant, *i.e.* with all *possible* arguments, that ϕx is asserted when we assert "$(x).\phi x$." Thus a convenient way to read "$(x).\phi x$" is "ϕx is true with all possible values of x." This is, however, a less accurate reading than "ϕx always," because the notion of *truth* is not part of the content of what is judged. When we judge "all men are mortal," we judge truly, but the notion of truth is not necessarily in our minds, any more than it need be when we judge "Socrates is mortal."

III. *Definition and Systematic Ambiguity of Truth and Falsehood.*

Since "$(x).\phi x$" involves the function $\phi\hat{x}$, it must, according to our principle, be impossible as an argument to ϕ. That is to say, the symbol "$\phi\{(x).\phi x\}$" must be meaningless. This principle would seem, at first sight, to have certain exceptions. Take, for example, the function "$\hat{p}$ is false," and consider the proposition "$(p).p$ is false." This should be a proposition asserting all propositions of the form "p is false." Such a proposition, we

* We use "always" as meaning "in all cases," not "at all times." Similarly "sometimes" will mean "in some cases."

should be inclined to say, must be false, because "p is false" is not always true. Hence we should be led to the proposition

"$\{(p) \,.\, p$ is false$\}$ is false,"

i.e. we should be led to a proposition in which "$(p) \,.\, p$ is false" is the argument to the function "$\hat{p}$ is false," which we had declared to be impossible. Now it will be seen that "$(p) \,.\, p$ is false," in the above, purports to be a proposition about all propositions, and that, by the general form of the vicious-circle principle, there must be no propositions about *all* propositions. Nevertheless, it seems plain that, given any function, there is a proposition (true or false) asserting all its values. Hence we are led to the conclusion that "p is false" and "q is false" must not always be the values, with the arguments p and q, for a single function "$\hat{p}$ is false." This, however, is only possible if the word "false" really has many different meanings, appropriate to propositions of different kinds.

That the words "true" and "false" have many different meanings, according to the kind of proposition to which they are applied, is not difficult to see. Let us take any function $\phi\hat{x}$, and let ϕa be one of its values. Let us call the sort of truth which is applicable to ϕa "*first* truth." (This is not to assume that this would be first truth in another context: it is merely to indicate that it is the first sort of truth in our context.) Consider now the proposition $(x) \,.\, \phi x$. If this has truth of the sort appropriate to it, that will mean that every value ϕx has "first truth." Thus if we call the sort of truth that is appropriate to $(x) \,.\, \phi x$ "*second* truth," we may define "$\{(x) \,.\, \phi x\}$ has second truth" as meaning "every value for $\phi\hat{x}$ has first truth," *i.e.* "$(x) \,.\, (\phi x$ has first truth)." Similarly, if we denote by "$(\exists x) \,.\, \phi x$" the proposition "ϕx sometimes," *i.e.* as we may less accurately express it, "ϕx with some value of x," we find that $(\exists x) \,.\, \phi x$ has second truth if there is an x with which ϕx has first truth; thus we may define "$\{(\exists x) \,.\, \phi x\}$ has second truth" as meaning "some value for $\phi\hat{x}$ has first truth," *i.e.* "$(\exists x) \,.\, (\phi x$ has first truth)." Similar remarks apply to falsehood. Thus "$\{(x) \,.\, \phi x\}$ has second falsehood" will mean "some value for $\phi\hat{x}$ has first falsehood," *i.e.* "$(\exists x) \,.\, (\phi x$ has first falsehood)," while "$\{(\exists x) \,.\, \phi x\}$ has second falsehood" will mean "all values for $\phi\hat{x}$ have first falsehood," *i.e.* "$(x) \,.\, (\phi x$ has first falsehood)." Thus the sort of falsehood that can belong to a general proposition is different from the sort that can belong to a particular proposition.

Applying these considerations to the proposition "$(p) \,.\, p$ is false," we see that the kind of falsehood in question must be specified. If, for example, first falsehood is meant, the function "$\hat{p}$ has first falsehood" is only significant when p is the sort of proposition which has first falsehood or first truth. Hence "$(p) \,.\, p$ is false" will be replaced by a statement which is equivalent to "all propositions having either first truth or first falsehood have first falsehood." This proposition has *second* falsehood, and is not a possible argument to the function "$\hat{p}$ has *first* falsehood." Thus the apparent exception to the principle that "$\phi\{(x) . \phi x\}$" must be meaningless disappears.

Similar considerations will enable us to deal with "not-p" and with "p or q." It might seem as if these were functions in which *any* proposition might appear as argument. But this is due to a systematic ambiguity in the meanings of "not" and "or," by which they adapt themselves to propositions of any order. To explain fully how this occurs, it will be well to begin with a definition of the simplest kind of *truth* and *falsehood.*

The universe consists of objects having various qualities and standing in various relations. Some of the objects which occur in the universe are complex. When an object is complex, it consists of interrelated parts. Let us consider a complex object composed of two parts a and b standing to each other in the relation R. The complex object "a-in-the-relation-R-to-b" may be capable of being *perceived*; when perceived, it is perceived as one object. Attention may show that it is complex; we then *judge* that a and b stand in the relation R. Such a judgment, being derived from perception by mere attention, may be called a "judgment of perception." This judgment of perception, considered as an actual occurrence, is a relation of four terms, namely a and b and R and the percipient. The perception, on the contrary, is a relation of two terms, namely "a-in-the-relation-R-to-b," and the percipient. Since an object of perception cannot be nothing, we cannot perceive "a-in-the-relation-R-to-b" unless a is in the relation R to b. Hence a judgment of perception, according to the above definition, must be true. This does not mean that, in a judgment which *appears* to us to be one of perception, we are sure of not being in error, since we may err in thinking that our judgment has really been derived merely by analysis of what was perceived. But if our judgment has been so derived, it must be true. In fact, we may define *truth*, where such judgments are concerned, as consisting in the fact that there is a complex *corresponding* to the discursive thought which is the judgment. That is, when we judge "a has the relation R to b," our judgment is said to be *true* when there is a complex "a-in-the-relation-R-to-b," and is said to be *false* when this is not the case. This is a definition of truth and falsehood in relation to judgments of this kind.

It will be seen that, according to the above account, a judgment does not have a single object, namely the proposition, but has several interrelated objects. That is to say, the relation which constitutes judgment is not a relation of two terms, namely the judging mind and the proposition, but is a relation of several terms, namely the mind and what are called the constituents of the proposition. That is, when we judge (say) "this is red," what occurs is a relation of three terms, the mind, and "this," and red. On the other hand, when we *perceive* "the redness of this," there is a relation of two terms, namely the mind and the complex object "the redness of this." When a judgment occurs, there is a certain complex entity, composed of the mind and the various objects of the judgment. When the judgment is *true*, in the case of the kind of judgments we have been considering, there is a corresponding complex of the *objects* of the judgment alone. Falsehood, in regard to our present class of judgments, consists in the absence of a corresponding complex composed of the objects alone. It follows from the above theory that a "proposition," in the sense in which a proposition is supposed to be *the* object of a judgment, is a false abstraction, because a judgment has several objects, not one. It is the severalness of the objects in judgment (as opposed to perception) which has led people to speak of thought as "discursive," though they do not appear to have realized clearly what was meant by this epithet.

Owing to the plurality of the objects of a single judgment, it follows that what we call a "proposition" (in the sense in which this is distinguished from the phrase expressing it) is not a single entity at all. That is to say, the phrase which expresses a proposition is what we call an "incomplete" symbol*; it does not have meaning in itself, but requires some supplementation in order to acquire a complete meaning. This fact is somewhat concealed by the circumstance that judgment in itself supplies a sufficient supplement, and that judgment in itself makes no *verbal* addition to the proposition. Thus "the proposition 'Socrates is human'" uses "Socrates is human" in a way which requires a supplement of some kind before it acquires a complete meaning; but when I judge "Socrates is human," the meaning is completed by the act of judging, and we no longer have an incomplete symbol. The fact that propositions are "incomplete symbols" is important philosophically, and is relevant at certain points in symbolic logic.

The judgments we have been dealing with hitherto are such as are of the same form as judgments of perception, *i.e.* their subjects are always particular and definite. But there are many judgments which are not of this form. Such are "all men are mortal," "I met a man," "some men are Greeks." Before dealing with such judgments, we will introduce some technical terms.

We will give the name of "a *complex*" to any such object as "a in the relation R to b" or "a having the quality q," or "a and b and c standing in the relation S." Broadly speaking, a *complex* is anything which occurs in the universe and is not simple. We will call a judgment *elementary* when it merely asserts such things as "a has the relation R to b," "a has the quality q" or "a and b and c stand in the relation S." Then an *elementary* judgment is true when there is a corresponding complex, and false when there is no corresponding complex.

But take now such a proposition as "all men are mortal." Here the judgment does not correspond to *one* complex, but to many, namely "Socrates

* See Chapter III.

is mortal," "Plato is mortal," "Aristotle is mortal," etc. (For the moment, it is unnecessary to inquire whether each of these does not require further treatment before we reach the ultimate complexes involved. For purposes of illustration, "Socrates is mortal" is here treated as an elementary judgment, though it is in fact not one, as will be explained later. Truly elementary judgments are not very easily found.) We do not mean to deny that there may be some relation of the concept *man* to the concept *mortal* which may be *equivalent* to "all men are mortal," but in any case this relation is not the same thing as what we affirm when we say that all men are mortal. Our judgment that all men are mortal collects together a number of elementary judgments. It is not, however, composed of these, since (*e.g.*) the fact that Socrates is mortal is no part of what we assert, as may be seen by considering the fact that our assertion can be understood by a person who has never heard of Socrates. In order to understand the judgment "all men are mortal," it is not necessary to know what men there are. We must admit, therefore, as a radically new kind of judgment, such general assertions as "all men are mortal." We assert that, given that x is human, x is always mortal. That is, we assert "x is mortal" of *every* x which is human. Thus we are able to judge (whether truly or falsely) that *all* the objects which have some assigned property also have some other assigned property. That is, given any propositional functions $\phi\hat{x}$ and $\psi\hat{x}$, there is a judgment asserting ψx with every x for which we have ϕx. Such judgments we will call *general judgments*.

It is evident (as explained above) that the definition of *truth* is different in the case of general judgments from what it was in the case of elementary judgments. Let us call the meaning of *truth* which we gave for elementary judgments "elementary truth." Then when we assert that it is true that all men are mortal, we shall mean that all judgments of the form "x is mortal," where x is a man, have elementary truth. We may define this as "truth of the second order" or "second-order truth." Then if we express the proposition "all men are mortal" in the form

"$(x) . x$ is mortal, where x is a man,"

and call this judgment p, then "p is true" must be taken to mean "p has second-order truth," which in turn means

"(x) . 'x is mortal' has elementary truth, where x is a man."

In order to avoid the necessity for stating explicitly the limitation to which our variable is subject, it is convenient to replace the above interpretation of "all men are mortal" by a slightly different interpretation. The proposition "all men are mortal" is equivalent to "'x is a man' implies 'x is mortal,' with all possible values of x." Here x is not restricted to such values as are men, but may have any value with which "'x is a man' implies 'x is mortal'" is *significant*, *i.e.* either true or false. Such a proposition is called a "formal implication." The advantage of this form is that the values which the variable may take are given by the function to which it is the argument: the values which the variable may take are all those with which the function is significant.

We use the symbol "$(x) . \phi x$" to express the general judgment which asserts all judgments of the form "ϕx." Then the judgment "all men are mortal" is equivalent to

"(x). 'x is a man' implies 'x is a mortal,'"

i.e. (in virtue of the definition of implication) to

"$(x) . x$ is not a man or x is mortal."

As we have just seen, the meaning of *truth* which is applicable to this proposition is not the same as the meaning of *truth* which is applicable to "x is a man" or to "x is mortal." And generally, in any judgment $(x) . \phi x$, the sense in which this judgment is or may be true is not the same as that in which ϕx is or may be true. If ϕx is an elementary judgment, it is true when it *points to* a corresponding complex. But $(x) . \phi x$ does not point to a single corresponding complex: the corresponding complexes are as numerous as the possible values of x.

It follows from the above that such a proposition as "all the judgments made by Epimenides are true" will only be prima facie capable of truth if all his judgments are of the same order. If they are of varying orders, of which the nth is the highest, we may make n assertions of the form "all the judgments of order m made by Epimenides are true," where m has all values up to n. But no such judgment can include itself in its own scope, since such a judgment is always of higher order than the judgments to which it refers.

Let us consider next what is meant by the negation of a proposition of the form "$(x) . \phi x$." We observe, to begin with, that "ϕx in some cases," or "ϕx sometimes," is a judgment which is on a par with "ϕx in all cases," or "ϕx always." The judgment "ϕx sometimes" is true if one or more values of x exist for which ϕx is true. We will express the proposition "ϕx sometimes" by the notation "$(\exists x) . \phi x$," where "$\exists$" stands for "there exists," and the whole symbol may be read "there exists an x such that ϕx." We take the two kinds of judgment expressed by "$(x) . \phi x$" and "$(\exists x) . \phi x$" as primitive ideas. We also take as a primitive idea the negation of an *elementary* proposition. We can then define the negations of $(x) . \phi x$ and $(\exists x) . \phi x$. The negation of any proposition p will be denoted by the symbol "$\sim p$." Then the negation of $(x) . \phi x$ will be *defined* as meaning

"$(\exists x) . \sim \phi x$,"

and the negation of $(\exists x) . \phi x$ will be *defined* as meaning "$(x) . \sim \phi x$." Thus, in the traditional language of formal logic, the negation of a universal affirmative is to be defined as the particular negative, and the negation of the particular affirmative is to be defined as the universal negative. Hence the meaning of negation for such propositions is different from the meaning of negation for elementary propositions.

An analogous explanation will apply to disjunction. Consider the statement "either p, or ϕx always." We will denote the disjunction of two propositions p, q by "$p \lor q$." Then our statement is "$p . \lor . (x) . \phi x$." We will suppose that p is an elementary proposition, and that ϕx is always an elementary proposition. We take the disjunction of two elementary propositions as a primitive idea, and we wish to *define* the disjunction

$$\text{"}p . \lor . (x) . \phi x.\text{"}$$

This may be defined as "$(x) . p \lor \phi x$," *i.e.* "either p is true, or ϕx is always true" is to mean "'p or ϕx' is always true." Similarly we will define

$$\text{"}p . \lor . (\exists x) . \phi x\text{"}$$

as meaning "$(\exists x) . p \lor \phi x$," *i.e.* we define "either p is true or there is an x for which ϕx is true" as meaning "there is an x for which either p or ϕx is true." Similarly we can define a disjunction of two universal propositions: "$(x) . \phi x . \lor . (y) . \psi y$" will be defined as meaning "$(x, y) . \phi x \lor \psi y$," *i.e.* "either ϕx is always true or ψy is always true" is to mean "'ϕx or ψy' is always true." By this method we obtain definitions of disjunctions containing propositions of the form $(x) . \phi x$ or $(\exists x) . \phi x$ in terms of disjunctions of elementary propositions; but the meaning of "disjunction" is not the same for propositions of the forms $(x) . \phi x$, $(\exists x) . \phi x$, as it was for elementary propositions.

Similar explanations could be given for implication and conjunction, but this is unnecessary, since these can be defined in terms of negation and disjunction.

IV. *Why a Given Function requires Arguments of a Certain Type.*

The considerations so far adduced in favour of the view that a function cannot significantly have as argument anything defined in terms of the function itself have been more or less indirect. But a direct consideration of the kinds of functions which have functions as arguments and the kinds of functions which have arguments other than functions will show, if we are not mistaken, that not only is it impossible for a function $\phi \hat{z}$ to have itself or anything derived from it as argument, but that, if $\psi \hat{z}$ is another function such that there are arguments a with which both "ϕa" and "ψa" are significant, then $\psi \hat{z}$ and anything derived from it cannot significantly be argument to $\phi \hat{z}$. This arises from the fact that a function is essentially an ambiguity, and that, if it is to occur in a definite proposition, it must occur in such a way that the ambiguity has disappeared, and a wholly unambiguous statement has resulted. A few illustrations will make this clear. Thus "$(x) . \phi x$," which we have already considered, is a function of $\phi \hat{x}$; as soon as $\phi \hat{x}$ is assigned, we have a definite proposition, wholly free from ambiguity. But it is obvious that we cannot substitute for the function something which is not a function: "$(x) . \phi x$" means "ϕx in all cases," and depends for its significance upon the fact that there are "cases" of ϕx, *i.e.* upon the ambiguity which is characteristic of a function. This instance illustrates the fact that, when a function can occur significantly as argument, something which is not a function cannot occur significantly as argument. But conversely, when something which is not a function can occur significantly as argument, a function cannot occur significantly. Take, *e.g.* "x is a man," and consider "$\phi\hat{x}$ is a man." Here there is nothing to eliminate the ambiguity which constitutes $\phi\hat{x}$; there is thus nothing definite which is said to be a man. A function, in fact, is not a definite object, which could be or not be a man; it is a mere ambiguity awaiting determination, and in order that it may occur significantly it must receive the necessary determination, which it obviously does not receive if it is merely substituted for something determinate in a proposition*. This argument does not, however, apply directly as against such a statement as "$\{(x) . \phi x\}$ is a man." Common sense would pronounce such a statement to be meaningless, but it cannot be condemned on the ground of ambiguity in its subject. We need here a new objection, namely the following: A proposition is not a single entity, but a relation of several; hence a statement in which a proposition appears as subject will only be significant if it can be reduced to a statement about the terms which appear in the proposition. A proposition, like such phrases as "the so-and-so," where grammatically it appears as subject, must be broken up into its constituents if we are to find the true subject or subjects†. But in such a statement as "p is a man," where p is a proposition, this is not possible. Hence "$\{(x) . \phi x\}$ is a man" is meaningless.

V. *The Hierarchy of Functions and Propositions.*

We are thus led to the conclusion, both from the vicious-circle principle and from direct inspection, that the functions to which a given object a can be an argument are incapable of being arguments to each other, and that they have no term in common with the functions to which they can be arguments. We are thus led to construct a hierarchy. Beginning with a and the other terms which can be arguments to the same functions to which a can be argument, we come next to functions to which a is a possible argument, and then to functions to which such functions are possible arguments, and so on. But the hierarchy which has to be constructed is not so simple as might at first appear. The functions which can take a as argument form an illegitimate totality, and themselves require division into a hierarchy of functions. This is easily seen as follows. Let $f(\phi\hat{z}, x)$ be a function of the two variables $\phi\hat{z}$ and x. Then if, keeping x fixed for the moment, we assert this with all possible values of ϕ, we obtain a proposition:

$$(\phi) . f(\phi\hat{z}, x).$$

* Note that statements concerning the significance of a phrase containing "$\phi\hat{z}$" concern the *symbol* "$\phi\hat{z}$," and therefore do not fall under the rule that the elimination of the functional ambiguity is necessary to significance. Significance is a property of signs. Cf. pp. 40, 41.

† Cf. Chapter III.

Here, if x is variable, we have a function of x; but as this function involves a totality of values of $\phi\hat{z}$*, it cannot itself be one of the values included in the totality, by the vicious-circle principle. It follows that the totality of values of $\phi\hat{z}$ concerned in $(\phi) . f(\phi\hat{z}, x)$ is not the totality of all functions in which x can occur as argument, and that there is no such totality as that of all functions in which x can occur as argument.

It follows from the above that a function in which $\phi\hat{z}$ appears as argument requires that "$\phi\hat{z}$" should not stand for *any* function which is capable of a given argument, but must be restricted in such a way that none of the functions which are possible values of "$\phi\hat{z}$" should involve any reference to the totality of such functions. Let us take as an illustration the definition of identity. We might attempt to define "x is identical with y" as meaning "whatever is true of x is true of y," *i.e.* "ϕx always implies ϕy." But here, since we are concerned to assert all values of "ϕx implies ϕy" regarded as a function of ϕ, we shall be compelled to impose upon ϕ some limitation which will prevent us from including among values of ϕ values in which "all possible values of ϕ" are referred to. Thus for example "x is identical with a" is a function of x; hence, if it is a legitimate value of ϕ in "ϕx always implies ϕy," we shall be able to infer, by means of the above definition, that if x is identical with a, and x is identical with y, then y is identical with a. Although the conclusion is sound, the reasoning embodies a vicious-circle fallacy, since we have taken "$(\phi) . \phi x$ implies ϕa" as a possible value of ϕx, which it cannot be. If, however, we impose any limitation upon ϕ, it may happen, so far as appears at present, that with other values of ϕ we might have ϕx true and ϕy false, so that our proposed definition of identity would plainly be wrong. This difficulty is avoided by the "axiom of reducibility," to be explained later. For the present, it is only mentioned in order to illustrate the necessity and the relevance of the hierarchy of functions of a given argument.

Let us give the name "a-functions" to functions that are significant for a given argument a. Then suppose we take any selection of a-functions, and consider the proposition "a satisfies all the functions belonging to the selection in question." If we here replace a by a variable, we obtain an a-function; but by the vicious-circle principle this a-function cannot be a member of our selection, since it refers to the whole of the selection. Let the selection consist of all those functions which satisfy $f(\phi\hat{z})$. Then our new function is

$$(\phi) . \{f(\phi\hat{z}) \text{ implies } \phi x\},$$

where x is the argument. It thus appears that, whatever selection of a-functions we may make, there will be other a-functions that lie outside our

* When we speak of "values *of* $\phi\hat{z}$" it is ϕ, not z, that is to be assigned. This follows from the explanation in the note on p. 40. When the function itself is the variable, it is possible and simpler to write ϕ rather than $\phi\hat{z}$, except in positions where it is necessary to emphasize that an argument must be supplied to secure significance.

selection. Such a-functions, as the above instance illustrates, will always arise through taking a function of two arguments, $\phi\hat{z}$ and x, and asserting all or some of the values resulting from varying ϕ. What is necessary, therefore, in order to avoid vicious-circle fallacies, is to divide our a-functions into "types," each of which contains no functions which refer to the whole of that type.

When something is asserted or denied about all possible values or about some (undetermined) possible values of a variable, that variable is called *apparent*, after Peano. The presence of the words *all* or *some* in a proposition indicates the presence of an apparent variable; but often an apparent variable is really present where language does not at once indicate its presence. Thus for example "A is mortal" means "there is a time at which A will die." Thus a variable time occurs as apparent variable.

The clearest instances of propositions not containing apparent variables are such as express immediate judgments of perception, such as "this is red" or "this is painful," where "this" is something immediately given. In other judgments, even where at first sight no variable appears to be present, it often happens that there really is one. Take (say) "Socrates is human." To Socrates himself, the word "Socrates" no doubt stood for an object of which he was immediately aware, and the judgment "Socrates is human" contained no apparent variable. But to us, who only know Socrates by description, the word "Socrates" cannot mean what it meant to him; it means rather "the person having such-and-such properties," (say) "the Athenian philosopher who drank the hemlock." Now in all propositions about "the so-and-so" there is an apparent variable, as will be shown in Chapter III. Thus in what *we* have in mind when we say "Socrates is human" there is an apparent variable, though there was no apparent variable in the corresponding judgment as made by Socrates, provided we assume that there is such a thing as immediate awareness of oneself.

Whatever may be the instances of propositions not containing apparent variables, it is obvious that propositional functions whose values do not contain apparent variables are the source of propositions containing apparent variables, in the sense in which the function $\phi\hat{x}$ is the source of the proposition $(x) . \phi x$. For the values for $\phi\hat{x}$ do not contain the apparent variable x, which appears in $(x) . \phi x$; if they contain an apparent variable y, this can be similarly eliminated, and so on. This process must come to an end, since no proposition which we can apprehend can contain more than a finite number of apparent variables, on the ground that whatever we can apprehend must be of finite complexity. Thus we must arrive at last at a function of as many variables as there have been stages in reaching it from our original proposition, and this function will be such that its values contain no apparent variables. We may call this function the *matrix* of our original proposition and of any other propositions and functions to be obtained by turning some of the arguments to the function into apparent variables. Thus for example, if we have a matrix-function whose values are $\phi(x, y)$, we shall derive from it

$(y) . \phi(x, y)$, which is a function of x,

$(x) . \phi(x, y)$, which is a function of y,

$(x, y) . \phi(x, y)$, meaning "$\phi(x, y)$ is true with all possible values of x and y."

This last is a proposition containing no *real* variable, *i.e.* no variable except apparent variables.

It is thus plain that all possible propositions and functions are obtainable from matrices by the process of turning the arguments to the matrices into apparent variables. In order to divide our propositions and functions into types, we shall, therefore, start from matrices, and consider how they are to be divided with a view to the avoidance of vicious-circle fallacies in the definitions of the functions concerned. For this purpose, we will use such letters as a, b, c, x, y, z, w, to denote objects which are neither propositions nor functions. Such objects we shall call *individuals*. Such objects will be constituents of propositions or functions, and will be *genuine* constituents, in the sense that they do not disappear on analysis, as (for example) classes do, or phrases of the form "the so-and-so."

The first matrices that occur are those whose values are of the forms

$$\phi x, \ \psi(x, y), \ \chi(x, y, z \ldots),$$

i.e. where the arguments, however many there may be, are all individuals. The functions ϕ, ψ, $\chi \ldots$, since (by definition) they contain no apparent variables, and have no arguments except individuals, do not presuppose any totality of functions. From the functions ψ, $\chi \ldots$ we may proceed to form other functions of x, such as $(y) . \psi(x, y)$, $(\exists y) . \psi(x, y)$, $(y, z) . \chi(x, y, z)$, $(y) : (\exists z) . \chi(x, y, z)$, and so on. All these presuppose no totality except that of individuals. We thus arrive at a certain collection of functions of x, characterized by the fact that they involve no variables except individuals. Such functions we will call "*first-order* functions."

We may now introduce a notation to express "any first-order function." We will denote any first-order function by "$\phi ! \hat{x}$" and any value for such a function by "$\phi ! x$." Thus "$\phi ! x$" stands for any value for any function which involves no variables except individuals. It will be seen that "$\phi ! x$" is itself a function of *two* variables, namely $\phi ! \hat{z}$ and x. Thus $\phi ! x$ involves a variable which is not an individual, namely $\phi ! \hat{z}$. Similarly "$(x) . \phi ! x$" is a function of the variable $\phi ! \hat{z}$, and thus involves a variable other than an individual. Again, if a is a given individual,

"$\phi ! x$ implies $\phi ! a$ with all possible values of ϕ"

is a function of x, but it is not a function of the form $\phi ! x$, because it involves an (apparent) variable ϕ which is not an individual. Let us give the name "predicate" to any first-order function $\phi ! \hat{x}$. (This use of the word "predicate" is only proposed for the purposes of the present discussion.) Then the statement "$\phi ! x$ implies $\phi ! a$ with all possible values of ϕ" may be read "all the predicates of x are predicates of a." This makes a statement about x, but does not attribute to x a *predicate* in the special sense just defined.

Owing to the introduction of the variable first-order function $\phi ! \hat{z}$, we now have a new set of matrices. Thus "$\phi ! x$" is a function which contains no apparent variables, but contains the two real variables $\phi ! \hat{z}$ and x. (It should be observed that when ϕ is assigned, we may obtain a function whose values do involve individuals as apparent variables, for example if $\phi ! x$ is $(y) . \psi (x, y)$. But so long as ϕ is variable, $\phi ! x$ contains no apparent variables.) Again, if a is a definite individual, $\phi ! a$ is a function of the one variable $\phi ! \hat{z}$. If a and b are definite individuals, "$\phi ! a$ implies $\psi ! b$" is a function of the two variables $\phi ! \hat{z}$, $\psi ! \hat{z}$, and so on. We are thus led to a whole set of new matrices,

$$f(\phi ! \hat{z}),\ g(\phi ! \hat{z}, \psi ! \hat{z}),\ F(\phi ! \hat{z}, x), \text{ and so on.}$$

These matrices contain individuals and first-order functions as arguments, but (like all matrices) they contain no apparent variables. Any such matrix, if it contains more than one variable, gives rise to new functions of one variable by turning all its arguments except one into apparent variables. Thus we obtain the functions

$$(\phi) . g(\phi ! \hat{z}, \psi ! \hat{z}), \text{ which is a function of } \psi ! \hat{z}.$$

$$(x) . F(\phi ! \hat{z}, x), \text{ which is a function of } \phi ! \hat{z}.$$

$$(\phi) . F(\phi ! \hat{z}, x), \text{ which is a function of } x.$$

We will give the name of *second-order matrices* to such matrices as have first-order functions among their arguments, and have no arguments except first-order functions and individuals. (It is not *necessary* that they should have individuals among their arguments.) We will give the name of *second-order functions* to such as either are second-order matrices or are derived from such matrices by turning some of the arguments into apparent variables. It will be seen that either an individual or a first-order function may appear as argument to a second-order function. Second-order functions are such as contain variables which are first-order functions, but contain no other variables except (possibly) individuals.

We now have various new classes of functions at our command. In the first place, we have second-order functions which have one argument which is a first-order function. We will denote a variable function of this kind by the notation $f ! (\hat{\phi} ! \hat{z})$, and any value of such a function by $f ! (\phi ! \hat{z})$. Like $\phi ! x$, $f ! (\phi ! \hat{z})$ is a function of two variables, namely $f ! (\hat{\phi} ! \hat{z})$ and $\phi ! \hat{z}$. Among possible values of $f ! (\phi ! \hat{z})$ will be $\phi ! a$ (where a is constant), $(x) . \phi ! x$, $(\exists x) . \phi ! x$, and so on. (These result from assigning a value to f, leaving ϕ to be assigned.) We will call such functions "predicative functions of first-order functions."

In the second place, we have second-order functions of two arguments, one of which is a first-order function while the other is an individual. Let us denote undetermined values of such functions by the notation

$$f\,!\,(\phi\,!\,\hat{z}, x).$$

As soon as x is assigned, we shall have a predicative function of $\phi\,!\,\hat{z}$. If our function contains no first-order function as apparent variable, we shall obtain a predicative function of x if we assign a value to $\phi\,!\,\hat{z}$. Thus, to take the simplest possible case, if $f\,!\,(\phi\,!\,\hat{z}, x)$ is $\phi\,!\,x$, the assignment of a value to ϕ gives us a predicative function of x, in virtue of the definition of "$\phi\,!\,x$." But if $f\,!\,(\phi\,!\,\hat{z}, x)$ contains a first-order function as apparent variable, the assignment of a value to $\phi\,!\,\hat{z}$ gives us a second-order function of x.

In the third place, we have second-order functions of individuals. These will all be derived from functions of the form $f\,!\,(\phi\,!\,\hat{z}, x)$ by turning ϕ into an apparent variable. We do not, therefore, need a new notation for them.

We have also second-order functions of two first-order functions, or of two such functions and an individual, and so on.

We may now proceed in exactly the same way to third-order matrices, which will be functions containing second-order functions as arguments, and containing no apparent variables, and no arguments except individuals and first-order functions and second-order functions. Thence we shall proceed, as before, to third-order functions; and so we can proceed indefinitely. If the highest order of variable occurring in a function, whether as argument or as apparent variable, is a function of the nth order, then the function in which it occurs is of the $n+1$th order. We do not arrive at functions of an infinite order, because the number of arguments and of apparent variables in a function must be finite, and therefore every function must be of a finite order. Since the orders of functions are only defined step by step, there can be no process of "proceeding to the limit," and functions of an infinite order cannot occur.

We will define a function of one variable as *predicative* when it is of the next order above that of its argument, *i.e.* of the lowest order compatible with its having that argument. If a function has several arguments, and the highest order of function occurring among the arguments is the nth, we call the function predicative if it is of the $n+1$th order, *i.e.* again, if it is of the lowest order compatible with its having the arguments it has. A function of several arguments is predicative if there is one of its arguments such that, when the other arguments have values assigned to them, we obtain a predicative function of the one undetermined argument.

It is important to observe that all possible functions in the above hierarchy can be obtained by means of predicative functions and apparent variables. Thus, as we saw, second-order functions of an individual x are of the form

$$(\phi)\,.\,f\,!\,(\phi\,!\,\hat{z}, x) \text{ or } (\exists\phi)\,.\,f\,!\,(\phi\,!\,\hat{z}, x) \text{ or } (\phi, \psi)\,.\,f\,!\,(\phi\,!\,\hat{z}, \psi\,!\,\hat{z}, x) \text{ or etc.,}$$

where f is a second-order predicative function. And speaking generally, a non-predicative function of the nth order is obtained from a predicative function of the nth order by turning all the arguments of the $n-1$th order into apparent variables. (Other arguments also may be turned into apparent variables.) Thus we need not introduce as variables any functions except predicative functions. Moreover, to obtain any function of one variable x, we need not go beyond predicative functions of *two* variables. For the function $(\psi) . f!(\phi ! \hat{z}, \psi ! \hat{z}, x)$, where f is given, is a function of $\phi ! \hat{z}$ and x, and is predicative. Thus it is of the form $F!(\phi ! \hat{z}, x)$, and therefore $(\phi, \psi) . f!(\phi ! \hat{z}, \psi ! \hat{z}, x)$ is of the form $(\phi) . F!(\phi ! \hat{z}, x)$. Thus speaking generally, by a succession of steps we find that, if $\phi ! \hat{u}$ is a predicative function of a sufficiently high order, any assigned non-predicative function of x will be of one of the two forms

$$(\phi) . F!(\phi ! \hat{u}, x), (\exists \phi) . F!(\phi ! \hat{u}, x),$$

where F is a predicative function of $\phi ! \hat{u}$ and x.

The nature of the above hierarchy of functions may be restated as follows. A function, as we saw at an earlier stage, presupposes as part of its meaning the totality of its values, or, what comes to the same thing, the totality of its possible arguments. The arguments to a function may be functions or propositions or individuals. (It will be remembered that individuals were defined as whatever is neither a proposition nor a function.) For the present we neglect the case in which the argument to a function is a proposition. Consider a function whose argument is an individual. This function presupposes the totality of individuals; but unless it contains functions as apparent variables, it does not presuppose any totality of functions. If, however, it does contain a function as apparent variable, then it cannot be defined until some totality of functions has been defined. It follows that we must first define the totality of those functions that have individuals as arguments and contain no functions as apparent variables. These are the *predicative* functions of individuals. Generally, a predicative function of a variable argument is one which involves no totality except that of the possible values of the argument, and those that are presupposed by any one of the possible arguments. Thus a predicative function of a variable argument is any function which can be specified without introducing new kinds of variables not necessarily presupposed by the variable which is the argument.

A closely analogous treatment can be developed for propositions. Propositions which contain no functions and no apparent variables may be called *elementary propositions*. Propositions which are not elementary, which contain no functions, and no apparent variables except individuals, may be called *first-order propositions*. (It should be observed that no variables except *apparent* variables can occur in a proposition, since whatever contains a *real* variable is a function, not a proposition.) Thus elementary and first-order propositions will be values of first-order functions. (It should be remembered that a function is not a constituent in one of its values: thus for example the function "$\hat{x}$ is human" is not a constituent of the proposition "Socrates is human.") Elementary and first-order propositions presuppose no totality except (at most) the totality of individuals. They are of one or other of the three forms

$$\phi ! x;\ (x) . \phi ! x;\ (\exists x) . \phi ! x,$$

where $\phi ! x$ is a predicative function of an individual. If follows that, if p represents a variable elementary proposition or a variable first-order proposition, a function fp is either $f(\phi ! x)$ or $f\{(x) . \phi ! x\}$ or $f\{(\exists x) . \phi ! x\}$. Thus a function of an elementary or a first-order proposition may always be reduced to a function of a first-order function. It follows that a proposition involving the totality of first-order propositions may be reduced to one involving the totality of first-order functions; and this obviously applies equally to higher orders. The propositional hierarchy can, therefore, be derived from the functional hierarchy, and we may define a proposition of the nth order as one which involves an apparent variable of the $n-1$th order in the functional hierarchy. The propositional hierarchy is never required in practice, and is only relevant for the solution of paradoxes; hence it is unnecessary to go into further detail as to the types of propositions.

VI. *The Axiom of Reducibility.*

It remains to consider the "axiom of reducibility." It will be seen that, according to the above hierarchy, no statement can be made significantly about "all a-functions," where a is some given object. Thus such a notion as "all properties of a," meaning "all functions which are true with the argument a," will be illegitimate. We shall have to distinguish the order of function concerned. We can speak of "all predicative properties of a," "all second-order properties of a," and so on. (If a is not an individual, but an object of order n, "second-order properties of a" will mean "functions of order $n+2$ satisfied by a.") But we cannot speak of "all properties of a." In some cases, we can see that some statement will hold of "all nth-order properties of a," whatever value n may have. In such cases, no practical harm results from regarding the statement as being about "all properties of a," provided we remember that it is really a number of statements, and not a single statement which could be regarded as assigning another property to a, over and above all properties. Such cases will always involve some systematic ambiguity, such as that involved in the meaning of the word "truth," as explained above. Owing to this systematic ambiguity, it will be possible, sometimes, to combine into a single verbal statement what are really a number of different statements, corresponding to different orders in the hierarchy. This is illustrated in the case of the liar, where the statement "all A's statements are false" should be broken up into different statements referring to his statements of various orders, and attributing to each the appropriate kind of falsehood.

The axiom of reducibility is introduced in order to legitimate a great mass of reasoning, in which, prima facie, we are concerned with such notions as "all properties of a" or "all a-functions," and in which, nevertheless, it seems scarcely possible to suspect any substantial error. In order to state the axiom, we must first define what is meant by "formal equivalence." Two functions $\phi\hat{x}$, $\psi\hat{x}$ are said to be "formally equivalent" when, with every possible argument x, ϕx is equivalent to ψx, *i.e.* ϕx and ψx are either both true or both false. Thus two functions are formally equivalent when they are satisfied by the same set of arguments. The axiom of reducibility is the assumption that, given any function $\phi\hat{x}$, there is a formally equivalent *predicative* function, *i.e.* there is a predicative function which is true when ϕx is true and false when ϕx is false. In symbols, the axiom is:

$$\vdash : (\exists\psi) : \phi x . \equiv_x . \psi ! x.$$

For two variables, we require a similar axiom, namely: Given any function $\phi(\hat{x}, \hat{y})$, there is a formally equivalent *predicative* function, *i.e.*

$$\vdash : (\exists\psi) : \phi(x, y) . \equiv_{x,y} . \psi ! (x, y).$$

In order to explain the purposes of the axiom of reducibility, and the nature of the grounds for supposing it true, we shall first illustrate it by applying it to some particular cases.

If we call a *predicate* of an object a predicative function which is true of that object, then the predicates of an object are only some among its properties. Take for example such a proposition as "Napoleon had all the qualities that make a great general." We may interpret this as meaning "Napoleon had all the predicates that make a great general." Here there is a predicate which is an apparent variable. If we put "$f(\phi ! \hat{z})$" for "$\phi ! \hat{z}$ is a predicate required in a great general," our proposition is

$$(\phi) : f(\phi ! \hat{z}) \text{ implies } \phi ! (\text{Napoleon}).$$

Since this refers to a totality of predicates, it is not itself a predicate of Napoleon. It by no means follows, however, that there is not some one predicate common and peculiar to great generals. In fact, it is certain that there is such a predicate. For the number of great generals is finite, and each of them certainly possessed some predicate not possessed by any other human being—for example, the exact instant of his birth. The disjunction of such predicates will constitute a predicate common and peculiar to great generals*. If we call this predicate $\psi ! \hat{z}$, the statement we made about Napoleon was equivalent to $\psi !$ (Napoleon). And this equivalence holds equally if we substitute any other individual for Napoleon. Thus we have arrived at a predicate which is always equivalent to the property we ascribed to Napoleon, *i.e.* it belongs to those objects which have this property, and to no others. The axiom of reducibility states that such a predicate always exists, *i.e.* that any property

* When a (finite) set of predicates is given by actual enumeration, their disjunction is a predicate, because no predicate occurs as apparent variable in the disjunction.

of an object belongs to the same collection of objects as those that possess some predicate.

We may next illustrate our principle by its application to *identity*. In this connection, it has a certain affinity with Leibniz's identity of indiscernibles. It is plain that, if x and y are identical, and ϕx is true, then ϕy is true. Here it cannot matter what sort of function $\phi \hat{x}$ may be: the statement must hold for *any* function. But we cannot say, conversely: "If, with all values of ϕ, ϕx implies ϕy, then x and y are identical"; because "all values of ϕ" is inadmissible. If we wish to speak of "all values of ϕ," we must confine ourselves to functions of one order. We may confine ϕ to predicates, or to second-order functions, or to functions of any order we please. But we must necessarily leave out functions of all but one order. Thus we shall obtain, so to speak, a hierarchy of different degrees of identity. We may say "all the predicates of x belong to y," "all second-order properties of x belong to y," and so on. Each of these statements implies all its predecessors: for example, if all second-order properties of x belong to y, then all predicates of x belong to y, for to have all the predicates of x is a second-order property, and this property belongs to x. But we cannot, without the help of an axiom, argue conversely that if all the predicates of x belong to y, all the second-order properties of x must also belong to y. Thus we cannot, without the help of an axiom, be sure that x and y are identical if they have the same predicates. Leibniz's identity of indiscernibles supplied this axiom. It should be observed that by "indiscernibles" he cannot have meant two objects which agree as to *all* their properties, for one of the properties of x is to be identical with x, and therefore this property would necessarily belong to y if x and y agreed in *all* their properties. Some limitation of the common properties necessary to make things indiscernible is therefore implied by the necessity of an axiom. For purposes of illustration (not of interpreting Leibniz) we may suppose the common properties required for indiscernibility to be limited to predicates. Then the identity of indiscernibles will state that if x and y agree as to all their predicates, they are identical. This can be proved if we assume the axiom of reducibility. For, in that case, every property belongs to the same collection of objects as is defined by some predicate. Hence there is some predicate common and peculiar to the objects which are identical with x. This predicate belongs to x, since x is identical with itself; hence it belongs to y, since y has all the predicates of x; hence y is identical with x. It follows that we may *define* x and y as identical when all the predicates of x belong to y, *i.e.* when $(\phi) : \phi ! x . \supset . \phi ! y$. We therefore adopt the following definition of identity*:

$$x = y . = : (\phi) : \phi ! x . \supset . \phi ! y \quad \text{Df.}$$

* Note that in this definition the second sign of equality is to be regarded as combining with "Df" to form one symbol; what is defined is the sign of equality *not* followed by the letters "Df."

But apart from the axiom of reducibility, or some axiom equivalent in this connection, we should be compelled to regard identity as indefinable, and to admit (what seems impossible) that two objects may agree in all their predicates without being identical.

The axiom of reducibility is even more essential in the theory of classes. It should be observed, in the first place, that if we assume the existence of classes, the axiom of reducibility can be proved. For in that case, given any function $\phi\hat{z}$ of whatever order, there is a class α consisting of just those objects which satisfy $\phi\hat{z}$. Hence "ϕx" is equivalent to "x belongs to α." But "x belongs to α" is a statement containing no apparent variable, and is therefore a predicative function of x. Hence if we assume the existence of classes, the axiom of reducibility becomes unnecessary. The assumption of the axiom of reducibility is therefore a smaller assumption than the assumption that there are classes. This latter assumption has hitherto been made unhesitatingly. However, both on the ground of the contradictions, which require a more complicated treatment if classes are assumed, and on the ground that it is always well to make the smallest assumption required for proving our theorems, we prefer to assume the axiom of reducibility rather than the existence of classes. But in order to explain the use of the axiom in dealing with classes, it is necessary first to explain the theory of classes, which is a topic belonging to Chapter III. We therefore postpone to that Chapter the explanation of the use of our axiom in dealing with classes.

It is worth while to note that all the purposes served by the axiom of reducibility are equally well served if we assume that there is always a function of the nth order (where n is fixed) which is formally equivalent to $\phi\hat{x}$, whatever may be the order of $\phi\hat{x}$. Here we shall mean by "a function of the nth order" a function of the nth order relative to the arguments to $\phi\hat{x}$; thus if these arguments are absolutely of the mth order, we assume the existence of a function formally equivalent to $\phi\hat{x}$ whose absolute order is the $m+n$th. The axiom of reducibility in the form assumed above takes $n=1$, but this is not necessary to the use of the axiom. It is also unnecessary that n should be the same for different values of m; what is necessary is that n should be constant so long as m is constant. What is needed is that, where extensional functions of functions are concerned, we should be able to deal with any a-function by means of some formally equivalent function of a given type, so as to be able to obtain results which would otherwise require the illegitimate notion of "all a-functions"; but it does not matter what the given type is. It does not appear, however, that the axiom of reducibility is rendered appreciably more plausible by being put in the above more general but more complicated form.

The axiom of reducibility is equivalent to the assumption that "any combination or disjunction of predicates* is equivalent to a single predicate," *i.e.* to the assumption that, if we assert that x has all the predicates that satisfy a function $f(\phi ! \hat{z})$, there is some one predicate which x will have whenever our assertion is true, and will not have whenever it is false, and similarly if we assert that x has some one of the predicates that satisfy a function $f(\phi ! \hat{z})$. For by means of this assumption, the order of a non-predicative function can be lowered by one; hence, after some finite number of steps, we shall be able to get from any non-predicative function to a formally equivalent predicative function. It does not seem probable that the above assumption could be substituted for the axiom of reducibility in symbolic deductions, since its use would require the explicit introduction of the further assumption that by a finite number of downward steps we can pass from any function to a predicative function, and this assumption could not well be made without developments that are scarcely possible at an early stage. But on the above grounds it seems plain that in fact, if the above alternative axiom is true, so is the axiom of reducibility. The converse, which completes the proof of equivalence, is of course evident.

VII. *Reasons for Accepting the Axiom of Reducibility.*

That the axiom of reducibility is self-evident is a proposition which can hardly be maintained. But in fact self-evidence is never more than a part of the reason for accepting an axiom, and is never indispensable. The reason for accepting an axiom, as for accepting any other proposition, is always largely inductive, namely that many propositions which are nearly indubitable can be deduced from it, and that no equally plausible way is known by which these propositions could be true if the axiom were false, and nothing which is probably false can be deduced from it. If the axiom is apparently self-evident, that only means, practically, that it is nearly indubitable; for things have been thought to be self-evident and have yet turned out to be false. And if the axiom itself is nearly indubitable, that merely adds to the inductive evidence derived from the fact that its consequences are nearly indubitable: it does not provide new evidence of a radically different kind. Infallibility is never attainable, and therefore some element of doubt should always attach to every axiom and to all its consequences. In formal logic, the element of doubt is less than in most sciences, but it is not absent, as appears from the fact that the paradoxes followed from premisses which were not previously known to require limitations. In the case of the axiom of reducibility, the inductive evidence in its favour is very strong, since the reasonings which it permits and the results to which it leads are all such as appear valid. But although it seems very improbable that the axiom should turn out to be false,

* Here the combination or disjunction is supposed to be given intensionally. If given extensionally (*i.e.* by enumeration), no assumption is required; but in this case the number of predicates concerned must be finite.

it is by no means improbable that it should be found to be deducible from some other more fundamental and more evident axiom. It is possible that the use of the vicious-circle principle, as embodied in the above hierarchy of types, is more drastic than it need be, and that by a less drastic use the necessity for the axiom might be avoided. Such changes, however, would not render anything false which had been asserted on the basis of the principles explained above: they would merely provide easier proofs of the same theorems. There would seem, therefore, to be but the slenderest ground for fearing that the use of the axiom of reducibility may lead us into error.

VIII. *The Contradictions.*

We are now in a position to show how the theory of types affects the solution of the contradictions which have beset mathematical logic. For this purpose, we shall begin by an enumeration of some of the more important and illustrative of these contradictions, and shall then show how they all embody vicious-circle fallacies, and are therefore all avoided by the theory of types. It will be noticed that these paradoxes do not relate exclusively to the ideas of number and quantity. Accordingly no solution can be adequate which seeks to explain them merely as the result of some illegitimate use of these ideas. The solution must be sought in some such scrutiny of fundamental logical ideas as has been attempted in the foregoing pages.

(1) The oldest contradiction of the kind in question is the *Epimenides.* Epimenides the Cretan said that all Cretans were liars, and all other statements made by Cretans were certainly lies. Was this a lie? The simplest form of this contradiction is afforded by the man who says "I am lying"; if he is lying, he is speaking the truth, and vice versa.

(2) Let w be the class of all those classes which are not members of themselves. Then, whatever class x may be, "x is a w" is equivalent to "x is not an x." Hence, giving to x the value w, "w is a w" is equivalent to "w is not a w."

(3) Let T be the relation which subsists between two relations R and S whenever R does not have the relation R to S. Then, whatever relations R and S may be, "R has the relation T to S" is equivalent to "R does not have the relation R to S." Hence, giving the value T to both R and S, "T has the relation T to T" is equivalent to "T does not have the relation T to T."

(4) Burali-Forti's contradiction* may be stated as follows: It can be shown that every well-ordered series has an ordinal number, that the series of ordinals up to and including any given ordinal exceeds the given ordinal by one, and (on certain very natural assumptions) that the series of all ordinals (in order of magnitude) is well-ordered. It follows that the series of all

* "Una questione sui numeri transfiniti," *Rendiconti del circolo matematico di Palermo*, Vol. XI. (1897). See *256.

ordinals has an ordinal number, Ω say. But in that case the series of all ordinals including Ω has the ordinal number $\Omega+1$, which must be greater than Ω. Hence Ω is not the ordinal number of all ordinals.

(5) The number of syllables in the English names of finite integers tends to increase as the integers grow larger, and must gradually increase indefinitely, since only a finite number of names can be made with a given finite number of syllables. Hence the names of some integers must consist of at least nineteen syllables, and among these there must be a least. Hence "the least integer not nameable in fewer than nineteen syllables" must denote a definite integer; in fact, it denotes 111,777. But "the least integer not nameable in fewer than nineteen syllables" is itself a name consisting of eighteen syllables; hence the least integer not nameable in fewer than nineteen syllables can be named in eighteen syllables, which is a contradiction*.

(6) Among transfinite ordinals some can be defined, while others can not; for the total number of possible definitions is $\aleph_0$†, while the number of transfinite ordinals exceeds $\aleph_0$. Hence there must be indefinable ordinals, and among these there must be a least. But this is defined as "the least indefinable ordinal," which is a contradiction‡.

(7) Richard's paradox§ is akin to that of the least indefinable ordinal. It is as follows: Consider all decimals that can be defined by means of a finite number of words; let E be the class of such decimals. Then E has $\aleph_0$ terms; hence its members can be ordered as the 1st, 2nd, 3rd, ... Let N be a number defined as follows: If the nth figure in the nth decimal is p, let the nth figure in N be $p+1$ (or 0, if $p=9$). Then N is different from all the members of E, since, whatever finite value n may have, the nth figure in N is different from the nth figure in the nth of the decimals composing E, and therefore N is different from the nth decimal. Nevertheless we have defined N in a finite number of words, and therefore N ought to be a member of E. Thus N both is and is not a member of E.

In all the above contradictions (which are merely selections from an indefinite number) there is a common characteristic, which we may describe as self-reference or reflexiveness. The remark of Epimenides must include itself in its own scope. If *all* classes, provided they are not members of themselves, are members of w, this must also apply to w; and similarly for the

* This contradiction was suggested to us by Mr G. G. Berry of the Bodleian Library.

† $\aleph_0$ is the number of finite integers. See *123.

‡ Cf. König, "Ueber die Grundlagen der Mengenlehre und das Kontinuumproblem," *Math. Annalen*, Vol. LXI. (1905); A. C. Dixon, "On 'well-ordered' aggregates," *Proc. London Math. Soc.* Series 2, Vol. IV. Part I. (1906); and E. W. Hobson, "On the Arithmetic Continuum," *ibid.* The solution offered in the last of these papers depends upon the variation of the "apparatus of definition," and is thus in outline in agreement with the solution adopted here. But it does not invalidate the statement in the text, if "definition" is given a constant meaning.

§ Cf. Poincaré, "Les mathématiques et la logique," *Revue de Métaphysique et de Morale*, Mai 1906, especially sections VII. and IX.; also Peano, *Revista de Mathematica*, Vol. VIII. No. 5 (1906), p. 149 ff.

analogous relational contradiction. In the cases of names and definitions, the paradoxes result from considering non-nameability and indefinability as elements in names and definitions. In the case of Burali-Forti's paradox, the series whose ordinal number causes the difficulty is the series of all ordinal numbers. In each contradiction something is said about *all* cases of some kind, and from what is said a new case seems to be generated, which both is and is not of the same kind as the cases of which *all* were concerned in what was said. But this is the characteristic of illegitimate totalities, as we defined them in stating the vicious-circle principle. Hence all our contradictions are illustrations of vicious-circle fallacies. It only remains to show, therefore, that the illegitimate totalities involved are excluded by the hierarchy of types which we have constructed.

(1) When a man says "I am lying," we may interpret his statement as: "There is a proposition which I am affirming and which is false." That is to say, he is asserting the truth of some value of the function "I assert p, and p is false." But we saw that the word "false" is ambiguous, and that, in order to make it unambiguous, we must specify the order of falsehood, or, what comes to the same thing, the order of the proposition to which falsehood is ascribed. We saw also that, if p is a proposition of the nth order, a proposition in which p occurs as an apparent variable is not of the nth order, but of a higher order. Hence the kind of truth or falsehood which can belong to the statement "there is a proposition p which I am affirming and which has falsehood of the nth order" is truth or falsehood of a higher order than the nth. Hence the statement of Epimenides does not fall within its own scope, and therefore no contradiction emerges.

If we regard the statement "I am lying" as a compact way of simultaneously making all the following statements: "I am asserting a false proposition of the first order," "I am asserting a false proposition of the second order," and so on, we find the following curious state of things: As no proposition of the first order is being asserted, the statement "I am asserting a false proposition of the first order" is false. This statement is of the second order, hence the statement "I am making a false statement of the second order" is true. This is a statement of the third order, and is the only statement of the third order which is being made. Hence the statement "I am making a false statement of the third order" is false. Thus we see that the statement "I am making a false statement of order $2n+1$" is false, while the statement "I am making a false statement of order $2n$" is true. But in this state of things there is no contradiction.

(2) In order to solve the contradiction about the class of classes which are not members of themselves, we shall assume, what will be explained in the next Chapter, that a proposition about a class is always to be reduced to a statement about a function which defines the class, *i.e.* about a function which

is satisfied by the members of the class and by no other arguments. Thus a class is an object derived from a function and presupposing the function, just as, for example, $(x) . \phi x$ presupposes the function $\phi \hat{x}$. Hence a class cannot, by the vicious-circle principle, significantly be the argument to its defining function, that is to say, if we denote by "$\hat{z}(\phi z)$" the class defined by $\phi \hat{z}$, the symbol "$\phi\{\hat{z}(\phi z)\}$" must be meaningless. Hence a class neither satisfies nor does not satisfy its defining function, and therefore (as will appear more fully in Chapter III) is neither a member of itself nor not a member of itself. This is an immediate consequence of the limitation to the possible arguments to a function which was explained at the beginning of the present Chapter. Thus if α is a class, the statement "α is not a member of α" is always meaningless, and there is therefore no sense in the phrase "the class of those classes which are not members of themselves." Hence the contradiction which results from supposing that there is such a class disappears.

(3) Exactly similar remarks apply to "the relation which holds between R and S whenever R does not have the relation R to S." Suppose the relation R is defined by a function $\phi(x, y)$, *i.e.* R holds between x and y whenever $\phi(x, y)$ is true, but not otherwise. Then in order to interpret "R has the relation R to S," we shall have to suppose that R and S can significantly be the arguments to ϕ. But (assuming, as will appear in Chapter III, that R presupposes its defining function) this would require that ϕ should be able to take as argument an object which is defined in terms of ϕ, and this no function can do, as we saw at the beginning of this Chapter. Hence "R has the relation R to S" is meaningless, and the contradiction ceases.

(4) The solution of Burali-Forti's contradiction requires some further developments for its solution. At this stage, it must suffice to observe that a series is a relation, and an ordinal number is a class of series. (These statements are justified in the body of the work.) Hence a series of ordinal numbers is a relation between classes of relations, and is of higher type than any of the series which are members of the ordinal numbers in question. Burali-Forti's "ordinal number of all ordinals" must be the ordinal number of all ordinals of a given type, and must therefore be of higher type than any of these ordinals. Hence it is not one of these ordinals, and there is no contradiction in its being greater than any of them*.

(5) The paradox about "the least integer not nameable in fewer than nineteen syllables" embodies, as is at once obvious, a vicious-circle fallacy. For the word "nameable" refers to the totality of names, and yet is allowed to occur in what professes to be one among names. Hence there can be no such thing as a totality of names, in the sense in which the paradox speaks

* The solution of Burali-Forti's paradox by means of the theory of types is given in detail in *256.

of "names." It is easy to see that, in virtue of the hierarchy of functions, the theory of types renders a totality of "names" impossible. We may, in fact, distinguish names of different orders as follows: (*a*) Elementary names will be such as are true "proper names," *i.e.* conventional appellations not involving any description. (*b*) First-order names will be such as involve a description by means of a first-order function; that is to say, if $\phi ! \hat{x}$ is a first-order function, "the term which satisfies $\phi ! \hat{x}$" will be a first-order name, though there will not always be an object named by this name. (*c*) Second-order names will be such as involve a description by means of a second-order function; among such names will be those involving a reference to the totality of first-order names. And so we can proceed through a whole hierarchy. But at no stage can we give a meaning to the word "nameable" unless we specify the order of names to be employed; and any name in which the phrase "nameable by names of order n" occurs is necessarily of a higher order than the nth. Thus the paradox disappears.

The solutions of the paradox about the least indefinable ordinal and of Richard's paradox are closely analogous to the above. The notion of "definable," which occurs in both, is nearly the same as "nameable," which occurs in our fifth paradox: "definable" is what "nameable" becomes when elementary names are excluded, *i.e.* "definable" means "nameable by a name which is not elementary." But here there is the same ambiguity as to type as there was before, and the same need for the addition of words which specify the type to which the definition is to belong. And however the type may be specified, "the least ordinal not definable by definitions of this type" is a definition of a higher type; and in Richard's paradox, when we confine ourselves, as we must, to decimals that have a definition of a given type, the number N, which causes the paradox, is found to have a definition which belongs to a higher type, and thus not to come within the scope of our previous definitions.

An indefinite number of other contradictions, of similar nature to the above seven, can easily be manufactured. In all of them, the solution is of the same kind. In all of them, the appearance of contradiction is produced by the presence of some word which has systematic ambiguity of type, such as *truth, falsehood, function, property, class, relation, cardinal, ordinal, name, definition*. Any such word, if its typical ambiguity is overlooked, will apparently generate a totality containing members defined in terms of itself, and will thus give rise to vicious-circle fallacies. In most cases, the conclusions of arguments which involve vicious-circle fallacies will not be self-contradictory, but wherever we have an illegitimate totality, a little ingenuity will enable us to construct a vicious-circle fallacy leading to a contradiction, which disappears as soon as the typically ambiguous words are rendered typically definite, *i.e.* are determined as belonging to this or that type.

Thus the appearance of contradiction is always due to the presence of words embodying a concealed typical ambiguity, and the solution of the apparent contradiction lies in bringing the concealed ambiguity to light.

In spite of the contradictions which result from unnoticed typical ambiguity, it is not desirable to avoid words and symbols which have typical ambiguity. Such words and symbols embrace practically all the ideas with which mathematics and mathematical logic are concerned: the systematic ambiguity is the result of a systematic analogy. That is to say, in almost all the reasonings which constitute mathematics and mathematical logic, we are using ideas which may receive any one of an infinite number of different typical determinations, any one of which leaves the reasoning valid. Thus by employing typically ambiguous words and symbols, we are able to make one chain of reasoning applicable to any one of an infinite number of different cases, which would not be possible if we were to forego the use of typically ambiguous words and symbols.

Among propositions wholly expressed in terms of typically ambiguous notions practically the only ones which may differ, in respect of truth or falsehood, according to the typical determination which they receive, are existence-theorems. If we assume that the total number of individuals is n, then the total number of classes of individuals is 2^n, the total number of classes of classes of individuals is 2^{2^n}, and so on. Here n may be either finite or infinite, and in either case $2^n > n$. Thus cardinals greater than n but not greater than 2^n exist as applied to classes of classes, but not as applied to classes of individuals, so that whatever may be supposed to be the number of individuals, there will be existence-theorems which hold for higher types but not for lower types. Even here, however, so long as the number of individuals is not asserted, but is merely assumed hypothetically, we may replace the type of individuals by any other type, provided we make a corresponding change in all the other types occurring in the same context. That is, we may give the name "relative individuals" to the members of an arbitrarily chosen type τ, and the name "relative classes of individuals" to classes of "relative individuals," and so on. Thus so long as only hypotheticals are concerned, in which existence-theorems for one type are shown to be implied by existence-theorems for another, only *relative* types are relevant even in existence-theorems. This applies also to cases where the hypothesis (and therefore the conclusion) is *asserted*, provided the assertion holds for any type, however chosen. For example, any type has at least one member; hence any type which consists of classes, of whatever order, has at least two members. But the further pursuit of these topics must be left to the body of the work.

CHAPTER III

INCOMPLETE SYMBOLS

(1) *Descriptions.* By an "incomplete" symbol we mean a symbol which is not supposed to have any meaning in isolation, but is only defined in certain contexts. In ordinary mathematics, for example, $\frac{d}{dx}$ and $\int_a^b$ are incomplete symbols: something has to be supplied before we have anything significant. Such symbols have what may be called a "definition in use." Thus if we put

$$\nabla^2 = \frac{\partial^2}{\partial x^2} + \frac{\partial^2}{\partial y^2} + \frac{\partial^2}{\partial z^2} \quad \text{Df},$$

we define the *use* of ∇^2, but ∇^2 by itself remains without meaning. This distinguishes such symbols from what (in a generalized sense) we may call *proper names*: "Socrates," for example, stands for a certain man, and therefore has a meaning by itself, without the need of any context. If we supply a context, as in "Socrates is mortal," these words express a fact of which Socrates himself is a constituent: there is a certain object, namely Socrates, which does have the property of mortality, and this object is a constituent of the complex fact which we assert when we say "Socrates is mortal." But in other cases, this simple analysis fails us. Suppose we say: "The round square does not exist." It seems plain that this is a true proposition, yet we cannot regard it as denying the existence of a certain object called "the round square." For if there were such an object, it would exist: we cannot first assume that there is a certain object, and then proceed to deny that there is such an object. Whenever the grammatical subject of a proposition can be supposed not to exist without rendering the proposition meaningless, it is plain that the grammatical subject is not a proper name, *i.e.* not a name directly representing some object. Thus in all such cases, the proposition must be capable of being so analysed that what was the grammatical subject shall have disappeared. Thus when we say "the round square does not exist," we may, as a first attempt at such analysis, substitute "it is false that there is an object x which is both round and square." Generally, when "the so-and-so" is said not to exist, we have a proposition of the form*

$$\text{“}\sim \mathrm{E}\,!\,(\iota x)(\phi x),\text{”}$$

i.e.

$$\sim\{(\exists c) : \phi x \,.\equiv_x .\, x = c\},$$

or some equivalent. Here the apparent grammatical subject $(\iota x)(\phi x)$ has completely disappeared; thus in "$\sim \mathrm{E}\,!\,(\iota x)(\phi x)$," $(\iota x)(\phi x)$ is an *incomplete* symbol.

* Cf. pp. 30, 31.

By an extension of the above argument, it can easily be shown that $(\imath x)(\phi x)$ is *always* an incomplete symbol. Take, for example, the following proposition: "Scott is the author of Waverley." [Here "the author of Waverley" is $(\imath x)$ (x wrote Waverley).] This proposition expresses an identity; thus if "the author of Waverley" could be taken as a proper name, and supposed to stand for some object c, the proposition would be "Scott is c." But if c is any one except Scott, this proposition is false; while if c *is* Scott, the proposition is "Scott is Scott," which is trivial, and plainly different from "Scott is the author of Waverley." Generalizing, we see that the proposition

$$a = (\imath x)(\phi x)$$

is one which may be true or may be false, but is never merely trivial, like $a = a$; whereas, if $(\imath x)(\phi x)$ were a proper name, $a = (\imath x)(\phi x)$ would necessarily be either false or the same as the trivial proposition $a = a$. We may express this by saying that $a = (\imath x)(\phi x)$ is not a value of the propositional function $a = y$, from which it follows that $(\imath x)(\phi x)$ is not a value of y. But since y may be anything, it follows that $(\imath x)(\phi x)$ is nothing. Hence, since in use it has meaning, it must be an incomplete symbol.

It might be suggested that "Scott is the author of Waverley" asserts that "Scott" and "the author of Waverley" are two names for the same object. But a little reflection will show that this would be a mistake. For if that were the meaning of "Scott is the author of Waverley," what would be required for its truth would be that Scott should have been *called* the author of Waverley: if he had been so called, the proposition would be true, even if some one else had written Waverley; while if no one called him so, the proposition would be false, even if he had written Waverley. But in fact he was the author of Waverley at a time when no one called him so, and he would not have been the author if every one had called him so but some one else had written Waverley. Thus the proposition "Scott is the author of Waverley" is not a proposition about names, like "Napoleon is Bonaparte"; and this illustrates the sense in which "the author of Waverley" differs from a true proper name.

Thus all phrases (other than propositions) containing the word *the* (in the singular) are incomplete symbols: they have a meaning in use, but not in isolation. For "the author of Waverley" cannot mean the same as "Scott," or "Scott is the author of Waverley" would mean the same as "Scott is Scott," which it plainly does not; nor can "the author of Waverley" mean anything other than "Scott," or "Scott is the author of Waverley" would be false. Hence "the author of Waverley" means nothing.

It follows from the above that we must not attempt to define "$(\imath x)(\phi x)$," but must define the *uses* of this symbol, *i.e.* the propositions in whose symbolic expression it occurs. Now in seeking to define the uses of this symbol, it is important to observe the import of propositions in which it occurs. Take as an illustration: "The author of Waverley was a poet." This implies (1) that Waverley was written, (2) that it was written by one man, and not in collaboration, (3) that the one man who wrote it was a poet. If any one of these fails, the proposition is false. Thus "the author of 'Slawkenburgius on Noses' was a poet" is false, because no such book was ever written; "the author of 'The Maid's Tragedy' was a poet" is false, because this play was written by Beaumont and Fletcher jointly. These two possibilities of falsehood do not arise if we say "Scott was a poet." Thus our interpretation of the uses of $(\imath x)(\phi x)$ must be such as to allow for them. Now taking ϕx to replace "x wrote Waverley," it is plain that any statement apparently about $(\imath x)(\phi x)$ requires (1) $(\exists x) . (\phi x)$ and (2) $\phi x . \phi y . \supset_{x,y} . x = y$; here (1) states that *at least* one object satisfies ϕx, while (2) states that *at most* one object satisfies ϕx. The two together are equivalent to

$$(\exists c) : \phi x . \equiv_x . x = c,$$

which we defined as $\quad E ! (\imath x)(\phi x)$.

Thus "$E ! (\imath x)(\phi x)$" must be part of what is affirmed by any proposition about $(\imath x)(\phi x)$. If our proposition is $f\{(\imath x)(\phi x)\}$, what is further affirmed is fc, if $\phi x . \equiv_x . x = c$. Thus we have

$$f\{(\imath x)(\phi x)\} . = : (\exists c) : \phi x . \equiv_x . x = c : fc \quad \text{Df},$$

i.e. "the x satisfying ϕx satisfies fx" is to mean: "There is an object c such that ϕx is true when, and only when, x is c, and fc is true," or, more exactly: "There is a c such that 'ϕx' is always equivalent to 'x is c,' and fc." In this, "$(\imath x)(\phi x)$" has completely disappeared; thus "$(\imath x)(\phi x)$" is merely symbolic, and does not directly represent an object, as single small Latin letters are assumed to do*.

The proposition "$a = (\imath x)(\phi x)$" is easily shown to be equivalent to "$\phi x . \equiv_x . x = a$." For, by the definition, it is

$$(\exists c) : \phi x . \equiv_x . x = c : a = c,$$

i.e. "there is a c for which $\phi x . \equiv_x . x = c$, and this c is a," which is equivalent to "$\phi x . \equiv_x . x = a$." Thus "Scott is the author of Waverley" is equivalent to:

"'x wrote Waverley' is always equivalent to 'x is Scott,'"

i.e. "x wrote Waverley" is true when x is Scott and false when x is not Scott.

Thus although "$(\imath x)(\phi x)$" has no meaning by itself, it may be substituted for y in any propositional function fy, and we get a significant proposition, though not a value of fy.

When $f\{(\imath x)(\phi x)\}$, as above defined, forms part of some other proposition, we shall say that $(\imath x)(\phi x)$ has a *secondary* occurrence. When $(\imath x)(\phi x)$ has a secondary occurrence, a proposition in which it occurs may be true even when $(\imath x)(\phi x)$ does not exist. This applies, *e.g.* to the proposition: "There

* We shall generally write "$f(\imath x)(\phi x)$" rather than "$f\{(\imath x)(\phi x)\}$" in future.

is no such person as the King of France." We may interpret this as

$$\sim \{E\,!\,(\imath x)(\phi x)\},$$

or as

$$\sim \{(\exists c) . c = (\imath x)(\phi x)\},$$

if "ϕx" stands for "x is King of France." In either case, what is asserted is that a proposition p in which $(\imath x)(\phi x)$ occurs is false, and this proposition p is thus part of a larger proposition. The same applies to such a proposition as the following: "If France were a monarchy, the King of France would be of the House of Orleans."

It should be observed that such a proposition as

$$\sim f\,\{(\imath x)(\phi x)\}$$

is ambiguous; it may deny $f\{(\imath x)(\phi x)\}$, in which case it will be true if $(\imath x)(\phi x)$ does not exist, or it may mean

$$(\exists c) : \phi x . \equiv_x . x = c : \sim fc,$$

in which case it can only be true if $(\imath x)(\phi x)$ exists. In ordinary language, the latter interpretation would usually be adopted. For example, the proposition "the King of France is not bald" would usually be rejected as false, being held to mean "the King of France exists and is not bald," rather than "it is false that the King of France exists and is bald." When $(\imath x)(\phi x)$ exists, the two interpretations of the ambiguity give equivalent results; but when $(\imath x)(\phi x)$ does not exist, one interpretation is true and one is false. It is necessary to be able to distinguish these in our notation; and generally, if we have such propositions as

$$\psi(\imath x)(\phi x) . \supset . p,$$
$$p . \supset . \psi(\imath x)(\phi x),$$
$$\psi(\imath x)(\phi x) . \supset . \chi(\imath x)(\phi x),$$

and so on, we must be able by our notation to distinguish whether the whole or only part of the proposition concerned is to be treated as the "$f(\imath x)(\phi x)$" of our definition. For this purpose, we will put "$[(\imath x)(\phi x)]$" followed by dots at the beginning of the part (or whole) which is to be taken as $f(\imath x)(\phi x)$, the dots being sufficiently numerous to bracket off the $f(\imath x)(\phi x)$; *i.e.* $f(\imath x)(\phi x)$ is to be everything following the dots until we reach an equal number of dots not signifying a logical product, or a greater number signifying a logical product, or the end of the sentence, or the end of a bracket enclosing "$[(\imath x)(\phi x)]$." Thus

$$[(\imath x)(\phi x)] . \psi(\imath x)(\phi x) . \supset . p$$

will mean

$$(\exists c) : \phi x . \equiv_x . x = c : \psi c : \supset . p,$$

but

$$[(\imath x)(\phi x)] : \psi(\imath x)(\phi x) . \supset . p$$

will mean

$$(\exists c) : \phi x . \equiv_x . x = c : \psi c . \supset . p.$$

It is important to distinguish these two, for if $(\imath x)(\phi x)$ does not exist, the first is true and the second false. Again

$$[(\imath x)(\phi x)] . \sim \psi(\imath x)(\phi x)$$

will mean $(\exists c) : \phi x . \equiv_x . x = c : \sim \psi c,$

while $\sim \{[(\imath x)(\phi x)] . \psi(\imath x)(\phi x)\}$

will mean $\sim \{(\exists c) : \phi x . \equiv_x . x = c : \psi c\}.$

Here again, when $(\imath x)(\phi x)$ does not exist, the first is false and the second true.

In order to avoid this ambiguity in propositions containing $(\imath x)(\phi x)$, we amend our definition, or rather our notation, putting

$$[(\imath x)(\phi x)] . f(\imath x)(\phi x) . = : (\exists c) : \phi x . \equiv_x . x = c : fc \quad \text{Df.}$$

By means of this definition, we avoid any doubt as to the portion of our whole asserted proposition which is to be treated as the "$f(\imath x)(\phi x)$" of the definition. This portion will be called the *scope* of $(\imath x)(\phi x)$. Thus in

$$[(\imath x)(\phi x)] . f(\imath x)(\phi x) . \supset . p$$

the scope of $(\imath x)(\phi x)$ is $f(\imath x)(\phi x)$; but in

$$[(\imath x)(\phi x)] : f(\imath x)(\phi x) . \supset . p$$

the scope is $f(\imath x)(\phi x) . \supset . p;$

in $\sim \{[(\imath x)(\phi x)] . f(\imath x)(\phi x)\}$

the scope is $f(\imath x)(\phi x)$; but in

$$[(\imath x)(\phi x)] . \sim f(\imath x)(\phi x)$$

the scope is $\sim f(\imath x)(\phi x).$

It will be seen that when $(\imath x)(\phi x)$ has the whole of the proposition concerned for its scope, the proposition concerned cannot be true unless $E!(\imath x)(\phi x)$; but when $(\imath x)(\phi x)$ has only part of the proposition concerned for its scope, it may often be true even when $(\imath x)(\phi x)$ does not exist. It will be seen further that when $E!(\imath x)(\phi x)$, we may enlarge or diminish the scope of $(\imath x)(\phi x)$ as much as we please without altering the truth-value of any proposition in which it occurs.

If a proposition contains two descriptions, say $(\imath x)(\phi x)$ and $(\imath x)(\psi x)$, we have to distinguish which of them has the larger scope, *i.e.* we have to distinguish

(1) $[(\imath x)(\phi x)] : [(\imath x)(\psi x)] . f\{(\imath x)(\phi x), (\imath x)(\psi x)\},$

(2) $[(\imath x)(\psi x)] : [(\imath x)(\phi x)] . f\{(\imath x)(\phi x), (\imath x)(\psi x)\}.$

The first of these, eliminating $(\imath x)(\phi x)$, becomes

(3) $(\exists c) : \phi x . \equiv_x . x = c : [(\imath x)(\psi x)] . f\{c, (\imath x)(\psi x)\},$

which, eliminating $(\imath x)(\psi x)$, becomes

(4) $(\exists c) :. \phi x . \equiv_x . x = c :. (\exists d) : \psi x . \equiv_x . x = d : f(c, d),$

and the same proposition results if, in (1), we eliminate first $(\imath x)(\psi x)$ and then $(\imath x)(\phi x)$. Similarly (2) becomes, when $(\imath x)(\phi x)$ and $(\imath x)(\psi x)$ are eliminated,

(5) $(\exists d) :. \psi x . \equiv_x . x = d :. (\exists c) : \phi x . \equiv_x . x = c : f(c, d).$

(4) and (5) are equivalent, so that the truth-value of a proposition containing two descriptions is independent of the question which has the larger scope.

It will be found that, in most cases in which descriptions occur, their scope is, in practice, the smallest proposition enclosed in dots or other brackets in which they are contained. Thus for example

$$[(\imath x)(\phi x)] . \psi(\imath x)(\phi x) . \supset . [(\imath x)(\phi x)] . \chi(\imath x)(\phi x)$$

will occur much more frequently than

$$[(\imath x)(\phi x)] : \psi(\imath x)(\phi x) . \supset . \chi(\imath x)(\phi x).$$

For this reason it is convenient to decide that, when the scope of an occurrence of $(\imath x)(\phi x)$ is the smallest proposition, enclosed in dots or other brackets, in which the occurrence in question is contained, the scope need not be indicated by "$[(\imath x)(\phi x)]$." Thus *e.g.*

$$p . \supset . a = (\imath x)(\phi x)$$

will mean $\quad p . \supset . [(\imath x)(\phi x)] . a = (\imath x)(\phi x);$

and $\quad p . \supset . (\exists a) . a = (\imath x)(\phi x)$

will mean $\quad p . \supset . (\exists a) . [(\imath x)(\phi x)] . a = (\imath x)(\phi x);$

and $\quad p . \supset . a \neq (\imath x)(\phi x)$

will mean $\quad p . \supset . [(\imath x)(\phi x)] . \sim \{a = (\imath x)(\phi x)\};$

but $\quad p . \supset . \sim \{a = (\imath x)(\phi x)\}$

will mean $\quad p . \supset . \sim \{[(\imath x)(\phi x)] . a = (\imath x)(\phi x)\}.$

This convention enables us, in the vast majority of cases that actually occur, to dispense with the explicit indication of the scope of a descriptive symbol; and it will be found that the convention agrees very closely with the tacit conventions of ordinary language on this subject. Thus for example, if "$(\imath x)(\phi x)$" is "the so-and-so," "$a \neq (\imath x)(\phi x)$" is to be read "$a$ is not the so-and-so," which would ordinarily be regarded as implying that "the so-and-so" exists; but "$\sim \{a = (\imath x)(\phi x)\}$" is to be read "it is not true that a is the so-and-so," which would generally be allowed to hold if "the so-and-so" does not exist. Ordinary language is, of course, rather loose and fluctuating in its implications on this matter; but subject to the requirement of definiteness, our convention seems to keep as near to ordinary language as possible.

In the case when the smallest proposition enclosed in dots or other brackets contains two or more descriptions, we shall assume, in the absence of any indication to the contrary, that one which typographically occurs earlier has a larger scope than one which typographically occurs later. Thus

$$(\imath x)(\phi x) = (\imath x)(\psi x)$$

will mean $\quad (\exists c) : \phi x . \equiv_x . x = c : [(\imath x)(\psi x)] . c = (\imath x)(\psi x),$

while $\quad (\imath x)(\psi x) = (\imath x)(\phi x)$

will mean $\quad (\exists d) : \psi x . \equiv_x . x = d : [(\imath x)(\phi x)] . (\imath x)(\phi x) = d.$

These two propositions are easily shown to be equivalent.

(2) *Classes.* The symbols for classes, like those for descriptions, are, in our system, incomplete symbols: their *uses* are defined, but they themselves are not assumed to mean anything at all. That is to say, the uses of such

symbols are so defined that, when the *definiens* is substituted for the *definiendum*, there no longer remains any symbol which could be supposed to represent a class. Thus classes, so far as we introduce them, are merely symbolic or linguistic conveniences, not genuine objects as their members are if they are individuals.

It is an old dispute whether formal logic should concern itself mainly with intensions or with extensions. In general, logicians whose training was mainly philosophical have decided for intensions, while those whose training was mainly mathematical have decided for extensions. The facts seem to be that, while mathematical logic requires extensions, philosophical logic refuses to supply anything except intensions. Our theory of classes recognizes and reconciles these two apparently opposite facts, by showing that an extension (which is the same as a class) is an incomplete symbol, whose use always acquires its meaning through a reference to intension.

In the case of descriptions, it was possible to *prove* that they are incomplete symbols. In the case of classes, we do not know of any equally definite proof, though arguments of more or less cogency can be elicited from the ancient problem of the One and the Many*. It is not necessary for our purposes, however, to assert dogmatically that there are no such things as classes. It is only necessary for us to show that the incomplete symbols which we introduce as representatives of classes yield all the propositions for the sake of which classes might be thought essential. When this has been shown, the mere principle of economy of primitive ideas leads to the non-introduction of classes except as incomplete symbols.

To explain the theory of classes, it is necessary first to explain the distinction between *extensional* and *intensional* functions. This is effected by the following definitions:

The *truth-value* of a proposition is truth if it is true, and falsehood if it is false. (This expression is due to Frege.)

Two propositions are said to be *equivalent* when they have the same truth-value, *i.e.* when they are both true or both false.

Two propositional functions are said to be *formally equivalent* when they are equivalent with every possible argument, *i.e.* when any argument which satisfies the one satisfies the other, and vice versa. Thus "$\hat{x}$ is a man" is formally equivalent to "$\hat{x}$ is a featherless biped"; "$\hat{x}$ is an even prime" is formally equivalent to "$\hat{x}$ is identical with 2."

A function of a function is called *extensional* when its truth-value with any argument is the same as with any formally equivalent argument. That is to

* Briefly, these arguments reduce to the following: If there is such an object as a class, it must be in some sense *one* object. Yet it is only of classes that *many* can be predicated. Hence, if we admit classes as objects, we must suppose that the same object can be both one and many, which seems impossible.

say, $f(\phi\hat{z})$ is an extensional function of $\phi\hat{z}$ if, provided $\psi\hat{z}$ is formally equivalent to $\phi\hat{z}$, $f(\phi\hat{z})$ is equivalent to $f(\psi\hat{z})$. Here the apparent variables ϕ and ψ are necessarily of the type from which arguments can significantly be supplied to f. We find no need to use as apparent variables any functions of non-predicative types; accordingly in the sequel all extensional functions considered are in fact functions of predicative functions*.

A function of a function is called *intensional* when it is not extensional.

The nature and importance of the distinction between intensional and extensional functions will be made clearer by some illustrations. The proposition "'x is a man' always implies 'x is a mortal'" is an extensional function of the function "$\hat{x}$ is a man," because we may substitute, for "x is a man," "x is a featherless biped," or any other statement which applies to the same objects to which "x is a man" applies, and to no others. But the proposition "A believes that 'x is a man' always implies 'x is a mortal'" is an intensional function of "$\hat{x}$ is a man," because A may never have considered the question whether featherless bipeds are mortal, or may believe wrongly that there are featherless bipeds which are not mortal. Thus even if "x is a featherless biped" is formally equivalent to "x is a man," it by no means follows that a person who believes that all men are mortal must believe that all featherless bipeds are mortal, since he may have never thought about featherless bipeds, or have supposed that featherless bipeds were not always men. Again the proposition "the number of arguments that satisfy the function $\phi ! \hat{z}$ is n" is an extensional function of $\phi ! \hat{z}$, because its truth or falsehood is unchanged if we substitute for $\phi ! \hat{z}$ any other function which is true whenever $\phi ! \hat{z}$ is true, and false whenever $\phi ! \hat{z}$ is false. But the proposition "A asserts that the number of arguments satisfying $\phi ! \hat{z}$ is n" is an intensional function of $\phi ! \hat{z}$, since, if A asserts this concerning $\phi ! \hat{z}$, he certainly cannot assert it concerning all predicative functions that are equivalent to $\phi ! \hat{z}$, because life is too short. Again, consider the proposition "two white men claim to have reached the North Pole." This proposition states "two arguments satisfy the function '$\hat{x}$ is a white man who claims to have reached the North Pole.'" The truth or falsehood of this proposition is unaffected if we substitute for "$\hat{x}$ is a white man who claims to have reached the North Pole" any other statement which holds of the same arguments, and of no others. Hence it is an extensional function. But the proposition "it is a strange coincidence that two white men should claim to have reached the North Pole," which states "it is a strange coincidence that two arguments should satisfy the function '$\hat{x}$ is a white man who claims to have reached the North Pole,'" is not equivalent to "it is a strange coincidence that two arguments should satisfy the function '$\hat{x}$ is Dr Cook or Commander Peary.'" Thus "it is a strange coincidence that $\phi ! \hat{x}$ should be satisfied by two arguments" is an intensional function of $\phi ! \hat{x}$.

* Cf. p. 53.

The above instances illustrate the fact that the functions of functions with which mathematics is specially concerned are extensional, and that intensional functions of functions only occur where non-mathematical ideas are introduced, such as what somebody believes or affirms, or the emotions aroused by some fact. Hence it is natural, in a mathematical logic, to lay special stress on *extensional* functions of functions.

When two functions are formally equivalent, we may say that they *have the same extension*. In this definition, we are in close agreement with usage. We do not assume that there is such a thing as an extension: we merely define the whole phrase "having the same extension." We may now say that an extensional function of a function is one whose truth or falsehood depends only upon the extension of its argument. In such a case, it is convenient to regard the statement concerned as being about the extension. Since extensional functions are many and important, it is natural to regard the extension as an object, called a *class*, which is supposed to be the subject of all the equivalent statements about various formally equivalent functions. Thus *e.g.* if we say "there were twelve Apostles," it is natural to regard this statement as attributing the property of being twelve to a certain collection of men, namely those who were Apostles, rather than as attributing the property of being satisfied by twelve arguments to the function "$\hat{x}$ was an Apostle." This view is encouraged by the feeling that there is something which is identical in the case of two functions which "have the same extension." And if we take such simple problems as "how many combinations can be made of n things?" it seems at first sight necessary that each "combination" should be a single object which can be counted as one. This, however, is certainly not necessary technically, and we see no reason to suppose that it is true philosophically. The technical procedure by which the apparent difficulty is overcome is as follows.

We have seen that an extensional function of a function may be regarded as a function of the class determined by the argument-function, but that an intensional function cannot be so regarded. In order to obviate the necessity of giving different treatment to intensional and extensional functions of functions, we construct an extensional function derived from any function of a predicative function $\psi ! \hat{z}$, and having the property of being equivalent to the function from which it is derived, provided this function is extensional, as well as the property of being significant (by the help of the systematic ambiguity of equivalence) with any argument $\phi\hat{z}$ whose arguments are of the same type as those of $\psi ! \hat{z}$. The derived function, written "$f\{\hat{z}(\phi z)\}$," is defined as follows: Given a function $f(\psi ! \hat{z})$, our derived function is to be "there is a predicative function which is formally equivalent to $\phi\hat{z}$ and satisfies f." If $\phi\hat{z}$ is a predicative function, our derived function will be true whenever $f(\phi\hat{z})$ is true. If $f(\phi\hat{z})$ is an extensional function, and $\phi\hat{z}$ is a predicative function, our derived function will not be true unless $f(\phi\hat{z})$ is true; thus in this case, our derived function is equivalent to $f(\phi\hat{z})$. If $f(\phi\hat{z})$ is not an extensional function, and if $\phi\hat{z}$ is a predicative function, our derived function may sometimes be true when the original function is false. But in any case the derived function is always extensional.

In order that the derived function should be significant for any function $\phi\hat{z}$, of whatever order, provided it takes arguments of the right type, it is necessary and sufficient that $f(\psi ! \hat{z})$ should be significant, where $\psi ! \hat{z}$ is any *predicative* function. The reason of this is that we only require, concerning an argument $\phi\hat{z}$, the hypothesis that it is formally equivalent to some predicative function $\psi ! \hat{z}$, and formal equivalence has the same kind of systematic ambiguity as to type that belongs to truth and falsehood, and can therefore hold between functions of any two different orders, provided the functions take arguments of the same type. Thus by means of our derived function we have not merely provided extensional functions everywhere in place of intensional functions, but we have *practically* removed the necessity for considering differences of type among functions whose arguments are of the same type. This effects the same kind of simplification in our hierarchy as would result from never considering any but predicative functions.

If $f(\psi ! \hat{z})$ can be built up by means of the primitive ideas of disjunction, negation, $(x) . \phi x$, and $(\exists x) . \phi x$, as is the case with all the functions of functions that explicitly occur in the present work, it will be found that, in virtue of the systematic ambiguity of the above primitive ideas, any function $\phi\hat{z}$ whose arguments are of the same type as those of $\psi ! \hat{z}$ can significantly be substituted for $\psi ! \hat{z}$ in f without any other symbolic change. Thus in such a case what is symbolically, though not really, the same function f can receive as arguments functions of various different types. If, with a given argument $\phi\hat{z}$, the function $f(\phi\hat{z})$, so interpreted, is equivalent to $f(\psi ! \hat{z})$ whenever $\psi ! \hat{z}$ is formally equivalent to $\phi\hat{z}$, then $f\{\hat{z}(\phi z)\}$ is equivalent to $f(\phi\hat{z})$ provided there is any predicative function formally equivalent to $\phi\hat{z}$. At this point, we make use of the axiom of reducibility, according to which there always is a predicative function formally equivalent to $\phi\hat{z}$.

As was explained above, it is convenient to regard an extensional function of a function as having for its argument not the function, but the class determined by the function. Now we have seen that our derived function is always extensional. Hence if our original function was $f(\psi ! \hat{z})$, we write the derived function $f\{\hat{z}(\phi z)\}$, where "$\hat{z}(\phi z)$" may be read "the class of arguments which satisfy $\phi\hat{z}$," or more simply "the class determined by $\phi\hat{z}$." Thus "$f\{\hat{z}(\phi z)\}$" will mean: "There is a predicative function $\psi ! \hat{z}$ which is formally equivalent to $\phi\hat{z}$ and is such that $f(\psi ! \hat{z})$ is true." This is in reality a function of $\phi\hat{z}$, but we treat it symbolically as if it had an argument $\hat{z}(\phi z)$. By the help of the axiom of reducibility, we find that the usual properties of classes result. For example, two formally equivalent functions determine the same class, and conversely, two functions which determine the same class are formally equivalent. Also to say that x is a member of $\hat{z}(\phi z)$, *i.e.* of the class determined by $\phi\hat{z}$, is true when ϕx is true, and false when ϕx is false. Thus all the mathematical purposes for which classes might seem to be required are fulfilled by the purely symbolic objects $\hat{z}(\phi z)$, provided we assume the axiom of reducibility.

In virtue of the axiom of reducibility, if $\phi\hat{z}$ is any function, there is a formally equivalent predicative function $\psi ! \hat{z}$; then the class $\hat{z}(\phi z)$ is identical with the class $\hat{z}(\psi ! z)$, so that every class can be defined by a *predicative* function. Hence the totality of the *classes* to which a given term can be significantly said to belong or not to belong is a legitimate totality, although the totality of *functions* which a given term can be significantly said to satisfy or not to satisfy is not a legitimate totality. The classes to which a given term a belongs or does not belong are the classes defined by a-functions; they are also the classes defined by *predicative* a-functions. Let us call them a-classes. Then "a-classes" form a legitimate totality, derived from that of predicative a-functions. Hence many kinds of general statements become possible which would otherwise involve vicious-circle paradoxes. These general statements are none of them such as lead to contradictions, and many of them such as it is very hard to suppose illegitimate. The fact that they are rendered possible by the axiom of reducibility, and that they would otherwise be excluded by the vicious-circle principle, is to be regarded as an argument in favour of the axiom of reducibility.

The above definition of "the class defined by the function $\phi\hat{z}$," or rather, of any proposition in which this phrase occurs, is, in symbols, as follows:

$$f\{\hat{z}(\phi z)\} . = : (\exists \psi) : \phi x . \equiv_x . \psi ! x : f\{\psi ! \hat{z}\} \quad \text{Df.}$$

In order to recommend this definition, we shall enumerate five requisites which a definition of classes must satisfy, and we shall then show that the above definition satisfies these five requisites.

We require of classes, if they are to serve the purposes for which they are commonly employed, that they shall have certain properties, which may be enumerated as follows. (1) Every propositional function must determine a class, which may be regarded as the collection of all the arguments satisfying the function in question. This principle must hold when the function is satisfied by an infinite number of arguments as well as when it is satisfied by a finite number. It must hold also when no arguments satisfy the function; *i.e.* the "null-class" must be just as good a class as any other. (2) Two propositional functions which are formally equivalent, *i.e.* such that any argument which satisfies either satisfies the other, must determine the same class; that is to say, a class must be something wholly determined by its membership, so that *e.g.* the class "featherless bipeds" is identical with the class "men," and the class "even primes" is identical with the class "numbers identical with 2." (3) Conversely, two propositional functions which determine the same class must be formally equivalent; in other words, when the class is given, the membership is determinate: two different sets of objects cannot yield the same class. (4) In the same sense in which there are classes (whatever this sense may be), or in some closely analogous sense, there must also be classes of classes. Thus for example "the combinations of n things m at a time," where the n things form a given class, is a class of classes; each combination of m things is a class, and each such class is a member of the specified set of combinations, which set is therefore a class whose members are classes. Again, the class of unit classes, or of couples, is absolutely indispensable; the former is the number 1, the latter the number 2. Thus without classes of classes, arithmetic becomes impossible. (5) It must under all circumstances be meaningless to suppose a class identical with one of its own members. For if such a supposition had any meaning "$\alpha \epsilon \alpha$" would be a significant propositional function*, and so would "$\alpha \sim \epsilon \alpha$." Hence, by (1) and (4), there would be a class of all classes satisfying the function "$\alpha \sim \epsilon \alpha$." If we call this class κ, we shall have

$$\alpha \epsilon \kappa . \equiv_\alpha . \alpha \sim \epsilon \alpha.$$

Since, by our hypothesis, "$\kappa \epsilon \kappa$" is supposed significant, the above equivalence, which holds with all possible values of α, holds with the value κ, *i.e.*

$$\kappa \epsilon \kappa . \equiv . \kappa \sim \epsilon \kappa.$$

But this is a contradiction†. Hence "$\alpha \epsilon \alpha$" and "$\alpha \sim \epsilon \alpha$" must always be meaningless. In general, there is nothing surprising about this conclusion, but it has two consequences which deserve special notice. In the first place, a class consisting of only one member must not be identical with that one member, *i.e.* we must not have $\iota ' x = x$. For we have $x \epsilon \iota ' x$, and therefore, if $x = \iota ' x$, we have $\iota ' x \epsilon \iota ' x$, which, we saw, must be meaningless. It follows that "$x = \iota ' x$" must be absolutely meaningless, not simply false. In the second place, it might appear as if the class of all classes were a class, *i.e.* as if (writing "Cls" for "class") "Cls ϵ Cls" were a true proposition. But this combination of symbols must be meaningless; unless, indeed, an ambiguity exists in the meaning of "Cls," so that, in "Cls ϵ Cls," the first "Cls" can be supposed to have a different meaning from the second.

As regards the above requisites, it is plain, to begin with, that, in accordance with our definition, every propositional function $\phi \hat{z}$ determines a class $\hat{z}(\phi z)$. Assuming the axiom of reducibility, there must always be true propositions about $\hat{z}(\phi z)$, *i.e.* true propositions of the form $f\{\hat{z}(\phi z)\}$. For suppose $\phi \hat{z}$ is formally equivalent to $\psi ! \hat{z}$, and suppose $\psi ! \hat{z}$ satisfies some function f. Then

* As explained in Chapter I (p. 25), "$x \epsilon \alpha$" means "x is a member of the class α," or, more shortly, "x is an α." The definition of this expression in terms of our theory of classes will be given shortly.

† This is the second of the contradictions discussed at the end of Chapter II.

$\hat{z}(\phi z)$ also satisfies f. Hence, given any function $\phi\hat{z}$, there are true propositions of the form $f\{\hat{z}(\phi z)\}$, *i.e.* true propositions in which "the class determined by $\phi\hat{z}$" is grammatically the subject. This shows that our definition fulfils the first of our five requisites.

The second and third requisites together demand that the classes $\hat{z}(\phi z)$ and $\hat{z}(\psi z)$ should be identical when, and only when, their defining functions are formally equivalent, *i.e.* that we should have

$$\hat{z}(\phi z) = \hat{z}(\psi z) . \equiv : \phi x . \equiv_x . \psi x.$$

Here the meaning of "$\hat{z}(\phi z) = \hat{z}(\psi z)$" is to be derived, by means of a two-fold application of the definition of $f\{\hat{z}(\phi z)\}$, from the definition of

$$\text{“}\chi ! \hat{z} = \theta ! \hat{z}\text{,”}$$

which is $\chi ! \hat{z} = \theta ! \hat{z} . = : (f) : f ! \chi ! \hat{z} . \supset . f ! \theta ! \hat{z}$ Df

by the general definition of identity.

In interpreting "$\hat{z}(\phi z) = \hat{z}(\psi z)$," we will adopt the convention which we adopted in regard to $(\imath x)(\phi x)$ and $(\imath x)(\psi x)$, namely that the incomplete symbol which occurs first is to have the larger scope. Thus $\hat{z}(\phi z) = \hat{z}(\psi z)$ becomes, by our definition,

$$(\exists \chi) : \phi x . \equiv_x . \chi ! x : \chi ! \hat{z} = \hat{z}(\psi z),$$

which, by eliminating $\hat{z}(\psi z)$, becomes

$$(\exists \chi) :. \phi x . \equiv_x . \chi ! x :. (\exists \theta) : \psi x . \equiv_x . \theta ! x : \chi ! \hat{z} = \theta ! \hat{z},$$

which is equivalent to

$$(\exists \chi, \theta) : \phi x . \equiv_x . \chi ! x : \psi x . \equiv_x . \theta ! x : \chi ! \hat{z} = \theta ! \hat{z},$$

which, again, is equivalent to

$$(\exists \chi) : \phi x . \equiv_x . \chi ! x : \psi x . \equiv_x . \chi ! x,$$

which, in virtue of the axiom of reducibility, is equivalent to

$$\phi x . \equiv_x . \psi x.$$

Thus our definition of the use of $\hat{z}(\phi z)$ is such as to satisfy the conditions (2) and (3) which we laid down for classes, *i.e.* we have

$$\vdash :. \hat{z}(\phi z) = \hat{z}(\psi z) . \equiv : \phi x . \equiv_x . \psi x.$$

Before considering classes of classes, it will be well to define membership of a class, *i.e.* to define the symbol "$x \epsilon \hat{z}(\phi z)$," which may be read "$x$ is a member of the class determined by $\phi\hat{z}$." Since this is a function of the form $f\{\hat{z}(\phi z)\}$, it must be derived, by means of our general definition of such functions, from the corresponding function $f\{\psi ! \hat{z}\}$. We therefore put

$$x \epsilon \psi ! \hat{z} . = . \psi ! x \quad \text{Df.}$$

This definition is only needed in order to give a meaning to "$x \epsilon \hat{z}(\phi z)$"; the meaning it gives is, in virtue of the definition of $f\{\hat{z}(\phi z)\}$,

$$(\exists \psi) : \phi y . \equiv_y . \psi ! y : \psi ! x.$$

It thus appears that "$x \epsilon \hat{z}(\phi z)$" implies ϕx, since it implies $\psi ! x$, and $\psi ! x$ is equivalent to ϕx; also, in virtue of the axiom of reducibility, ϕx implies "$x \epsilon \hat{z}(\phi z)$," since there is a predicative function ψ formally equivalent to ϕ, and x must satisfy ψ, since x (*ex hypothesi*) satisfies ϕ. Thus in virtue of the axiom of reducibility we have

$$\vdash : x \,\epsilon\, \hat{z}(\phi z) \,.\equiv.\, \phi x,$$

i.e. x is a member of the class $\hat{z}(\phi z)$ when, and only when, x satisfies the function ϕ which defines the class.

We have next to consider how to interpret a class of classes. As we have defined $f\{\hat{z}(\phi z)\}$, we shall naturally regard a class of classes as consisting of those values of $\hat{z}(\phi z)$ which satisfy $f\{\hat{z}(\phi z)\}$. Let us write α for $\hat{z}(\phi z)$; then we may write $\hat{\alpha}(f\alpha)$ for the class of values of α which satisfy $f\alpha$*. We shall apply the same definition, and put

$$F\{\hat{\alpha}(f\alpha)\} \,.=:\, (\exists g) : f\beta \,.\equiv_\beta.\, g!\beta : F\{g!\hat{\alpha}\} \quad \text{Df},$$

where "β" stands for any expression of the form $\hat{z}(\psi! z)$.

Let us take "$\gamma \,\epsilon\, \hat{\alpha}(f\alpha)$" as an instance of $F\{\hat{\alpha}(f\alpha)\}$. Then

$$\vdash :. \gamma \,\epsilon\, \hat{\alpha}(f\alpha) \,.\equiv:\, (\exists g) : f\beta \,.\equiv_\beta.\, g!\beta : \gamma \,\epsilon\, g!\hat{\alpha}.$$

Just as we put $\quad x \,\epsilon\, \psi!\hat{z} \,.=.\, \psi! x \quad$ Df,

so we put $\quad \gamma \,\epsilon\, g!\hat{\alpha} \,.=.\, g!\gamma \quad$ Df.

Thus we find

$$\vdash :. \gamma \,\epsilon\, \hat{\alpha}(f\alpha) \,.\equiv:\, (\exists g) : f\beta \,.\equiv_\beta.\, g!\beta : g!\gamma.$$

If we now extend the axiom of reducibility so as to apply to functions of functions, *i.e.* if we assume

$$(\exists g) : f(\psi!\hat{z}) \,.\equiv_\psi.\, g!(\psi!\hat{z}),$$

we easily deduce

$$\vdash : (\exists g) : f\{\hat{z}(\psi! z)\} \,.\equiv_\psi.\, g!\{\hat{z}(\psi! z)\},$$

i.e. $\quad \vdash : (\exists g) : f\beta \,.\equiv_\beta.\, g!\beta.$

Thus $\quad \vdash : \gamma \,\epsilon\, \hat{\alpha}(f\alpha) \,.\equiv.\, f\gamma.$

Thus every function which can take classes as arguments, *i.e.* every function of functions, determines a class of classes, whose members are those classes which satisfy the determining function. Thus the theory of classes of classes offers no difficulty.

We have next to consider our fifth requisite, namely that "$\hat{z}(\phi z) \,\epsilon\, \hat{z}(\phi z)$" is to be meaningless. Applying our definition of $f\{\hat{z}(\phi z)\}$, we find that if this collection of symbols had a meaning, it would mean

$$(\exists \psi) : \phi x \,.\equiv_x.\, \psi! x : \psi!\hat{z} \,\epsilon\, \psi!\hat{z},$$

i.e. in virtue of the definition

$$x \,\epsilon\, \psi!\hat{z} \,.=.\, \psi! x \quad \text{Df},$$

it would mean $\quad (\exists \psi) : \phi x \,.\equiv_x.\, \psi! x : \psi!(\psi!\hat{z}).$

But here the symbol "$\psi!(\psi!\hat{z})$" occurs, which assigns a function as argument to itself. Such a symbol is always meaningless, for the reasons explained at the beginning of Chapter II (pp. 38—41). Hence "$\hat{z}(\phi z) \,\epsilon\, \hat{z}(\phi z)$" is meaningless, and our fifth and last requisite is fulfilled.

* The use of a single letter, such as α or β, to represent a variable class, will be further explained shortly.

As in the case of $f(\imath x)(\phi x)$, so in that of $f\{\hat{z}(\phi z)\}$, there is an ambiguity as to the scope of $\hat{z}(\phi z)$ if it occurs in a proposition which itself is part of a larger proposition. But in the case of classes, since we always have the axiom of reducibility, namely

$$(\exists\psi) : \phi x . \equiv_x . \psi ! x,$$

which takes the place of $E!(\imath x)(\phi x)$, it follows that the truth-value of any proposition in which $\hat{z}(\phi z)$ occurs is the same whatever scope we may give to $\hat{z}(\phi z)$, provided the proposition is an extensional function of whatever functions it may contain. Hence we may adopt the convention that the scope is to be always the smallest proposition enclosed in dots or brackets in which $\hat{z}(\phi z)$ occurs. If at any time a larger scope is required, we may indicate it by "$[\hat{z}(\phi z)]$" followed by dots, in the same way as we did for $[(\imath x)(\phi x)]$.

Similarly when two class symbols occur, *e.g.* in a proposition of the form $f\{\hat{z}(\phi z), \hat{z}(\psi z)\}$, we need not remember rules for the scopes of the two symbols, since all choices give equivalent results, as it is easy to prove. For the preliminary propositions a rule is desirable, so we can decide that the class symbol which occurs first in the order of writing is to have the larger scope.

The representation of a class by a single letter α can now be understood. For the denotation of α is ambiguous, in so far as it is undecided as to which of the symbols $\hat{z}(\phi z)$, $\hat{z}(\psi z)$, $\hat{z}(\chi z)$, etc. it is to stand for, where $\phi\hat{z}$, $\psi\hat{z}$, $\chi\hat{z}$, etc. are the various determining functions of the class. According to the choice made, different propositions result. But all the resulting propositions are equivalent by virtue of the easily proved proposition:

$$\text{"}\vdash : \phi x \equiv_x \psi x . \supset . f\{\hat{z}(\phi z)\} \equiv f\{\hat{z}(\psi z)\}.\text{"}$$

Hence unless we wish to discuss the determining function itself, so that the notion of a class is really not properly present, the ambiguity in the denotation of α is entirely immaterial, though, as we shall see immediately, we are led to limit ourselves to predicative determining functions. Thus "$f(\alpha)$," where α is a variable class, is really "$f\{\hat{z}(\phi z)\}$," where ϕ is a variable function, that is, it is

$$\text{"}(\exists\psi) . \phi x \equiv_x \psi ! x . f\{\psi ! \hat{z}\},\text{"}$$

where ϕ is a variable function. But here a difficulty arises which is removed by a limitation to our practice and by the axiom of reducibility. For the determining functions $\phi\hat{z}$, $\psi\hat{z}$, etc. will be of different types, though the axiom of reducibility secures that some are predicative functions. Then, in interpreting α as a variable in terms of the variation of any determining function, we shall be led into errors unless we confine ourselves to predicative determining functions. These errors especially arise in the transition to total variation (cf. pp. 15, 16). Accordingly

$$f\alpha = . (\exists\psi) . \phi ! x \equiv_x \psi ! x . f\{\psi ! \hat{z}\} \quad \text{Df.}$$

It is the peculiarity of a definition of the use of a single letter [viz. α] for a variable incomplete symbol that it, though in a sense a real variable, occurs only in the *definiendum*, while "ϕ," though a real variable, occurs only in the *definiens*.

Thus "$f\hat{\alpha}$" stands for

$$\text{"}(\exists\psi) \,.\, \hat{\phi} ! x \equiv_x \psi ! x \,.\, f\{\psi ! \hat{z}\},\text{"}$$

and "$(\alpha) \,.\, f\alpha$" stands for

$$\text{"}(\phi) : (\exists\psi) \,.\, \phi ! x \equiv_x \psi ! x \,.\, f\{\psi ! \hat{z}\}.\text{"}$$

Accordingly, in mathematical reasoning, we can dismiss the whole apparatus of functions and think only of classes as "quasi-things," capable of immediate representation by a single name. The advantages are two-fold: (1) classes are determined by their membership, so that to one set of members there is one class, (2) the "type" of a class is entirely defined by the type of its members.

Also a predicative function of a class can be defined thus

$$f ! \alpha \,. = .\, (\exists\psi) \,.\, \phi ! x \equiv_x \psi ! x \,.\, f ! \{\psi ! \hat{z}\} \quad \text{Df.}$$

Thus a predicative function of a class is always a predicative function of any predicative determining function of the class, though the converse does not hold.

(3) *Relations.* With regard to relations, we have a theory strictly analogous to that which we have just explained as regards classes. Relations in extension, like classes, are incomplete symbols. We require a division of functions of two variables into predicative and non-predicative functions, again for reasons which have been explained in Chapter II. We use the notation "$\phi ! (x, y)$" for a *predicative* function of x and y.

We use "$\phi ! (\hat{x}, \hat{y})$" for the function as opposed to its values; and we use "$\hat{x}\hat{y}\phi(x, y)$" for the relation (in extension) determined by $\phi(x, y)$. We put

$$f\{\hat{x}\hat{y}\phi(x, y)\} \,. = : (\exists\psi) : \phi(x, y) \,.\, \equiv_{x,y} .\, \psi ! (x, y) : f\{\psi ! (\hat{x}, \hat{y})\} \quad \text{Df.}$$

Thus even when $f\{\psi ! (\hat{x}, \hat{y})\}$ is not an extensional function of ψ, $f\{\hat{x}\hat{y}\phi(x, y)\}$ *is* an extensional function of ϕ. Hence, just as in the case of classes, we deduce

$$\vdash :.\, \hat{x}\hat{y}\phi(x, y) = \hat{x}\hat{y}\psi(x, y) \,. \equiv : \phi(x, y) \,.\, \equiv_{x,y} .\, \psi(x, y),$$

i.e. a relation is determined by its extension, and vice versa.

On the analogy of the definition of "$x \,\epsilon\, \psi ! \hat{z}$," we put

$$x\{\psi ! (\hat{x}, \hat{y})\} y \,. = .\, \psi ! (x, y) \quad \text{Df*.}$$

This definition, like that of "$x \,\epsilon\, \psi ! \hat{z}$," is not introduced for its own sake, but in order to give a meaning to

$$x\{\hat{x}\hat{y}\phi(x, y)\} y.$$

This meaning, in virtue of our definitions, is

$$(\exists\psi) : \phi(x, y) \,.\, \equiv_{x,y} .\, \psi ! (x, y) : x\{\psi ! (\hat{x}, \hat{y})\} y,$$

i.e.

$$(\exists\psi) : \phi(x, y) \,.\, \equiv_{x,y} .\, \psi ! (x, y) : \psi ! (x, y),$$

and this, in virtue of the axiom of reducibility

$$\text{"}(\exists\psi) : \phi(x, y) \,.\, \equiv_{x,y} .\, \psi ! (x, y),\text{"}$$

is equivalent to

$$\phi(x, y).$$

Thus we have always

$$\vdash : x\{\hat{x}\hat{y}\phi(x, y)\} y \,. \equiv .\, \phi(x, y).$$

* This definition raises certain questions as to the two senses of a relation, which are dealt with in *21.

Whenever the determining function of a relation is not relevant, we may replace $\hat{x}\hat{y}\,\phi(x, y)$ by a single capital letter. In virtue of the propositions given above,

$$\vdash :. R = S . \equiv : xRy . \equiv_{x,y} . xSy,$$

$$\vdash :. R = \hat{x}\hat{y}\,\phi(x, y) . \equiv : xRy . \equiv_{x,y} . \phi(x, y),$$

and $$\vdash . R = \hat{x}\hat{y}\,(xRy).$$

Classes of relations, and relations of relations, can be dealt with as classes of classes were dealt with above.

Just as a class must not be capable of being or not being a member of itself, so a relation must neither be nor not be referent or relatum with respect to itself. This turns out to be equivalent to the assertion that $\phi!(\hat{x}, \hat{y})$ cannot significantly be either of the arguments x or y in $\phi!(x, y)$. This principle, again, results from the limitation to the possible arguments to a function explained at the beginning of Chapter II.

We may sum up this whole discussion on incomplete symbols as follows.

The use of the symbol "$(\imath x)(\phi x)$" as if in "$f(\imath x)(\phi x)$" it *directly* represented an argument to the function $f\hat{z}$ is rendered possible by the theorems

$$\vdash :. E!(\imath x)(\phi x) . \supset : (x) . fx . \supset . f(\imath x)(\phi x),$$

$$\vdash : (\imath x)(\phi x) = (\imath x)(\psi x) . \supset . f(\imath x)(\phi x) \equiv f(\imath x)(\psi x),$$

$$\vdash : E!(\imath x)(\phi x) . \supset . (\imath x)(\phi x) = (\imath x)(\phi x),$$

$$\vdash : (\imath x)(\phi x) = (\imath x)(\psi x) . \equiv . (\imath x)(\psi x) = (\imath x)(\phi x),$$

$$\vdash : (\imath x)(\phi x) = (\imath x)(\psi x) . (\imath x)(\psi x) = (\imath x)(\chi x) . \supset . (\imath x)(\phi x) = (\imath x)(\chi x).$$

The use of the symbol "$\hat{x}(\phi x)$" (or of a single letter, such as α, to represent such a symbol) as if, in "$f\{\hat{x}(\phi x)\}$," it *directly* represented an argument α to a function $f\hat{\alpha}$, is rendered possible by the theorems

$$\vdash : (\alpha) . f\alpha . \supset . f\{\hat{x}(\phi x)\},$$

$$\vdash : \hat{x}(\phi x) = \hat{x}(\psi x) . \supset . f\{\hat{x}(\phi x)\} \equiv f\{\hat{x}(\psi x)\},$$

$$\vdash . \hat{x}(\phi x) = \hat{x}(\phi x),$$

$$\vdash : \hat{x}(\phi x) = \hat{x}(\psi x) . \equiv . \hat{x}(\psi x) = \hat{x}(\phi x),$$

$$\vdash : \hat{x}(\phi x) = \hat{x}(\psi x) . \hat{x}(\psi x) = \hat{x}(\chi x) . \supset . \hat{x}(\phi x) = \hat{x}(\chi x).$$

Throughout these propositions the types must be supposed to be properly adjusted, where ambiguity is possible.

The use of the symbol "$\hat{x}\hat{y}\,\{\phi(x, y)\}$" (or of a single letter, such as R, to represent such a symbol) as if, in "$f\{\hat{x}\hat{y}\,\phi(x, y)\}$," it *directly* represented an argument R to a function $f\hat{R}$, is rendered possible by the theorems

$$\vdash : (R) . fR . \supset . f\{\hat{x}\hat{y}\,\phi(x, y)\},$$

$$\vdash : \hat{x}\hat{y}\,\phi(x, y) = \hat{x}\hat{y}\,\psi(x, y) . \supset . f\{\hat{x}\hat{y}\,\phi(x, y)\} \equiv f\{\hat{x}\hat{y}\,\psi(x, y)\},$$

$$\vdash . \hat{x}\hat{y}\,\phi(x, y) = \hat{x}\hat{y}\,\phi(x, y),$$

$$\vdash : \hat{x}\hat{y}\,\phi(x, y) = \hat{x}\hat{y}\,\psi(x, y) . \equiv . \hat{x}\hat{y}\,\psi(x, y) = \hat{x}\hat{y}\,\phi(x, y),$$

$$\vdash : \hat{x}\hat{y}\,\phi(x, y) = \hat{x}\hat{y}\,\psi(x, y) . \hat{x}\hat{y}\,\psi(x, y) = \hat{x}\hat{y}\,\chi(x, y) . \supset . \hat{x}\hat{y}\,\phi(x, y) = \hat{x}\hat{y}\,\chi(x, y).$$

Throughout these propositions the types must be supposed to be properly adjusted where ambiguity is possible.

It follows from these three groups of theorems that these incomplete symbols are obedient to the same formal rules of identity as symbols which directly represent objects, so long as we only consider the *equivalence* of the resulting variable (or constant) values of propositional functions and not their identity. This consideration of the *identity* of propositions never enters into our formal reasoning.

Similarly the *limitations* to the use of these symbols can be summed up as follows. In the case of $(\imath x)(\phi x)$, the chief way in which its incompleteness is relevant is that we do not always have

$$(x) . fx . \supset . f(\imath x)(\phi x),$$

i.e. a function which is always true may nevertheless not be true of $(\imath x)(\phi x)$. This is possible because $f(\imath x)(\phi x)$ is not a value of $f\hat{x}$, so that even when all values of $f\hat{x}$ are true, $f(\imath x)(\phi x)$ may not be true. This happens when $(\imath x)(\phi x)$ does not exist. Thus for example we have $(x) . x = x$, but we do not have

the round square = the round square.

The inference $$(x) . fx . \supset . f(\imath x)(\phi x)$$

is only valid when $E ! (\imath x)(\phi x)$. As soon as we know $E ! (\imath x)(\phi x)$, the fact that $(\imath x)(\phi x)$ is an incomplete symbol becomes irrelevant so long as we confine ourselves to truth-functions* of whatever proposition is its scope. But even when $E ! (\imath x)(\phi x)$, the incompleteness of $(\imath x)(\phi x)$ may be relevant when we pass outside truth-functions. For example, George IV wished to know whether Scott was the author of Waverley, *i.e.* he wished to know whether a proposition of the form "$c = (\imath x)(\phi x)$" was true. But there was no proposition of the form "$c = y$" concerning which he wished to know if it was true.

In regard to classes, the relevance of their incompleteness is somewhat different. It may be illustrated by the fact that we may have

$$\hat{z}(\phi z) = \psi ! \hat{z} . \hat{z}(\phi z) = \chi ! \hat{z}$$

without having $$\psi ! \hat{z} = \chi ! \hat{z}.$$

For, by a direct application of the definitions, we find that

$$\vdash : \hat{z}(\phi z) = \psi ! \hat{z} . \equiv . \phi x \equiv_x \psi ! x.$$

Thus we shall have

$$\vdash : \phi x \equiv_x \psi ! x . \phi x \equiv_x \chi ! x . \supset . \hat{z}(\phi z) = \psi ! \hat{z} . \hat{z}(\phi z) = \chi ! \hat{z},$$

but we shall not necessarily have $\psi ! \hat{z} = \chi ! \hat{z}$ under these circumstances, for two functions may well be formally equivalent without being identical; for example,

$$x = \text{Scott} . \equiv_x . x = \text{the author of Waverley},$$

but the function "$\hat{z}$ = the author of Waverley" has the property that George IV wished to know whether its value with the argument "Scott" was true, whereas

* Cf. p. 8.

the function "$\hat{z} = \text{Scott}$" has no such property, and therefore the two functions are not identical. Hence there is a propositional function, namely

$$x = y \,.\, x = z \,.\, \supset \,.\, y = z,$$

which holds without any exception, and yet does not hold when for x we substitute a class, and for y and z we substitute functions. This is only possible because a class is an incomplete symbol, and therefore "$\hat{z}(\phi z) = \psi ! \hat{z}$" is not a value of "$x = y$."

It will be observed that "$\theta ! \hat{z} = \psi ! \hat{z}$" is not an extensional function of $\psi ! \hat{z}$. Thus the scope of $\hat{z}(\phi z)$ is relevant in interpreting the product

$$\hat{z}(\phi z) = \psi ! \hat{z} \,.\, \hat{z}(\phi z) = \chi ! \hat{z}.$$

If we take the whole of the product as the scope of $\hat{z}(\phi z)$, the product is equivalent to

$$(\exists \theta) : \phi x \equiv_x \theta ! x \,.\, \theta ! \hat{z} = \psi ! \hat{z} \,.\, \theta ! \hat{z} = \chi ! \hat{z},$$

and this *does* imply $\psi ! \hat{z} = \chi ! \hat{z}$.

We may say generally that the fact that $\hat{z}(\phi z)$ is an incomplete symbol is not relevant so long as we confine ourselves to extensional functions of functions, but is apt to become relevant for other functions of functions.

PART I

MATHEMATICAL LOGIC

SUMMARY OF PART I

In this Part, we shall deal with such topics as belong traditionally to symbolic logic, or deserve to belong to it in virtue of their generality. We shall, that is to say, establish such properties of propositions, propositional functions, classes and relations as are likely to be required in any mathematical reasoning, and not merely in this or that branch of mathematics.

The subjects treated in Part I may be viewed in two aspects: (1) as a deductive chain depending on the primitive propositions, (2) as a formal calculus. Taking the first view first: We begin, in ∗1, with certain axioms as to deduction of one proposition or asserted propositional function from another. From these primitive propositions, in Section A, we deduce various propositions which are all concerned with four ways of obtaining new propositions from given propositions, namely negation, disjunction, joint assertion and implication, of which the last two can be defined in terms of the first two. Throughout this first section, although, as will be shown at the beginning of Section B, our propositions, symbolically unchanged, will apply to any propositions as values of our variables, yet it will be supposed that our variable propositions are all what we shall call *elementary* propositions, *i.e.* such as contain no reference, explicit or implicit, to any totality. This restriction is imposed on account of the distinction between different *types* of propositions, explained in Chapter II of the Introduction. Its importance and purpose, however, are purely philosophical, and so long as only mathematical purposes are considered, it is unnecessary to remember this preliminary restriction to elementary propositions, which is symbolically removed at the beginning of the next section.

Section B deals, to begin with, with the relations of propositions containing apparent variables (*i.e.* involving the notions of "all" or "some") to each other and to propositions not containing apparent variables. We show that, where propositions containing apparent variables are concerned, we can define negation, disjunction, joint assertion and implication in such a way that their properties shall be exactly analogous to the properties of the corresponding ideas as applied to elementary propositions. We show also that *formal implication*, *i.e.* "$(x) . \phi x \supset \psi x$" considered as a relation of $\phi \hat{x}$ to $\psi \hat{x}$, has many properties analogous to those of *material implication*, *i.e.* "$p \supset q$" considered as a relation of p and q. We then consider *predicative* functions and the *axiom of reducibility*, which are vital in the employment of *functions* as apparent variables. An example of such employment is afforded by *identity*, which is the next topic considered in Section B. Finally, this section deals with *descriptions*, *i.e.* phrases of the form "the so-and-so" (in the singular). It is shown that the appearance of a grammatical subject "the so-and-so" is deceptive, and that such propositions, fully stated, contain no such subject, but contain instead an apparent variable.

Section C deals with classes, and with relations in so far as they are analogous to classes. Classes and relations, like descriptions, are shown to be "incomplete symbols" (cf. Introduction, Chapter III), and it is shown that a proposition which is grammatically about a class is to be regarded as really concerned with a propositional function and an apparent variable whose values are *predicative* propositional functions (with a similar result for relations). The remainder of Section C deals with the calculus of classes, and with the calculus of relations in so far as it is analogous to that of classes.

Section D deals with those properties of relations which have no analogues for classes. In this section, a number of ideas and notations are introduced which are constantly needed throughout the rest of the work. Most of the properties of relations which have analogues in the theory of classes are comparatively unimportant, while those that have no such analogues are of the very greatest utility. It is partly for this reason that emphasis on the calculus-aspect of symbolic logic has proved a hindrance, hitherto, to the proper development of the theory of relations.

Section E, finally, extends the notions of the addition and multiplication of classes or relations to cases where the summands or factors are not individually given, but are given as the members of some class. The advantage obtained by this extension is that it enables us to deal with an infinite number of summands or factors.

Considered as a formal calculus, mathematical logic has three analogous branches, namely (1) the calculus of propositions, (2) the calculus of classes, (3) the calculus of relations. Of these, (1) is dealt with in Section A, while (2) and (3), in so far as they are analogous, are dealt with in Section C. We have, for each of the three, the four analogous ideas of negation, addition, multiplication, and implication or inclusion. Of these, negation is analogous to the negative in ordinary algebra, and implication or inclusion is analogous to the relation "less than or equal to" in ordinary algebra. But the analogy must not be pressed, as it has important limitations. The sum of two propositions is their disjunction, the sum of two classes is the class of terms belonging to one or other, the sum of two relations is the relation consisting in the fact that one or other of the two relations holds. The sum of a class of classes is the class of all terms belonging to some one or other of the classes, and the sum of a class of relations is the relation consisting in the fact that some one relation of the class holds. The product of two propositions is their joint assertion, the product of two classes is their common part, the product of two relations is the relation consisting in the fact that both the relations hold. The product of a class of classes is the part common to all of them, and the product of a class of relations is the relation consisting in the fact that all relations of the class in question hold. The inclusion of one class in another consists in the fact that all members of the one are members of the other, while the inclusion of one relation in another consists in the fact that every pair of terms which has the one relation also has the other relation. It is then shown that the properties of negation, addition, multiplication and inclusion are exactly analogous for classes and relations, and are, with certain exceptions, analogous to the properties of negation, addition, multiplication and implication for propositions. (The exceptions arise chiefly from the fact that "p implies q" is itself a proposition, and can therefore imply and be implied, while "α is contained in β," where α and β are classes, is not a class, and can therefore neither contain nor be contained in another class γ.) But classes have certain properties not possessed by propositions: these arise from the fact that classes have not a *two-fold* division corresponding to the division of propositions into true and false, but a *three-fold* division, namely into (1) the universal class, which contains the whole of a certain type, (2) the null-class, which has no members, (3) all other classes, which neither contain nothing nor contain everything of the appropriate type. The resulting properties of classes, which are not analogous to properties of propositions, are dealt with in ∗24. And just as classes have properties not analogous to any properties of propositions, so relations have properties not analogous to any properties of classes, though all the properties of classes have analogues among relations. The special properties of relations are much more numerous and important than the properties belonging to classes but not to propositions. These special properties of relations therefore occupy a whole section, namely Section D.

SECTION A

THE THEORY OF DEDUCTION

THE purpose of the present section is to set forth the first stage of the deduction of pure mathematics from its logical foundations. This first stage is necessarily concerned with deduction itself, *i.e.* with the principles by which conclusions are inferred from premisses. If it is our purpose to make all our assumptions explicit, and to effect the deduction of all our other propositions from these assumptions, it is obvious that the first assumptions we need are those that are required to make deduction possible. Symbolic logic is often regarded as consisting of two coordinate parts, the theory of classes and the theory of propositions. But from our point of view these two parts are not coordinate; for in the theory of classes we deduce one proposition from another by means of principles belonging to the theory of propositions, whereas in the theory of propositions we nowhere require the theory of classes. Hence, in a deductive system, the theory of propositions necessarily precedes the theory of classes.

But the subject to be treated in what follows is not quite properly described as the theory of *propositions*. It is in fact the theory of how one proposition can be inferred from another. Now in order that one proposition may be inferred from another, it is necessary that the two should have that relation which makes the one a consequence of the other. When a proposition q is a consequence of a proposition p, we say that p *implies* q. Thus deduction depends upon the relation of *implication*, and every deductive system must contain among its premisses as many of the properties of implication as are necessary to legitimate the ordinary procedure of deduction. In the present section, certain propositions will be stated as premisses, and it will be shown that they are sufficient for all common forms of inference. It will not be shown that they are all *necessary*, and it is possible that the number of them might be diminished. All that is affirmed concerning the premisses is (1) that they are true, (2) that they are sufficient for the theory of deduction, (3) that we do not know how to diminish their number. But with regard to (2), there must always be some element of doubt, since it is hard to be sure that one never uses some principle unconsciously. The habit of being rigidly guided by formal symbolic rules is a safeguard against unconscious assumptions; but even this safeguard is not always adequate.

✱1. PRIMITIVE IDEAS AND PROPOSITIONS

Since all definitions of terms are effected by means of other terms, every system of definitions which is not circular must start from a certain apparatus of undefined terms. It is to some extent optional what ideas we take as undefined in mathematics; the motives guiding our choice will be (1) to make the number of undefined ideas as small as possible, (2) as between two systems in which the number is equal, to choose the one which seems the simpler and easier. We know no way of proving that such and such a system of undefined ideas contains as few as will give such and such results*. Hence we can only say that such and such ideas are undefined in such and such a system, not that they are indefinable. Following Peano, we shall call the undefined ideas and the undemonstrated propositions *primitive* ideas and *primitive* propositions respectively. The primitive ideas are *explained* by means of descriptions intended to point out to the reader what is meant; but the explanations do not constitute definitions, because they really involve the ideas they explain.

In the present number, we shall first enumerate the primitive ideas required in this section; then we shall define *implication*; and then we shall enunciate the primitive propositions required in this section. Every definition or proposition in the work has a number, for purposes of reference. Following Peano, we use numbers having a decimal as well as an integral part, in order to be able to insert new propositions between any two. A change in the integral part of the number will be used to correspond to a new chapter. Definitions will generally have numbers whose decimal part is less than ·1, and will be usually put at the beginning of chapters. In references, the integral parts of the numbers of propositions will be distinguished by being preceded by a star; thus "✱1·01" will mean the definition or proposition so numbered, and "✱1" will mean the chapter in which propositions have numbers whose integral part is 1, *i.e.* the present chapter. Chapters will generally be called "numbers."

PRIMITIVE IDEAS.

(1) *Elementary propositions.* By an "elementary" proposition we mean one which does not involve any variables, or, in other language, one which does not involve such words as "all," "some," "the" or equivalents for such words. A proposition such as "this is red," where "this" is something given in sensation, will be elementary. Any combination of given elementary propositions by means of negation, disjunction or conjunction (see below) will

* The recognized methods of proving independence are not applicable, without reserve, to fundamentals. Cf. *Principles of Mathematics*, § 17. What is there said concerning primitive propositions applies with even greater force to primitive ideas.

be elementary. In the primitive propositions of the present number, and therefore in the deductions from these primitive propositions in ∗2—∗5, the letters p, q, r, s will be used to denote elementary propositions.

(2) *Elementary propositional functions.* By an "elementary propositional function" we shall mean an expression containing an undetermined constituent, *i.e.* a variable, or several such constituents, and such that, when the undetermined constituent or constituents are determined, *i.e.* when values are assigned to the variable or variables, the resulting value of the expression in question is an elementary proposition. Thus if p is an undetermined elementary proposition, "not-p" is an elementary propositional function.

We shall show in ∗9 how to extend the results of this and the following numbers (∗1—∗5) to propositions which are not elementary.

(3) *Assertion.* Any proposition may be either asserted or merely considered. If I say "Caesar died," I assert the proposition "Caesar died," if I say "'Caesar died' is a proposition," I make a different assertion, and "Caesar died" is no longer asserted, but merely considered. Similarly in a hypothetical proposition, *e.g.* "if $a = b$, then $b = a$," we have two unasserted propositions, namely "$a = b$" and "$b = a$," while what *is* asserted is that the first of these implies the second. In language, we indicate when a proposition is merely considered by "*if* so-and-so" or "*that* so-and-so" or merely by inverted commas. In symbols, if p is a proposition, p by itself will stand for the unasserted proposition, while the asserted proposition will be designated by

$$\text{"} \vdash . p. \text{"}$$

The sign "$\vdash$" is called the assertion-sign*; it may be read "it is true that" (although philosophically this is not exactly what it means). The dots after the assertion-sign indicate its range; that is to say, everything following is asserted until we reach either an equal number of dots preceding a sign of implication or the end of the sentence. Thus "$\vdash : p . \supset . q$" means "it is true that p implies q," whereas "$\vdash . p . \supset \vdash . q$" means "$p$ is true; therefore q is true†." The first of these does not necessarily involve the truth either of p or of q, while the second involves the truth of both.

(4) *Assertion of a propositional function.* Besides the assertion of definite propositions, we need what we shall call "assertion of a propositional function." The general notion of asserting *any* propositional function is not used until ∗9, but we use at once the notion of asserting various special *elementary* propositional functions. Let ϕx be a propositional function whose argument is x; then we may assert ϕx without assigning a value to x. This is done, for example, when the law of identity is asserted in the form "A is A." Here A is left undetermined, because, however A may be deter-

* We have adopted both the idea and the symbol of assertion from Frege.

† Cf. *Principles of Mathematics*, § 38.

mined, the result will be true. Thus when we assert ϕx, leaving x undetermined, we are asserting an ambiguous value of our function. This is only legitimate if, however the ambiguity may be determined, the result will be true. Thus take, as an illustration, the primitive proposition *1·2 below, namely

$$\text{“}\vdash : p \vee p . \supset . p,\text{”}$$

i.e. "'p or p' implies p." Here p may be *any* elementary proposition: by leaving p undetermined, we obtain an assertion which can be applied to any particular elementary proposition. Such assertions are like the particular enunciations in Euclid: when it is said "let ABC be an isosceles triangle; then the angles at the base will be equal," what is said applies to *any* isosceles triangle; it is stated concerning *one* triangle, but not concerning a definite one. All the assertions in the present work, with a very few exceptions, assert propositional functions, not definite propositions.

As a matter of fact, no constant elementary proposition will occur in the present work, or can occur in any work which employs only logical ideas. The ideas and propositions of logic are all *general*: an assertion (for example) which is true of Socrates but not of Plato, will not belong to logic*, and if an assertion which is true of both is to occur in logic, it must not be made concerning either, but concerning a variable x. In order to obtain, in logic, a definite proposition instead of a propositional function, it is necessary to take some propositional function and assert that it is true always or sometimes, *i.e.* with all possible values of the variable or with some possible value. Thus, giving the name "individual" to whatever there is that is neither a proposition nor a function, the proposition "every individual is identical with itself" or the proposition "there are individuals" will be a proposition belonging to logic. But these propositions are not elementary.

(5) *Negation.* If p is any proposition, the proposition "not-p," or "p is false," will be represented by "$\sim p$." For the present, p must be an *elementary* proposition.

(6) *Disjunction.* If p and q are any propositions, the proposition "p or q," *i.e.* "either p is true or q is true," where the alternatives are to be not mutually exclusive, will be represented by

$$\text{“}p \vee q.\text{”}$$

This is called the *disjunction* or the *logical sum* of p and q. Thus "$\sim p \vee q$" will mean "p is false or q is true"; "$\sim (p \vee q)$" will mean "it is false that either p or q is true," which is equivalent to "p and q are both false"; "$\sim (\sim p \vee \sim q)$" will mean "it is false that either p is false or q is false," which is equivalent to "p and q are both true"; and so on. For the present, p and q must be elementary propositions.

* When we say that a proposition "belongs to logic," we mean that it can be expressed in terms of the primitive ideas of logic. We do not mean that logic *applies* to it, for that would of course be true of any proposition.

The above are all the primitive ideas required in the theory of deduction. Other primitive ideas will be introduced in Section B.

Definition of Implication. When a proposition q follows from a proposition p, so that if p is true, q must also be true, we say that p *implies* q. The idea of implication, in the form in which we require it, can be defined. The meaning to be given to implication in what follows may at first sight appear somewhat artificial; but although there are other legitimate meanings, the one here adopted is very much more convenient for our purposes than any of its rivals. The essential property that we require of implication is this: "What is implied by a true proposition is true." It is in virtue of this property that implication yields proofs. But this property by no means determines whether anything, and if so what, is implied by a false proposition. What it does determine is that, if p implies q, then it cannot be the case that p is true and q is false, *i.e.* it must be the case that either p is false or q is true. The most convenient interpretation of implication is to say, conversely, that if either p is false or q is true, then "p implies q" is to be true. Hence "p implies q" is to be defined to mean: "Either p is false or q is true." Hence we put:

***1·01.** $p \supset q \,.\!=\!.\, \sim p \vee q$ Df.

Here the letters "Df" stand for "definition." They and the sign of equality together are to be regarded as forming one symbol, standing for "is defined to mean*." Whatever comes to the left of the sign of equality is defined to mean the same as what comes to the right of it. Definition is not among the primitive ideas, because definitions are concerned solely with the symbolism, not with what is symbolised; they are introduced for practical convenience, and are theoretically unnecessary.

In virtue of the above definition, when "$p \supset q$" holds, then either p is false or q is true; hence if p is true, q must be true. Thus the above definition preserves the essential characteristic of implication; it gives, in fact, the most general meaning compatible with the preservation of this characteristic.

PRIMITIVE PROPOSITIONS.

***1·1.** Anything implied by a true elementary proposition is true. Pp†.

The above principle will be extended in *9 to propositions which are not elementary. It is not the same as "*if* p is true, then *if* p implies q, q is true." This is a true proposition, but it holds equally when p is not true and when p does not imply q. It does not, like the principle we are concerned with, enable us to assert q simply, without any hypothesis. We cannot express the principle symbolically, partly because any symbolism in which p is variable only gives the *hypothesis* that p is true, not the *fact* that it is true‡.

* The sign of equality not followed by the letters "Df" will have a different meaning, to be defined later.

† The letters "Pp" stand for "primitive proposition," as with Peano.

‡ For further remarks on this principle, cf. *Principles of Mathematics*, § 38.

The above principle is used whenever we have to deduce a *proposition* from a *proposition*. But the immense majority of the assertions in the present work are assertions of propositional functions, *i.e.* they contain an undetermined variable. Since the assertion of a propositional function is a different primitive idea from the assertion of a proposition, we require a primitive proposition different from ∗1·1, though allied to it, to enable us to deduce the assertion of a propositional function "ψx" from the assertions of the two propositional functions "ϕx" and "$\phi x \supset \psi x$." This primitive proposition is as follows:

∗1·11. When ϕx can be asserted, where x is a real variable, and $\phi x \supset \psi x$ can be asserted, where x is a real variable, then ψx can be asserted, where x is a real variable. Pp.

This principle is also to be assumed for functions of several variables.

Part of the importance of the above primitive proposition is due to the fact that it expresses in the symbolism a result following from the theory of types, which requires symbolic recognition. Suppose we have the two assertions of *propositional functions* "$\vdash . \phi x$" and "$\vdash . \phi x \supset \psi x$"; then the "$x$" in ϕx is not absolutely anything, but anything for which as argument the function "ϕx" is significant; similarly in "$\phi x \supset \psi x$" the x is anything for which "$\phi x \supset \psi x$" is significant. Apart from some axiom, we do not know that the x's for which "$\phi x \supset \psi x$" is significant are the same as those for which "ϕx" is significant. The primitive proposition ∗1·11, by securing that, as the result of the assertions of the *propositional functions* "ϕx" and "$\phi x \supset \psi x$," the propositional function "ψx" can also be asserted, secures partial symbolic recognition, in the form most useful in actual deductions, of an important principle which follows from the theory of types, namely that, if there is any one argument a for which both "ϕa" and "ψa" are significant, then the range of arguments for which "ϕx" is significant is the same as the range of arguments for which "ψx" is significant. It is obvious that, if the propositional function "$\phi x \supset \psi x$" can be asserted, there must be arguments a for which "$\phi a \supset \psi a$" is significant, and for which, therefore, "ϕa" and "ψa" must be significant. Hence, by our principle, the values of x for which "ϕx" is significant are the same as those for which "ψx" is significant, *i.e.* the type of possible arguments for $\phi \hat{x}$ (cf. p. 15) is the same as that of possible arguments for $\psi \hat{x}$. The primitive proposition ∗1·11, since it states a practically important consequence of this fact, is called the "axiom of identification of type."

Another consequence of the principle that, if there is an argument a for which both ϕa and ψa are significant, then ϕx is significant whenever ψx is significant, and vice versa, will be given in the "axiom of identification of real variables," introduced in ∗1·72. These two propositions, ∗1·11 and ∗1·72, give what is symbolically essential to the conduct of demonstrations in accordance with the theory of types.

The above proposition $*1\cdot 11$ is used in every inference from one asserted propositional function to another. We will illustrate the use of this proposition by setting forth at length the way in which it is first used, in the proof of $*2\cdot 06$. That proposition is

$$\text{"}\vdash :. p \supset q . \supset : q \supset r . \supset . p \supset r.\text{"}$$

We have already proved, in $*2\cdot 05$, the proposition

$$\vdash :. q \supset r . \supset : p \supset q . \supset . p \supset r.$$

It is obvious that $*2\cdot 06$ results from $*2\cdot 05$ by means of $*2\cdot 04$, which is

$$\vdash :. p . \supset . q \supset r : \supset : q . \supset . p \supset r.$$

For if, in this proposition, we replace p by $q \supset r$, q by $p \supset q$, and r by $p \supset r$, we obtain, as an instance of $*2\cdot 04$, the proposition

$$\vdash :: q \supset r . \supset : p \supset q . \supset . p \supset r :. \supset :. p \supset q . \supset : q \supset r . \supset . p \supset r \qquad (1),$$

and here the hypothesis is asserted by $*2\cdot 05$. Thus our primitive proposition $*1\cdot 11$ enables us to assert the conclusion.

$*1\cdot 2$. $\vdash : p \vee p . \supset . p$ Pp.

This proposition states: "If either p is true or p is true, then p is true." It is called the "principle of tautology," and will be quoted by the abbreviated title of "Taut." It is convenient, for purposes of reference, to give names to a few of the more important propositions; in general, propositions will be referred to by their numbers.

$*1\cdot 3$. $\vdash : q . \supset . p \vee q$ Pp.

This principle states: "If q is true, then 'p or q' is true." Thus *e.g.* if q is "to-day is Wednesday" and p is "to-day is Tuesday," the principle states: "If to-day is Wednesday, then to-day is either Tuesday or Wednesday." It is called the "principle of addition," because it states that if a proposition is true, any alternative may be added without making it false. The principle will be referred to as "Add."

$*1\cdot 4$. $\vdash : p \vee q . \supset . q \vee p$ Pp.

This principle states that "p or q" implies "q or p." It states the permutative law for logical addition of propositions, and will be called the "principle of permutation." It will be referred to as "Perm."

$*1\cdot 5$. $\vdash : p \vee (q \vee r) . \supset . q \vee (p \vee r)$ Pp.

This principle states: "If either p is true, or 'q or r' is true, then either q is true, or 'p or r' is true." It is a form of the associative law for logical addition, and will be called the "associative principle." It will be referred to as "Assoc." The proposition

$$p \vee (q \vee r) . \supset . (p \vee q) \vee r,$$

which would be the natural form for the associative law, has less deductive power, and is therefore not taken as a primitive proposition.

***1·6.** $\vdash :. q \supset r . \supset : p \vee q . \supset . p \vee r$ Pp.

This principle states: "If q implies r, then 'p or q' implies 'p or r.'" In other words, in an implication, an alternative may be added to both premiss and conclusion without impairing the truth of the implication. The principle will be called the "principle of summation," and will be referred to as "Sum."

***1·7.** If p is an elementary proposition, $\sim p$ is an elementary proposition. Pp.

***1·71.** If p and q are elementary propositions, $p \vee q$ is an elementary proposition. Pp.

***1·72.** If ϕp and ψp are elementary propositional functions which take elementary propositions as arguments, $\phi p \vee \psi p$ is an elementary propositional function. Pp.

This axiom is to apply also to functions of two or more variables. It is called the "axiom of identification of real variables." It will be observed that if ϕ and ψ are functions which take arguments of different types, there is no such function as "$\phi x \vee \psi x$," because ϕ and ψ cannot significantly have the same argument. A more general form of the above axiom will be given in *9.

The use of the above axioms *1·7·71·72 will generally be tacit. It is only through them and the axioms of *9 that the theory of types explained in the Introduction becomes relevant, and any view of logic which justifies these axioms justifies such subsequent reasoning as employs the theory of types.

This completes the list of primitive propositions required for the theory of deduction as applied to elementary propositions.

∗2. IMMEDIATE CONSEQUENCES OF THE PRIMITIVE PROPOSITIONS

Summary of ∗2.

The proofs of the earlier of the propositions of this number consist simply in noticing that they are instances of the general rules given in ∗1. In such cases, these rules are not premisses, since they assert any instance of themselves, not something other than their instances. Hence when a general rule is adduced in early proofs, it will be adduced in brackets*, with indications, when required, as to the changes of letters from those given in the rule to those in the case considered. Thus "Taut $\frac{\sim p}{p}$" will mean what "Taut" becomes when $\sim p$ is written in place of p. If "Taut $\frac{\sim p}{p}$" is enclosed in square brackets before an asserted proposition, that means that, in accordance with "Taut," we are asserting what "Taut" becomes when $\sim p$ is written in place of p. The recognition that a certain proposition is an instance of some general proposition previously proved or assumed is essential to the process of deduction from general rules, but cannot itself be erected into a general rule, since the application required is particular, and no general rule can *explicitly* include a particular application.

Again, when two different sets of symbols express the same proposition in virtue of a definition, say ∗1·01, and one of these, which we will call (1), has been asserted, the assertion of the other is made by writing "[(1).(∗1·01)]" before it, meaning that, in virtue of ∗1·01, the new set of symbols asserts the same proposition as was asserted in (1). A reference to a definition is distinguished from a reference to a previous proposition by being enclosed in round brackets.

The propositions in this number are all, or nearly all, actually needed in deducing mathematics from our primitive propositions. Although certain abbreviating processes will be gradually introduced, proofs will be given very fully, because the importance of the present subject lies, not in the propositions themselves, but (1) in the fact that they follow from the primitive propositions, (2) in the fact that the subject is the easiest, simplest, and most elementary example of the symbolic method of dealing with the principles of mathematics generally. Later portions—the theories of classes, relations, cardinal numbers, series, ordinal numbers, geometry, etc.—all employ the same method, but with an increasing complexity in the entities and functions considered.

* Later on we shall cease to mark the distinction between a premiss and a rule according to which an inference is conducted. It is only in early proofs that this distinction is important.

The most important propositions proved in the present number are the following:

***2·02.** $\vdash : q . \supset . p \supset q$

I.e. q implies that p implies q, *i.e.* a true proposition is implied by any proposition. This proposition is called the "principle of simplification" (referred to as "Simp"), because, as will appear later, it enables us to pass from the joint assertion of q and p to the assertion of q simply. When the special meaning which we have given to implication is remembered, it will be seen that this proposition is obvious.

***2·03.** $\vdash : p \supset \sim q . \supset . q \supset \sim p$

***2·15.** $\vdash : \sim p \supset q . \supset . \sim q \supset p$

***2·16.** $\vdash : p \supset q . \supset . \sim q \supset \sim p$

***2·17.** $\vdash : \sim q \supset \sim p . \supset . p \supset q$

These four analogous propositions constitute the "principle of transposition," referred to as "Transp." They lead to the rule that in an implication the two sides may be interchanged by turning negative into positive and positive into negative. They are thus analogous to the algebraical rule that the two sides of an equation may be interchanged by changing the signs.

***2·04.** $\vdash :. p . \supset . q \supset r : \supset : q . \supset . p \supset r$

This is called the "commutative principle" and referred to as "Comm." It states that, if r follows from q provided p is true, then r follows from p provided q is true.

***2·05.** $\vdash :. q \supset r . \supset : p \supset q . \supset . p \supset r$

***2·06.** $\vdash :. p \supset q . \supset : q \supset r . \supset . p \supset r$

These two propositions are the source of the syllogism in Barbara (as will be shown later) and are therefore called the "principle of the syllogism" (referred to as "Syll"). The first states that, if r follows from q, then if q follows from p, r follows from p. The second states the same thing with the premisses interchanged.

***2·08.** $\vdash . p \supset p$

I.e. any proposition implies itself. This is called the "principle of identity" and referred to as "Id." It is not the same as the "law of identity" ("x is identical with x"), but the law of identity is inferred from it (cf. *13·15).

***2·21.** $\vdash : \sim p . \supset . p \supset q$

I.e. a false proposition implies any proposition.

The later propositions of the present number are mostly subsumed under propositions in *3 or *4, which give the same results in more compendious forms. We now proceed to formal deductions.

***2·01.** $\vdash : p \supset \sim p . \supset . \sim p$

This proposition states that, if p implies its own falsehood, then p is false. It is called the "principle of the *reductio ad absurdum*," and will be referred to as 'Abs."* The proof is as follows (where "*Dem.*" is short for "demonstration"):

Dem.

$$\left[\text{Taut}\ \frac{\sim p}{p}\right] \quad \vdash : \sim p \vee \sim p . \supset . \sim p \qquad (1)$$

$$[(1).(*1\cdot01)] \quad \vdash : p \supset \sim p . \supset . \sim p$$

***2·02.** $\vdash : q . \supset . p \supset q$

Dem.

$$\left[\text{Add}\ \frac{\sim p}{p}\right] \quad \vdash : q . \supset . \sim p \vee q \qquad (1)$$

$$[(1).(*1\cdot01)] \quad \vdash : q . \supset . p \supset q$$

***2·03.** $\vdash : p \supset \sim q . \supset . q \supset \sim p$

Dem.

$$\left[\text{Perm}\ \frac{\sim p, \sim q}{p, \quad q}\right] \quad \vdash : \sim p \vee \sim q . \supset . \sim q \vee \sim p \qquad (1)$$

$$[(1).(*1\cdot01)] \quad \vdash : p \supset \sim q . \supset . q \supset \sim p$$

***2·04.** $\vdash :. p . \supset . q \supset r : \supset : q . \supset . p \supset r$

Dem.

$$\left[\text{Assoc}\ \frac{\sim p, \sim q}{p, \quad q}\right] \quad \vdash :. \sim p \vee (\sim q \vee r) . \supset . \sim q \vee (\sim p \vee r) \qquad (1)$$

$$[(1).(*1\cdot01)] \quad \vdash :. p . \supset . q \supset r : \supset : q . \supset . p \supset r$$

***2·05.** $\vdash :. q \supset r . \supset : p \supset q . \supset . p \supset r$

Dem.

$$\left[\text{Sum}\ \frac{\sim p}{p}\right] \quad \vdash :. q \supset r . \supset : \sim p \vee q . \supset . \sim p \vee r \qquad (1)$$

$$[(1).(*1\cdot01)] \quad \vdash :. q \supset r . \supset : p \supset q . \supset . p \supset r$$

***2·06.** $\vdash :. p \supset q . \supset : q \supset r . \supset . p \supset r$

Dem.

$$\left[\text{Comm}\ \frac{q \supset r, p \supset q, p \supset r}{p, \quad q, \quad r}\right] \vdash :: q \supset r . \supset : p \supset q . \supset . p \supset r :.$$

$$\supset :. p \supset q . \supset : q \supset r . \supset . p \supset r \qquad (1)$$

$$[*2\cdot05] \quad \vdash :. q \supset r . \supset : p \supset q . \supset . p \supset r \qquad (2)$$

$$[(1).(2).*1\cdot11] \quad \vdash :. p \supset q . \supset : q \supset r . \supset . p \supset r$$

In the last line of this proof, "(1) . (2) . *1·11" means that we are inferring in accordance with *1·11, having before us a proposition, namely $p \supset q . \supset : q \supset r . \supset . p \supset r$, which, by (1), is implied by $q \supset r . \supset : p \supset q . \supset . p \supset r$, which, by (2), is true. In general, in such cases, we shall omit the reference to *1·11.

* There is an interesting historical article on this principle by Vailati, "A proposito d' un passo del Teeteto e di una dimostrazione di Euclide," *Rivista di Filosofia e scienze affine*, 1904.

The above two propositions will both be referred to as the "principle of the syllogism" (shortened to "Syll"), because, as will appear later, the syllogism in Barbara is derived from them.

$*2\cdot07$. $\vdash : p \,.\, \supset . \, p \vee p \quad \left[*1\cdot3 \frac{p}{q}\right]$

Here we put nothing beyond "$*1\cdot3 \frac{p}{q}$," because the proposition to be proved is what $*1\cdot3$ becomes when p is written in place of q.

$*2\cdot08$. $\vdash . \, p \supset p$

Dem.

$\left[*2\cdot05 \frac{p \vee p,\, p}{q,\quad r}\right]$	$\vdash :: p \vee p \,.\, \supset . \, p : \supset :. \, p \,.\, \supset . \, p \vee p : \supset . \, p \supset p$	(1)
[Taut]	$\vdash : p \vee p \,.\, \supset . \, p$	(2)
$[(1).(2).*1\cdot11]$	$\vdash :. \, p \,.\, \supset . \, p \vee p : \supset . \, p \supset p$	(3)
$[*2\cdot07]$	$\vdash : p \,.\, \supset . \, p \vee p$	(4)
$[(3).(4).*1\cdot11]$	$\vdash . \, p \supset p$	

$*2\cdot1$. $\vdash . \sim p \vee p \quad [*2\cdot08 \,.\, (*1\cdot01)]$

$*2\cdot11$. $\vdash . \, p \vee \sim p$

Dem.

$\left[\text{Perm} \frac{\sim p,\, p}{p,\quad q}\right]$	$\vdash : \sim p \vee p \,.\, \supset . \, p \vee \sim p$	(1)
$[(1).*2\cdot1.*1\cdot11]$	$\vdash . \, p \vee \sim p$	

This is the law of excluded middle.

$*2\cdot12$. $\vdash . \, p \supset \sim(\sim p)$

Dem.

$\left[*2\cdot11 \frac{\sim p}{p}\right]$	$\vdash . \sim p \vee \sim(\sim p)$	(1)
$[(1).(*1\cdot01)]$	$\vdash . \, p \supset \sim(\sim p)$	

$*2\cdot13$. $\vdash . \, p \vee \sim\{\sim(\sim p)\}$

This proposition is a lemma for $*2\cdot14$, which, with $*2\cdot12$, constitutes the principle of double negation.

Dem.

$\left[\text{Sum} \frac{\sim p,\, \sim\{\sim(\sim p)\}}{q,\qquad r}\right]$	$\vdash :. \sim p \,.\, \supset . \sim\{\sim(\sim p)\} \,.\, \supset :$ $p \vee \sim p \,.\, \supset . \, p \vee \sim\{\sim(\sim p)\}$	(1)
$\left[*2\cdot12 \frac{\sim p}{p}\right]$	$\vdash : \sim p \,.\, \supset . \sim\{\sim(\sim p)\}$	(2)
$[(1).(2).*1\cdot11]$	$\vdash : p \vee \sim p \,.\, \supset . \, p \vee \sim\{\sim(\sim p)\}$	(3)
$[(3).*2\cdot11.*1\cdot11]$	$\vdash . \, p \vee \sim\{\sim(\sim p)\}$	

$*2{\cdot}14.\ \vdash .\sim(\sim p)\supset p$

Dem.

$\left[\text{Perm}\ \frac{\sim\{\sim(\sim p)\}}{q}\right]\ \vdash : p \vee \sim\{\sim(\sim p)\} . \supset . \sim\{\sim(\sim p)\} \vee p$ (1)

$[(1).*2{\cdot}13.*1{\cdot}11]\quad \vdash . \sim\{\sim(\sim p)\} \vee p$ (2)

$[(2).(*1{\cdot}01)]\quad \vdash . \sim(\sim p)\supset p$

$*2{\cdot}15.\ \vdash : \sim p\supset q . \supset . \sim q\supset p$

Dem.

$\left[*2{\cdot}05\ \frac{\sim p, \sim(\sim q)}{p,\quad r}\right]\quad \vdash :. q\supset\sim(\sim q) . \supset : \sim p\supset q . \supset . \sim p\supset\sim(\sim q)$ (1)

$\left[*2{\cdot}12\ \frac{q}{p}\right]\quad \vdash . q\supset\sim(\sim q)$ (2)

$[(1).(2).*1{\cdot}11]\quad \vdash : \sim p\supset q . \supset . \sim p\supset\sim(\sim q)$ (3)

$\left[*2{\cdot}03\ \frac{\sim p, \sim q}{p,\quad q}\right]\quad \vdash : \sim p\supset\sim(\sim q) . \supset . \sim q\supset\sim(\sim p)$ (4)

$\left[*2{\cdot}05\ \frac{\sim q, \sim(\sim p), p}{p,\quad q,\quad r}\right]\ \vdash :. \sim(\sim p)\supset p . \supset : \sim q\supset\sim(\sim p) . \supset . \sim q\supset p$ (5)

$[(5).*2{\cdot}14.*1{\cdot}11]\quad \vdash : \sim q\supset\sim(\sim p) . \supset . \sim q\supset p$ (6)

$\left[*2{\cdot}05\ \frac{\sim p\supset q, \sim p\supset\sim(\sim q), \sim q\supset\sim(\sim p)}{p,\quad q,\quad r}\right]\ \vdash ::$

$\sim p\supset\sim(\sim q) . \supset . \sim q\supset\sim(\sim p) : \supset :.$

$\sim p\supset q . \supset . \sim p\supset\sim(\sim q) : \supset : \sim p\supset q . \supset . \sim q\supset\sim(\sim p)$ (7)

$[(4).(7).*1{\cdot}11]\ \vdash :. \sim p\supset q . \supset . \sim p\supset\sim(\sim q) : \supset :$

$\sim p\supset q . \supset . \sim q\supset\sim(\sim p)$ (8)

$[(3).(8).*1{\cdot}11]\ \vdash : \sim p\supset q . \supset . \sim q\supset\sim(\sim p)$ (9)

$\left[*2{\cdot}05\ \frac{\sim p\supset q, \sim q\supset\sim(\sim p), \sim q\supset p}{p,\quad q,\quad r}\right]\ \vdash :: \sim q\supset\sim(\sim p) . \supset . \sim q\supset p :$

$\supset :. \sim p\supset q . \supset . \sim q\supset\sim(\sim p) : \supset : \sim p\supset q . \supset . \sim q\supset p$ (10)

$[(6).(10).*1{\cdot}11]\ \vdash :. \sim p\supset q . \supset . \sim q\supset \sim(\sim p) : \supset :$

$\sim p\supset q . \supset . \sim q\supset p$ (11)

$[(9).(11).*1{\cdot}11]\ \vdash : \sim p\supset q . \supset . \sim q\supset p$

Note on the proof of $*2{\cdot}15$. In the above proof, it will be seen that (3), (4), (6) are respectively of the forms $p_1\supset p_2$, $p_2\supset p_3$, $p_3\supset p_4$, where $p_1\supset p_4$ is the proposition to be proved. From $p_1\supset p_2$, $p_2\supset p_3$, $p_3\supset p_4$ the proposition $p_1\supset p_4$ results by repeated applications of $*2{\cdot}05$ or $*2{\cdot}06$ (both of which are called "Syll"). It is tedious and unnecessary to repeat this process every time it is used; it will therefore be abbreviated into

"[Syll] $\vdash . (a) . (b) . (c) . \supset \vdash . (d)$,"

where (a) is of the form $p_1\supset p_2$, (b) of the form $p_2\supset p_3$, (c) of the form $p_3\supset p_4$, and (d) of the form $p_1\supset p_4$. The same abbreviation will be applied to a sorites of any length.

Also where we have "$\vdash . p_1$" and "$\vdash . p_1 \supset p_2$," and p_2 is the proposition to be proved, it is convenient to write simply

$$\text{"}\vdash . p_1 . \supset$$

$$[\text{etc.}] \qquad \vdash . p_2 \text{,"}$$

where "etc." will be a reference to the previous propositions in virtue of which the implication "$p_1 \supset p_2$" holds. This form embodies the use of ∗1·11 or ∗1·1, and makes many proofs at once shorter and easier to follow. It is used in the first two lines of the following proof.

∗2·16. $\vdash : p \supset q . \supset . \sim q \supset \sim p$

Dem.

$$
\begin{array}{lll}
[∗2·12] & \vdash . q \supset \sim(\sim q) . \supset & \\
[∗2·05] & \vdash : p \supset q . \supset . p \supset \sim(\sim q) & (1) \\
\left[∗2·03 \dfrac{\sim q}{q}\right] & \vdash : p \supset \sim(\sim q) . \supset . \sim q \supset \sim p & (2) \\
[\text{Syll}] & \vdash . (1) . (2) . \supset \vdash : p \supset q . \supset . \sim q \supset \sim p &
\end{array}
$$

Note. The proposition to be proved will be called "Prop," and when a proof ends, like that of ∗2·16, by an implication between asserted propositions, of which the consequent is the proposition to be proved, we shall write "$\vdash$. etc. $\supset \vdash$. Prop". Thus "$\supset \vdash$. Prop" ends a proof, and more or less corresponds to "Q.E.D."

∗2·17. $\vdash : \sim q \supset \sim p . \supset . p \supset q$

Dem.

$$
\begin{array}{lll}
\left[∗2·03 \dfrac{\sim q, p}{p, q}\right] & \vdash : \sim q \supset \sim p . \supset . p \supset \sim(\sim q) & (1) \\
[∗2·14] & \vdash : \sim(\sim q) \supset q : \supset & \\
[∗2·05] & \vdash : p \supset \sim(\sim q) . \supset . p \supset q & (2) \\
[\text{Syll}] & \vdash . (1) . (2) . \supset \vdash . \text{Prop} &
\end{array}
$$

∗2·15, ∗2·16 and ∗2·17 are forms of the principle of transposition, and will be all referred to as "Transp."

∗2·18. $\vdash : \sim p \supset p . \supset . p$

Dem.

$$
\begin{array}{lll}
[∗2·12] & \vdash . p \supset \sim(\sim p) . \supset & \\
[∗2·05] & \vdash . \sim p \supset p . \supset . \sim p \supset \sim(\sim p) & (1) \\
\left[∗2·01 \dfrac{\sim p}{p}\right] & \vdash : \sim p \supset \sim(\sim p) . \supset . \sim(\sim p) & (2) \\
[\text{Syll}] & \vdash . (1) . (2) . \supset \vdash : \sim p \supset p . \supset . \sim(\sim p) & (3) \\
[∗2·14] & \vdash . \sim(\sim p) \supset p & (4) \\
[\text{Syll}] & \vdash . (3) . (4) . \supset \vdash . \text{Prop} &
\end{array}
$$

This is the complement of the principle of the *reductio ad absurdum*. It states that a proposition which follows from the hypothesis of its own falsehood is true.

***2·2.** $\vdash : p . \supset . p \vee q$

Dem.

$$\vdash . \text{Add} . \supset \vdash : p . \supset . q \vee p \qquad (1)$$
$$[\text{Perm}] \vdash : q \vee p . \supset . p \vee q \qquad (2)$$
$$[\text{Syll}] \vdash . (1) . (2) . \supset \vdash . \text{Prop}$$

***2·21.** $\vdash : \sim p . \supset . p \supset q \quad \left[*2 \cdot 2 \frac{\sim p}{p} \right]$

The above two propositions are very frequently used.

***2·24.** $\vdash : p . \supset . \sim p \supset q$ [*2·21 . Comm]

***2·25.** $\vdash :. p : \vee : p \vee q . \supset . q$

Dem.

$$\vdash . *2 \cdot 1 . \supset \vdash : \sim (p \vee q) . \vee . (p \vee q) :$$
$$[\text{Assoc}] \supset \vdash : p . \vee . \{ \sim (p \vee q) . \vee . q \} : \supset \vdash . \text{Prop}$$

***2·26.** $\vdash :. \sim p : \vee : p \supset q . \supset . q \quad \left[*2 \cdot 25 \frac{\sim p}{p} \right]$

***2·27.** $\vdash :. p . \supset : p \supset q . \supset . q$ [*2·26]

***2·3.** $\vdash : p \vee (q \vee r) . \supset . p \vee (r \vee q)$

Dem.

$$\left[\text{Perm} \frac{q, r}{p, q} \right] \quad \vdash : q \vee r . \supset . r \vee q :$$
$$\left[\text{Sum} \frac{q \vee r, r \vee q}{q, \quad r} \right] \supset \vdash : p \vee (q \vee r) . \supset . p \vee (r \vee q)$$

***2·31.** $\vdash : p \vee (q \vee r) . \supset . (p \vee q) \vee r$

This proposition and *2·32 together constitute the associative law for logical addition of propositions. In the proof, the following abbreviation (constantly used hereafter) will be employed*: When we have a series of propositions of the form $a \supset b$, $b \supset c$, $c \supset d$, all asserted, and "$a \supset d$" is the proposition to be proved, the proof in full is as follows:

$$[\text{Syll}] \quad \vdash :. a \supset b . \supset : b \supset c . \supset . a \supset c \qquad (1)$$
$$\vdash : a . \supset . b \qquad (2)$$
$$[(1).(2).*1 \cdot 11] \quad \vdash : b \supset c . \supset . a \supset c \qquad (3)$$
$$\vdash : b . \supset . c \qquad (4)$$
$$[(3).(4).*1 \cdot 11] \quad \vdash : a . \supset . c \qquad (5)$$
$$[\text{Syll}] \quad \vdash :. a \supset c . \supset : c \supset d . \supset . a \supset d \qquad (6)$$
$$[(5).(6).*1 \cdot 11] \quad \vdash : c \supset d . \supset . a \supset d \qquad (7)$$
$$\vdash : c . \supset . d \qquad (8)$$
$$[(7).(8).*1 \cdot 11] \quad \vdash : a . \supset . d$$

* This abbreviation applies to the same type of cases as those concerned in the note to *2·15, but is often more convenient than the abbreviation explained in that note.

It is tedious to write out this process in full; we therefore write simply

$\vdash : a . \supset . b .$
[etc.] $\supset . c .$
[etc.] $\supset . d : \supset \vdash .$ Prop,

where "$a \supset d$" is the proposition to be proved. We indicate on the left by references in square brackets the propositions in virtue of which the successive implications hold. We put one dot (not two) after "b," to show that it is b, not "$a \supset b$," that implies c. But we put two dots after d, to show that now the whole proposition "$a \supset d$" is concerned. If "$a \supset d$" is not the proposition to be proved, but is to be used subsequently in the proof, we put

$\vdash : a . \supset . b .$
[etc.] $\supset . c .$
[etc.] $\supset . d$ (1),

and then "(1)" means "$a \supset d$." The proof of ∗2·31 is as follows:

Dem.

[∗2·3] $\vdash : p \vee (q \vee r) . \supset . p \vee (r \vee q) .$
$\left[\text{Assoc} \dfrac{r, q}{q, r}\right]$ $\supset . r \vee (p \vee q) .$
$\left[\text{Perm} \dfrac{r, p \vee q}{p, \quad q}\right]$ $\supset . (p \vee q) \vee r : \supset \vdash .$ Prop

∗2·32. $\vdash : (p \vee q) \vee r . \supset . p \vee (q \vee r)$

Dem.

$\left[\text{Perm} \dfrac{p \vee q, r}{p, \quad q}\right]$ $\vdash : (p \vee q) \vee r . \supset . r \vee (p \vee q)$
$\left[\text{Assoc} \dfrac{r, p, q}{p, q, r}\right]$ $\supset . p \vee (r \vee q)$
[∗2·3] $\supset . p \vee (q \vee r) : \supset \vdash .$ Prop

∗2·33. $p \vee q \vee r . = . (p \vee q) \vee r$ Df

This definition serves only for the avoidance of brackets.

∗2·36. $\vdash :. q \supset r . \supset : p \vee q . \supset . r \vee p$

Dem.

[Perm] $\vdash : p \vee r . \supset . r \vee p :$
$\left[\text{Syll} \dfrac{p \vee q, p \vee r, r \vee p}{p, \quad q, \quad r}\right]$ $\supset \vdash :. p \vee q . \supset . p \vee r : \supset : p \vee q . \supset . r \vee p$ (1)
[Sum] $\vdash :. q \supset r . \supset : p \vee q . \supset . p \vee r$ (2)
$\vdash . (1) . (2) .$ Syll $. \supset \vdash .$ Prop

∗2·37. $\vdash :. q \supset r . \supset : q \vee p . \supset . p \vee r$
[Syll . Perm . Sum]

∗2·38. $\vdash :. q \supset r . \supset : q \vee p . \supset . r \vee p$
[Syll . Perm . Sum]

The proofs of $*2{\cdot}37{\cdot}38$ are exactly analogous to that of $*2{\cdot}36$. (We use "$*2{\cdot}37{\cdot}38$" as an abbreviation for "$*2{\cdot}37$ and $*2{\cdot}38$." Such abbreviations will be used throughout.)

The use of a general principle of deduction, such as either form of "Syll," in a proof, is different from the use of the particular premisses to which the principle of deduction is applied. The principle of deduction gives the general rule according to which the inference is made, but is not itself a premiss in the inference. If we treated it as a premiss, we should need either it or some other general rule to enable us to infer the desired conclusion, and thus we should gradually acquire an increasing accumulation of premisses without ever being able to make any inference. Thus when a general rule is adduced in drawing an inference, as when we write "[Syll] $\vdash . (1) . (2) . \supset \vdash .$ Prop," the mention of "Syll" is only required in order to remind the reader how the inference is drawn.

The rule of inference may, however, also occur as one of the ordinary premisses, that is to say, in the case of "Syll" for example, the proposition "$p \supset q . \supset : q \supset r . \supset . p \supset r$" may be one of those to which our rules of deduction are applied, and it is then an ordinary premiss. The distinction between the two uses of principles of deduction is of some philosophical importance, and in the above proofs we have indicated it by putting the rule of inference in square brackets. It is, however, practically inconvenient to continue to distinguish in the manner of the reference. We shall therefore henceforth both adduce ordinary premisses in square brackets where convenient, and adduce rules of inference, along with other propositions, in asserted premisses, *i.e.* we shall write *e.g.*

"$\vdash . (1) . (2) .$ Syll $. \supset \vdash .$ Prop"

rather than "[Syll] $\vdash . (1) . (2) . \supset \vdash .$ Prop"

$*2{\cdot}4$. $\vdash :. p . \vee . p \vee q : \supset . p \vee q$

Dem.

$\vdash . *2{\cdot}31 . \supset \vdash :. p . \vee . p \vee q : \supset : p \vee p . \vee . q :$

[Taut.$*2{\cdot}38$] $\supset : p \vee q :. \supset \vdash .$ Prop

$*2{\cdot}41$. $\vdash :. q . \vee . p \vee q : \supset . p \vee q$

Dem.

$\left[\text{Assoc}\,\frac{q, p, q}{p, q, r}\right] \vdash :. q . \vee . p \vee q : \supset : p . \vee . q \vee q :$

[Taut.Sum] $\supset : p \vee q :. \supset \vdash .$ Prop

$*2{\cdot}42$. $\vdash :. \sim p . \vee . p \supset q : \supset . p \supset q$ $\left[*2{\cdot}4 \frac{\sim p}{p}\right]$

$*2{\cdot}43$. $\vdash :. p . \supset . p \supset q : \supset . p \supset q$ [$*2{\cdot}42$]

$*2{\cdot}45$. $\vdash : \sim(p \vee q) . \supset . \sim p$ [$*2{\cdot}2$. Transp]

$*2{\cdot}46$. $\vdash : \sim(p \vee q) . \supset . \sim q$ [$*1{\cdot}3$. Transp]

*2·47. $\vdash:\sim(p\vee q).\supset.\sim p\vee q$ $\left[*2\cdot45\,.\,*2\cdot2\,\dfrac{\sim p}{p}\,.\,\mathrm{Syll}\right]$

*2·48. $\vdash:\sim(p\vee q).\supset.p\vee\sim q$ $\left[*2\cdot46\,.\,*1\cdot3\,\dfrac{\sim q}{q}\,.\,\mathrm{Syll}\right]$

*2·49. $\vdash:\sim(p\vee q).\supset.\sim p\vee\sim q$ $\left[*2\cdot45\,.\,*2\cdot2\,\dfrac{\sim p,\ \sim q}{p,\ \ q}\,.\,\mathrm{Syll}\right]$

*2·5. $\vdash:\sim(p\supset q).\supset.\sim p\supset q$ $\left[*2\cdot47\,\dfrac{\sim p}{p}\right]$

*2·51. $\vdash:\sim(p\supset q).\supset.p\supset\sim q$ $\left[*2\cdot48\,\dfrac{\sim p}{p}\right]$

*2·52. $\vdash:\sim(p\supset q).\supset.\sim p\supset\sim q$ $\left[*2\cdot49\,\dfrac{\sim p}{p}\right]$

*2·521. $\vdash:\sim(p\supset q).\supset.q\supset p$ [*2·52·17]

*2·53. $\vdash:p\vee q.\supset.\sim p\supset q$

Dem.

$\vdash.*2\cdot12\cdot38.\supset\vdash:p\vee q.\supset.\sim(\sim p)\vee q:\supset\vdash.\mathrm{Prop}$

*2·54. $\vdash:\sim p\supset q.\supset.p\vee q$ [*2·14·38]

*2·55. $\vdash:.\sim p.\supset:p\vee q.\supset.q$ [*2·53 . Comm]

*2·56. $\vdash:.\sim q.\supset:p\vee q.\supset.p$ $\left[*2\cdot55\,\dfrac{q,\ p}{p,\ q}\,.\,\mathrm{Perm}\right]$

*2·6. $\vdash:.\sim p\supset q.\supset:p\supset q.\supset.q$

Dem.

[*2·38] $\vdash:.\sim p\supset q.\supset:\sim p\vee q.\supset.q\vee q$ (1)

[Taut . Syll] $\vdash:.\sim p\vee q.\supset.q\vee q:\supset:\sim p\vee q.\supset.q$ (2)

$\vdash.(1).(2).\mathrm{Syll}.\supset\vdash:.\sim p\supset q.\supset:\sim p\vee q.\supset.q:.\supset\vdash.\mathrm{Prop}$

*2·61. $\vdash:.p\supset q.\supset:\sim p\supset q.\supset.q$ [*2·6 . Comm]

*2·62. $\vdash:.p\vee q.\supset:p\supset q.\supset.q$ [*2·53·6 . Syll]

*2·621. $\vdash:.p\supset q.\supset:p\vee q.\supset.q$ [*2·62 . Comm]

*2·63. $\vdash:.p\vee q.\supset:\sim p\vee q.\supset.q$ [*2·62]

*2·64. $\vdash:.p\vee q.\supset:p\vee\sim q.\supset.p$ $\left[*2\cdot63\,\dfrac{q,\ p}{p,\ q}\,.\,\mathrm{Perm}\right]$

*2·65. $\vdash:.p\supset q.\supset:p\supset\sim q.\supset.\sim p$ $\left[*2\cdot64\,\dfrac{\sim p}{p}\right]$

*2·67. $\vdash:.p\vee q.\supset.q:\supset.p\supset q$

Dem.

[*2·54.Syll] $\vdash:.p\vee q.\supset.q:\supset:\sim p\supset q.\supset.q$ (1)

[*2·24.Syll] $\vdash:.\sim p\supset q.\supset.q:\supset.p\supset q$ (2)

$\vdash.(1).(2).\mathrm{Syll}.\supset\vdash.\mathrm{Prop}$

***2·68.** $\vdash :. p \supset q . \supset . q : \supset . p \lor q$

Dem.

$$\left[*2{\cdot}67 \frac{\sim p}{p}\right] \vdash :. p \supset q . \supset . q : \supset . \sim p \supset q \quad (1)$$

$\vdash . (1) . *2{\cdot}54 . \supset \vdash . \mathrm{Prop}$

***2·69.** $\vdash :. p \supset q . \supset . q : \supset : q \supset p . \supset . p$ $\quad \left[*2{\cdot}68 . \mathrm{Perm} . *2{\cdot}62 \frac{q, p}{p, q}\right]$

***2·73.** $\vdash :. p \supset q . \supset : p \lor q \lor r . \supset . q \lor r$ $\quad [*2{\cdot}621{\cdot}38]$

***2·74.** $\vdash :. q \supset p . \supset : p \lor q \lor r . \supset . p \lor r$ $\quad \left[*2{\cdot}73 \frac{q, p}{p, q} . \mathrm{Assoc} . \mathrm{Syll}\right]$

***2·75.** $\vdash :: p \lor q . \supset :. p . \lor . q \supset r : \supset . p \lor r$ $\quad \left[*2{\cdot}74 \frac{\sim q}{q} . *2{\cdot}53{\cdot}31\right]$

***2·76.** $\vdash :. p . \lor . q \supset r : \supset : p \lor q . \supset . p \lor r$ $\quad [*2{\cdot}75 . \mathrm{Comm}]$

***2·77.** $\vdash :. p . \supset . q \supset r : \supset : p \supset q . \supset . p \supset r$ $\quad \left[*2{\cdot}76 \frac{\sim p}{p}\right]$

***2·8.** $\vdash :. q \lor r . \supset : \sim r \lor s . \supset . q \lor s$

Dem.

$\vdash . *2{\cdot}53 . \mathrm{Perm} . \supset \vdash :. q \lor r . \supset : \sim r \supset q :$

$[*2{\cdot}38] \qquad \supset : \sim r \lor s . \supset . q \lor s :. \supset \vdash . \mathrm{Prop}$

***2·81.** $\vdash :: q . \supset . r \supset s : \supset :. p \lor q . \supset : p \lor r . \supset . p \lor s$

Dem.

$\vdash . \mathrm{Sum} . \supset \vdash :: q . \supset . r \supset s : \supset :. p \lor q . \supset : p . \lor . r \supset s \quad (1)$

$\vdash . *2{\cdot}76 . \mathrm{Syll} . \supset \vdash :: p \lor q . \supset : p . \lor . r \supset s :. \supset :.$

$\qquad p \lor q . \supset : p \lor r . \supset . p \lor s \quad (2)$

$\vdash . (1) . (2) . \supset \vdash . \mathrm{Prop}$

***2·82.** $\vdash :. p \lor q \lor r . \supset : p \lor \sim r \lor s . \supset . p \lor q \lor s$

$$\left[*2{\cdot}8 . *2{\cdot}81 \frac{q \lor r, \sim r \lor s, q \lor s}{q, \quad r, \quad s}\right]$$

***2·83.** $\vdash :: p . \supset . q \supset r : \supset :. p . \supset . r \supset s : \supset : p . \supset . q \supset s$

$$\left[*2{\cdot}82 \frac{\sim p, \sim q}{p, \quad q}\right]$$

***2·85.** $\vdash :. p \lor q . \supset . p \lor r : \supset : p . \lor . q \supset r$

Dem.

$[\mathrm{Add}.\mathrm{Syll}] \vdash :. p \lor q . \supset . r : \supset . q \supset r \quad (1)$

$\vdash . *2{\cdot}55 . \supset \vdash :: \sim p . \supset :. p \lor r . \supset . r :.$

$[\mathrm{Syll}] \qquad \supset :. p \lor q . \supset . p \lor r : \supset : p \lor q . \supset . r :.$

$[(1) . *2{\cdot}83] \qquad \supset :. p \lor q . \supset . p \lor r : \supset : q \supset r \quad (2)$

$\vdash . (2) . \mathrm{Comm} . \supset \vdash :. \quad p \lor q . \supset . p \lor r : \supset : \sim p . \supset . q \supset r :$

$[*2{\cdot}54] \qquad \supset : p . \lor . q \supset r :. \supset \vdash . \mathrm{Prop}$

***2·86.** $\vdash :. p \supset q . \supset . p \supset r : \supset : p . \supset . q \supset r$ $\quad \left[*2{\cdot}85 \frac{\sim p}{p}\right]$

$*3$. THE LOGICAL PRODUCT OF TWO PROPOSITIONS

Summary of $*3$.

The logical product of two propositions p and q is practically the proposition "p and q are both true." But this as it stands would have to be a new primitive idea. We therefore take as the logical product the proposition $\sim(\sim p \lor \sim q)$, *i.e.* "it is false that either p is false or q is false," which is obviously true when and only when p and q are both true. Thus we put

$*3{\cdot}01$. $p \,.\, q \,.\!=.\, \sim(\sim p \lor \sim q)$ Df

where "$p \,.\, q$" is the logical product of p and q.

$*3{\cdot}02$. $p \supset q \supset r \,.\!=.\, p \supset q \,.\, q \supset r$ Df

This definition serves merely to abbreviate proofs.

When we are given two asserted propositional functions "$\vdash . \phi x$" and "$\vdash . \psi x$," we shall have "$\vdash . \phi x . \psi x$" whenever ϕ and ψ take arguments of the same type. This will be proved for any functions in $*9$; for the present, we are confined to *elementary* propositional functions of elementary propositions. In this case, the result is proved as follows:

By $*1{\cdot}7$, $\sim \phi p$ and $\sim \psi p$ are elementary propositional functions, and therefore, by $*1{\cdot}72$, $\sim \phi p \lor \sim \psi p$ is an elementary propositional function. Hence by $*2{\cdot}11$,

$$\vdash : \sim \phi p \lor \sim \psi p \,.\!\lor.\, \sim(\sim \phi p \lor \sim \psi p).$$

Hence by $*2{\cdot}32$ and $*1{\cdot}01$,

$$\vdash :. \phi p \,.\!\supset\!: \psi p \,.\!\supset.\, \sim(\sim \phi p \lor \sim \psi p),$$

i.e. by $*3{\cdot}01$,

$$\vdash :. \phi p \,.\!\supset\!: \psi p \,.\!\supset.\, \phi p \,.\, \psi p.$$

Hence by $*1{\cdot}11$, when we have "$\vdash . \phi p$" and "$\vdash . \psi p$" we have "$\vdash . \phi p . \psi p$." This proposition is $*3{\cdot}03$. It is to be understood, like $*1{\cdot}72$, as applying also to functions of two or more variables.

The above is the practically most useful form of the axiom of identification of real variables (cf. $*1{\cdot}72$). In practice, when the restriction to *elementary* propositions and propositional functions has been removed, a convenient means by which two functions can often be recognized as taking arguments of the same type is the following:

If ϕx contains, in any way, a constituent $\chi(x, y, z, \ldots)$ and ψx contains, in any way, a constituent $\chi(x, u, v, \ldots)$, then both ϕx and ψx take arguments of the type of the argument x in $\chi(x, y, z, \ldots)$, and therefore both ϕx and ψx take arguments of the same type. Hence, in such a case, if both ϕx and ψx can be asserted, so can $\phi x \,.\, \psi x$.

As an example of the use of this proposition, take the proof of ∗3·47. We there prove

$$\vdash :. p \supset r . q \supset s . \supset : p . q . \supset . q . r \qquad (1)$$

and

$$\vdash :. p \supset r . q \supset s . \supset : q . r . \supset . r . s \qquad (2)$$

and what we wish to prove is

$$p \supset r . q \supset s . \supset : p . q . \supset . r . s,$$

which is ∗3·47. Now in (1) and (2), p, q, r, s are elementary propositions (as everywhere in Section A); hence by ∗1·7·71, applied repeatedly, "$p \supset r . q \supset s . \supset : p . q . \supset . q . r$" and "$p \supset r . q \supset s . \supset : q . r . \supset . r . s$" are elementary propositional functions. Hence by ∗3·03, we have

$$\vdash :: p \supset r . q \supset s . \supset : p . q . \supset . q . r :. p \supset r . q \supset s . \supset : q . r . \supset . r . s,$$

whence the result follows by ∗3·43 and ∗3·33.

The principal propositions of the present number are the following:

∗3·2. $\vdash :. p . \supset : q . \supset . p . q$

I.e. "p implies that q implies $p . q$," *i.e.* if each of two propositions is true, so is their logical product.

∗3·26. $\vdash : p . q . \supset . p$

∗3·27. $\vdash : p . q . \supset . q$

I.e. if the logical product of two propositions is true, then each of the two propositions severally is true.

∗3·3. $\vdash :. p . q . \supset . r : \supset : p . \supset . q \supset r$

I.e. if p and q jointly imply r, then p implies that q implies r. This principle (following Peano) will be called "exportation," because q is "exported" from the hypothesis. It will be referred to as "Exp."

∗3·31. $\vdash :. p . \supset . q \supset r : \supset : p . q . \supset . r$

This is the correlative of the above, and will be called (following Peano) "importation" (referred to as "Imp").

∗3·35. $\vdash : p . p \supset q . \supset . q$

I.e. "if p is true, and q follows from it, then q is true." This will be called the "principle of assertion" (referred to as "Ass"). It differs from ∗1·1 by the fact that it does not apply only when p really is true, but requires merely the *hypothesis* that p is true.

∗3·43. $\vdash :. p \supset q . p \supset r . \supset : p . \supset . q . r$

I.e. if a proposition implies each of two propositions, then it implies their logical product. This is called by Peano the "principle of composition." It will be referred to as "Comp."

∗3·45. $\vdash :. p \supset q . \supset : p . r . \supset . q . r$

I.e. both sides of an implication may be multiplied by a common factor. This is called by Peano the "principle of the factor." It will be referred to as "Fact."

***3·47.** $\vdash :. p \supset r . q \supset s . \supset : p . q . \supset . r . s$

I.e. if p implies q and r implies s, then p and q jointly imply r and s jointly. The law of contradiction, "$\vdash . \sim (p . \sim p)$," is proved in this number (*3·24); but in spite of its fame we have found few occasions for its use.

***3·01.** $p . q . = . \sim (\sim p \vee \sim q)$ Df

***3·02.** $p \supset q \supset r . = . p \supset q . q \supset r$ Df

***3·03.** Given two asserted elementary propositional functions "$\vdash . \phi p$" and "$\vdash . \psi p$" whose arguments are elementary propositions, we have $\vdash . \phi p . \psi p$.

Dem.

$\vdash . *1·7·72 . *2·11 . \supset \vdash : \sim \phi p \vee \sim \psi p . \vee . \sim (\sim \phi p \vee \sim \psi p)$ (1)

$\vdash . (1) . *2·32 . (*1·01) . \supset \vdash :. \phi p . \supset : \psi p . \supset . \sim (\sim \phi p \vee \sim \psi p)$ (2)

$\vdash . (2) . (*3·01) . \supset \vdash :. \phi p . \supset : \psi p . \supset . \phi p . \psi p$ (3)

$\vdash . (3) . *1·11 . \supset \vdash .$ Prop

***3·1.** $\vdash : p . q . \supset . \sim (\sim p \vee \sim q)$ [Id . (*3·01)]

***3·11.** $\vdash : \sim (\sim p \vee \sim q) . \supset . p . q$ [Id . (*3·01)]

***3·12.** $\vdash : \sim p . \vee . \sim q . \vee . p . q$ $\left[*2·11 \frac{\sim p \vee \sim q}{p}\right]$

***3·13.** $\vdash : \sim (p . q) . \supset . \sim p \vee \sim q$ [*3·11 . Transp]

***3·14.** $\vdash : \sim p \vee \sim q . \supset . \sim (p . q)$ [*3·1 . Transp]

***3·2.** $\vdash :. p . \supset : q . \supset . p . q$ [*3·12]

***3·21.** $\vdash :. q . \supset : p . \supset . p . q$ [*3·2 . Comm]

***3·22.** $\vdash : p . q . \supset . q . p$

This is one form of the commutative law for logical multiplication. A more complete form is given in *4·3.

Dem.

$\left[*3·13 \frac{q, p}{p, q}\right]$ $\vdash : \sim (q . p) . \supset . \sim q \vee \sim p .$

[Perm] $\supset . \sim p \vee \sim q .$

[*3·14] $\supset . \sim (p . q)$ (1)

$\vdash . (1) .$ Transp $. \supset \vdash .$ Prop

Note that, in the above proof, "(1)" stands for the proposition

"$\sim (q . p) . \supset . \sim (p . q)$,"

as was explained in the proof of *2·31.

***3·24.** $\vdash . \sim (p . \sim p)$

Dem.

$\left[*2·11 \frac{\sim p}{p}\right]$ $\vdash . \sim p \vee \sim (\sim p) . \supset$

$\left[*3·14 \frac{\sim p}{q}\right]$ $\vdash . \sim (p . \sim p)$

The above is the law of contradiction.

***3·26.** $\vdash : p\,.\,q\,.\supset .\,p$

Dem.

$\left[*2\cdot02\,\frac{q,\,p}{p,\,q}\right] \quad \vdash : p\,.\supset .\,q\supset p \qquad (1)$

$[(1).(*1\cdot01)] \quad \vdash : \sim p\,.\vee .\sim q\vee p :$

$[*2\cdot31] \quad \supset\vdash : \sim p\vee\sim q\,.\vee .\,p :$

$\left[*2\cdot53\,\frac{\sim p\vee\sim q,\,p}{p,\qquad q}\right] \quad \supset\vdash : \sim(\sim p\vee\sim q)\,.\supset .\,p \qquad (2)$

$[(2).(*3\cdot01)] \quad \vdash : p\,.\,q\,.\supset .\,p$

***3·27.** $\vdash : p\,.\,q\,.\supset .\,q$

Dem.

$[*3\cdot22] \quad \vdash : p\,.\,q\,.\supset .\,q\,.\,p\,.$

$\left[*3\cdot26\,\frac{q,\,p}{p,\,q}\right] \quad \supset .\,q : \supset\vdash .$ Prop

*3·26·27 will both be called the "principle of simplification," like *2·02, from which they are deduced. They will be referred to as "Simp."

***3·3.** $\vdash :.\, p\,.\,q\,.\supset .\,r : \supset : p\,.\supset .\,q\supset r$

Dem.

$[\text{Id}.(*3\cdot01)] \quad \vdash :.\, p\,.\,q\,.\supset .\,r : \supset : \sim(\sim p\vee\sim q)\,.\supset .\,r :$

$[\text{Transp}] \quad \supset : \sim r\,.\supset .\sim p\vee\sim q :$

$[\text{Id}.(*1\cdot01)] \quad \supset : \sim r\,.\supset .\,p\supset\sim q :$

$[\text{Comm}] \quad \supset : p\,.\supset .\sim r\supset\sim q :$

$[\text{Transp}.\text{Syll}] \quad \supset : p\,.\supset .\,q\supset r :.\supset\vdash .$ Prop

***3·31.** $\vdash :.\, p\,.\supset .\,q\supset r : \supset : p\,.\,q\,.\supset .\,r$

Dem.

$[\text{Id}.(*1\cdot01)] \quad \vdash :.\, p\,.\supset .\,q\supset r : \supset : \sim p\,.\vee .\sim q\vee r :$

$[*2\cdot31] \quad \supset : \sim p\vee\sim q\,.\vee .\,r :$

$\left[*2\cdot53\,\frac{\sim p\vee\sim q,\,r}{p,\qquad q}\right] \quad \supset : \sim(\sim p\vee\sim q)\,.\supset .\,r :$

$[\text{Id}.(*3\cdot01)] \quad \supset : p\,.\,q\,.\supset .\,r :.\supset\vdash .$ Prop

***3·33.** $\vdash : p\supset q\,.\,q\supset r\,.\supset .\,p\supset r$ [Syll . Imp]

***3·34.** $\vdash : q\supset r\,.\,p\supset q\,.\supset .\,p\supset r$ [Syll . Imp]

These two propositions will hereafter be referred to as "Syll"; they are usually more convenient than either *2·05 or *2·06.

***3·35.** $\vdash : p\,.\,p\supset q\,.\supset .\,q$ [*2·27 . Imp]

***3·37.** $\vdash :.\, p\,.\,q\,.\supset .\,r : \supset : p\,.\sim r\,.\supset .\sim q$

Dem.

$\vdash .\,\text{Transp}\,.\supset\vdash : q\supset r\,.\supset .\sim r\supset\sim q :$

$[\text{Syll}] \quad \supset\vdash :.\, p\,.\supset :\!.\, q\supset r : \supset : p\,.\supset .\sim r\supset\sim q \qquad (1)$

$\vdash .\,\text{Exp}\,. \quad \supset\vdash :.\, p\,.\,q\,.\supset .\,r : \supset : p\,.\supset .\,q\supset r \qquad (2)$

$\vdash .\,\text{Imp}\,. \quad \supset\vdash :.\, p\,.\supset .\sim r\supset\sim q : \supset : p\,.\sim r\,.\supset .\sim q \qquad (3)$

$\vdash .\,(2)\,.\,(1)\,.\,(3)\,.\,\text{Syll}\,.\supset\vdash .$ Prop

This is another form of transposition.

***3·4.** $\vdash : p . q . \supset . p \supset q$ [*2·51 . Transp . (*1·01 . *3·01)]

***3·41.** $\vdash :. p \supset r . \supset : p . q . \supset . r$ [*3·26 . Syll]

***3·42.** $\vdash :. q \supset r . \supset : p . q . \supset . r$ [*3·27 . Syll]

***3·43.** $\vdash :. p \supset q . p \supset r . \supset : p . \supset . q . r$

Dem.

$\vdash . *3{\cdot}2 . \supset \vdash :. q . \supset : r . \supset . q . r$ (1)

$\vdash . (1) . \text{Syll} . \supset \vdash :: p \supset q . \supset :. p . \supset : r . \supset . q . r :.$

[*2·77] $\supset :. p \supset r . \supset : p . \supset . q . r$ (2)

$\vdash . (2) . \text{Imp} . \supset \vdash . \text{Prop}$

***3·44.** $\vdash :. q \supset p . r \supset p . \supset : q \vee r . \supset . p$

This principle is analogous to *3·43. The analogy between *3·43 and *3·44 is of a sort which generally subsists between formulae concerning products and formulae concerning sums.

Dem.

$\vdash . \text{Syll} . \supset \vdash :. \sim q \supset r . r \supset p . \supset : \sim q \supset p :$

[*2·6] $\supset : q \supset p . \supset . p$ (1)

$\vdash . (1) . \text{Exp} . \supset \vdash :: \sim q \supset r . \supset :. r \supset p . \supset : q \supset p . \supset . p :.$

[Comm.Imp] $\supset :. q \supset p . r \supset p . \supset . p$ (2)

$\vdash . (2) . \text{Comm} . \supset \vdash :. q \supset p . r \supset p . \supset : \sim q \supset r . \supset . p :.$

[*2·53.Syll] $\supset \vdash . \text{Prop}$

***3·45.** $\vdash :. p \supset q . \supset : p . r . \supset . q . r$

This principle shows that we may multiply both sides of an implication by a common factor; hence it is called by Peano the "principle of the factor." We shall refer to it as "Fact." It is the analogue, for multiplication, of the primitive proposition *1·6.

Dem.

$\vdash . \text{Syll} \frac{\sim r}{r} . \supset \vdash :. p \supset q . \supset : q \supset \sim r . \supset . p \supset \sim r :$

[Transp] $\supset : \sim (p \supset \sim r) . \supset . \sim (q \supset \sim r) :.$

[Id.(*1·01.*3·01)] $\supset \vdash . \text{Prop}$

***3·47.** $\vdash :. p \supset r . q \supset s . \supset : p . q . \supset . r . s$

This proposition, or rather its analogue for classes, was proved by Leibniz, and evidently pleased him, since he calls it "præclarum theorema*."

Dem.

$\vdash . *3{\cdot}26 . \supset \vdash :. p \supset r . q \supset s . \supset : p \supset r :$

[Fact] $\supset : p . q . \supset . r . q :$

[*3·22] $\supset : p . q . \supset . q . r$ (1)

* *Philosophical works*, Gerhardt's edition, Vol. VII. p. 223.

$\vdash . *3{\cdot}27 . \supset \vdash :. p \supset r . q \supset s . \supset : q \supset s :$

[Fact] $\supset : q . r . \supset . s . r :$

[$*3{\cdot}22$] $\supset : q . r . \supset . r . s$ (2)

$\vdash . (1) . (2) . *3{\cdot}03 . *2{\cdot}83 . \supset$

$\vdash :. p \supset r . q \supset s . \supset : p . q . \supset . r . s :. \supset \vdash .$ Prop

$*3{\cdot}48$. $\vdash :. p \supset r . q \supset s . \supset : p \vee q . \supset . r \vee s$

This theorem is the analogue of $*3{\cdot}47$.

Dem.

$\vdash . *3{\cdot}26 . \supset \vdash :. p \supset r . q \supset s . \supset : p \supset r :$

[Sum] $\supset : p \vee q . \supset . r \vee q :$

[Perm] $\supset : p \vee q . \supset . q \vee r$ (1)

$\vdash . *3{\cdot}27 . \supset \vdash :. p \supset r . q \supset s . \supset : q \supset s :$

[Sum] $\supset : q \vee r . \supset . s \vee r :$

[Perm] $\supset : q \vee r . \supset . r \vee s$ (2)

$\vdash . (1) . (2) . *2{\cdot}83 . \supset$

$\vdash :. p \supset r . q \supset s . \supset : p \vee q . \supset . r \vee s :. \supset \vdash .$ Prop

*4. EQUIVALENCE AND FORMAL RULES

*Summary of *4.*

In this number, we shall be concerned with rules analogous, more or less, to those of ordinary algebra. It is from these rules that the usual "calculus of formal logic" starts. Treated as a "calculus," the rules of deduction are capable of many other interpretations. But all other interpretations depend upon the one here considered, since in all of them we deduce consequences from our rules, and thus presuppose the theory of deduction. One very simple interpretation of the "calculus" is as follows: The entities considered are to be numbers which are all either 0 or 1; "$p \supset q$" is to have the value 0 if p is 1 and q is 0; otherwise it is to have the value 1; $\sim p$ is to be 1 if p is 0, and 0 if p is 1; $p \,.\, q$ is to be 1 if p and q are both 1, and is to be 0 in any other case; $p \vee q$ is to be 0 if p and q are both 0, and is to be 1 in any other case; and the assertion-sign is to mean that what follows has the value 1. Symbolic logic considered as a calculus has undoubtedly much interest on its own account; but in our opinion this aspect has hitherto been too much emphasized, at the expense of the aspect in which symbolic logic is merely the most elementary part of mathematics, and the logical pre-requisite of all the rest. For this reason, we shall only deal briefly with what is required for the algebra of symbolic logic.

When each of two propositions implies the other, we say that the two are *equivalent*, which we write "$p \equiv q$." We put

***4·01.** $p \equiv q \,.=.\, p \supset q \,.\, q \supset p$ Df

It is obvious that two propositions are equivalent when, and only when, both are true or both are false. Following Frege, we shall call the *truth-value of a proposition* truth if it is true, and falsehood if it is false. Thus two propositions are equivalent when they have the same truth-value.

It should be observed that, if $p \equiv q$, q may be substituted for p without altering the truth-value of any function of p which involves no primitive ideas except those enumerated in *1. This can be proved in each separate case, but not generally, because we have no means of specifying (with our apparatus of primitive ideas) that a function is one which can be built up out of these ideas alone. We shall give the name of a *truth-function* to a function $f(p)$ whose argument is a proposition, and whose truth-value depends only upon the truth-value of its argument. All the functions of propositions with which we shall be specially concerned will be truth-functions, *i.e.* we shall have

$$p \equiv q \,.\supset.\, f(p) \equiv f(q).$$

The reason of this is, that the functions of propositions with which we deal are all built up by means of the primitive ideas of ∗1. But it is not a universal characteristic of functions of propositions to be truth-functions. For example, "A believes p" may be true for one true value of p and false for another.

The principal propositions of this number are the following:

∗4·1. $\vdash : p \supset q . \equiv . \sim q \supset \sim p$

∗4·11. $\vdash : p \equiv q . \equiv . \sim p \equiv \sim q$

These are both forms of the "principle of transposition."

∗4·13. $\vdash . p \equiv \sim(\sim p)$

This is the principle of double negation, *i.e.* a proposition is equivalent to the falsehood of its negation.

∗4·2. $\vdash . p \equiv p$

∗4·21. $\vdash : p \equiv q . \equiv . q \equiv p$

∗4·22. $\vdash : p \equiv q . q \equiv r . \supset . p \equiv r$

These propositions assert that equivalence is *reflexive, symmetrical* and *transitive*.

∗4·24. $\vdash : p . \equiv . p . p$

∗4·25. $\vdash : p . \equiv . p \vee p$

I.e. p is equivalent to "p and p" and to "p or p," which are two forms of the *law of tautology*, and are the source of the principal differences between the algebra of symbolic logic and ordinary algebra.

∗4·3. $\vdash : p . q . \equiv . q . p$

This is the commutative law for the product of propositions.

∗4·31. $\vdash : p \vee q . \equiv . q \vee p$

This is the commutative law for the sum of propositions.

The associative laws for multiplication and addition of propositions, namely

∗4·32. $\vdash : (p . q) . r . \equiv . p . (q . r)$

∗4·33. $\vdash : (p \vee q) \vee r . \equiv . p \vee (q \vee r)$

The distributive law in the two forms

∗4·4. $\vdash :. p . q \vee r . \equiv : p . q . \vee . p . r$

∗4·41. $\vdash :. p . \vee . q . r : \equiv . p \vee q . p \vee r$

The second of these forms has no analogue in ordinary algebra.

∗4·71. $\vdash :. p \supset q . \equiv : p . \equiv . p . q$

I.e. p implies q when, and only when, p is equivalent to $p . q$. This proposition is used constantly; it enables us to replace any implication by an equivalence.

∗4·73. $\vdash :. q . \supset : p . \equiv . p . q$

I.e. a true factor may be dropped from or added to a proposition without altering the truth-value of the proposition.

✱4·01. $p \equiv q . = . p \supset q . q \supset p$ Df

✱4·02. $p \equiv q \equiv r . = . p \equiv q . q \equiv r$ Df

This definition serves merely to provide a convenient abbreviation.

✱4·1. $\vdash : p \supset q . \equiv . \sim q \supset \sim p$ [✱2·16·17]

✱4·11. $\vdash : p \equiv q . \equiv . \sim p \equiv \sim q$ [✱2·16·17 . ✱3·47·22]

✱4·12. $\vdash : p \equiv \sim q . \equiv . q \equiv \sim p$ [✱2·03·15]

✱4·13. $\vdash . p \equiv \sim (\sim p)$ [✱2·12·14]

✱4·14. $\vdash :. p . q . \supset . r : \equiv : p . \sim r . \supset . \sim q$ [✱3·37 . ✱4·13]

✱4·15. $\vdash :. p . q . \supset . \sim r : \equiv : q . r . \supset . \sim p$ [✱3·22 . ✱4·13·14]

✱4·2. $\vdash . p \equiv p$ [Id . ✱3·2]

✱4·21. $\vdash : p \equiv q . \equiv . q \equiv p$ [✱3·22]

✱4·22. $\vdash : p \equiv q . q \equiv r . \supset . p \equiv r$

Dem.

$\vdash .$ ✱3·26 . $\supset \vdash : p \equiv q . q \equiv r . \supset . p \equiv q .$
[✱3·26] $\supset . p \supset q$ (1)
$\vdash .$ ✱3·27 . $\supset \vdash : p \equiv q . q \equiv r . \supset . q \equiv r .$
[✱3·26] $\supset . q \supset r$ (2)
$\vdash . (1) . (2) .$ ✱2·83 . $\supset \vdash : p \equiv q . q \equiv r . \supset . p \supset r$ (3)
$\vdash .$ ✱3·27 . $\supset \vdash : p \equiv q . q \equiv r . \supset . q \equiv r .$
[✱3·27] $\supset . r \supset q$ (4)
$\vdash .$ ✱3·26 . $\supset \vdash : p \equiv q . q \equiv r . \supset . p \equiv q .$
[✱3·27] $\supset . q \supset p$ (5)
$\vdash . (4) . (5) .$ ✱2·83 . $\supset \vdash : p \equiv q . q \equiv r . \supset . r \supset p$ (6)
$\vdash . (3) . (6) .$ Comp . $\supset \vdash .$ Prop

Note. The above three propositions show that the relation of equivalence is reflexive (✱4·2), symmetrical (✱4·21), and transitive (✱4·22). Implication is reflexive and transitive, but not symmetrical. The properties of being symmetrical, transitive, and (at least within a certain field) reflexive are essential to any relation which is to have the formal characters of equality.

✱4·24. $\vdash : p . \equiv . p . p$

Dem.

$\vdash .$ ✱3·26 . $\supset \vdash : p . p . \supset . p$ (1)
$\vdash .$ ✱3·2 . $\supset \vdash :. p . \supset : p . \supset . p . p :.$
[✱2·43] $\supset \vdash : p . \supset . p . p$ (2)
$\vdash . (1) . (2) .$ ✱3·2 . $\supset \vdash .$ Prop

✱4·25. $\vdash : p . \equiv . p \vee p$ $\left[\text{Taut . Add } \frac{p}{q}\right]$

Note. ✱4·24·25 are two forms of the *law of tautology*, which is what chiefly distinguishes the algebra of symbolic logic from ordinary algebra.

***4·3.** $\vdash : p . q . \equiv . q . p$ [*3·22]

Note. Whenever we have, whatever values p and q may have,

$$\phi(p, q) . \supset . \phi(q, p),$$

we have also

$$\phi(p, q) . \equiv . \phi(q, p).$$

For $$\{\phi(p, q) . \supset . \phi(q, p)\} \frac{q, p}{p, q} . \supset : \phi(q, p) . \supset . \phi(p, q).$$

***4·31.** $\vdash : p \vee q . \equiv . q \vee p$ [Perm]

***4·32.** $\vdash : (p . q) . r . \equiv . p . (q . r)$

Dem.

$\vdash . *4{\cdot}15 . \quad \supset \vdash :. p . q . \supset . \sim r : \equiv : q . r . \supset . \sim p :$

$[*4{\cdot}12] \quad \equiv : p . \supset . \sim(q . r)$ (1)

$\vdash . (1) . *4{\cdot}11 . \supset \vdash : \sim(p . q . \supset . \sim r) . \equiv . \sim\{p . \supset . \sim(q . r)\} :$

$[(*1{\cdot}01 . *3{\cdot}01)] \supset \vdash .$ Prop

Note. Here "(1)" stands for "$\vdash :. p . q . \supset . \sim r : \equiv : p . \supset . \sim(q . r)$," which is obtained from the above steps by *4·22. The use of *4·22 will often be tacit, as above. The principle is the same as that explained in respect of implication in *2·31.

***4·33.** $\vdash : (p \vee q) \vee r . \equiv . p \vee (q \vee r)$ [*2·31·32]

The above are the associative laws for multiplication and addition. To avoid brackets, we introduce the following definition:

***4·34.** $p . q . r . = . (p . q) . r$ Df

***4·36.** $\vdash :. p \equiv q . \supset : p . r . \equiv . q . r$ [Fact . *3·47]

***4·37.** $\vdash :. p \equiv q . \supset : p \vee r . \equiv . q \vee r$ [Sum . *3·47]

***4·38.** $\vdash :. p \equiv r . q \equiv s . \supset : p . q . \equiv . r . s$ [*3·47 . *4·32 . *3·22]

***4·39.** $\vdash :. p \equiv r . q \equiv s . \supset : p \vee q . \equiv . r \vee s$ [*3·48·47 . *4·32 . *3·22]

***4·4.** $\vdash :. p . q \vee r . \equiv : p . q . \vee . p . r$

This is the first form of the distributive law.

Dem.

$\vdash . *3{\cdot}2 . \quad \supset \vdash :: p . \supset : q . \supset . p . q :. p . \supset : r . \supset . p . r ::$

$[\text{Comp}] \quad \supset \vdash :: p . \supset :. q . \supset . p . q : r . \supset . p . r :.$

$[*3{\cdot}48] \quad \supset :. q \vee r . \supset : p . q . \vee . p . r$ (1)

$\vdash . (1) . \text{Imp} . \quad \supset \vdash :. p . q \vee r . \supset : p . q . \vee . p . r$ (2)

$\vdash . *3{\cdot}26 . \quad \supset \vdash :. p . q . \supset . p : p . r . \supset . p :.$

$[*3{\cdot}44] \quad \supset \vdash :. p . q . \vee . p . r : \supset . p$ (3)

$\vdash . *3{\cdot}27 . \quad \supset \vdash :. p . q . \supset . q : p . r . \supset . r :.$

$[*3{\cdot}48] \quad \supset \vdash :. p . q . \vee . p . r : \supset . q \vee r$ (4)

$\vdash . (3) . (4) . \text{Comp} . \supset \vdash :. p . q . \vee . p . r : \supset . p . q \vee r$ (5)

$\vdash . (2) . (5) . \quad \supset \vdash .$ Prop

***4·41.** $\vdash :. p . \vee . q . r : \equiv . p \vee q . p \vee r$

This is the second form of the distributive law—a form to which there is nothing analogous in ordinary algebra. By the conventions as to dots, "$p . \vee . q . r$" means "$p \vee (q . r)$."

Dem.

$\vdash . *3{\cdot}26 . \text{Sum} . \quad \supset \vdash :. p . \vee . q . r : \supset . p \vee q$ (1)
$\vdash . *3{\cdot}27 . \text{Sum} . \quad \supset \vdash :. p . \vee . q . r : \supset . p \vee r$ (2)
$\vdash . (1) . (2) . \text{Comp} . \supset \vdash :. p . \vee . q . r : \supset . p \vee q . p \vee r$ (3)
$\vdash . *2{\cdot}53 . *3{\cdot}47 . \quad \supset \vdash :. p \vee q . p \vee r . \supset : \sim p \supset q . \sim p \supset r :$
$[\text{Comp}] \quad \supset : \sim p . \supset . q . r :$
$[*2{\cdot}54] \quad \supset : p . \vee . q . r$ (4)
$\vdash . (3) . (4) . \quad \supset \vdash . \text{Prop}$

***4·42.** $\vdash :. p . \equiv : p . q . \vee . p . \sim q$

Dem.

$\vdash . *3{\cdot}21 . \quad \supset \vdash :. q \vee \sim q . \supset : p . \supset . p . q \vee \sim q :.$
$[*2{\cdot}11] \quad \supset \vdash : p . \supset . p . q \vee \sim q$ (1)
$\vdash . *3{\cdot}26 . \quad \supset \vdash : p . q \vee \sim q . \supset . p$ (2)
$\vdash . (1) . (2) . \supset \vdash :. p . \equiv : p . q \vee \sim q :$
$[*4{\cdot}4] \quad \equiv : p . q . \vee . p . \sim q :. \supset \vdash . \text{Prop}$

***4·43.** $\vdash :. p . \equiv : p \vee q . p \vee \sim q$

Dem.

$\vdash . *2{\cdot}2 . \quad \supset \vdash : p . \supset . p \vee q : p . \supset . p \vee \sim q :$
$[\text{Comp}] \quad \supset \vdash : p . \supset . p \vee q . p \vee \sim q$ (1)
$\vdash . *2{\cdot}65 \frac{\sim p}{p} . \supset \vdash :. \sim p \supset q . \supset : \sim p \supset \sim q . \supset . p :.$
$[\text{Imp}] \quad \supset \vdash :. \sim p \supset q . \sim p \supset \sim q . \supset . p :.$
$[*2{\cdot}53 . *3{\cdot}47] \quad \supset \vdash :. p \vee q . p \vee \sim q . \supset . p$ (2)
$\vdash . (1) . (2) . \quad \supset \vdash . \text{Prop}$

***4·44.** $\vdash :. p . \equiv : p . \vee . p . q$

Dem.

$\vdash . *2{\cdot}2 . \quad \supset \vdash :. p . \supset : p . \vee . p . q$ (1)
$\vdash . \text{Id} . *3{\cdot}26 . \supset \vdash :. p \supset p : p . q . \supset . p :.$
$[*3{\cdot}44] \quad \supset \vdash :. p . \vee . p . q : \supset . p$ (2)
$\vdash . (1) . (2) . \quad \supset \vdash . \text{Prop}$

***4·45.** $\vdash : p . \equiv . p . p \vee q$ [*3·26 . *2·2]

The following formulae are due to De Morgan, or rather, are the propositional analogues of formulae given by De Morgan for classes. The first of them, it will be observed, merely embodies our definition of the logical product.

*4·5. $\vdash : \quad p . q . \equiv . \sim(\sim p \vee \sim q)$ [*4·2 . (*3·01)]

*4·51. $\vdash : \quad \sim(p . q) . \equiv . \sim p \vee \sim q$ [*4·5·12]

*4·52. $\vdash : \quad p . \sim q . \equiv . \sim(\sim p \vee q)$ $\left[*4\cdot5 \frac{\sim q}{q} . *4\cdot13\right]$

*4·53. $\vdash : \quad \sim(p . \sim q) . \equiv . \sim p \vee q$ [*4·52·12]

*4·54. $\vdash : \quad \sim p . q . \equiv . \sim(p \vee \sim q)$ $\left[*4\cdot5 \frac{\sim p}{p} . *4\cdot13\right]$

*4·55. $\vdash : \quad \sim(\sim p . q) . \equiv . p \vee \sim q$ [*4·54·12]

*4·56. $\vdash : \quad \sim p . \sim q . \equiv . \sim(p \vee q)$ $\left[*4\cdot54 \frac{\sim q}{q} . *4\cdot13\right]$

*4·57. $\vdash : \sim(\sim p . \sim q) . \equiv . p \vee q$ [*4·56·12]

The following formulae are obtained immediately from the above. They are important as showing how to transform implications into sums or into denials of products, and vice versa. It will be observed that the first of them merely embodies the definition *1·01.

*4·6. $\vdash : \quad p \supset q . \equiv . \sim p \vee q$ [*4·2 . (*1·01)]

*4·61. $\vdash : \quad \sim(p \supset q) . \equiv . p . \sim q$ [*4·6·11·52]

*4·62. $\vdash : \quad p \supset \sim q . \equiv . \sim p \vee \sim q$ $\left[*4\cdot6 \frac{\sim q}{q}\right]$

*4·63. $\vdash : \quad \sim(p \supset \sim q) . \equiv . p . q$ [*4·62·11·5]

*4·64. $\vdash : \quad \sim p \supset q . \equiv . p \vee q$ [*2·53·54]

*4·65. $\vdash : \quad \sim(\sim p \supset q) . \equiv . \sim p . \sim q$ [*4·64·11·56]

*4·66. $\vdash : \quad \sim p \supset \sim q . \equiv . p \vee \sim q$ $\left[*4\cdot64 \frac{\sim q}{q}\right]$

*4·67. $\vdash : \sim(\sim p \supset \sim q) . \equiv . \sim p . q$ [*4·66·11·54]

*4·7. $\vdash :. p \supset q . \equiv : p . \supset . p . q$

Dem.

$\vdash . *3\cdot27 . \text{Syll} . \supset \vdash :. p . \supset . p . q : \supset . p \supset q$ (1)
$\vdash . \text{Comp} . \quad \supset \vdash :. p \supset p . p \supset q . \supset : p . \supset . p . q :.$
$[\text{Exp}] \quad \supset \vdash :: p \supset p . \supset :. p \supset q . \supset : p . \supset . p . q ::$
$[\text{Id}] \quad \supset \vdash :. p \supset q . \supset : p . \supset . p . q$ (2)
$\vdash . (1) . (2) . \quad \supset \vdash . \text{Prop}$

*4·71. $\vdash :. p \supset q . \equiv : p . \equiv . p . q$

Dem.

$\vdash . *3\cdot21 . \quad \supset \vdash :: p . q . \supset . p : \supset :. p . \supset . p . q : \supset : p . \equiv . p . q ::$
$[*3\cdot26] \quad \supset \vdash :. p . \supset . p . q : \supset : p . \equiv . p . q$ (1)
$\vdash . *3\cdot26 . \quad \supset \vdash :. p . \equiv . p . q : \supset : p . \supset . p . q$ (2)
$\vdash . (1) . (2) . \quad \supset \vdash :. p . \supset . p . q : \equiv : p . \equiv . p . q$ (3)
$\vdash . (3) . *4\cdot7\cdot22 . \supset \vdash . \text{Prop}$

The above proposition is constantly used. It enables us to transform every implication into an equivalence, which is an advantage if we wish to assimilate symbolic logic as far as possible to ordinary algebra. But when symbolic logic is regarded as an instrument of proof, we need implications, and it is usually inconvenient to substitute equivalences. Similar remarks apply to the following proposition.

***4·72.** $\vdash :. p \supset q . \equiv : q . \equiv . p \vee q$

Dem.

$\vdash . *4·1 . \supset \vdash :. p \supset q . \equiv : \sim q \supset \sim p :$

$\left[*4·71 \frac{\sim q, \sim p}{p, \quad q}\right] \quad \equiv : \sim q . \equiv . \sim q . \sim p :$

$[*4·12] \quad \equiv : q . \equiv . \sim(\sim q . \sim p) :$

$[*4·57] \quad \equiv : q . \equiv . q \vee p :$

$[*4·31] \quad \equiv : q . \equiv . p \vee q :. \supset \vdash . \text{Prop}$

***4·73.** $\vdash :. q . \supset : p . \equiv . p . q$ [Simp . *4·71]

This proposition is very useful, since it shows that a true factor may be omitted from a product without altering its truth or falsehood, just as a true hypothesis may be omitted from an implication.

***4·74.** $\vdash :. \sim p . \supset : q . \equiv . p \vee q$ [*2·21 . *4·72]

***4·76.** $\vdash :. p \supset q . p \supset r . \equiv : p . \supset . q . r$ $\left[*4·41 \frac{\sim p}{p} . (*1·01)\right]$

***4·77.** $\vdash :. q \supset p . r \supset p . \equiv : q \vee r . \supset . p$ [*3·44 . Add . *2·2]

***4·78.** $\vdash :. p \supset q . \vee . p \supset r : \equiv : p . \supset . q \vee r$

Dem.

$\vdash . *4·2 . (*1·01) . \supset \vdash :. p \supset q . \vee . p \supset r : \equiv : \sim p \vee q . \vee . \sim p \vee r :$

$[*4·33] \quad \equiv . \sim p . \vee . q \vee \sim p \vee r :$

$[*4·31·37] \quad \equiv : \sim p . \vee . \sim p \vee q \vee r :$

$[*4·33] \quad \equiv : \sim p \vee \sim p . \vee . q \vee r :$

$[*4·25·37] \quad \equiv : \sim p . \vee . q \vee r :$

$[*4·2 . (*1·01)] \quad \equiv : p . \supset . q \vee r :. \supset \vdash . \text{Prop}$

***4·79.** $\vdash :. q \supset p . \vee . r \supset p : \equiv : q . r . \supset . p$

Dem.

$\vdash . *4·1·39 . \supset \vdash :. q \supset p . \vee . r \supset p : \equiv : \sim p \supset \sim q . \vee . \sim p \supset \sim r :$

$[*4·78] \quad \equiv : \sim p . \supset . \sim q \vee \sim r :$

$[*2·15] \quad \equiv : \sim(\sim q \vee \sim r) . \supset . p :$

$[*4·2 . (*3·01)] \quad \equiv : q . r . \supset . p :. \supset \vdash . \text{Prop}$

Note. The analogues, for classes, of *4·78·79 are false. Take, *e.g.* *4·78, and put $p =$ English people, $q =$ men, $r =$ women. Then p is contained in q or r, but is not contained in q and is not contained in r.

***4·8.** $\vdash : p \supset \sim p . \equiv . \sim p$ [*2·01 . Simp]

***4·81.** $\vdash : \sim p \supset p . \equiv . p$ [*2·18 . Simp]

***4·82.** $\vdash : p \supset q . p \supset \sim q . \equiv . \sim p$ [*2·65 . Imp . *2·21 . Comp]

***4·83.** $\vdash : p \supset q . \sim p \supset q . \equiv . q$ [*2·61 . Imp . Simp . Comp]

Note. *4·82·83 may also be obtained from *4·43, of which they are virtually other forms.

***4·84.** $\vdash :. p \equiv q . \supset : p \supset r . \equiv . q \supset r$ [*2·06 . *3·47]

***4·85.** $\vdash :. p \equiv q . \supset : r \supset p . \equiv . r \supset q$ [*2·05 . *3·47]

***4·86.** $\vdash :. p \equiv q . \supset : p \equiv r . \equiv . q \equiv r$ [*4·21·22]

***4·87.** $\vdash :. p . q . \supset . r : \equiv : p . \supset . q \supset r : \equiv : q . \supset . p \supset r : \equiv : q . p . \supset . r$

[Exp . Comm . Imp]

*4·87 embodies in one proposition the principles of exportation and importation and the commutative principle.

$*5$. MISCELLANEOUS PROPOSITIONS

Summary of $*5$.

The present number consists chiefly of propositions of two sorts: (1) those which will be required as lemmas in one or more subsequent proofs, (2) those which are on their own account illustrative, or would be important in other developments than those that we wish to make. A few of the propositions of this number, however, will be used very frequently. These are:

$*5 \cdot 1$. $\vdash : p . q . \supset . p \equiv q$

I.e. two propositions are equivalent if they are both true. (The statement that two propositions are equivalent if they are both false is $*5 \cdot 21$.)

$*5 \cdot 32$. $\vdash :. p . \supset . q \equiv r : \equiv : p . q . \equiv . p . r$

I.e. to say that, on the hypothesis p, q and r are equivalent, is equivalent to saying that the joint assertion of p and q is equivalent to the joint assertion of p and r. This is a very useful rule in inference.

$*5 \cdot 6$. $\vdash :. p . \sim q . \supset . r : \equiv : p . \supset . q \vee r$

I.e. "p and not-q imply r" is equivalent to "p implies q or r."

Among propositions never subsequently referred to, but inserted for their intrinsic interest, are the following: $*5 \cdot 11 \cdot 12 \cdot 13 \cdot 14$, which state that, given any two propositions p, q, either p or $\sim p$ must imply q, and p must imply either q or not-q, and either p implies q or q implies p; and given any third proposition r, either p implies q or q implies r*.

Other propositions not subsequently referred to are $*5 \cdot 22 \cdot 23 \cdot 24$; in these it is shown that two propositions are not equivalent when, and only when, one is true and the other false, and that two propositions are equivalent when, and only when, both are true or both false. It follows ($*5 \cdot 24$) that the negation of "$p . q . \vee . \sim p . \sim q$" is equivalent to "$p . \sim q . \vee . q . \sim p$." $*5 \cdot 54 \cdot 55$ state that both the product and the sum of p and q are equivalent, respectively, either to p or to q.

The proofs of the following propositions are all easy, and we shall therefore often merely indicate the propositions used in the proofs.

$*5 \cdot 1$. $\vdash : p . q . \supset . p \equiv q$ [$*3 \cdot 4 \cdot 22$]

$*5 \cdot 11$. $\vdash : p \supset q . \vee . \sim p \supset q$ [$*2 \cdot 5 \cdot 54$]

$*5 \cdot 12$. $\vdash : p \supset q . \vee . p \supset \sim q$ [$*2 \cdot 51 \cdot 54$]

$*5 \cdot 13$. $\vdash : p \supset q . \vee . q \supset p$ [$*2 \cdot 521$]

$*5 \cdot 14$. $\vdash : p \supset q . \vee . q \supset r$ [Simp . Transp . $*2 \cdot 21$]

* Cf. Schröder, *Vorlesungen über Algebra der Logik*, Zweiter Band (Leipzig, 1891), pp. 270—271, where the apparent oddity of the above proposition is explained.

***5·15.** $\vdash : p \equiv q . \vee . p \equiv \sim q$

Dem.

$\vdash . *4·61 . \supset \vdash : \sim(p \supset q) . \supset . p . \sim q .$
$[*5·1] \quad \supset . p \equiv \sim q :$
$[*2·54] \quad \supset \vdash : p \supset q . \vee . p \equiv \sim q \quad (1)$
$\vdash . *4·61 . \supset \vdash : \sim(q \supset p) . \supset . q . \sim p .$
$[*5·1] \quad \supset . q \equiv \sim p .$
$[*4·12] \quad \supset . p \equiv \sim q :$
$[*2·54] \quad \supset \vdash : q \supset p . \vee . p \equiv \sim q \quad (2)$
$\vdash . (1) . (2) . *4·41 . \supset \vdash .$ Prop

***5·16.** $\vdash . \sim(p \equiv q . p \equiv \sim q)$

Dem.

$\vdash . *3·26 . \supset \vdash : p \equiv q . p \supset \sim q . \supset . p \supset q . p \supset \sim q .$
$[*4·82] \quad \supset . \sim p \quad (1)$
$\vdash . *3·27 . \supset \vdash : p \equiv q . p \supset \sim q . \supset . q \supset p . p \supset \sim q .$
$[\text{Syll}] \quad \supset . q \supset \sim q .$
$[\text{Abs}] \quad \supset . \sim q \quad (2)$
$\vdash . (1) . (2) . \text{Comp} . \supset \vdash : p \equiv q . p \supset \sim q . \supset . \sim p . \sim q .$
$\left[*4·65 \frac{q, p}{p, q}\right] \quad \supset . \sim(\sim q \supset p) \quad (3)$
$\vdash . (3) . \text{Exp} . \supset \vdash :. p \equiv q . \supset : p \supset \sim q . \supset . \sim(\sim q \supset p) :$
$[\text{Id}.(*1·01)] \quad \supset : \sim(p \supset \sim q) . \vee . \sim(\sim q \supset p) :$
$[*4·51.(*4·01)] \quad \supset : \sim(p \equiv \sim q) :. \supset \vdash .$ Prop

***5·17.** $\vdash : p \vee q . \sim(p . q) . \equiv . p \equiv \sim q$

Dem.

$\vdash . *4·64·21 . \quad \supset \vdash : p \vee q . \equiv . \sim q \supset p \quad (1)$
$\vdash . *4·63 . \text{Transp} . \quad \supset \vdash : \sim(p . q) . \equiv . p \supset \sim q \quad (2)$
$\vdash . (1) . (2) . *4·38·21 . \supset \vdash .$ Prop

***5·18.** $\vdash : p \equiv q . \equiv . \sim(p \equiv \sim q)$ $\quad \left[*5·15·16 . *5·17 \frac{p \equiv q, \; p \equiv \sim q}{p, \quad q}\right]$

***5·19.** $\vdash . \sim(p \equiv \sim p)$ $\quad \left[*5·18 \frac{p}{q} . *4·2\right]$

***5·21.** $\vdash : \sim p . \sim q . \supset . p \equiv q$ $\quad [*5·1 . *4·11]$

***5·22.** $\vdash :. \sim(p \equiv q) . \equiv : p . \sim q . \vee . q . \sim p$ $\quad [*4·61·51·39]$

***5·23.** $\vdash :. p \equiv q . \equiv : p . q . \vee . \sim p . \sim q$ $\quad \left[*5·18 . *5·22 \frac{\sim q}{q} . *4·13·36\right]$

***5·24.** $\vdash :. \sim(p . q . \vee . \sim p . \sim q) . \equiv : p . \sim q . \vee . q . \sim p$ $\quad [*5·22·23]$

***5·25.** $\vdash :. p \vee q . \equiv : p \supset q . \supset . q$ $\quad [*2·62·68]$

From *5·25 it appears that we might have taken implication, instead of disjunction, as a primitive idea, and have defined "$p \vee q$" as meaning "$p \supset q . \supset . q$." This course, however, requires more primitive propositions than are required by the method we have adopted.

***5·3.** $\vdash :. p . q . \supset . r : \equiv : p . q . \supset . p . r$ [Simp . Comp . Syll]

***5·31.** $\vdash :. r . p \supset q : \supset : p . \supset . q . r$ [Simp . Comp]

***5·32.** $\vdash :. p . \supset . q \equiv r : \equiv : p . q . \equiv . p . r$ [*4·76 . *3·3·31 . *5·3]

This proposition is constantly required in subsequent proofs.

***5·33.** $\vdash :. p . q \supset r . \equiv : p : p . q . \supset . r$ [*4·73·84 . *5·32]

***5·35.** $\vdash :. p \supset q . p \supset r . \supset : p . \supset . q \equiv r$ [Comp . *5·1]

***5·36.** $\vdash : p . p \equiv q . \equiv . q . p \equiv q$ [Ass . *4·38]

***5·4.** $\vdash :. p . \supset . p \supset q : \equiv . p \supset q$ [Simp . *2·43]

***5·41.** $\vdash :. p \supset q . \supset . p \supset r : \equiv : p . \supset . q \supset r$ [*2·77·86]

***5·42.** $\vdash :: p . \supset . q \supset r : \equiv :. p . \supset : q . \supset . p . r$ [*5·3 . *4·87]

***5·44.** $\vdash :: p \supset q . \supset :. p \supset r . \equiv : p . \supset . q . r$ [*4·76 . *5·3·32]

***5·5.** $\vdash :. p . \supset : p \supset q . \equiv . q$ [Ass . Exp . Simp]

***5·501.** $\vdash :. p . \supset : q . \equiv . p \equiv q$ [*5·1 . Exp . Ass]

***5·53.** $\vdash :. p \vee q \vee r . \supset . s : \equiv : p \supset s . q \supset s . r \supset s$ [*4·77]

***5·54.** $\vdash :. p . q . \equiv . p : \vee : p . q . \equiv . q$ [*4·73 . *4·44 . Transp . *5·1]

***5·55.** $\vdash :. p \vee q . \equiv . p : \vee : p \vee q . \equiv . q$ [*1·3 . *5·1 . *4·74]

***5·6.** $\vdash :. p . \sim q . \supset . r : \equiv : p . \supset . q \vee r$ $\left[*4{\cdot}87 \dfrac{\sim q}{q} . *4{\cdot}64{\cdot}85\right]$

***5·61.** $\vdash : p \vee q . \sim q . \equiv . p . \sim q$ [*4·74 . *5·32]

***5·62.** $\vdash :. p . q . \vee . \sim q : \equiv . p \vee \sim q$ $\left[*4{\cdot}7 \dfrac{q, p}{p, q}\right]$

***5·63.** $\vdash :. p \vee q . \equiv : p . \vee . \sim p . q$ $\left[*5{\cdot}62 \dfrac{\sim p, q}{q, p}\right]$

***5·7.** $\vdash :. p \vee r . \equiv . q \vee r : \equiv : r . \vee . p \equiv q$ [*4·74 . *1·3 . *5·1 . *4·37]

***5·71.** $\vdash :. q \supset \sim r . \supset : p \vee q . r . \equiv . p . r$

In the following proof, as always henceforth, "Hp" means the hypothesis of the proposition to be proved.

Dem.

$\vdash . *4{\cdot}4 . \quad \supset \vdash :. p \vee q . r . \equiv : p . r . \vee . q . r$ (1)

$\vdash . *4{\cdot}62{\cdot}51 . \supset \vdash :: \text{Hp} . \supset :. \sim (q . r) :.$

[*4·74] $\supset :. p . r . \vee . q . r : \equiv : p . r$ (2)

$\vdash . (1) . (2) . *4{\cdot}22 . \supset \vdash . \text{Prop}$

***5·74.** $\vdash :. p . \supset . q \equiv r : \equiv : p \supset q . \equiv . p \supset r$

Dem.

$\vdash . *5·41 . \supset \vdash :: p \supset q . \supset . p \supset r : \equiv : p . \supset . q \supset r :.$
$p \supset r . \supset . p \supset q : \equiv : p . \supset . r \supset q$ (1)

$\vdash . (1) . *4·38 . \supset \vdash :: p \supset q . \equiv . p \supset r . \equiv :. p . \supset . q \supset r : p . \supset . r \supset q :.$
[*4·76] $\equiv :. p . \supset . q \equiv r :: \supset \vdash .$ Prop

***5·75.** $\vdash :. r \supset \sim q : p . \equiv . q \vee r : \supset : p . \sim q . \equiv . r$

Dem.

$\vdash . *5·6 . \supset \vdash :.$ Hp $. \supset : p . \sim q . \supset . r$ (1)

$\vdash . *3·27 . \supset \vdash :.$ Hp $. \supset : q \vee r . \supset . p :$
[*4·77] $\supset : r \supset p$ (2)

$\vdash . *3·26 . \supset \vdash :.$ Hp $. \supset : r \supset \sim q$ (3)

$\vdash . (2) . (3) .$ Comp $. \supset \vdash :.$ Hp $. \supset : r \supset p . r \supset \sim q :$
[Comp] $\supset : r . \supset . p . \sim q$ (4)

$\vdash . (1) . (4) .$ Comp $. \supset \vdash :.$ Hp $. \supset : p . \sim q . \equiv . r :. \supset \vdash .$ Prop

SECTION B

THEORY OF APPARENT VARIABLES

✱9. EXTENSION OF THE THEORY OF DEDUCTION FROM LOWER TO HIGHER TYPES OF PROPOSITIONS

Summary of ✱9.

In the present number, we introduce two new primitive ideas, which may be expressed as "ϕx is always* true" and "ϕx is sometimes* true," or, more correctly, as "ϕx always" and "ϕx sometimes." When we assert "ϕx always," we are asserting all values of $\phi \hat{x}$, where "$\phi \hat{x}$" means the function itself, as opposed to an ambiguous value of the function (cf. pp. 15, 40); we are not asserting that ϕx is true for all values of x, because, in accordance with the theory of types, there are values of x for which "ϕx" is meaningless; for example, the function $\phi \hat{x}$ itself must be such a value. We shall denote "ϕx always" by the notation

$$(x) \,.\, \phi x,$$

where the "(x)" will be followed by a sufficiently large number of dots to cover the function of which "all values" are concerned. The form in which such propositions most frequently occur is the "formal implication," *i.e.* such a proposition as

$$(x) : \phi x \,.\, \supset \,.\, \psi x,$$

i.e. "ϕx always implies ψx." This is the form in which we express the universal affirmative "all objects having the property ϕ have the property ψ."

We shall denote "ϕx sometimes" by the notation

$$(\exists x) \,.\, \phi x.$$

Here "$\exists$" stands for "there exists," and the whole symbol may be read "there exists an x such that ϕx."

In a proposition of either of the two forms $(x) \,.\, \phi x$, $(\exists x) \,.\, \phi x$, the x is called an *apparent variable.* A proposition which contains no apparent variables is called "elementary," and a function, all whose values are elementary propositions, is called an elementary function. For reasons explained in Chapter II of the Introduction, it would seem that negation and disjunction and their derivatives must have a different meaning when applied to elementary propositions from that which they have when applied to such propositions as $(x) \,.\, \phi x$ or $(\exists x) \,.\, \phi x$. If $\phi \hat{x}$ is an elementary function, we will in this number call $(x) \,.\, \phi x$ and $(\exists x) \,.\, \phi x$ "first-order propositions." Then in virtue of the fact

* We use "always" as meaning "in all cases," not "at all times." A similar remark applies to "sometimes."

that disjunction and negation do not have the same meanings as applied to elementary or to first-order propositions, it follows that, in asserting the primitive propositions of ∗1, we must either confine them, in their application, to propositions of a single type, or we must regard them as the simultaneous assertion of a number of different primitive propositions, corresponding to the different meanings of "disjunction" and "negation." Likewise in regard to the primitive ideas of disjunction and negation, we must either, in the primitive propositions of ∗1, confine them to disjunctions and negations of elementary propositions, or we must regard them as really each multiple, so that in regard to each type of propositions we shall need a new primitive idea of negation and a new primitive idea of disjunction. In the present number, we shall show how, when the primitive ideas of negation and disjunction are restricted to elementary propositions, and the p, q, r of ∗1—∗5 are therefore necessarily elementary propositions, it is possible to obtain definitions of the negation and disjunction of first-order propositions, and proofs of the analogues, for first-order propositions, of the primitive propositions ∗1·2—·6. (∗1·1 and ∗1·11 have to be assumed afresh for first-order propositions, and the analogues of ∗1·7·71·72 require a fresh treatment.) It follows that the analogues of the propositions of ∗2—∗5 follow by merely repeating previous proofs. It follows also that the theory of deduction can be extended from first-order propositions to such as contain two apparent variables, by merely repeating the process which extends the theory of deduction from elementary to first-order propositions. Thus by merely repeating the process set forth in the present number, propositions of any order can be reached. Hence negation and disjunction may be treated in practice as if there were no difference in these ideas as applied to different types; that is to say, when "$\sim p$" or "$p \vee q$" occurs, it is unnecessary in practice to know what is the type of p or q, since the properties of negation and disjunction assumed in ∗1 (which are alone used in proving other properties) can be asserted, without formal change, of propositions of any order or, in the case of $p \vee q$, of any two orders. The limitation, in practice, to the treatment of negation or disjunction as single ideas, the same in all types, would only arise if we ever wished to assume that there is some one function of p whose value is always $\sim p$, whatever may be the order of p, or that there is some one function of p and q whose value is always $p \vee q$, whatever may be the orders of p and q. Such an assumption is not involved so long as p (and q) remain *real* variables, since, in that case, there is no need to give the same meaning to negation and disjunction for different values of p (and q), when these different values are of different types. But if p (or q) is going to be turned into an apparent variable, then since our two primitive ideas $(x) . \phi x$ and $(\exists x) . \phi x$ both demand some definite function ϕ, and restrict the apparent variable to possible arguments for ϕ, it follows that negation and disjunction must, wherever they occur in the expression in which p (or q) is an apparent variable, be restricted to the kind of negation or disjunction appropriate to a given type or pair of types. Thus, to take an instance, if we assert the law of excluded middle in the form

$$\text{“}\vdash . p \vee \sim p\text{”}$$

there is no need to place any restriction upon p: we may give to p a value of any order, and then give to the negation and disjunction involved those meanings which are appropriate to that order. But if we assert

$$\text{“}\vdash . (p) . p \vee \sim p\text{”}$$

it is necessary, if our symbol is to be significant, that "$p \vee \sim p$" should be the value, for the argument p, of a function ϕp; and this is only possible if the negation and disjunction involved have meanings fixed in advance, and if, therefore, p is limited to one type. Thus the assertion of the law of excluded middle in the form involving a real variable is more general than in the form involving an apparent variable. Similar remarks apply generally where the variable is the argument to a typically ambiguous function.

In what follows the single letters p and q will represent *elementary* propositions, and so will "ϕx," "ψx," etc. We shall show how, assuming the primitive ideas and propositions of $*1$ as applied to elementary propositions, we can define and prove analogous ideas and propositions as applied to propositions of the forms $(x) . \phi x$ and $(\exists x) . \phi x$. By mere repetition of the analogous process, it will then follow that analogous ideas and propositions can be defined and proved for propositions of any order; whence, further, it follows that, in all that concerns disjunction and negation, so long as propositions do not appear as apparent variables, we may wholly ignore the distinction between different types of propositions and between different meanings of negation and disjunction. Since we never have occasion, in practice, to consider propositions as apparent variables, it follows that the hierarchy of propositions (as opposed to the hierarchy of functions) will never be relevant in practice after the present number.

The purpose and interest of the present number are purely philosophical, namely to show how, by means of certain primitive propositions, we can deduce the theory of deduction for propositions containing apparent variables from the theory of deduction for elementary propositions. From the purely technical point of view, the distinction between elementary and other propositions may be ignored, so long as propositions do not appear as apparent variables; we may then regard the primitive propositions of $*1$ as applying to propositions of any type, and proceed as in $*10$, where the purely technical development is resumed.

It should be observed that although, in the present number, we prove that the analogues of the primitive propositions of $*1$, if they hold for propositions containing n apparent variables, also hold for such as contain $n+1$, yet we must not suppose that mathematical induction may be used to infer that the analogues of the primitive propositions of $*1$ hold for propositions containing any number of apparent variables. Mathematical induction is a method of proof which is not yet applicable, and is (as will appear) incapable of being used freely until the theory of propositions containing apparent variables has been established. What we are enabled to do, by means of the propositions in the present number, is to prove our desired result for any assigned number of apparent variables—say ten—by ten applications of the same proof. Thus we can prove, concerning any assigned proposition, that it obeys the analogues of the primitive propositions of ✱1, but we can only do this by proceeding step by step, not by any such compendious method as mathematical induction would afford. The fact that higher types can only be reached step by step is essential, since to proceed otherwise we should need an apparent variable which would wander from type to type, which would contradict the principle upon which types are built up.

Definition of Negation. We have first to define the negations of $(x) . \phi x$ and $(\exists x) . \phi x$. We define the negation of $(x) . \phi x$ as $(\exists x) . \sim \phi x$, *i.e.* "it is not the case that ϕx is always true" is to mean "it is the case that not-ϕx is sometimes true." Similarly the negation of $(\exists x) . \phi x$ is to be defined as $(x) . \sim \phi x$. Thus we put

✱9·01. $\sim \{(x) . \phi x\} . = . (\exists x) . \sim \phi x$ Df

✱9·02. $\sim \{(\exists x) . \phi x\} . = . (x) . \sim \phi x$ Df

To avoid brackets, we shall write $\sim (x) . \phi x$ in place of $\sim \{(x) . \phi x\}$, and $\sim (\exists x) . \phi x$ in place of $\sim \{(\exists x) . \phi x\}$. Thus:

✱9·011. $\sim (x) . \phi x . = . \sim \{(x) . \phi x\}$ Df

✱9·021. $\sim (\exists x) . \phi x . = . \sim \{(\exists x) . \phi x\}$ Df

Definition of Disjunction. To define disjunction when one or both of the propositions concerned is of the first order, we have to distinguish six cases, as follows:

✱9·03. $(x) . \phi x . \vee . p : = . (x) . \phi x \vee p$ Df

✱9·04. $p . \vee . (x) . \phi x : = . (x) . p \vee \phi x$ Df

✱9·05. $(\exists x) . \phi x . \vee . p : = . (\exists x) . \phi x \vee p$ Df

✱9·06. $p . \vee . (\exists x) . \phi x : = . (\exists x) . p \vee \phi x$ Df

✱9·07. $(x) . \phi x . \vee . (\exists y) . \psi y : = : (x) : (\exists y) . \phi x \vee \psi y$ Df

✱9·08. $(\exists y) . \psi y . \vee . (x) . \phi x : = : (x) : (\exists y) . \psi y \vee \phi x$ Df

(The definitions ✱9·07·08 are to apply also when ϕ and ψ are not both elementary functions.)

In virtue of these definitions, the true scope of an apparent variable is always the whole of the asserted proposition in which it occurs, even when, typographically, its scope appears to be only part of the asserted proposition. Thus when $(\exists x) . \phi x$ or $(x) . \phi x$ *appears* as *part* of an asserted proposition, it does not really occur, since the scope of the apparent variable really extends to the whole asserted proposition. It will be shown, however, that, so far as the theory of deduction is concerned, $(\exists x) . \phi x$ and $(x) . \phi x$ behave like propositions not containing apparent variables.

The definitions of implication, the logical product, and equivalence are to be transferred unchanged to $(x) . \phi x$ and $(\exists x) . \phi x$.

The above definitions can be repeated for successive types, and thus reach propositions of any type.

Primitive Propositions. The primitive propositions required are six in number, and may be divided into three sets of two. We have first two propositions, which effect the passage from elementary to first-order propositions, namely

***9·1.** $\vdash : \phi x . \supset . (\exists z) . \phi z$ Pp

***9·11.** $\vdash : \phi x \vee \phi y . \supset . (\exists z) . \phi z$ Pp

Of these, the first states that, if ϕx is true, then there is a value of $\phi \hat{z}$ which is true; *i.e.* if we can find an instance of a function which is true, then the function is "sometimes true." (When we speak of a function as "sometimes" true, we do not mean to assert that there is *more* than one argument for which it is true, but only that there is *at least* one.) Practically, the above primitive proposition gives the only method of proving "existence-theorems": in order to prove such theorems, it is necessary (and sufficient) to find some instance in which an object possesses the property in question. If we were to assume what may be called "existence-axioms," *i.e.* axioms stating $(\exists z) . \phi z$ for some particular ϕ, these axioms would give other methods of proving existence. Instances of such axioms are the multiplicative axiom (*88) and the axiom of infinity (defined in *120·03). But we have not assumed any such axioms in the present work.

The second of the above primitive propositions is only used once, in proving $(\exists z) . \phi z . \vee . (\exists z) . \phi z : \supset . (\exists z) . \phi z$, which is the analogue of *1·2 (namely $p \vee p . \supset . p$) when p is replaced by $(\exists z) . \phi z$. The effect of this primitive proposition is to emphasize the ambiguity of the z required in order to secure $(\exists z) . \phi z$. We have, of course, in virtue of *9·1,

$$\phi x . \supset . (\exists z) . \phi z \text{ and } \phi y . \supset . (\exists z) . \phi z.$$

But if we try to infer from these that $\phi x \vee \phi y . \supset . (\exists z) . \phi z$, we must use the proposition $q \supset p . r \supset p . \supset . q \vee r \supset p$, where p is $(\exists z) . \phi z$. Now it will be found, on referring to *4·77 and the propositions used in its proof, that this proposition depends upon *1·2, *i.e.* $p \vee p . \supset . p$. Hence it cannot be used by us to prove $(\exists x) . \phi x . \vee . (\exists x) . \phi x : \supset . (\exists x) . \phi x$, and thus we are compelled to assume the primitive proposition *9·11.

We have next two propositions concerned with *inference* to or from propositions containing apparent variables, as opposed to implication. First, we have, for the new meaning of implication resulting from the above definitions of negation and disjunction, the analogue of ∗1·1, namely

∗9·12. What is implied by a true premiss is true. Pp.

That is to say, given "$\vdash . p$" and "$\vdash . p \supset q$," we may proceed to "$\vdash . q$," even when the propositions p and q are not elementary. Also, as in ∗1·11, we may proceed from "$\vdash . \phi x$" and "$\vdash . \phi x \supset \psi x$" to "$\vdash . \psi x$," where x is a real variable, and ϕ and ψ are not necessarily elementary functions. It is in this latter form that the axiom is usually needed. It is to be assumed for functions of several variables as well as for functions of one variable.

We have next the primitive proposition which permits the passage from a real to an apparent variable, namely "when ϕy may be asserted, where y may be any possible argument, then $(x) . \phi x$ may be asserted." In other words, when ϕy is true however y may be chosen among possible arguments, then $(x) . \phi x$ is true, *i.e.* all values of ϕ are true. That is to say, if we can assert a wholly ambiguous value ϕy, that must be because all values are true. We may express this primitive proposition by the words: "What is true in *any* case, however the case may be selected, is true in *all* cases." We cannot symbolise this proposition, because if we put

$$\text{"}\vdash : \phi y . \supset . (x) . \phi x\text{"}$$

that means: "However y may be chosen, ϕy implies $(x) . \phi x$," which is in general false. What we mean is: "If ϕy is true however y may be chosen, then $(x) . \phi x$ is true." But we have not supplied a symbol for the mere *hypothesis* of what is *asserted* in "$\vdash . \phi y$," where y is a real variable, and it is not worth while to supply such a symbol, because it would be very rarely required. If, for the moment, we use the symbol $[\phi y]$ to express this hypothesis, then our primitive proposition is

$$\vdash : [\phi y] . \supset . (x) . \phi x \quad \text{Pp.}$$

In practice, this primitive proposition is only used for *inference*, not for implication; that is to say, when we actually have an assertion containing a real variable, it enables us to turn this real variable into an apparent variable by placing it in brackets immediately after the assertion-sign, followed by enough dots to reach to the end of the assertion. This process will be called "turning a real variable into an apparent variable." Thus we may assert our primitive proposition, for technical use, in the form:

∗9·13. In any assertion containing a real variable, this real variable may be turned into an apparent variable of which all possible values are asserted to satisfy the function in question. Pp.

We have next two primitive propositions concerned with types. These require some preliminary explanations.

Primitive Idea: Individual. We say that x is "individual" if x is neither a proposition nor a function (cf. p. 51).

***9·131.** *Definition of "being of the same type."* The following is a step-by-step definition, the definition for higher types presupposing that for lower types. We say that u and v "are of the same type" if (1) both are individuals, (2) both are elementary functions taking arguments of the same type, (3) u is a function and v is its negation, (4) u is $\phi\hat{x}$ or $\psi\hat{x}$, and v is $\phi\hat{x} \vee \psi\hat{x}$, where $\phi\hat{x}$ and $\psi\hat{x}$ are elementary functions, (5) u is $(y) . \phi(\hat{x}, y)$ and v is $(z) . \psi(\hat{x}, z)$, where $\phi(\hat{x}, \hat{y})$, $\psi(\hat{x}, \hat{y})$ are of the same type, (6) both are elementary propositions, (7) u is a proposition and v is $\sim u$, or (8) u is $(x) . \phi x$ and v is $(y) . \psi y$, where $\phi\hat{x}$ and $\psi\hat{x}$ are of the same type.

Our primitive propositions are:

***9·14.** If "ϕx" is significant, then if x is of the same type as a, "ϕa" is significant, and vice versa. Pp. (Cf. note on *10·121, p. 140.)

***9·15.** If, for some a, there is a proposition ϕa, then there is a function $\phi\hat{x}$, and vice versa. Pp.

It will be seen that, in virtue of the definitions,

$$(x) . \phi x . \supset . p \text{ means } \sim(x) . \phi x . \vee . p, \textit{ i.e. } (\exists x) . \sim\phi x . \vee . p,$$
$$\textit{i.e. } (\exists x) . \sim\phi x \vee p, \textit{ i.e. } (\exists x) . \phi x \supset p$$
$$(\exists x) . \phi x . \supset . p \text{ means } \sim(\exists x) . \phi x . \vee . p, \textit{ i.e. } (x) . \sim\phi x . \vee . p,$$
$$\textit{i.e. } (x) . \sim\phi x \vee p, \textit{ i.e. } (x) . \phi x \supset p$$

In order to prove that $(x) . \phi x$ and $(\exists x) . \phi x$ obey the same rules of deduction as ϕx, we have to prove that propositions of the forms $(x) . \phi x$ and $(\exists x) . \phi x$ may replace one or more of the propositions p, q, r in *1·2—·6. When this has been proved, the previous proofs of subsequent propositions in *2—*5 become applicable. These proofs are given below. Certain other propositions, required in the proofs, are also proved.

***9·2.** $\vdash : (x) . \phi x . \supset . \phi y$

The above proposition states the principle of deduction from the general to the particular, *i.e.* "what holds in all cases, holds in any one case."

Dem.

$\vdash . *2·1 . \supset \vdash . \sim\phi y \vee \phi y$		(1)
$\vdash . *9·1 . \supset \vdash : \sim\phi y \vee \phi y . \supset . (\exists x) . \sim\phi x \vee \phi y$		(2)
$\vdash . (1) . (2) . *1·11 . \supset \vdash . (\exists x) . \sim\phi x \vee \phi y$		(3)
[(3).(*9·05)]	$\vdash : (\exists x) . \sim\phi x . \vee . \phi y$	(4)
[(4).(*9·01.*1·01)]	$\vdash : (x) . \phi x . \supset . \phi y$	

In the second line of the above proof, "$\sim\phi y \vee \phi y$" is taken as the value, for the argument y, of the function "$\sim\phi x \vee \phi y$," where x is the argument. A similar method of using *9·1 is employed in most of the following proofs.

*1·11 is used, as in the third line of the above proof, in almost all steps except such as are mere applications of definitions. Hence it will not be further referred to, unless in cases where its employment is obscure or specially important.

***9·21.** $\vdash :. (x) . \phi x \supset \psi x . \supset : (x) . \phi x . \supset . (x) . \psi x$

I.e. if ϕx always implies ψx, then "ϕx always" implies "ψx always." The use of this proposition is constant throughout the remainder of this work.

Dem.

$\vdash . *2·08 . \quad \supset \vdash : \phi z \supset \psi z . \supset . \phi z \supset \psi z \quad (1)$

$\vdash . (1) . *9·1 . \quad \supset \vdash : (\exists y) : \phi z \supset \psi z . \supset . \phi y \supset \psi z \quad (2)$

$\vdash . (2) . *9·1 . \quad \supset \vdash :. (\exists x) :. (\exists y) : \phi x \supset \psi x . \supset . \phi y \supset \psi z \quad (3)$

$\vdash . (3) . *9·13 . \quad \supset \vdash :: (z) :: (\exists x) :. (\exists y) : \phi x \supset \psi x . \supset . \phi y \supset \psi z \quad (4)$

$[(4).(*9·06)] \quad \vdash :: (z) :: (\exists x) :. \phi x \supset \psi x . \supset : (\exists y) . \phi y \supset \psi z \quad (5)$

$[(5).(*1·01.*9·08)] \quad \vdash :. (\exists x) . \sim(\phi x \supset \psi x) : \vee : (z) : (\exists y) . \sim \phi y \vee \psi z \quad (6)$

$[(6).(*9·08)] \quad \vdash :. (\exists x) . \sim(\phi x \supset \psi x) : \vee : (\exists y) . \sim \phi y . \vee . (z) . \psi z \quad (7)$

$[(7).(*1·01)] \quad \vdash :. (x) . \phi x \supset \psi x . \supset : (y) . \phi y . \supset . (z) . \psi z$

This is the proposition to be proved, since "$(y) . \phi y$" is the same proposition as "$(x) . \phi x$," and "$(z) . \psi z$" is the same proposition as "$(x) . \psi x$."

***9·22.** $\vdash :. (x) . \phi x \supset \psi x . \supset : (\exists x) . \phi x . \supset . (\exists x) . \psi x$

I.e. if ϕx always implies ψx, then if ϕx is sometimes true, so is ψx. This proposition, like *9·21, is constantly used in the sequel.

Dem.

$\vdash . *2·08 . \quad \supset \vdash : \phi y \supset \psi y . \supset . \phi y \supset \psi y \quad (1)$

$\vdash . (1) . *9·1 . \quad \supset \vdash : (\exists z) : \phi y \supset \psi y . \supset . \phi y \supset \psi z \quad (2)$

$\vdash . (2) . *9·1 . \quad \supset \vdash :. (\exists x) :. (\exists z) : \phi x \supset \psi x . \supset . \phi y \supset \psi z \quad (3)$

$\vdash . (3) . *9·13 . \quad \supset \vdash :: (y) :: (\exists x) :. (\exists z) : \phi x \supset \psi x . \supset . \phi y \supset \psi z \quad (4)$

$[(4).(*9·06)] \quad \vdash :: (y) :: (\exists x) :. \phi x \supset \psi x . \supset : (\exists z) . \phi y \supset \psi z \quad (5)$

$[(5).(*1·01.*9·08)] \quad \vdash :: (\exists x) . \sim(\phi x \supset \psi x) : \vee : (y) : (\exists z) . \phi y \supset \psi z \quad (6)$

$[(6).(*1·01.*9·07)] \quad \vdash :: (\exists x) . \sim(\phi x \supset \psi x) : \vee : (y) . \sim \phi y . \vee . (\exists z) . \psi z \quad (7)$

$[(7).(*1·01.*9·01·02)] \vdash :. (x) . \phi x \supset \psi x . \supset : (\exists y) . \phi y . \supset . (\exists z) . \psi z$

This is the proposition to be proved, because $(\exists y) . \phi y$ is the same proposition as $(\exists x) . \phi x$, and $(\exists z) . \psi z$ is the same proposition as $(\exists x) . \psi x$.

***9·23.** $\vdash : (x) . \phi x . \supset . (x) . \phi x \quad$ [Id . *9·13·21]

***9·24.** $\vdash : (\exists x) . \phi x . \supset . (\exists x) . \phi x \quad$ [Id . *9·13·22]

***9·25.** $\vdash :. (x) . p \vee \phi x . \supset : p . \vee . (x) . \phi x \quad$ [*9·23 . (*9·04)]

We are now in a position to prove the analogues of *1·2—·6, replacing one of the letters p, q, r in those propositions by $(x) . \phi x$ or $(\exists x) . \phi x$. The proofs are given below.

*9·3. $\vdash :. (x) . \phi x . \vee . (x) . \phi x : \supset . (x) . \phi x$

Dem.

$\vdash . *1\cdot 2 . \quad \supset \vdash . \phi x \vee \phi x . \supset . \phi x$ (1)

$\vdash . (1) . *9\cdot 1 . \quad \supset \vdash : (\exists y) : \phi x \vee \phi y . \supset . \phi x$ (2)

$\vdash . (2) . *9\cdot 13 . \quad \supset \vdash :. (x) :. (\exists y) : \phi x \vee \phi y . \supset . \phi x$ (3)

$[(3).(*9\cdot 05\cdot 01\cdot 04)] \quad \vdash :. (x) :. \phi x . \vee . (y) . \phi y : \supset . \phi x$ (4)

$\vdash . (4) . *9\cdot 21 . \quad \supset \vdash :. (x) : \phi x . \vee . (y) . \phi y : \supset . (x) . \phi x$ (5)

$[(5).(*9\cdot 03)] \quad \vdash :. (x) . \phi x . \vee . (y) . \phi y : \supset . (x) . \phi x :. \supset \vdash . \mathrm{Prop}$

*9·31. $\vdash :. (\exists x) . \phi x . \vee . (\exists x) . \phi x : \supset . (\exists x) . \phi x$

This is the only proposition which employs *9·11.

Dem.

$\vdash . *9\cdot 11\cdot 13 . \quad \supset \vdash : (y) : \phi x \vee \phi y . \supset . (\exists z) . \phi z$ (1)

$[(1).(*9\cdot 03\cdot 02)] \quad \vdash : (\exists y) . \phi x \vee \phi y . \supset . (\exists z) . \phi z$ (2)

$\vdash . (2) . *9\cdot 13 . \supset \vdash : (x) : (\exists y) . \phi x \vee \phi y . \supset . (\exists z) . \phi z$ (3)

$[(3).(*9\cdot 03\cdot 02)] \quad \vdash :. (\exists x) : (\exists y) . \phi x \vee \phi y : \supset . (\exists z) . \phi z$ (4)

$[(4).(*9\cdot 05\cdot 06)] \quad \vdash :. (\exists x) . \phi x . \vee . (\exists y) . \phi y : \supset . (\exists z) . \phi z$

*9·32. $\vdash :. q . \supset : (x) . \phi x . \vee . q$

Dem.

$\vdash . *1\cdot 3 . \quad \supset \vdash :. q . \supset : \phi x . \vee . q$ (1)

$\vdash . (1) . *9\cdot 13 . \supset \vdash :. (x) :. q . \supset : \phi x . \vee . q$

$[*9\cdot 25] \quad \supset \vdash :. q . \supset : (x) : \phi x . \vee . q$ (2)

$[(2).(*9\cdot 03)] \quad \vdash :. q . \supset : (x) . \phi x . \vee . q$

*9·33. $\vdash :. q . \supset : (\exists x) . \phi x . \vee . q$ [Proof as above]

*9·34. $\vdash :. (x) . \phi x . \supset : p . \vee . (x) . \phi x$

Dem.

$\vdash . *1\cdot 3 . \quad \supset \vdash : \phi x . \supset . p \vee \phi x$ (1)

$\vdash . (1) . *9\cdot 13 . \quad \supset \vdash : (x) : \phi x . \supset . p \vee \phi x$ (2)

$\vdash . (2) . *9\cdot 21 . \quad \supset \vdash : (x) . \phi x . \supset . (x) . p \vee \phi x$ (3)

$\vdash . (3) . (*9\cdot 04) . \supset \vdash . \mathrm{Prop}$

*9·35. $\vdash :. (\exists x) . \phi x . \supset : p . \vee . (\exists x) . \phi x$ [Proof as above]

*9·36. $\vdash :. p . \vee . (x) . \phi x : \supset : (x) . \phi x . \vee . p$

Dem.

$\vdash . *1\cdot 4 . \quad \supset \vdash : p \vee \phi x . \supset . \phi x \vee p$ (1)

$\vdash . (1) . *9\cdot 13\cdot 21 . \quad \supset \vdash : (x) . p \vee \phi x . \supset . (x) . \phi x \vee p$ (2)

$\vdash . (2) . (*9\cdot 03\cdot 04) . \supset \vdash . \mathrm{Prop}$

*9·361. $\vdash :. (x) . \phi x . \vee . p : \supset : p . \vee . (x) . \phi x$ [Similar proof]

*9·37. $\vdash :. p . \vee . (\exists x) . \phi x : \supset : (\exists x) . \phi x . \vee . p$ [Similar proof]

*9·371. $\vdash :. (\exists x) . \phi x . \vee . p : \supset : p . \vee . (\exists x) . \phi x$ [Similar proof]

***9·4.** $\vdash :: p : \lor : q . \lor . (x) . \phi x :. \supset :. q : \lor : p . \lor . (x) . \phi x$

Dem.

$\vdash . *1{\cdot}5 . *9{\cdot}21 . \supset \vdash :. (x) : p . \lor . q \lor \phi x : \supset : (x) : q . \lor . p \lor \phi x$ (1)

$\vdash . (1) . (*9{\cdot}04) . \supset \vdash . \text{Prop}$

***9·401.** $\vdash :: p : \lor : q . \lor . (\exists x) . \phi x :. \supset :. q : \lor : p . \lor . (\exists x) . \phi x$ [As above]

***9·41.** $\vdash :: p : \lor : (x) . \phi x . \lor . r :. \supset :. (x) . \phi x : \lor : p \lor r$ [As above]

***9·411.** $\vdash :: p : \lor : (\exists x) . \phi x . \lor . r :. \supset :. (\exists x) . \phi x : \lor : p \lor r$ [As above]

***9·42.** $\vdash :: (x) . \phi x : \lor : q \lor r :. \supset :. q : \lor : (x) . \phi x . \lor . r$ [As above]

***9·421.** $\vdash :: (\exists x) . \phi x : \lor : q \lor r :. \supset :. q : \lor : (\exists x) . \phi x . \lor . r$ [As above]

***9·5.** $\vdash :: p \supset q . \supset :. p . \lor . (x) . \phi x : \supset : q . \lor . (x) . \phi x$

Dem.

$\vdash . *1{\cdot}6 . \qquad \supset \vdash :. p \supset q . \supset : p \lor \phi y . \supset . q \lor \phi y$ (1)

$\vdash . (1) . *9{\cdot}1 . (*9{\cdot}06) . \supset \vdash :. p \supset q . \supset : (\exists x) : p \lor \phi x . \supset . q \lor \phi y$ (2)

$\vdash . (2) . *9{\cdot}13 . (*9{\cdot}04) . \supset \vdash :: p \supset q . \supset :. (y) :. (\exists x) : p \lor \phi x . \supset . q \lor \phi y$ (3)

$[(3).(*9{\cdot}08)] \qquad \vdash :: p \supset q . \supset :. (\exists x) . \sim(p \lor \phi x) . \lor . (y) . q \lor \phi y$ (4)

$[(4).(*9{\cdot}01)] \qquad \vdash :: p \supset q . \supset :. (x) . p \lor \phi x . \supset . (y) . q \lor \phi y$ (5)

$[(5).(*9{\cdot}04)] \qquad \vdash :: p \supset q . \supset :. p . \lor . (x) . \phi x : \supset : q . \lor . (y) . \phi y$

***9·501.** $\vdash :: p \supset q . \supset :. p . \lor . (\exists x) . \phi x : \supset : q . \lor . (\exists x) . \phi x$ [As above]

***9·51.** $\vdash :: p . \supset . (x) . \phi x : \supset :. p \lor r . \supset : (x) . \phi x . \lor . r$

Dem.

$\vdash . *1{\cdot}6 . \qquad \supset \vdash :. p \supset \phi x . \supset : p \lor r . \supset . \phi x \lor r$ (1)

$\vdash . (1) . *9{\cdot}13{\cdot}21 . \supset \vdash :: (x) . p \supset \phi x . \supset :. (x) : p \lor r . \supset . \phi x \lor r$ (2)

$\vdash . (2) . (*9{\cdot}03{\cdot}04) . \supset \vdash . \text{Prop}$

***9·511.** $\vdash :: p . \supset . (\exists x) . \phi x : \supset :. p \lor r . \supset : (\exists x) . \phi x . \lor . r$ [As above]

***9·52.** $\vdash :: (x) . \phi x . \supset . q : \supset :. (x) . \phi x . \lor . r : \supset . q \lor r$

Dem.

$\vdash . *1{\cdot}6 . \qquad \supset \vdash :. \phi x \supset q . \supset : \phi x \lor r . \supset . q \lor r$ (1)

$\vdash . (1) . *9{\cdot}13{\cdot}22 . \supset \vdash :: (\exists x) . \phi x \supset q . \supset :. (\exists x) : \phi x \lor r . \supset . q \lor r$ (2)

$\vdash . (2) . (*9{\cdot}05{\cdot}01) . \supset \vdash :: (x) . \phi x . \supset . q : \supset :. (x) . \phi x \lor r . \supset . q \lor r$ (3)

$\vdash . (3) . (*9{\cdot}03) . \qquad \supset \vdash . \text{Prop}$

***9·521.** $\vdash :: (\exists x) . \phi x . \supset . q : \supset :. (\exists x) . \phi x . \lor . r : \supset . q \lor r$ [As above]

***9·6.** $(x) . \phi x$, $\sim(x) . \phi x$, $(\exists x) . \phi x$ and $\sim(\exists x) . \phi x$ are of the same type. [*9·131, (7) and (8)]

***9·61.** If $\phi \hat{x}$ and $\psi \hat{x}$ are elementary functions of the same type, there is a function $\phi \hat{x} \lor \psi \hat{x}$.

Dem.

By *9·14·15, there is an a for which "ψa," and therefore "ϕa," are significant, and therefore so is "$\phi a \lor \psi a$," by the primitive idea of disjunction. Hence the result by *9·15.

The same proof holds for functions of any number of variables.

***9·62.** If $\phi(\hat{x}, \hat{y})$ and $\psi\hat{z}$ are elementary functions, and the x-argument to ϕ is of the same type as the argument to ψ, there are functions

$$(y) \,.\, \phi(\hat{x}, y) \,.\, \mathsf{v} \,.\, \psi\hat{x}, \quad (\exists y) \,.\, \phi(\hat{x}, y) \,.\, \mathsf{v} \,.\, \psi\hat{x}.$$

Dem.

By *9·15, there are propositions $\phi(x, b)$ and ψa, where by hypothesis x and a are of the same type. Hence by *9·14 there is a proposition $\phi(a, b)$, and therefore, by the primitive idea of disjunction, there is a proposition $\phi(a, b) \mathsf{v} \psi a$, and therefore, by *9·15 and *9·03, there is a proposition $(y) \,.\, \phi(a, y) \,.\, \mathsf{v} \,.\, \psi a$. Similarly there is a proposition $(\exists y) \,.\, \phi(a, y) \,.\, \mathsf{v} \,.\, \psi a$. Hence the result, by *9·15.

***9·63.** If $\phi(\hat{x}, \hat{y})$, $\psi(\hat{x}, \hat{y})$ are elementary functions of the same type, there are functions $(y) \,.\, \phi(\hat{x}, y) \,.\, \mathsf{v} \,.\, (z) \,.\, \psi(\hat{x}, z)$, etc. [Proof as above]

We have now completed the proof that, in the primitive propositions of *1, any *one* of the propositions that occur may be replaced by $(x) \,.\, \phi x$ or $(\exists x) \,.\, \phi x$. It follows that, by merely repeating the proofs, we can show that any other of the propositions that occur in these propositions can be simultaneously replaced by $(x) \,.\, \psi x$ or $(\exists x) \,.\, \psi x$. Thus all the primitive propositions of *1, and therefore all the propositions of *2—*5, hold equally when some or all of the propositions concerned are of one of the forms $(x) \,.\, \phi x$, $(\exists x) \,.\, \phi x$, which was to be proved.

It follows, by mere repetition of the proofs, that the propositions of *1—*5 hold when p, q, r are replaced by propositions containing any number of apparent variables.

*10. THEORY OF PROPOSITIONS CONTAINING ONE APPARENT VARIABLE

Summary of *10.

The chief purpose of the propositions of this number is to extend to formal implications (*i.e.* to propositions of the form $(x) . \phi x \supset \psi x$) as many as possible of the propositions proved previously for material implications, *i.e.* for propositions of the form $p \supset q$. Thus *e.g.* we have proved in *3·33 that

$$p \supset q . q \supset r . \supset . p \supset r.$$

Put $p =$ Socrates is a Greek,

$q =$ Socrates is a man,

$r =$ Socrates is a mortal.

Then we have "if 'Socrates is a Greek' implies 'Socrates is a man,' and 'Socrates is a man' implies 'Socrates is a mortal,' it follows that 'Socrates is a Greek' implies 'Socrates is a mortal.'" But this does not of itself prove that if all Greeks are men, and all men are mortals, then all Greeks are mortals.

Putting $\phi x . = . x$ is a Greek,

$\psi x . = . x$ is a man,

$\chi x . = . x$ is a mortal,

we have to prove

$$(x) . \phi x \supset \psi x : (x) . \psi x \supset \chi x : \supset : (x) . \phi x \supset \chi x.$$

It is such propositions that have to be proved in the present number. It will be seen that formal implication $((x) . \phi x \supset \psi x)$ is a relation of two functions $\phi \hat{x}$ and $\psi \hat{x}$. Many of the formal properties of this relation are analogous to properties of the relation "$p \supset q$" which expresses material implication; it is such analogues that are to be proved in this number.

We shall assume in this number, what has been proved in *9, that the propositions of *1—*5 can be applied to such propositions as $(x) . \phi x$ and $(\exists x) . \phi x$. Instead of the method adopted in *9, it is possible to take negation and disjunction as new primitive ideas, as applied to propositions containing apparent variables, and to assume that, with the new meanings of negation and disjunction, the primitive propositions of *1 still hold. If this method is adopted, we need not take $(\exists x) . \phi x$ as a primitive idea, but may put

***10·01.** $(\exists x) . \phi x . = . \sim (x) . \sim \phi x$ Df

In order to make it clear how this alternative method can be developed, we shall, in the present number, assume nothing of what has been proved in *9 except certain propositions which, in the alternative method, will be primitive propositions, and (what in part characterizes the alternative method)

the applicability to propositions containing apparent variables of analogues of the primitive ideas and propositions of ∗1, and therefore of their consequences as set forth in ∗2—∗5.

The two following definitions merely serve to introduce a notation which is often more convenient than the notation $(x) . \phi x \supset \psi x$ or $(x) . \phi x \equiv \psi x$.

∗10·02. $\phi x \supset_x \psi x . = . (x) . \phi x \supset \psi x$ Df

∗10·03. $\phi x \equiv_x \psi x . = . (x) . \phi x \equiv \psi x$ Df

The first of these notations is due to Peano, who, however, has no notation for $(x) . \phi x$ except in the special case of a formal implication.

The following propositions (∗10·1·11·12·121·122) have already been given in ∗9. ∗10·1 is ∗9·2, ∗10·11 is ∗9·13, ∗10·12 is ∗9·25, ∗10·121 is ∗9·14, and ∗10·122 is ∗9·15. These five propositions must all be taken as primitive propositions in the alternative method; on the other hand, ∗9·1 and ∗9·11 are not required as primitive propositions in the alternative method.

The propositions of the present number are very much used throughout the rest of the work. The propositions most used are the following:

∗10·1. $\vdash : (x) . \phi x . \supset . \phi y$

I.e. what is true in all cases is true in any one case.

∗10·11. If ϕy is true whatever possible argument y may be, then $(x) . \phi x$ is true. In other words, whenever the propositional function ϕy can be asserted, so can the proposition $(x) . \phi x$.

∗10·21. $\vdash :. (x) . p \supset \phi x . \equiv : p . \supset . (x) . \phi x$

∗10·22. $\vdash :. (x) . \phi x . \psi x . \equiv : (x) . \phi x : (x) . \psi x$

The conditions of significance in this proposition demand that ϕ and ψ should take arguments of the same type.

∗10·23. $\vdash :. (x) . \phi x \supset p . \equiv : (\exists x) . \phi x . \supset . p$

I.e. if ϕx always implies p, then if ϕx is ever true, p is true.

∗10·24. $\vdash : \phi y . \supset . (\exists x) . \phi x$

I.e. if ϕy is true, then there is an x for which ϕx is true. This is the sole method of proving existence-theorems.

∗10·27. $\vdash :. (z) . \phi z \supset \psi z . \supset : (z) . \phi z . \supset . (z) . \psi z$

I.e. if ϕz always implies ψz, then "ϕz always" implies "ψz always." The three following propositions, which are equally useful, are analogous to ∗10·27.

∗10·271. $\vdash :. (z) . \phi z \equiv \psi z . \supset : (z) . \phi z . \equiv . (z) . \psi z$

∗10·28. $\vdash :. (x) . \phi x \supset \psi x . \supset : (\exists x) . \phi x . \supset . (\exists x) . \psi x$

∗10·281. $\vdash :. (x) . \phi x \equiv \psi x . \supset : (\exists x) . \phi x . \equiv . (\exists x) . \psi x$

∗10·35. $\vdash :. (\exists x) . p . \phi x . \equiv : p : (\exists x) . \phi x$

∗10·42. $\vdash :. (\exists x) . \phi x . \vee . (\exists x) . \psi x : \equiv . (\exists x) . \phi x \vee \psi x$

∗10·5. $\vdash :. (\exists x) . \phi x . \psi x . \supset : (\exists x) . \phi x : (\exists x) . \psi x$

It should be noticed that whereas ✱10·42 expresses an equivalence, ✱10·5 only expresses an implication. This is the source of many subsequent differences between formulae concerning addition and formulae concerning multiplication.

✱10·51. $\vdash :. \sim\{(\exists x) . \phi x . \psi x\} . \equiv : \phi x . \supset_x . \sim \psi x$

This proposition is analogous to

$$\vdash : \sim(p . q) . \equiv . p \supset \sim q$$

which results from ✱4·63 by transposition.

Of the remaining propositions of this number, some are employed fairly often, while others are lemmas which are used only once or twice, sometimes at a much later stage.

✱10·01. $(\exists x) . \phi x . = . \sim(x) . \sim\phi x$ Df

This definition is only to be used when we discard the method of ✱9 in favour of the alternative method already explained. In either case we have

$$\vdash : (\exists x) . \phi x . \equiv . \sim(x) . \sim\phi x.$$

✱10·02. $\phi x \supset_x \psi x . = . (x) . \phi x \supset \psi x$ Df

✱10·03. $\phi x \equiv_x \psi x . = . (x) . \phi x \equiv \psi x$ Df

✱10·1. $\vdash : (x) . \phi x . \supset . \phi y$ [✱9·2]

✱10·11. If ϕy is true whatever possible argument y may be, then $(x) . \phi x$ is true. [✱9·13]

This proposition is, in a sense, the converse of ✱10·1. ✱10·1 may be stated: "What is true of all is true of any," while ✱10·11 may be stated: "What is true of any, however chosen, is true of all."

✱10·12. $\vdash :. (x) . p \lor \phi x . \supset : p . \lor . (x) . \phi x$ [✱9·25]

According to the definitions in ✱9, this proposition is a mere example of "$q \supset q$," since by definition the two sides of the implication are different symbols for the same proposition. According to the alternative method, on the contrary, ✱10·12 is a substantial proposition.

✱10·121. If "ϕx" is significant, then if a is of the same type as x, "ϕa" is significant, and vice versa. [✱9·14]

It follows from this proposition that two arguments to the same function must be of the same type; for if x and a are arguments to $\phi \hat{x}$, "ϕx" and "ϕa" are significant, and therefore x and a are of the same type. Thus the above primitive proposition embodies the outcome of our discussion of the vicious-circle paradoxes in Chapter II of the Introduction.

✱10·122. If, for some a, there is a proposition ϕa, then there is a function $\phi \hat{x}$, and vice versa. [✱9·15]

✱10·13. If $\phi \hat{x}$ and $\psi \hat{x}$ take arguments of the same type, and we have "$\vdash . \phi x$" and "$\vdash . \psi x$," we shall have "$\vdash . \phi x . \psi x$."

Dem.

By repeated use of $*9{\cdot}61{\cdot}62{\cdot}63{\cdot}131$ (3), there is a function $\sim\phi\hat{x} \vee \sim\psi\hat{x}$. Hence by $*2{\cdot}11$ and $*3{\cdot}01$,

$\vdash : \sim\phi x \vee \sim\psi x . \vee . \phi x . \psi x$ (1)

$\vdash . (1) . *2{\cdot}32 . (*1{\cdot}01) . \supset \vdash :. \phi x . \supset : \psi x . \supset . \phi x . \psi x$ (2)

$\vdash . (2) . *9{\cdot}12 . \supset \vdash .$ Prop

$*10{\cdot}14$. $\vdash :. (x) . \phi x : (x) . \psi x : \supset . \phi y . \psi y$

This proposition is true whenever it is significant, but it is not always significant when its hypothesis is significant. For the thesis demands that ϕ and ψ should take arguments of the same type, while the hypothesis does not demand this. Hence, if it is to be applied when ϕ and ψ are given, or when ψ is given as a function of ϕ or vice versa, we must not argue from the hypothesis to the thesis unless, in the supposed case, ϕ and ψ take arguments of the same type.

Dem.

$\vdash . *10{\cdot}1 . \quad \supset \vdash : (x) . \phi x . \supset . \phi y$ (1)

$\vdash . *10{\cdot}1 . \quad \supset \vdash : (x) . \psi x . \supset . \psi y$ (2)

$\vdash . (1) . (2) . *10{\cdot}13 . \supset \vdash : (x) . \phi x . \supset . \phi y : (x) . \psi x . \supset . \psi y :$

$[*3{\cdot}47] \quad \supset \vdash :. (x) . \phi x : (x) . \psi x : \supset . \phi y . \psi y :. \supset \vdash .$ Prop

$*10{\cdot}2$. $\vdash :. (x) . p \vee \phi x . \equiv : p . \vee . (x) . \phi x$

Dem.

$\vdash . *10{\cdot}1 . *1{\cdot}6 . \supset \vdash :. p . \vee . (x) . \phi x : \supset . p \vee \phi y :.$

$[*10{\cdot}11] \quad \supset \vdash :. (y) :. p . \vee . (x) . \phi x : \supset . p \vee \phi y :.$

$[*10{\cdot}12] \quad \supset \vdash :. p . \vee . (x) . \phi x : \supset . (y) . p \vee \phi y$ (1)

$\vdash . *10{\cdot}12 . \quad \supset \vdash :. (y) . p \vee \phi y . \supset : p . \vee . (x) . \phi x$ (2)

$\vdash . (1) . (2) . \quad \supset \vdash .$ Prop

$*10{\cdot}21$. $\vdash :. (x) . p \supset \phi x . \equiv : p . \supset . (x) . \phi x \quad \left[*10{\cdot}2 \dfrac{\sim p}{p}\right]$

This proposition is much more used than $*10{\cdot}2$.

$*10{\cdot}22$. $\vdash :. (x) . \phi x . \psi x . \equiv : (x) . \phi x : (x) . \psi x$

Dem.

$\vdash . *10{\cdot}1 . \quad \supset \vdash : (x) . \phi x . \psi x . \supset . \phi y . \psi y .$ (1)

$[*3{\cdot}26] \quad \supset . \phi y :$

$[*10{\cdot}11] \quad \supset \vdash :. (y) : (x) . \phi x . \psi x . \supset . \phi y :.$

$[*10{\cdot}21] \quad \supset \vdash :. (x) . \phi x . \psi x . \supset . (y) . \phi y$ (2)

$\vdash . (1) . *3{\cdot}27 . \quad \supset \vdash :. (x) . \phi x . \psi x . \supset . \psi z :.$

$[*10{\cdot}11] \quad \supset \vdash :. (z) : (x) . \phi x . \psi x . \supset . \psi z :.$

$[*10{\cdot}21] \quad \supset \vdash :. (x) . \phi x . \psi x . \supset . (z) . \psi z$ (3)

$\vdash . (2) . (3) . \text{Comp} . \supset \vdash :. (x) . \phi x . \psi x . \supset : (y) . \phi y : (z) . \psi z$ (4)

$\vdash . *10{\cdot}14{\cdot}11 . \quad \supset \vdash :. (y) :. (x) . \phi x : (x) . \psi x : \supset . \phi y . \psi y :.$

$[*10{\cdot}21] \quad \supset \vdash :. (x) . \phi x : (x) . \psi x : \supset . (y) . \phi y . \psi y$ (5)

$\vdash . (4) . (5) . \quad \supset \vdash .$ Prop

The above proposition is true whenever it is significant; but, as was pointed out in connexion with ∗10·14, it is not always significant when "$(x) . \phi x : (x) . \psi x$" is significant.

∗10·221. If ϕx contains a constituent $\chi(x, y, z, \ldots)$ and ψx contains a constituent $\chi(x, u, v, \ldots)$, where χ is an elementary function and $y, z, \ldots u, v, \ldots$ are either constants or apparent variables, then $\phi\hat{x}$ and $\psi\hat{x}$ take arguments of the same type. This can be proved in each particular case, though not generally, provided that, in obtaining ϕ and ψ from χ, χ is only submitted to negations, disjunctions and generalizations. The process may be illustrated by an example. Suppose ϕx is $(y) . \chi(x, y) . \supset . \theta x$, and ψx is $fx . \supset . (y) . \chi(x, y)$. By the definitions of ∗9, ϕx is $(\exists y) . \sim\chi(x, y) \vee \theta x$, and ψx is $(y) . \sim fx \vee \chi(x, y)$. Hence since the primitive ideas $(x) . Fx$ and $(\exists x) . Fx$ only apply to functions, there are functions $\sim\chi(\hat{x}, \hat{y}) \vee \theta\hat{x}$, $\sim f\hat{x} \vee \chi(\hat{x}, \hat{y})$. Hence there is a proposition $\sim\chi(a, b) \vee \theta a$. Hence, since "$p \vee q$" and "$\sim p$" are only significant when p and q are propositions, there is a proposition $\chi(a, b)$. Similarly, for some u and v, there are propositions $\sim fu \vee \chi(u, v)$ and $\chi(u, v)$. Hence by ∗9·14, u and a, v and b are respectively of the same type, and (again by ∗9·14) there is a proposition $\sim fa \vee \chi(a, b)$. Hence (∗9·15) there are functions $\sim\chi(a, \hat{y}) \vee \theta a$, $\sim fa \vee \chi(a, \hat{y})$, and therefore there are propositions

$$(\exists y) . \sim\chi(a, y) \vee \theta a, \quad (y) . \sim fa \vee \chi(a, y),$$

i.e. there are propositions ϕa, ψa, which was to be proved. This process can be applied similarly in any other instance.

∗10·23. $\vdash :. (x) . \phi x \supset p . \equiv : (\exists x) . \phi x . \supset . p$

Dem.

$\vdash . ∗4·2 . (∗9·03) . \supset \vdash :. (x) . \sim\phi x \vee p . \equiv : (x) . \sim\phi x . \vee . p :$

$[(∗9·02)] \qquad \equiv . (\exists x) . \phi x . \supset . p \qquad (1)$

$\vdash . (1) . (∗1·01) . \supset \vdash . \text{Prop}$

In the above proof, we employ the definitions of ∗9. In the alternative method, in which $(\exists x) . \phi x$ is defined in accordance with ∗10·01, the proof proceeds as follows.

∗10·23. $\vdash :. (x) . \phi x \supset p . \equiv : (\exists x) . \phi x . \supset . p$

Dem.

$\vdash . \text{Transp} . (∗10·01) . \supset \vdash :. (\exists x) . \phi x . \supset . p : \equiv : \sim p . \supset . (x) . \sim\phi x :$

$[∗10·21] \qquad \equiv : (x) : \sim p . \supset . \sim\phi x : \qquad (1)$

$[∗10·1] \qquad \supset : \sim p . \supset . \sim\phi x :$

$[\text{Transp}] \qquad \supset : \phi x . \supset . p :.$

$[∗10·11] \qquad \supset \vdash :. (x) :. (\exists x) . \phi x . \supset . p : \supset : \phi x . \supset . p :.$

[$*10\cdot21$] $\supset\vdash:.(\exists x).\phi x.\supset.p:\supset:(x):\phi x.\supset.p$ (2)

$\vdash.*10\cdot1.$ $\supset\vdash:.(x):\phi x.\supset.p:\supset:\phi x\supset p:$

[Transp] $\supset:\sim p.\supset.\sim\phi x:.$

[$*10\cdot11\cdot21$] $\supset\vdash:.(x):\phi x.\supset.p:\supset:(x):\sim p.\supset.\sim\phi x:$

[(1)] $\supset:(\exists x).\phi x.\supset.p$ (3)

$\vdash.(2).(3).$ $\supset\vdash.$ Prop

Whenever we have an asserted proposition of the form $p\supset\phi x$, we can pass by $*10\cdot11\cdot21$ to an asserted proposition $p.\supset.(x).\phi x$. This passage is constantly required, as in the last line but one of the above proof. It will be indicated merely by the reference "$*10\cdot11\cdot21$," and the two steps which it requires will not be separately put down.

$*10\cdot24$. $\vdash:\phi y.\supset.(\exists x).\phi x$

This is $*9\cdot1$. In the alternative method, the proof is as follows.

Dem.

$\vdash.*10\cdot1.\supset\vdash:(x).\sim\phi x.\supset.\sim\phi y:$

[Transp] $\supset\vdash:\phi y.\supset.\sim(x).\sim\phi x:$

[($*10\cdot01$)] $\supset\vdash.$ Prop

$*10\cdot25$. $\vdash:(x).\phi x.\supset.(\exists x).\phi x$ [$*10\cdot1\cdot24$]

$*10\cdot251$. $\vdash:(x).\sim\phi x.\supset.\sim\{(x).\phi x\}$ [$*10\cdot25$. Transp]

$*10\cdot252$. $\vdash:\sim\{(\exists x).\phi x\}.\equiv.(x).\sim\phi x$ [$*4\cdot2$. ($*9\cdot02$)]

$*10\cdot253$. $\vdash:\sim\{(x).\phi x\}.\equiv.(\exists x).\sim\phi x$ [$*4\cdot2$. ($*9\cdot01$)]

In the alternative method, in which $(\exists x).\phi x$ is defined as in $*10\cdot01$, the proofs of $*10\cdot252\cdot253$ are as follows.

$*10\cdot252$. $\vdash:\sim\{(\exists x).\phi x\}.\equiv.(x).\sim\phi x$ [$*4\cdot13$. ($*10\cdot01$)]

$*10\cdot253$. $\vdash:\sim\{(x).\phi x\}.\equiv.(\exists x).\sim\phi x$

Dem.

$\vdash.*10\cdot1.$ $\supset\vdash:(x).\phi x.\supset.\phi y.$

[$*2\cdot12$] $\supset.\sim(\sim\phi y):$

[$*10\cdot11\cdot21$] $\supset\vdash:(x).\phi x.\supset.(y).\sim(\sim\phi y):$

[Transp] $\supset\vdash:\sim\{(y).\sim(\sim\phi y)\}.\supset.\sim\{(x).\phi x\}:$

[($*10\cdot01$)] $\supset\vdash:(\exists y).\sim\phi y.$ $\supset.\sim\{(x).\phi x\}$ (1)

$\vdash.*10\cdot1.$ $\supset\vdash:(y).\sim(\sim\phi y).$ $\supset.\sim(\sim\phi x).$

[$*2\cdot14$] $\supset.\phi x:$

[$*10\cdot11\cdot21$] $\supset\vdash:(y).\sim(\sim\phi y).$ $\supset.(x).\phi x:$

[Transp] $\supset\vdash:\sim\{(x).\phi x\}.$ $\supset.\sim\{(y).\sim(\sim\phi y)\}.$

[($*10\cdot01$)] $\supset.(\exists y).\sim\phi y$ (2)

$\vdash.(1).(2).\supset\vdash.$ Prop

***10·26.** $\vdash :. (z) . \phi z \supset \psi z : \phi x : \supset . \psi x$ [*10·1 . Imp]

This is one form of the syllogism in Barbara. *E.g.* put $\phi z . = . z$ is a man, $\psi z . = . z$ is mortal, $x =$ Socrates. Then the proposition becomes:

"If all men are mortal, and Socrates is a man, then Socrates is mortal."

Another form of the syllogism in Barbara is given in *10·3. The two forms, formerly wrongly identified, were first distinguished by Peano and Frege.

***10·27.** $\vdash :. (z) . \phi z \supset \psi z . \supset : (z) . \phi z . \supset . (z) . \psi z$

This is *9·21. In the alternative method, the proof is as follows.

Dem.

$\vdash . *10·14 . \quad \supset \vdash :. (z) . \phi z \supset \psi z : (z) . \phi z : \supset . \phi y \supset \psi y . \phi y .$
[Ass] $\quad \supset . \psi y :.$
[*10·1] $\quad \supset \vdash :. (y) :. (z) . \phi z \supset \psi z : (z) . \phi z : \supset . \psi y :.$
[*10·21] $\quad \supset \vdash :. (z) . \phi z \supset \psi z : (z) . \phi z : \supset . (y) . \psi y$ (1)
$\vdash . (1) . \text{Exp} . \supset \vdash . \text{Prop}$

***10·271.** $\vdash :. (z) . \phi z \equiv \psi z . \supset : (z) . \phi z . \equiv . (z) . \psi z$

Dem.

$\vdash . *10·22 . \quad \supset \vdash :. \text{Hp} . \supset : (z) . \phi z \supset \psi z :$
[*10·27] $\quad \supset : (z) . \phi z . \supset . (z) . \psi z$ (1)
$\vdash . *10·22 . \quad \supset \vdash :. \text{Hp} . \supset : (z) . \psi z \supset \phi z :$
[*10·27] $\quad \supset : (z) . \psi z . \supset . (z) . \phi z$ (2)
$\vdash . (1) . (2) . \text{Comp} . \supset \vdash . \text{Prop}$

***10·28.** $\vdash :. (x) . \phi x \supset \psi x . \supset : (\exists x) . \phi x . \supset . (\exists x) . \psi x$

This is *9·22. In the alternative method, the proof is as follows.

Dem.

$\vdash . *10·1 . \quad \supset \vdash :. (x) . \phi x \supset \psi x . \supset . \phi y \supset \psi y .$
[Transp] $\quad \supset . \sim\psi y \supset \sim\phi y :.$
[*10·11·21] $\supset \vdash :. (x) . \phi x \supset \psi x . \supset : (y) . \sim\psi y \supset \sim\phi y :$
[*10·27] $\quad \supset : (y) . \sim\psi y . \supset . (y) . \sim\phi y :$
[Transp] $\quad \supset : (\exists y) . \phi y . \supset . (\exists y) . \psi y :. \supset \vdash . \text{Prop}$

***10·281.** $\vdash :. (x) . \phi x \equiv \psi x . \supset : (\exists x) . \phi x . \equiv . (\exists x) . \psi x$ [*10·22·28 . Comp]

***10·29.** $\vdash :. (x) . \phi x \supset \psi x : (x) . \phi x \supset \chi x : \equiv : (x) : \phi x . \supset . \psi x . \chi x.$

Dem.

$\vdash . *10·22 . \quad \supset \vdash :. (x) . \phi x \supset \psi x : (x) . \phi x \supset \chi x :$
$\quad \equiv : (x) : \phi x \supset \psi x . \phi x \supset \chi x$ (1)
$\vdash . *4·76 . \quad \supset \vdash :. \phi x \supset \psi x . \phi x \supset \chi x . \equiv : \phi x . \supset . \psi x . \chi x :.$
[*10·11] $\quad \supset \vdash :. (x) :. \phi x \supset \psi x . \phi x \supset \chi x . \equiv : \phi x . \supset . \psi x . \chi x :.$
[*10·271] $\quad \supset \vdash :. (x) : \phi x \supset \psi x . \phi x \supset \chi x : \equiv : (x) : \phi x . \supset . \psi x . \chi x$ (2)
$\vdash . (1) . (2) . \supset \vdash . \text{Prop}$

This is an extension of the principle of composition.

✱10·3. $\vdash :. (x) . \phi x \supset \psi x : (x) . \psi x \supset \chi x : \supset . (x) . \phi x \supset \chi x$

This is the second form of the syllogism in Barbara.

Dem.

$\vdash . ✱10·22·221 . \supset \vdash : \mathrm{Hp} . \supset . (x) . \phi x \supset \psi x . \psi x \supset \chi x .$

$[\mathrm{Syll}.✱10·27] \qquad \supset . (x) . \phi x \supset \chi x : \supset \vdash . \mathrm{Prop}$

✱10·301. $\vdash :. (x) . \phi x \equiv \psi x : (x) . \psi x \equiv \chi x : \supset . (x) . \phi x \equiv \chi x$

Dem.

$\vdash . ✱10·22·221 . \supset \vdash :. \mathrm{Hp} . \supset : (x) . \phi x \equiv \psi x . \psi x = \chi x :$

$[✱4·22 . ✱10·27] \qquad \supset : (x) . \phi x \equiv \chi x :. \supset \vdash . \mathrm{Prop}$

In the second line of the proofs of ✱10·3 and ✱10·301, we abbreviate the process of proof in a way which is often convenient. In ✱10·3, the full process would be as follows:

$\vdash . \mathrm{Syll} . \supset \vdash : \phi x \supset \psi x . \psi x \supset \chi x . \supset . \phi x \supset \chi x :$

$[✱10·11] \supset \vdash : (x) : \phi x \supset \psi x . \psi x \supset \chi x . \supset . \phi x \supset \chi x :$

$[✱10·27] \supset \vdash : (x) . \phi x \supset \psi x . \psi x \supset \chi x . \supset . (x) . \phi x \supset \chi x$

The above two propositions show that formal implication and formal equivalence are transitive relations between functions.

✱10·31. $\vdash :. (x) . \phi x \supset \psi x . \supset : (x) : \phi x . \chi x . \supset . \psi x . \chi x$

Dem.

$\vdash . \mathrm{Fact} . ✱10·11 . \supset \vdash :. (x) :. \phi x \supset \psi x . \supset : \phi x . \chi x . \supset . \psi x . \chi x \qquad (1)$

$\vdash . (1) . ✱10·27 . \quad \supset \vdash . \mathrm{Prop}$

✱10·311. $\vdash :. (x) . \phi x \equiv \psi x . \supset : (x) : \phi x . \chi x . \equiv . \psi x . \chi x$

Dem.

$\vdash . ✱4·36 . ✱10·11 . \supset \vdash :. (x) :. \phi x \equiv \psi x . \supset : \phi x . \chi x . \equiv . \psi x . \chi x \qquad (1)$

$\vdash . (1) . ✱10·27 . \quad \supset \vdash . \mathrm{Prop}$

The above two propositions are extensions of the principle of the factor.

✱10·32. $\vdash : \phi x \equiv_x \psi x . \equiv . \psi x \equiv_x \phi x$

Dem.

$\vdash . ✱10·22 . \supset \vdash : \phi x \equiv_x \psi x . \equiv . \phi x \supset_x \psi x . \psi x \supset_x \phi x .$

$[✱4·3] \qquad \equiv . \psi x \supset_x \phi x . \phi x \supset_x \psi x .$

$[✱10·22] \qquad \equiv . \psi x \equiv_x \phi x : \supset \vdash . \mathrm{Prop}$

This proposition shows that formal equivalence is symmetrical.

✱10·321. $\vdash : \phi x \equiv_x \psi x . \phi x \equiv_x \chi x . \supset . \psi x \equiv_x \chi x$

Dem.

$\vdash . ✱10·32 . \mathrm{Fact} . \supset \vdash : \mathrm{Hp} . \supset . \psi x \equiv_x \phi x . \phi x \equiv_x \chi x .$

$[✱10·301] \qquad \supset . \psi x \equiv_x \chi x : \supset \vdash . \mathrm{Prop}$

✱10·322. $\vdash : \psi x \equiv_x \phi x . \chi x \equiv_x \phi x . \supset . \psi x \equiv_x \chi x$

Dem.

$\vdash . ✱10·32 . \supset \vdash : \mathrm{Hp} . \supset . \psi x \equiv_x \phi x . \phi x \equiv_x \chi x .$

$[✱10·301] \qquad \supset . \psi x \equiv_x \chi x : \supset \vdash . \mathrm{Prop}$

***10·33.** $\vdash :. (x) : \phi x . p : \equiv : (x) . \phi x : p$

Dem.

$\vdash . *10 \cdot 1 . \quad \supset \vdash :. (x) : \phi x . p : \supset . \phi y . p .$ (1)

$[*3 \cdot 27] \quad \supset . p$ (2)

$\vdash . (1) . *3 \cdot 26 . \supset \vdash :. (x) : \phi x . p : \supset . \phi y :$

$[*10 \cdot 11 \cdot 21] \quad \supset \vdash :. (x) : \phi x . p : \supset . (y) . \phi y$ (3)

$\vdash . (2) . (3) . \quad \supset \vdash :. (x) : \phi x . p : \supset : (y) . \phi y : p$ (4)

$\vdash . *10 \cdot 1 . \quad \supset \vdash :. (y) . \phi y . \quad \supset . \phi x :.$

$[\text{Fact}] \quad \supset \vdash :. (y) . \phi y : p : \supset . \phi x . p :.$

$[*10 \cdot 11 \cdot 21] \quad \supset \vdash :. (y) . \phi y : p : \supset : (x) : \phi x . p$ (5)

$\vdash . (4) . (5) . \quad \supset \vdash . \text{Prop}$

***10·34.** $\vdash :. (\exists x) . \phi x \supset p . \equiv : (x) . \phi x . \supset . p$

This follows immediately from *9·05·01 and *1·01. In the alternative method, the proof is as follows.

Dem.

$\vdash . *4 \cdot 2 . (*10 \cdot 01) . \supset$

$\vdash :. (\exists x) . \phi x \supset p . \quad \equiv : \sim \{(x) . \sim (\phi x \supset p)\} :$

$[*4 \cdot 61 . *10 \cdot 271] \quad \equiv : \sim \{(x) : \phi x . \sim p\} :$

$[*10 \cdot 33] \quad \equiv : \sim \{(x) . \phi x : \sim p\} :$

$[*4 \cdot 53] \quad \equiv : \sim \{(x) . \phi x\} . \vee . p :$

$[*4 \cdot 6] \quad \equiv : (x) . \phi x . \supset . p$

***10·35.** $\vdash :. (\exists x) . p . \phi x . \equiv : p : (\exists x) . \phi x$

Dem.

$\vdash . *3 \cdot 26 . \quad \supset \vdash : p . \phi x . \supset . p :$

$[*10 \cdot 11] \quad \supset \vdash : (x) : p . \phi x . \supset . p :$

$[*10 \cdot 23] \quad \supset \vdash : (\exists x) . p . \phi x . \supset . p$ (1)

$\vdash . *3 \cdot 27 . \quad \supset \vdash : p . \phi x . \supset . \phi x :$

$[*10 \cdot 11] \quad \supset \vdash : (x) : p . \phi x . \supset . \phi x :$

$[*10 \cdot 28] \quad \supset \vdash : (\exists x) . p . \phi x . \supset . (\exists x) . \phi x$ (2)

$\vdash . *3 \cdot 2 . \quad \supset \vdash :. p . \supset : \phi x . \supset . p . \phi x .$

$[*10 \cdot 11 \cdot 21] \supset \vdash :. p . \supset : (x) : \phi x . \supset . p . \phi x :$

$[*10 \cdot 28] \quad \supset : (\exists x) . \phi x . \supset . (\exists x) . p . \phi x$ (3)

$\vdash . (1) . (2) . (3) . \text{Imp} . \supset \vdash . \text{Prop}$

***10·36.** $\vdash :. (\exists x) . \phi x \vee p . \equiv : (\exists x) . \phi x . \vee . p$

This follows immediately from *9·05. In the alternative method, the proof is as follows.

Dem.

$\vdash . *4 \cdot 64 . \quad \supset \vdash : \phi x \vee p . \equiv . \sim \phi x \supset p :$

$[*10 \cdot 11] \quad \supset \vdash : (x) : \phi x \vee p . \equiv . \sim \phi x \supset p :$

$[*10 \cdot 281] \quad \supset \vdash :. (\exists x) . \phi x \vee p . \equiv : (\exists x) . \sim \phi x \supset p :$

$[*10 \cdot 34] \quad \equiv : (x) . \sim \phi x . \supset . p :$

$[*4 \cdot 6 . (*10 \cdot 01)] \quad \equiv : (\exists x) . \phi x . \vee . p :. \supset \vdash . \text{Prop}$

The above proposition is only required in order to lead to the following:

***10·37.** $\vdash :. (\exists x) . p \supset \phi x . \equiv : p . \supset . (\exists x) . \phi x$ $\left[*10\cdot36 \frac{\sim p}{p}\right]$

***10·39.** $\vdash :. \phi x \supset_x \chi x : \psi x \supset_x \theta x : \supset : \phi x . \psi x . \supset_x . \chi x . \theta x$

Dem.

$\vdash . *10\cdot22 . \supset \vdash :. \text{Hp} . \supset : (x) : \phi x \supset \chi x . \psi x \supset \theta x :$

$[*3\cdot47 . *10\cdot27] \quad \supset : (x) : \phi x . \psi x . \supset . \chi x . \theta x :. \supset \vdash . \text{Prop}$

This proposition is only true when the conclusion is significant; the significance of the hypothesis does not insure that of the conclusion. On the conditions of significance, see the remarks on *10·4, below.

***10·4.** $\vdash :. \phi x \equiv_x \chi x . \psi x \equiv_x \theta x . \supset : \phi x . \psi x . \equiv_x . \chi x . \theta x$

Dem.

$\vdash . *10\cdot22 . \quad \supset \vdash :. \text{Hp} . \supset : \phi x \supset_x \chi x . \psi x \supset_x \theta x :$

$[*10\cdot39] \quad \supset : \phi x . \psi x . \supset_x . \chi x . \theta x \quad (1)$

Similarly $\quad \vdash :. \text{Hp} . \supset : \chi x . \theta x . \supset_x . \phi x . \psi x \quad (2)$

$\vdash . (1) . (2) . \text{Comp} . \supset \vdash :. \text{Hp} . \supset : \phi x . \psi x . \supset_x . \chi x . \theta x : \chi x . \theta x . \supset_x . \phi x . \psi x :$

$[*10\cdot22] \quad \supset : \phi x . \psi x . \equiv_x . \chi x . \theta x :. \supset \vdash . \text{Prop}$

In *10·4 and many later propositions, as in *10·39, the conclusion may be not significant when the hypothesis is true. Hence, in order that it may be legitimate to use *10·4 in *inference*, *i.e.* to pass from the *assertion* of the hypothesis to the *assertion* of the conclusion, the functions ϕ, ψ, χ, θ must be such as to have overlapping ranges of significance. In virtue of *10·221, this is secured if they are of the forms $F\{x, \chi(x, \hat{y}, \hat{z}, \ldots)\}$, $f\{x, \chi(x, \hat{y}, \hat{z}, \ldots)\}$, $G\{x, \chi(x, \hat{y}, \hat{z}, \ldots)\}$, $g\{x, \chi(x, \hat{y}, \hat{z}, \ldots)\}$. It is also secured if ϕ and ψ or ϕ and θ or χ and ψ or χ and θ are of such forms, for ϕ and χ must have overlapping ranges of significance if the hypothesis is to be significant, and so must ψ and θ.

***10·41.** $\vdash :. (x) . \phi x . \vee . (x) . \psi x : \supset . (x) . \phi x \vee \psi x$

Dem.

$\vdash . *10\cdot1 . \quad \supset \vdash : (x) . \phi x . \supset . \phi y .$

$[*2\cdot2] \quad \supset . \phi y \vee \psi y \quad (1)$

$\vdash . *10\cdot1 . \quad \supset \vdash : (x) . \psi x . \supset . \psi y .$

$[*1\cdot3] \quad \supset . \phi y \vee \psi y \quad (2)$

$\vdash . (1) . (2) . *10\cdot13 . \supset \vdash :. (x) . \phi x . \supset . \phi y \vee \psi y : (x) . \psi x . \supset . \phi y \vee \psi y :.$

$[*3\cdot44] \quad \supset \vdash :. (x) . \phi x . \vee . (x) . \psi x : \supset . \phi y \vee \psi y$

$[*10\cdot11\cdot21] \quad \supset \vdash :. (x) . \phi x . \vee . (x) . \psi x : \supset . (y) . \phi y \vee \psi y :. \supset \vdash . \text{Prop}$

Observe that in the above proof the uses of *2·2 and *1·3 are only legitimate if ϕy and ψy have overlapping ranges of significance, for otherwise, if y is such that there is a proposition ϕy, it is such that there is no proposition ψy, and conversely.

***10·411.** $\vdash :. \phi x \equiv_x \chi x . \psi x \equiv_x \theta x . \supset : \phi x \lor \psi x . \equiv_x . \chi x \lor \theta x$

Dem.

$\vdash . *10·14 . \supset \vdash :. \text{Hp} . \supset : \phi x \equiv \chi x . \psi x \equiv \theta x :$

$[*4·39] \quad \supset : \phi x \lor \psi x . \equiv . \chi x \lor \theta x \quad (1)$

$\vdash . (1) . *10·11·21 . \supset \vdash . \text{Prop}$

***10·412.** $\vdash : \phi x \equiv_x \psi x . \equiv . \sim\phi x \equiv_x \sim\psi x$ [*4·11 . *10·11·271]

***10·413.** $\vdash :. \phi x \equiv_x \chi x . \psi x \equiv_x \theta x . \supset : \phi x \supset \psi x . \equiv_x . \chi x \supset \theta x$

Dem.

$\vdash . *10·411·412 . \supset \vdash :. \text{Hp} . \supset : \sim\phi x \lor \psi x . \equiv_x . \sim\chi x \lor \theta x$

$[(*1·01)] \quad \supset : \phi x \supset \psi x . \equiv_x . \chi x \supset \theta x :. \supset \vdash . \text{Prop}$

***10·414.** $\vdash :. \phi x \equiv_x \chi x . \psi x \equiv_x \theta x . \supset : \phi x \equiv \psi x . \equiv_x . \chi x \equiv \theta x$

Dem.

$\vdash . *10·413 \dfrac{\psi, \phi, \theta, \chi}{\phi, \psi, \chi, \theta} . *10·32 . \supset \vdash :. \text{Hp} . \supset : \psi x \supset \phi x . \equiv_x . \theta x \supset \chi x \quad (1)$

$\vdash . *10·413 . (1) . *10·4 . \quad \supset \vdash . \text{Prop}$

The propositions *10·413·414 are chiefly used in cases where either χ is replaced by ϕ or θ is replaced by ψ, in which case half the hypothesis becomes superfluous, being true by *4·2.

***10·42.** $\vdash :. (\exists x) . \phi x . \lor . (\exists x) . \psi x : \equiv . (\exists x) . \phi x \lor \psi x$

Dem.

$\vdash . *10·22 . \quad \supset \vdash :. (x) . \sim\phi x : (x) . \sim\psi x : \equiv . (x) . \sim\phi x . \sim\psi x :.$

$[*4·11] \quad \supset \vdash :. \sim\{(x) . \sim\phi x : (x) . \sim\psi x\} . \equiv . \sim\{(x) . \sim\phi x . \sim\psi x\} :.$

$[*4·51·56 . *10·271] \quad \supset \vdash :. \sim\{(x) . \sim\phi x\} . \lor . \sim\{(x) . \sim\psi x\} :$

$\equiv . \sim\{(x) . \sim(\phi x \lor \psi x)\} :.$

$[*10·253] \quad \supset \vdash :. (\exists x) . \phi x . \lor . (\exists x) . \psi x : \equiv . (\exists x) . \phi x \lor \psi x :.$

$\supset \vdash . \text{Prop}$

This proposition is very frequently used. It should be contrasted with *10·5, in which we have only an implication, not an equivalence.

***10·43.** $\vdash : \phi z \equiv_z \psi z . \phi x . \equiv . \phi z \equiv_z \psi z . \psi x$

Dem.

$\vdash . *10·1 . \quad \supset \vdash : \phi z \equiv_z \psi z . \supset . \phi x \equiv \psi x \quad (1)$

$\vdash . (1) . *5·32 . \supset \vdash . \text{Prop}$

***10·5.** $\vdash :. (\exists x) . \phi x . \psi x . \supset : (\exists x) . \phi x : (\exists x) . \psi x$

Dem.

$\vdash . *3·26 . *10·11 . \supset \vdash : (x) : \phi x . \psi x . \supset . \phi x :$

$[*10·28] \quad \supset \vdash : (\exists x) . \phi x . \psi x . \supset . (\exists x) . \phi x \quad (1)$

$\vdash . *3·27 . *10·11 . \supset \vdash :. (x) : \phi x . \psi x . \supset . \psi x :$

$[*10·28] \quad \supset \vdash : (\exists x) . \phi x . \psi x . \supset . (\exists x) . \psi x \quad (2)$

$\vdash . (1) . (2) . \text{Comp} . \supset \vdash :. \text{Prop}$

The converse of the above proposition is false. The fact that this proposition states an implication, while ∗10·42 states an equivalence, is the source of many subsequent differences between formulae concerning logical addition and formulae concerning logical multiplication.

∗10·51. $\vdash :. \sim\{(\exists x) . \phi x . \psi x\} . \equiv : \phi x . \supset_x . \sim\psi x$

Dem.

$\vdash . *10\cdot252 . \supset \vdash :. \sim\{(\exists x) . \phi x . \psi x\} . \equiv : (x) . \sim(\phi x . \psi x) :$

$[*4\cdot51\cdot62 . *10\cdot271] \quad \equiv : (x) : \phi x . \supset . \sim\psi x :. \supset \vdash . \text{Prop}$

∗10·52. $\vdash :. (\exists x) . \phi x . \supset : (x) . \phi x \supset p . \equiv . p$

Dem.

$\vdash . *5\cdot5 . \supset \vdash :: \text{Hp} . \supset :. p . \equiv : (\exists x) . \phi x . \supset . p :$

$[*10\cdot23] \quad \equiv : (x) . \phi x \supset p :: \supset \vdash . \text{Prop}$

∗10·53. $\vdash :. \sim(\exists x) . \phi x . \supset : \phi x . \supset_x . \psi x$

Dem.

$\vdash . *2\cdot21 . *10\cdot11 . \supset$

$\vdash :. (x) :. \sim\phi x . \supset : \phi x . \supset . \psi x :.$

$[*10\cdot27] \supset \vdash :. (x) . \sim\phi x . \supset : (x) : \phi x . \supset . \psi x :.$

$[*10\cdot252] \supset \vdash :. \sim(\exists x) . \phi x . \supset : (x) : \phi x . \supset . \psi x :. \supset \vdash . \text{Prop}$

∗10·541. $\vdash :. \phi y . \supset_y . p \vee \psi y : \equiv : p . \vee . \phi y \supset_y \psi y$

Dem.

$\vdash . *4\cdot2 . (*1\cdot01) . \supset \vdash :. \phi y . \supset_y . p \vee \psi y : \equiv : (y) . \sim\phi y \vee p \vee \psi y :$

$[\text{Assoc} . *10\cdot271] \quad \equiv : (y) . p \vee \sim\phi y \vee \psi y :$

$[*10\cdot2] \quad \equiv : p . \vee . (y) . \sim\phi y \vee \psi y :$

$[(*1\cdot01)] \quad \equiv : p . \vee . \phi y \supset_y \psi y :. \supset \vdash . \text{Prop}$

The above proposition is only needed in order to lead to the following:

10·542. $\vdash :. \phi y . \supset_y . p \supset \psi y : \equiv : p . \supset . \phi y \supset_y \psi y \quad \left[*10\cdot541 \frac{\sim p}{p}\right]$

This proposition is a lemma for ∗84·43.

∗10·55. $\vdash :. (\exists x) . \phi x . \psi x : \phi x \supset_x \psi x : \equiv : (\exists x) . \phi x : \phi x \supset_x \psi x$

Dem.

$\vdash . *4\cdot71 . \supset \vdash :. \phi x \supset \psi x . \supset : \phi x . \psi x . \equiv . \phi x \quad (1)$

$\vdash . (1) . *10\cdot11\cdot27 . \supset$

$\vdash :. \phi x \supset_x \psi x . \supset : (x) : \phi x . \psi x . \equiv . \phi x :$

$[*10\cdot281] \quad \supset : (\exists x) . \phi x . \psi x . \equiv . (\exists x) . \phi x \quad (2)$

$\vdash . (2) . *5\cdot32 . \supset \vdash . \text{Prop}$

This proposition is a lemma for ∗117·12·121.

$*10\cdot56$. $\vdash :. \phi x . \supset_x . \psi x : (\exists x) . \phi x . \chi x : \supset . (\exists x) . \psi x . \chi x$

Dem.

$\vdash . *10\cdot31 . \quad \supset \vdash :. \phi x . \supset_x . \psi x : \supset : \phi x . \chi x . \supset_x . \psi x . \chi x :$

$[*10\cdot28] \qquad \supset : (\exists x) . \phi x . \chi x . \supset . (\exists x) . \psi x . \chi x \quad (1)$

$\vdash . (1) . \text{Imp} . \supset \vdash . \text{Prop}$

This proposition and $*10\cdot57$ are used in the theory of series (Part V).

$*10\cdot57$. $\vdash :. \phi x . \supset_x . \psi x \vee \chi x : \supset : \phi x \supset_x \psi x . \vee . (\exists x) . \phi x . \chi x$

Dem.

$\vdash . *10\cdot51 . \text{Fact} . \supset$

$\vdash :. \phi x . \supset_x . \psi x \vee \chi x : \sim(\exists x) . \phi x . \chi x : \supset : \phi x . \supset_x . \psi x \vee \chi x : \phi x . \supset_x . \sim\chi x :$

$[*10\cdot29] \qquad \supset : \phi x . \supset_x . \psi x \vee \chi x . \sim\chi x :$

$[*5\cdot61] \qquad \supset : \phi x . \supset_x . \psi x \quad (1)$

$\vdash . (1) . *5\cdot6 . \supset \vdash . \text{Prop}$

$*11$. THEORY OF TWO APPARENT VARIABLES

Summary of $*11$.

In this number, the propositions proved for one variable in $*10$ are to be extended to two variables, with the addition of a few propositions having no analogues for one variable, such as $*11\cdot2\cdot21\cdot23\cdot24$ and $*11\cdot53\cdot55\cdot6\cdot7$. "$\phi(x, y)$" stands for a proposition containing x and containing y; when x and y are unassigned, $\phi(x, y)$ is a propositional function of x and y. The definition $*11\cdot01$ shows that "the truth of all values of $\phi(x, y)$" does not need to be taken as a new primitive idea, but is definable in terms of "the truth of all values of ψx." The reason is that, when x is assigned, $\phi(x, y)$ becomes a function of one variable, namely y, whence it follows that, for every possible value of x, "$(y) . \phi(x, y)$" embodies merely the primitive idea introduced in $*9$. But "$(y) . \phi(x, y)$" is again only a function of one variable, namely x, since y has here become an apparent variable. Hence the definition $*11\cdot01$ below is legitimate. We put:

$*11\cdot01$. $(x, y) . \phi(x, y) . = : (x) : (y) . \phi(x, y)$ Df

$*11\cdot02$. $(x, y, z) . \phi(x, y, z) . = : (x) : (y, z) . \phi(x, y, z)$ Df

$*11\cdot03$. $(\exists x, y) . \phi(x, y) . = : (\exists x) : (\exists y) . \phi(x, y)$ Df

$*11\cdot04$. $(\exists x, y, z) . \phi(x, y, z) . = : (\exists x) : (\exists y, z) . \phi(x, y, z)$ Df

$*11\cdot05$. $\phi(x, y) . \supset_{x,y} . \psi(x, y) : = : (x, y) : \phi(x, y) . \supset . \psi(x, y)$ Df

$*11\cdot06$. $\phi(x, y) . \equiv_{x,y} . \psi(x, y) : = : (x, y) : \phi(x, y) . \equiv . \psi(x, y)$ Df

All the above definitions are supposed extended to any number of variables that may occur.

The propositions of this section can all be extended to any finite number of variables; as the analogy is exact, it is not necessary to carry the process beyond two variables in our proofs.

In addition to the definition $*11\cdot01$, we need the primitive proposition that "whatever possible argument x may be, $\phi(x, y)$ is true whatever possible argument y may be" implies the corresponding statement with x and y interchanged except in "$\phi(x, y)$". Either may be taken as the meaning of "$\phi(x, y)$ is true whatever possible arguments x and y may be."

The propositions of the present number are somewhat less used than those of $*10$, but some of them are used frequently. Such are the following:

$*11\cdot1$. $\vdash : (x, y) . \phi(x, y) . \supset . \phi(z, w)$

$*11\cdot11$. If $\phi(z, w)$ is true whatever possible arguments z and w may be, then $(x, y) . \phi(x, y)$ is true

These two propositions are the analogues of $*10\cdot1\cdot11$.

***11·2.** $\vdash : (x, y) . \phi(x, y) . \equiv . (y, x) . \phi(x, y)$

I.e. to say that "for all possible values of x, $\phi(x, y)$ is true for all possible values of y" is equivalent to saying "for all possible values of y, $\phi(x, y)$ is true for all possible values of x."

***11·3.** $\vdash :. p . \supset . (x, y) . \phi(x, y) : \equiv : (x, y) : p . \supset . \phi(x, y)$

This is the analogue of *10·21.

***11·32.** $\vdash :. (x, y) : \phi(x, y) . \supset . \psi(x, y) : \supset : (x, y) . \phi(x, y) . \supset . (x, y) . \psi(x, y)$

I.e. "if $\phi(x, y)$ always implies $\psi(x, y)$, then '$\phi(x, y)$ always' implies '$\psi(x, y)$ always.'" This is the analogue of *10·27. *11·33·34·341 are respectively the analogues of *10·271·28·281, and are also much used.

***11·35.** $\vdash :. (x, y) : \phi(x, y) . \supset . p : \equiv : (\exists x, y) . \phi(x, y) . \supset . p$

I.e. if $\phi(x, y)$ always implies p, then if $\phi(x, y)$ is ever true, p is true, and vice versa. This is the analogue of *10·23.

***11·45.** $\vdash :. (\exists x, y) : p . \phi(x, y) : \equiv : p : (\exists x, y) . \phi(x, y)$

This is the analogue of *10·35.

***11·54.** $\vdash :. (\exists x, y) . \phi x . \psi y . \equiv : (\exists x) . \phi x : (\exists y) . \psi y$

This proposition is useful because it analyses a proposition containing two apparent variables into two propositions which each contain only one. "$\phi x . \psi y$" is a function of two variables, but is compounded of two functions of one variable each. Such a function is like a conic which is two straight lines: it may be called an "analysable" function.

***11·55.** $\vdash :. (\exists x, y) . \phi x . \psi(x, y) . \equiv : (\exists x) : \phi x : (\exists y) . \psi(x, y)$

I.e. to say "there are values of x and y for which $\phi x . \psi(x, y)$ is true" is equivalent to saying "there is a value of x for which ϕx is true and for which there is a value of y such that $\psi(x, y)$ is true."

***11·6.** $\vdash :: (\exists x) :. (\exists y) . \phi(x, y) . \psi y : \chi x :. \equiv :. (\exists y) :. (\exists x) . \phi(x, y) . \chi x : \psi y$

This gives a transformation which is useful in many proofs.

***11·62.** $\vdash :: \phi x . \psi(x, y) . \supset_{x,y} . \chi(x, y) : \equiv :. \phi x . \supset_x : \psi(x, y) . \supset_y . \chi(x, y)$

This transformation also is often useful.

***11·01.** $(x, y) . \phi(x, y) . = : (x) : (y) . \phi(x, y)$ Df

***11·02.** $(x, y, z) . \phi(x, y, z) . = : (x) : (y, z) . \phi(x, y, z)$ Df

***11·03.** $(\exists x, y) . \phi(x, y) . = : (\exists x) : (\exists y) . \phi(x, y)$ Df

***11·04.** $(\exists x, y, z) . \phi(x, y, z) . = : (\exists x) : (\exists y, z) . \phi(x, y, z)$ Df

***11·05.** $\phi(x, y) . \supset_{x,y} . \psi(x, y) : = : (x, y) : \phi(x, y) . \supset . \psi(x, y)$ Df

***11·06.** $\phi(x, y) . \equiv_{x,y} . \psi(x, y) : = : (x, y) : \phi(x, y) . \equiv . \psi(x, y)$ Df

with similar definitions for any number of variables.

***11·07.** "Whatever possible argument x may be, $\phi(x, y)$ is true whatever possible argument y may be" implies the corresponding statement with x and y interchanged except in "$\phi(x, y)$". Pp.

∗11·1. $\vdash : (x, y) . \phi(x, y) . \supset . \phi(z, w)$

Dem.

$$\begin{aligned} &\vdash . ∗10·1 . \supset \vdash : \text{Hp} . \supset . (y) . \phi(z, y) . \\ &[∗10·1] \qquad \supset . \phi(z, w) : \supset \vdash . \text{Prop} \end{aligned}$$

∗11·11. If $\phi(z, w)$ is true whatever possible arguments z and w may be, then $(x, y) . \phi(x, y)$ is true.

Dem.

By ∗10·11, the hypothesis implies that $(y) . \phi(z, y)$ is true whatever possible argument z may be; and this, by ∗10·11, implies $(x, y) . \phi(x, y)$.

∗11·12. $\vdash :. (x, y) . p \lor \phi(x, y) . \supset : p . \lor . (x, y) . \phi(x, y)$

Dem.

$$\begin{aligned} &\vdash . ∗10·12 . \supset \vdash :. (y) . p \lor \phi(x, y) . \quad \supset : p . \lor . (y) . \phi(x, y) :. \\ &[∗10·11·27] \supset \vdash :. (x, y) . p \lor \phi(x, y) . \supset : (x) : p . \lor . (y) . \phi(x, y) : \\ &[∗10·12] \qquad \supset : p . \lor . (x, y) . \phi(x, y) :. \supset \vdash . \text{Prop} \end{aligned}$$

This proposition is only used for proving ∗11·2.

∗11·13. If $\phi(\hat{x}, \hat{y})$, $\psi(\hat{x}, \hat{y})$ take their first and second arguments respectively of the same type, and we have "$\vdash . \phi(x, y)$" and "$\vdash . \psi(x, y)$," we shall have "$\vdash . \phi(x, y) . \psi(x, y)$." [Proof as in ∗10·13]

∗11·14. $\vdash :. (x, y) . \phi(x, y) : (x, y) . \psi(x, y) : \supset : \phi(z, w) . \psi(z, w)$

Dem.

$$\begin{aligned} &\vdash . ∗10·14 . \supset \vdash :. \text{Hp} . \supset : (y) . \phi(z, y) : (y) . \psi(z, y) \\ &[∗10·14] \qquad \supset : \phi(z, w) . \psi(z, w) :. \supset \vdash . \text{Prop} \end{aligned}$$

This proposition, like ∗10·14, is not always significant when its hypothesis is true. ∗11·13, on the contrary, is always significant when its hypothesis is true. For this reason, ∗11·13 may always be safely used in *inference*, whereas ∗11·14 can only be used in *inference* (*i.e.* for the actual assertion of the conclusion when the hypothesis is asserted) if it is known that the conclusion is significant.

∗11·2. $\vdash : (x, y) . \phi(x, y) . \equiv . (y, x) . \phi(x, y)$

Dem.

$$\begin{aligned} &\vdash . ∗11·1 . \qquad \supset \vdash : (x, y) . \phi(x, y) . \supset . \phi(z, w) \qquad (1) \\ &\vdash . (1) . ∗11·07·11 . \supset \vdash :. (w, z) : (x, y) . \phi(x, y) . \supset . \phi(z, w) \qquad (2) \\ &\vdash . (2) . ∗11·12 \frac{\sim \{(x, y) . \phi(x, y)\}}{p} . \supset \\ &\qquad \vdash :. (x, y) . \phi(x, y) . \supset . (w, z) . \phi(z, w) \qquad (3) \\ &\text{Similarly} \quad \vdash :. (w, z) . \phi(z, w) . \supset . (x, y) . \phi(x, y) \qquad (4) \\ &\vdash . (3) . (4) . \quad \supset \vdash . \text{Prop} \end{aligned}$$

Note that "$(w, z) . \phi(z, w)$" is the same proposition as "$(y, x) . \phi(x, y)$"; a proposition is not a function of any apparent variable which occurs in it.

*11·21. $\vdash : (x, y, z) . \phi(x, y, z) . \equiv . (y, z, x) . \phi(x, y, z)$

Dem.

[(*11·01·02)] $\vdash :: (x, y, z) . \phi(x, y, z) . \equiv :. (x) :. (y) : (z) . \phi(x, y, z) :.$

[*11·2] $\equiv :. (y) :. (x) : (z) . \phi(x, y, z) :.$

[*11·2.*10·271] $\equiv :. (y) :. (z) : (x) . \phi(x, y, z) :.$

[(*11·01·02)] $\equiv :. (y, z, x) . \phi(x, y, z) :: \supset \vdash .$ Prop

*11·22. $\vdash : (\exists x, y) . \phi(x, y) . \equiv . \sim \{(x, y) . \sim \phi(x, y)\}$

Dem.

$\vdash .$ *10·252 . Transp . (*11·03) . $\supset$

$\vdash : (\exists x, y) . \phi(x, y) . \equiv . \sim \{(x) : \sim (\exists y) . \phi(x, y)\} .$

[*10·252·271] $\equiv . \sim \{(x) : (y) . \sim \phi(x, y)\} .$

[(*11·01)] $\equiv . \sim \{(x, y) . \sim \phi(x, y)\} : \supset \vdash .$ Prop

*11·23. $\vdash : (\exists x, y) . \phi(x, y) . \equiv . (\exists y, x) . \phi(x, y)$

Dem.

$\vdash .$ *11·22 . $\supset \vdash : (\exists x, y) . \phi(x, y) . \equiv . \sim \{(x, y) . \sim \phi(x, y)\} .$

[*11·2.Transp] $\equiv . \sim \{(y, x) . \sim \phi(x, y)\} .$

[*11·22] $\equiv . (\exists y, x) . \phi(x, y) : \supset \vdash .$ Prop

*11·24. $\vdash : (\exists x, y, z) . \phi(x, y, z) . \equiv . (\exists y, z, x) . \phi(x, y, z)$

Dem.

[(*11·03·04)] $\vdash :: (\exists x, y, z) . \phi(x, y, z) . \equiv :. (\exists x) :. (\exists y) : (\exists z) . \phi(x, y, z) :.$

[*11·23] $\equiv :. (\exists y) :. (\exists x) : (\exists z) . \phi(x, y, z) :.$

[*11·23.*10·281] $\equiv :. (\exists y) :. (\exists z) : (\exists x) . \phi(x, y, z) :.$

[(*11·03·04)] $\equiv :. (\exists y, z, x) . \phi(x, y, z) :: \supset \vdash .$ Prop

*11·25. $\vdash : \sim \{(\exists x, y) . \phi(x, y)\} . \equiv . (x, y) . \sim \phi(x, y)$ [*11·22 . Transp]

*11·26. $\vdash :. (\exists x) : (y) . \phi(x, y) : \supset : (y) : (\exists x) . \phi(x, y)$

Dem.

$\vdash .$ *10·1·28 . $\supset \vdash :. (\exists x) : (y) . \phi(x, y) : \supset : (\exists x) . \phi(x, y)$ (1)

$\vdash .$ (1) . *10·11·21 . $\supset \vdash .$ Prop

Note that the converse of this proposition is false. *E.g.* let $\phi(x, y)$ be the propositional function "if y is a proper fraction, then x is a proper fraction greater than y." Then for all values of y we have $(\exists x) . \phi(x, y)$, so that $(y) : (\exists x) . \phi(x, y)$ is satisfied. In fact "$(y) : (\exists x) . \phi(x, y)$" expresses the proposition: "If y is a proper fraction, then there is always a proper fraction greater than y." But "$(\exists x) : (y) . \phi(x, y)$" expresses the proposition: "There is a proper fraction which is greater than any proper fraction," which is false.

*11·27. $\vdash :. (\exists x, y) : (\exists z) . \phi(x, y, z) : \equiv : (\exists x) : (\exists y, z) . \phi(x, y, z) :$

$\equiv : (\exists x, y, z) . \phi(x, y, z)$

Dem.

$\vdash . *4{\cdot}2 . (*11{\cdot}03) . \supset$

$\vdash :: (\exists x, y) : (\exists z) . \phi(x, y, z) : \equiv :. (\exists x) :. (\exists y) : (\exists z) . \phi(x, y, z)$ (1)

$\vdash . *4{\cdot}2 . (*11{\cdot}03) . \supset$

$\vdash :. (\exists y) : (\exists z) . \phi(x, y, z) : \equiv : (\exists y, z) . \phi(x, y, z)$ (2)

$\vdash . (2) . *10{\cdot}11{\cdot}281 . \supset$

$\vdash :: (\exists x) :. (\exists y) : (\exists z) . \phi(x, y, z) :. \equiv :. (\exists x) : (\exists y, z) . \phi(x, y, z)$ (3)

$\vdash . (1) . (3) . (*11{\cdot}04) . \supset \vdash . \text{Prop}$

All the propositions of $*10$ have analogues which hold for two or more variables. The more important of these are proved in what follows.

$*11{\cdot}3$. $\vdash :. p . \supset . (x, y) . \phi(x, y) : \equiv : (x, y) : p . \supset . \phi(x, y)$

Dem.

$\vdash . *10{\cdot}21 . \supset \vdash :. p . \supset . (x, y) . \phi(x, y) : \equiv : (x) : p . \supset . (y) . \phi(x, y) :$

$[*10{\cdot}21{\cdot}271]$ $\equiv : (x, y) : p . \supset . \phi(x, y) :. \supset \vdash . \text{Prop}$

$*11{\cdot}31$. $\vdash :. (x, y) . \phi(x, y) : (x, y) . \psi(x, y) : \equiv : (x, y) : \phi(x, y) . \psi(x, y)$

Here the conditions of significance on the right-hand side require that ϕ and ψ should take arguments of the same types.

Dem.

$\vdash . *10{\cdot}22 . \supset \vdash :: (x, y) . \phi(x, y) : (x, y) . \psi(x, y) :$

$\equiv :. (x) :. (y) . \phi(x, y) : (y) . \psi(x, y) :.$

$[*10{\cdot}22{\cdot}271]$ $\equiv :. (x, y) : \phi(x, y) . \psi(x, y) :: \supset \vdash . \text{Prop}$

The proofs of most of the following propositions are conducted exactly as those of $*11{\cdot}3{\cdot}31$ are conducted: the analogous proposition in $*10$ is used twice, together with $*10{\cdot}27$ or $*10{\cdot}271$ or $*10{\cdot}28$ or $*10{\cdot}281$ as the case may be. When proofs conform to this pattern we shall merely give references to the propositions used.

$*11{\cdot}311$. If $\phi(\hat{x}, \hat{y})$, $\psi(\hat{x}, \hat{y})$ take arguments of the same type, and we have "$\vdash . \phi(x, y)$" and "$\vdash . \psi(x, y)$," we shall have "$\vdash . \phi(x, y) . \psi(x, y)$." [Proof as in $*10{\cdot}13$.]

$*11{\cdot}32$. $\vdash :. (x, y) : \phi(x, y) . \supset . \psi(x, y) : \supset : (x, y) . \phi(x, y) . \supset . (x, y) . \psi(x, y)$ $[*10{\cdot}27]$

$*11{\cdot}33$. $\vdash :. (x, y) : \phi(x, y) . \equiv . \psi(x, y) : \supset : (x, y) . \phi(x, y) . \equiv . (x, y) . \psi(x, y)$ $[*10{\cdot}271]$

$*11{\cdot}34$. $\vdash :. (x, y) : \phi(x, y) . \supset . \psi(x, y) : \supset :$

$(\exists x, y) . \phi(x, y) . \supset . (\exists x, y) . \psi(x, y)$ $[*10{\cdot}27{\cdot}28]$

$*11{\cdot}341$. $\vdash :. (x, y) : \phi(x, y) . \equiv . \psi(x, y) : \supset :$

$(\exists x, y) . \phi(x, y) . \equiv . (\exists x, y) . \psi(x, y)$ $[*10{\cdot}271{\cdot}281]$

$*11{\cdot}35$. $\vdash :. (x, y) : \phi(x, y) . \supset . p : \equiv : (\exists x, y) . \phi(x, y) . \supset . p$ $[*10{\cdot}23{\cdot}271]$

$*11{\cdot}36$. $\vdash : \phi(z, w) . \supset . (\exists x, y) . \phi(x, y)$

Dem.

$\vdash . *11{\cdot}1 . \supset \vdash : (x, y) . \sim\phi(x, y) . \supset . \sim\phi(z, w)$ (1)

$\vdash . (1) . \text{Transp} . \supset \vdash . \text{Prop}$

***11·37.** $\vdash :: (x, y) : \phi(x, y) . \supset . \psi(x, y) :. (x, y) : \psi(x, y) . \supset . \chi(x, y) :.$
$\supset : (x, y) : \phi(x, y) . \supset . \chi(x, y)$

Dem.

In the following demonstration, "Hp" means the hypothesis of the proposition to be proved. We shall employ this abbreviation, whenever convenient, in all cases where the proposition to be proved is a hypothetical, *i.e.* is of the form "$p \supset q$." Similarly "Hp (1)" will mean "the hypothesis of (1)," and so on.

$\vdash . *11\cdot31 . \supset \vdash :: \text{Hp} . \supset :. (x, y) :. \phi(x, y) . \supset . \psi(x, y) : \psi(x, y) . \supset . \chi(x, y)$ (1)

$\vdash . \text{Syll} . *11\cdot11 . \supset \vdash :. (x, y) :. \phi(x, y) . \supset . \psi(x, y) : \psi(x, y) . \supset . \chi(x, y) :$
$\supset : \phi(x, y) . \supset . \chi(x, y) :.$

[*11·32] $\supset \vdash :. (x, y) : \phi(x, y) . \supset . \psi(x, y) : \psi(x, y) . \supset . \chi(x, y) :$
$\supset : (x, y) : \phi(x, y) . \supset . \chi(x, y)$ (2)

$\vdash . (1) . (2) . \text{Syll} . \supset \vdash . \text{Prop}$

The above is a type of proof which recurs frequently in what follows. Proofs conforming to this pattern will be indicated only by the numbers of the propositions used.

***11·371.** $\vdash :: (x, y) : \phi(x, y) . \equiv . \psi(x, y) :. (x, y) : \psi(x, y) . \equiv . \chi(x, y) :.$
$\supset :. (x, y) : \phi(x, y) . \equiv . \chi(x, y)$ [*11·31·11·33]

***11·38.** $\vdash :: (x, y) : \phi(x, y) . \supset . \psi(x, y) :. \supset :.$
$(x, y) : \phi(x, y) . \chi(x, y) . \supset . \psi(x, y) . \chi(x, y)$ [Fact . *11·11·32]

***11·39.** $\vdash :: (x, y) : \phi(x, y) . \supset . \psi(x, y) :. (x, y) : \chi(x, y) . \supset . \theta(x, y) :. \supset :.$
$(x, y) : \phi(x, y) . \chi(x, y) . \supset . \psi(x, y) . \theta(x, y)$ [*3·47 . *11·11·32]

***11·391.** $\vdash :: (x, y) : \phi(x, y) . \supset . \psi(x, y) :. (x, y) : \phi(x, y) . \supset . \chi(x, y) :.$
$\equiv : (x, y) : \phi(x, y) . \supset . \psi(x, y) . \chi(x, y)$

Dem.

$\vdash . *4\cdot76 . \quad \supset \vdash :. \phi(x, y) . \supset . \psi(x, y) : \phi(x, y) . \supset . \chi(x, y) :$
$\equiv : \phi(x, y) . \supset . \psi(x, y) . \chi(x, y) :.$

[*11·11·33] $\supset \vdash :. (x, y) : \phi(x, y) . \supset . \psi(x, y) : \phi(x, y) . \supset . \chi(x, y) :$
$\equiv : (x, y) : \phi(x, y) . \supset . \psi(x, y) . \chi(x, y) ::$

[*11·31] $\supset \vdash :: (x, y) : \phi(x, y) . \supset . \psi(x, y) :. (x, y) : \phi(x, y) . \supset . \chi(x, y) :.$
$\equiv : (x, y) : \phi(x, y) . \supset . \psi(x, y) . \chi(x, y) ::$
$\supset \vdash . \text{Prop}$

***11·4.** $\vdash :: (x, y) : \phi(x, y) . \equiv . \psi(x, y) :. (x, y) : \chi(x, y) . \equiv . \theta(x, y) :. \supset :.$
$(x, y) : \phi(x, y) . \chi(x, y) . \equiv . \psi(x, y) . \theta(x, y)$

Dem.

$\vdash . *11\cdot31 . \supset \vdash :: \text{Hp} . \supset :. (x, y) :. \phi(x, y) . \equiv . \psi(x, y) : \chi(x, y) . \equiv . \theta(x, y) :.$

[*4·38 . *11·11·32] $\supset :. (x, y) : \phi(x, y) . \chi(x, y) . \equiv . \psi(x, y) . \theta(x, y) ::$
$\supset \vdash . \text{Prop}$

∗11·401. $\vdash :: (x, y) : \phi(x, y) . \equiv . \psi(x, y) : \supset :.$
$(x, y) : \phi(x, y) . \chi(x, y) . \equiv . \psi(x, y) . \chi(x, y)$ $\left[∗11·4 \frac{\chi}{\theta} . \mathrm{Id}\right]$

∗11·41. $\vdash :. (\exists x, y) . \phi(x, y) : \mathbf{v} : (\exists x, y) . \psi(x, y) :$
$\equiv : (\exists x, y) : \phi(x, y) . \mathbf{v} . \psi(x, y)$ [∗10·42·281]

∗11·42. $\vdash :. (\exists x, y) . \phi(x, y) . \psi(x, y) . \supset : (\exists x, y) . \phi(x, y) : (\exists x, y) . \psi(x, y)$
[∗10·5]

∗11·421. $\vdash :. (x, y) . \phi(x, y) . \mathbf{v} . (x, y) . \psi(x, y) : \supset : (x, y) : \phi(x, y) . \mathbf{v} . \psi(x, y)$
$\left[∗11·42 \frac{\sim\phi, \sim\psi}{\phi, \psi} . \text{Transp} . ∗4·56\right]$

∗11·43. $\vdash :. (\exists x, y) : \phi(x, y) . \supset . p : \equiv : (x, y) . \phi(x, y) . \supset . p$ [∗10·34·281]

∗11·44. $\vdash :. (x, y) : \phi(x, y) . \mathbf{v} . p : \equiv : (x, y) . \phi(x, y) . \mathbf{v} . p$ [∗10·2·271]

∗11·45. $\vdash :. (\exists x, y) : p . \phi(x, y) : \equiv : p : (\exists x, y) . \phi(x, y)$ [∗10·35·281]

∗11·46. $\vdash :. (\exists x, y) : p . \supset . \phi(x, y) : \equiv : p . \supset . (\exists x, y) . \phi(x, y)$ [∗10·37·281]

∗11·47. $\vdash :. (x, y) : p . \phi(x, y) : \equiv : p : (x, y) . \phi(x, y)$ [∗10·33·271]

∗11·5. $\vdash :. (\exists x) : \sim\{(y) . \phi(x, y)\} : \equiv : \sim\{(x, y) . \phi(x, y)\} : \equiv : (\exists x, y) . \sim\phi(x, y)$

Dem.

$\vdash . ∗10·253 . \supset \vdash :. (\exists x) : \sim\{(y) . \phi(x, y)\} : \equiv : \sim\{(x) : (y) . \phi(x, y)\} :$
$[(∗11·01)]$ $\equiv : \sim\{(x, y) . \phi(x, y)\}$ (1)
$\vdash . ∗10·253 . \supset \vdash : \sim\{(y) . \phi(x, y)\} .$ $\equiv . (\exists y) . \sim\phi(x, y) :$
$[∗10·11·281] \supset \vdash :. (\exists x) : \sim\{(y) . \phi(x, y)\} : \equiv : (\exists x) : (\exists y) . \sim\phi(x, y) :$
$[(∗11·03)]$ $\equiv : (\exists x, y) . \sim\phi(x, y)$ (2)
$\vdash . (1) . (2) . \supset \vdash . \text{Prop}$

∗11·51. $\vdash :. (\exists x) : (y) . \phi(x, y) : \equiv : \sim\{(x) : (\exists y) . \sim\phi(x, y)\}$

Dem.

$\vdash . ∗10·252 . \text{Transp} . \supset \vdash :. (\exists x) : (y) . \phi(x, y) : \equiv : \sim[(x) : \sim(y) . \phi(x, y)]$ (1)
$\vdash . ∗10·253 . \supset \vdash :. \sim(y) . \phi(x, y) .$ $\equiv : (\exists y) . \sim\phi(x, y) :.$
$[∗10·11·271] \supset \vdash :. (x) : \sim(y) . \phi(x, y) :$ $\equiv : (x) : (\exists y) . \sim\phi(x, y) :.$
$[\text{Transp}] \quad \supset \vdash :. \sim[(x) : \sim\{(y) . \phi(x, y)\}] .$ $\equiv : \sim\{(x) : (\exists y) . \sim\phi(x, y)\}$ (2)
$\vdash . (1) . (2) . \supset \vdash . \text{Prop}$

∗11·52. $\vdash :. (\exists x, y) . \phi(x, y) . \psi(x, y) . \equiv . \sim\{(x, y) : \phi(x, y) . \supset . \sim\psi(x, y)\}$

Dem.

$\vdash . ∗4·51·62 . \supset$
$\vdash :. \sim\{\phi(x, y) . \psi(x, y)\} .$ $\equiv : \phi(x, y) . \supset . \sim\psi(x, y)$ (1)
$\vdash . (1) . ∗11·11·33 . \supset$
$\vdash :. (x, y) . \sim\{\phi(x, y) . \psi(x, y)\} : \equiv : (x, y) : \phi(x, y) . \supset . \sim\psi(x, y)$ (2)
$\vdash : (2) . \text{Transp} . ∗11·22 . \supset \vdash . \text{Prop}$

∗11·521. $\vdash :. \sim(\exists x, y) . \phi(x, y) . \sim\psi(x, y) . \equiv : (x, y) : \phi(x, y) . \supset . \psi(x, y)$
$\left[∗11·52 . \text{Transp} . \frac{\sim\psi(x, y)}{\psi(x, y)}\right]$

***11·53.** $\vdash :. (x, y) . \phi x \supset \psi y . \equiv : (\exists x) . \phi x . \supset . (y) . \psi y$

Dem.

$\vdash . *10·21·271 . \supset \vdash :. (x, y) . \phi x \supset \psi y . \equiv : (x) : \phi x . \supset . (y) . \psi y :$

$[*10·23] \qquad \equiv : (\exists x) . \phi x . \supset . (y) . \psi y :. \supset \vdash . \text{Prop}$

***11·54.** $\vdash :. (\exists x, y) . \phi x . \psi y . \equiv : (\exists x) . \phi x : (\exists y) . \psi y$

Dem.

$\vdash . *10·35 . \supset \vdash :. (\exists y) . \phi x . \psi y . \quad \equiv : \phi x : (\exists y) . \psi y :.$

$[*10·11·281] \supset \vdash :. (\exists x, y) . \phi x . \psi y . \equiv : (\exists x) : \phi x : (\exists y) . \psi y :$

$[*10·35] \qquad \equiv : (\exists x) . \phi x : (\exists y) . \psi y :. \supset \vdash . \text{Prop}$

This proposition is very often used.

***11·55.** $\vdash :. (\exists x, y) . \phi x . \psi (x, y) . \equiv : (\exists x) : \phi x : (\exists y) . \psi (x, y)$

Dem.

$\vdash . *10·35 . \supset \vdash :. (\exists y) . \phi x . \psi (x, y) . \qquad \equiv : \phi x : (\exists y) . \psi (x, y) :.$

$[*10·11] \quad \supset \vdash :. (x) :. (\exists y) . \phi x . \psi (x, y) . \equiv : \phi x : (\exists y) . \psi (x, y) :.$

$[*10·281] \quad \supset \vdash :. (\exists x) : (\exists y) . \phi x . \psi (x, y) . \equiv : (\exists x) : \phi x : (\exists y) . \psi (x, y) :. \supset \vdash . \text{Prop}$

This proposition is very often used.

***11·56.** $\vdash :. (x) . \phi x : (y) . \psi y : \equiv : (x, y) . \phi x . \psi y$

Dem.

$\vdash . *10·33 . \supset \vdash :: (x) . \phi x : (y) . \psi y : \equiv :. (x) :. \phi x : (y) . \psi y \qquad (1)$

$\vdash . *10·33 . \supset \vdash :. \quad \phi x : (y) . \psi y : \equiv : (y) . \phi x . \psi y :.$

$[*10·11] \quad \supset \vdash :. (x) :. \phi x : (y) . \psi y : \equiv : (y) . \phi x . \psi y :.$

$[*10·271] \quad \supset \vdash :: (x) :. \phi x : (y) . \psi y :. \equiv : (x) : (y) . \phi x . \psi y :$

$[(*11·01)] \qquad \equiv : (x, y) . \phi x . \psi y \qquad (2)$

$\vdash . (1) . (2) . \supset \vdash . \text{Prop}$

***11·57.** $\vdash : (x) . \phi x . \equiv . (x, y) . \phi x . \phi y \quad [*11·56 . *4·24]$

The use of *4·24 here depends upon the fact that $(x) . \phi x$ and $(y) . \phi y$ are the same proposition.

***11·58.** $\vdash : (\exists x) . \phi x . \equiv . (\exists x, y) . \phi x . \phi y \quad [*11·54 . *4·24]$

***11·59.** $\vdash :. \phi x . \supset_x . \psi x : \equiv : \phi x . \phi y . \supset_{x, y} . \psi x . \psi y$

Dem.

$\vdash . *11·57 . \supset \vdash :. \phi x . \supset_x . \psi x : \equiv : (x, y) : \phi x . \supset . \psi x : \phi y . \supset . \psi y :$

$[*3·47 . *11·32] \qquad \supset : (x, y) : \phi x . \phi y . \supset . \psi x . \psi y \qquad (1)$

$\vdash . *11·1 . \supset \vdash :. (x, y) : \phi x . \phi y . \supset . \psi x . \psi y : \supset : \phi x . \phi y . \supset . \psi x . \psi y \quad (2)$

$\vdash . (2) \frac{x}{y} . *4·24 . \supset \vdash :. \text{Hp} (2) . \supset : \phi x . \supset . \psi x \qquad (3)$

$\vdash . (3) . *10·11·21 . \supset$

$\vdash :. (x, y) : \phi x . \phi y . \supset . \psi x . \psi y : \supset : \phi x . \supset_x . \psi x \qquad (4)$

$\vdash . (1) . (4) . \supset \vdash . \text{Prop}$

***11·6.** $\vdash :: (\exists x) :. (\exists y) . \phi(x,y) . \psi y : \chi x :. \equiv :. (\exists y) :. (\exists x) . \phi(x,y) . \chi x : \psi y$

This proposition is very frequently employed in subsequent proofs.

Dem.

$\vdash .$ *10·35 . $\supset \vdash :. (\exists y) . \phi(x,y) . \psi y : \chi x : \equiv : (\exists y) : \phi(x,y) . \psi y . \chi x :.$

[*10·11·281] $\supset \vdash :: (\exists x) :. (\exists y) . \phi(x,y) . \psi y : \chi x :$

$\equiv :. (\exists x) :. (\exists y) . \phi(x,y) . \psi y . \chi x :.$

[*11·23] $\equiv :. (\exists y) :. (\exists x) . \phi(x,y) . \psi y . \chi x :.$

[*11·341.Perm] $\equiv :. (\exists y) :. (\exists x) . \phi(x,y) . \chi x . \psi y :.$

[*10·35·281] $\equiv :. (\exists y) :. (\exists x) . \phi(x,y) . \chi x : \psi y :: \supset \vdash .$ Prop

***11·61.** $\vdash :. (\exists y) : \phi x . \supset_x . \psi(x,y) : \supset : \phi x . \supset_x . (\exists y) . \psi(x,y)$

Dem.

$\vdash .$ *11·26 . $\supset \vdash ::$ Hp . $\supset :. (x) :. (\exists y) : \phi x . \supset . \psi(x,y)$ (1)

$\vdash .$ *10·37 . $\supset \vdash :. (\exists y) : \phi x . \supset . \psi(x,y) : \supset : \phi x . \supset . (\exists y) . \psi(x,y) :.$

[*10·11·27] $\supset \vdash :: (x) :. (\exists y) : \phi x . \supset . \psi(x,y) :. \supset :. (x) : \phi x . \supset . (\exists y) . \psi(x,y)$ (2)

$\vdash . (1) . (2) . \supset \vdash .$ Prop

***11·62.** $\vdash :: \phi x . \psi(x,y) . \supset_{x,y} . \chi(x,y) : \equiv :. \phi x . \supset_x : \psi(x,y) . \supset_y . \chi(x,y)$

Dem.

$\vdash .$ *4·87 . *11·11·33 . $\supset$

$\vdash :: \phi x . \psi(x,y) . \supset_{x,y} . \chi(x,y) : \equiv :. (x,y) :. \phi x . \supset : \psi(x,y) . \supset . \chi(x,y)$

[*10·21·11·271] $\equiv :. (x) :. \phi x . \supset : (y) : \psi(x,y) . \supset . \chi(x,y) ::$

$\supset \vdash .$ Prop

***11·63.** $\vdash :. \sim(\exists x,y) . \phi(x,y) . \supset : \phi(x,y) . \supset_{x,y} . \psi(x,y)$

Dem.

$\vdash .$ *2·21 . *11·11 . $\supset \vdash :. (x,y) :. \sim\phi(x,y) . \supset : \phi(x,y) . \supset . \psi(x,y) :.$

[*11·32] $\supset \vdash :. (x,y) . \sim\phi(x,y) . \supset : (x,y) : \phi(x,y) . \supset . \psi(x,y) :.$

[*11·25] $\supset \vdash :. \sim(\exists x,y) . \phi(x,y) . \supset : (x,y) : \phi(x,y) . \supset . \psi(x,y) :.$

$\supset \vdash .$ Prop

***11·7.** $\vdash :. (\exists x,y) : \phi(x,y) . \vee . \phi(y,x) : \equiv . (\exists x,y) . \phi(x,y)$

Dem.

$\vdash .$ *11·41 . $\supset \vdash :. (\exists x,y) : \phi(x,y) . \vee . \phi(y,x) :$

$\equiv : (\exists x,y) . \phi(x,y) . \vee . (\exists x,y) . \phi(y,x) :$

[*11·23] $\equiv : (\exists x,y) . \phi(x,y) . \vee . (\exists y,x) . \phi(y,x) :$

[*4·25] $\equiv : (\exists x,y) . \phi(x,y) :. \supset \vdash .$ Prop

In the last line of the above proof, use is made of the fact that

$$(\exists x,y) . \phi(x,y) \text{ and } (\exists y,x) . \phi(y,x)$$

are the same proposition.

The first use of the following proposition occurs in the proof of *234·12. Its utility lies in its enabling us to pass from a hypothesis

$$\phi z . \chi w . \supset_{z,w} . \psi z . \theta w,$$

containing two apparent variables, to the product of two hypotheses each containing only one.

***11·71.** $\vdash :: (\exists z) . \phi z : (\exists w) . \chi w : \supset :.$

$$\phi z . \supset_z . \psi z : \chi w . \supset_w . \theta w : \equiv : \phi z . \chi w . \supset_{z,w} . \psi z . \theta w$$

Dem.

$\vdash . *10·1 . *3·47 . \supset \vdash :. \phi z . \supset_z . \psi z : \chi w . \supset_w . \theta w :$

$\supset : \phi z . \chi w . \supset . \psi z . \theta w$ (1)

$\vdash . (1) . *11·11·3 . \supset \vdash :. \phi z . \supset_z . \psi z : \chi w . \supset_w . \theta w :$

$\supset : \phi z . \chi w . \supset_{z,w} . \psi z . \theta w$ (2)

$\vdash . *10·1 . \supset \vdash :: \phi z . \chi w . \supset_{z,w} . \psi z . \theta w : \supset :. \phi z . \chi w . \supset_w . \psi z . \theta w :.$

[*10·28] $\supset :. (\exists w) . \phi z . \chi w . \supset . (\exists w) . \psi z . \theta w :.$

[*10·35] $\supset :. \phi z : (\exists w) . \chi w : \supset : \psi z : (\exists w) . \theta w$ (3)

$\vdash . (3) . \text{Comm} . *3·26 . \supset \vdash :: (\exists w) . \chi w : \supset :. \phi z . \chi w . \supset_{z,w} . \psi z . \theta w :$

$\supset : \phi z . \supset . \psi z$ (4)

$\vdash . (4) . *10·11·21 . \supset \vdash :: (\exists w) . \chi w . \supset :. \phi z . \chi w . \supset_{z,w} . \psi z . \theta w :$

$\supset : \phi z . \supset_z . \psi z$ (5)

Similarly $\vdash :: (\exists z) . \phi z . \supset :. \phi z . \chi w . \supset_{z,w} . \psi z . \theta w :$

$\supset : \chi w . \supset_w . \theta w$ (6)

$\vdash . (5) . (6) . *3·47 . \text{Comp} . \supset$

$\vdash :: \text{Hp} . \supset :. \phi z . \chi w . \supset_{z,w} . \psi z . \theta w : \supset : \phi z . \supset_z . \psi z : \chi w . \supset_w . \theta w$ (7)

$\vdash . (2) . (7) . \supset \vdash . \text{Prop}$

∗12. THE HIERARCHY OF TYPES AND THE AXIOM OF REDUCIBILITY

The primitive idea "$(x) \,.\, \phi x$" has been explained to mean "ϕx is always true," *i.e.* "all values of ϕx are true." But whatever function ϕ may be, there will be arguments x with which ϕx is meaningless, *i.e.* with which as arguments ϕ does not have any value. The arguments with which ϕx has values form what we will call the "range of significance" of ϕx. A "*type*" is defined as the range of significance of some function. In virtue of ∗9·14, if ϕx, ϕy, and ψx are significant, *i.e.* either true or false, so is ψy. From this it follows that two types which have a common member coincide, and that two different types are mutually exclusive. Any proposition of the form $(x) \,.\, \phi x$, *i.e.* any proposition containing an apparent variable, determines some type as the range of the apparent variable, the type being fixed by the function ϕ.

The division of objects into types is necessitated by the vicious-circle fallacies which otherwise arise*. These fallacies show that there must be no totalities which, if legitimate, would contain members defined in terms of themselves. Hence any expression containing an apparent variable must not be in the range of that variable, *i.e.* must belong to a different type. Thus the apparent variables contained or presupposed in an expression are what determines its type. This is the guiding principle in what follows.

As explained in ∗9, propositions containing variables are generated from propositional functions which do not contain these apparent variables, by the process of asserting all or some values of such functions. Suppose ϕa is a proposition containing a; we will give the name of *generalization* to the process which turns ϕa into $(x) \,.\, \phi x$ or $(\exists x) \,.\, \phi x$, and we will give the name of *generalized propositions* to all such as contain apparent variables. It is plain that propositions containing apparent variables presuppose others not containing apparent variables, from which they can be derived by generalization. Propositions which contain no apparent variables we call *elementary propositions*†, and the terms of such propositions, other than functions, we call *individuals*. Then individuals form the first type.

It is unnecessary, in practice, to know what objects belong to the lowest type, or even whether the lowest type of variable occurring in a given context is that of individuals or some other. For in practice only the *relative* types of variables are relevant; thus the lowest type occurring in a given context may be called that of individuals, so far as that context is concerned. Accordingly the above account of individuals is not essential to the truth of what

* Cf. Introduction, Chapter II.

† Cf. pp. 91, 92.

follows; all that is essential is the way in which other types are generated from individuals, however the type of individuals may be constituted.

By applying the process of generalization to individuals occurring in elementary propositions, we obtain new propositions. The legitimacy of this process requires only that no individuals should be propositions. That this is so, is to be secured by the meaning we give to the word *individual*. We may explain an individual as something which exists on its own account; it is then obviously not a proposition, since propositions, as explained in Chapter II of the Introduction (p. 43), are incomplete symbols, having no meaning except in use. Hence in applying the process of generalization to individuals we run no risk of incurring reflexive fallacies. We will give the name of *first-order propositions* to such as contain one or more apparent variables whose possible values are individuals, but contain no other apparent variables. First-order propositions are not all of the same type, since, as was explained in ∗9, two propositions which do not contain the same number of apparent variables cannot be of the same type. But owing to the systematic ambiguity of negation and disjunction, their differences of type may usually be ignored in practice. No reflexive fallacies will result, since no first-order proposition involves any totality except that of individuals.

Let us denote by "$\phi ! \hat{x}$" or "$\phi ! (\hat{x}, \hat{y})$" or etc. an elementary function whose argument or arguments are individual. We will call such a function a *predicative function of an individual*. Such functions, together with those derived from them by generalization, will be called *first-order functions*. In practice we may without risk of reflexive fallacies treat first-order functions as a type, since the only totality they involve is that of individuals, and, by means of the systematic ambiguity of negation and disjunction, any function of a first-order function which will concern us will be significant whatever first-order function is taken as argument, provided the right meanings are given to the negations and disjunctions involved.

For the sake of clearness, we will repeat in somewhat different terms our account of what is meant by a first-order function. Let us give the name of *matrix* to any function, of however many variables, which does not involve any apparent variables. Then any possible function other than a matrix is derived from a matrix by means of generalization, *i.e.* by considering the proposition which asserts that the function in question is true with all possible values or with some value of one of the arguments, the other argument or arguments remaining undetermined. Thus *e.g.* from the function $\phi(x, y)$ we shall be able to derive the four functions

$$(x) . \phi(x, y), \quad (\exists x) . \phi(x, y), \quad (y) . \phi(x, y), \quad (\exists y) . \phi(x, y),$$

of which the two first are functions of y, while the two last are functions of x. (All *propositions*, with the exception of such as are values of matrices, are also derived from matrices by the above process of generalization. In order to obtain a proposition from a matrix containing n variables, without assigning values to any of the variables, it is necessary to turn all the variables into apparent variables. Thus if $\phi(x, y)$ is a matrix, $(x, y) . \phi(x, y)$ is a proposition.) We will give the name *first-order matrices* to such as have only individuals for their arguments, and we will give the name of *first-order functions* (of any number of variables) to such as either are first-order matrices or are derived from first-order matrices by generalization applied to some (not all) of the arguments to such matrices. First-order *propositions* will be such as result from applying generalization to *all* the arguments to a first-order matrix.

As we have already stated, the notation "$\phi ! \hat{z}$" is used for any elementary function of one variable. Thus "$\phi ! x$" represents any value of any elementary function of one variable. It will be seen that "$\phi ! x$" is a function of two variables, namely $\phi ! \hat{z}$ and x. Since it contains no apparent variable, it is a matrix, but since it contains a variable (namely $\phi ! \hat{z}$) which is not an individual, it is not a first-order matrix. The same applies to $\phi ! a$, where a is some definite constant. We can build up a number of new matrices, such as

$$\sim \phi ! a, \quad \sim \phi ! x, \quad \phi ! x \vee \phi ! y, \quad \phi ! x \vee \psi ! x, \quad \phi ! x \vee \psi ! y,$$
$$\phi ! x . \supset . \psi ! x, \quad \phi ! x . \psi ! x, \quad \phi ! x \vee \psi ! y \vee \chi ! z, \text{ and so on.}$$

All these are matrices which involve first-order functions among their arguments. Such matrices we will call *second-order matrices*. From these matrices, by applying generalization to their arguments, whether to such as are functions or to such (if any) as are individuals, we obtain new functions and propositions. Such functions (together with second-order matrices) will be called *second-order functions*, and such propositions will be called *second-order propositions*. Thus we are led to the following definitions:

A *second-order matrix* is one which has at least one first-order matrix among its arguments, but has no arguments other than first-order matrices and individuals.

A *second-order function* is one which either is a second-order matrix or results from one by applying generalization to some (not all) of the arguments to a second-order matrix.

A *second-order proposition* is one which results from a second-order matrix by applying generalization to all its arguments.

In addition to the above illustrations of second-order matrices, we may give the following examples of second-order functions:

(1) Functions in which the argument is $\phi ! \hat{z}$: $(x) . \phi ! x$, $(\exists x) . \phi ! x$, $\phi ! a . \supset . \phi ! b$, where a and b are constants, $\phi ! x . \supset_x . g ! x$, where $g ! \hat{z}$ is a constant function, and so on.

(2) Functions in which the arguments are $\phi ! \hat{z}$ and $\psi ! \hat{z}$:

$$\phi ! x . \supset_x . \psi ! x, \quad \phi ! x . \equiv_x . \psi ! x, \quad (\exists x) . \phi x . \psi x, \quad \phi ! a . \supset . \psi ! b,$$

where a and b are constants, and so on.

(3) Functions in which the argument is an individual x: $(\phi) . \phi ! x$, $(\exists \phi) . \phi ! x$, $\phi ! x . \supset_\phi . \phi ! a$, where a is constant, and so on.

(4) Functions in which the arguments are $\phi ! \hat{z}$ and x: $\phi ! x$, $\phi ! x . \supset . \phi ! a$, where a is constant, $(\exists \psi) : \phi ! x . \equiv . \psi ! x$, and so on.

Examples of second-order functions might, of course, be multiplied indefinitely, but the above seem sufficient for purposes of illustration.

A second-order matrix of one variable will be called a *predicative second-order function of one variable* or a *predicative function of a first-order matrix*. Thus $\phi ! a$, $\sim \phi ! a$ and $\phi ! a \supset \phi ! b$ are predicative functions of $\phi ! \hat{z}$. Similarly a function of several variables of which at least one is a first-order matrix, while the rest are either individuals or first-order matrices, will be called *predicative* if it is a matrix.

It will be seen, however, that a second-order function may have only individuals for its arguments; instances were given just now under the heading (3). Such functions we shall not call predicative, since predicative functions of individuals have already been defined as being such as are of the first order. Thus the order of a function is not determined by the order of its argument or arguments; indeed, the function may be of any order superior to the order or orders of its arguments.

A variable matrix whose argument is $\phi ! \hat{z}$ will be denoted by $f ! \phi ! \hat{z}$, and generally, a matrix whose arguments are $\phi ! \hat{z}, \psi ! \hat{z}, \ldots x, y, \ldots$ (where there is at least one function among the arguments) will be denoted by

$$f ! (\phi ! \hat{z}, \psi ! \hat{z}, \ldots x, y, \ldots).$$

Such a matrix is not of the first or second order, since it contains the new variable f, whose values are second-order matrices. We proceed to construct new matrices as we did with the matrix $\phi ! \hat{x}$; these constitute *third-order matrices*. These together with the functions derived from them by generalization are called *third-order functions*, and the propositions derived from third-order matrices by generalization are called *third-order propositions*.

In this way we can proceed indefinitely to matrices, functions and propositions of higher and higher orders. We introduce the following definition:

A function is said to be *predicative* when it is a matrix. It will be observed that, in a hierarchy in which all the variables are individuals or matrices, a matrix is the same thing as an elementary function (cf. pp. 127, 128).

"Matrix" or "predicative function" is a primitive idea.

The fact that a function is predicative is indicated, as above, by a note of exclamation after the functional letter.

The variables occurring in the present work, from this point onwards, will all be either individuals or matrices of some order in the above hierarchy. Propositions, which have occurred hitherto as variables, will no longer do so except in a few isolated cases of which no subsequent use is made. In practice, for the reasons explained on p. 162, a function of a matrix may be regarded as capable of any argument which is a function of the same order and takes arguments of the same type.

In practice, we never need to know the absolute types of our variables, but only their *relative* types. That is to say, if we prove any proposition on the assumption that one of our variables is an individual, and another is a function of order n, the proof will still hold if, in place of an individual, we take a function of order m, and in place of our function of order n we take a function of order $n + m$, with corresponding changes for any other variables that may be involved. This results from the assumption that our primitive propositions are to apply to variables of any order.

We shall use small Latin letters (other than p, q, r, s) for variables of the lowest type concerned in any context. For functions, we shall use the letters ϕ, ψ, χ, θ, f, g, F (except that, at a later stage, F will be defined as a constant relation, and θ will be defined as the order-type of the continuum).

We shall explain later a different hierarchy, that of classes and relations, which is derived from the functional hierarchy explained above, but is more convenient in practice.

When any predicative function, say $\phi ! \hat{z}$, occurs as apparent variable, it would be strictly more correct to indicate the fact by placing "$(\phi ! \hat{z})$" before what follows, as thus: "$(\phi ! \hat{z}) . f(\phi ! \hat{z})$." But for the sake of brevity we write simply "(ϕ)" instead of "$(\phi ! \hat{z})$." Since what follows the ϕ in brackets must always contain ϕ with arguments supplied, no confusion can result from this practice.

It should be observed that, in virtue of the manner in which our hierarchy of functions was generated, non-predicative functions always result from such as are predicative by means of generalization. Hence it is unnecessary to introduce a special notation for non-predicative functions of a given order and taking arguments of a given order. For example, second-order functions of an individual x are always derived by generalization from a matrix

$$f ! (\phi ! \hat{z}, \psi ! \hat{z}, \ldots x, y, z, \ldots),$$

where the functions f, ϕ, ψ, ... are predicative. It is possible, therefore, without loss of generality, to use no apparent variables except such as are predicative.

We require, however, a means of symbolizing a function whose order is not assigned. We shall use "ϕx" or "$f(\chi ! \hat{z})$" or etc. to express a function (ϕ or f) whose order, relatively to its argument, is not given. Such a function cannot be made into an apparent variable, unless we suppose its order previously fixed. As the only purpose of the notation is to avoid the necessity of fixing the order, such a function will not be used as an apparent variable; the only functions which will be so used will be predicative functions, because, as we have just seen, this restriction involves no loss of generality.

We have now to state and explain the *axiom of reducibility*.

It is important to observe that, since there are various types of propositions and functions, and since generalization can only be applied within some one type (or, by means of systematic ambiguity, within some well-defined and completed set of types), all phrases referring to "all propositions" or "all functions," or to "some (undetermined) proposition" or "some (undetermined) function," are *prima facie* meaningless, though in certain cases they are capable of an unobjectionable interpretation. Contradictions arise from the use of such phrases in cases where no innocent meaning can be found.

If mathematics is to be possible, it is absolutely necessary (as explained in the Introduction, Chapter II) that we should have some method of making statements which will usually be equivalent to what we have in mind when we (inaccurately) speak of "all properties of x." (A "property of x" may be defined as a propositional function satisfied by x.) Hence we must find, if possible, some method of reducing the order of a propositional function without affecting the truth or falsehood of its values. This seems to be what common-sense effects by the admission of *classes*. Given any propositional function ψx, of whatever order, this is assumed to be equivalent, for all values of x, to a statement of the form "x belongs to the class α." Now assuming that there is such an entity as the class α, this statement is of the first order, since it involves no allusion to a variable function. Indeed its only practical advantage over the original statement ψx is that it is of the first order. There is no advantage in assuming that there really are such things as classes, and the contradiction about the classes which are not members of themselves shows that, if there are classes, they must be something radically different from individuals. It would seem that the sole purpose which classes serve, and one main reason which makes them linguistically convenient, is that they provide a method of reducing the order of a propositional function. We shall, therefore, not assume anything of what may seem to be involved in the common-sense admission of classes, except this, that every propositional function is equivalent, for all its values, to some predicative function of the same argument or arguments.

This assumption with regard to functions is to be made whatever may be the type of their arguments. Let fu be a function, of any order, of an argument u, which may itself be either an individual or a function of any order. If f is a matrix, we write the function in the form $f!u$; in such a case we call f a *predicative* function. Thus a predicative function of an individual is a first-order function; and for higher types of arguments, predicative functions take the place that first-order functions take in respect of individuals. We assume, then, that every function of one variable is equivalent, for all its values, to some predicative function of the same argument. This assumption seems to be the essence of the usual assumption of classes; at any rate, it retains as much of classes as we have any use for, and little enough to avoid the contradictions which a less grudging admission of classes is apt to entail. We will call this assumption the *axiom of classes*, or the *axiom of reducibility*.

We shall assume similarly that every function of two variables is equivalent, for all its values, to a predicative function of those variables, *i.e.* to a matrix. This assumption is what seems to be meant by saying that any statement about two variables defines a relation between them. We will call this assumption the *axiom of relations* or (like the previous axiom) the *axiom of reducibility*.

In dealing with relations between more than two terms, similar assumptions would be needed for three, four, ... variables. But these assumptions are not indispensable for our purpose, and are therefore not made in this work.

Stated in symbols, the two forms of the axiom of reducibility are as follows:

***12·1.** $\vdash : (\exists f) : \phi x \,.\equiv_x .\, f!x$ Pp

***12·11.** $\vdash : (\exists f) : \phi(x, y) \,.\equiv_{x,y} .\, f!(x, y)$ Pp

We call two functions $\phi\hat{x}$, $\psi\hat{x}$ *formally equivalent* when $\phi x \,.\equiv_x .\, \psi x$, and similarly we call $\phi(\hat{x}, \hat{y})$ and $\psi(\hat{x}, \hat{y})$ formally equivalent when

$$\phi(x, y) \,.\equiv_{x,y} .\, \psi(x, y).$$

Thus the above axioms state that any function of one or two variables is formally equivalent to some *predicative* function of one or two variables, as the case may be.

Of the above two axioms, the first is chiefly needed in the theory of classes (*20), and the second in the theory of relations (*21). But the first is also essential to the theory of identity, if identity is to be defined (as we have done, in *13·01); its use in the theory of identity is embodied in the proof of *13·101, below.

We may sum up what has been said in the present number as follows:

(1) A function of the first order is one which involves no variables except individuals, whether as apparent variables or as arguments.

(2) A function of the $(n+1)$th order is one which has at least one argument or apparent variable of order n, and contains no argument or apparent variable which is not either an individual or a first-order function or a second-order function or ... or a function of order n.

(3) A predicative function is one which contains no apparent variables, *i.e.* is a matrix. It is possible, without loss of generality, to use no variables except matrices and individuals, so long as variable *propositions* are not required.

(4) Any function of one argument or of two is formally equivalent to a predicative function of the same argument or arguments.

$*13$. IDENTITY

Summary of $*13$.

The propositional function "x is identical with y" will be written "$x = y$." We shall find that this use of the sign of equality covers all the common uses of equality that occur in mathematics. The definition is as follows:

$*13{\cdot}01$. $x = y \,.=:\, (\phi) : \phi ! x \,.\supset.\, \phi ! y$ Df

This definition states that x and y are to be called identical when every predicative function satisfied by x is also satisfied by y. We cannot state that *every* function satisfied by x is to be satisfied by y, because x satisfies functions of various orders, and these cannot all be covered by one apparent variable. But in virtue of the axiom of reducibility it follows that, if $x = y$ and x satisfies ψx, where ψ is any function, predicative or non-predicative, then y also satisfies ψy (cf. $*13{\cdot}101$, below). Hence in effect the definition is as powerful as it would be if it could be extended to cover *all* functions of x.

Note that the second sign of equality in the above definition is combined with "Df," and thus is not really the same symbol as the sign of equality which is defined. Thus the definition is not circular, although at first sight it appears so.

The propositions of the present number are constantly referred to. Most of them are self-evident, and the proofs offer no difficulty. The most important of the propositions of this number are the following:

$*13{\cdot}101$. $\vdash : x = y \,.\supset.\, \psi x \supset \psi y$

I.e. if x and y are identical, any property of x is a property of y.

$*13{\cdot}12$. $\vdash : x = y \,.\supset.\, \psi x \equiv \psi y$

This includes $*13{\cdot}101$ together with the fact that if x and y are identical any property of y is a property of x.

$*13{\cdot}15{\cdot}16{\cdot}17$, which state that identity is reflexive, symmetrical and transitive.

$*13{\cdot}191$. $\vdash :. y = x \,.\supset_y.\, \phi y : \equiv . \phi x$

I.e. to state that everything that is identical with x has a certain property is equivalent to stating that x has that property.

$*13{\cdot}195$. $\vdash : (\exists y) . y = x . \phi y . \equiv . \phi x$

I.e. to state that something identical with x has a certain property is equivalent to saying that x has that property.

$*13{\cdot}22$. $\vdash : (\exists z, w) . z = x . w = y . \phi (z, w) . \equiv . \phi (x, y)$

This is the analogue of $*13{\cdot}195$ for two variables.

***13·01.** $x = y . = : (\phi) : \phi ! x . \supset . \phi ! y$ Df

The following definitions embody abbreviations which are often convenient.

***13·02.** $x \neq y . = . \sim(x = y)$ Df

***13·03.** $x = y = z . = . x = y . y = z$ Df

***13·1.** $\vdash :. x = y . \equiv : \phi ! x . \supset_{\phi} . \phi ! y$ [*4·2 . (*13·01) . (*10·02)]

***13·101.** $\vdash : x = y . \supset . \psi x \supset \psi y$

Dem.

$\vdash . *12 \cdot 1 . \supset \vdash :. (\exists \phi) :. \psi x . \equiv . \phi ! x : \psi y . \equiv . \phi ! y$ (1)

$\vdash . *13 \cdot 1 . \supset \vdash :: \mathrm{Hp} . \supset :. \phi ! x . \supset_{\phi} . \phi ! y :.$

[*4·84·85.*10·27] $\supset :. \psi x . \equiv . \phi ! x : \psi y . \equiv . \phi ! y : \supset_{\phi} : \psi x . \supset . \psi y :.$

[*10·23] $\supset :. (\exists \phi) : \psi x . \equiv . \phi ! x : \psi y . \equiv . \phi ! y : \supset : \psi x . \supset . \psi y$ (2)

$\vdash . (1) . (2) . \supset \vdash .$ Prop

In virtue of this proposition, if $x = y$, y satisfies any function, whether predicative or non-predicative, which is satisfied by x. It will be observed that the proof uses the axiom of reducibility (*12·1). But for this axiom, two terms x and y might agree in respect of all *predicative* functions, but not in respect of all non-predicative functions. We should thus be led to identities of different degrees, according to the degree of the functions in respect of which x and y agreed. Strict identity would, in this case, have to be taken as a primitive idea, and *13·101 would have to be a primitive proposition, as would also *13·15·16·17.

***13·11.** $\vdash :. x = y . \equiv : \phi ! x . \equiv_{\phi} . \phi ! y$

Dem.

$\vdash . *10 \cdot 22 . \quad \supset \vdash :. \phi ! x . \equiv_{\phi} . \phi ! y : \supset : \phi ! x . \supset_{\phi} . \phi ! y :$

[*13·1] $\supset : x = y$ (1)

$\vdash . *13 \cdot 101 . \quad \supset \vdash :. x = y . \supset . \phi ! x \supset \phi ! y$ (2)

$\vdash . *13 \cdot 101 . *1 \cdot 7 . \quad \supset \vdash :. x = y . \supset . \sim \phi ! x \supset \sim \phi ! y .$

[Transp] $\supset . \phi ! y \supset \phi ! x$ (3)

$\vdash . (2) . (3) . \mathrm{Comp} . \supset \vdash : x = y . \supset . \phi ! x \equiv \phi ! y :$

[*10·11·21] $\supset \vdash :. x = y . \supset : \phi ! x . \equiv_{\phi} . \phi ! y$ (4)

$\vdash . (1) . (4) . \quad \supset \vdash .$ Prop

***13·12.** $\vdash : x = y . \supset . \psi x \equiv \psi y$

Dem.

$\vdash . *13 \cdot 101 . \mathrm{Comp} . \supset \vdash : x = y . \supset . \psi x \supset \psi y . \sim \psi x \supset \sim \psi y .$

[Transp] $\supset . \psi x \equiv \psi y : \supset \vdash .$ Prop

***13·13.** $\vdash : \psi x . x = y . \supset . \psi y$ [*13·101 . Comm . Imp]

***13·14.** $\vdash : \psi x . \sim \psi y . \supset . x \neq y$ [*13·13 . *4·14]

***13·15.** $\vdash . x = x$ [Id . *10·11 . *13·1]

***13·16.** $\vdash : x = y . \equiv . y = x$ [*13·11 . *10·32]

***13·17.** $\vdash : x = y \,.\, y = z \,.\, \supset \,.\, x = z$

Dem.

$\vdash \,.\,$ *13·1 $\,.\, \supset \vdash :: \mathrm{Hp} \,.\, \supset :. \phi ! x \,.\, \supset_\phi \,.\, \phi ! y : \phi ! y \,.\, \supset_\phi \,.\, \phi ! z :.$

[*10·3] $\supset :. \phi ! x \,.\, \supset_\phi \,.\, \phi ! z :: \supset \vdash \,.\,$ Prop

In the above use of *10·3, $\phi ! x$, $\phi ! y$, $\phi ! z$ are regarded as three different functions of ϕ, and ϕ replaces the x of *10·3.

The above three propositions show that identity is reflexive (*13·15), symmetrical (*13·16), and transitive (*13·17). These are the three marks of relations having the formal properties which we associate commonly with the sign of equality.

***13·171.** $\vdash : x = y \,.\, x = z \,.\, \supset \,.\, y = z$ [*13·16·17]

***13·172.** $\vdash : y = x \,.\, z = x \,.\, \supset \,.\, y = z$ [*13·16·17]

***13·18.** $\vdash : x = y \,.\, x \neq z \,.\, \supset \,.\, y \neq z$ [*13·17 . *4·14]

***13·181.** $\vdash : x = y \,.\, y \neq z \,.\, \supset \,.\, x \neq z$ [*13·171 . *4·14]

***13·182.** $\vdash :. x = y \,.\, \supset : z = x \,.\, \equiv \,.\, z = y$ [*13·17·172 . Exp . Comp]

***13·183.** $\vdash :. x = y \,.\, \equiv : z = x \,.\, \equiv_z \,.\, z = y$

Dem.

$\vdash \,.\,$ *13·182 . *10·11·21 $\,.\, \supset \vdash :. x = y \,.\, \supset : z = x \,.\, \equiv_z \,.\, z = y$ (1)

$\vdash \,.\,$ *10·1 $\,.\, \supset \vdash :. z = x \,.\, \equiv_z \,.\, z = y : \supset : x = x \,.\, \supset \,.\, x = y :$

[*13·15] $\supset : x = y$ (2)

$\vdash \,.\, (1) \,.\, (2) \,.\, \supset \vdash \,.\,$ Prop

***13·19.** $\vdash \,.\, (\exists y) \,.\, y = x$ [*13·15 . *10·24]

***13·191.** $\vdash :. y = x \,.\, \supset_y \,.\, \phi y : \equiv \,.\, \phi x$

Dem.

$\vdash \,.\,$ *10·1 $\,.\, \supset \vdash :. y = x \,.\, \supset_y \,.\, \phi y : \supset : x = x \,.\, \supset \,.\, \phi x :$

[*13·15] $\supset : \phi x$ (1)

$\vdash \,.\,$ *13·12 $\,.\, \supset \vdash :. y = x \,.\, \supset : \phi x \,.\, \supset \,.\, \phi y :.$

[Comm] $\supset \vdash :. \phi x \,.\, \supset : y = x \,.\, \supset \,.\, \phi y :.$

[*10·11·21] $\supset \vdash :. \phi x \,.\, \supset : y = x \,.\, \supset_y \,.\, \phi y$ (2)

$\vdash \,.\, (1) \,.\, (2) \,.\, \supset \vdash \,.\,$ Prop

This proposition is constantly used in subsequent proofs.

***13·192.** $\vdash :. (\exists c) : x = b \,.\, \equiv_x \,.\, x = c : \psi c : \equiv \,.\, \psi b$

Dem.

$\vdash \,.\,$ *4·2 . *3·2 $\,.\, \supset \vdash :: \psi b \,.\, \supset :. x = b \,.\, \equiv_x \,.\, x = b : \psi b :.$

[*10·24] $\supset :. (\exists c) : x = b \,.\, \equiv_x \,.\, x = c : \psi c$ (1)

$\vdash \,.\,$ *10·1 $\,.\, \supset \vdash :. x = b \,.\, \equiv_x \,.\, x = c : \psi c : \supset : b = b \,.\, \equiv \,.\, b = c : \psi c :$

[*5·501.*13·15] $\supset : b = c \,.\, \psi c :$

[*13·13] $\supset : \psi b$ (2)

$\vdash \,.\, (2) \,.\,$ *10·11·23 $\,.\, \supset \vdash :. (\exists c) : x = b \,.\, \equiv_x \,.\, x = c : \psi c : \supset \,.\, \psi b$ (3)

$\vdash \,.\, (1) \,.\, (3) \,.\, \supset \vdash \,.\,$ Prop

This proposition is useful in the theory of descriptions (*14).

***13·193.** $\vdash : \phi x . x = y . \equiv . \phi y . x = y$

Dem.

$\vdash$. Simp . $\supset \vdash : \phi x . x = y . \supset . x = y$ (1)

$\vdash$. *13·13 . $\supset \vdash : \phi x . x = y . \supset . \phi y$ (2)

$\vdash$. (1) . (2) . Comp . $\supset \vdash : \phi x . x = y . \supset . \phi y . x = y$ (3)

$\vdash$. *13·16 . Fact . $\supset \vdash : \phi y . x = y . \supset . \phi y . y = x .$

$\left[(3)\dfrac{y, x}{x, y}\right]$ $\supset . \phi x . y = x .$

[*13·16.Fact] $\supset . \phi x . x = y$ (4)

$\vdash$. (3) . (4) . $\supset \vdash$. Prop

This proposition is very often used.

***13·194.** $\vdash : \phi x . x = y . \equiv . \phi x . \phi y . x = y$ [*13·13 . *4·71]

This proposition is used in *37·65 and *101·14.

***13·195.** $\vdash : (\exists y) . y = x . \phi y . \equiv . \phi x$

Dem.

$\vdash$. *3·2 . *13·15 . $\supset \vdash : \phi x . \supset . x = x . \phi x .$

[*10·24] $\supset . (\exists y) . y = x . \phi y$ (1)

$\vdash$. *13·13 . *10·11 . $\supset \vdash :. (y) : y = x . \phi y . \supset . \phi x :$

[*10·23] $\supset \vdash :. (\exists y) . y = x . \phi y . \supset . \phi x$ (2)

$\vdash$. (1) . (2) . $\supset \vdash$. Prop

The use of this proposition in subsequent proofs is very frequent.

***13·196.** $\vdash :. \sim\phi x . \equiv : \phi y . \supset_y . y \neq x$ [*13·195 . Transp . *10·51]

***13·21.** $\vdash :. z = x . w = y . \supset_{z, w} . \phi(z, w) : \equiv . \phi(x, y)$

Dem.

$\vdash$. *11·62 . $\supset$

$\vdash :: z = x . w = y . \supset_{z, w} . \phi(z, w) : \equiv :. z = x . \supset_z : w = y . \supset_w . \phi(z, w) :.$

[*13·191] $\equiv :. w = y . \supset_w . \phi(x, w) :.$

[*13·191] $\equiv :. \phi(x, y) :: \supset \vdash$. Prop

This proposition is the analogue, for two variables, of *13·191.

***13·22.** $\vdash : (\exists z, w) . z = x . w = y . \phi(z, w) . \equiv . \phi(x, y)$

Dem.

$\vdash$. *11·55 . $\supset \vdash :. (\exists z, w) . z = x . w = y . \phi(z, w) .$

$\equiv : (\exists z) : z = x : (\exists w) . w = y . \phi(z, w) :$

[*13·195] $\equiv : (\exists w) . w = y . \phi(x, w) :$

[*13·195] $\equiv : \phi(x, y) :. \supset \vdash$. Prop

This proposition is the analogue, for two variables, of *13·195. It is frequently used, especially in the theory of couples (*54, *55, *56).

The following proposition is useful in the theory of types. Its purpose is to show that, if a is any argument for which "ϕa" is significant, *i.e.* for which we have $\phi a \lor \sim\phi a$, then "ϕx" is significant when, and only when, x is either identical with a or not identical with a. It follows (as will be proved in *20·81) that, if "ϕa" and "ψa" are both significant, the class of values of x for which "ϕx" is significant is the same as the class of those for which "ψx" is significant, *i.e.* two types which have a common member are identical.

In the following proof, the chief point to observe is the use of *10·221. There are two variables, a and x, to be identified. In the first use, we depend upon the fact that ϕa and $x = a$ both occur in both (4) and (5): the occurrence of ϕa in both justifies the identification of the two a's, and when these have been identified, the occurrence of $x = a$ in both justifies the identification of the two x's. (Unless the a's had been already identified, this would not be legitimate, because "$x = a$" is typically ambiguous if neither x nor a is of given type.) The second use of *10·221 is justified by the fact that both ϕa and ϕx occur in both (2) and (6).

***13·3.** $\vdash :: \phi a \vee \sim \phi a . \supset :. \phi x \vee \sim \phi x . \equiv : x = a . \vee . x \neq a$

Dem.

$\vdash . *2 \cdot 11 . \quad \supset \vdash . \phi x \vee \sim \phi x$ (1)

$\vdash . (1) . \text{Simp} . \quad \supset \vdash : \phi a \vee \sim \phi a . \supset . \phi x \vee \sim \phi x$ (2)

$\vdash . *2 \cdot 11 . \quad \supset \vdash : x = a . \vee . x \neq a$ (3)

$\vdash . (3) . \text{Simp} . \quad \supset \vdash :. \phi a \vee \sim \phi a . \supset : x = a . \vee . x \neq a$ (4)

$\vdash . *13 \cdot 101 . \text{Comm} . \supset \vdash :. \phi a \vee \sim \phi a . \supset : x = a . \supset . \phi x \vee \sim \phi x$ (5)

$\vdash . (4) . (5) . *10 \cdot 13 \cdot 221 . \supset$

$\vdash :: \phi a \vee \sim \phi a . \supset : x = a . \vee . x \neq a :. \phi a \vee \sim \phi a . \supset : x = a . \supset . \phi x \vee \sim \phi x$ (6)

$\vdash . (2) . (6) . *10 \cdot 13 \cdot 221 . \supset$

$\vdash :: \phi a \vee \sim \phi a . \supset . \phi x \vee \sim \phi x :. \phi a \vee \sim \phi a . \supset : x = a . \vee . x \neq a :.$
$\phi a \vee \sim \phi a . \supset : x = a . \supset . \phi x \vee \sim \phi x$ (7)

$\vdash . (7) . \text{Simp} . \supset$

$\vdash :: \phi a \vee \sim \phi a . \supset . \phi x \vee \sim \phi x :. \phi a \vee \sim \phi a . \supset : x = a . \vee . x \neq a$ (8)

$\vdash . (8) . *5 \cdot 35 . \quad \supset \vdash :: \phi a \vee \sim \phi a . \supset :. \phi x \vee \sim \phi x . \equiv : x = a . \vee . x \neq a ::$
$\supset \vdash . \text{Prop}$

∗14. DESCRIPTIONS

Summary of ∗14.

A *description* is a phrase of the form "the term which etc.," or, more explicitly, "the term x which satisfies $\phi\hat{x}$," where $\phi\hat{x}$ is some function satisfied by one and only one argument. For reasons explained in the Introduction (Chapter III), we do not define "the x which satisfies $\phi\hat{x}$," but we define any proposition in which this phrase occurs. Thus when we say: "The term x which satisfies $\phi\hat{x}$ satisfies $\psi\hat{x}$," we shall mean: "There is a term b such that ϕx is true when, and only when, x is b, and ψb is true." That is, writing "$(\iota x)(\phi x)$" for "the term x which satisfies ϕx," $\psi(\iota x)(\phi x)$ is to mean

$$(\exists b) : \phi x \, . \equiv_x . \, x = b : \psi b.$$

This, however, is not yet quite adequate as a definition, for when $(\iota x)(\phi x)$ occurs in a proposition which is part of a larger proposition, there is doubt whether the smaller or the larger proposition is to be taken as the "$\psi(\iota x)(\phi x)$." Take, for example, $\psi(\iota x)(\phi x) \, . \supset . \, p$. This may be either

$$(\exists b) : \phi x \, . \equiv_x . \, x = b : \psi b : \supset . \, p$$

or

$$(\exists b) :. \phi x \, . \equiv_x . \, x = b : \psi b \, . \supset . \, p.$$

If "$(\exists b) : \phi x \, . \equiv_x . \, x = b$" is false, the first of these must be true, while the second must be false. Thus it is very necessary to distinguish them.

The proposition which is to be treated as the "$\psi(\iota x)(\phi x)$" will be called the *scope* of $(\iota x)(\phi x)$. Thus in the first of the above two propositions, the scope of $(\iota x)(\phi x)$ is $\psi(\iota x)(\phi x)$, while in the second it is $\psi(\iota x)(\phi x) \, . \supset . \, p$. In order to avoid ambiguities as to scope, we shall indicate the scope by writing "$[(\iota x)(\phi x)]$" at the beginning of the scope, followed by enough dots to extend to the end of the scope. Thus of the above two propositions the first is

$$[(\iota x)(\phi x)] \, . \, \psi(\iota x)(\phi x) \, . \supset . \, p,$$

while the second is

$$[(\iota x)(\phi x)] : \psi(\iota x)(\phi x) \, . \supset . \, p.$$

Thus we arrive at the following definition:

∗14·01. $[(\iota x)(\phi x)] \, . \, \psi(\iota x)(\phi x) \, . = : (\exists b) : \phi x \, . \equiv_x . \, x = b : \psi b$ Df

It will be found in practice that the scope usually required is the smallest proposition enclosed in dots or brackets in which "$(\iota x)(\phi x)$" occurs. Hence when this scope is to be given to $(\iota x)(\phi x)$, we shall usually omit explicit mention of the scope. Thus *e.g.* we shall have

$$a \neq (\iota x)(\phi x) \, . = : \quad (\exists b) : \phi x \, . \equiv_x . \, x = b : a \neq b,$$

$$\sim\{a = (\iota x)(\phi x)\} \, . = . \sim\{(\exists b) : \phi x \, . \equiv_x . \, x = b : a = b\}.$$

Of these the first necessarily implies $(\exists b) : \phi x . \equiv_x . x = b$, while the second does not. We put

***14·02.** $E ! (\imath x) (\phi x) . = : (\exists b) : \phi x . \equiv_x . x = b$ Df

This defines: "The x satisfying $\phi \hat{x}$ exists," which holds when, and only when, $\phi \hat{x}$ is satisfied by one value of x and by no other value.

When two or more descriptions occur in the same proposition, there is need of avoiding ambiguity as to which has the larger scope. For this purpose, we put

***14·03.** $[(\imath x)(\phi x), (\imath x)(\psi x)] . f\{(\imath x)(\phi x), (\imath x)(\psi x)\} . = :$
$$[(\imath x)(\phi x)] : [(\imath x)(\psi x)] . f\{(\imath x)(\phi x), (\imath x)(\psi x)\} \quad \text{Df}$$

It will be shown (*14·113) that the truth-value of a proposition containing two descriptions is unaffected by the question which has the larger scope. Hence we shall in general adopt the convention that the description occurring first typographically is to have the larger scope, unless the contrary is expressly indicated. Thus *e.g.*

$$(\imath x)(\phi x) = (\imath x)(\psi x)$$

will mean $\quad (\exists b) : \phi x . \equiv_x . x = b : b = (\imath x)(\psi x)$,

i.e. $\quad (\exists b) :. \phi x . \equiv_x . x = b :. (\exists c) : \psi x . \equiv_x . x = c : b = c$.

By this convention we are able almost always to avoid explicit indication of the order of elimination of two or more descriptions. If, however, we require a larger scope for the later description, we put

***14·04.** $[(\imath x)(\psi x)] . f\{(\imath x)(\phi x), (\imath x)(\psi x)\} . = .$
$$[(\imath x)(\psi x), (\imath x)(\phi x)] . f\{(\imath x)(\phi x), (\imath x)(\psi x)\} \quad \text{Df}$$

Whenever we have $E ! (\imath x)(\phi x)$, $(\imath x)(\phi x)$ behaves, formally, like an ordinary argument to any function in which it may occur. This fact is embodied in the following proposition:

***14·18.** $\vdash :. E ! (\imath x)(\phi x) . \supset : (x) . \psi x . \supset . \psi (\imath x)(\phi x)$

That is to say, when $(\imath x)(\phi x)$ exists, it has any property which belongs to everything. This does not hold when $(\imath x)(\phi x)$ does not exist; for example, the present King of France does not have the property of being either bald or not bald.

If $(\imath x)(\phi x)$ has any property whatever, it must exist. This fact is stated in the proposition:

***14·21.** $\vdash : \psi (\imath x)(\phi x) . \supset . E ! (\imath x)(\phi x)$

This proposition is obvious, since "$E ! (\imath x)(\phi x)$" is, by the definitions, part of "$\psi (\imath x)(\phi x)$." When, in ordinary language or in philosophy, something is said to "exist," it is always something *described*, *i.e.* it is not something immediately presented, like a taste or a patch of colour, but something like "matter" or "mind" or "Homer" (meaning "the author of the Homeric poems"), which is known by description as "the so-and-so," and is thus of the form $(\imath x)(\phi x)$. Thus in all such cases, the existence of the (grammatical) subject $(\imath x)(\phi x)$ can be analytically inferred from any true proposition having this grammatical subject. It would seem that the word "existence" cannot be significantly applied to subjects immediately given; *i.e.* not only does our definition give no meaning to "$E ! x$," but there is no reason, in philosophy, to suppose that a meaning of existence could be found which would be applicable to immediately given subjects.

Besides the above, the following are among the more useful propositions of the present number.

***14·202.** $\vdash :. \phi x . \equiv_x . x = b : \equiv : (\imath x)(\phi x) = b : \equiv : \phi x . \equiv_x . b = x : \equiv : b = (\imath x)(\phi x)$

From the first equivalence in the above, it follows that

***14·204.** $\vdash : E ! (\imath x)(\phi x) . \equiv . (\exists b) . (\imath x)(\phi x) = b$

I.e. $(\imath x)(\phi x)$ exists when there is something which $(\imath x)(\phi x)$ is.

We have

***14·205.** $\vdash : \psi (\imath x)(\phi x) . \equiv . (\exists b) . b = (\imath x)(\phi x) . \psi b$

I.e. $(\imath x)(\phi x)$ has the property ψ when there is something which is $(\imath x)(\phi x)$ and which has the property ψ.

We have to prove that such symbols as "$(\imath x)(\phi x)$" obey the same rules with regard to identity as symbols which directly represent objects. To this, however, there is one partial exception, for instead of having

$$(\imath x)(\phi x) = (\imath x)(\phi x),$$

we only have

***14·28.** $\vdash : E ! (\imath x)(\phi x) . \equiv . (\imath x)(\phi x) = (\imath x)(\phi x)$

I.e. "$(\imath x)(\phi x)$" only satisfies the reflexive property of identity if $(\imath x)(\phi x)$ exists.

The symmetrical property of identity holds for such symbols as $(\imath x)(\phi x)$, without the need of assuming existence, *i.e.* we have

***14·13.** $\vdash : a = (\imath x)(\phi x) . \equiv . (\imath x)(\phi x) = a$

***14·131.** $\vdash : (\imath x)(\phi x) = (\imath x)(\psi x) . \equiv . (\imath x)(\psi x) = (\imath x)(\phi x)$

Similarly the transitive property of identity holds without the need of assuming existence. This is proved in *14·14·142·144.

***14·01.** $[(\imath x)(\phi x)] . \psi (\imath x)(\phi x) . = : (\exists b) : \phi x . \equiv_x . x = b : \psi b$ Df

***14·02.** $E ! (\imath x)(\phi x) . = : (\exists b) : \phi x . \equiv_x . x = b$ Df

***14·03.** $[(\imath x)(\phi x), (\imath x)(\psi x)] . f\{(\imath x)(\phi x), (\imath x)(\psi x)\} . = :$
$[(\imath x)(\phi x)] : [(\imath x)(\psi x)] . f\{(\imath x)(\phi x), (\imath x)(\psi x)\}$ Df

***14·04.** $[(\imath x)(\psi x)] . f\{(\imath x)(\phi x), (\imath x)(\psi x)\} . = .$
$[(\imath x)(\psi x), (\imath x)(\phi x)] . f\{(\imath x)(\phi x), (\imath x)(\psi x)\}$ Df

∗14·1. $\vdash :. [(\imath x)(\phi x)] . \psi(\imath x)(\phi x) . \equiv : (\exists b) : \phi x . \equiv_x . x = b : \psi b$ [∗4·2 . (∗14·01)]

In virtue of our conventions as to the scope intended when no scope is explicitly indicated, the above proposition is the same as the following:·

∗14·101. $\vdash :. \psi(\imath x)(\phi x) . \equiv : (\exists b) : \phi x . \equiv_x . x = b : \psi b$ [∗14·1]

∗14·11. $\vdash :. E!(\imath x)(\phi x) . \equiv : (\exists b) : \phi x . \equiv_x . x = b$ [∗4·2 . (∗14·02)]

∗14·111. $\vdash :. [(\imath x)(\psi x)] . f\{(\imath x)(\phi x), (\imath x)(\psi x)\} . \equiv :$

$(\exists b, c) : \phi x . \equiv_x . x = b : \psi x . \equiv_x . x = c : f(b, c)$

Dem.

$\vdash . ∗4·2 . (∗14·04·03) . \supset$

$\vdash :: [(\imath x)(\psi x)] . f\{(\imath x)(\phi x), (\imath x)(\psi x)\} . \equiv :.$

$[(\imath x)(\psi x)] : [(\imath x)(\phi x)] . f\{(\imath x)(\phi x), (\imath x)(\psi x)\} :.$

[∗14·1] $\equiv :. [(\imath x)(\psi x)] :. (\exists b) : \phi x . \equiv_x . x = b : f\{b, (\imath x)(\psi x)\} :.$

[∗14·1] $\equiv :. (\exists c) :. \psi x . \equiv_x . x = c :. (\exists b) : \phi x . \equiv_x . x = b : f(b, c) :.$

[∗11·55] $\equiv :. (\exists b, c) : \phi x . \equiv_x . x = c : \psi x . \equiv_x . x = b : f(b, c) :: \supset \vdash . \text{Prop}$

∗14·112. $\vdash :. f\{(\imath x)(\phi x), (\imath x)(\psi x)\} . \equiv :$

$(\exists b, c) : \phi x . \equiv_x . x = b : \psi x . \equiv_x . x = c : f(b, c)$

[Proof as in ∗14·111]

In the above proposition, we assume the convention explained on p. 174, after the statement of ∗14·03.

∗14·113. $\vdash : [(\imath x)(\psi x)] . f\{(\imath x)(\phi x), (\imath x)(\psi x)\} . \equiv . f\{(\imath x)(\phi x), (\imath x)(\psi x)\}$

[∗14·111·112]

This proposition shows that when two descriptions occur in the same proposition, the truth-value of the proposition is unaffected by the question which has the larger scope.

∗14·12. $\vdash :. E!(\imath x)(\phi x) . \supset : \phi x . \phi y . \supset_{x,y} . x = y$

Dem.

$\vdash . ∗14·11 \qquad \supset \vdash :. \text{Hp} . \supset : (\exists b) : \phi x . \equiv_x . x = b \qquad (1)$

$\vdash . ∗4·38 . ∗10·1 . ∗11·11·3 . \supset$

$\vdash :. \phi x . \equiv_x . x = b : \supset : \phi x . \phi y . \equiv_{x,y} . x = b . y = b .$

[∗13·172] $\qquad \supset_{x,y} . x = y \qquad (2)$

$\vdash . (2) . ∗10·11·23 . \qquad \supset \vdash :. (\exists b) : \phi x . \equiv_x . x = b : \supset : \phi x . \phi y . \supset_{x,y} . x = y \; (3)$

$\vdash . (1) . (3) . \qquad \supset \vdash . \text{Prop}$

∗14·121. $\vdash :. \phi x . \equiv_x . x = b : \phi x . \equiv_x . x = c : \supset . b = c$

Dem.

$\vdash . ∗10·1 . \supset \vdash :. \text{Hp} . \supset : \phi b . \equiv . b = b : \phi b . \equiv . b = c :$

[∗13·15] $\qquad \supset : \phi b : \phi b . \equiv . b = c :$

[Ass] $\qquad \supset : b = c :. \supset \vdash . \text{Prop}$

∗14·122. $\vdash :. \phi x . \equiv_x . x = b : \equiv : \phi x . \supset_x . x = b : \phi b :$

$\equiv : \phi x . \supset_x . x = b : (\exists x) . \phi x$

Dem.

$\vdash . *10\cdot22 . \quad \supset \vdash :. \phi x . \equiv_x . x = b : \equiv : \phi x . \supset_x . x = b : x = b . \supset_x . \phi x :$

[*13·191] $\qquad \equiv : \phi x . \supset_x . x = b : \phi b \qquad (1)$

$\vdash . *4\cdot71 . \quad \supset \vdash :. \phi x . \supset . x = b : \supset : \phi x . \equiv . \phi x . x = b :.$

[*10·11·27] $\supset \vdash :. \phi x . \supset_x . x = b : \supset : \phi x . \equiv_x . \phi x . x = b :$

[*10·281] $\qquad \supset : (\exists x) . \phi x . \equiv . (\exists x) . \phi x . x = b .$

[*13·195] $\qquad \equiv . \phi b \qquad (2)$

$\vdash . (2) . *5\cdot32 . \supset \vdash :. \phi x . \supset_x . x = b : (\exists x) . \phi x : \equiv : \phi x . \supset_x . x = b : \phi b \qquad (3)$

$\vdash . (1) . (3) . \quad \supset \vdash .$ Prop

The two following propositions (*14·123·124) are placed here because of the analogy with *14·122, but they are not used until we come to the theory of couples (*55 and *56).

***14·123.** $\vdash :. \phi (z, w) . \equiv_{z,w} . z = x . w = y :$

$\qquad \equiv : \phi (z, w) . \supset_{z,w} . z = x . w = y : \phi (x, y) :$

$\qquad \equiv : \phi (z, w) . \supset_{z,w} . z = x . w = y : (\exists z, w) . \phi (z, w)$

Dem.

$\vdash . *11\cdot31 . \quad \supset \vdash :. \phi (z, w) . \equiv_{z,w} . z = x . w = y :$

$\qquad \equiv : \phi (z, w) . \supset_{z,w} . z = x . w = y : z = x . w = y . \supset_{z,w} . \phi (z, w) :$

[*13·21] $\qquad \equiv : \phi (z, w) . \supset_{z,w} . z = x . w = y : \phi (x, y) \qquad (1)$

$\vdash . *4\cdot71 . \quad \supset \vdash :. \phi (z, w) . \supset . z = x . w = y :$

$\qquad \supset : \phi (z, w) . \equiv . \phi (z, w) . z = x . w = y :.$

[*11·11·32] $\supset \vdash :. \phi (z, w) . \supset_{z,w} . z = x . w = y :$

$\qquad \supset : \phi (z, w) . \equiv_{z,w} . \phi (z, w) . z = x . w = y :$

[*11·341] $\qquad \supset : (\exists z, w) . \phi (z, w) . \equiv . (\exists z, w) . \phi (z, w) . z = x . w = y .$

[*13·22] $\qquad \equiv . \phi (x, y) \qquad (2)$

$\vdash . (2) . *5\cdot32 . \supset \vdash :. \phi (z, w) . \supset_{z,w} . z = x . w = y : (\exists z, w) . \phi (z, w) :$

$\qquad \equiv : \phi (z, w) . \supset_{z,w} . z = x . w = y : \phi (x, y) \qquad (3)$

$\vdash . (1) . (3) . \quad \supset \vdash .$ Prop

***14·124.** $\vdash :. (\exists x, y) : \phi (z, w) . \equiv_{z,w} . z = x . w = y :$

$\qquad \equiv : (\exists x, y) . \phi (x, y) : \phi (z, w) . \phi (u, v) . \supset_{z,w,u,v} . z = u . w = v$

Dem.

$\vdash . *14\cdot123 . *3\cdot27 . \supset$

$\qquad \vdash :. (\exists x, y) : \phi (z, w) . \equiv_{z,w} . z = x . w = y : \supset . (\exists x, y) . \phi (x, y) \qquad (1)$

$\vdash . *11\cdot1 . *3\cdot47 . \supset \vdash :. \phi (z, w) . \equiv_{z,w} . z = x . w = y :$

$\qquad \supset : \phi (z, w) . \phi (u, v) . \supset . z = x . w = y . u = x . v = y .$

[*13·172] $\qquad \supset . z = u . w = v \qquad (2)$

$\vdash . (2) . *11\cdot11\cdot35 . \supset$

$\vdash :. (\exists x, y) : \phi (z, w) . \equiv_{z,w} . z = x . w = y :$

$\qquad \supset : \phi (z, w) . \phi (u, v) . \supset . z = u . w = v \qquad (3)$

$\vdash . (3) . *11\cdot 11\cdot 3 . \supset$

$\vdash :. (\exists x, y) : \phi(z, w) . \equiv_{z,w} . z = x . w = y :$

$\supset : \phi(z, w) . \phi(u, v) . \supset_{z,w,u,v} . z = u . w = v$ (4)

$\vdash . *11\cdot 1 . \supset \vdash :. \phi(x, y) : \phi(z, w) . \phi(u, v) . \supset_{z,w,u,v} . z = u . w = v :$

$\supset : \phi(x, y) : \phi(z, w) . \phi(x, y) . \supset_{z,w} . z = x . w = y :$

[*5·33] $\supset : \phi(x, y) : \phi(z, w) . \supset_{z,w} . z = x . w = y :$

[*14·123] $\supset : \phi(z, w) . \equiv_{z,w} . z = x . w = y$ (5)

$\vdash . (5) . *11\cdot 11\cdot 34\cdot 45 . \supset$

$\vdash :. (\exists x, y) . \phi(x, y) : \phi(z, w) . \phi(u, v) . \supset_{z,w,u,v} . z = u . w = v :$

$\supset : (\exists x, y) : \phi(z, w) . \equiv_{z,w} . z = x . w = y$ (6)

$\vdash . (1) . (4) . (6) . \supset \vdash .$ Prop

***14·13.** $\vdash : a = (\imath x)(\phi x) . \equiv . (\imath x)(\phi x) = a$

Dem.

$\vdash . *14\cdot 1 . \quad \supset \vdash :. a = (\imath x)(\phi x) . \equiv : (\exists b) : \phi x . \equiv_x . x = b : a = b$ (1)

$\vdash . *13\cdot 16 . *4\cdot 36 . \supset \vdash :. \phi x . \equiv_x . x = b : a = b : \equiv : \phi x . \equiv_x . x = b : b = a :$

[*10·11·281] $\supset \vdash :. (\exists b) : \phi x . \equiv_x . x = b : a = b :$

$\equiv : (\exists b) : \phi x . \equiv_x . x = b : b = a :$

[*14·1] $\equiv : (\imath x)(\phi x) = a$ (2)

$\vdash . (1) . (2) . \quad \supset \vdash .$ Prop

This proposition is not an *immediate* consequence of *13·16, because "$a = (\imath x)(\phi x)$" is not a value of the function "$x = y$." Similar remarks apply to the following propositions.

***14·131.** $\vdash : (\imath x)(\phi x) = (\imath x)(\psi x) . \equiv . (\imath x)(\psi x) = (\imath x)(\phi x)$

Dem.

$\vdash . *14\cdot 1 . \supset \vdash :: (\imath x)(\phi x) = (\imath x)(\psi x) . \equiv :. (\exists b) : \phi x . \equiv_x . x = b : b = (\imath x)(\psi x) :.$

[*14·1] $\equiv :. (\exists b) :. \phi x . \equiv_x . x = b :. (\exists c) : \psi x . \equiv_x . x = c : b = c :.$

[*11·6] $\equiv :. (\exists c) :. \psi x . \equiv_x . x = c :. (\exists b) : \phi x . \equiv_x . x = b : b = c :.$

[*14·1] $\equiv :. (\exists c) :. \psi x . \equiv_x . x = c : (\imath x)(\phi x) = c :.$

[*14·13] $\equiv :. (\exists c) :. \psi x . \equiv_x . x = c : c = (\imath x)(\phi x) :.$

[*14·1] $\equiv :. (\imath x)(\psi x) = (\imath x)(\phi x) :: \supset \vdash .$ Prop

In the above proposition, in accordance with our convention, the descriptive expression $(\imath x)(\phi x)$ is eliminated before $(\imath x)(\psi x)$, because it occurs first in "$(\imath x)(\phi x) = (\imath x)(\psi x)$"; but in "$(\imath x)(\psi x) = (\imath x)(\phi x)$," $(\imath x)(\psi x)$ is to be first eliminated. The order of elimination makes no difference to the truth-value, as was proved in *14·113.

The above proposition may also be proved as follows:

$\vdash . *14\cdot 111 . \supset \vdash :. (\imath x)(\phi x) = (\imath x)(\psi x) .$

$\equiv : (\exists b, c) : \phi x . \equiv_x . x = b : \psi x . \equiv_x . x = c : b = c :$

[*4·3.*13·16.*11·11·341] $\equiv : (\exists b, c) : \psi x . \equiv_x . x = c : \phi x . \equiv_x . x = b : c = b :$

[*11·2.*14·111] $\equiv : (\imath x)(\psi x) = (\imath x)(\phi x) :. \supset \vdash .$ Prop

***14·14.** $\vdash : a = b \,.\, b = (\imath x)(\phi x) \,.\, \supset \,.\, a = (\imath x)(\phi x)$ [*13·13]

***14·142.** $\vdash : a = (\imath x)(\phi x) \,.\, (\imath x)(\phi x) = (\imath x)(\psi x) \,.\, \supset \,.\, a = (\imath x)(\psi x)$

Dem.

$\vdash \,.\, *14\cdot1 \,.\, \supset \vdash :: \mathrm{Hp} \,.\, \supset :. (\exists b) : \phi x \,.\, \equiv_x \,.\, x = b : a = b :.$

$(\exists c) : \phi x \,.\, \equiv_x \,.\, x = c : c = (\imath x)(\psi x) :.$

[*13·195] $\supset :. \phi x \,.\, \equiv_x \,.\, x = a :. (\exists c) : \phi x \,.\, \equiv_x \,.\, x = c : c = (\imath x)(\psi x) :.$

[*10·35] $\supset :. (\exists c) :. \phi x \,.\, \equiv_x \,.\, x = a : \phi x \,.\, \equiv_x \,.\, x = c : c = (\imath x)(\psi x) :.$

[*14·121] $\supset :. (\exists c) :. \phi x \,.\, \equiv_x \,.\, x = a : a = c : c = (\imath x)(\psi x) :.$

[*3·27.*13·195] $\supset :. a = (\imath x)(\psi x) :: \supset \vdash \,.\, \mathrm{Prop}$

***14·144.** $\vdash : (\imath x)(\phi x) = (\imath x)(\psi x) \,.\, (\imath x)(\psi x) = (\imath x)(\chi x) \,.\, \supset \,.\, (\imath x)(\phi x) = (\imath x)(\chi x)$

Dem.

$\vdash \,.\, *14\cdot111 \,.\, \supset \vdash :: \mathrm{Hp} \,.\, \supset :. (\exists a, b) : \phi x \,.\, \equiv_x \,.\, x = a : \psi x \,.\, \equiv_x \,.\, x = b : a = b :.$

$(\exists c, d) : \psi x \,.\, \equiv_x \,.\, x = c : \chi x \,.\, \equiv_x \,.\, x = d : c = d :.$

[*13·195] $\supset :. (\exists a) : \phi x \,.\, \equiv_x \,.\, x = a : \psi x \,.\, \equiv_x \,.\, x = a :.$

$(\exists c) : \psi x \,.\, \equiv_x \,.\, x = c : \chi x \,.\, \equiv_x \,.\, x = c :.$

[*11·54] $\supset :. (\exists a, c) : \phi x \,.\, \equiv_x \,.\, x = a : \psi x \,.\, \equiv_x \,.\, x = a :$

$\psi x \,.\, \equiv_x \,.\, x = c : \chi x \,.\, \equiv_x \,.\, x = c :.$

[*14·121.*11·42] $\supset :. (\exists a, c) : \phi x \,.\, \equiv_x \,.\, x = a : \chi x \,.\, \equiv_x \,.\, x = c : a = c :.$

[*14·111] $\supset :. (\imath x)(\phi x) = (\imath x)(\chi x) :: \supset \vdash \,.\, \mathrm{Prop}$

***14·145.** $\vdash : a = (\imath x)(\phi x) \,.\, a = (\imath x)(\psi x) \,.\, \supset \,.\, (\imath x)(\phi x) = (\imath x)(\psi x)$

Dem.

$\vdash \,.\, *14\cdot1 \,.\, \quad \supset \vdash :. a = (\imath x)(\phi x) \,.\, \equiv : (\exists b) : \phi x \,.\, \equiv_x \,.\, x = b : a = b :$

[*13·195] . $\equiv : \phi x \,.\, \equiv_x \,.\, x = a$ (1)

$\vdash \,.\, (1) \,.\, *14\cdot1 \,.\, \supset \vdash :: \mathrm{Hp} \,.\, \equiv :. \phi x \,.\, \equiv_x \,.\, x = a :. (\exists b) : \psi x \,.\, \equiv_x \,.\, x = b : a = b :.$

[*10·35] $\equiv :. (\exists b) :. \phi x \,.\, \equiv_x \,.\, x = a : \psi x \,.\, \equiv_x \,.\, x = b : a = b :.$

[*14·111] $\supset :. (\imath x)(\phi x) = (\imath x)(\psi x) :: \supset \vdash \,.\, \mathrm{Prop}$

***14·15.** $\vdash :. (\imath x)(\phi x) = b \,.\, \supset : \psi \{(\imath x)(\phi x)\} \,.\, \equiv .\, \psi b$

Dem.

$\vdash \,.\, *14\cdot1 \,.\, \supset$

$\vdash :: \mathrm{Hp} \,.\, \supset :. (\exists c) : \phi x \,.\, \equiv_x \,.\, x = c : c = b :.$

[*13·195] $\supset :. \phi x \,.\, \equiv_x \,.\, x = b$ (1)

$\vdash \,.\, (1) \,.\, *14\cdot1 \,.\, \supset$

$\vdash :: \mathrm{Hp} \,.\, \supset :. \psi \{(\imath x)(\phi x)\} \,.\, \equiv : (\exists c) : x = b \,.\, \equiv_x \,.\, x = c : \psi c :$

[*13·192] $\equiv : \psi b :: \supset \vdash \,.\, \mathrm{Prop}$

***14·16.** $\vdash :. (\imath x)(\phi x) = (\imath x)(\psi x) \,.\, \supset : \chi \{(\imath x)(\phi x)\} \,.\, \equiv \,.\, \chi \{(\imath x)(\psi x)\}$

Dem.

$\vdash \,.\, *14\cdot1 \,.\, \supset \vdash :. \mathrm{Hp} \,.\, \supset : (\exists b) : \phi x \,.\, \equiv_x \,.\, x = b : b = (\imath x)(\psi x)$ (1)

$\vdash \,.\, *14\cdot1 \,.\, \supset \vdash :: \phi x \,.\, \equiv_x \,.\, x = b : \supset :.$

$\chi \{(\imath x)(\phi x)\} \,.\, \equiv : (\exists c) : x = b \,.\, \equiv_x \,.\, x = c : \chi c :$

[*13·192] $\equiv : \chi b$ (2)

$\vdash . *14\cdot13\cdot15 . \supset \vdash :. b = (\iota x)(\psi x) . \supset : \chi b . \equiv . \chi\{(\iota x)(\psi x)\}$ (3)

$\vdash . (2) . (3) . \quad \supset \vdash :. \phi x . \equiv_x . x = b : b = (\iota x)(\psi x) :$

$\supset : \chi\{(\iota x)(\phi x)\} . \equiv . \chi\{(\iota x)(\psi x)\}$ (4)

$\vdash . (1) . (4) . *10\cdot1\cdot23 . \supset \vdash$. Prop

***14·17.** $\vdash :. (\iota x)(\phi x) = b . \equiv : \psi ! (\iota x)(\phi x) . \equiv_\psi . \psi ! b$

Dem.

$\vdash . *14\cdot15 . *10\cdot11\cdot21 . \supset$

$\vdash :. (\iota x)(\phi x) = b . \supset : \psi ! (\iota x)(\phi x) . \equiv_\psi . \psi ! b$ (1)

$\vdash . *10\cdot1 . *4\cdot22 . \supset \vdash :: \chi ! x . \equiv_x . x = b : \psi ! (\iota x)(\phi x) . \equiv_\psi . \psi ! b :$

$\supset : (\iota x)(\phi x) = b . \equiv . b = b :$

[*13·15] $\supset : (\iota x)(\phi x) = b$ (2)

$\vdash . (2) .$ Exp $. *10\cdot11\cdot23 . \supset$

$\vdash :: (\exists \chi) : \chi ! x . \equiv_x . x = b : \supset :. \psi ! (\iota x)(\phi x) . \equiv_\psi . \psi ! b : \supset . (\iota x)(\phi x) = b$ (3)

$\vdash . *12\cdot1 . \quad \supset \vdash : (\exists \chi) : \chi ! x . \equiv_x . x = b$ (4)

$\vdash . (3) . (4) . \supset \vdash :. \psi ! (\iota x)(\phi x) . \equiv_\psi . \psi ! b : \supset . (\iota x)(\phi x) = b$ (5)

$\vdash . (1) . (5) . \supset \vdash$. Prop

It should be observed that we do *not* have

$$(\iota x)(\phi x) = b . \equiv : \psi ! (\iota x)(\phi x) . \supset_\psi . \psi ! b$$

for, if $\sim E ! (\iota x)(\phi x)$, $\psi ! (\iota x)(\phi x)$ is always false, and therefore

$$\psi ! (\iota x)(\phi x) . \supset_\psi . \psi ! b$$

holds for all values of b. But we do have

***14·171.** $\vdash :. (\iota x)(\phi x) = b . \equiv : \psi ! b . \supset_\psi . \psi ! (\iota x)(\phi x)$

Dem.

$\vdash . *14\cdot17 . \quad \supset \vdash :. (\iota x)(\phi x) = b . \supset : \psi ! b . \supset_\psi . \psi ! (\iota x)(\phi x)$ (1)

$\vdash . *10\cdot1 . *12\cdot1 . \supset \vdash :. \psi ! b . \supset_\psi . \psi ! (\iota x)(\phi x) : \supset : b = b . \supset . (\iota x)(\phi x) = b :$

[*13·15] $\supset : (\iota x)(\phi x) = b$ (2)

$\vdash . (1) . (2) . \quad \supset \vdash$. Prop

***14·18.** $\vdash :. E ! (\iota x)(\phi x) . \supset : (x) . \psi x . \supset . \psi (\iota x)(\phi x)$

Dem.

$\vdash . *10\cdot1 . \quad \supset \vdash : (x) . \psi x . \supset . \psi b :$

[Fact] $\supset \vdash :. \phi x . \equiv_x . x = b : (x) . \psi x : \supset : \phi x . \equiv_x . x = b : \psi b :$

[*10·11·28] $\supset \vdash :. (\exists b) : \phi x . \equiv_x . x = b : (x) . \psi x : \supset : (\exists b) : \phi x . \equiv_x . x = b : \psi b :.$

[*10·35] $\supset \vdash :: (\exists b) : \phi x . \equiv_x . x = b :. (x) . \psi x :. \supset : (\exists b) : \phi x . \equiv_x . x = b : \psi b :.$

[*14·1·11] $\supset \vdash :. E ! (\iota x)(\phi x) : (x) . \psi x : \supset : \psi (\iota x)(\phi x) :. \supset \vdash$. Prop

The above proposition shows that, provided $(\iota x)(\phi x)$ exists, it has (speaking formally) all the logical properties of symbols which directly represent objects. Hence when $(\iota x)(\phi x)$ exists, the fact that it is an incomplete symbol becomes irrelevant to the truth-values of logical propositions in which it occurs.

***14·2.** $\vdash . (\imath x)(x = a) = a$

Dem.

$\vdash . *14\cdot101 . \supset \vdash :. (\imath x)(x = a) = a . \equiv : (\exists b) : x = a . \equiv_x . x = b : b = a :$

$[*13\cdot195] \quad \equiv : x = a . \equiv_x . x = a \qquad (1)$

$\vdash . (1) . \mathrm{Id} . \supset \vdash . \mathrm{Prop}$

***14·201.** $\vdash : \mathrm{E} ! (\imath x)(\phi x) . \supset . (\exists x) . \phi x$

Dem.

$\vdash . *14\cdot11 . \supset \vdash :. \mathrm{Hp} . \supset : (\exists b) : \phi x . \equiv_x . x = b :$

$[*10\cdot1] \quad \supset : (\exists b) : \phi b . \equiv . b = b :$

$[*13\cdot15] \quad \supset : (\exists b) . \phi b :. \supset \vdash . \mathrm{Prop}$

***14·202.** $\vdash :. \phi x . \equiv_x . x = b : \equiv : (\imath x)(\phi x) = b : \equiv : \phi x . \equiv_x . b = x : \equiv : b = (\imath x)(\phi x)$

Dem.

$\vdash . *14\cdot1 . \supset \vdash :. (\imath x)(\phi x) = b . \equiv : (\exists c) : \phi x . \equiv_x . x = c : c = b :$

$[*13\cdot195] \quad \equiv : \phi x . \equiv_x . x = b :. \supset \vdash . \mathrm{Prop}$

[The second half is proved in the same way as the first half.]

***14·203.** $\vdash :. \mathrm{E} ! (\imath x)(\phi x) . \equiv : (\exists x) . \phi x : \phi x . \phi y . \supset_{x,y} . x = y$

Dem.

$\vdash . *14\cdot12\cdot201 . \quad \supset \vdash :. \mathrm{E} ! (\imath x)(\phi x) . \supset : (\exists x) . \phi x : \phi x . \phi y . \supset_{x,y} . x = y \qquad (1)$

$\vdash . *10\cdot1 . \quad \supset \vdash :. \phi b : \phi x . \phi y . \supset_{x,y} . x = y : \supset : \phi b : \phi x . \phi b . \supset_x . x = b :$

$[*5\cdot33] \quad \supset : \phi b : \phi x . \supset_x . x = b :$

$[*13\cdot191] \quad \supset : x = b . \supset_x . \phi x :$

$\phi x . \supset_x . x = b :$

$[*10\cdot22] \quad \supset : \phi x . \equiv_x . x = b \qquad (2)$

$\vdash . (2) . *10\cdot1\cdot28 . \supset \vdash :. (\exists b) : \phi b : \phi x . \phi y . \supset_{x,y} . x = y : \supset : (\exists b) : \phi x . \equiv_x . x = b :.$

$[*10\cdot35] \quad \supset \vdash :. (\exists b) . \phi b : \phi x . \phi y . \supset_{x,y} . x = y : \supset : (\exists b) : \phi x . \equiv_x . x = b :$

$[*14\cdot11] \quad \supset : \mathrm{E} ! (\imath x)(\phi x) \qquad (3)$

$\vdash . (1) . (3) . \quad \supset \vdash . \mathrm{Prop}$

***14·204.** $\vdash :. \mathrm{E} ! (\imath x)(\phi x) . \equiv : (\exists b) . (\imath x)(\phi x) = b$

Dem.

$\vdash . *14\cdot202 . *10\cdot11 . \supset$

$\vdash :. (b) :. \phi x . \equiv_x . x = b : \equiv : (\imath x)(\phi x) = b :. \supset$

$[*10\cdot281] \vdash :. (\exists b) : \phi x . \equiv_x . x = b : \equiv : (\exists b) . (\imath x)(\phi x) = b \qquad (1)$

$\vdash . (1) . *14\cdot11 . \supset \vdash . \mathrm{Prop}$

***14·205.** $\vdash : \psi (\imath x)(\phi x) . \equiv . (\exists b) . b = (\imath x)(\phi x) . \psi b$ [*14·202·1]

***14·21.** $\vdash : \psi (\imath x)(\phi x) . \supset . \mathrm{E} ! (\imath x)(\phi x)$

Dem.

$\vdash . *14\cdot1 . \supset$

$\vdash :. \psi \{(\imath x)(\phi x)\} . \supset : (\exists b) : \phi x . \equiv_x . x = b : \psi b :$

$[*10\cdot5] \quad \supset : (\exists b) : \phi x . \equiv_x . x = b :$

$[*14\cdot11] \quad \supset : \mathrm{E} ! (\imath x)(\phi x) :. \supset \vdash . \mathrm{Prop}$

This proposition shows that if any true statement can be made about $(\imath x)(\phi x)$, then $(\imath x)(\phi x)$ must exist. Its use throughout the remainder of the work will be very frequent.

When $(\imath x)(\phi x)$ does not exist, there are still true propositions in which "$(\imath x)(\phi x)$" occurs, but it has, in such propositions, a *secondary* occurrence, in the sense explained in Chapter III of the Introduction, *i.e.* the asserted proposition concerned is not of the form $\psi(\imath x)(\phi x)$, but of the form $f\{\psi(\imath x)(\phi x)\}$, in other words, the proposition which is the scope of $(\imath x)(\phi x)$ is only part of the whole asserted proposition.

***14·22.** $\vdash : E!(\imath x)(\phi x) . \equiv . \phi(\imath x)(\phi x)$

Dem.

$\vdash . *14\cdot122 . \supset \vdash :. \phi x . \equiv_x . x = b : \supset . \phi b$ (1)

$\vdash . (1) . *4\cdot71 . \supset \vdash :. \phi x . \equiv_x . x = b : \equiv : \phi x . \equiv_x . x = b : \phi b :.$

$[*10\cdot11\cdot281] \quad \supset \vdash :. (\exists b) : \phi x . \equiv_x . x = b : \equiv : (\exists b) : \phi x . \equiv_x . x = b : \phi b :.$

$[*14\cdot11\cdot101] \quad \supset \vdash : E!(\imath x)(\phi x) . \equiv . \phi(\imath x)(\phi x) : \supset \vdash . \text{Prop}$

As an instance of the above proposition, we may take the following: "The proposition 'the author of Waverley existed' is equivalent to 'the man who wrote Waverley wrote Waverley.'" Thus such a proposition as "the man who wrote Waverley wrote Waverley" does not embody a logically necessary truth, since it would be false if Waverley had not been written, or had been written by two men in collaboration. For example, "the man who squared the circle squared the circle" is a false proposition.

***14·23.** $\vdash : E!(\imath x)(\phi x . \psi x) . \equiv . \phi\{(\imath x)(\phi x . \psi x)\}$

Dem.

$\vdash . *14\cdot22 . \supset \vdash :. E!(\imath x)(\phi x . \psi x) .$

$\equiv : [(\imath x)(\phi x . \psi x)] : \phi\{(\imath x)(\phi x . \psi x)\} . \psi\{(\imath x)(\phi x . \psi x)\}$

$[*10\cdot5 . *3\cdot26] \quad \supset : \phi\{(\imath x)(\phi x . \psi x)\}$ (1)

$\vdash . *14\cdot21 . \supset \vdash : \phi\{(\imath x)(\phi x . \psi x)\} . \supset . E!(\imath x)(\phi x . \psi x)$ (2)

$\vdash . (1) . (2) . \supset \vdash . \text{Prop}$

Note that in the second line of the above proof *10·5, not only *3·26, is required. For the scope of the descriptive symbol $(\imath x)(\phi x . \psi x)$ is the whole product $\phi\{(\imath x)(\phi x . \psi x)\} . \psi\{(\imath x)(\phi x . \psi x)\}$, so that, applying *14·1, the proposition on the right in the first line becomes

$$(\exists b) : \phi x . \psi x . \equiv_x . x = b : \phi b . \psi b$$

which, by *10·5 and *3·26, implies

$$(\exists b) : \phi x . \psi x . \equiv_x . x = b : \phi b,$$

i.e. $\qquad \phi\{(\imath x)(\phi x . \psi x)\}.$

***14·24.** $\vdash :. E!(\imath x)(\phi x) . \equiv : [(\imath x)(\phi x)] : \phi y . \equiv_y . y = (\imath x)(\phi x)$

Dem.

$\vdash . *14\cdot1 . \supset \vdash :. [(\imath x)(\phi x)] : \phi y . \equiv_y . y = (\imath x)(\phi x) :$

$\equiv : (\exists b) : \phi y . \equiv_y . y = b : \phi y . \equiv_y . y = b :$

$[*4\cdot24.*10\cdot281] \qquad \equiv : (\exists b) : \phi y \,.\, \equiv_y \,.\, y = b :$

$[*14\cdot11] \qquad \equiv : E\,!\,(\imath x)(\phi x) :. \supset \vdash . \text{Prop}$

This proposition should be compared with $*14\cdot241$, where, in virtue of the smaller scope of $(\imath x)(\phi x)$, we get an implication instead of an equivalence.

$*14\cdot241$. $\vdash :. E\,!\,(\imath x)(\phi x) \,.\supset: \phi y \,.\, \equiv_y \,.\, y = (\imath x)(\phi x)$

Dem.

$\vdash . *14\cdot203 . \supset \vdash :: \text{Hp} . \supset :. \phi y . \phi x . \supset . y = x :.$

$[\text{Exp}] \qquad \supset :. \phi y . \supset : \phi x . \supset . y = x ::$

$[*10\cdot11\cdot21] \quad \supset \vdash :: \text{Hp} . \supset :. \phi y . \supset : \phi x . \supset_x . y = x :.$

$[*4\cdot71] \qquad \supset :. \phi y . \equiv : \phi y : \phi x . \supset_x . y = x :$

$[*13\cdot191] \qquad \equiv : y = x . \supset_x . \phi x : \phi x . \supset_x . y = x :$

$[*10\cdot22] \qquad \equiv : \phi x . \equiv_x . y = x :$

$[*14\cdot202] \qquad \equiv : y = (\imath x)(\phi x) :: \supset \vdash . \text{Prop}$

$*14\cdot242$. $\vdash :. \phi x . \equiv_x . x = b : \supset : \psi b . \equiv . \psi (\imath x)(\phi x) \quad [*14\cdot202\cdot15]$

$*14\cdot25$. $\vdash :. E\,!\,(\imath x)(\phi x) . \supset : \phi x \supset_x \psi x . \equiv . \psi (\imath x)(\phi x)$

Dem.

$\vdash . *4\cdot84 . *10\cdot27\cdot271 . \supset \vdash :: \phi x . \equiv_x . x = b : \supset :. \phi x \supset_x \psi x . \equiv : x = b . \supset_x . \psi x :$

$[*13\cdot191] \qquad \equiv : \psi b :$

$[*14\cdot242] \qquad \equiv . \psi (\imath x)(\phi x) \quad (1)$

$\vdash . (1) . *10\cdot11\cdot23 . \supset \vdash :. (\exists b) : \phi x . \equiv_x . x = b :$

$\supset : \phi x \supset_x \psi x . \equiv . \psi (\imath x)(\phi x) \qquad (2)$

$\vdash . (2) . *14\cdot11 . \quad \supset \vdash . \text{Prop}$

$*14\cdot26$. $\vdash :. E\,!\,(\imath x)(\phi x) . \supset : (\exists x) . \phi x . \psi x . \equiv . \psi \{(\imath x)(\phi x)\} . \equiv . \phi x \supset_x \psi x$

Dem.

$\vdash . *14\cdot11 . \supset$

$\vdash :. \text{Hp} . \supset : (\exists b) : \phi x . \equiv_x . x = b \qquad (1)$

$\vdash . *10\cdot311 . \supset \vdash :: \phi x . \equiv_x . x = b : \supset :. \phi x . \psi x . \equiv_x . x = b . \psi x :.$

$[*10\cdot281] \qquad \supset :. (\exists x) . \phi x . \psi x . \equiv . (\exists x) . x = b . \psi x .$

$[*13\cdot195] \qquad \equiv . \psi b .$

$[*14\cdot242] \qquad \equiv . \psi \{(\imath x)(\phi x)\} \qquad (2)$

$\vdash . (2) . *10\cdot11\cdot23 . \supset$

$\vdash :. (\exists b) : \phi x . \equiv_x . x = b : \supset : (\exists x) . \phi x . \psi x . \equiv . \psi \{(\imath x)(\phi x)\} \qquad (3)$

$\vdash . (1) . (3) . *14\cdot25 . \supset \vdash . \text{Prop}$

$*14\cdot27$. $\vdash :. E\,!\,(\imath x)(\phi x) . \supset : \phi x \equiv_x \psi x . \equiv . (\imath x)(\phi x) = (\imath x)(\psi x)$

Dem.

$\vdash . *4\cdot86\cdot21 . \qquad \supset \vdash :: \phi x . \equiv \ . x = b : \supset :. \phi x . \equiv . \psi x : \equiv : \psi x . \equiv . x = b \ (1)$

$\vdash . (1) . *10\cdot11\cdot27 . \supset \vdash :: \phi x . \equiv_x . x = b : \supset :. (x) :. \phi x . \equiv . \psi x : \equiv : \psi x . \equiv . x = b :.$

$[*10\cdot271] \qquad \supset :. \phi x . \equiv_x . \psi x : \equiv : \psi x . \equiv_x . x = b :$

$[*14\cdot202] \qquad \equiv : b = (\imath x)(\psi x) :$

$[*14\cdot242] \qquad \equiv : (\imath x)(\phi x) = (\imath x)(\psi x) \quad (2)$

$\vdash . (2) . *10\cdot11\cdot23 . *14\cdot11 . \supset \vdash . \text{Prop}$

***14·271.** $\vdash :. \phi x . \equiv_x . \psi x : \supset : E ! (\imath x)(\phi x) . \equiv . E ! (\imath x)(\psi x)$

Dem.

$\vdash . *4·86 . \supset \vdash :: \phi x \equiv \psi x . \supset :. \phi x . \equiv . x = b : \equiv : \psi x . \equiv . x = b ::$

$[*10·11·27] \supset \vdash :: \text{Hp} . \quad \supset :. (x) :. \phi x . \equiv . x = b : \equiv : \psi x . \equiv . x = b :.$

$[*10·271] \quad \supset :. (x) : \phi x . \equiv . x = b : \equiv : (x) : \psi x . \equiv . x = b ::$

$[*10·11·21] \supset \vdash :: \text{Hp} . \quad \supset :. (b) :. \phi x . \equiv_x . x = b : \equiv : \psi x . \equiv_x . x = b :.$

$[*10·281] \quad \supset :. (\exists b) : \phi x . \equiv_x . x = b : \equiv : (\exists b) : \psi x . \equiv_x . x = b ::$

$\supset \vdash . \text{Prop}$

***14·272.** $\vdash :. \phi x . \equiv_x . \psi x : \supset : \chi (\imath x)(\phi x) . \equiv . \chi (\imath x)(\psi x)$

Dem.

$\vdash . *4·86 . \quad \supset \vdash :: \phi x \equiv \psi x . \supset :. \phi x . \equiv . x = b : \equiv : \psi x . \equiv . x = b :.$

$[*10·11·414] \supset \vdash :: \text{Hp} . \quad \supset :. \phi x . \equiv_x . x = b : \equiv : \psi x . \equiv_x . x = b :.$

$[\text{Fact}] \quad \supset :. \phi x . \equiv_x . x = b : \chi b : \equiv : \psi x . \equiv_x . x = b : \chi b :.$

$[*10·11·21] \quad \supset \vdash :: \text{Hp} . \quad \supset :. (b) :. \phi x . \equiv_x . x = b : \chi b : \equiv : \psi x . \equiv_x . x = b : \chi b :.$

$[*10·281] \quad \supset :. (\exists b) : \phi x . \equiv_x . x = b : \chi b : \equiv$

$: (\exists b) : \psi x . \equiv_x . x = b : \chi b :.$

$[*14·101] \quad \supset :. \chi (\imath x)(\phi x) . \equiv . \chi (\imath x)(\psi x) :: \supset \vdash . \text{Prop}$

The above two propositions show that $E ! (\imath x)(\phi x)$ and $\chi (\imath x)(\phi x)$ are "extensional" properties of $\phi \hat{x}$, *i.e.* their truth-value is unchanged by the substitution, for $\phi \hat{x}$, of any formally equivalent function $\psi \hat{x}$.

***14·28.** $\vdash : E ! (\imath x)(\phi x) . \equiv . (\imath x)(\phi x) = (\imath x)(\phi x)$

Dem.

$\vdash . *13·15 . *4·73 . \supset \vdash :. \phi x . \equiv_x . x = b : \equiv : \phi x . \equiv_x . x = b : b = b \quad (1)$

$\vdash . (1) . *10·11·281 . \supset$

$\vdash :. (\exists b) : \phi x . \equiv_x . x = b : \equiv : (\exists b) : \phi x . \equiv_x . x = b : b = b \quad (2)$

$\vdash . (2) . *14·1·11 . \supset \vdash . \text{Prop}$

This proposition states that $(\imath x)(\phi x)$ is identical with itself whenever it exists, but not otherwise. Thus for example the proposition "the present King of France is the present King of France" is false.

The purpose of the following propositions is to show that, when $E ! (\imath x)(\phi x)$, the scope of $(\imath x)(\phi x)$ does not matter to the truth-value of any proposition in which $(\imath x)(\phi x)$ occurs. This proposition cannot be proved generally, but it can be proved in each particular case. The following propositions show the method, which proceeds always by means of *14·242, *10·23 and *14·11. The proposition can be proved generally when $(\imath x)(\phi x)$ occurs in the form $\chi (\imath x)(\phi x)$, and $\chi (\imath x)(\phi x)$ occurs in what we may call a "truth-function," *i.e.* a function whose truth or falsehood depends only upon the truth or falsehood of its argument or arguments. This covers all the cases with which we are ever concerned. That is to say, if $\chi (\imath x)(\phi x)$ occurs in any of the ways which can be generated by the processes of *1—*11, then, provided $E ! (\imath x)(\phi x)$, the truth-value of $f \{[(\imath x)(\phi x)] . \chi (\imath x)(\phi x)\}$ is the same as that of

$$[(\imath x)(\phi x)] . f \{\chi (\imath x)(\phi x)\}.$$

This is proved in the following proposition. In this proposition, however, the use of propositions as apparent variables involves an apparatus not required elsewhere, and we have therefore not used this proposition in subsequent proofs.

***14·3.** $\vdash :. p \equiv q . \supset_{p,q} . f(p) \equiv f(q) : \mathrm{E}!(\imath x)(\phi x) : \supset :$

$$f\{[(\imath x)(\phi x)] . \chi(\imath x)(\phi x)\} . \equiv . [(\imath x)(\phi x)] . f\{\chi(\imath x)(\phi x)\}$$

Dem.

$\vdash . *14·242 . \supset$

$\vdash :. \phi x . \equiv_x . x = b : \supset : [(\imath x)(\phi x)] . \chi(\imath x)(\phi x) . \equiv . \chi b$ (1)

$\vdash . (1) . \supset \vdash :. p \equiv q . \supset_{p,q} . f(p) \equiv f(q) : \phi x . \equiv_x . x = b : \supset :$

$f\{[(\imath x)(\phi x)] . \chi(\imath x)(\phi x)\} . \equiv . f(\chi b)$ (2)

$\vdash . *14·242 . \supset$

$\vdash :. \phi x . \equiv_x . x = b : \supset : [(\imath x)(\phi x)] . f\{\chi(\imath x)(\phi x)\} . \equiv . f(\chi b)$ (3)

$\vdash . (2) . (3) . \supset$

$\vdash :. p \equiv q . \supset_{p,q} . f(p) \equiv f(q) : \phi x . \equiv_x . x = b : \supset :$

$f\{[(\imath x)(\phi x)] . \chi(\imath x)(\phi x)\} . \equiv . [(\imath x)(\phi x)] . f\{\chi(\imath x)(\phi x)\}$ (4)

$\vdash . (4) . *10·23 . *14·11 . \supset \vdash .$ Prop

The following propositions are immediate applications of the above. They are, however, independently proved, because *14·3 introduces propositions (p, q namely) as apparent variables, which we have not done elsewhere, and cannot do legitimately without the explicit introduction of the hierarchy of propositions with a reducibility-axiom such as *12·1.

***14·31.** $\vdash :: \mathrm{E}!(\imath x)(\phi x) . \supset :. [(\imath x)(\phi x)] . p \vee \chi(\imath x)(\phi x) .$

$$\equiv : p . \vee . [(\imath x)(\phi x)] . \chi(\imath x)(\phi x)$$

Dem.

$\vdash . *14·242 . \supset \vdash :. \phi x . \equiv_x . x = b : \supset : [(\imath x)(\phi x)] . p \vee \chi(\imath x)(\phi x) . \equiv . p \vee \chi b$ (1)

$\vdash . *14·242 . \supset \vdash :. \phi x . \equiv_x . x = b : \supset : [(\imath x)(\phi x)] . \chi(\imath x)(\phi x) . \equiv . \chi b :$

[*4·37] $\supset : p \vee [(\imath x)(\phi x)] \chi(\imath x)(\phi x) . \equiv . p \vee \chi b$ (2)

$\vdash . (1) . (2) . \supset \vdash :. \phi x . \equiv_x . x = b : \supset : [(\imath x)(\phi x)] . p \vee \chi(\imath x)(\phi x) .$

$\equiv . p \vee [(\imath x)(\phi x)] \chi(\imath x)(\phi x)$ (3)

$\vdash . (3) . *10·23 . *14·11 . \supset \vdash .$ Prop

The following propositions are proved in precisely the same way as *14·31; hence we shall merely give references to the propositions used in the proofs.

***14·32.** $\vdash :. \mathrm{E}!(\imath x)(\phi x) . \equiv : [(\imath x)(\phi x)] . \sim\chi(\imath x)(\phi x) .$

$$\equiv . \sim\{[(\imath x)(\phi x)] . \chi(\imath x)(\phi x)\}$$

[*14·242 . *4·11 . *10·23 . *14·11]

The equivalence asserted here fails when $\sim\mathrm{E}!(\imath x)(\phi x)$. Thus, for example, let ϕy be "y is King of France." Then $(\imath x)(\phi x) =$ the King of France. Let χy be "y is bald." Then $[(\imath x)(\phi x)] . \sim\chi(\imath x)(\phi x) . = .$ the King of France exists and is not bald; but $\sim\{[(\imath x)(\phi x)] . \chi(\imath x)(\phi x)\} . = .$ it is false that the King of France exists and is bald. Of these the first is false, the second true.

Either might be meant by "the King of France is not bald," which is ambiguous; but it would be more natural to take the first (false) interpretation as the meaning of the words. If the King of France existed, the two would be equivalent; thus as applied to the King of England, both are true or both false.

***14·33.** $\vdash :: E ! (\imath x)(\phi x) . \supset :. [(\imath x)(\phi x)] . p \supset \chi (\imath x)(\phi x) .$
$\equiv : p . \supset . [(\imath x)(\phi x)] . \chi (\imath x)(\phi x)$
[*14·242 . *4·85 . *10·23 . *14·11]

***14·331.** $\vdash :: E ! (\imath x)(\phi x) . \supset :. [(\imath x)(\phi x)] . \chi (\imath x)(\phi x) \supset p .$
$\equiv : [(\imath x)(\phi x)] . \chi (\imath x)(\phi x) . \supset . p$
[*4·84 . *14·242 . *10·23 . *14·11]

***14·332.** $\vdash :: E ! (\imath x)(\phi x) . \supset :. [(\imath x)(\phi x)] . p \equiv \chi (\imath x)(\phi x) . \equiv$
$: p . \equiv . [(\imath x)(\phi x)] . \chi (\imath x)(\phi x)$
[*4·86 . *14·242 . *10·23 . *14·11]

***14·34.** $\vdash :. p : [(\imath x)(\phi x)] . \chi (\imath x)(\phi x) : \equiv : [(\imath x)(\phi x)] : p . \chi (\imath x)(\phi x)$

This proposition does not require the hypothesis $E ! (\imath x)(\phi x)$.

Dem.

$\vdash . *14·1 . \supset$
$\vdash :. p : [(\imath x)(\phi x)] . \chi (\imath x)(\phi x) : \equiv : p : (\exists b) : \phi x . \equiv_x . x = b : \chi b :$
[*10·35] $\equiv : (\exists b) : p : \phi x . \equiv_x . x = b : \chi b :$
[*14·1] $\equiv : [(\imath x)(\phi x)] : p . \chi (\imath x)(\phi x) :. \supset \vdash .$ Prop

Propositions of the above type might be continued indefinitely, but as they are proved on a uniform plan, it is unnecessary to go beyond the fundamental cases of $p \vee q$, $\sim p$, $p \supset q$ and $p . q$.

It should be observed that the proposition in which $(\imath x)(\phi x)$ has the larger scope always implies the corresponding one in which it has the smaller scope, but the converse implication only holds if either (a) we have $E ! (\imath x)(\phi x)$ or (b) the proposition in which $(\imath x)(\phi x)$ has the smaller scope implies $E ! (\imath x)(\phi x)$. The second case occurs in *14·34, and is the reason why we get an equivalence without the hypothesis $E ! (\imath x)(\phi x)$. The proposition in which $(\imath x)(\phi x)$ has the larger scope always implies $E ! (\imath x)(\phi x)$, in virtue of *14·21.

SECTION C

CLASSES AND RELATIONS

*20. GENERAL THEORY OF CLASSES

Summary of *20.

The following theory of classes, although it provides a notation to represent them, avoids the assumption that there are such things as classes. This it does by merely defining propositions in whose expression the symbols representing classes occur, just as, in *14, we defined propositions containing descriptions.

The characteristics of a class are that it consists of all the terms satisfying some propositional function, so that every propositional function determines a class, and two functions which are formally equivalent (*i.e.* such that whenever either is true, the other is true also) determine the same class, while conversely two functions which determine the same class are formally equivalent. When two functions are formally equivalent, we shall say that they have the same *extension*. The incomplete symbols which take the place of classes serve the purpose of technically providing something identical in the case of two functions having the same extension; without something to represent classes, we cannot, for example, count the combinations that can be formed out of a given set of objects.

Propositions in which a function ϕ occurs may depend, for their truth-value, upon the particular function ϕ, or they may depend only upon the *extension* of ϕ. In the former case, we will call the proposition concerned an *intensional* function of ϕ; in the latter case, an *extensional* function of ϕ. Thus, for example, $(x) . \phi x$ or $(\exists x) . \phi x$ is an extensional function of ϕ, because, if ϕ is formally equivalent to ψ, *i.e.* if $\phi x . \equiv_x . \psi x$, we have $(x) . \phi x . \equiv . (x) . \psi x$ and $(\exists x) . \phi x . \equiv . (\exists x) . \psi x$. But on the other hand "I believe $(x) . \phi x$" is an *intensional* function, because, even if $\phi x . \equiv_x . \psi x$, it by no means follows that I believe $(x) . \psi x$ provided I believe $(x) . \phi x$. The mark of an extensional function f of a function $\phi ! \hat{z}$ is

$$\phi ! x . \equiv_x . \psi ! x : \supset_{\phi, \psi} : f(\phi ! \hat{z}) . \equiv . f(\psi ! \hat{z}).$$

(We write "$\phi ! \hat{z}$" when we wish to speak of the function itself as opposed to its argument.) The functions of functions with which mathematics is specially concerned are all extensional.

When a function of $\phi ! \hat{z}$ is extensional, it may be regarded as being about the class determined by $\phi ! \hat{z}$, since its truth-value remains unchanged so long as the class is unchanged. Hence we require, for the theory of classes, a method of obtaining an extensional function from any given function of a function. This is effected by the following definition:

***20·01.** $f\{\hat{z}(\psi z)\} . = : (\exists \phi) : \phi ! x . \equiv_x . \psi x : f\{\phi ! \hat{z}\}$ Df

Here $f\{\hat{z}(\psi z)\}$ is in reality a function of $\psi\hat{z}$, which is defined whenever $f\{\phi ! \hat{z}\}$ is significant for predicative functions $\phi ! \hat{z}$. But it is convenient to regard $f\{\hat{z}(\psi z)\}$ as though it had an argument $\hat{z}(\psi z)$, which we will call "the class determined by the function $\psi\hat{z}$." It will be proved shortly that $f\{\hat{z}(\psi z)\}$ is always an *extensional* function of $\psi\hat{z}$, and that, applying the definition of identity (*13·01) to the fictitious objects $\hat{z}(\phi z)$ and $\hat{z}(\psi z)$, we have

$$\hat{z}(\phi z) = \hat{z}(\psi z) . \equiv : (x) : \phi x . \equiv . \psi x.$$

This last is the distinguishing characteristic of classes, and justifies us in treating $\hat{z}(\psi z)$ as the class determined by $\psi\hat{z}$.

With regard to the scope of $\hat{z}(\psi z)$, and to the order of elimination of two such expressions, we shall adopt the same conventions as were explained in *14 for $(\iota x)(\phi x)$. The condition corresponding to

$$\text{E} ! (\iota x)(\psi x) \text{ is } (\exists \phi) : \phi ! x . \equiv_x . \psi x,$$

which is always satisfied because of *12·1.

Following Peano, we shall use the notation

$$x \in \hat{z}(\psi z)$$

to express "x is a member of the class determined by $\psi\hat{z}$." We therefore introduce the following definition:

***20·02.** $x \in (\phi ! \hat{z}) . = . \phi ! x$ Df

In this form, the definition is never used; it is introduced for the sake of the proposition

$$\vdash :. x \in \hat{z}(\psi z) . \equiv : (\exists \phi) : \psi y . \equiv_y . \phi ! y : \phi ! x$$

which results from *20·02 and *20·01, and leads to

$$\vdash : x \in \hat{z}(\psi z) . \equiv . \psi x$$

by the help of *12·1.

We shall use small Greek letters (other than $\epsilon, \iota, \pi, \phi, \psi, \chi, \theta$) to represent classes, *i.e.* to stand for symbols of the form $\hat{z}(\phi z)$ or $\hat{z}(\phi ! z)$. When a small Greek letter occurs as apparent variable, it is to be understood to stand for a symbol of the form $\hat{z}(\phi ! z)$, where ϕ is properly the apparent variable concerned. The use of single letters in place of such symbols as $\hat{z}(\phi z)$ or $\hat{z}(\phi ! z)$ is practically almost indispensable, since otherwise the notation rapidly becomes intolerably cumbrous. Thus "$x \in \alpha$" will mean "x is a member of the class α," and may be used wherever no special defining function of the class α is in question.

The following definition defines what is meant by a *class*.

***20·03.** $\text{Cls} = \hat{\alpha}\{(\exists \phi) . \alpha = \hat{z}(\phi ! z)\}$ Df

Note that the expression "$\hat{\alpha}\{(\exists \phi) . \alpha = \hat{z}(\phi ! z)\}$" has no meaning in isolation: we have merely defined (in *20·01) certain *uses* of such expressions. What the above definition decides is that the symbol "Cls" may replace the symbol "$\hat{\alpha}\{(\exists \phi) . \alpha = \hat{z}(\phi ! z)\}$," wherever the latter occurs, and that the meaning of the combination of symbols concerned is to be unchanged thereby. Thus "Cls," also, has no meaning in isolation, but merely in certain uses.

The above definition, like many future definitions, is ambiguous as to type. The Latin letter z, according to our conventions, is to represent the lowest type concerned; thus ϕ is of the type next above this. It is convenient to speak of a class as being of the same type as its defining function; thus α is of the type next above that of z, and "Cls" is of the type next above that of α. Thus the type of "Cls" is fixed relatively to the lowest type concerned; but if, in two different contexts, different types are the lowest concerned, the meaning of "Cls" will be different in these two contexts. The meaning of "Cls" only becomes definite when the lowest type concerned is specified.

Equality between classes is defined by applying *13·01, symbolically unchanged, to their defining functions, and then using *20·01.

The propositions of the present number may be divided into three sets. First, we have those that deal with the fundamental properties of classes; these end with *20·43. Then we have a set of propositions dealing with both classes and descriptions; these extend from *20·5 to *20·59 (with the exception of *20·53·54). Lastly, we have a set of propositions designed to prove that classes of classes have all the same formal properties as classes of individuals.

In the first set, the principal propositions are the following.

***20·15.** $\vdash :. \psi x . \equiv_x . \chi x : \equiv . \hat{z}(\psi z) = \hat{z}(\chi z)$

I.e. two classes are identical when, and only when, their defining functions are formally equivalent. This is the principal property of classes.

***20·31.** $\vdash :. \hat{z}(\psi z) = \hat{z}(\chi z) . \equiv : x \epsilon \hat{z}(\psi z) . \equiv_x . x \epsilon \hat{z}(\chi z)$

I.e. two classes are identical when, and only when, they have the same members.

***20·43.** $\vdash :. \alpha = \beta . \equiv : x \epsilon \alpha . \equiv_x . x \epsilon \beta$

This is the same proposition as *20·31, merely employing Greek letters in place of $\hat{z}(\psi z)$ and $\hat{z}(\chi z)$.

***20·18.** $\vdash :. \hat{z}(\phi z) = \hat{z}(\psi z) . \supset : f\{\hat{z}(\phi z)\} . \equiv . f\{\hat{z}(\psi z)\}$

I.e. if two classes are identical, any property of either belongs also to the other. This is the analogue of *13·12.

***20·2·21·22**, which prove that identity between classes is reflexive, symmetrical and transitive.

***20·3.** $\vdash : x \epsilon \hat{z}(\psi z) . \equiv . \psi x$

I.e. a term belongs to a class when, and only when, it satisfies the defining function of the class.

In the second set of propositions (*20·5—·59), we show that, under suitable circumstances, expressions such as $(\imath x)(\phi x)$ may be substituted for x in *20·3 and various other propositions of the first set, and we prove a few properties of such expressions as "$(\imath\alpha)(f\alpha)$," *i.e.* "the class which satisfies the function f." Here it is to be remembered that "α" stands for "$\hat{z}(\phi z)$," and that "$f\alpha$" therefore stands for "$f\{\hat{z}(\phi z)\}$." This is, in reality, a function of $\phi\hat{z}$, namely the extensional function associated with $f(\psi!\hat{z})$ by means of $*20{\cdot}01$. Thus an expression containing a variable class is always an abbreviation for an expression containing a variable function.

In the third set of propositions, we prove that variable classes satisfy all the primitive propositions assumed for variable individuals or functions, whence it follows, by merely repeating the proofs of the first set of propositions ($*20{\cdot}1$—$\cdot43$), that classes of classes have all the formal properties of classes of individuals or functions. We shall never have occasion explicitly to consider classes of functions, but classes of classes will occur constantly—for example, every cardinal number will be defined as a class of classes. Classes of relations, which will also frequently occur, will be considered in $*21$.

$*20{\cdot}01$. $f\{\hat{z}(\psi z)\} . = : (\exists\phi) : \phi!x . \equiv_x . \psi x : f\{\phi!\hat{z}\}$ Df

$*20{\cdot}02$. $x \epsilon (\phi!\hat{z}) . = . \phi!x$ Df

$*20{\cdot}03$. $\text{Cls} = \hat{\alpha}\{(\exists\phi) . \alpha = \hat{z}(\phi!z)\}$ Df

The three following definitions serve merely for purposes of abbreviation.

$*20{\cdot}04$. $x, y \epsilon \alpha . = . x \epsilon \alpha . y \epsilon \alpha$ Df

$*20{\cdot}05$. $x, y, z \epsilon \alpha . = . x, y \epsilon \alpha . z \epsilon \alpha$ Df

$*20{\cdot}06$. $x \sim\epsilon \alpha . = . \sim(x \epsilon \alpha)$ Df

The following definitions merely extend to symbols representing classes the definitions which have already been given for other symbols, with the smallest possible modifications.

$*20{\cdot}07$. $(\alpha) . f\alpha . = . (\phi) . f\{\hat{z}(\phi!z)\}$ Df

$*20{\cdot}071$. $(\exists\alpha) . f\alpha . = . (\exists\phi) . f\{\hat{z}(\phi!z)\}$ Df

$*20{\cdot}072$. $[(\imath\alpha)(\phi\alpha)] . f(\imath\alpha)(\phi\alpha) . = : (\exists\gamma) : \phi\alpha . \equiv_\alpha . \alpha = \gamma : f\gamma$ Df

$*20{\cdot}08$. $f\{\hat{\alpha}(\psi\alpha)\} . = : (\exists\phi) : \psi\alpha . \equiv_\alpha . \phi!\alpha : f(\phi!\hat{\alpha})$ Df

$*20{\cdot}081$. $\alpha \epsilon \psi!\hat{\alpha} . = . \psi!\alpha$ Df

The propositions which follow give the most general properties of classes.

$*20{\cdot}1$. $\vdash :. f\{\hat{z}(\psi z)\} . \equiv : (\exists\phi) : \phi!x . \equiv_x . \psi x : f\{\phi!\hat{z}\}$ [$*4{\cdot}2 . (*20{\cdot}01)$]

$*20{\cdot}11$. $\vdash :. \psi x . \equiv_x . \chi x : \supset : f\{\hat{z}(\psi z)\} . \equiv . f\{\hat{z}(\chi z)\}$

Dem.

$\vdash . *4{\cdot}86 . \supset \vdash :: \text{Hp} . \supset :. \phi!x . \equiv_x . \psi x : \equiv_\phi : \phi!x . \equiv_x . \chi x :.$

[$*4{\cdot}36$] $\supset :. \phi!x . \equiv_x . \psi x : f\{\phi!\hat{z}\} : \equiv_\phi : \phi!x . \equiv_x . \chi x : f\{\phi!\hat{z}\} :.$

[$*10{\cdot}281$] $\supset :. (\exists\phi) : \phi!x . \equiv_x . \psi x : f\{\phi!\hat{z}\} :$

$\equiv : (\exists\phi) : \phi!x . \equiv_x . \chi x : f\{\phi!\hat{z}\} :.$

[$*20{\cdot}1$] $\supset :. f\{\hat{z}(\psi z)\} . \equiv . f\{\hat{z}(\chi z)\} :: \supset \vdash . \text{Prop}$

This proves that every proposition about a class expresses an extensional property of the determining function of the class, and therefore does not depend for its truth or falsehood upon the particular function selected for determining the class, but only upon the extension of the determining function.

***20·111.** $\vdash :. f(\phi ! \hat{z}) \,.\equiv_\phi.\, g(\phi ! \hat{z}) : \supset : f\{\hat{z}(\phi ! z)\} \,.\equiv_\phi.\, g\{\hat{z}(\phi ! z)\}$

Dem.

$\vdash .\, \text{Fact} \,.\, \supset \vdash :: \text{Hp} \,.\supset :.\, \phi ! x \,.\equiv_x.\, \psi ! x : f(\psi ! \hat{z}) : \equiv : \phi ! x \,.\equiv_x.\, \psi ! x : g(\psi ! \hat{z}) ::$

[*10·11·21] $\supset \vdash :: \text{Hp} \,.\supset :.\, \phi ! x \,.\equiv_x.\, \psi ! x : f(\psi ! \hat{z}) : \equiv_\psi : \phi ! x \,.\equiv_x.\, \psi ! x : g(\psi ! \hat{z}) :.$

[*10·281] $\supset :. (\exists\psi) : \phi ! x \,.\equiv_x.\, \psi ! x : f(\psi ! \hat{z}) : \equiv : (\exists\psi) : \phi ! x \,.\equiv_x.\, \psi ! x : g(\psi ! \hat{z}) :.$

[*20·1] $\supset :. f\{\hat{z}(\phi ! x)\} \,.\equiv.\, g\{\hat{z}(\phi ! x)\}$ (1)

$\vdash .\, (1) \,.\, *10{\cdot}11{\cdot}21 \,.\, \supset \vdash .\, \text{Prop}$

***20·112.** $\vdash :. (\exists g) :. f\{\hat{z}(\phi ! z)\} \,.\equiv_\phi.\, g ! \{\hat{z}(\phi ! z)\}$

Dem.

$\vdash .\, *12{\cdot}1 \,.\supset \vdash :. (\exists g) : f(\phi ! \hat{z}) \,.\equiv_\phi.\, g ! (\phi ! \hat{z})$ (1)

$\vdash .\, (1) \,.\, *20{\cdot}111 \,.\supset \vdash .\, \text{Prop}$

Thus the axiom of reducibility still holds for classes as arguments.

***20·12.** $\vdash : (\exists\phi) : \phi ! x \,.\equiv_x.\, \psi x : f\{\hat{z}(\psi z)\} \,.\equiv.\, f\{\hat{z}(\phi ! z)\}$ [*20·11 . *12·1]

***20·13.** $\vdash :. \psi x \,.\equiv_x.\, \chi x : \supset .\, \hat{z}(\psi z) = \hat{z}(\chi z)$

The meaning of "$\hat{z}(\psi z) = \hat{z}(\chi z)$" is obtained by a double application of *20·01 to *13·01, remembering the convention that $\hat{z}(\psi z)$ is to have a larger scope than $\hat{z}(\chi z)$ because it occurs first.

Dem.

$\vdash .\, *20{\cdot}1 \,.\supset \vdash :: \hat{z}(\psi z) = \hat{z}(\chi z) \,.\equiv :. (\exists\phi) : \psi x \,.\equiv_x.\, \phi ! x : \phi ! \hat{z} = \hat{z}(\chi z) :.$

[*20·1] $\equiv :. (\exists\phi, \theta) :. \psi x \,.\equiv_x.\, \phi ! x : \chi x \,.\equiv_x.\, \theta ! x : \phi ! \hat{z} = \theta ! \hat{z}$ (1)

$\vdash .\, *12{\cdot}1 \,.\, *10{\cdot}321 \,.\supset$

$\vdash :: \text{Hp} \,.\supset :. (\exists\phi) : \psi x \,.\equiv_x.\, \phi ! x : \chi x \,.\equiv_x.\, \phi ! x :.$

[*13·195] $\supset :. (\exists\phi, \theta) :. \psi x \,.\equiv_x.\, \phi ! x : \chi x \,.\equiv_x.\, \theta ! x : \phi ! \hat{z} = \theta ! \hat{z}$ (2)

$\vdash .\, (1) \,.\, (2) \,.\supset \vdash .\, \text{Prop}$

***20·14.** $\vdash :. \hat{z}(\psi z) = \hat{z}(\chi z) \,.\supset : \psi x \,.\equiv_x.\, \chi x$

Dem.

$\vdash .\, *20{\cdot}1 \,.\supset \vdash :: \hat{z}(\psi z) = \hat{z}(\chi z) \,.\equiv :. (\exists\phi) : \psi x \,.\equiv_x.\, \phi ! x : \phi ! \hat{z} = \hat{z}(\chi z) :.$

[*20·1] $\equiv :. (\exists\phi, \theta) :. \psi x \,.\equiv_x.\, \phi ! x : \chi x \,.\equiv_x.\, \theta ! x : \phi ! \hat{z} = \theta ! \hat{z} :.$

[*13·195] $\equiv :. (\exists\phi) :. \psi x \,.\equiv_x.\, \phi ! x : \chi x \,.\equiv_x.\, \phi ! x :.$

[*10·322] $\supset :. \psi x \,.\equiv_x.\, \chi x :: \supset \vdash .\, \text{Prop}$

This proposition is the converse of *20·13.

***20·15.** $\vdash :. \psi x \,.\equiv_x.\, \chi x : \equiv .\, \hat{z}(\psi z) = \hat{z}(\chi z)$ [*20·13·14]

This proposition states that two functions determine the same class when, and only when, they are formally equivalent, *i.e.* are satisfied by the same set of values. This is the essential property of classes, and gives the justification of the definition *20·01.

***20·151.** $\vdash . (\exists \phi) . \hat{z}(\psi z) = \hat{z}(\phi ! z)$

Dem.

$\vdash . *20{\cdot}15 . \quad \supset \vdash :. \psi x . \equiv_x . \phi ! x : \supset . \hat{z}(\psi z) = \hat{z}(\phi ! z) :.$

$[*10{\cdot}11{\cdot}28] \quad \supset \vdash :. (\exists \phi) : \psi x . \equiv_x . \phi ! x : \supset . (\exists \phi) . \hat{z}(\psi z) = \hat{z}(\phi ! z)$ (1)

$\vdash . (1) . *12{\cdot}1 . \supset \vdash .$ Prop

In virtue of this proposition, all classes can be obtained from predicative functions. This fact is especially important when classes are used as apparent variables. For in that case, according to the definitions *20·07·071, the apparent variable really involved is a predicative function. In virtue of *20·151, this places no limitation upon the classes concerned, except the limitation which inevitably results from the nature of their membership. A class, therefore, unlike a function, has its order completely determined by the order of its possible members, *i.e.* of the arguments which render its defining function significant.

***20·16.** $\vdash : (\exists \phi) : f\{\hat{z}(\psi z)\} . \equiv . f\{\hat{z}(\phi ! z)\}$ [*20·12]

***20·17.** $\vdash : (\phi) . f\{\hat{z}(\phi ! z)\} . \supset . f\{\hat{z}(\psi z)\}$ [*20·16 . *10·1]

***20·18.** $\vdash :. \hat{z}(\phi z) = \hat{z}(\psi z) . \supset : f\{\hat{z}(\phi z)\} . \equiv . f\{\hat{z}(\psi z)\}$ [*20·11·15]

***20·19.** $\vdash :. \hat{z}(\psi z) = \hat{z}(\chi z) . \equiv : (f) : f ! \hat{z}(\psi z) . \supset . f ! \hat{z}(\chi z)$

Dem.

$\vdash . *20{\cdot}18 . *10{\cdot}11{\cdot}21 . \supset \vdash :. \hat{z}(\psi z) = \hat{z}(\chi z) . \supset :$

$(f) : f ! \hat{z}(\psi z) . \supset . f ! \hat{z}(\chi z)$ (1)

$\vdash . *20{\cdot}18{\cdot}15 . \supset \vdash :: \phi ! x . \equiv_x . \psi x : \theta ! x . \equiv_x . \chi x : f ! \hat{z}(\psi z) . \supset . f ! \hat{z}(\chi z) : \supset :$

$f ! \hat{z}(\phi ! z) . \supset . f ! \hat{z}(\theta ! z)$ (2)

$\vdash . (2) . *10{\cdot}11{\cdot}27{\cdot}33 . \supset$

$\vdash :: \phi ! x . \equiv_x . \psi x : \theta ! x . \equiv_x . \chi x :. (f) : f ! \hat{z}(\psi z) . \supset . f ! \hat{z}(\chi z) :. \supset :.$

$(f) : f ! \hat{z}(\phi ! z) . \supset . f ! \hat{z}(\theta ! z) :.$

$[*20{\cdot}112 . *10{\cdot}1] \quad \supset :. \phi ! x . \equiv_x . \phi ! x : \supset : \phi ! x . \equiv_x . \theta ! x :.$

$[*4{\cdot}2] \quad \supset :. \phi ! x . \equiv_x . \theta ! x :.$

$[*10{\cdot}301{\cdot}32 . \mathrm{Hp}] \quad \supset :. \psi x . \equiv_x . \chi x :.$

$[*20{\cdot}15] \quad \supset :. \hat{z}(\psi z) = \hat{z}(\chi z)$ (3)

$\vdash . (3) . *10{\cdot}11{\cdot}23{\cdot}35 . \supset$

$\vdash :: (\exists \phi, \theta) : \phi ! x . \equiv_x . \psi x : \theta ! x . \equiv_x . \chi x :. (f) : f ! \hat{z}(\psi z) . \supset . f ! \hat{z}(\chi z) :.$

$\supset . \hat{z}(\psi z) = \hat{z}(\chi z)$ (4)

$\vdash . (4) . *12{\cdot}1 . \supset \vdash :. (f) : f ! \hat{z}(\psi z) . \supset . f ! \hat{z}(\chi z) : \supset . \hat{z}(\psi z) = \hat{z}(\chi z)$ (5)

$\vdash . (1) . (5) . \quad \supset \vdash .$ Prop

***20·191.** $\vdash :. \hat{z}(\psi z) = \hat{z}(\chi z) . \equiv : (f) : f ! \hat{z}(\psi z) . \equiv . f ! \hat{z}(\chi z)$

[*20·18·19 . *10·22]

***20·2.** $\vdash . \hat{z}(\phi z) = \hat{z}(\phi z)$

Dem.

$\vdash . *20{\cdot}15 . \supset \vdash :. \hat{z}(\phi z) = \hat{z}(\phi z) . \equiv : \phi x . \equiv_x . \phi x$ (1)

$\vdash . (1) . *4{\cdot}2 . *10{\cdot}11 . \supset \vdash .$ Prop

***20·21.** $\vdash : \hat{z}(\phi z) = \hat{z}(\psi z) . \equiv . \hat{z}(\psi z) = \hat{z}(\phi z)$ [*20·15 . *10·32]

***20·22.** $\vdash : \hat{z}(\phi z) = \hat{z}(\psi z) . \hat{z}(\psi z) = \hat{z}(\chi z) . \supset . \hat{z}(\phi z) = \hat{z}(\chi z)$
[*20·15 . *10·301]

The above propositions are not *immediate* consequences of *13·15·16·17, for a reason analogous to that explained in the note to *14·13, namely because $f\{\hat{z}(\phi z)\}$ is not a value of fx, and therefore in particular "$\hat{z}(\phi z) = \hat{z}(\psi z)$" is not a value of "$x = y$."

***20·23.** $\vdash : \hat{z}(\phi z) = \hat{z}(\psi z) . \hat{z}(\phi z) = \hat{z}(\chi z) . \supset . \hat{z}(\psi z) = \hat{z}(\chi z)$ [*20·21·22]

***20·24.** $\vdash : \hat{z}(\psi z) = \hat{z}(\phi z) . \hat{z}(\chi z) = \hat{z}(\phi z) . \supset . \hat{z}(\psi z) = \hat{z}(\chi z)$ [*20·21·22]

***20·25.** $\vdash :. \alpha = \hat{z}(\phi z) . \equiv_\alpha . \alpha = \hat{z}(\psi z) : \equiv . \hat{z}(\phi z) = \hat{z}(\psi z)$

Dem.

$\vdash . *10·1 . \quad \supset \vdash :. \alpha = \hat{z}(\phi z) . \equiv_\alpha . \alpha = \hat{z}(\psi z) : \supset :$
$\hat{z}(\phi z) = \hat{z}(\phi z) . \equiv . \hat{z}(\phi z) = \hat{z}(\psi z) :$
[*20·2] $\supset : \hat{z}(\phi z) = \hat{z}(\psi z)$ (1)
$\vdash . *20·22 . \quad \supset \vdash : \alpha = \hat{z}(\phi z) . \hat{z}(\phi z) = \hat{z}(\psi z) . \supset . \alpha = \hat{z}(\psi z) :$
[Exp.Comm] $\supset \vdash :. \hat{z}(\phi z) = \hat{z}(\psi z) . \supset : \alpha = \hat{z}(\phi z) . \supset . \alpha = \hat{z}(\psi z)$ (2)
$\vdash . *20·24 . \quad \supset \vdash :. \hat{z}(\phi z) = \hat{z}(\psi z) . \alpha = \hat{z}(\psi z) . \supset . \alpha = \hat{z}(\phi z) :.$
[Exp] $\supset \vdash :. \hat{z}(\phi z) = \hat{z}(\psi z) . \supset : \alpha = \hat{z}(\psi z) . \supset . \alpha = \hat{z}(\phi z)$ (3)
$\vdash . (2) . (3) . \quad \supset \vdash :. \hat{z}(\phi z) = \hat{z}(\psi z) . \supset : \alpha = \hat{z}(\phi z) . \equiv . \alpha = \hat{z}(\psi z) :.$
[*10·11·21] $\supset \vdash :. \hat{z}(\phi z) = \hat{z}(\psi z) . \supset : \alpha = \hat{z}(\phi z) . \equiv_\alpha . \alpha = \hat{z}(\psi z)$ (4)
$\vdash . (1) . (4) . \quad \supset \vdash .$ Prop

***20·3.** $\vdash : x \in \hat{z}(\psi z) . \equiv . \psi x$

Dem.

$\vdash . *20·1 . \supset$
$\vdash :: x \in \hat{z}(\psi z) . \equiv :. (\exists \phi) :. \psi y . \equiv_y . \phi ! y : x \in (\phi ! \hat{z}) :.$
[(*20·02)] $\equiv :. (\exists \phi) :. \psi y . \equiv_y . \phi ! y : \phi ! x :.$
[*10·43] $\equiv :. (\exists \phi) :. \psi y . \equiv_y . \phi ! y : \psi x :.$
[*10·35] $\equiv :. (\exists \phi) : \psi y . \equiv_y . \phi ! y :. \psi x :.$
[*12·1] $\equiv :. \psi x :: \supset \vdash .$ Prop

This proposition shows that x is a member of the class determined by ψ when, and only when, x satisfies ψ.

***20·31.** $\vdash :. \hat{z}(\psi z) = \hat{z}(\chi z) . \equiv : x \in \hat{z}(\psi z) . \equiv_x . x \in \hat{z}(\chi z)$ [*20·15·3]

***20·32.** $\vdash . \hat{x}\{x \in \hat{z}(\phi z)\} = \hat{z}(\phi z)$ [*20·3·15]

***20·33.** $\vdash :. \alpha = \hat{z}(\phi z) . \equiv : x \in \alpha . \equiv_x . \phi x$

Dem.

$\vdash . *20·31 . \quad \supset \vdash :. \alpha = \hat{z}(\phi z) . \equiv : x \in \alpha . \equiv_x . x \in \hat{z}(\phi z)$ (1)
$\vdash . (1) . *20·3 . \supset \vdash .$ Prop

Here α is written in place of some expression of the form $\hat{z}(\psi z)$. The use of the single Greek letter is more convenient whenever the determining function is irrelevant.

***20·34.** $\vdash :. x = y . \equiv : x \epsilon \alpha . \supset_\alpha . y \epsilon \alpha$

Dem.

$\vdash . *4\cdot2 . (*20\cdot07) . \supset \vdash :. x \epsilon \alpha . \supset_\alpha . y \epsilon \alpha : \equiv : x \epsilon \hat{z}(\phi ! z) . \supset_\phi . y \epsilon \hat{z}(\phi ! z) :$

[*20·3] $\equiv : \phi ! x . \supset_\phi . \phi ! y :$

[*13·1] $\equiv : x = y :. \supset \vdash . \text{Prop}$

The above proposition and *20·25 illustrate the use of Greek letters as apparent variables.

***20·35.** $\vdash :. x = y . \equiv : x \epsilon \alpha . \equiv_\alpha . y \epsilon \alpha$ [*20·3 . *13·11]

***20·4.** $\vdash : \alpha \epsilon \text{Cls} . \equiv . (\exists \phi) . \alpha = \hat{z}(\phi ! z)$ [*20·3 . (*20·03)]

***20·41.** $\vdash . \hat{z}(\psi z) \epsilon \text{Cls}$ [*20·4·151]

***20·42.** $\vdash . \hat{z}(z \epsilon \alpha) = \alpha$

A Greek letter, such as α, is merely an abbreviation for an expression of the form $\hat{z}(\phi z)$; thus this proposition is *20·32 repeated.

Dem.

$\vdash . *20\cdot3 . *10\cdot11 . \supset \vdash : x \epsilon \hat{z}(\psi z) . \equiv_x . \psi x :$

[*20·15] $\supset \vdash . \hat{x}\{x \epsilon \hat{z}(\psi z)\} = \hat{x}(\psi x) . \supset \vdash . \text{Prop}$

***20·43.** $\vdash :. \alpha = \beta . \equiv : x \epsilon \alpha . \equiv_x . x \epsilon \beta$ [*20·31]

The following propositions deal with cases in which both classes and descriptions occur. In such cases, we shall, in the absence of any indication to the contrary, adopt the convention that the descriptions are to have a larger scope than the classes, in applying the definitions *14·01 and *20·01.

***20·5.** $\vdash : (\imath x)(\phi x) \epsilon \hat{z}(\psi z) . \equiv . \psi\{(\imath x)(\phi x)\}$

Dem.

$\vdash . *14\cdot1 . \supset \vdash :: (\imath x)(\phi x) \epsilon \hat{z}(\psi z) . \equiv :. (\exists c) : \phi x . \equiv_x . x = c : c \epsilon \hat{z}(\psi z) :.$

[*20·3] $\equiv :. (\exists c) : \phi x . \equiv_x . x = c : \psi c :.$

[*14·1] $\equiv :. \psi\{(\imath x)(\phi x)\} :: \supset \vdash . \text{Prop}$

***20·51.** $\vdash :. (\imath x)(\phi x) = b . \equiv : (\imath x)(\phi x) \epsilon \alpha . \equiv_\alpha . b \epsilon \alpha$

Dem.

$\vdash . *20\cdot5\cdot3 . \supset$

$\vdash :. (\imath x)(\phi x) \epsilon \hat{z}(\psi ! z) . \equiv . b \epsilon \hat{z}(\psi ! z) : \equiv : \psi ! (\imath x)(\phi x) . \equiv . \psi ! b :. \supset$

[*10·11] $\vdash :. (\imath x)(\phi x) \epsilon \alpha . \equiv_\alpha . b \epsilon \alpha : \equiv : \psi ! (\imath x)(\phi x) . \equiv_\psi . \psi ! b :$

[*14·17] $\equiv : (\imath x)(\phi x) = b :. \supset \vdash . \text{Prop}$

***20·52.** $\vdash :. E ! (\imath x)(\phi x) . \equiv : (\exists b) : (\imath x)(\phi x) \epsilon \alpha . \equiv_\alpha . b \epsilon \alpha$

Dem.

$\vdash . *20\cdot51 . *10\cdot11\cdot281 . \supset$

$\vdash :. (\exists b) . (\imath x)(\phi x) = b . \equiv : (\exists b) : (\imath x)(\phi x) \epsilon \alpha . \equiv_\alpha . b \epsilon \alpha$ (1)

$\vdash . (1) . *14\cdot204 . \supset \vdash . \text{Prop}$

***20·53.** $\vdash :. \beta = \alpha . \supset_\beta . \phi\beta : \equiv . \phi\alpha$

This is the analogue of *13·191.

Dem.

$\vdash . *10\cdot1 . \quad \supset \vdash :. \beta = \alpha . \supset_\beta . \phi\beta : \supset : \alpha = \alpha . \supset . \phi\alpha :$

[*20·2] $\quad \supset : \phi\alpha \quad (1)$

$\vdash . *20\cdot18\cdot21 . \supset \vdash :. \beta = \alpha . \supset : \phi\alpha . \supset . \phi\beta :.$

[Comm] $\quad \supset \vdash :. \phi\alpha . \supset : \beta = \alpha . \supset . \phi\beta :.$

[*10·11·21] $\quad \supset \vdash :. \phi\alpha . \supset : \beta = \alpha . \supset_\beta . \phi\beta \quad (2)$

$\vdash . (1) . (2) . \quad \supset \vdash .$ Prop

***20·54.** $\vdash : (\exists\beta) . \beta = \alpha . \phi\beta . \equiv . \phi\alpha$

This proposition is the analogue of *13·195.

Dem.

$\vdash . *20\cdot18 . *10\cdot11 . \supset \vdash : \beta = \alpha . \phi\beta . \supset_\beta . \phi\alpha :$

[*10·23] $\quad \supset \vdash : (\exists\beta) . \beta = \alpha . \phi\beta . \supset . \phi\alpha \quad (1)$

$\vdash . *20\cdot2 . *3\cdot2 . \quad \supset \vdash : \phi\alpha . \supset . \alpha = \alpha . \phi\alpha .$

[*10·24] $\quad \supset . (\exists\beta) . \beta = \alpha . \phi\beta \quad (2)$

$\vdash . (1) . (2) . \quad \supset \vdash .$ Prop

***20·55.** $\vdash . \hat{z}(\phi z) = (\imath\alpha)(x \in \alpha . \equiv_x . \phi x)$

Dem.

$\vdash . *20\cdot33 . \supset \vdash :: x \in \alpha . \equiv_x . \phi x : \equiv_\alpha . \alpha = \hat{z}(\phi z) :.$

[*20·54] $\quad \supset \vdash :. (\exists\beta) :. x \in \alpha . \equiv_x . \phi x : \equiv_\alpha . \alpha = \beta :. \hat{z}(\phi z) = \beta :.$

[*14·1] $\quad \supset \vdash . \hat{z}(\phi z) = (\imath\alpha)(x \in \alpha . \equiv_x . \phi x) . \supset \vdash .$ Prop

***20·56.** $\vdash . E!(\imath\alpha)(x \in \alpha . \equiv_x . \phi x)$ [*20·55 . *14·21]

***20·57.** $\vdash :. \hat{z}(\phi z) = (\imath\alpha)(f\alpha) . \supset : g\{\hat{z}(\phi z)\} . \equiv . g\{(\imath\alpha)(f\alpha)\}$

Dem.

$\vdash . *14\cdot1 . \quad \supset \vdash :: \text{Hp} . \equiv :. (\exists\beta) : f\alpha . \equiv_\alpha . \alpha = \beta : \hat{z}(\phi z) = \beta :.$

[*20·54] $\quad \equiv :. f\alpha . \equiv_\alpha . \alpha = \hat{z}(\phi z) \quad (1)$

$\vdash . *14\cdot1 . \quad \supset \vdash :. g\{(\imath\alpha)(f\alpha)\} . \equiv : (\exists\beta) : f\alpha . \equiv_\alpha . \alpha = \beta : g\beta \quad (2)$

$\vdash . (1) . (2) . \supset \vdash :: \text{Hp} . \supset :. g\{(\imath\alpha)(f\alpha)\} . \equiv : (\exists\beta) : \alpha = \hat{z}(\phi z) . \equiv_\alpha . \alpha = \beta : g\beta :$

[*13·183] $\quad \equiv : (\exists\beta) . \hat{z}(\phi z) = \beta . g\beta :$

[*20·54] $\quad \equiv : g\{\hat{z}(\phi z)\} :: \supset \vdash .$ Prop

***20·58.** $\vdash . \hat{z}(\phi z) = (\imath\alpha)\{\alpha = \hat{z}(\phi z)\}$

Dem.

$\vdash . *4\cdot2 . *10\cdot11 . \supset \vdash : \alpha = \hat{z}(\phi z) . \equiv_\alpha . \alpha = \hat{z}(\phi z) :$

[*20·54] $\quad \supset \vdash :. (\exists\beta) :. \alpha = \hat{z}(\phi z) . \equiv_\alpha . \alpha = \beta : \hat{z}(\phi z) = \beta :.$

[*14·1] $\quad \supset \vdash . \hat{z}(\phi z) = (\imath\alpha)\{\alpha = \hat{z}(\phi z)\} . \supset \vdash .$ Prop

***20·59.** $\vdash : \hat{z}(\phi z) = (\imath\alpha)(f\alpha) . \equiv . (\imath\alpha)(f\alpha) = \hat{z}(\phi z)$

Dem.

$\vdash . *20\cdot1 . \supset \vdash :. \hat{z}(\phi z) = (\imath\alpha)(f\alpha) . \equiv : (\exists\psi) : \phi x . \equiv_x . \psi ! x : \psi ! \hat{z} = (\imath\alpha)(f\alpha) :$

[*14·13] $\quad \equiv : (\exists\psi) : \phi x . \equiv_x . \psi ! x : (\imath\alpha)(f\alpha) = \psi ! \hat{z} :$

[*20·1] $\quad \equiv : (\imath\alpha)(f\alpha) = \hat{z}(\phi z) :. \supset \vdash .$ Prop

In the following propositions, we shall prove that classes have all the formal properties of individuals, and have the same relations to classes of classes as individuals have to classes of individuals. It is only necessary to prove the analogues of our primitive propositions, and of our definitions in cases where their analogues are not themselves definitions. We shall take the propositions ∗10·1·11·12·121·122, rather than those of ∗9, and we shall prove the analogue of ∗10·01. As was pointed out in ∗10, we shall thus have proved everything upon which subsequent proofs depend. The analogues of ∗20·01·02 and of ∗14·01 remain definitions, but those of ∗10·01 and ∗13·01 become propositions to be proved. ∗9·131 must be extended by the definition: Two classes are "of the same type" when they have predicative defining functions of the same type. In addition to these, we have to prove the analogues of ∗10·1·11·12·121·122, ∗11·07 and ∗12·1·11. When these have been proved, the analogues of other propositions follow by merely repeating previous proofs. These analogues will, therefore, be quoted by the numbers of the original propositions whose analogues they are.

∗20·6. $\vdash : (\exists \alpha) . f\alpha . \equiv . \sim \{(\alpha) . \sim f\alpha\}$

Dem.

$\vdash . *4\cdot 2 . (*20\cdot 071) . \supset$

$\vdash : (\exists \alpha) . f\alpha . \equiv . (\exists \phi) . f\{\hat{z}(\phi ! z)\} .$

$[(*10\cdot 01)] \quad \equiv . \sim [(\phi) . \sim f\{\hat{z}(\phi ! z)\}] .$

$[(*20\cdot 07)] \quad \equiv . \sim \{(\alpha) . \sim f\alpha\} : \supset \vdash . \text{Prop}$

This is the analogue of ∗10·01.

∗20·61. $\vdash : (\alpha) . f\alpha . \supset . f\beta$

Dem.

$\vdash . *10\cdot 1 . (*20\cdot 07) . \supset \vdash : (\alpha) . f\alpha . \supset . f\{\hat{z}(\phi ! z)\} : \supset \vdash . \text{Prop}$

This is the analogue of ∗10·1.

In practice we also need

$$\vdash : (\alpha) . f\alpha . \supset . f\{\hat{z}(\psi z)\}.$$

This is ∗20·17.

We need further $\quad \vdash . (\exists \alpha) . \hat{z}(\psi z) = \alpha.$

This is ∗20·41.

∗20·62. When $f\beta$ is true, whatever possible argument of the form $\hat{z}(\phi ! z)$ β may be, then $(\alpha) . f\alpha$ is true.

This is the analogue of ∗10·11.

Dem.

$\vdash . *10\cdot 11 . \supset .$ when $f\{\hat{z}(\phi ! z)\}$ is true, whatever possible argument ϕ may be, then $(\phi) . f\{\hat{z}(\phi ! z)\}$ is true, *i.e.* (by ∗20·07), $(\alpha) . f\alpha$ is true.

∗20·63. $\vdash :. (\alpha) . p \vee f\alpha . \supset : p . \vee . (\alpha) . f\alpha$

This is the analogue of ∗10·12.

Dem.

$\vdash . *4\cdot2 . (*20\cdot07) . \supset$

$\vdash :. (\alpha) . p \vee f\alpha . \equiv : (\phi) . p \vee f\{\hat{z}(\phi ! z)\} :$

[*10·12] $\equiv : p . \vee . (\phi) . f\{\hat{z}(\phi ! z)\} :$

[(*20·07)] $\equiv : p . \vee . (\alpha) . f\alpha :. \supset \vdash . \text{Prop}$

***20·631.** If "$f\alpha$" is significant, then if β is of the same type as α, "$f\beta$" is significant, and vice versa.

This is the analogue of *10·121.

Dem.

By *20·151, α is of the form $\hat{z}(\phi ! z)$, and therefore, by *20·01, $f\alpha$ is a function of $\phi ! \hat{z}$. Similarly β is of the form $\hat{z}(\psi ! z)$, and $f\beta$ is a function of $\psi ! \hat{z}$. Hence by applying *10·121 to $\phi ! \hat{z}$ and $\psi ! \hat{z}$ the result follows.

***20·632.** If, for some α, there is a proposition $f\alpha$, then there is a function $f\hat{\alpha}$, and vice versa.

Dem.

By the definition in *20·01, $f\{\hat{z}(\psi ! z)\}$ is a function of $\psi ! \hat{z}$. Hence the proposition follows from *10·122.

***20·633.** "Whatever possible class α may be, $f(\alpha, \beta)$ is true whatever possible class β may be" implies the corresponding statement with α and β interchanged except in "$f(\alpha, \beta)$."

This is the analogue of *11·07, and follows at once from *11·07 because $f(\alpha, \beta)$ is a function of the defining functions of α and β.

***20·64.** $\vdash :. (\alpha) . f\alpha : (\alpha) . g\alpha : \supset . f\beta . g\beta$

Dem.

$\vdash . *4\cdot2 . (*20\cdot07) . \supset$

$\vdash :. (\alpha) . f\alpha : (\alpha) . g\alpha : \equiv : (\phi) . f\{\hat{z}(\phi ! z)\} : (\phi) . g\{\hat{z}(\phi ! z)\} :$

[*10·14] $\supset : f\{\hat{z}(\psi ! z)\} . g\{\hat{z}(\psi ! z)\} :. \supset \vdash . \text{Prop}$

Observe that "β" is merely an abbreviation for any symbol of the form $\hat{z}(\psi ! z)$. This is why nothing further is required in the above proof.

The above proposition is the analogue of *10·14. Like that proposition, it requires, for the significance of the conclusion, that f and g should be functions which take arguments of the same type. This is not required for the significance of the hypothesis. Hence, though the above proposition is true whenever it is significant, it is not true whenever its hypothesis is significant.

***20·7.** $\vdash : (\exists g) : f\alpha . \equiv_\alpha . g ! \alpha$ [*20·112]

This is the analogue of *12·1.

***20·701.** $\vdash : (\exists g) : f\{\hat{z}(\phi ! z), x\} . \equiv_{\phi, x} . g ! \{\hat{z}(\phi ! z), x\}$

[The proof proceeds as in *20·112, using *12·11 instead of *12·1.]

***20·702.** $\vdash : (\exists g) : f\{x, \hat{z}(\phi ! z)\} . \equiv_{\phi, x} . g ! \{x, \hat{z}(\phi ! z)\}$

[Proof as in *20·701.]

***20·703.** $\vdash : (\exists g) : f\{\hat{z}(\phi ! z), \hat{z}(\psi ! z)\} . \equiv_{\phi, \psi} . g ! \{\hat{z}(\phi ! z), \hat{z}(\psi ! z)\}$

Dem.

$\vdash . *10\cdot311 . \supset \vdash :. f\{\chi ! \hat{z}, \theta ! \hat{z}\} . \equiv_{\chi, \theta} . g ! \{\chi ! \hat{z}, \theta ! \hat{z}\} : \supset :$
$\phi ! x \equiv_x \chi ! x . \psi ! x \equiv_x \theta ! x . f\{\chi ! \hat{z}, \theta ! \hat{z}\} . \equiv_{\chi, \theta} .$
$\phi ! x \equiv_x \chi ! x . \psi ! x \equiv_x \theta ! x . g ! \{\chi ! \hat{z}, \theta ! \hat{z}\}$ (1)

$\vdash . (1) . *11\cdot11\cdot3\cdot341 . \supset$

$\vdash :. \text{Hp}(1) . \quad \supset : (\exists \chi, \theta) . \phi ! x \equiv_x \chi ! x . \psi ! x \equiv_x \theta ! x . f\{\chi ! \hat{z}, \theta ! \hat{z}\} . \equiv_{\phi, \psi} .$
$(\exists \chi, \theta) . \phi ! x \equiv_x \chi ! x . \psi ! x \equiv_x \theta ! x . g ! \{\chi ! \hat{z}, \theta ! \hat{z}\} :$

$[*20\cdot1 . *10\cdot35] \supset : f\{\hat{z}(\phi ! z), \hat{z}(\psi ! z)\} . \equiv_{\phi, \psi} . g ! \{\phi ! \hat{z}, \psi ! \hat{z}\}$ (2)

$\vdash . (2) . *10\cdot11\cdot281 . \supset$

$\vdash :. (\exists g) : f\{\chi ! \hat{z}, \theta ! \hat{z}\} . \equiv_{\chi, \theta} . g ! \{\chi ! \hat{z}, \theta ! \hat{z}\} : \supset :$
$(\exists g) : f\{\hat{z}(\phi ! z), \hat{z}(\psi ! z)\} . \equiv_{\phi, \psi} . g ! \{\hat{z}(\phi ! z), \hat{z}(\psi ! z)\}$ (3)

$\vdash . (3) . *12\cdot11 . \supset \vdash . \text{Prop}$

*20·701·702·703 give the analogues, for classes, of *12·11.

***20·71.** $\vdash :. \alpha = \beta . \equiv : g ! \alpha . \supset_g . g ! \beta$ [*20·19]

This is the analogue of *13·01.

This completes the proof that all propositions hitherto given apply to classes as well as to individuals. Precisely similar reasoning extends this result to classes of classes, classes of classes of classes, etc.

From the above propositions it appears that, although expressions such as $\hat{z}(\phi z)$ have no meaning in isolation, yet those of their formal properties with which we have been hitherto concerned are the same as the corresponding properties of symbols which have a meaning in isolation. Hence nothing in the apparatus hitherto introduced requires us to determine whether a given symbol stands for a class or not, unless the symbol occurs in a way in which only a class can occur significantly. This is an important result, which enables us to give much greater generality to our propositions than would otherwise be possible.

The two following propositions (*20·8·81) are consequences of *13·3. The "type" of any object x will be defined in *63 as the class of terms either identical with x or not identical with x. We may define the "type of the arguments to $\phi\hat{z}$" as the class of arguments x for which "ϕx" is significant, *i.e.* the class $\hat{x}(\phi x \vee \sim \phi x)$. Then the first of the following propositions shows that if "ϕa" is significant, the type of the arguments to $\phi\hat{z}$ is the type of a; the second proposition shows that, if "ϕa" and "ψa" are both significant, the type of the arguments to $\phi\hat{z}$ is the same as the type of the arguments to $\psi\hat{z}$, because each is the type of a. *20·8 will be used in *63·11, which is a fundamental proposition in the theory of relative types.

***20·8.** $\vdash : \phi a \lor \sim \phi a \,.\supset .\, \hat{x}(\phi x \lor \sim \phi x) = \hat{x}(x = a \,.\lor .\, x \neq a)$

Dem.

$\vdash . *13\cdot3 . *10\cdot11\cdot21 . \supset$

$\vdash :: \mathrm{Hp} . \supset :. \phi x \lor \sim \phi x \,.\equiv_x :\, x = a \,.\lor .\, x \neq a :.$

$[*20\cdot15] \supset :. \hat{x}(\phi x \lor \sim \phi x) = \hat{x}(x = a \,.\lor .\, x \neq a) :: \supset \vdash . \mathrm{Prop}$

***20·81.** $\vdash : \phi a \lor \sim \phi a \,.\, \psi a \lor \sim \psi a \,.\supset .\, \hat{x}(\phi x \lor \sim \phi x) = \hat{x}(\psi x \lor \sim \psi x)$

Dem.

$\vdash . *20\cdot8 . \supset \vdash : \mathrm{Hp} . \supset . \hat{x}(\phi x \lor \sim \phi x) = \hat{x}(x = a \,.\lor .\, x \neq a)$ (1)

$\vdash . *20\cdot8 . \supset \vdash : \mathrm{Hp} . \supset . \hat{x}(\psi x \lor \sim \psi x) = \hat{x}(x = a \,.\lor .\, x \neq a)$ (2)

$\vdash . (1) . (2) . *10\cdot121\cdot13 . \mathrm{Comp} . \supset$

$\vdash : \mathrm{Hp} . \supset . \hat{x}(\phi x \lor \sim \phi x) = \hat{x}(x = a . \lor . x \neq a) . \hat{x}(\psi x \lor \sim \psi x) = \hat{x}(x = a . \lor . x \neq a) .$

$[*20\cdot24] \supset . \hat{x}(\phi x \lor \sim \phi x) = \hat{x}(\psi x \lor \sim \psi x) : \supset \vdash . \mathrm{Prop}$

In the third line of the above proof, the use of *10·121 depends upon the fact that the "a" in both (1) and (2) must be such as to render the hypothesis significant, *i.e.* such as to render

$$\text{“}\phi a \lor \sim \phi a \,.\, \psi a \lor \sim \psi a\text{”}$$

significant. Hence the "a" in (1) and the "a" in (2) must be of the same type, by *10·121, and hence by *10·13 we can assert the product of (1) and (2), identifying the two "a's."

Since a type is the range of significance of a function, if ϕx is a function which is always true, $\hat{z}(\phi z)$ must be a type. For if a function is always true, the arguments for which it is true are the same as the arguments for which it is significant; hence $\hat{z}(\phi z)$ is the range of significance of ϕx, if $(x) . \phi x$ holds. Thus any class α is a type if $(x) . x \in \alpha$. It follows that, whatever function ϕ may be, $\hat{x}(\phi x \lor \sim \phi x)$ is a type; and in particular, $\hat{x}(x = a \,.\lor .\, x \neq a)$ is a type. Since a is a member of this class, this class is the type to which a belongs. In virtue of *20·8, if ϕa is significant, the type to which a belongs is the class of arguments for which ϕx is significant, *i.e.* $\hat{x}(\phi x \lor \sim \phi x)$. And if there is any argument a for which ϕa and ψa are both significant, then $\phi \hat{x}$ and $\psi \hat{x}$ have the same range of significance, in virtue of *20·81.

*21. GENERAL THEORY OF RELATIONS

Summary of *21.

The definitions and propositions of this number are exactly analogous to those of *20, from which they differ by being concerned with functions of two variables instead of one. A *relation*, as we shall use the word, will be understood in extension: it may be regarded as the class of couples (x, y) for which some given function $\psi(x, y)$ is true. Its relation to the function $\psi(\hat{x}, \hat{y})$ is just like that of the class to its determining function. We put

***21·01.** $f\{\hat{x}\hat{y}\psi(x, y)\} \,.\!=:\, (\exists\phi) : \phi\,!\,(x, y) \,.\equiv_{x,y}.\, \psi(x, y) : f\{\phi\,!\,(\hat{u}, \hat{v})\}$ Df

Here "$\hat{x}\hat{y}\psi(x, y)$" has no meaning in isolation, but only in certain of its uses. In *21·01 the *alphabetical* order of u and v corresponds to the *typographical* order of $\hat{x}$ and $\hat{y}$ in $f\{\hat{x}\hat{y}\psi(x, y)\}$, so that

$$f\{\hat{y}\hat{x}\psi(x, y)\} \,.\!=:\, (\exists\phi) : \phi\,!\,(x, y) \,.\equiv_{x,y}.\, \psi(x, y) : f\{\phi\,!\,(\hat{v}, \hat{u})\} \quad \text{Df}$$

This is important in relation to the substitution-convention below.

It will be shown that

$$\hat{x}\hat{y}\psi(x, y) = \hat{x}\hat{y}\chi(x, y) \,.\equiv:\, \psi(x, y) \,.\equiv_{x,y}.\, \chi(x, y),$$

i.e. that two relations, as above defined, are identical when, and only when, they are satisfied by the same pair of arguments.

For substitution in $\phi\,!\,(\hat{x}, \hat{y})$ and $\phi\,!\,(\hat{y}, \hat{x})$, we adopt the convention that when a function (as opposed to its values) is represented in a form involving $\hat{x}$ and $\hat{y}$, or any other two letters of the alphabet, the value of this function for the arguments a and b is to be found by substituting a for $\hat{x}$ and b for $\hat{y}$, while the value for the arguments b and a is to be found by substituting b for $\hat{x}$ and a for $\hat{y}$. That is, the argument mentioned first is to be substituted for the letter which comes first in the alphabet, and the argument mentioned second for the later letter; thus the mode of substitution depends upon the *alphabetical* order of the letters which have circumflexes and the *typographical* order of the other letters.

The above convention as to order is presupposed in the following definition, where a is the first argument mentioned and b the second:

***21·02.** $a\{\phi\,!\,(\hat{x}, \hat{y})\}\,b \,.=.\, \phi\,!\,(a, b)$ Df

Hence, following the convention,

$$b\{\phi\,!\,(\hat{x}, \hat{y})\}\,a \,.=.\, \phi\,!\,(b, a) \quad \text{Df}$$
$$a\{\phi\,!\,(\hat{y}, \hat{x})\}\,b \,.=.\, \phi\,!\,(b, a) \quad \text{Df}$$
$$b\{\phi\,!\,(\hat{y}, \hat{x})\}\,a \,.=.\, \phi\,!\,(a, b) \quad \text{Df}$$

This definition is not used as it stands, but is introduced for the sake of

$$a\{\hat{x}\hat{y}\psi(x, y)\}\,b \,.\equiv:\, (\exists\phi) : \phi\,!\,(x, y) \,.\equiv_{x,y}.\, \psi(x, y) : \phi\,!\,(a, b)$$

which results from ∗21·01·02. We shall use capital Latin letters to represent variable expressions of the form $\hat{x}\hat{y}\phi\,!\,(x, y)$, just as we used Greek letters for variable expressions of the form $\hat{z}\,(\phi\,!\,z)$. If a capital Latin letter, say R, is used as an apparent variable, it is supposed that the R which occurs in the form "(R)" or "$(\exists R)$" is to be replaced by "(ϕ)" or "$(\exists\phi)$," while the R which occurs later is to be replaced by "$\hat{x}\hat{y}\phi\,!\,(x, y)$." In fact we put

$$(R)\,.\,fR\,.\,=\,.\,(\phi)\,.\,f\{\hat{x}\hat{y}\phi\,!\,(x, y)\}\quad \text{Df.}$$

The use of single letters for such expressions as $\hat{x}\hat{y}\phi\,(x, y)$ is a practically indispensable convenience.

The following is the definition of the class of relations:

∗21·03. $\text{Rel} = \hat{R}\,\{(\exists\phi)\,.\,R = \hat{x}\hat{y}\phi\,!\,(x, y)\}$ Df

Similar remarks apply to it as to the definition of "Cls" (∗20·03).

In virtue of the definitions ∗21·01·02 and the convention as to capital Latin letters, the notation "xRy" will mean "x has the relation R to y." This notation is practically convenient, and will, after the preliminaries, wholly replace the cumbrous notation $x\,\{\hat{x}\hat{y}\phi\,(x, y)\}\,y$.

The proofs of the propositions of this number are usually omitted, since they are exactly analogous to those of ∗20, merely substituting ∗12·11 for ∗12·1, and propositions in ∗11 for propositions in ∗10.

The propositions of this number, like those of ∗20, fall into three sections. Those of the second section are seldom referred to. Those of the third section, extending to relations the formal properties hitherto assumed or proved for individuals and functions, are not explicitly referred to in the sequel, but are constantly relevant, namely whenever a proposition which has been assumed or proved for individuals and functions is applied to relations. The principal propositions of the first section are the following.

∗21·15. $\vdash :.\ \psi\,(x, y)\,.\,\equiv_{x,y}\,.\,\chi\,(x, y) : \equiv .\ \hat{x}\hat{y}\psi\,(x, y) = \hat{x}\hat{y}\chi\,(x, y)$

I.e. two relations are identical when, and only when, their defining functions are formally equivalent.

∗21·31. $\vdash :.\ \hat{x}\hat{y}\psi\,(x, y) = \hat{x}\hat{y}\chi\,(x, y)\,.\,\equiv : x\,\{\hat{x}\hat{y}\psi\,(x, y)\}\,y\,.\,\equiv_{x,y}\,.\,x\,\{\hat{x}\hat{y}\chi\,(x, y)\}\,y$

I.e. two relations are identical when, and only when, they hold between the same pairs of terms. The same fact is expressed by the following proposition:

∗21·43. $\vdash :.\ R = S\,.\,\equiv : xRy\,.\,\equiv_{x,y}\,.\,xSy$

∗21·2·21·22 show that identity of relations is reflexive, symmetrical and transitive.

∗21·3. $\vdash : x\,\{\hat{x}\hat{y}\psi\,(x, y)\}\,y\,.\,\equiv .\ \psi\,(x, y)$

I.e. two terms have a given relation when, and only when, they satisfy its defining function.

***21·151.** $\vdash . (\exists\phi) . \hat{x}\hat{y}\psi(x, y) = \hat{x}\hat{y}\phi ! (x, y)$

I.e. every relation can be defined by a predicative function. Hence when, using *21·07 or *21·071, we have a relation as apparent variable, and are therefore confined to predicative defining functions, there is no loss of generality.

***21·01.** $f\{\hat{x}\hat{y}\psi(x, y)\} . = : (\exists\phi) : \phi ! (x, y) . \equiv_{x,y} . \psi(x, y) : f\{\phi ! (\hat{u}, \hat{v})\}$ Df

On the convention as to order in *21·01·02, cf. p. 200, and thus relate $\hat{u}, \hat{v}$ to $\hat{x}, \hat{y}$ so that

$$f\{\hat{y}\hat{x}\psi(x, y)\} . = : (\exists\phi) : \phi ! (x, y) . \equiv_{x,y} . \psi(x, y) : f\{\phi ! (\hat{v}, \hat{u})\} \quad \text{Df}$$

***21·02.** $a\{\phi ! (\hat{x}, \hat{y})\} b . = . \phi ! (a, b)$ Df

***21·03.** $\text{Rel} = \hat{R}\{(\exists\phi) . R = \hat{x}\hat{y}\phi ! (x, y)\}$ Df

The following definitions merely extend to relations, with as little modification as possible, the definitions already given for other symbols.

***21·07.** $(R) . fR . = . (\phi) . f\{\hat{x}\hat{y}\phi ! (x, y)\}$ Df

***21·071.** $(\exists R) . fR . = . (\exists\phi) . f\{\hat{x}\hat{y}\phi ! (x, y)\}$ Df

***21·072.** $[(\imath R)(\phi R)] . f(\imath R)(\phi R) . = : (\exists S) : \phi R . \equiv_R . R = S : fS$ Df

***21·08.** $f\{\hat{R}\hat{S}\psi(R, S)\} . = : (\exists\phi) : \psi(R, S) . \equiv_{R,S} . \phi ! (R, S) : f\{\phi ! (\hat{R}, \hat{S})\}$ Df

***21·081.** $P\{\phi ! (\hat{R}, \hat{S})\} Q . = . \phi ! (P, Q)$ Df

The convention as to typographic and alphabetic order is here retained.

***21·082.** $f\{\hat{R}(\psi R)\} . = : (\exists\phi) : \psi R . \equiv_R . \phi ! R : f(\phi ! \hat{R})$ Df

***21·083.** $R \epsilon \phi ! \hat{R} . = . \phi ! R$ Df

***21·1.** $\vdash :. f\{\hat{x}\hat{y}\psi(x, y)\} . \equiv : (\exists\phi) : \phi ! (x, y) . \equiv_{x,y} . \psi(x, y) : f\{\phi ! (\hat{u}, \hat{v})\}$
[*4·2 . (*21·01)]

***21·11.** $\vdash :. \psi(x, y) . \equiv_{x,y} . \chi(x, y) : \supset : f\{\hat{x}\hat{y}\psi(x, y)\} . \equiv . f\{\hat{x}\hat{y}\chi(x, y)\}$
[*4·86·36 . *10·281 . *21·1]

This proposition proves that every proposition about a relation expresses an extensional property of the determining function.

***21·111.** $\vdash :. f\{\phi ! (\hat{x}, \hat{y})\} . \equiv_\phi . g\{\phi ! (x, y)\} : \supset : f\{\hat{x}\hat{y}\phi ! (x, y)\} . \equiv_\phi . g\{\hat{x}\hat{y}\phi ! (x, y)\}$
[Fact . *11·11·3 . *10·281 . *21·1]

***21·112.** $\vdash :. (\exists g) :. f\{\hat{x}\hat{y}\phi ! (x, y)\} . \equiv_\phi . g ! \{\hat{x}\hat{y}\phi ! (x, y)\}$ [*12·1 . *21·111]

It is *12·1, not *12·11, which is required in this proposition, because we are concerned with a function (f) of *one* variable, namely ϕ, although that one variable is itself a function of two variables.

***21·12.** $\vdash :. (\exists\phi) :. \phi ! (x, y) . \equiv_{x,y} . \psi(x, y) : f\{\hat{x}\hat{y}\psi(x, y)\} . \equiv . f\{\hat{x}\hat{y}\phi ! (x, y)\}$
[*21·11 . *12·11]

This is the first use of the primitive proposition *12·11, except in *20·701·702·703.

***21·13.** $\vdash :. \psi(x, y) . \equiv_{x,y} . \chi(x, y) : \supset . \hat{x}\hat{y}\psi(x, y) = \hat{x}\hat{y}\chi(x, y)$
[*21·1 . *12·11 . *13·195]

*21·14. $\vdash :. \hat{x}\hat{y}\psi(x,y) = \hat{x}\hat{y}\chi(x,y) . \supset : \psi(x,y) . \equiv_{x,y} . \chi(x,y)$
[Proof as in *20·14]

*21·15. $\vdash :. \psi(x,y) . \equiv_{x,y} . \chi(x,y) : \equiv . \hat{x}\hat{y}\psi(x,y) = \hat{x}\hat{y}\chi(x,y)$ [*21·13·14]

This proposition states that two double functions determine the same relation when, and only when, they are formally equivalent, *i.e.* are satisfied by the same pairs of arguments. This is a fundamental property of relations as defined above (*21·01).

*21·151. $\vdash . (\exists\phi) . \hat{x}\hat{y}\psi(x,y) = \hat{x}\hat{y}\phi!(x,y)$ [*21·15 . *12·11]

*21·16. $\vdash : (\exists\phi) : f\{\hat{x}\hat{y}\psi(x,y)\} . \equiv . f\{\hat{x}\hat{y}\phi!(x,y)\}$ [*21·12]

*21·17. $\vdash : (\phi) . f\{\hat{x}\hat{y}\phi!(x,y)\} . \supset . f\{\hat{x}\hat{y}\psi(x,y)\}$ [*21·16 . *10·1]

*21·18. $\vdash :. \hat{x}\hat{y}\phi(x,y) = \hat{x}\hat{y}\psi(x,y) . \supset : f\{\hat{x}\hat{y}\phi(x,y)\} . \equiv . f\{\hat{x}\hat{y}\psi(x,y)\}$
[*21·11·15]

*21·19. $\vdash :. \hat{x}\hat{y}\psi(x,y) = \hat{x}\hat{y}\chi(x,y) . \equiv : (f) : f!\hat{x}\hat{y}\psi(x,y) . \supset . f!\hat{x}\hat{y}\chi(x,y)$
[*21·18 . *10·11·21 . *21·1 . *10·35 . (*13·01) . *21·112 . *10·301]

*21·191. $\vdash :. \hat{x}\hat{y}\psi(x,y) = \hat{x}\hat{y}\chi(x,y) . \equiv : (f) : f!\hat{x}\hat{y}\psi(x,y) . \equiv . f!\hat{x}\hat{y}\chi(x,y)$
[*21·18·19]

*21·2. $\vdash . \hat{x}\hat{y}\phi(x,y) = \hat{x}\hat{y}\phi(x,y)$ [*21·15 . *4·2]

*21·21. $\vdash : \hat{x}\hat{y}\phi(x,y) = \hat{x}\hat{y}\psi(x,y) . \equiv . \hat{x}\hat{y}\psi(x,y) = \hat{x}\hat{y}\phi(x,y)$ [*21·15 . *10·32]

*21·22. $\vdash : \hat{x}\hat{y}\phi(x,y) = \hat{x}\hat{y}\psi(x,y) . \hat{x}\hat{y}\psi(x,y) = \hat{x}\hat{y}\chi(x,y) . \supset .$
$\hat{x}\hat{y}\phi(x,y) = \hat{x}\hat{y}\chi(x,y)$ [*21·15 . *10·301]

*21·23. $\vdash : \hat{x}\hat{y}\phi(x,y) = \hat{x}\hat{y}\psi(x,y) . \hat{x}\hat{y}\phi(x,y) = \hat{x}\hat{y}\chi(x,y) . \supset .$
$\hat{x}\hat{y}\psi(x,y) = \hat{x}\hat{y}\chi(x,y)$ [*21·21·22]

*21·24. $\vdash : \hat{x}\hat{y}\psi(x,y) = \hat{x}\hat{y}\phi(x,y) . \hat{x}\hat{y}\chi(x,y) = \hat{x}\hat{y}\phi(x,y) . \supset .$
$\hat{x}\hat{y}\psi(x,y) = \hat{x}\hat{y}\chi(x,y)$ [*21·21·22]

*21·3. $\vdash : x\{\hat{x}\hat{y}\psi(x,y)\}y . \equiv . \psi(x,y)$ [*21·1·02 . *10·43·35 . *12·11]

This shows that x has to y the relation determined by ψ when, and only when, x and y satisfy $\psi(x,y)$.

Note that the primitive proposition *12·11 is again required here.

*21·31. $\vdash :. \hat{x}\hat{y}\psi(x,y) = \hat{x}\hat{y}\chi(x,y) . \equiv : x\{\hat{x}\hat{y}\psi(x,y)\}y . \equiv_{x,y} . x\{\hat{x}\hat{y}\chi(x,y)\}y$
[*21·15·3]

*21·32. $\vdash . \hat{x}\hat{y}[x\{\hat{x}\hat{y}\phi(x,y)\}y] = \hat{x}\hat{y}\phi(x,y)$ [*21·3·15]

*21·33. $\vdash :. R = \hat{x}\hat{y}\phi(x,y) . \equiv : xRy . \equiv_{x,y} . \phi(x,y)$ [*21·31·3]

Here R is written for some expression of the form $\hat{x}\hat{y}\psi(x,y)$. The use of a single capital letter for a relation is convenient whenever the determining function is irrelevant.

*21·4. $\vdash : R \epsilon \mathrm{Rel} . \equiv . (\exists\phi) . R = \hat{x}\hat{y}\phi!(x,y)$ [*20·3 . (*21·03)]

*21·41. $\vdash . \hat{x}\hat{y}\phi(x,y) \epsilon \mathrm{Rel}$ [*21·4·151]

*21·42. $\vdash . \hat{x}\hat{y}(xRy) = R$ [*21·3·15]

*21·43. $\vdash :. R = S . \equiv : xRy . \equiv_{x,y} . xSy$ [*21·15·3]

*20·5·51·52 have no analogues in the theory of relations.

***21·53.** $\vdash :. S = R \,.\, \supset_S . \, \phi S : \equiv . \, \phi R$ [*10·1 . *21·2·18·21 . Comm . *10·11·21]

***21·54.** $\vdash :. (\exists S) . S = R . \phi S . \equiv . \phi R$ [*21·18 . *10·11·23 . *21·2 . *10·24]

***21·55.** $\vdash . \hat{x}\hat{y}\phi(x, y) = (\imath R)\{xRy . \equiv_{x,y} . \phi(x, y)\}$ [*21·33·54 . *14·1]

***21·56.** $\vdash . E!(\imath R)\{xRy . \equiv_{x,y} . \phi(x, y)\}$ [*21·55 . *14·21]

***21·57.** $\vdash :. \hat{x}\hat{y}\phi(x, y) = (\imath R)(fR) . \supset : g\{\hat{x}\hat{y}\phi(x, y)\} . \equiv . g\{(\imath R)(fR)\}$
[*14·1 . *21·54 . *13·183]

***21·58.** $\vdash : \hat{x}\hat{y}\phi(x, y) = (\imath R)\{R = \hat{x}\hat{y}\phi(x, y)\}$ [*4·2 . *10·11 . *21·54 . *14·1]

The following propositions are the analogues of *20·6 ff., and have a similar purpose.

***21·6.** $\vdash : (\exists R) . fR . \equiv . \sim\{(R) . \sim fR\}$ [Proof as in *20·6]

***21·61.** $\vdash : (R) . fR . \supset . fS$ [Proof as in *20·61]

***21·62.** When fR is true, whatever possible argument of the form $\hat{x}\hat{y}\phi!(x, y)$ R may be, $(R) . fR$ is true. [Proof as in *20·62]

***21·63.** $\vdash :. (R) . p \vee fR . \supset : p . \vee . (R) . fR$ [Proof as in *20·63]

***21·631.** If "fR" is significant, then if S is of the same type as R, "fS" is significant, and vice versa. [Proof as in *20·631]

***21·632.** If, for some R, there is a proposition fR, then there is a function $f\hat{R}$, and vice versa. [Proof as in *20·632]

***21·633.** "Whatever possible relation R may be, $f(R, S)$ is true whatever possible relation S may be" implies "whatever possible relation S may be, $f(R, S)$ is true whatever possible relation R may be."
[Proof as in *20·633]

***21·64.** $\vdash :. (R) . fR : (R) . gR : \supset . fS . gS$ [Proof as in *20·64]

***21·7.** $\vdash : (\exists g) : fR . \equiv_R . g!R$ [Proof as in *20·7]

***21·701** $\vdash : (\exists g) : f(R, x) . \equiv_{R,x} . g!(R, x)$ [Proof as in *20·701]

***21·702** $\vdash : (\exists g) : f(x, R) . \equiv_{R,x} . g!(R, x)$ [Proof as in *20·702]

***21·703** $\vdash : (\exists g) : f(R, S) . \equiv_{R,S} . g!(R, S)$ [Proof as in *20·703]

***21·704.** $\vdash : (\exists g) : f(R, \alpha) . \equiv_{R,\alpha} . g!(R, \alpha)$ [Proof as in *20·703]

***21·705.** $\vdash : (\exists g) : f(\alpha, R) . \equiv_{\alpha,R} . g!(\alpha, R)$ [Proof as in *20·703]

***21·71.** $\vdash :. R = S . \equiv : g!R . \supset_g . g!S$ [Proof as in *20·71]

From the above propositions it appears that relations, like classes, have all the formal properties which they would have if they were symbols having a meaning in isolation. Hence unless a symbol occurs in a way in which only a relation can occur significantly, we do not need to decide whether it stands for a relation or not. This result, like the corresponding result for classes mentioned at the end of *20, is important as giving greater generality to our propositions than they would otherwise possess. The results obtained in *20 and *21 for classes and relations whose members or terms are neither classes nor relations can be extended, by mere repetition of the proofs, to classes of classes, classes of relations, relations of classes, relations of relations, and so on.

*22. CALCULUS OF CLASSES

Summary of *22.

In this number we reach what was historically the starting-point of symbolic logic. The Greek letters used (except ϕ, ψ, χ, θ) are always to stand for expressions of the form $\hat{x}(\phi ! x)$, or, where the Greek letters are not apparent variables, $\hat{x}(\phi x)$. The small Latin letters may either be such as have a meaning in isolation, or may represent classes or relations; this is possible in virtue of the notes at the ends of *20 and *21. We put:

***22·01.** $\alpha \subset \beta \,.=:\, x \epsilon \alpha \,.\supset_x.\, x \epsilon \beta$ Df

This defines "the class α is contained in the class β," or "all α's are β's."

***22·02.** $\alpha \cap \beta = \hat{x}(x \epsilon \alpha \,.\, x \epsilon \beta)$ Df

This defines the logical product or common part of two classes α and β.

***22·03.** $\alpha \cup \beta = \hat{x}(x \epsilon \alpha \,.\vee.\, x \epsilon \beta)$ Df

This defines the logical sum of two classes; it is the class consisting of all the members of one together with all the members of the other.

***22·04.** $-\alpha = \hat{x}(x \sim\epsilon\, \alpha)$ Df

This defines the negation of a class. It is read "not-α." It does not contain every object x concerning which "$x \epsilon \alpha$" is *not true*, but only those objects concerning which "$x \epsilon \alpha$" is *false*; *i.e.* it excludes those objects for which "$x \epsilon \alpha$" is meaningless. Thus it consists of all objects, of the type next below α, which are not members of α; but it does not contain objects of any other type but this.

***22·05.** $\alpha - \beta = \alpha \cap -\beta$ Df

This definition gives an abbreviation which is often convenient.

The postulates required for the algebra of logic have been enumerated by Huntington*. In our notation, they are as follows.

We assume a class K, with two rules of combination, namely $\cup$ and $\cap$; and we then require the following ten postulates:

I *a*. $a \cup b$ is in the class whenever a and b are in the class.

I *b*. $a \cap b$ is in the class whenever a and b are in the class.

II *a*. There is an element Λ such that $a \cup \Lambda = a$ for every element a.

II *b*. There is an element V such that $a \cap \mathrm{V} = a$ for every element a.

III *a*. $a \cup b = b \cup a$ whenever a, b, $a \cup b$ and $b \cup a$ are in the class.

III *b*. $a \cap b = b \cap a$ whenever a, b, $a \cap b$ and $b \cap a$ are in the class.

* *Trans. Amer. Math. Soc.* Vol. 5, July 1904, p. 292.

IV a. $a \cup (b \cap c) = (a \cup b) \cap (a \cup c)$ whenever $a, b, c, a \cup b, a \cup c, b \cap c, a \cup (b \cap c)$, and $(a \cup b) \cap (a \cup c)$ are in the class.

IV b. $a \cap (b \cup c) = (a \cap b) \cup (a \cap c)$ whenever $a, b, c, a \cap b, a \cap c, b \cup c, a \cap (b \cup c)$, and $(a \cap b) \cup (a \cap c)$ are in the class.

V. If the elements Λ and V in postulates II a and II b exist and are unique, then for every element a there is an element $-a$ such that $a \cup -a = V$ and $a \cap -a = \Lambda$.

VI. There are at least two elements, x and y, in the class, such that $x \neq y$.

The form of the above postulates is such that they are mutually independent, *i.e.* any nine of them are satisfied by interpretations of the symbols which do not satisfy the remaining one.

For our purposes, "K" must be replaced by "Cls." Λ and V will be the null-class and the universal class, which are defined in ∗24. Then the above ten postulates are proved below, as follows:

I a, in ∗22·37, namely "$\vdash . \alpha \cup \beta \in \text{Cls}$"
I b, in ∗22·36, namely "$\vdash . \alpha \cap \beta \in \text{Cls}$"
II a, in ∗24·24, namely "$\vdash . \alpha \cup \Lambda = \alpha$"
II b, in ∗24·26, namely "$\vdash . \alpha \cap V = \alpha$"
III a, in ∗22·57, namely "$\vdash . \alpha \cup \beta = \beta \cup \alpha$"
III b, in ∗22·51, namely "$\vdash . \alpha \cap \beta = \beta \cap \alpha$"
IV a, in ∗22·69, namely "$\vdash . (\alpha \cup \beta) \cap (\alpha \cup \gamma) = \alpha \cup (\beta \cap \gamma)$"
IV b, in ∗22·68, namely "$\vdash . (\alpha \cap \beta) \cup (\alpha \cap \gamma) = \alpha \cap (\beta \cup \gamma)$"
V, in ∗24·21·22, namely "$\vdash . \alpha \cap -\alpha = \Lambda$" and "$\vdash . \alpha \cup -\alpha = V$"
VI, in ∗24·1, namely "$\vdash . \Lambda \neq V$"

Hence, assuming Huntington's analysis of the postulates for the formal algebra of logic, the propositions proved in what follows suffice to establish that this algebra holds for classes. The corresponding propositions of ∗23 and ∗25 prove that it holds for relations, substituting Rel, $\dot{\cup}$, $\dot{\cap}$, $\dot{\Lambda}$, $\dot{V}$ for Cls, $\cup$, $\cap$, Λ, V.

The principal propositions of the present number are the following:

(1) Those embodying the formal rules:

∗22·51. $\vdash . \alpha \cap \beta = \beta \cap \alpha$

∗22·57. $\vdash . \alpha \cup \beta = \beta \cup \alpha$

These embody the commutative law.

∗22·52. $\vdash . (\alpha \cap \beta) \cap \gamma = \alpha \cap (\beta \cap \gamma)$

∗22·7. $\vdash . (\alpha \cup \beta) \cup \gamma = \alpha \cup (\beta \cup \gamma)$

These embody the associative law.

∗22·5. $\vdash . \alpha \cap \alpha = \alpha$

∗22·56. $\vdash . \alpha \cup \alpha = \alpha$

These embody the law of tautology.

***22·68.** $\vdash . (\alpha \cap \beta) \cup (\alpha \cap \gamma) = \alpha \cap (\beta \cup \gamma)$

***22·69.** $\vdash . (\alpha \cup \beta) \cap (\alpha \cup \gamma) = \alpha \cup (\beta \cap \gamma)$

These embody the distributive law. It will be seen that the second results from the first by everywhere interchanging the signs of addition and multiplication.

***22·8.** $\vdash . -(-\alpha) = \alpha$

This is the principle of double negation.

***22·81.** $\vdash : \alpha \subset \beta . \equiv . -\beta \subset -\alpha$

This is the principle of transposition.

(2) Other useful propositions:

***22·44.** $\vdash : \alpha \subset \beta . \beta \subset \gamma . \supset . \alpha \subset \gamma$

***22·441.** $\vdash : \alpha \subset \beta . x \epsilon \alpha . \supset . x \epsilon \beta$

These embody the two forms of the syllogism in Barbara.

***22·62.** $\vdash : \alpha \subset \beta . \equiv . \alpha \cup \beta = \beta$

***22·621.** $\vdash : \alpha \subset \beta . \equiv . \alpha \cap \beta = \alpha$

These two propositions enable us to transform any inclusion ($\alpha \subset \beta$) into an equation.

***22·91.** $\vdash . \alpha \cup \beta = \alpha \cup (\beta - \alpha)$

I.e. "α or β" is identical with "α or the part of β which is excluded from α."

***22·01.** $\alpha \subset \beta . = : x \epsilon \alpha . \supset_x . x \epsilon \beta$ Df

***22·02.** $\alpha \cap \beta = \hat{x}(x \epsilon \alpha . x \epsilon \beta)$ Df

***22·03.** $\alpha \cup \beta = \hat{x}(x \epsilon \alpha . \vee . x \epsilon \beta)$ Df

***22·04.** $-\alpha = \hat{x}(x \sim\epsilon \alpha)$ Df

***22·05.** $\alpha - \beta = \alpha \cap -\beta$ Df

***22·1.** $\vdash :. \alpha \subset \beta . \equiv : x \epsilon \alpha . \supset_x . x \epsilon \beta$ [*4·2 . (*22·01)]

***22·2.** $\vdash . \alpha \cap \beta = \hat{x}(x \epsilon \alpha . x \epsilon \beta)$ [*20·2 . (*22·02)]

***22·3.** $\vdash . \alpha \cup \beta = \hat{x}(x \epsilon \alpha . \vee . x \epsilon \beta)$ [*20·2 . (*22·03)]

***22·31.** $\vdash . -\alpha = \hat{x}(x \sim\epsilon \alpha)$ [*20·2 . (*22·04)]

***22·32.** $\vdash . \alpha - \beta = \hat{x}(x \epsilon \alpha . x \sim\epsilon \beta)$ [*20·2 . (*22·05) . *22·2 . *20·32]

***22·33.** $\vdash : x \epsilon \alpha \cap \beta . \equiv . x \epsilon \alpha . x \epsilon \beta$ [*20·3 . *22·2]

***22·34.** $\vdash :. x \epsilon \alpha \cup \beta . \equiv : x \epsilon \alpha . \vee . x \epsilon \beta$ [*20·3 . *22·3]

***22·35.** $\vdash : x \epsilon -\alpha . \equiv . x \sim\epsilon \alpha$ [*20·3 . *22·31]

***22·351.** $\vdash . -\alpha \neq \alpha$

Dem.

$\vdash$. *22·35 . *5·19 . $\supset \vdash : \sim\{x \epsilon -\alpha . \equiv . x \epsilon \alpha\} :$

[*10·11] $\supset \vdash : (x) : \sim\{x \epsilon -\alpha . \equiv . x \epsilon \alpha\} :$

[*10·251] $\supset \vdash : \sim\{(x) : x \epsilon -\alpha . \equiv . x \epsilon \alpha\} :$

[*20·43.Transp] $\supset \vdash : \sim(-\alpha = \alpha) : \supset \vdash$. Prop

This proposition is used in proving that the null-class is not identical with the class containing everything (*24·1), which is used to show that at least two classes exist. Our axioms do not suffice to prove that more than one *individual* exists, but they prove the existence of at least two *classes* and at least two *relations*.

***22·36.** $\vdash . \alpha \cap \beta \epsilon \mathrm{Cls}$ [*20·41]

***22·37.** $\vdash . \alpha \cup \beta \epsilon \mathrm{Cls}$ [*20·41]

***22·38.** $\vdash . - \alpha \epsilon \mathrm{Cls}$ [*20·41]

***22·39.** $\vdash . \hat{z}(\phi z) \cap \hat{z}(\psi z) = \hat{z}(\phi z . \psi z)$

Dem.

$\vdash . *22·33 . \supset \vdash : x \epsilon \hat{z}(\phi z) \cap \hat{z}(\psi z) . \equiv . x \epsilon \hat{z}(\phi z) . x \epsilon \hat{z}(\psi z) .$

[*20·3] $\equiv . \phi x . \psi x$ (1)

$\vdash . (1) . *20·33 . \supset \vdash . \mathrm{Prop}$

***22·391.** $\vdash . \hat{z}(\phi z) \cup \hat{z}(\psi z) = \hat{z}(\phi z \vee \psi z)$ [Similar proof]

***22·392.** $\vdash . - \hat{z}(\phi z) = \hat{z}(\sim \phi z)$ [Similar proof]

***22·4.** $\vdash :. \alpha \subset \beta . \beta \subset \alpha . \equiv : x \epsilon \alpha . \equiv_x . x \epsilon \beta$

Dem.

$\vdash . *22·1 . \supset \vdash :: \alpha \subset \beta . \equiv : x \epsilon \alpha . \supset_x . x \epsilon \beta :. \beta \subset \alpha . \equiv : x \epsilon \beta . \supset_x . x \epsilon \alpha :.$

[*4·38] $\supset \vdash :: \alpha \subset \beta . \beta \subset \alpha . \equiv :. x \epsilon \alpha . \supset_x . x \epsilon \beta : x \epsilon \beta . \supset_x . x \epsilon \alpha :.$

[*10·22] $\equiv :. x \epsilon \alpha . \equiv_x . x \epsilon \beta :: \supset \vdash . \mathrm{Prop}$

***22·41.** $\vdash : \alpha \subset \beta . \beta \subset \alpha . \equiv . \alpha = \beta$ [*22·4 . *20·43]

***22·42.** $\vdash . \alpha \subset \alpha$ [Id . *10·11]

***22·43.** $\vdash : \alpha \cap \beta \subset \alpha$ [*3·26 . *10·11]

***22·44.** $\vdash : \alpha \subset \beta . \beta \subset \gamma . \supset . \alpha \subset \gamma$ [*10·3]

This is one form of the syllogism in Barbara. Another form is the following:

***22·441.** $\vdash : \alpha \subset \beta . x \epsilon \alpha . \supset . x \epsilon \beta$ [*10·1 . Imp]

***22·45.** $\vdash : \alpha \subset \beta . \alpha \subset \gamma . \equiv . \alpha \subset \beta \cap \gamma$

Dem.

$\vdash . *22·1 . \supset \vdash :. \alpha \subset \beta . \alpha \subset \gamma . \equiv : x \epsilon \alpha . \supset_x . x \epsilon \beta : x \epsilon \alpha . \supset_x . x \epsilon \gamma :$

[*10·29] $\equiv : x \epsilon \alpha . \supset_x . x \epsilon \beta . x \epsilon \gamma :$

[*22·33.*10·413] $\equiv : x \epsilon \alpha . \supset_x . x \epsilon \beta \cap \gamma :. \supset \vdash . \mathrm{Prop}$

***22·46.** $\vdash : x \epsilon \alpha . \alpha \subset \beta . \supset . x \epsilon \beta$ [*22·441 . Perm]

***22·47.** $\vdash : \alpha \subset \gamma . \supset . \alpha \cap \beta \subset \gamma$ [22·43.44]

***22·48.** $\vdash : \alpha \subset \beta . \supset . \alpha \cap \gamma \subset \beta \cap \gamma$ [*10·31]

***22·481.** $\vdash : \alpha = \beta . \supset . \alpha \cap \gamma = \beta \cap \gamma$

Dem.

$\vdash . *22·41 . \supset :. \mathrm{Hp} . \supset : \alpha \subset \beta . \beta \subset \alpha :$

[*22·48] $\supset : \alpha \cap \gamma \subset \beta \cap \gamma . \beta \cap \gamma \subset \alpha \cap \gamma :$

[*22·41] $\supset : \alpha \cap \gamma = \beta \cap \gamma :. \supset \vdash . \mathrm{Prop}$

***22·49.** $\vdash : \alpha \subset \beta \,.\, \gamma \subset \delta \,.\, \supset \,.\, \alpha \cap \gamma \subset \beta \cap \delta$ [*10·39]

***22·5.** $\vdash \,.\, \alpha \cap \alpha = \alpha$

Dem.

$$\begin{aligned} &\vdash \,.\, *22{\cdot}33 \,.\, \supset \vdash :.\, x \in \alpha \cap \alpha \,.\, \equiv : x \in \alpha \,.\, x \in \alpha : \\ &[*4{\cdot}24] \qquad\qquad \equiv : x \in \alpha \qquad (1) \\ &\vdash \,.\, (1) \,.\, *10{\cdot}11 \,.\, *20{\cdot}43 \,.\, \supset \vdash \,.\, \text{Prop} \end{aligned}$$

The above is the law of tautology for the logical multiplication of classes.

***22·51.** $\vdash \,.\, \alpha \cap \beta = \beta \cap \alpha$ [*22·33 . *4·3 . *10·11 . *20·43]

***22·52.** $\vdash \,.\, (\alpha \cap \beta) \cap \gamma = \alpha \cap (\beta \cap \gamma)$ [*22·33 . *4·32 . *10·11 . *20·43]

Thus logical multiplication of classes obeys the commutative and associative laws. References to *22·33·34·35 and to *20·43 will in future often be omitted.

***22·53.** $\alpha \cap \beta \cap \gamma = (\alpha \cap \beta) \cap \gamma$ Df

This definition serves merely for the avoidance of brackets.

***22·54.** $\vdash :.\, \alpha = \beta \,.\, \supset : \alpha \subset \gamma \,.\, \equiv \,.\, \beta \subset \gamma$ [*20·18]

***22·55.** $\vdash :.\, \alpha = \beta \,.\, \supset : \gamma \subset \alpha \,.\, \equiv \,.\, \gamma \subset \beta$ [*20·18]

***22·551.** $\vdash : \alpha = \beta \,.\, \supset \,.\, \alpha \cup \gamma = \beta \cup \gamma$ [*10·411]

***22·56.** $\vdash \,.\, \alpha \cup \alpha = \alpha$ [*4·25 . *10·11]

The above is the law of tautology for the logical addition of classes.

***22·57.** $\vdash \,.\, \alpha \cup \beta = \beta \cup \alpha$ [*4·31 . *10·11]

***22·58.** $\vdash \,.\, \alpha \subset \alpha \cup \beta \,.\, \beta \subset \alpha \cup \beta$ [*1·3 . *2·2]

***22·59.** $\vdash : \alpha \subset \gamma \,.\, \beta \subset \gamma \,.\, \equiv \,.\, \alpha \cup \beta \subset \gamma$

Dem.

$$\begin{aligned} &\vdash \,.\, *22{\cdot}1 \,.\, \supset \vdash :: \text{Hp} \,.\, \equiv :.\, x \in \alpha \,.\, \supset_x \,.\, x \in \gamma : x \in \beta \,.\, \supset_x \,.\, x \in \gamma :. \\ &[*10{\cdot}22] \qquad \equiv :.\, (x) :.\, x \in \alpha \,.\, \supset \,.\, x \in \gamma : x \in \beta \,.\, \supset \,.\, x \in \gamma :. \\ &[*4{\cdot}77 . *10{\cdot}271] \qquad \equiv :.\, (x) :.\, x \in \alpha \,.\, \vee \,.\, x \in \beta : \supset \,.\, x \in \gamma :. \\ &[*22{\cdot}34 . *10{\cdot}413] \qquad \equiv :.\, (x) : x \in \alpha \cup \beta \,.\, \supset \,.\, x \in \gamma :: \supset \vdash \,.\, \text{Prop} \end{aligned}$$

The analogue of *4·78, *i.e.*

$$\alpha \subset \beta \,.\, \vee \,.\, \alpha \subset \gamma : \equiv \,.\, \alpha \subset \beta \cup \gamma$$

is false. We have only

$$\alpha \subset \beta \,.\, \vee \,.\, \alpha \subset \gamma : \supset \,.\, \alpha \subset \beta \cup \gamma.$$

A similar remark applies to the analogue of *4·79. Cf. *22·64·65.

***22·6.** $\vdash :.\, x \in \alpha \cup \beta \,.\, \equiv : \alpha \subset \gamma \,.\, \beta \subset \gamma \,.\, \supset_\gamma \,.\, x \in \gamma$

Dem.

$$\begin{aligned} &\vdash \,.\, *22{\cdot}59 \,.\, \supset \vdash :.\, \alpha \subset \gamma \,.\, \beta \subset \gamma \,.\, \supset : x \in \alpha \cup \beta \,.\, \supset \,.\, x \in \gamma :. \\ &[\text{Comm}] \quad \supset \vdash :.\, x \in \alpha \cup \beta \,.\, \supset : \alpha \subset \gamma \,.\, \beta \subset \gamma \,.\, \supset \,.\, x \in \gamma :. \\ &[*10{\cdot}11{\cdot}21] \supset \vdash :.\, x \in \alpha \cup \beta \,.\, \supset : \alpha \subset \gamma \,.\, \beta \subset \gamma \,.\, \supset_\gamma \,.\, x \in \gamma \qquad (1) \\ &\vdash \,.\, *10{\cdot}1 \,.\, \supset \vdash :.\, \alpha \subset \gamma \,.\, \beta \subset \gamma \,.\, \supset_\gamma \,.\, x \in \gamma : \supset : \alpha \subset \alpha \cup \beta \,.\, \beta \subset \alpha \cup \beta \,.\, \supset \,.\, x \in \alpha \cup \beta : \\ &[*22{\cdot}58] \qquad\qquad \supset : x \in \alpha \cup \beta \qquad (2) \\ &\vdash \,.\, (1) \,.\, (2) \,.\, \supset \vdash \,.\, \text{Prop} \end{aligned}$$

***22·61.** $\vdash : \alpha \subset \beta . \supset . \alpha \subset \beta \cup \gamma$ [*22·44·58]

***22·62.** $\vdash : \alpha \subset \beta . \equiv . \alpha \cup \beta = \beta$

Dem.

$\vdash . *4·72 . \supset \vdash :: x \in \alpha . \supset . x \in \beta : \equiv :. x \in \alpha . \vee . x \in \beta : \equiv . x \in \beta :.$

[*22·34] $\equiv :. x \in \alpha \cup \beta . \equiv . x \in \beta$ (1)

$\vdash . (1) . *10·271 . \supset \vdash :: \alpha \subset \beta . \equiv :. x \in \alpha \cup \beta . \equiv_x . x \in \beta :.$

[*20·43] $\equiv :. \alpha \cup \beta = \beta :: \supset \vdash .$ Prop

***22·621.** $\vdash : \alpha \subset \beta . \equiv . \alpha \cap \beta = \alpha$ [*4·71]

The proof proceeds as in *22·62. The proposition *22·621 is one of the most useful propositions in the present number.

***22·63.** $\vdash : \alpha \cup (\alpha \cap \beta) = \alpha$ [*4·44]

The process of obtaining *22·63 from *4·44 is of the same kind as the process employed in the proofs that have been written out in this number. Hence only *4·44 is referred to. We shall similarly restrict references for later propositions in this number. The process is always roughly as follows: p, q, r are replaced by $x \in \alpha$, $x \in \beta$, $x \in \gamma$; then *10·11 is applied, and such further propositions of *10 as may be required, together with *22·33·34·35.

***22·631.** $\vdash . \alpha \cap (\alpha \cup \beta) = \alpha$ [*22·58·621]

***22·632.** $\vdash : \alpha = \beta . \supset . \alpha = \alpha \cap \beta$ [*22·42·621]

***22·633.** $\vdash : \alpha \subset \beta . \supset . \alpha \cup \gamma = (\alpha \cap \beta) \cup \gamma$ [*22·551·621]

***22·64.** $\vdash :. \alpha \subset \gamma . \vee . \beta \subset \gamma : \supset . \alpha \cap \beta \subset \gamma$

Dem.

$\vdash . *22·47·51 . \supset \vdash : \alpha \subset \gamma . \supset . \alpha \cap \beta \subset \gamma : \beta \subset \gamma . \supset . \alpha \cap \beta \subset \gamma$ (1)

$\vdash . (1) . *4·77 . \supset \vdash .$ Prop

The converse of this proposition does not hold, because the converse of *10·41 does not hold.

***22·65.** $\vdash :. \alpha \subset \beta . \vee . \alpha \subset \gamma : \supset . \alpha \subset \beta \cup \gamma$ [*22·61·57 . *4·77]

Here again the converse is untrue.

***22·66.** $\vdash : \alpha \subset \beta . \supset . \alpha \cup \gamma \subset \beta \cup \gamma$ [*2·38]

***22·68.** $\vdash . (\alpha \cap \beta) \cup (\alpha \cap \gamma) = \alpha \cap (\beta \cup \gamma)$

Dem.

$\vdash . *22·34 . \supset \vdash :: x \in \{(\alpha \cap \beta) \cup (\alpha \cap \gamma)\} . \equiv :. x \in \alpha \cap \beta . \vee . x \in \alpha \cap \gamma :.$

[*22·33] $\equiv :. x \in \alpha . x \in \beta . \vee . x \in \alpha . x \in \gamma :.$

[*4·4] $\equiv :. x \in \alpha : x \in \beta . \vee . x \in \gamma :.$

[*22·34] $\equiv :. x \in \alpha . x \in \beta \cup \gamma :.$

[*22·33] $\equiv :. x \in \alpha \cap (\beta \cup \gamma)$ (1)

$\vdash . (1) . *10·11 . *20·43 . \supset \vdash .$ Prop

***22·69.** $\vdash . (\alpha \cup \beta) \cap (\alpha \cup \gamma) = \alpha \cup (\beta \cap \gamma)$ [Similar proof, by *4·41]

The above propositions *22·68·69 are the two forms of the distributive law. Note that either results from the other by interchanging the signs of addition and multiplication.

***22·7.** $\vdash . (\alpha \cup \beta) \cup \gamma = \alpha \cup (\beta \cup \gamma)$ [*4·33]

***22·71.** $\alpha \cup \beta \cup \gamma = (\alpha \cup \beta) \cup \gamma$ Df

***22·72.** $\vdash : \alpha \subset \gamma . \beta \subset \delta . \supset . \alpha \cup \beta \subset \gamma \cup \delta$ [*3·48]

***22·73.** $\vdash : \alpha = \gamma . \beta = \delta . \supset . \alpha \cup \beta = \gamma \cup \delta$ [*10·411]

***22·74.** $\vdash : \alpha \cap \beta \subset \gamma . \alpha \cap \gamma \subset \beta . \equiv . \alpha \cap \beta = \alpha \cap \gamma$

Dem.

$\vdash . *22·43 . *4·73 . \supset \vdash : \alpha \cap \beta \subset \gamma . \equiv . \alpha \cap \beta \subset \alpha . \alpha \cap \beta \subset \gamma .$

[*22·45] $\equiv . \alpha \cap \beta \subset \alpha \cap \gamma$ (1)

$\vdash . (1) \frac{\gamma, \beta}{\beta, \gamma} .$ $\supset \vdash : \alpha \cap \gamma \subset \beta . \equiv . \alpha \cap \gamma \subset \alpha \cap \beta$ (2)

$\vdash . (1) . (2) . *4·38 . \supset \vdash : \alpha \cap \beta \subset \gamma . \alpha \cap \gamma \subset \beta . \equiv . \alpha \cap \beta \subset \alpha \cap \gamma . \alpha \cap \gamma \subset \alpha \cap \beta .$

[*22·41] $\equiv . \alpha \cap \beta = \alpha \cap \gamma : \supset \vdash .$ Prop

***22·8.** $\vdash . -(-\alpha) = \alpha$ [*4·13]

***22·81.** $\vdash : \alpha \subset \beta . \equiv . -\beta \subset -\alpha$ [*4·1]

***22·811.** $\vdash : \alpha \subset -\beta . \equiv . \beta \subset -\alpha$ [*4·1 . *22·8]

***22·82.** $\vdash : \alpha \cap \beta \subset \gamma . \equiv . \alpha - \gamma \subset -\beta$ [*4·14]

***22·83.** $\vdash : \alpha = \beta . \equiv . -\alpha = -\beta$ [*4·11]

***22·831.** $\vdash : \alpha = -\beta . \equiv . \beta = -\alpha$ [*4·12]

***22·84.** $\vdash . -(\alpha \cap \beta) = -\alpha \vee -\beta$ [*4·51]

***22·85.** $\vdash . \alpha \cap \beta = -(-\alpha \vee -\beta)$ [*22·84·831]

***22·86.** $\vdash . -(-\alpha \cap -\beta) = \alpha \cup \beta$ [*4·57]

***22·87.** $\vdash . -\alpha \cap -\beta = -(\alpha \cup \beta)$ [*22·86·831]

*22·84·85·86·87 are De Morgan's formulae.

***22·88.** $\vdash . (x) . x \in (\alpha \cup -\alpha)$ [*2·11]

This is a form of the law of excluded middle.

***22·89.** $\vdash . (x) . x \sim\in (\alpha - \alpha)$ [*3·24]

This is a form of the law of contradiction.

***22·9.** $\vdash . (\alpha \cup \beta) - \beta = \alpha - \beta$ [*5·61]

***22·91.** $\vdash . \alpha \cup \beta = \alpha \cup (\beta - \alpha)$

Dem.

$\vdash . *5·63 .$ $\supset \vdash :. x \in \alpha . \vee . x \in \beta : \equiv : x \in \alpha . \vee . x \in \beta . x \sim\in \alpha :.$

[*22·33·34·35] $\supset \vdash :. x \in \alpha \cup \beta . \equiv : x \in \alpha . \vee . x \in (\beta - \alpha) :$

[*22·34] $\equiv : x \in \alpha \cup (\beta - \alpha)$ (1)

$\vdash . (1) . *10·11 . *20·43 . \supset \vdash .$ Prop

***22·92.** $\vdash : \alpha \subset \beta . \supset . \beta = \alpha \cup (\beta - \alpha)$ [*22·91·62]

***22·93.** $\vdash . \alpha - \beta = \alpha - (\alpha \cap \beta)$

Dem.

$\vdash$. *4·73 . Transp . $\supset \vdash :. x \epsilon \alpha . \supset : x \sim\epsilon \beta . \equiv . \sim (x \epsilon \alpha . x \epsilon \beta) .$

[*22·33] $\equiv . x \sim\epsilon (\alpha \cap \beta) :.$

[*5·32] $\supset \vdash :. x \epsilon \alpha . x \sim\epsilon \beta . \equiv . x \epsilon \alpha . x \sim\epsilon (\alpha \cap \beta) :.$

[*22·35·33] $\supset \vdash : x \epsilon \alpha - \beta . \equiv . x \epsilon \alpha - (\alpha \cap \beta) :$

[*10·11.*20·43] $\supset \vdash . \alpha - \beta = \alpha - (\alpha \cap \beta) . \supset \vdash$. Prop

***22·94.** $\vdash : (\alpha) . f\alpha . \equiv . (\alpha) . f(-\alpha)$

Dem.

$\vdash$. *10·1 . $\supset \vdash : (\alpha) . f\alpha . \supset . f(-\alpha) :$

[*10·11·21] $\supset \vdash : (\alpha) . f\alpha . \supset . (\alpha) . f(-\alpha)$ (1)

$\vdash$. *10·1 . $\supset \vdash : (\alpha) . f(-\alpha) . \supset . f\{-(-\alpha)\} .$

[*22·8.*20·18] $\supset . f\alpha :$

[*10·11·21] $\supset \vdash : (\alpha) . f(-\alpha) . \supset . (\alpha) . f\alpha$ (2)

$\vdash$. (1) . (2) . $\supset \vdash$. Prop

This proposition is used in connection with mathematical induction, in *90·102, which is required for the proof of *90·132, which is one of the fundamental propositions in the theory of mathematical induction.

***22·95.** $\vdash : (\exists\alpha) . f\alpha . \equiv . (\exists\alpha) . f(-\alpha)$

Dem.

$\vdash$. *22·94 . $\supset \vdash : (\alpha) . \sim f\alpha . \equiv . (\alpha) . \sim f(-\alpha)$ (1)

$\vdash$. (1) . Transp . *20·6 . $\supset \vdash$. Prop

$*23$. CALCULUS OF RELATIONS

Summary of $*23$.

The definitions and propositions of this number are to be exact analogues of those of $*22$. Properties of relations which have no analogues for classes will not be dealt with till Section D. Proofs will be omitted in the present number, as they are precisely analogous to those of analogous propositions in $*22$. In this number, as always in future, capital Latin letters stand for expressions of the form $\hat{x}\hat{y}\phi!(x, y)$, or, where they are not being used as apparent variables, for $\hat{x}\hat{y}\phi(x, y)$. The principal propositions of this number are the analogues of those of $*22$.

$*23\cdot01$. $R \mathrel{\dot{\subset}} S \,.=:\, xRy \,.\supset_{x,y}.\, xSy$ Df

$*23\cdot02$. $R \mathbin{\dot{\cap}} S = \hat{x}\hat{y}\,(xRy \,.\, xSy)$ Df

$*23\cdot03$. $R \mathbin{\dot{\cup}} S = \hat{x}\hat{y}\,(xRy \,.\vee.\, xSy)$ Df

$*23\cdot04$. $\dot{-}R = \hat{x}\hat{y}\,\{\sim(xRy)\}$ Df

$*23\cdot05$. $R \mathbin{\dot{-}} S = R \mathbin{\dot{\cap}} \dot{-}S$ Df

Similar remarks apply to these definitions as to those of $*22$.

$*23\cdot1$. $\vdash :.\, R \mathrel{\dot{\subset}} S \,.\equiv:\, xRy \,.\supset_{x,y}.\, xSy$

$*23\cdot2$. $\vdash .\, R \mathbin{\dot{\cap}} S = \hat{x}\hat{y}\,(xRy \,.\, xSy)$

$*23\cdot3$. $\vdash .\, R \mathbin{\dot{\cup}} S = \hat{x}\hat{y}\,(xRy \,.\vee.\, xSy)$

$*23\cdot31$. $\vdash .\, \dot{-}R = \hat{x}\hat{y}\,\{\sim(xRy)\}$

$*23\cdot32$. $\vdash .\, R \mathbin{\dot{-}} S = \hat{x}\hat{y}\,\{xRy \,.\sim(xSy)\}$

$*23\cdot33$. $\vdash :\, x\,(R \mathbin{\dot{\cap}} S)\,y \,.\equiv.\, xRy \,.\, xSy$

$*23\cdot34$. $\vdash :.\, x\,(R \mathbin{\dot{\cup}} S)\,y \,.\equiv:\, xRy \,.\vee.\, xSy$

$*23\cdot35$. $\vdash :\, x \mathbin{\dot{-}} Ry \,.\equiv.\sim(xRy)$

$*23\cdot351$. $\vdash .\, \dot{-}R \neq R$

$*23\cdot36$. $\vdash .\, R \mathbin{\dot{\cap}} S \in \mathrm{Rel}$

$*23\cdot37$. $\vdash .\, R \mathbin{\dot{\cup}} S \in \mathrm{Rel}$

$*23\cdot38$. $\vdash .\, \dot{-}R \in \mathrm{Rel}$

$*23\cdot39$. $\vdash .\, \hat{x}\hat{y}\phi(x, y) \mathbin{\dot{\cap}} \hat{x}\hat{y}\psi(x, y) = \hat{x}\hat{y}\,\{\phi(x, y) \,.\, \psi(x, y)\}$

$*23\cdot391$. $\vdash .\, \hat{x}\hat{y}\phi(x, y) \mathbin{\dot{\cup}} \hat{x}\hat{y}\psi(x, y) = \hat{x}\hat{y}\,\{\phi(x, y) \,.\vee.\, \psi(x, y)\}$

$*23\cdot392$. $\vdash .\, \dot{-}\hat{x}\hat{y}\phi(x, y) = \hat{x}\hat{y}\,\{\sim\phi(x, y)\}$

$*23\cdot4$. $\vdash :.\, R \mathrel{\dot{\subset}} S \,.\, S \mathrel{\dot{\subset}} R \,.\equiv:\, xRy \,.\equiv_{x,y}.\, xSy$

$*23\cdot41$. $\vdash :\, R \mathrel{\dot{\subset}} S \,.\, S \mathrel{\dot{\subset}} R \,.\equiv.\, R = S$

$*23\cdot42$. $\vdash .\, R \mathrel{\dot{\subset}} R$

$*23\cdot43$. $\vdash .\, R \mathbin{\dot{\cap}} S \mathrel{\dot{\subset}} R$

$*23\cdot44$. $\vdash :\, R \mathrel{\dot{\subset}} S \,.\, S \mathrel{\dot{\subset}} T \,.\supset.\, R \mathrel{\dot{\subset}} T$

$*23\cdot441$. $\vdash :\, R \mathrel{\dot{\subset}} S \,.\, xRy \,.\supset.\, xSy$

***23·45.** $\vdash : R \dot{\subset} S . R \dot{\subset} T . \supset . R \dot{\subset} S \dot{\cap} T$

***23·46.** $\vdash : xRy . R \dot{\subset} S . \supset . xSy$

***23·47.** $\vdash : R \dot{\subset} T . \supset . R \dot{\cap} S \dot{\subset} T$

***23·48.** $\vdash : R \dot{\subset} S . \supset . R \dot{\cap} T \dot{\subset} S \dot{\cap} T$

***23·481.** $\vdash : R = S . \supset . R \dot{\cap} T = S \dot{\cap} T$

***23·49.** $\vdash : P \dot{\subset} Q . R \dot{\subset} S . \supset . P \dot{\cap} R \dot{\subset} Q \dot{\cap} S$

***23·5.** $\vdash . R \dot{\cap} R = R$

***23·51.** $\vdash . R \dot{\cap} S = S \dot{\cap} R$

***23·52.** $\vdash . (R \dot{\cap} S) \dot{\cap} T = R \dot{\cap} (S \dot{\cap} T)$

***23·53.** $R \dot{\cap} S \dot{\cap} T = (R \dot{\cap} S) \dot{\cap} T$ Df

***23·54.** $\vdash :. R = S . \supset : R \dot{\subset} T . \equiv . S \dot{\subset} T$

***23·55.** $\vdash :. R = S . \supset : T \dot{\subset} R . \equiv . T \dot{\subset} S$

***23·551.** $\vdash : R = S . \supset . R \dot{\cup} T = S \dot{\cup} T$

***23·56.** $\vdash . R \dot{\cup} R = R$

***23·57.** $\vdash . R \dot{\cup} S = S \dot{\cup} R$

***23·58.** $\vdash . R \dot{\subset} R \dot{\cup} S . S \dot{\subset} R \dot{\cup} S$

***23·59.** $\vdash : R \dot{\subset} T . S \dot{\subset} T . \equiv . R \dot{\cup} S \dot{\subset} T$

***23·6.** $\vdash :. x (R \dot{\cup} S) y . \equiv : R \dot{\subset} T . S \dot{\subset} T . \supset_T . xTy$

***23·61.** $\vdash : R \dot{\subset} S . \supset . R \dot{\subset} S \dot{\cup} T$

***23·62.** $\vdash : R \dot{\subset} S . \equiv . R \dot{\cup} S = S$

***23·621.** $\vdash : R \dot{\subset} S . \equiv . R \dot{\cap} S = R$

***23·63.** $\vdash . R \dot{\cup} (R \dot{\cap} S) = R$

***23·631.** $\vdash . R \dot{\cap} (R \dot{\cup} S) = R$

***23·632.** $\vdash : R = S . \supset . R = R \dot{\cap} S$

***23·633.** $\vdash : R \dot{\subset} S . \supset . R \dot{\cup} T = (R \dot{\cap} S) \dot{\cup} T$

***23·64.** $\vdash :. R \dot{\subset} T . \vee . S \dot{\subset} T : \supset . R \dot{\cap} S \dot{\subset} T$

***23·65.** $\vdash :. R \dot{\subset} S . \vee . R \dot{\subset} T : \supset . R \dot{\subset} S \dot{\cup} T$

***23·66.** $\vdash : R \dot{\subset} S . \supset . R \dot{\cup} T \dot{\subset} S \dot{\cup} T$

***23·68.** $\vdash . (R \dot{\cap} S) \dot{\cup} (R \dot{\cap} T) = R \dot{\cap} (S \dot{\cup} T)$

***23·69.** $\vdash . (R \dot{\cup} S) \dot{\cap} (R \dot{\cup} T) = R \dot{\cup} (S \dot{\cap} T)$

***23·7.** $\vdash . (R \dot{\cup} S) \dot{\cup} T = R \dot{\cup} (S \dot{\cup} T)$

***23·71.** $R \dot{\cup} S \dot{\cup} T = (R \dot{\cup} S) \dot{\cup} T$ Df

***23·72.** $\vdash : P \dot{\subset} R . Q \dot{\subset} S . \supset . P \dot{\cup} Q \dot{\subset} R \dot{\cup} S$

***23·73.** $\vdash : P = R . Q = S . \supset . P \dot{\cup} Q = R \dot{\cup} S$

***23·74.** $\vdash : P \dot{\cap} Q \dot{\subset} R . P \dot{\cap} R \dot{\subset} Q . \equiv . P \dot{\cap} Q = P \dot{\cap} R$

***23·8.** $\vdash . \dot{-} (\dot{-} R) = R$

***23·81.** $\vdash : R \dot{\subset} S . \equiv . \dot{-} S \dot{\subset} \dot{-} R$

***23·811.** $\vdash : R \dot{\subset} \dot{-} S . \equiv . S \dot{\subset} \dot{-} R$

***23·82.** $\vdash : R \dot{\cap} S \dot{\subset} T . \equiv . R \dot{-} T \dot{\subset} \dot{-} S$

***23·83.** $\vdash : R = S . \equiv . \dot{-} R = \dot{-} S$

***23·831.** $\vdash : R = \dot{-} S \,.\equiv .\, S = \dot{-} R$

***23·84.** $\vdash .\, \dot{-}(R \mathbin{\dot{\cap}} S) = \dot{-} R \mathbin{\dot{\cup}} \dot{-} S$

***23·85.** $\vdash .\, R \mathbin{\dot{\cap}} S = \dot{-}(\dot{-} R \mathbin{\dot{\cup}} \dot{-} S)$

***23·86.** $\vdash .\, \dot{-}(\dot{-} R \mathbin{\dot{\cap}} \dot{-} S) = R \mathbin{\dot{\cup}} S$

***23·87.** $\vdash .\, \dot{-} R \mathbin{\dot{\cap}} \dot{-} S = \dot{-}(R \mathbin{\dot{\cup}} S)$

***23·88.** $\vdash .\, (x, y) \,.\, x (R \mathbin{\dot{\cup}} \dot{-} R)\, y$

***23·89.** $\vdash .\, (x, y) \,.\sim \{x (R \dot{-} R)\, y\}$

***23·9.** $\vdash .\, (R \mathbin{\dot{\cup}} S) \dot{-} S = R \dot{-} S$

***23·91.** $\vdash .\, R \mathbin{\dot{\cup}} S = R \mathbin{\dot{\cup}} (S \dot{-} R)$

***23·92.** $\vdash : R \mathrel{\dot{\subset}} S \,.\supset .\, S = R \mathbin{\dot{\cup}} (S \dot{-} R)$

***23·93.** $\vdash .\, R \dot{-} S = R \dot{-} (R \mathbin{\dot{\cap}} S)$

***23·94.** $\vdash : (R) \,.\, fR \,.\equiv .\, (R) \,.\, f(\dot{-} R)$

***23·95.** $\vdash : (\exists R) \,.\, fR \,.\equiv .\, (\exists R) \,.\, f(\dot{-} R)$

$*24$. THE UNIVERSAL CLASS, THE NULL-CLASS, AND THE EXISTENCE OF CLASSES

Summary of $*24$.

The universal class, denoted by V, is the class of all objects of the type which, in the given context, is being denoted by small Latin letters, *i.e.* of the lowest type concerned. Thus V, like "Cls," is ambiguous as to type. Its definition is as follows:

$*24{\cdot}01$. $V = \hat{x}(x = x)$ Df

Any other property possessed by everything would do as well as "$x = x$," but this is the only such property which we have hitherto studied.

The null-class, denoted by Λ, is the class which has no members. Like V, it is ambiguous as to type. We use the same symbol, Λ, for null-classes of various types; but these null-classes differ. The type of Λ is determined by that of the terms x concerning which "$x \epsilon \Lambda$" is false: whatever x may be, "$x \epsilon \Lambda$" will not represent a *true* proposition, but unless x is of the appropriate type, "$x \epsilon \Lambda$" will be meaningless, not false. Thus Λ is of the type next above that of an x concerning which "$x \epsilon \Lambda$" is significant and false. The definition of Λ is

$*24{\cdot}02$. $\Lambda = -V$ Df

When a class α is not null, so that it has one or more members, it is said to *exist*. (This sense of "existence" must not be confused with that defined in $*14{\cdot}02$.) We write "$\exists ! \alpha$" for "α exists." The definition is

$*24{\cdot}03$. $\exists ! \alpha . = . (\exists x) . x \epsilon \alpha$ Df

In the present number, we shall deal first with the properties of Λ and V, then with those of existence. In comparing the algebra of symbolic logic with ordinary algebra, Λ takes the place of 0, while V combines the properties of 1 and of ∞.

Among the more important properties of Λ and V which are proved in this number are the following:

$*24{\cdot}1$. $\vdash . \Lambda \neq V$

I.e. "nothing is not everything." This is useful as giving us the existence of at least two classes. If the monistic philosophers were right in maintaining that only one individual exists, there would be only two classes, Λ and V, V being (in that case) the class whose only member is the one individual. Our primitive propositions do not require the existence of more than one individual.

***24·102·103** show that any function which is always true determines the universal class, and any function which is always false determines the null-class.

***24·21·22** give forms of the laws of contradiction and excluded middle, namely "nothing is both α and not-α" ($\alpha \cap -\alpha = \Lambda$) and "everything is either α or not-α" ($\alpha \cup -\alpha = V$).

***24·23·24·26·27** give the properties of Λ and V with respect to addition and multiplication, namely: multiplication by V and addition of Λ make no change in a class (*24·26·24); addition of V gives V, and multiplication by Λ gives Λ (*24·27·23). It will be observed that the properties of Λ and V result from each other by interchanging addition and multiplication.

***24·3.** $\vdash : \alpha \subset \beta . \equiv . \alpha - \beta = \Lambda$

I.e. "α is contained in β" is equivalent to "nothing is α but not β."

***24·311.** $\vdash : \alpha \subset -\beta . \equiv . \alpha \cap \beta = \Lambda$

I.e. "no α is a β" is equivalent to "nothing is both α and β."

***24·411.** $\vdash : \beta \subset \alpha . \supset . \alpha = \beta \cup (\alpha - \beta)$

***24·43.** $\vdash : \alpha - \beta \subset \gamma . \equiv . \alpha \subset \beta \cup \gamma$

As a rule, propositions concerning V are much less used than the correlative propositions concerning Λ.

The properties of the existence of classes result from those of Λ, owing to the fact that $\exists ! \alpha$ is the contradictory of $\alpha = \Lambda$, as is proved in *24·54. Thus we have, in virtue of *24·3,

***24·55.** $\vdash : \sim(\alpha \subset \beta) . \equiv . \exists ! \alpha - \beta$

I.e. "not all α's are β's" is equivalent to "there are α's which are not β's." This is the familiar proposition of formal logic, that the contradictory of the universal affirmative is the particular negative.

We have

***24·56.** $\vdash :. \exists ! (\alpha \cup \beta) . \equiv : \exists ! \alpha . \vee . \exists ! \beta$

***24·561.** $\vdash : \exists ! (\alpha \cap \beta) . \supset . \exists ! \alpha . \exists ! \beta$

I.e. if a sum exists, then one of the summands exists, and vice versa; and if a product exists, both the factors exist (but not vice versa).

The proofs of propositions in the present number offer no difficulty.

***24·01.** $V = \hat{x}(x = x)$ Df

***24·02.** $\Lambda = -V$ Df

***24·03.** $\exists ! \alpha . = . (\exists x) . x \in \alpha$ Df

***24·1.** $\vdash . \Lambda \neq V$ [*22·351 . (*24·02)]

***24·101.** $\vdash . V = -\Lambda$ [*22·831 . (*24·02)]

***24·102.** $\vdash : (x) . \phi x . \equiv . \hat{z}(\phi z) = V$

Dem.

$\vdash . *13\cdot15 . *5\cdot501 . \supset \vdash :. \phi x . \equiv : \phi x . \equiv . x = x :.$

$[*10\cdot11\cdot271] \quad \supset \vdash :. (x) . \phi x . \equiv : (x) : \phi x . \equiv . x = x :$

$[*20\cdot15] \quad \equiv : \hat{z}(\phi z) = \hat{z}(x = x) :$

$[(*24\cdot01)] \quad \equiv : \hat{z}(\phi z) = V :. \supset \vdash . \text{Prop}$

Thus any function which is always true determines the universal class, and vice versa.

***24·103.** $\vdash : (x) . \sim\phi x . \equiv . \hat{z}(\phi z) = \Lambda$

Dem.

$\vdash . *24\cdot102 . \supset \vdash :. (x) . \sim\phi x . \equiv : \hat{z}(\sim\phi z) = V :$

$[*22\cdot392] \quad \equiv : -\hat{z}(\phi z) = V :$

$[*22\cdot831] \quad \equiv : \hat{z}(\phi z) = -V :$

$[(*24\cdot02)] \quad \equiv : \hat{z}(\phi z) = \Lambda :. \supset \vdash . \text{Prop}$

***24·104.** $\vdash . (x) . x \in V$

Dem.

$\vdash . *20\cdot3 . \supset \vdash : x \in V . \equiv . x = x \quad (1)$

$\vdash . (1) . *13\cdot15 . *10\cdot11\cdot271 . \supset \vdash . \text{Prop}$

***24·105.** $\vdash . (x) . x \sim\in \Lambda$

Dem.

$\vdash . *22\cdot35 . \supset \vdash : x \in \Lambda . \equiv . x \sim\in V :$

$[*4\cdot12] \quad \supset \vdash : x \sim\in \Lambda . \equiv . x \in V \quad (1)$

$\vdash . (1) . *10\cdot11\cdot271 . *24\cdot104 . \supset \vdash . \text{Prop}$

***24·11.** $\vdash . (\alpha) . \alpha \subset V$

Dem.

$\vdash . *24\cdot104 . *10\cdot1 . \supset \vdash . x \in V .$

$[\text{Simp}] \quad \supset \vdash : x \in \alpha . \supset . x \in V :$

$[*10\cdot11 . *22\cdot1] \quad \supset \vdash : \alpha \subset V :$

$[*10\cdot11] \quad \supset \vdash : (\alpha) . \alpha \subset V : \supset \vdash . \text{Prop}$

***24·12.** $\vdash . (\alpha) . \Lambda \subset \alpha$

Dem.

$\vdash . *24\cdot105 . *10\cdot1 . \quad \supset \vdash . x \sim\in \Lambda .$

$[*2\cdot21] \quad \supset \vdash : x \in \Lambda . \supset . x \in \alpha \quad (1)$

$\vdash . (1) . *10\cdot11 . *22\cdot1 . \supset \vdash . \text{Prop}$

***24·13.** $\vdash : \alpha = \Lambda . \equiv . \alpha \subset \Lambda$

Dem.

$\vdash . *24\cdot12 . *4\cdot73 . \supset \vdash : \alpha \subset \Lambda . \equiv . \alpha \subset \Lambda . \Lambda \subset \alpha .$

$[*22\cdot41] \quad \equiv . \alpha = \Lambda : \supset \vdash . \text{Prop}$

***24·14.** $\vdash : (x) . x \in \alpha . \equiv . \alpha = V$

Dem.

$\vdash . *24\cdot102 . \supset \vdash : (x) . x \in \alpha . \equiv . \hat{x}(x \in \alpha) = V .$

$[*20\cdot32] \quad \equiv . \alpha = V : \supset \vdash . \text{Prop}$

***24·141.** $\vdash : \mathrm{V} \subset \alpha . \equiv . \mathrm{V} = \alpha$

Dem.

$\vdash . *24·11 . *4·73 . \supset \vdash : \mathrm{V} \subset \alpha . \equiv . \alpha \subset \mathrm{V} . \mathrm{V} \subset \alpha .$

[*22·41] $\equiv . \alpha = \mathrm{V} : \supset \vdash . \mathrm{Prop}$

***24·15.** $\vdash : (x) . x \sim \epsilon \alpha . \equiv . \alpha = \Lambda$

Dem.

$\vdash . *24·103 . \supset \vdash : (x) . x \sim \epsilon \alpha . \equiv . \hat{x}(x \epsilon \alpha) = \Lambda .$

[*20·32] $\equiv . \alpha = \Lambda : \supset \vdash . \mathrm{Prop}$

***24·17.** $\vdash : \alpha = \mathrm{V} . \equiv . -\alpha = \Lambda$ [*22·83 . (*24·02)]

***24·21.** $\vdash . \alpha \cap -\alpha = \Lambda$ [*24·103 . *22·89]

***24·22.** $\vdash . \alpha \cup -\alpha = \mathrm{V}$ [*22·88 . *24·102]

***24·23.** $\vdash . \alpha \cap \Lambda = \Lambda$ [*24·12 . *22·621]

***24·24.** $\vdash . \alpha \cup \Lambda = \alpha$ [*24·12 . *22·62]

The above two propositions (*24·23·24) exhibit the algebraical analogy of Λ to zero.

***24·26.** $\vdash . \alpha \cap \mathrm{V} = \alpha$ [*22·621 . *24·11]

This exhibits the analogy of V to 1.

***24·27.** $\vdash . \alpha \cup \mathrm{V} = \mathrm{V}$ [*22·62 . *24·11]

This exhibits the analogy of V to ∞.

***24·3.** $\vdash : \alpha \subset \beta . \equiv . \alpha - \beta = \Lambda$

Dem.

$\vdash . *4·53·6 . \supset$

$\vdash :. x \epsilon \alpha . \supset . x \epsilon \beta : \equiv : \sim(x \epsilon \alpha . x \sim \epsilon \beta) :$

[*22·35] $\equiv : \sim(x \epsilon \alpha . x \epsilon -\beta) :$

[*22·33] $\equiv : \sim(x \epsilon \alpha - \beta)$ (1)

$\vdash . (1) . *10·11·271 . \supset$

$\vdash : \alpha \subset \beta . \equiv . (x) . \sim(x \epsilon \alpha - \beta) .$

[*24·15] $\equiv . \alpha - \beta = \Lambda : \supset \vdash . \mathrm{Prop}$

The above proposition is very frequently used.

***24·31.** $\vdash : \alpha \subset \beta . \equiv . -\alpha \cup \beta = \mathrm{V}$

Dem.

$\vdash . *4·6 . \supset \vdash :. x \epsilon \alpha . \supset . x \epsilon \beta : \equiv : x \sim \epsilon \alpha . \vee . x \epsilon \beta :.$

[*10·11·271] $\supset \vdash :. \alpha \subset \beta . \equiv : (x) : x \sim \epsilon \alpha . \vee . x \epsilon \beta :$

[*22·35] $\equiv : (x) : x \epsilon -\alpha . \vee . x \epsilon \beta :$

[*22·34] $\equiv : (x) . x \epsilon (-\alpha \cup \beta) :$

[*24·14] $\equiv : -\alpha \cup \beta = \mathrm{V} :. \supset \vdash . \mathrm{Prop}$

This proposition is the correlative of *24·3, but, unlike that proposition, it is not useful in the sequel. Every proposition concerning Λ has a correlative concerning V, but we shall often not give these correlatives, since they are seldom required for subsequent proofs.

***24·311.** $\vdash : \alpha \subset - \beta . \equiv . \alpha \cap \beta = \Lambda$

Dem.

$\vdash . *22{\cdot}35 . \supset \vdash :. x \epsilon \alpha . \supset . x \epsilon - \beta : \equiv : x \epsilon \alpha . \supset . x \sim\epsilon \beta :$

[*4·51·62] $\equiv : \sim(x \epsilon \alpha . x \epsilon \beta) :$

[*22·33] $\equiv : \sim(x \epsilon \alpha \cap \beta)$ (1)

$\vdash . (1) . *10{\cdot}11{\cdot}271 . \supset \vdash : \alpha \subset - \beta . \equiv . (x) . x \sim\epsilon \alpha \cap \beta .$

[*24·15] $\equiv . \alpha \cap \beta = \Lambda : \supset \vdash . \text{Prop}$

***24·312.** $\vdash : - \alpha \subset \beta . \equiv . \alpha \cup \beta = \text{V}$

Dem.

$\vdash . *22{\cdot}35 . \supset \vdash :. - \alpha \subset \beta . \equiv : x \sim\epsilon \alpha . \supset_x . x \epsilon \beta :$

[*4·64] $\equiv : (x) : x \epsilon \alpha . \vee . x \epsilon \beta :$

[*22·34] $\equiv : (x) . x \epsilon \alpha \cup \beta :$

[*24·14] $\equiv : \alpha \cup \beta = \text{V} :. \supset \vdash . \text{Prop}$

***24·313.** $\vdash : \alpha \cap \beta = \Lambda . \equiv . \alpha = \alpha - \beta$ [*24·311 . *22·621]

***24·32.** $\vdash :. \alpha \cup \beta = \Lambda . \equiv . \alpha = \Lambda . \beta = \Lambda$

Dem.

$\vdash . *24{\cdot}13 . \supset \vdash :. \alpha \cup \beta = \Lambda . \equiv : \alpha \cup \beta \subset \Lambda :$

[*22·59] $\equiv : \alpha \subset \Lambda . \beta \subset \Lambda :$

[*24·13] $\equiv : \alpha = \Lambda . \beta = \Lambda :. \supset \vdash . \text{Prop}$

***24·33.** $\vdash : \alpha = \text{V} . \supset . \alpha \cup \beta = \text{V}$

Dem.

$\vdash . *22{\cdot}551 . \supset \vdash : \text{Hp} . \supset . \alpha \cup \beta = \text{V} \cup \beta$

[*24·27.*22·57] $= \text{V} : \supset \vdash . \text{Prop}$

***24·34.** $\vdash : \alpha = \Lambda . \supset . \alpha \cap \beta = \Lambda$ [*22·481 . *24·23]

***24·35.** $\vdash : \alpha = \text{V} . \supset . \alpha \cap \beta = \beta$ [*22·481 . *24·26]

***24·36.** $\vdash : \alpha = \Lambda . \supset . \alpha \cup \beta = \beta$ [*22·551 . *24·24]

***24·37.** $\vdash :. \alpha \cap \beta = \Lambda . \equiv : x \epsilon \alpha . y \epsilon \beta . \supset_{x,y} . x \neq y$

Dem.

$\vdash . *24{\cdot}15 . \supset \vdash :. \alpha \cap \beta = \Lambda . \equiv : (x) . x \sim\epsilon (\alpha \cap \beta) :$

[*22·33] $\equiv : (x) . \sim(x \epsilon \alpha . x \epsilon \beta) :$

[*13·191] $\equiv : (x, y) : x = y . \supset . \sim(x \epsilon \alpha . y \epsilon \beta) :$

[Transp] $\equiv : (x, y) : x \epsilon \alpha . y \epsilon \beta . \supset . x \neq y :. \supset \vdash . \text{Prop}$

***24·38.** $\vdash :. \alpha \cap \beta = \Lambda . \supset : \alpha \neq \beta . \vee . \alpha = \Lambda . \beta = \Lambda$

Dem.

$\vdash . *22{\cdot}481 . \supset \vdash : \alpha \cap \beta = \Lambda . \alpha = \beta . \supset . \alpha \cap \alpha = \Lambda .$

[*22·5] $\supset . \alpha = \Lambda .$

[*20·23] $\supset . \alpha = \Lambda . \beta = \Lambda$ (1)

$\vdash . (1) . \text{Exp} . \supset \vdash :. \alpha \cap \beta = \Lambda . \supset : \alpha = \beta . \supset . \alpha = \Lambda . \beta = \Lambda :$

[*4·6] $\supset : \alpha \neq \beta . \vee . \alpha = \Lambda . \beta = \Lambda :. \supset \vdash . \text{Prop}$

***24·39.** $\vdash :. \alpha \cap \beta = \Lambda . \equiv : x \in \alpha . \supset_x . x \sim \in \beta$ [*24·311 . *22·35]

***24·4.** $\vdash : \alpha \cap \beta = \Lambda . \equiv . (\alpha \cup \beta) - \alpha = \beta . \equiv . (\alpha \cup \beta) - \beta = \alpha$

Dem.

$\vdash . *24{\cdot}311 . \supset \vdash : \alpha \cap \beta = \Lambda . \equiv . \beta \subset -\alpha .$
[*22·621] $\equiv . \beta - \alpha = \beta .$
[*22·9] $\equiv . (\alpha \cup \beta) - \alpha = \beta$ (1)
$\vdash . (1) \dfrac{\beta, \alpha}{\alpha, \beta} . \supset \vdash : \beta \cap \alpha = \Lambda . \equiv . (\beta \cup \alpha) - \beta = \alpha :$
[*22·51·57] $\supset \vdash : \alpha \cap \beta = \Lambda . \equiv . (\alpha \cup \beta) - \beta = \alpha$ (2)
$\vdash . (1) . (2) . \supset \vdash .$ Prop

***24·401.** $\vdash : \beta \subset \alpha . \supset . (\beta \cup \gamma) - \alpha = \gamma - \alpha$

Dem.

$\vdash . *22{\cdot}68 . \supset \vdash . (\beta \cup \gamma) - \alpha = (\beta - \alpha) \cup (\gamma - \alpha)$ (1)
$\vdash . *24{\cdot}3 . \supset \vdash : \mathrm{Hp} . \supset . \beta - \alpha = \Lambda$ (2)
$\vdash . (1) . (2) . \supset \vdash : \mathrm{Hp} . \supset . (\beta \cup \gamma) - \alpha = \Lambda \cup (\gamma - \alpha)$
[*24·24] $= \gamma - \alpha : \supset \vdash .$ Prop

***24·402.** $\vdash : \alpha \cap \beta = \Lambda . \xi \subset \alpha . \eta \subset \beta . \supset . \xi \cap \eta = \Lambda$

Dem.

$\vdash . *22{\cdot}49 . \supset \vdash : \mathrm{Hp} . \supset . \xi \cap \eta \subset \alpha \cap \beta .$
[*22·55] $\supset . \xi \cap \eta \subset \Lambda .$
[*24·13] $\supset . \xi \cap \eta = \Lambda : \supset \vdash .$ Prop

***24·41.** $\vdash . \alpha = (\alpha \cap \beta) \cup (\alpha - \beta)$

Dem.

$\vdash . *22{\cdot}68 . \supset \vdash . (\alpha \cap \beta) \cup (\alpha - \beta) = \alpha \cap (\beta \cup -\beta)$
[*24·22] $= \alpha \cap V$
[*24·26] $= \alpha . \supset \vdash .$ Prop

***24·411.** $\vdash : \beta \subset \alpha . \supset . \alpha = \beta \cup (\alpha - \beta)$

Dem.

$\vdash . *22{\cdot}633 \dfrac{\beta, \alpha, \alpha - \beta}{\alpha, \beta, \gamma} . \supset \vdash : \beta \subset \alpha . \supset . \beta \cup (\alpha - \beta) = (\alpha \cap \beta) \cup (\alpha - \beta)$
[*24·41] $= \alpha : \supset \vdash .$ Prop

***24·412.** $\vdash : \beta \subset \alpha . \gamma \subset \beta . \supset . (\alpha - \beta) \cup (\beta - \gamma) = \alpha - \gamma$

Dem.

$\vdash . *24{\cdot}41 . \supset \vdash : \mathrm{Hp} . \supset . (\alpha - \beta) \cup (\beta - \gamma) = (\alpha - \beta \cap \gamma) \cup (\alpha - \beta - \gamma) \cup (\beta - \gamma)$
[*24·3·23] $= (\alpha - \beta - \gamma) \cup (\beta - \gamma)$
[*22·68] $= \{(\alpha - \beta) \cup \beta\} - \gamma$
[*24·411] $= \alpha - \gamma : \supset \vdash .$ Prop

This proposition is used in *234·181, in the theory of continuous functions.

***24·42.** $\vdash : \alpha \cap \beta \subset \gamma . \alpha - \beta \subset \gamma . \equiv . \alpha \subset \gamma$

Dem.

$\vdash . *22{\cdot}59 . \supset \vdash : \alpha \cap \beta \subset \gamma . \alpha - \beta \subset \gamma . \equiv . (\alpha \cap \beta) \cup (\alpha - \beta) \subset \gamma .$
[*24·41] $\equiv . \alpha \subset \gamma : \supset \vdash .$ Prop

***24·43.** $\vdash : \alpha - \beta \subset \gamma . \equiv . \alpha \subset \beta \cup \gamma$

Dem.

$\vdash . *5\cdot6 . \quad \supset \vdash :: x \epsilon \alpha . x \sim\epsilon \beta . \supset . x \epsilon \gamma : \equiv :. x \epsilon \alpha . \supset : x \epsilon \beta . \vee . x \epsilon \gamma :.$

$[*22\cdot35\cdot33] \supset \vdash :: x \epsilon \alpha - \beta . \supset . x \epsilon \gamma : \equiv :. x \epsilon \alpha . \supset : x \epsilon \beta . \vee . x \epsilon \gamma :.$

$[*22\cdot34] \qquad \equiv :. x \epsilon \alpha . \supset . x \epsilon (\beta \cup \gamma) \qquad (1)$

$\vdash . (1) . *10\cdot11\cdot271 . \supset \vdash . \text{Prop}$

***24·431.** $\vdash . (\alpha \cup \gamma) \cap (\beta \cup -\gamma) = (\alpha \cap \beta) \cup (\alpha - \gamma) \cup (\beta \cap \gamma)$

This and the following proposition are lemmas for *24·44.

Dem.

$\vdash . *22\cdot68 . \supset \vdash . (\alpha \cup \gamma) \cap (\beta \cup -\gamma) = \{(\alpha \cup \gamma) \cap \beta\} \cup \{(\alpha \cup \gamma) \cap -\gamma\}$

$[*22\cdot68] \qquad = (\alpha \cap \beta) \cup (\gamma \cap \beta) \cup (\alpha - \gamma) \cup (\gamma - \gamma)$

$[*24\cdot21] \qquad = (\alpha \cap \beta) \cup (\gamma \cap \beta) \cup (\alpha - \gamma) \cup \Lambda$

$[*24\cdot24] \qquad = (\alpha \cap \beta) \cup (\gamma \cap \beta) \cup (\alpha - \gamma)$

$[*22\cdot51\cdot57] \qquad = (\alpha \cap \beta) \cup (\alpha - \gamma) \cup (\beta \cap \gamma) . \supset \vdash . \text{Prop}$

***24·432.** $\vdash . (\alpha - \gamma) \cup (\beta \cap \gamma) = (\alpha \cap \beta) \cup (\alpha - \gamma) \cup (\beta \cap \gamma)$

Dem.

$\vdash . *24\cdot22\cdot35 . \supset \vdash . \alpha \cap \beta = (\alpha \cap \beta) \cap (\gamma \cup -\gamma)$

$[*22\cdot68] \qquad = (\alpha \cap \beta \cap \gamma) \cup (\alpha \cap \beta - \gamma)$

$[*22\cdot51] \qquad = (\alpha \cap \beta \cap \gamma) \cup (\alpha \cap -\gamma \cap \beta) .$

$[*22\cdot551] \quad \supset \vdash . (\alpha \cap \beta) \cup (\alpha - \gamma) = (\alpha \cap \beta \cap \gamma) \cup (\alpha \cap -\gamma \cap \beta) \cup (\alpha - \gamma)$

$[*22\cdot63] \qquad = (\alpha \cap \beta \cap \gamma) \cup (\alpha - \gamma)$

$[*22\cdot57] \qquad = (\alpha - \gamma) \cup (\alpha \cap \beta \cap \gamma) .$

$[*22\cdot551] \quad \supset \vdash . (\alpha \cap \beta) \cup (\alpha - \gamma) \cup (\beta \cap \gamma) = (\alpha - \gamma) \cup (\alpha \cap \beta \cap \gamma) \cup (\beta \cap \gamma)$

$[*22\cdot63] \qquad = (\alpha - \gamma) \cup (\beta \cap \gamma) . \supset \vdash . \text{Prop}$

***24·44.** $\vdash . (\alpha \cup \gamma) \cap (\beta \cup -\gamma) = (\alpha \cap -\gamma) \cup (\beta \cap \gamma) \quad$ [*24·431·432]

***24·45.** $\vdash : (\alpha \cap \gamma) \cup (\beta - \gamma) = \Lambda . \equiv . \beta \subset \gamma . \gamma \subset -\alpha$

Dem.

$\vdash . *24\cdot32 . \supset \vdash : (\alpha \cap \gamma) \cup (\beta - \gamma) = \Lambda . \equiv . \alpha \cap \gamma = \Lambda . \beta - \gamma = \Lambda .$

$[*24\cdot3\cdot311] \qquad \equiv . \gamma \subset -\alpha . \beta \subset \gamma : \supset \vdash . \text{Prop}$

***24·46.** $\vdash : (\alpha \cap \gamma) \cup (\beta - \gamma) = \Lambda . \supset . \alpha \cap \beta = \Lambda$

Dem.

$\vdash . *24\cdot45 . *22\cdot44 . \supset \vdash : \text{Hp} . \supset . \beta \subset -\alpha .$

$[*22\cdot811] \qquad \supset . \alpha \subset -\beta .$

$[*24\cdot311] \qquad \supset . \alpha \cap \beta = \Lambda : \supset \vdash . \text{Prop}$

The following propositions, down to *24·495 inclusive, are lemmas inserted for use in much later propositions, most of them being only used a few times.

***24·47.** $\vdash : \alpha \cap \beta = \Lambda . \alpha \cup \beta = \gamma . \equiv . \alpha \subset \gamma . \beta = \gamma - \alpha$

Dem.

$\vdash . *24\cdot311 . \supset \vdash : \alpha \cap \beta = \Lambda . \equiv . \beta \subset - \alpha$ (1)

$\vdash . *22\cdot41 . \supset \vdash : \alpha \cup \beta = \gamma . \equiv . \alpha \cup \beta \subset \gamma . \gamma \subset \alpha \cup \beta .$

[*22·59.*24·43] $\equiv . \alpha \subset \gamma . \beta \subset \gamma . \gamma - \alpha \subset \beta$ (2)

$\vdash . (1) . (2) . \supset \vdash : \alpha \cap \beta = \Lambda . \alpha \cup \beta = \gamma . \equiv . \beta \subset - \alpha . \alpha \subset \gamma . \beta \subset \gamma . \gamma - \alpha \subset \beta .$

[*4·3] $\equiv . \alpha \subset \gamma . \beta \subset \gamma . \beta \subset - \alpha . \gamma - \alpha \subset \beta .$

[*22·45] $\equiv . \alpha \subset \gamma . \beta \subset \gamma - \alpha . \gamma - \alpha \subset \beta .$

[*22·41] $\equiv . \alpha \subset \gamma . \beta = \gamma - \alpha : \supset \vdash . \text{Prop}$

***24·48.** $\vdash :. \xi \subset \alpha . \xi' \subset \alpha . \eta \subset \beta . \eta' \subset \beta . \alpha \cap \beta = \Lambda . \supset :$

$\xi \cup \eta = \xi' \cup \eta' . \equiv . \xi = \xi' . \eta = \eta'$

Dem.

$\vdash . *22\cdot73 . \quad \supset \vdash : \xi = \xi' . \eta = \eta' . \supset . \xi \cup \eta = \xi' \cup \eta'$ (1)

$\vdash . *22\cdot481 . \quad \supset \vdash :. \xi \cup \eta = \xi' \cup \eta' . \supset : (\xi \cup \eta) \cap \alpha = (\xi' \cup \eta') \cap \alpha :$

[*22·68] $\supset : (\xi \cap \alpha) \cup (\eta \cap \alpha) = (\xi' \cap \alpha) \cup (\eta' \cap \alpha)$ (2)

$\vdash . *22\cdot621 . \quad \supset \vdash : \xi \subset \alpha . \supset . \xi \cap \alpha = \xi : \xi' \subset \alpha . \supset . \xi' \cap \alpha = \xi' :$

[*3·47] $\supset \vdash : \xi \subset \alpha . \xi' \subset \alpha . \supset . \xi \cap \alpha = \xi . \xi' \cap \alpha = \xi'$ (3)

$\vdash . *22\cdot48 . \quad \supset \vdash : \eta \subset \beta . \supset . \eta \cap \alpha \subset \alpha \cap \beta :$

[*22·55] $\supset \vdash : \eta \subset \beta . \alpha \cap \beta = \Lambda . \supset . \eta \cap \alpha \subset \Lambda .$

[*24·13] $\supset . \eta \cap \alpha = \Lambda$ (4)

Similarly $\vdash : \eta' \subset \beta . \alpha \cap \beta = \Lambda . \supset . \eta' \cap \alpha = \Lambda$ (5)

$\vdash . (3) . (4) . \quad \supset \vdash :. \text{Hp} . \supset : (\xi \cap \alpha) \cup (\eta \cap \alpha) = \xi \cup \Lambda$

[*24·24] $= \xi$ (6)

$\vdash . (3) . (5) . \quad \supset \vdash :. \text{Hp} . \supset : (\xi' \cap \alpha) \cup (\eta' \cap \alpha) = \xi' \cup \Lambda$

[*24·24] $= \xi'$ (7)

$\vdash . (2) . (6) . (7) . \supset \vdash :. \text{Hp} . \supset : \xi \cup \eta = \xi' \cup \eta' . \supset . \xi = \xi'$ (8)

Similarly $\vdash :. \text{Hp} . \supset : \xi \cup \eta = \xi' \cup \eta' . \supset . \eta = \eta'$ (9)

$\vdash . (1) . (8) . (9) . \supset \vdash . \text{Prop}$

The above proposition, besides being used in the next two, is used in the theory of couples (*54·6), in the theory of greater and less (*117·632), and in the chapter on the ordering of classes by the principle of first differences (*170·68).

***24·481.** $\vdash :. \alpha \cap \beta = \Lambda . \alpha \cap \gamma = \Lambda . \supset : \alpha \cup \beta = \alpha \cup \gamma . \equiv . \beta = \gamma$

Dem.

$\vdash . *24\cdot48 \dfrac{\alpha, -\alpha, \alpha, \alpha, \beta, \gamma}{\alpha, \beta, \xi, \xi', \eta, \eta'} . \supset$

$\vdash :. \alpha \subset \alpha . \alpha \subset \alpha . \beta \subset - \alpha . \gamma \subset - \alpha . \alpha - \alpha = \Lambda . \supset :$

$\alpha \cup \beta = \alpha \cup \gamma . \equiv . \alpha = \alpha . \beta = \gamma$ (1)

$\vdash . *22\cdot42 . *24\cdot21 . \supset$

$\vdash :. \alpha \subset \alpha . \alpha \subset \alpha . \beta \subset - \alpha . \gamma \subset - \alpha . \alpha - \alpha = \Lambda . \equiv . \beta \subset - \alpha . \gamma \subset - \alpha .$

$[*24\cdot311] \qquad \equiv . \alpha \cap \beta = \Lambda . \alpha \cap \gamma = \Lambda \quad (2)$

$\vdash . *20\cdot2 . *4\cdot73 . \supset \vdash : \alpha = \alpha . \beta = \gamma . \equiv . \beta = \gamma \quad (3)$

$\vdash . (1) . (2) . (3) . \supset \vdash .$ Prop

The above proposition is used in the theory of selections (*83·74), in the theory of greater and less (*117·582), and in the theory of transfinite induction (*257).

***24·482.** $\vdash :. \xi \subset \alpha . \eta \subset \beta . \alpha \cap \beta = \Lambda . \supset : \xi \cup \eta = \alpha \cup \beta . \equiv . \xi = \alpha . \eta = \beta$

$\left[*24\cdot48 \dfrac{\alpha, \beta}{\xi', \eta'} . *22\cdot42 \right]$

The above proposition is used in the theory of convergence (*232·34).

***24·49.** $\vdash :. \alpha \cap \beta = \Lambda . \supset : \alpha \subset \beta \cup \gamma . \equiv . \alpha \subset \gamma$

Dem.

$\vdash . *22\cdot621 . \supset \vdash : \alpha \subset \beta \cup \gamma . \equiv . \alpha = \alpha \cap (\beta \cup \gamma)$

$[*22\cdot68] \qquad = (\alpha \cap \beta) \cup (\alpha \cap \gamma) \quad (1)$

$\vdash . *24\cdot24 . \supset \vdash : \alpha \cap \beta = \Lambda . \supset . (\alpha \cap \beta) \cup (\alpha \cap \gamma) = \alpha \cap \gamma \quad (2)$

$\vdash . (1) . (2) . \supset \vdash :. \mathrm{Hp} . \supset : \alpha \subset \beta \cup \gamma . \equiv . \alpha = \alpha \cap \gamma .$

$[*22\cdot621] \qquad \equiv . \alpha \subset \gamma : \supset \vdash .$ Prop

***24·491.** $\vdash : \beta \cap \gamma = \Lambda . \alpha \subset \beta \cup \gamma .$

$\supset . \alpha - \beta = \alpha \cap \gamma . \alpha - \gamma = \alpha \cap \beta . \alpha = (\alpha - \beta) \cup (\alpha - \gamma)$

Dem.

$\vdash . *22\cdot621 . \quad \supset \vdash : \mathrm{Hp} . \supset . \alpha = \alpha \cap (\beta \cup \gamma) .$

$[*22\cdot481] \qquad \supset . \alpha - \gamma = \alpha \cap (\beta \cup \gamma) - \gamma$

$[*24\cdot4] \qquad = \alpha \cap \beta \quad (1)$

Similarly $\quad \vdash : \mathrm{Hp} . \supset . \alpha - \beta = \alpha \cap \gamma \quad (2)$

$\vdash . (1) . (2) . \quad \supset \vdash : \mathrm{Hp} . \supset . (\alpha - \beta) \cup (\alpha - \gamma) = (\alpha \cap \gamma) \cup (\alpha \cap \beta)$

$[*22\cdot68] \qquad = \alpha \cap (\gamma \cup \beta)$

$[*22\cdot621] \qquad = \alpha \quad (3)$

$\vdash . (1) . (2) . (3) . \supset \vdash .$ Prop

The above proposition is used in the theory of selections (*83·63·65) and in the theory of segments of a series (*211·84).

***24·492.** $\vdash : \beta \subset \alpha . \alpha - \beta = \gamma . \supset . \alpha - \gamma = \beta$

Dem.

$\vdash . *22\cdot481 . \supset \vdash : \mathrm{Hp} . \supset . \alpha - \gamma = \alpha - (\alpha - \beta)$

$[*22\cdot8\cdot86] \qquad = \alpha \cap (-\alpha \cup \beta)$

$[*22\cdot8\cdot9] \qquad = \alpha \cap \beta$

$[*22\cdot621] \qquad = \beta : \supset \vdash .$ Prop

The above proposition is used fairly frequently, especially in the theory of series. It is first used in *93·273, in the theory of "generations."

***24·493.** $\vdash : \beta \cap \gamma = \Lambda . \supset . \alpha = (\alpha - \beta) \cup (\alpha - \gamma)$

Dem.

$\vdash . *22·84 . *24·17 . \supset \vdash : \text{Hp} . \supset . -\beta \cup -\gamma = V .$

$[*24·26] \quad \supset . \alpha = \alpha \cap (-\beta \cup -\gamma)$

$[*22·68] \quad = (\alpha - \beta) \cup (\alpha - \gamma) : \supset \vdash . \text{Prop}$

***24·494.** $\vdash : \xi \subset \alpha . \eta \subset \beta . \alpha \cap \beta = \Lambda . \supset . (\xi \cup \eta) - \alpha = \eta . (\xi \cup \eta) - \beta = \xi$

Dem.

$\vdash . *24·3 . \quad \supset \vdash : \text{Hp} . \supset . \xi - \alpha = \Lambda \quad (1)$

$\vdash . *24·311 . \quad \supset \vdash : \text{Hp} . \supset . \beta \subset -\alpha .$

$[*22·44] \quad \supset . \eta \subset -\alpha .$

$[*22·621] \quad \supset . \eta - \alpha = \eta \quad (2)$

$\vdash . *22·68 . \quad \supset \vdash . (\xi \cup \eta) - \alpha = (\xi - \alpha) \cup (\eta - \alpha) \quad (3)$

$\vdash . (1) . (2) . (3) . *24·24 . \supset \vdash : \text{Hp} . \supset . (\xi \cup \eta) - \alpha = \eta \quad (4)$

Similarly $\quad \vdash : \text{Hp} . \supset . (\xi \cup \eta) - \beta = \xi \quad (5)$

$\vdash . (4) . (5) . \quad \supset \vdash . \text{Prop}$

This proposition is used in the theory of selections (*83·63 and *88·45).

***24·495.** $\vdash : \alpha \cap \gamma = \Lambda . \supset . (\alpha \cup \gamma) - (\beta \cup \gamma) = \alpha - \beta$

Dem.

$\vdash . *22·87·68 . \supset$

$\vdash . (\alpha \cup \gamma) - (\beta \cup \gamma) = (\alpha - \beta - \gamma) \cup (\gamma - \beta - \gamma)$

$[*24·21] \quad = \alpha - \beta - \gamma \quad (1)$

$\vdash . *24·311 . *22·621 . \supset \vdash : \text{Hp} . \supset . \alpha - \gamma = \alpha \quad (2)$

$\vdash . (1) . (2) . \quad \supset \vdash . \text{Prop}$

The above proposition is used in the theory of minimum points (*205·83·832·84).

In the remainder of this number we shall be concerned with the existence of classes. Many of the properties of the existence of classes follow from the fact that to say a class exists is equivalent to saying that the class is not equal to the null-class. This is proved in *24·54.

***24·5.** $\vdash : \exists ! \alpha . \equiv . (\exists x) . x \in \alpha \quad$ [*4·2 . (*24·03)]

***24·51.** $\vdash : \sim \exists ! \alpha . \equiv . \alpha = \Lambda$

Dem.

$\vdash . *24·5 . \supset \vdash : \sim \exists ! \alpha . \equiv . \sim \{(\exists x) . x \in \alpha\} .$

$[*10·252] \quad \equiv . (x) . x \sim\in \alpha .$

$[*24·15] \quad \equiv . \alpha = \Lambda : \supset \vdash . \text{Prop}$

***24·52.** $\vdash . \exists ! V \quad$ [*24·51·1 . Transp]

This proposition states that the class of all objects of the type in question is not null, but has at least one member. The assumption that there is something, which is equivalent to this proposition, is implicit in the proposition ∗10·1, that what is true always is true in any instance. This would not hold if there were no instances of anything; hence it implies the existence of something. It will be observed that the above proposition (∗24·52) depends on ∗24·1, which depends on ∗22·351, which depends on ∗10·251, which depends on ∗10·24, which depends on ∗10·1 or on ∗9·1. The assumption that there is something is involved in the use of the real variable, which would otherwise be meaningless. This is made explicit in ∗9·1, and in the proof of ∗9·2, which is the same proposition as ∗10·1.

∗24·53. $\vdash . \sim \exists ! \Lambda$ [∗24·51 . ∗20·2]

∗24·54. $\vdash : \exists ! \alpha . \equiv . \alpha \neq \Lambda$ [∗24·51 . Transp]

∗24·55. $\vdash : \sim(\alpha \subset \beta) . \equiv . \exists ! \alpha - \beta$ [∗24·3 . Transp . ∗24·54]

∗24·56. $\vdash :. \exists ! (\alpha \cup \beta) . \equiv : \exists ! \alpha . \vee . \exists ! \beta$ [∗10·42 . ∗22·34]

∗24·561. $\vdash : \exists ! (\alpha \cap \beta) . \supset . \exists ! \alpha . \exists ! \beta$ [∗10·5 . ∗22·33]

∗24·57. $\vdash :. \alpha \cap \beta = \Lambda . \supset : \exists ! \alpha . \supset . \alpha \neq \beta$

Dem.

$\vdash .$ ∗22·481 $. \supset \vdash : \alpha \cap \beta = \Lambda . \alpha = \beta . \supset . \alpha \cap \alpha = \Lambda .$
[∗22·5] $\supset . \alpha = \Lambda .$
[∗24·51] $\supset . \sim \exists ! \alpha$ (1)
$\vdash . (1) .$ Exp . Transp $. \supset \vdash .$ Prop

∗24·571. $\vdash : \exists ! \alpha . \alpha = \beta . \supset . \exists ! (\alpha \cap \beta)$

Dem.

$\vdash .$ ∗24·57 . Comm $. \supset \vdash :. \exists ! \alpha . \supset : \alpha \cap \beta = \Lambda . \supset . \alpha \neq \beta :$
[Transp] $\supset : \alpha = \beta . \supset . \alpha \cap \beta \neq \Lambda .$
[∗24·54] $\supset . \exists ! (\alpha \cap \beta)$ (1)
$\vdash . (1) .$ Imp $. \supset \vdash .$ Prop

∗24·58. $\vdash :. \alpha \subset \beta . \supset : \exists ! \alpha . \supset . \exists ! \beta$ [∗10·28]

∗24·6. $\vdash :. \alpha \subset \beta \supset : \alpha \neq \beta . \equiv . \exists ! \beta - \alpha$

Dem.

$\vdash .$ ∗22·41 . Transp $. \supset \vdash :.$ Hp $. \supset : \alpha \neq \beta . \supset . \sim(\beta \subset \alpha) .$
[∗24·55] $\supset . \exists ! \beta - \alpha$ (1)
$\vdash .$ ∗24·21 $. \supset \vdash : \alpha = \beta . \supset . \beta - \alpha = \Lambda$ (2)
$\vdash . (2) .$ Transp . ∗24·54 $. \supset \vdash : \exists ! \beta - \alpha . \supset . \alpha \neq \beta$ (3)
$\vdash . (1) . (3) . \supset \vdash .$ Prop

∗24·61. $\vdash : \sim \exists ! \beta . \supset . \alpha \cup \beta = \alpha$ [∗24·51·24]

∗24·62. $\vdash : \sim \exists ! \beta . \supset . \alpha \cap \beta = \Lambda$ [∗24·51·23]

***24·63.** $\vdash :. \Lambda \sim \epsilon \kappa . \equiv : \alpha \epsilon \kappa . \supset_\alpha . \exists ! \alpha$

In this proposition, the conditions of significance require that κ should be a class of classes. The condition "$\alpha \epsilon \kappa . \supset_\alpha . \exists ! \alpha$" is one required as hypothesis in many propositions. In virtue of the above proposition, this hypothesis may be replaced by "$\Lambda \sim \epsilon \kappa$."

Dem.

$\vdash . *13\cdot191 . \supset \vdash :. \Lambda \sim \epsilon \kappa . \equiv : \alpha = \Lambda . \supset_\alpha . \alpha \sim \epsilon \kappa :$

[Transp] $\equiv : \alpha \epsilon \kappa . \supset_\alpha . \alpha \neq \Lambda :$

[*24·54] $\equiv : \alpha \epsilon \kappa . \supset_\alpha . \exists ! \alpha :. \supset \vdash .$ Prop

This proposition is frequently used in later parts of the work. We often have to deal with classes of existent classes, and the most convenient form in which to state that all the members of a class of classes exist is "$\Lambda \sim \epsilon \kappa$."

*25. THE UNIVERSAL RELATION, THE NULL RELATION, AND THE EXISTENCE OF RELATIONS

Summary of *25.

This number contains the analogues, for relations, of the definitions and propositions of *24. Proofs will not be given, as they proceed precisely as in *24.

The universal relation, denoted by $\dot{\mathrm{V}}$, is the relation which holds between any two terms whatever of the appropriate types, whatever these may be in the given context. The null relation, $\dot{\Lambda}$, is the relation which does not hold between any pair of terms whatever, its type being fixed by the types of the terms concerning which the denial that it holds is significant. A relation R is said to *exist* when there is at least one pair of terms between which it holds; "R exists" is written "$\dot{\exists} ! R$."

The propositions of this number are much less often referred to than those of *24, but for the sake of uniformity we have given the analogues of all propositions in *24, with the same numeration (except for the integral part).

All the remarks made in *24 apply, *mutatis mutandis*, in the present number.

***25·01.** $\dot{\mathrm{V}} = \hat{x}\hat{y}\,(x = x \,.\, y = y)$ Df

***25·02.** $\dot{\Lambda} = \dot{-}\,\dot{\mathrm{V}}$ Df

***25·03.** $\dot{\exists} ! R \,.=.\, (\exists x, y) \,.\, xRy$ Df

***25·1.** $\vdash \,.\, \dot{\Lambda} \neq \dot{\mathrm{V}}$

***25·101.** $\vdash \,.\, \dot{\mathrm{V}} = \dot{-}\,\dot{\Lambda}$

***25·102.** $\vdash : (x, y) \,.\, \phi(x, y) \,.\equiv.\, \hat{x}\hat{y}\,\phi(x, y) = \dot{\mathrm{V}}$

***25·103.** $\vdash : (x, y) \,.\, {\sim}\phi(x, y) \,.\equiv.\, \hat{x}\hat{y}\,\phi(x, y) = \dot{\Lambda}$

***25·104.** $\vdash \,.\, (x, y) \,.\, x\dot{\mathrm{V}}y$

***25·105.** $\vdash \,.\, (x, y) \,.\, {\sim}(x\dot{\Lambda}y)$

***25·11.** $\vdash \,.\, (R) \,.\, R \mathrel{\dot{\subset}} \dot{\mathrm{V}}$

***25·12.** $\vdash \,.\, (R) \,.\, \dot{\Lambda} \mathrel{\dot{\subset}} R$

***25·13.** $\vdash : R = \dot{\Lambda} \,.\equiv.\, R \mathrel{\dot{\subset}} \dot{\Lambda}$

***25·14.** $\vdash : (x, y) \,.\, xRy \,.\equiv.\, R = \mathrm{V}$

***25·141.** $\vdash : \dot{\mathrm{V}} \mathrel{\dot{\subset}} R \,.\equiv.\, \mathrm{V} = R$

***25·15.** $\vdash : (x, y) \,.\, {\sim}(xRy) \,.\equiv.\, R = \dot{\Lambda}$

***25·17.** $\vdash : R = \dot{\mathrm{V}} \,.\equiv.\, \dot{-}R = \dot{\Lambda}$

***25·21.** $\vdash \,.\, R \mathbin{\dot{\cap}} \dot{-}R = \dot{\Lambda}$

***25·22.** $\vdash . R \mathbin{\dot\cup} \dot- R = \dot V$

***25·23.** $\vdash . R \mathbin{\dot\cap} \dot\Lambda = \dot\Lambda$

***25·24.** $\vdash . R \mathbin{\dot\cup} \dot\Lambda = R$

***25·26.** $\vdash . R \mathbin{\dot\cap} \dot V = R$

***25·27.** $\vdash . R \mathbin{\dot\cup} \dot V = \dot V$

***25·3.** $\vdash : R \mathrel{\dot\subset} S . \equiv . R \mathbin{\dot-} S = \dot\Lambda$

***25·31.** $\vdash : R \mathrel{\dot\subset} S . \equiv . \dot- R \mathbin{\dot\cup} S = V$

***25·311.** $\vdash : R \mathrel{\dot\subset} \dot- S . \equiv . R \mathbin{\dot\cap} S = \dot\Lambda$

***25·312.** $\vdash : \dot- R \mathrel{\dot\subset} S . \equiv . R \mathbin{\dot\cup} S = \dot V$

***25·313.** $\vdash : R \mathbin{\dot\cap} S = \dot\Lambda . \equiv . R \mathbin{\dot-} S = R$

***25·32.** $\vdash : R \mathbin{\dot\cup} S = \dot\Lambda . \equiv . R = \dot\Lambda . S = \dot\Lambda$

***25·33.** $\vdash : R = \dot V . \supset . R \mathbin{\dot\cup} S = \dot V$

***25·34.** $\vdash : R = \dot\Lambda . \supset . R \mathbin{\dot\cap} S = \dot\Lambda$

***25·35.** $\vdash : R = \dot V . \supset . R \mathbin{\dot\cap} S = S$

***25·36.** $\vdash : R = \dot\Lambda . \supset . R \mathbin{\dot\cup} S = S$

***25·37.** $\vdash :: R \mathbin{\dot\cap} S = \dot\Lambda . \equiv :. xRy . zSw . \supset_{x,y,z,w} : x \neq z . \vee . y \neq w$

***25·38.** $\vdash :. R \mathbin{\dot\cap} S = \dot\Lambda . \supset : R \neq S . \vee . R = \dot\Lambda . S = \dot\Lambda$

***25·39.** $\vdash :. R \mathbin{\dot\cap} S = \dot\Lambda . \equiv : xRy . \supset_{x,y} . \sim(xSy)$

***25·4.** $\vdash : P \mathbin{\dot\cap} Q = \dot\Lambda . \equiv . (P \mathbin{\dot\cup} Q) \mathbin{\dot-} P = Q . \equiv . (P \mathbin{\dot\cup} Q) \mathbin{\dot-} Q = P$

***25·401.** $\vdash : Q \mathrel{\dot\subset} P . \supset . (Q \mathbin{\dot\cup} R) \mathbin{\dot-} P = R \mathbin{\dot-} P$

***25·402.** $\vdash : P \mathbin{\dot\cap} Q = \dot\Lambda . R \mathrel{\dot\subset} P . S \mathrel{\dot\subset} Q . \supset . R \mathbin{\dot\cap} S = \dot\Lambda$

***25·41.** $\vdash . R = (R \mathbin{\dot\cap} S) \mathbin{\dot\cup} (R \mathbin{\dot-} S)$

***25·411.** $\vdash : S \mathrel{\dot\subset} R . \supset . R = S \mathbin{\dot\cup} (R \mathbin{\dot-} S)$

***25·412.** $\vdash : Q \mathrel{\dot\subset} P . S \mathrel{\dot\subset} Q . \supset . (P \mathbin{\dot-} Q) \mathbin{\dot\cup} (Q \mathbin{\dot-} S) = P \mathbin{\dot-} S$

***25·42.** $\vdash : P \mathbin{\dot\cap} Q \mathrel{\dot\subset} R . P \mathbin{\dot-} Q \mathrel{\dot\subset} R . \equiv . P \mathrel{\dot\subset} R$

***25·43.** $\vdash : P \mathbin{\dot-} Q \mathrel{\dot\subset} R . \equiv . P \mathrel{\dot\subset} Q \mathbin{\dot\cup} R$

***25·431.** $\vdash . (P \mathbin{\dot\cup} R) \mathbin{\dot\cap} (Q \mathbin{\dot\cup} \dot- R) = (P \mathbin{\dot\cap} Q) \mathbin{\dot\cup} (P \mathbin{\dot-} R) \mathbin{\dot\cup} (Q \mathbin{\dot\cap} R)$

***25·432.** $\vdash . (P \mathbin{\dot-} R) \mathbin{\dot\cup} (Q \mathbin{\dot\cap} R) = (P \mathbin{\dot\cap} Q) \mathbin{\dot\cup} (P \mathbin{\dot-} R) \mathbin{\dot\cup} (Q \mathbin{\dot\cap} R)$

***25·44.** $\vdash . (P \mathbin{\dot\cup} R) \mathbin{\dot\cap} (Q \mathbin{\dot\cup} \dot- R) = (P \mathbin{\dot\cap} \dot- R) \mathbin{\dot\cup} (Q \mathbin{\dot\cap} R)$

***25·45.** $\vdash : (P \mathbin{\dot\cap} R) \mathbin{\dot\cup} (Q \mathbin{\dot-} R) = \dot\Lambda . \equiv . Q \mathrel{\dot\subset} R . R \mathrel{\dot\subset} \dot- P$

***25·46.** $\vdash : (P \mathbin{\dot\cap} R) \mathbin{\dot\cup} (Q \mathbin{\dot-} R) = \dot\Lambda . \supset . P \mathbin{\dot\cap} Q = \dot\Lambda$

***25·47.** $\vdash : P \mathbin{\dot\cap} Q = \dot\Lambda . P \mathbin{\dot\cup} Q = R . \equiv . P \mathrel{\dot\subset} R . Q = R \mathbin{\dot-} P$

***25·48.** $\vdash :. R \mathrel{\dot\subset} P . R' \mathrel{\dot\subset} P . S \mathrel{\dot\subset} Q . S' \mathrel{\dot\subset} Q . P \mathbin{\dot\cap} Q = \dot\Lambda . \supset :$
$R \mathbin{\dot\cup} S = R' \mathbin{\dot\cup} S' . \equiv . R = R' . S = S'$

***25·481.** $\vdash :. P \mathbin{\dot\cap} Q = \dot\Lambda . P \mathbin{\dot\cap} R = \dot\Lambda . \supset : P \mathbin{\dot\cup} Q = P \mathbin{\dot\cup} R . \equiv . Q = R$

***25·482.** $\vdash :. R \mathrel{\dot\subset} P . S \mathrel{\dot\subset} Q . P \mathbin{\dot\cap} Q = \dot\Lambda . \supset : R \mathbin{\dot\cup} S = P \mathbin{\dot\cup} Q . \equiv . R = P . S = Q$

***25·49.** $\vdash :. P \mathbin{\dot\cap} Q = \dot\Lambda . \supset : P \mathrel{\dot\subset} Q \mathbin{\dot\cup} R . \equiv . P \mathrel{\dot\subset} R$

***25·491.** $\vdash : Q \dot{\frown} R = \dot{\Lambda} . P \mathrel{\dot{\subset}} Q \dot{\cup} R . \supset .$
$P \dot{-} Q = P \dot{\frown} R . P \dot{-} R = P \dot{\frown} Q . P = (P \dot{-} Q) \dot{\cup} (P \dot{-} R)$

***25·492.** $\vdash : Q \mathrel{\dot{\subset}} P . P \dot{-} Q = R . \supset . P \dot{-} R = Q$

***25·493.** $\vdash : Q \dot{\frown} R = \dot{\Lambda} . \supset . P = (P \dot{-} Q) \dot{\cup} (P \dot{-} R)$

***25·494.** $\vdash : R \mathrel{\dot{\subset}} P . S \mathrel{\dot{\subset}} Q . P \dot{\frown} Q = \dot{\Lambda} . \supset . (R \dot{\cup} S) \dot{-} P = S . (R \dot{\cup} S) \dot{-} Q = R$

***25·495.** $\vdash : P \dot{\frown} R = \dot{\Lambda} . \supset . (P \dot{\cup} R) \dot{-} (Q \dot{\cup} R) = P \dot{-} Q$

***25·5.** $\vdash : \dot{\exists} ! R . \equiv . (\exists x, y) . xRy$

***25·51.** $\vdash : \sim \dot{\exists} ! R . \equiv . R = \dot{\Lambda}$

***25·52.** $\vdash . \dot{\exists} ! \dot{V}$

***25·53.** $\vdash . \sim \dot{\exists} ! \dot{\Lambda}$

***25·54.** $\vdash : \dot{\exists} ! R . \equiv . R \neq \dot{\Lambda}$

***25·55.** $\vdash : \sim (R \mathrel{\dot{\subset}} S) . \equiv . \dot{\exists} ! R \dot{-} S$

***25·56.** $\vdash :. \dot{\exists} ! (R \dot{\cup} S) . \equiv : \dot{\exists} ! R . \vee . \dot{\exists} ! S$

***25·561.** $\vdash : \dot{\exists} ! (R \dot{\frown} S) . \supset . \dot{\exists} ! R . \dot{\exists} ! S$

***25·57.** $\vdash :. R \dot{\frown} S = \dot{\Lambda} . \supset : \dot{\exists} ! R . \supset . R \neq S$

***25·571.** $\vdash : \dot{\exists} ! R . R = S . \supset . \dot{\exists} ! (R \dot{\frown} S)$

***25·58.** $\vdash :. R \mathrel{\dot{\subset}} S . \supset : \dot{\exists} ! R . \supset . \dot{\exists} ! S$

***25·6.** $\vdash :. R \mathrel{\dot{\subset}} S . \supset : R \neq S . \equiv . \dot{\exists} ! S \dot{-} R$

***25·61.** $\vdash : \sim \dot{\exists} ! S . \supset . R \dot{\cup} S = R$

***25·62.** $\vdash : \sim \dot{\exists} ! S . \supset . R \dot{\frown} S = \dot{\Lambda}$

***25·63.** $\vdash :. \dot{\Lambda} \sim \epsilon \kappa . \equiv : R \epsilon \kappa . \supset_R . \dot{\exists} ! R$

$*30$. DESCRIPTIVE FUNCTIONS

Summary of $*30$.

The functions hitherto considered, with the exception of a few particular functions such as $\alpha \cap \beta$, have been propositional, *i.e.* have had propositions for their values. But the ordinary functions of mathematics, such as x^2, $\sin x$, $\log x$, are not propositional. Functions of this kind always mean "the term having such and such a relation to x." For this reason they may be called *descriptive* functions, because they *describe* a certain term by means of its relation to their argument. Thus "$\sin \pi/2$" describes the number 1; yet propositions in which $\sin \pi/2$ occurs are not the same as they would be if 1 were substituted for $\sin \pi/2$. This appears *e.g.* from the proposition "$\sin \pi/2 = 1$," which conveys valuable information, whereas "$1 = 1$" is trivial. Descriptive functions, like descriptions in general, have no meaning by themselves, but only as constituents of propositions*.

The general definition of a descriptive function is:

$*30 \cdot 01$. $R'y = (\imath x)(xRy)$ Df

That is, "$R'y$" is to mean "the term x which has the relation R to y." If there are several terms or none having the relation R to y, all propositions about $R'y$, *i.e.* all propositions of the form "$\phi(R'y)$," will be false. The apostrophe in "$R'y$" may be read "of." Thus if R is the relation of father to son, "$R'y$" means "the father of y." If R is the relation of son to father, "$R'y$" means "the son of y"; in this case, all propositions of the form "$\phi(R'y)$" will be false unless y has one son and no more.

All the functions that occur in ordinary mathematics are instances of the above definition; all are obtained in the above manner from some relation. Thus in our notation "$R'y$" takes the place of what would commonly be "fy," this latter notation being reserved for *propositional* functions. We should write "$\sin 'y$" in place of "$\sin y$," using "$\sin$" to express the relation of x to y when $x = \sin y$.

A definition such as $R'y = (\imath x)(xRy)$, where the meaning given to the term defined is a description, must be understood to mean that the term defined (in this case $R'y$) and the description assigned as its meaning (in this case $(\imath x)(xRy)$) are to be interchangeable in use: the definition is, in a sense, more purely symbolic than other definitions, since the description assigned as the meaning has itself no meaning except in use. It would perhaps be more formally correct to write

$$f(R'y) . = . f\{(\imath x)(xRy)\} \quad \text{Df.}$$

* Cf. $*14$, above.

SECTION D

LOGIC OF RELATIONS

In the present section we shall be concerned with such of the general properties of relations as have no analogues in the theory of classes. The notations introduced in this section will be used constantly throughout the rest of the work, and the ideas expressed in the definitions will be found to be of fundamental importance.

But even this definition would not be quite complete, because it omits mention of the *scope* of the two descriptions $R‘y$ and $(\iota x)(xRy)$. Thus the complete form would be

$$[R‘y] . f(R‘y) . = . [(\iota x)(xRy)] . f\{(\iota x)(xRy)\} \quad \text{Df.}$$

But it is unnecessary to adopt this form of definition, provided it is understood that the definition $*30·01$ means that "$R‘y$" may be written for "$(\iota x)(xRy)$" *everywhere, i.e.* in indications of scope as well as elsewhere. The *use* of the definition occurs always in accordance with the proposition:

$$\vdash : [R‘y] . f(R‘y) . \equiv . [(\iota x)(xRy)] . f(\iota x)(xRy),$$

which is $*30·1$, below.

It is to be observed that $*30·01$ does not necessarily involve

$$R‘y = (\iota x)(xRy).$$

For this, by the definition, is equivalent to

$$(\iota x)(xRy) = (\iota x)(xRy),$$

which, by $*14·28$, only holds when $E ! (\iota x)(xRy)$, *i.e.* when there is one term, and no more, which has the relation R to y.

All the conventions as to scope explained in $*14$ are to be transferred to $R‘x$, *i.e.*, in the absence of any contrary indication, the scope of $R‘x$ is to be the smallest proposition, enclosed in dots or other brackets, in which the $R‘x$ in question occurs.

We put

$*30·02$. $R‘S‘y = R‘(S‘y)$ Df

This definition serves merely for the avoidance of brackets. It is to be interpreted as meaning

$$[R‘S‘y] . f(R‘S‘y) . = . [R‘(S‘y)] . f\{R‘(S‘y)\} \quad \text{Df.}$$

In future, we shall often define a new expression as having a descriptive phrase for its meaning; in such a case, the definition is always to be interpreted as above. That is, any proposition in which the new expression occurs is to be the proposition which is obtained by substituting the old expression for the new one wherever the latter occurs.

$R‘(S‘y)$, in the above, is to be interpreted by first treating $S‘y$ as if it were not a descriptive symbol, and applying $*30·01$ and $*14·01$ or $*14·02$ to $R‘(S‘y)$, and by then applying $*30·01$ and $*14·01$ or $*14·02$ to $S‘y$.

The majority of the propositions of the present number are immediate consequences of the corresponding propositions in $*14$. Thus $*14·31$—$·34$ and $*14·113$ lead immediately to $*30·12$—$·16$, which show that, either always or when $R‘y$ exists, the "scope" of $R‘y$ or of $R‘y$ and $S‘y$ makes no difference to the truth-values of such propositions as we are concerned with. We have

$*30·18$. $\vdash :. E ! R‘y : (z) . \phi z : \supset . \phi(R‘y)$

so that what holds of everything holds of $R`y$, provided $R`y$ exists. This results immediately from ✱14·18, and shows that, provided $R`y$ exists, the fact that "$R`y$" is an incomplete symbol does not prevent its being substituted as a value of z whenever we have $(z) . \phi z$, or an assertion of the propositional function ϕz.

One of the most used propositions of this number is:

✱30·3. $\vdash :. x = R`y . \equiv : zRy . \equiv_z . z = x$

which results immediately from ✱14·202. The following analogous proposition results from the above by means of ✱14·122:

✱30·31. $\vdash :. x = R`y . \equiv : xRy : zRy . \supset_z . z = x$

I.e. "$x = R`y$" involves, in addition to "xRy," the statement that whatever has the relation R to y is identical with x.

A proposition constantly referred to is:

✱30·37. $\vdash : E ! R`y . y = z . \supset . R`y = R`z$

In the hypothesis, $E ! R`y$ might be replaced by $E ! R`z$, but one or other of them is essential. For, by ✱14·21, "$R`y = R`z$" implies $E ! R`y$ and $E ! R`z$ (these are equivalent when $y = z$), and therefore cannot be true when $R`y$ and $R`z$ do not exist.

The use of ✱30·37 is chiefly in cases where y or z or both are replaced by descriptive functions. Suppose, for example, that z is replaced by $S`w$. By ✱30·18, we may substitute $S`w$ for z if $S`w$ exists. By ✱14·21, both sides of the implication in ✱30·37 will become false if $S`w$ does not exist, and therefore the implication will still hold. Hence whether $S`w$ exists or not, we may substitute it for z and obtain

$$\vdash : E ! R`y . y = S`w . \supset . R`y = R`S`w.$$

In like manner, if we replace y by $T`v$, we obtain

$$\vdash : E ! R`T`v . T`v = S`w . \supset . R`T`v = R`S`w.$$

A very important proposition is:

✱30·4. $\vdash :. E ! R`y . \supset : a = R`y . \equiv . aRy$

This proposition states that, provided $R`y$ exists, to say that a is *the* term which has the relation R to y is equivalent to saying that a has the relation R to y. Thus for example "a is the occupier of the house y" is equivalent to "a occupies the house y," "a is the writer of Waverley" is equivalent to "a wrote Waverley," "a is the father of y" is equivalent to "a begot y." But we cannot argue from "John Smith inhabits London" to "John Smith is *the* inhabitant of London."

We shall introduce in this and subsequent sections many constant relations for which $E ! R`y$ is always true. When $R$ is such that $E ! R`y$ is always true, we have, in virtue of ✱30·4,

$$a = R`y . \equiv . aRy$$

for every possible value of y. The following proposition is useful in cases where both R and S are such that $R`y$ and $S`y$ always exist:

***30·41.** $\vdash :. (y) . R`y = S`y . \equiv : (y) . E ! R`y : R = S$

Thus if we know that $R`y$ and $S`y$ are always identical, we know not only that R and S are identical, but also that $R`y$ (and therefore $S`y$) always exists.

***30·01.** $R`y = (\iota x)(xRy)$ Df

***30·02.** $R`S`y = R`(S`y)$ Df

In interpreting $R`(S`y)$, $S`y$ is to be treated as an ordinary symbol until $R`(S`y)$ has been eliminated by *30·01 and *14·01 or *14·02, and then the above definitions are to be applied to $S`y$.

***30·1.** $\vdash : [R`y] . f(R`y) . \equiv . [(\iota x)(xRy)] . f(\iota x)(xRy)$ [*4·2 . (*30·01)]

***30·11.** $\vdash :. [R`y] . f(R`y) . \equiv : (\exists b) : xRy . \equiv_x . x = b : fb$ [*30·1 . *14·1]

The following propositions are immediate applications of *14·31 ff., made in accordance with *30·1.

***30·12.** $\vdash :: E ! R`y . \supset :. [R`y] . p \vee \chi(R`y) . \equiv : p . \vee . [R`y] . \chi(R`y)$ [*14·31]

***30·13.** $\vdash :: E ! R`y . \supset :. [R`y] . \sim\chi(R`y) . \equiv . \sim\{[R`y] . \chi(R`y)\}$ [*14·32]

***30·14.** $\vdash :: E ! R`y . \supset :. [R`y] . p \supset \chi(R`y) . \equiv : p . \supset . [R`y] . \chi(R`y)$ [*14·33]

***30·141.** $\vdash :: E ! R`y . \supset :. [R`y] . \chi(R`y) \supset p . \equiv : [R`y] . \chi(R`y) . \supset . p$ [*14·331]

***30·142.** $\vdash :: E ! R`y . \supset :. [R`y] . p \equiv \chi(R`y) . \equiv : p . \equiv . [R`y] . \chi(R`y)$ [*14·332]

***30·15.** $\vdash :. p : [R`y] . \chi(R`y) : \equiv : [R`y] . p . \chi(R`y)$ [*14·34]

The following two propositions are immediate consequences of *14·113·112.

***30·16.** $\vdash : [R`y] . f(R`y, S`z) . \equiv . [S`z] . f(R`y, S`z)$ [*14·113]

***30·17.** $\vdash :. [R`y] . f(R`y, S`z) . \equiv :$
$(\exists b, c) : xRy . \equiv_x . x = b : xSz . \equiv_x . x = c : f(b, c)$ [*14·112]

***30·18.** $\vdash :. E ! R`y : (z) . \phi z : \supset . \phi(R`y)$ [*14·18]

***30·19.** $\vdash :. R`y = b . \supset : \psi(R`y) . \equiv . \psi b$ [*14·15]

***30·2.** $\vdash :. E ! R`y . \equiv : (\exists b) : xRy . \equiv_x . x = b$ [*4·2 . *14·11 . (*30·01)]

In proving *30·2, we have to use the definition *30·01, not *30·1, because $E ! (\iota x)(\phi x)$ is not of the form $f(\iota x)(\phi x)$. This appears if we attempt to apply the definition *14·01 to $E ! (\iota x)(\phi x)$, which leads to an expression containing the meaningless constituent $E ! b$. But by the definition *30·01, every typographical occurrence of the symbol "$R`y$" means what results when this symbol is replaced by "$(\iota x)(xRy)$," hence "$E ! R`y$" means "$E ! (\iota x)(xRy)$."

***30·21.** $\vdash :: E ! R'y . \equiv :. (\exists x) . xRy : xRy . zRy . \supset_{x,z} . x = z$
[*14·203 . (*30·01)]

***30·22.** $\vdash : E ! R'y . \equiv . R'y = (\iota x)(xRy)$ [*14·28 . (*30·01)]

Note that we do not necessarily have

$$R'y = (\iota x)(xRy),$$

which is only true when $E ! R'y$.

***30·3.** $\vdash :. x = R'y . \equiv : zRy . \equiv_z . z = x$ [*14·202]

***30·31.** $\vdash :. x = R'y . \equiv : xRy : zRy . \supset_z . z = x$ [*14·122 . *30·3]

***30·32.** $\vdash : E ! R'y . \equiv . (R'y) Ry$ [*14·22]

***30·33.** $\vdash :: E ! R'y . \supset :. \psi(R'y) : \equiv : (\exists x) . xRy . \psi x : \equiv : xRy . \supset_x . \psi x$
[*14·26]

***30·34.** $\vdash :. xRy . \equiv_x . xSy : \supset : E ! R'y . \equiv . E ! S'y$ [*14·271]

***30·341.** $\vdash :. xRy . \equiv_x . xSy : \supset : E ! R'y . \equiv . R'y = S'y$

Dem.

$\vdash$. *14·21 . $\supset \vdash : R'y = S'y . \supset . E ! R'y$ (1)

$\vdash$. *14·27 . Comm . $\supset \vdash :. \text{Hp} . \supset : E ! R'y . \supset . R'y = S'y$ (2)

$\vdash$. (1) . (2) . $\supset \vdash$. Prop

***30·35.** $\vdash :. R = S . \supset : E ! R'y . \equiv . E ! S'y$ [*30·34 . *21·43]

***30·36.** $\vdash : E ! R'y . R = S . \supset . R'y = S'y$ [*14·27 . Imp . *21·43]

***30·37.** $\vdash : E ! R'y . y = z . \supset . R'y = R'z$

Dem.

$\vdash$. *14·28 . $\supset \vdash : E ! R'y . \supset . R'y = R'y$ (1)

$\vdash$. *13·12 . $\supset \vdash :. y = z . \supset : R'y = R'y . \equiv . R'y = R'z$ (2)

$\vdash$. (1) . (2) . Ass . $\supset \vdash$. Prop

This proposition is very frequently used.

***30·4.** $\vdash :. E ! R'y . \supset : a = R'y . \equiv . aRy$ [*14·241]

This is a very important proposition, of which the use is constant.

***30·41.** $\vdash :. (y) . R'y = S'y . \equiv : (y) . E ! R'y : R = S$

Dem.

$\vdash$. *14·21 . *10·11·27 . $\supset \vdash : (y) . R'y = S'y . \supset . (y) . E ! R'y$ (1)

$\vdash$. *14·13·142 . $\supset \vdash :. (y) . R'y = S'y . \supset : (x, y) : x = R'y . \equiv . x = S'y :$

[(1).*30·4] $\supset : (x, y) : xRy . \equiv . xSy :$

[*21·43] $\supset : R = S$ (2)

$\vdash$. *30·36 . $\supset \vdash : E ! R'y . R = S . \supset . R'y = S'y :$

[*10·11·27·35] $\supset \vdash :. (y) . E ! R'y : R = S : \supset . (y) . R'y = S'y$ (3)

$\vdash$. (1) . (2) . (3) . $\supset \vdash$. Prop

***30·42.** $\vdash :. (y) . E ! R‘y . \supset : (y) . R‘y = S‘y . \equiv . R = S$ [*30·41]

The hypothesis $(y) . E ! R‘y$ is fulfilled by a number of important special relations, of which examples will occur in the subsequent numbers of the present section.

***30·5.** $\vdash : E ! P‘Q‘z . \supset . E ! Q‘z$

Dem.

$\vdash . *30·2 . \supset \vdash :. E ! P‘Q‘z . \equiv : (\exists b) : xP(Q‘z) . \equiv_x . x = b :$

[*10·1] $\supset : (\exists b) : bP(Q‘z) . \equiv . b = b :$

[*13·15] $\supset : (\exists b) . bP(Q‘z) :$

[*14·21] $\supset : E ! Q‘z :. \supset \vdash . \text{Prop}$

***30·501.** $\vdash : \phi(P‘Q‘z) . \equiv . (\exists b, c) . c = Q‘z . b = P‘c . \phi b$

On the meaning of "$\phi(P‘Q‘z)$," see note to the definition *30·02.

Dem.

$\vdash . *14·1·122 . \supset \vdash :: \phi(P‘Q‘z) . \equiv :. (\exists b) : bP(Q‘z) : xP(Q‘z) . \supset_x . x = b : \phi b :.$

[*14·205] $\equiv :. (\exists b) :. (\exists c) : c = Q‘z : bPc : xPc . \supset_x . x = b : \phi b :.$

[*14·122·202] $\equiv :. (\exists b, c) . c = Q‘z . b = P‘c . \phi b :: \supset \vdash . \text{Prop}$

***30·51.** $\vdash : b = P‘Q‘z . \equiv . (\exists c) . b = P‘c . c = Q‘z$ [*30·501 . *13·195]

***30·52.** $\vdash : E ! P‘Q‘z . \equiv . (\exists b, c) . b = P‘c . c = Q‘z$ [*30·51 . *14·204]

*31. CONVERSES OF RELATIONS

Summary of *31.

If R is a relation, the relation which y has to x when xRy is called the *converse* of R. Thus *greater* is the converse of *less*, *before* of *after*, *husband* of *wife*. The converse of identity is identity, and the converse of diversity is diversity. The converse of R is written $\breve{R}$ (read "R-converse"). When $R=\breve{R}$, R is called a *symmetrical* relation, otherwise it is called *not-symmetrical*. When R is incompatible with $\breve{R}$, R is called *asymmetrical*. Thus "cousin" is symmetrical, "brother" is not-symmetrical (because when x is the brother of y, y may be either the brother or the sister of x), and "husband" is asymmetrical.

The relation of $\breve{R}$ to R is called "Cnv." It will be shown that every relation has one, and only one, converse; hence, applying the notation of *30, that one is $\text{Cnv}‘R$. Thus $\breve{R}=\text{Cnv}‘R$. We have thus two notations for the converse of R; the second is more convenient for the converse of a relation not denoted by a single letter.

The more important propositions of the present number are the following:

***31·13.** $\vdash . \text{E}\,!\,\text{Cnv}‘P$

I.e. any relation P has a converse. Hence the relation "Cnv" verifies the hypothesis $(y) . \text{E}\,!\,R‘y$, *i.e.* we have $(P) . \text{E}\,!\,\text{Cnv}‘P$.

***31·32.** $\vdash : P=Q . \equiv . \breve{P}=\breve{Q}$

I.e. two relations are identical when, and only when, their converses are identical.

***31·33.** $\vdash . \text{Cnv}‘\text{Cnv}‘P=P$

I.e. any relation is the converse of its converse.

Very many of the subsequent uses of the notion of the converse of a relation require only the propositions which embody the definitions of $\breve{P}$ and Cnv, namely

***31·11.** $\vdash : x\breve{P}y . \equiv . yPx$

and

***31·131.** $\vdash : x(\text{Cnv}‘P)y . \equiv . yPx$

***31·01.** $\mathrm{Cnv} = \hat{Q}\hat{P}\{xQy \,.\equiv_{x,y}.\, yPx\}$ Df

***31·02.** $\breve{P} = \hat{x}\hat{y}(yPx)$ Df

***31·1.** $\vdash :.\, Q \,\mathrm{Cnv}\, P \,.\equiv:\, xQy \,.\equiv_{x,y}.\, yPx$ [*21·3 . (*31·01)]

***31·101.** $\vdash : Q \,\mathrm{Cnv}\, P \,.\, R \,\mathrm{Cnv}\, P \,.\supset.\, Q = R$

Dem.

$\vdash .\, *31·1 \,.\supset\vdash :.\, \mathrm{Hp} \,.\supset:\, xQy \,.\equiv_{x,y}.\, yPx : xRy \,.\equiv_{x,y}.\, yPx :$

[*11·371] $\supset : xQy \,.\equiv_{x,y}.\, xRy :$

[*21·43] $\supset : Q = R :.\supset \vdash .\, \mathrm{Prop}$

***31·11.** $\vdash : x\breve{P}y \,.\equiv.\, yPx$ [*21·3 . (*31·02)]

***31·111.** $\vdash .\, \breve{P} \,\mathrm{Cnv}\, P$ [*31·1·11]

***31·12.** $\vdash .\, \breve{P} = \mathrm{Cnv}‘P$

Dem.

$\vdash .\, *31·101 \,.\supset \vdash : Q \,\mathrm{Cnv}\, P \,.\, \breve{P} \,\mathrm{Cnv}\, P \,.\supset.\, Q = \breve{P} :$

[*31·111] $\supset \vdash : Q \,\mathrm{Cnv}\, P \,.\supset.\, Q = \breve{P}$ (1)

$\vdash .\, (1) \,.\, *10·11 \,.\, *31·111 \,.\supset$

$\vdash : \breve{P} \,\mathrm{Cnv}\, P : Q \,\mathrm{Cnv}\, P \,.\supset_Q.\, Q = \breve{P} :$

[*30·31] $\supset \vdash .\, \breve{P} = \mathrm{Cnv}‘P$

***31·13.** $\vdash .\, \mathrm{E!}\, \mathrm{Cnv}‘P$ [*14·21 . *31·12]

***31·131.** $\vdash : x(\mathrm{Cnv}‘P)y \,.\equiv.\, yPx$ [*31·11·12 . *21·43]

***31·132.** $\vdash : Q \,\mathrm{Cnv}\, P \,.\equiv.\, Q = \mathrm{Cnv}‘P \,.\equiv.\, Q = \breve{P}$ [*30·4 . *31·13·12]

***31·14.** $\vdash .\, \mathrm{Cnv}‘(P \dot{\wedge} Q) = \mathrm{Cnv}‘P \dot{\wedge} \mathrm{Cnv}‘Q$

Dem.

$\vdash .\, *31·131 \,.\supset \vdash : x\{\mathrm{Cnv}‘(P \dot{\wedge} Q)\}y \,.\equiv.\, y(P \dot{\wedge} Q)x \,.$

[*21·33] $\equiv .\, yPx \,.\, yQx \,.$

[*31·131] $\equiv .\, x(\mathrm{Cnv}‘P)y \,.\, x(\mathrm{Cnv}‘Q)y \,.$

[*21·33] $\equiv .\, x\{\mathrm{Cnv}‘P \dot{\wedge} \mathrm{Cnv}‘Q\}y$ (1)

$\vdash .\, (1) \,.\, *11·11 \,.\, *21·43 \,.\supset \vdash .\, \mathrm{Prop}$

***31·15.** $\vdash .\, \mathrm{Cnv}‘(P \dot{\vee} Q) = \mathrm{Cnv}‘P \dot{\vee} \mathrm{Cnv}‘Q$ [Similar proof]

***31·16.** $\vdash .\, \mathrm{Cnv}‘\dot{-}P = \dot{-}(\mathrm{Cnv}‘P)$

Dem.

$\vdash .\, *31·131 \,.\supset \vdash : x(\mathrm{Cnv}‘\dot{-}P)y \,.\equiv.\, y \dot{-} Px \,.$

[*23·35] $\equiv .\, \sim(yPx) \,.$

[*31·131] $\equiv .\, \sim\{x(\mathrm{Cnv}‘P)y\} \,.$

[*23·35] $\equiv .\, x\{\dot{-}(\mathrm{Cnv}‘P)\}y$ (1)

$\vdash .\, (1) \,.\, *11·11 \,.\, *21·43 \,.\supset \vdash .\, \mathrm{Prop}$

$*31 \cdot 17$. $\vdash :. y = \breve{P}`x . \equiv : xPz . \equiv_z . z = y$ [$*30 \cdot 3 . *31 \cdot 11$]

$*31 \cdot 18$. $\vdash :. \mathrm{E} ! \breve{P}`x . \equiv : (\exists y) : xPz . \equiv_z . z = y$ [$*30 \cdot 2 . *31 \cdot 11$]

$*31 \cdot 21$. $\vdash . \mathrm{Cnv}`\dot{\Lambda} = \dot{\Lambda}$

Dem.

$\vdash . *31 \cdot 131 . \supset \vdash : x (\mathrm{Cnv}`\dot{\Lambda}) y . \equiv . y \dot{\Lambda} x :$
[$*25 \cdot 105$] $\supset \vdash . \sim x (\mathrm{Cnv}`\dot{\Lambda}) y$ (1)
$\vdash . (1) . *11 \cdot 11 . *25 \cdot 15 . \supset \vdash . \mathrm{Prop}$

$*31 \cdot 22$. $\vdash . \mathrm{Cnv}`\dot{V} = \dot{V}$ [Similar proof]

$*31 \cdot 23$. $\vdash : \breve{P} = \dot{V} . \equiv . P = \dot{V}$

Dem.

$\vdash . *25 \cdot 14 . \supset \vdash : \breve{P} = \dot{V} . \equiv . (x, y) . x \breve{P} y .$
[$*31 \cdot 11 . *11 \cdot 33$] $\equiv . (x, y) . yPx .$
[$*11 \cdot 2$] $\equiv . (y, x) . yPx .$
[$*25 \cdot 14$] $\equiv . P = \dot{V} : \supset \vdash . \mathrm{Prop}$

$*31 \cdot 24$. $\vdash : \breve{P} = \dot{\Lambda} . \equiv . P = \dot{\Lambda}$ [Similar proof]

$*31 \cdot 32$. $\vdash : P = Q . \equiv . \breve{P} = \breve{Q}$

Dem.

$\vdash . *21 \cdot 43 . \supset \vdash :. P = Q . \equiv : xPy . \equiv_{x,y} . xQy :$
[$*4 \cdot 86 \cdot 21 . *31 \cdot 11$] $\equiv : y\breve{P}x . \equiv_{x,y} . y\breve{Q}x :$
[$*11 \cdot 2$] $\equiv : y\breve{P}x . \equiv_{y,x} . y\breve{Q}x :$
[$*21 \cdot 43$] $\equiv : \breve{P} = \breve{Q} :. \supset \vdash . \mathrm{Prop}$

$*31 \cdot 33$. $\vdash . \mathrm{Cnv}`\mathrm{Cnv}`P = P$

Dem.

$\vdash . *31 \cdot 131 . \supset \vdash : x (\mathrm{Cnv}`\mathrm{Cnv}`P) y . \equiv . y (\mathrm{Cnv}`P) x .$
[$*31 \cdot 131$] $\equiv . xPy$ (1)
$\vdash . (1) . *11 \cdot 11 . *21 \cdot 43 . \supset \vdash . \mathrm{Prop}$

$*31 \cdot 34$. $\vdash : P = \breve{Q} . \equiv . Q = \breve{P}$

Dem.

$\vdash . *31 \cdot 32 . \supset \vdash : P = \breve{Q} . \equiv . \breve{P} = \mathrm{Cnv}`\breve{Q}$
[$*31 \cdot 12 \cdot 32$] $= \mathrm{Cnv}`\mathrm{Cnv}`Q$
[$*31 \cdot 33$] $= Q : \supset \vdash . \mathrm{Prop}$

$*31 \cdot 4$. $\vdash : P \mathrel{\dot{\subset}} Q . \equiv . \breve{P} \mathrel{\dot{\subset}} \breve{Q}$ [$*31 \cdot 11 . *11 \cdot 33$]

$*31 \cdot 41$. $\vdash : P \mathrel{\dot{\subset}} \breve{Q} . \equiv . \breve{P} \mathrel{\dot{\subset}} Q$ [$*31 \cdot 4 \cdot 33 \cdot 12$]

$*31 \cdot 5$. $\vdash : \dot{\exists} ! P . \equiv . \dot{\exists} ! \breve{P}$ [$*31 \cdot 24$. Transp . $*25 \cdot 54$]

∗31·51. $\vdash : (P) . f\breve{P} . \equiv . (P) . fP$

Dem.

$\vdash . *10\cdot1 . \supset \vdash : (P) . fP . \supset . f\breve{P} :$

$[*10\cdot11\cdot21] \supset \vdash : (P) . fP . \supset . (P) . f\breve{P}$ (1)

$\vdash . *10\cdot1 . *31\cdot12 . \supset$

$\vdash : (P) . f\breve{P} . \supset . f(\text{Cnv}`\breve{P}) .$

$[*31\cdot33\cdot12] \qquad \supset . fP :$

$[*10\cdot11\cdot21] \supset \vdash : (P) . f\breve{P} . \supset . (P) . fP$ (2)

$\vdash . (1) . (2) . \supset \vdash . \text{Prop}$

∗31·52. $\vdash : (\exists P) . f\breve{P} . \equiv . (\exists P) . fP$ $[*31\cdot51 . \text{Transp}]$

*32. REFERENTS AND RELATA OF A GIVEN TERM WITH RESPECT TO A GIVEN RELATION

Summary of *32.

Given any relation R, the class of terms which have the relation R to a given term y are called the *referents* of y, and the class of terms to which a given term x has the relation R are called the *relata* of x. We shall denote by $\overrightarrow{R}$ the relation of the class of referents of y to y, and by $\overleftarrow{R}$ the relation of the class of relata of x to x. It is convenient also to have a notation for the relations of $\overrightarrow{R}$ and $\overleftarrow{R}$ to R. We shall denote the relation of $\overrightarrow{R}$ to R by "sg," where "sg" stands for "sagitta." Similarly we shall denote by "gs" the relation of $\overleftarrow{R}$ to R, to suggest an arrow running from right to left instead of from left to right. $\overrightarrow{R}$ and $\overleftarrow{R}$ are chiefly useful for the sake of the descriptive functions to which they give rise; thus $\overrightarrow{R}`y = \hat{x}(xRy)$ and $\overleftarrow{R}`x = \hat{y}(xRy)$. Thus *e.g.* if R is the relation of parent to son, $\overrightarrow{R}`y =$ the parents of $y$, $\overleftarrow{R}`x =$ the sons of x. If R is the relation of less to greater among numbers of any kind, $\overrightarrow{R}`y =$ numbers less than $y$, and $\overleftarrow{R}`x =$ numbers greater than x. When $R`y$ exists, $\overrightarrow{R}`y$ is the class whose only member is $R`y$. But when there are many terms having the relation $R$ to $y$, $\overrightarrow{R}`y$, which is the class of those terms, supplies a notation which cannot be supplied by $R`y$. And similarly if there are many terms to which $x$ has the relation $R$, $\overleftarrow{R}`x$ supplies the notation for these terms. Thus for example let R be the relation "sin," *i.e.* the relation which x has to y when $x = \sin y$. Then "$\overleftarrow{\sin}`x$" represents all values of y such that $x = \sin y$, *i.e.* all values of $\sin^{-1} x$ or $\arcsin x$. Unlike the usual symbol, it is not ambiguous, since instead of representing some one of these values, it represents the class of them.

The definitions of $\overrightarrow{R}$, $\overleftarrow{R}$, sg, gs are as follows:

***32·01.** $\overrightarrow{R} = \hat{\alpha}\hat{y}\{\alpha = \hat{x}(xRy)\}$ Df

***32·02.** $\overleftarrow{R} = \hat{\beta}\hat{x}\{\beta = \hat{y}(xRy)\}$ Df

***32·03.** $\text{sg} = \hat{A}\hat{R}(A = \overrightarrow{R})$ Df

***32·04.** $\text{gs} = \hat{A}\hat{R}(A = \overleftarrow{R})$ Df

In virtue of the above definitions, we shall have $\text{sg}`R = \overrightarrow{R}$, $\text{gs}`R = \overleftarrow{R}$. This gives an alternative notation which is convenient in dealing with a relation not represented by a single letter.

It should be observed that if R is a homogeneous relation (*i.e.* one in which referents and relata are of the same type), then $\overrightarrow{R}$ and $\overleftarrow{R}$ are not homogeneous, but relate a class to objects of the type of its members.

In virtue of the definitions of $\overrightarrow{R}$ and $\overleftarrow{R}$, we shall have

***32·13.** $\vdash . \overrightarrow{R}\text{‘}y = \hat{x}(xRy)$

***32·131.** $\vdash . \overleftarrow{R}\text{‘}x = \hat{y}(xRy)$

Thus by *14·21, we always have $E ! \overrightarrow{R}\text{‘}y$ and $E ! \overleftarrow{R}\text{‘}x$. Thus whatever relation R may be, we have $(y) . E ! \overrightarrow{R}\text{‘}y$ and $(x) . E ! \overleftarrow{R}\text{‘}x$. We do not in general have $(y) . \exists ! \overrightarrow{R}\text{‘}y$ or $(x) . \exists ! \overleftarrow{R}\text{‘}x$. Thus taking R to be the relation of parent and child, $\overrightarrow{R}\text{‘}y =$ the parents of y and $\overleftarrow{R}\text{‘}x =$ the children of x. Thus $\overleftarrow{R}\text{‘}x = \Lambda$, *i.e* $\sim \exists ! \overleftarrow{R}\text{‘}x$, when x is childless, and $\overrightarrow{R}\text{‘}y = \Lambda$, *i.e.* $\sim \exists ! \overrightarrow{R}\text{‘}y$, when y is Adam or Eve. The two sorts of existence, $E ! \overrightarrow{R}\text{‘}y$ and $\exists ! \overrightarrow{R}\text{‘}y$, can both be *significantly* predicated of $\overrightarrow{R}\text{‘}y$, because "$\overrightarrow{R}\text{‘}y$" is a descriptive function whose value is a class; and the same applies to $\overleftarrow{R}\text{‘}x$. It will be seen that (by *14·21) $\exists ! \overrightarrow{R}\text{‘}y . \supset . E ! \overrightarrow{R}\text{‘}y$, but the converse implication does not hold in general.

We have

***32·16.** $\vdash : \overrightarrow{R} = \overrightarrow{S} . \equiv . \overleftarrow{R} = \overleftarrow{S} . \equiv . R = S$

Aso by *32·18·181,

$$\vdash : x \epsilon \overrightarrow{R}\text{‘}y . \equiv . xRy . \equiv . y \epsilon \overleftarrow{R}\text{‘}x.$$

Thus by the use of $\overrightarrow{R}\text{‘}y$ or $\overleftarrow{R}\text{‘}x$, every statement of the form "xRy" can be reduced to a statement asserting membership of a class. Since, however, the class in question is given by a descriptive function, and descriptive functions are defined by means of relations, we do not thus obtain a method of reducing the theory of relations to the theory of classes.

***32·01.** $\overrightarrow{R} = \hat{\alpha}\hat{y}\{\alpha = \hat{x}(xRy)\}$ Df

***32·02.** $\overleftarrow{R} = \hat{\beta}\hat{x}\{\beta = \hat{y}(xRy)\}$ Df

***32·03.** $\mathrm{sg} = \hat{A}\hat{R}(A = \overrightarrow{R})$ Df

***32·04.** $\mathrm{gs} = \hat{A}\hat{R}(A = \overleftarrow{R})$ Df

***32·1.** $\vdash : \alpha \overrightarrow{R} y . \equiv . \alpha = \hat{x}(xRy)$ [*21·3 . (*32·01)]

***32·101.** $\vdash : \beta \overleftarrow{R} x . \equiv . \beta = \hat{y}(xRy)$ [*21·3 . (*32·02)]

***32·11.** $\vdash . \hat{x}(xRy) = \overrightarrow{R}\text{‘}y$ [*32·1 . *30·3]

***32·111.** $\vdash . \hat{y}(xRy) = \overleftarrow{R}\text{‘}x$ [*32·101 . *30·3]

***32·12.** $\vdash . E ! \overrightarrow{R}\text{‘}y$ [*32·11 . *14·21]

***32·121.** $\vdash . E ! \overleftarrow{R}\text{‘}x$ [*32·111 . *14·21]

"$E ! \overrightarrow{R}\text{‘}y$" must not be confounded with "$\exists ! \overrightarrow{R}\text{‘}y$." The former means that there is such a class as $\overrightarrow{R}\text{‘}y$, which, as we have just seen, is always true; the latter means that $\overrightarrow{R}\text{‘}y$ is not null, which is only true if y is a term to which some other term has the relation R. Note that, by *14·21, both $\exists ! \overrightarrow{R}\text{‘}y$ and $\sim\exists ! \overrightarrow{R}\text{‘}y$ imply $E ! \overrightarrow{R}\text{‘}y$. The contradictory of $\exists ! \overrightarrow{R}\text{‘}y$ is not $\sim\exists ! \overrightarrow{R}\text{‘}y$, but $\sim\{[\overrightarrow{R}\text{‘}y] . \exists ! \overrightarrow{R}\text{‘}y\}$. This last would not imply $E ! \overrightarrow{R}\text{‘}y$, but for the fact that $E ! \overrightarrow{R}\text{‘}y$ is always true.

***32·13.** $\vdash . \overrightarrow{R}\text{‘}y = \hat{x}(xRy)$ [*32·11 . *20·59]

***32·131.** $\vdash . \overleftarrow{R}\text{‘}x = \hat{y}(xRy)$ [*32·111 . *20·59]

***32·132.** $\vdash : \alpha\overrightarrow{R}y . \equiv . \alpha = \overrightarrow{R}\text{‘}y . \equiv . \alpha = \hat{x}(xRy)$ [*32·1·13 . *20·57]

***32·133.** $\vdash : \beta\overleftarrow{R}x . \equiv . \beta = \overleftarrow{R}\text{‘}x . \equiv . \beta = \hat{y}(xRy)$ [*32·101·131 . *20·57]

The use of *20·57 will in general be tacit. It happens constantly that we have propositions such as *32·13, in which a descriptive expression is shown to be identical with a class. In such cases, whenever the properties of the class are asserted of the descriptive expression, *20·57 is relevant.

***32·14.** $\vdash : \overrightarrow{R} = \overrightarrow{S} . \equiv . R = S$

Dem.

$\vdash . *21·43 . \supset \vdash :: \overrightarrow{R} = \overrightarrow{S} . \equiv :. \alpha\overrightarrow{R}y . \equiv_{\alpha, y} . \alpha\overrightarrow{S}y :.$

[*32·1] $\equiv :. \alpha = \hat{x}(xRy) . \equiv_{\alpha, y} . \alpha = \hat{x}(xSy) :.$

[*11·2] $\equiv :. (y) :. \alpha = \hat{x}(xRy) . \equiv_{\alpha} . \alpha = \hat{x}(xSy) :.$

[*20·25] $\equiv :. (y) : \hat{x}(xRy) = \hat{x}(xSy) :.$

[*20·15] $\equiv :. (y) :. (x) : xRy . \equiv . xSy :.$

[*11·2] $\equiv :. (x, y) : xRy . \equiv . xSy :.$

[*21·43] $\equiv :. R = S :: \supset \vdash .$ Prop

***32·15.** $\vdash : \overleftarrow{R} = \overleftarrow{S} . \equiv . R = S$ [Similar proof]

***32·16.** $\vdash : \overrightarrow{R} = \overrightarrow{S} . \equiv . \overleftarrow{R} = \overleftarrow{S} . \equiv . R = S$ [*32·14·15]

***32·18.** $\vdash : x \in \overrightarrow{R}\text{‘}y . \equiv . xRy$ [*32·13 . *20·33]

***32·181.** $\vdash : y \in \overleftarrow{R}\text{‘}x . \equiv . xRy$ [*32·131 . *20·33]

***32·182.** $\vdash : x \in \overrightarrow{R}\text{‘}y . \equiv . y \in \overleftarrow{R}\text{‘}x$ [*32·18·181]

The transformation from "xRy" to "$x \epsilon \overrightarrow{R}‘y$" is one commonly effected in language. *E.g.* suppose "xRy" is "x loves y," then "$x \epsilon \overrightarrow{R}‘y$" is "$x$ is a lover of y."

***32·19.** $\vdash : R \Subset S . \supset . \overrightarrow{R}‘y \subset \overrightarrow{S}‘y . \overleftarrow{R}‘x \subset \overleftarrow{S}‘x$

Dem.

$\vdash . *32·18 . \supset \vdash :. \text{Hp} . \supset : x \epsilon \overrightarrow{R}‘y . \supset_x . x \epsilon \overrightarrow{S}‘y :$

[*22·1] $\supset : \overrightarrow{R}‘y \subset \overrightarrow{S}‘y$ (1)

$\vdash . *32·181 . \supset \vdash :. \text{Hp} . \supset : y \epsilon \overleftarrow{R}‘x . \supset_y . y \epsilon \overleftarrow{S}‘x :$

[*22·1] $\supset : \overleftarrow{R}‘x \subset \overleftarrow{S}‘x$ (2)

$\vdash . (1) . (2) . \supset \vdash . \text{Prop}$

***32·2.** $\vdash : A \text{ sg } R . \equiv . A = \overrightarrow{R}$ [*21·3 . (*32·03)]

***32·201.** $\vdash : A \text{ gs } R . \equiv . A = \overleftarrow{R}$ [*21·3 . (*32·04)]

***32·21.** $\vdash . \overrightarrow{R} = \text{sg}‘R$ [*32·2 . *30·3]

***32·211.** $\vdash . \overleftarrow{R} = \text{gs}‘R$ [*32·201 . *30·3]

***32·22.** $\vdash . E ! \text{ sg}‘R$ [*32·21 . *14·21]

***32·221.** $\vdash . E ! \text{ gs}‘R$ [*32·211 . *14·21]

***32·23.** $\vdash . \text{sg}‘R = \overrightarrow{R}$ [*32·21 . *21·2·57]

***32·231.** $\vdash . \text{gs}‘R = \overleftarrow{R}$ [*32·211 . *21·2·57]

***32·24.** $\vdash . \text{sg}‘\breve{R} = \text{gs}‘R$

Dem.

$\vdash . *32·23 . (*32·01) . \quad \supset \vdash . \text{sg}‘\breve{R} = \hat{\alpha}\hat{y} \{\alpha = \hat{x}(x\breve{R}y)\} .$

[*21·33] $\supset \vdash : \alpha (\text{sg}‘\breve{R}) y . \equiv . \alpha = \hat{x}(x\breve{R}y) .$

[*31·11.*20·15] $\equiv . \alpha = \hat{x}(yRx) .$

[*32·101] $\equiv . \alpha \overleftarrow{R} x .$

[*32·211] $\equiv . \alpha (\text{gs}‘R) x$ (1)

$\vdash . (1) . *11·11 . *21·43 . \supset \vdash . \text{Prop}$

***32·241.** $\vdash . \text{gs}‘\breve{R} = \text{sg}‘R$ [Similar proof]

***32·25.** $\vdash : A \text{ sg } R . \equiv . A = \text{sg}‘R$ [*30·4 . *32·22]

***32·251.** $\vdash : A \text{ gs } R . \equiv . A = \text{gs}‘R$ [*30·4 . *32·221]

***32·3.** $\vdash . \{\text{sg}‘(R \dot{\cap} S)\}‘y = \overrightarrow{R}‘y \cap \overrightarrow{S}‘y$

Note that we do *not* have

$$\text{sg}‘(R \dot{\cap} S) = \text{sg}‘R \dot{\cap} \text{sg}‘S.$$

Dem.

$\vdash . *32\cdot23\cdot13 . \supset \vdash . \{\text{sg}`(R \dot{\cap} S)\}`y = \hat{x}\{x(R \dot{\cap} S)y\}$

[$*23\cdot33$] $= \hat{x}(xRy . xSy)$

[$*22\cdot39$] $= \hat{x}(xRy) \cap \hat{x}(xSy)$

[$*32\cdot13$] $= \overrightarrow{R}`y \cap \overrightarrow{S}`y . \supset \vdash . \text{Prop}$

$*32\cdot31$. $\vdash . \{\text{gs}`(R \dot{\cap} S)\}`x = \overleftarrow{R}`x \cap \overleftarrow{S}`x$

$*32\cdot32$. $\vdash . \{\text{sg}`(R \dot{\cup} S)\}`y = \overrightarrow{R}`y \cup \overrightarrow{S}`y$

$*32\cdot33$. $\vdash . \{\text{gs}`(R \dot{\cup} S)\}`x = \overleftarrow{R}`x \cup \overleftarrow{S}`x$

$*32\cdot34$. $\vdash . \{\text{sg}`(\dot{-}R)\}`y = -\overrightarrow{R}`y$

$*32\cdot35$. $\vdash . \{\text{gs}`(\dot{-}R)\}`x = -\overleftarrow{R}`x$

The proofs of the above propositions are similar to that of $*32\cdot3$.

$*32\cdot4$. $\vdash :. \text{E}!\, R`z . \equiv : \exists !\, \overrightarrow{R}`z : x, y \in \overrightarrow{R}`z . \supset_{x,y} . x = y$ [$*30\cdot21 . *32\cdot18$]

$*32\cdot41$. $\vdash :. \text{E}!\, S`y . \supset : \overrightarrow{R}`y = \overrightarrow{S}`y . \equiv . R`y = S`y$

Dem.

$\vdash . *4\cdot86 . \quad \supset \vdash :: xSy . \equiv_x . x = b : \supset :.$
$xRy . \equiv_x . xSy : \equiv : xRy . \equiv_x . x = b$ (1)

$\vdash . (1) . *5\cdot32 . \supset \vdash :. xSy . \equiv_x . x = b : xRy . \equiv_x . xSy : \equiv :$
$xSy . \equiv_x . x = b : xRy . \equiv_x . x = b$ (2)

$\vdash . (2) . *10\cdot11\cdot281 . *32\cdot18\cdot181 . \supset$

$\vdash :. (\exists b) : xSy . \equiv_x . x = b : \overrightarrow{R}`y = \overrightarrow{S}`y : \equiv : (\exists b) : xSy . \equiv_x . x = b : xRy . \equiv_x . x = b :$

[$*30\cdot3 . *14\cdot13$] $\equiv : (\exists b) : xSy . \equiv_x . x = b : R`y = b :$

[$*14\cdot101$] $\equiv : R`y = S`y$ (3)

$\vdash . (3) . *30\cdot2 . \supset \vdash :. \text{E}!\, S`y . \overrightarrow{R}`y = \overrightarrow{S}`y . \equiv . R`y = S`y :. \supset \vdash . \text{Prop}$

$*32\cdot42$. $\vdash :. \overrightarrow{R}`y = \overrightarrow{S}`y . \supset : \text{E}!\, R`y . \equiv . \text{E}!\, S`y$ [$*30\cdot34 . *32\cdot18$]

$*33$. DOMAINS, CONVERSE DOMAINS, AND FIELDS OF RELATIONS

Summary of $*33$.

If R is any relation, the *domain* of R, which we denote by $D`R$, is the class of terms which have the relation $R$ to something or other; the *converse domain*, $\text{Ɑ}`R$, is the class of terms to which something or other has the relation R; and the *field*, $C`R$, is the sum of the domain and the converse domain. (Note that the field is only significant when R is a *homogeneous* relation.)

The above notations $D`R$, $\text{Ɑ}`R$, $C`R$ are derivative from the notations D, Ɑ, C for the relations, to a relation, of its domain, converse domain, and field respectively. We are to have

$$D`R = \hat{x}\{(\exists y) . xRy\}$$
$$\text{Ɑ}`R = \hat{y}\{(\exists x) . xRy\}$$
$$C`R = \hat{x}\{(\exists y) : xRy . \vee . yRx\};$$

hence we define D, Ɑ, C as follows:

$*33{\cdot}01$. $D = \hat{\alpha}\hat{R}\,[\alpha = \hat{x}\{(\exists y) . xRy\}]$ Df

$*33{\cdot}02$. $\text{Ɑ} = \hat{\beta}\hat{R}\,[\beta = \hat{y}\{(\exists x) . xRy\}]$ Df

$*33{\cdot}03$. $C = \hat{\gamma}\hat{R}\,[\gamma = \hat{x}\{(\exists y) : xRy . \vee . yRx\}]$ Df

The letter C is chosen as the initial of the word "campus." We require one other definition, namely of the relation of x to R when x is a member of the field of R. This relation, which we will call F, is defined as follows:

$*33{\cdot}04$. $F = \hat{x}\hat{R}\{(\exists y) : xRy . \vee . yRx\}$ Df

We shall find that $C = \overrightarrow{F}$. $\breve{D}$ will be the relation of a relation to its domain, $\overleftarrow{D}`\alpha$ will be the class of relations having α for their domain. Similar remarks apply to Ɑ and C. The *field* of a relation is specially important in connection with series.

The propositions of this number are constantly used throughout the remainder of the work. The ideas of the domain, converse domain, and field are very general, and have somewhat different uses for relations of different kinds. Consider first the sort of relation that gives rise to a descriptive function $R`y$. For this we require that $R`y$ should exist whenever there is anything having the relation R to y, *i.e.* that there should never be more than one term having the relation R to a given term y. In this case, the values of y for which $R`y$ exists will constitute the "converse domain" of $R$, *i.e.* $\text{Ɑ}`R$, and the values which $R`y$ assumes for various values of y will constitute the "domain" of R, *i.e.* $\text{D}\text{‘}R$. Thus the converse domain is the class of possible arguments for the descriptive function $R\text{‘}y$, and the domain is the class of all values of the function. Thus, for example, if R is the relation of the square of an integer y to y, then $R\text{‘}y =$ the square of y, provided y is an integer. In this case, $\text{Ɑ}\text{‘}R$ is the class of integers, and $\text{D}\text{‘}R$ is the class of perfect squares. Or again, suppose R is the relation of wife to husband; then $R\text{‘}y =$ the wife of y, $\text{Ɑ}\text{‘}R =$ married men, $\text{D}\text{‘}R =$ married women. In such cases, the *field* usually has little importance; and if the values of the function $R\text{‘}y$ are not of the same type as its arguments, *i.e.* if the relation R is not *homogeneous*, the field is meaningless. Thus, for example, if R is a homogeneous relation, $\overrightarrow{R}$ and $\overleftarrow{R}$ are not homogeneous, and therefore "$C\text{‘}\overrightarrow{R}$" and "$C\text{‘}\overleftarrow{R}$" are meaningless.

Let us next suppose that R is the sort of relation that generates a series, say the relation of less to greater among integers. Then $\text{D}\text{‘}R =$ all integers that are less than some other integer = all integers, $\text{Ɑ}\text{‘}R =$ all integers that are greater than some other integer = all integers except 0. In this case, $C\text{‘}R =$ all integers that are either greater or less than some other integer = all integers. Generally, if R generates a series, $\text{D}\text{‘}R =$ all members of the series except the last (if any), $\text{Ɑ}\text{‘}R =$ all members of the series except the first (if any), and $C\text{‘}R =$ all members of the series. In this case, "xFR" expresses the fact that x is a member of the series. Thus when R generates a series, $C\text{‘}R$ becomes important, and the relation F is likely to be useful.

We shall have occasion to deal with many relations having some of the properties of series, and with many propositions which, though only important in connection with serial relations, hold much more generally. In such cases, the field of a relation is likely to be important. Thus in the section on Induction (Part II, Section E), where we are preparing the way for the construction of serial relations by means of a certain kind of non-serial relation, and throughout relation-arithmetic (Part IV), the fields of relations will occur constantly. But in the earlier parts of the work, it is chiefly domains and converse domains that occur.

Among the more important properties of domains, converse domains and fields, which are proved in the present number, are the following.

We have always $\text{E} ! \text{D}\text{‘}R$, $\text{E} ! \text{Ɑ}\text{‘}R$, $\text{E} ! C\text{‘}R$ (∗33·12·121·122). (The last of these, however, is only significant when R is homogeneous.)

∗33·13. $\vdash : x \in \text{D}\text{‘}R . \equiv . (\exists y) . xRy$

∗33·131. $\vdash : y \in \text{Ɑ}\text{‘}R . \equiv . (\exists x) . xRy$

∗33·132. $\vdash :. x \in C\text{‘}R . \equiv : (\exists y) : xRy . \vee . yRx$

∗33·14. $\vdash : xRy . \supset . x \in \text{D}\text{‘}R . y \in \text{Ɑ}\text{‘}R$

∗33·16. $\vdash . C\text{‘}R = \text{D}\text{‘}R \cup \text{Ɑ}\text{‘}R$

***33·2·21·22.** The converse domain of a relation is the domain of its converse, the domain of a relation is the converse domain of its converse, and the field of a relation is the field of its converse.

***33·24.** $\vdash : \exists ! \text{D‘}R . \equiv . \exists ! \text{Ɑ‘}R . \equiv . \exists ! C\text{‘}R . \equiv . \dot{\exists} ! R$

***33·4.** $\vdash . \text{D‘}R = \hat{x}\{\exists ! \overleftarrow{R}\text{‘}x\}$

with corresponding propositions (*33·41·42) for $\text{Ɑ‘}R$ and $C\text{‘}R$.

***33·43.** $\vdash : \text{E} ! R\text{‘}y . \supset . y \in \text{Ɑ‘}R . R\text{‘}y \in \text{D‘}R$

***33·431.** $\vdash : (y) . \text{E} ! R\text{‘}y . \supset . (\beta) . \beta \subset \text{Ɑ‘}R$

***33·5.** $\vdash . C = \overrightarrow{F}$

***33·51.** $\vdash : x \in C\text{‘}R . \equiv . xFR$

The proofs of propositions concerning Ɑ and C are usually similar to those for D, and are therefore often omitted.

***33·01.** $\text{D} = \hat{\alpha}\hat{R}\,[\alpha = \hat{x}\,\{(\exists y) . xRy\}]$ Df

***33·02.** $\text{Ɑ} = \hat{\beta}\hat{R}\,[\beta = \hat{y}\,\{(\exists x) . xRy\}]$ Df

***33·03.** $C = \hat{\gamma}\hat{R}\,[\gamma = \hat{x}\,\{(\exists y) : xRy . \vee . yRx\}]$ Df

***33·04.** $F = \hat{x}\hat{R}\,\{(\exists y) : xRy . \vee . yRx\}$ Df

***33·1.** $\vdash : \alpha \text{D} R . \equiv . \alpha = \hat{x}\,\{(\exists y) . xRy\}$ [*21·3 . (*33·01)]

***33·101.** $\vdash : \beta \text{Ɑ} R . \equiv . \beta = \hat{y}\,\{(\exists x) . xRy\}$

***33·102.** $\vdash : \gamma C R . \equiv . \gamma = \hat{x}\,\{(\exists y) : xRy . \vee . yRx\}$

***33·103.** $\vdash :. xFR . \equiv : (\exists y) : xRy . \vee . yRx$

***33·11.** $\vdash . \text{D‘}R = \hat{x}\,\{(\exists y) . xRy\}$ [*33·1 . *30·3 . *20·59]

***33·111.** $\vdash . \text{Ɑ‘}R = \hat{y}\,\{(\exists x) . xRy\}$

***33·112.** $\vdash . C\text{‘}R = \hat{x}\,\{(\exists y) : xRy . \vee . yRx\}$

***33·12.** $\vdash . \text{E} ! \text{D‘}R$ [*33·11 . *14·21]

***33·121.** $\vdash . \text{E} ! \text{Ɑ‘}R$

***33·122.** $\vdash . \text{E} ! C\text{‘}R$

***33·123.** $\vdash : \alpha \text{D} R . \equiv . \alpha = \text{D‘}R$ [*30·4 . *33·12]

***33·124.** $\vdash : \beta \text{Ɑ} R . \equiv . \beta = \text{Ɑ‘}R$ [*30·4 . *33·121]

***33·125.** $\vdash : \gamma C R . \equiv . \gamma = C\text{‘}R$ [*30·4 . *32·123]

***33·13.** $\vdash : x \in \text{D‘}R . \equiv . (\exists y) . xRy$ [*33·11 . *20·3·57]

***33·131.** $\vdash : y \in \text{Ɑ‘}R . \equiv . (\exists x) . xRy$

***33·132.** $\vdash :. x \in C\text{‘}R . \equiv : (\exists y) : xRy . \vee . yRx$

***33·14.** $\vdash : xRy . \supset . x \in \text{D‘}R . y \in \text{Ɑ‘}R$

Dem.

$\vdash . *10{\cdot}24 . \supset \vdash :. \text{Hp} . \supset : (\exists y) . xRy : (\exists x) . xRy :$

[*33·13·131] $\supset : x \in \text{D‘}R . y \in \text{Ɑ‘}R :. \supset \vdash . \text{Prop}$

***33·15.** $\vdash . \overrightarrow{R}\text{‘}y \subset D\text{‘}R$

Dem.

$\vdash . *32·18 . \supset \vdash : x \in \overrightarrow{R}\text{‘}y . \supset_x . xRy .$
$[*10·24] \qquad \supset_x . (\exists y) . xRy .$
$[*33·13] \qquad \supset_x . x \in D\text{‘}R : \supset \vdash . \text{Prop}$

***33·151.** $\vdash . \overleftarrow{R}\text{‘}x \subset \text{Ꮧ}\text{‘}R$

***33·152.** $\vdash . \overrightarrow{R}\text{‘}x \cup \overleftarrow{R}\text{‘}x \subset C\text{‘}R$

***33·16.** $\vdash . C\text{‘}R = D\text{‘}R \cup \text{Ꮧ}\text{‘}R$

Dem.

$\vdash . *33·132 . *10·42 . \supset$
$\vdash :. x \in C\text{‘}R . \equiv : (\exists y) . xRy . \vee . (\exists y) . yRx :$
$[*33·13·131] \equiv : x \in D\text{‘}R . \vee . x \in \text{Ꮧ}\text{‘}R :$
$[*22·34] \qquad \equiv : x \in D\text{‘}R \cup \text{Ꮧ}\text{‘}R \qquad (1)$
$\vdash . (1) . *10·11 . *20·43 . \supset \vdash . \text{Prop}$

***33·161.** $\vdash . D\text{‘}R \subset C\text{‘}R . \text{Ꮧ}\text{‘}R \subset C\text{‘}R$ [*33·16 . *22·58]

***33·17.** $\vdash : xRy . \supset . x, y \in C\text{‘}R$ [*33·14·161]

***33·18.** $\vdash : D\text{‘}R = \text{Ꮧ}\text{‘}R . \supset . D\text{‘}R = C\text{‘}R$

Dem.

$\vdash . *22·56 . \supset \vdash : D\text{‘}R = \text{Ꮧ}\text{‘}R . \supset . D\text{‘}R = D\text{‘}R \cup \text{Ꮧ}\text{‘}R$
$[*33·16] \qquad = C\text{‘}R : \supset \vdash . \text{Prop}$

***33·181.** $\vdash : \text{Ꮧ}\text{‘}R \subset D\text{‘}R . \equiv . D\text{‘}R = C\text{‘}R$

Dem.

$\vdash . *22·62 . \supset \vdash : \text{Ꮧ}\text{‘}R \subset D\text{‘}R . \equiv . D\text{‘}R = D\text{‘}R \cup \text{Ꮧ}\text{‘}R$
$[*33·16] \qquad = C\text{‘}R : \supset \vdash . \text{Prop}$

***33·182.** $\vdash : D\text{‘}R \subset \text{Ꮧ}\text{‘}R . \equiv . \text{Ꮧ}\text{‘}R = C\text{‘}R$ [Similar proof]

If R is the sort of relation which generates a series, so that "xRy" may be read "x precedes y," then $\text{Ꮧ}\text{‘}R \subset D\text{‘}R$ is the condition that the series may have no last term, since it states that every term which follows some term precedes some other term, and is therefore not the last of the series.

***33·2.** $\vdash . \text{Ꮧ}\text{‘}R = D\text{‘}\breve{R}$

Dem.

$\vdash . *31·11 . *10·11 . \supset \vdash : xRy . \equiv_x . y\breve{R}x :$
$[*10·281] \qquad \supset \vdash : (\exists x) . xRy . \equiv . (\exists x) . y\breve{R}x :$
$[*33·13·131] \qquad \supset \vdash : y \in \text{Ꮧ}\text{‘}R . \equiv . y \in D\text{‘}\breve{R} \qquad (1)$
$\vdash . (1) . *10·11 . *20·43 . \supset \vdash . \text{Prop}$

***33·21.** $\vdash . D\text{‘}R = \text{Ꮧ}\text{‘}\breve{R}$ [Similar proof]

***33·22.** $\vdash . C\text{‘}R = C\text{‘}\breve{R}$

Dem.

$\vdash . *33·16·2·21 . \supset \vdash . C\text{‘}R = \text{ᗡ}\text{‘}\breve{R} \cup \text{D}\text{‘}\breve{R}$

$[*33·16] \qquad = C\text{‘}\breve{R} . \supset \vdash . \text{Prop}$

***33·24.** $\vdash : \exists ! \text{D}\text{‘}R . \equiv . \exists ! \text{ᗡ}\text{‘}R . \equiv . \exists ! C\text{‘}R . \equiv . \dot{\exists} ! R$

Dem.

$\vdash . *33·13 . \supset \vdash :. \exists ! \text{D}\text{‘}R . \equiv : (\exists x) : (\exists y) . xRy :$

$[*25·5.(*11·03)] \qquad \equiv : \dot{\exists} ! R \qquad (1)$

$\vdash . *33·131 . \supset \vdash :. \exists ! \text{ᗡ}\text{‘}R . \equiv : (\exists y) : (\exists x) . xRy :$

$[*11·2] \qquad \equiv : (\exists x, y) . xRy :$

$[*25·5] \qquad \equiv : \dot{\exists} ! R \qquad (2)$

$\vdash . *33·132 . \supset \vdash :: \exists ! C\text{‘}R . \equiv :. (\exists x) :. (\exists y) : xRy . \vee . yRx :.$

$[*11·7] \qquad \equiv :. (\exists x, y) . xRy :.$

$[*25·5] \qquad \equiv :. \dot{\exists} ! R \qquad (3)$

$\vdash . (1) . (2) . (3) . \supset \vdash . \text{Prop}$

***33·241.** $\vdash : \text{D}\text{‘}R = \Lambda . \equiv . \text{ᗡ}\text{‘}R = \Lambda . \equiv . C\text{‘}R = \Lambda . \equiv . R = \dot{\Lambda}$

$[*33·24 . \text{Transp} . *24·51 . *25·51]$

***33·25.** $\vdash . \text{D}\text{‘}(R \dot{\cap} S) \subset \text{D}\text{‘}R \cap \text{D}\text{‘}S$

Dem.

$\vdash . *33·13 . \supset \vdash :. x \in \text{D}\text{‘}(R \dot{\cap} S) . \equiv : (\exists y) . x (R \dot{\cap} S) y :$

$[*21·33.*10·281] \qquad \equiv : (\exists y) . xRy . xSy :$

$[*10·5] \qquad \supset : (\exists y) . xRy : (\exists y) . xSy :$

$[*33·13] \qquad \supset : x \in \text{D}\text{‘}R . x \in \text{D}\text{‘}S :$

$[*21·33] \qquad \supset : x \in \text{D}\text{‘}R \cap \text{D}\text{‘}S \qquad (1)$

$\vdash . (1) . *10·11 . \supset \vdash . \text{Prop}$

***33·251.** $\vdash . \text{ᗡ}\text{‘}(R \dot{\cap} S) \subset \text{ᗡ}\text{‘}R \cap \text{ᗡ}\text{‘}S$ [Similar proof]

***33·252.** $\vdash . C\text{‘}(R \dot{\cap} S) \subset C\text{‘}R \cap C\text{‘}S$ [Similar proof]

***33·26.** $\vdash . \text{D}\text{‘}(R \dot{\cup} S) = \text{D}\text{‘}R \cup \text{D}\text{‘}S$

Dem.

$\vdash . *33·13 . \supset \vdash :. x \in \text{D}\text{‘}(R \dot{\cup} S) . \equiv : (\exists y) . x (R \dot{\cup} S) y :$

$[*23·34.*10·281] \qquad \equiv : (\exists y) : xRy . \vee . xSy :$

$[*10·42] \qquad \equiv : (\exists y) . xRy : \vee : (\exists y) . xSy :$

$[*33·13] \qquad \equiv : x \in \text{D}\text{‘}R . \vee . x \in \text{D}\text{‘}S :$

$[*22·34] \qquad \equiv : x \in \text{D}\text{‘}R \cup \text{D}\text{‘}S \qquad (1)$

$\vdash . (1) . *10·11 . *20·43 . \supset \vdash . \text{Prop}$

***33·261.** $\vdash . \text{ᗡ}\text{‘}(R \dot{\cup} S) = \text{ᗡ}\text{‘}R \cup \text{ᗡ}\text{‘}S$ [Similar proof]

***33·262.** $\vdash . C\text{‘}(R \dot{\cup} S) = C\text{‘}R \cup C\text{‘}S$ [*33·26·261·16]

***33·263.** $\vdash : R \mathrel{\dot{\subset}} S . \supset . \mathrm{D}‘R \subset \mathrm{D}‘S$

Dem.

$\vdash . *23·1 . \supset \vdash :. \mathrm{Hp} . \supset : xRy . \supset_{x,y} . xSy :$

$[*10·28·27] \quad \supset : (x) : (\exists y) . xRy . \supset . (\exists y) . xSy :$

$[*33·13] \quad \supset : (x) : x \in \mathrm{D}‘R . \supset . x \in \mathrm{D}‘S :$

$[*23·1] \quad \supset : \mathrm{D}‘R \subset \mathrm{D}‘S :. \supset \vdash . \mathrm{Prop}$

***33·264.** $\vdash : R \mathrel{\dot{\subset}} S . \supset . \text{Ɑ}‘R \subset \text{Ɑ}‘S$ [Similar proof]

***33·265.** $\vdash : R \mathrel{\dot{\subset}} S . \supset . C‘R \subset C‘S$ [*33·263·264·16 . *22·72]

***33·27.** $\vdash . C‘R = \mathrm{D}‘(R \mathrel{\dot{\cup}} \breve{R})$

Dem.

$\vdash . *33·16·2 . \supset \vdash . C‘R = \mathrm{D}‘R \cup \mathrm{D}‘\breve{R}$

$[*33·26] \quad = \mathrm{D}‘(R \mathrel{\dot{\cup}} \breve{R}) . \supset \vdash . \mathrm{Prop}$

***33·271.** $\vdash . C‘R = \text{Ɑ}‘(R \mathrel{\dot{\cup}} \breve{R})$ [Similar proof]

***33·272.** $\vdash . \mathrm{D}‘(R \mathrel{\dot{\cup}} \breve{R}) = \text{Ɑ}‘(R \mathrel{\dot{\cup}} \breve{R}) = C‘(R \mathrel{\dot{\cup}} \breve{R}) = C‘R$ [*33·27·271·16]

***33·28.** $\vdash . \mathrm{D}‘\dot{\mathrm{V}} = \text{Ɑ}‘\dot{\mathrm{V}} = C‘\dot{\mathrm{V}} = \mathrm{V}$

Dem.

$\vdash . *10·25 . *25·104 . \supset \vdash :. (x) : (\exists y) . x \dot{\mathrm{V}} y :. (x) : (\exists y) . y \dot{\mathrm{V}} x :.$

$[*33·13·131] \quad \supset \vdash :. (x) . x \in \mathrm{D}‘\dot{\mathrm{V}} : (x) . x \in \text{Ɑ}‘\dot{\mathrm{V}} :.$

$[*24·14] \quad \supset \vdash : \mathrm{D}‘\dot{\mathrm{V}} = \mathrm{V} . \text{Ɑ}‘\dot{\mathrm{V}} = \mathrm{V}$ (1)

$[*33·16] \quad \supset \vdash . C‘\dot{\mathrm{V}} = \mathrm{V} \cup \mathrm{V}$

$[*22·56] \quad = \mathrm{V}$ (2)

$\vdash . (1) . (2) . \supset \vdash . \mathrm{Prop}$

***33·29.** $\vdash . \mathrm{D}‘\dot{\Lambda} = \text{Ɑ}‘\dot{\Lambda} = C‘\dot{\Lambda} = \Lambda$ [*33·241 . *21·2]

***33·3.** $\vdash :. \alpha \subset \mathrm{D}‘R . \equiv : x \in \alpha . \supset_x . \exists ! \overleftarrow{R}‘x$

Dem.

$\vdash . *32·181 . \supset \vdash :. x \in \alpha . \supset_x . \exists ! \overleftarrow{R}‘x : \equiv : x \in \alpha . \supset_x . (\exists y) . xRy :$

$[*33·13] \quad \equiv : x \in \alpha . \supset_x . x \in \mathrm{D}‘R :. \supset \vdash . \mathrm{Prop}$

***33·31.** $\vdash :. \beta \subset \text{Ɑ}‘R . \equiv : y \in \beta . \supset_y . \exists ! \overrightarrow{R}‘y$ [Proof as in *33·3]

The three following propositions are used in the theory of selections (*80, *83 and *85). The second of them is also used in the theory of greater and less (*117) and in the theory of transitive relations (*201).

***33·32.** $\vdash : \mathrm{D}‘R \cap \mathrm{D}‘S = \Lambda . \supset . R \mathrel{\dot{\cap}} S = \dot{\Lambda}$

The converse of this proposition is not true.

Dem.

$\vdash . *23·33 . \quad \supset \vdash : x (R \mathrel{\dot{\cap}} S) y . \supset . xRy . xSy .$

$[*33·14 . *22·33] \quad \supset . x \in \mathrm{D}‘R \cap \mathrm{D}‘S .$

[*10·24] $\supset . \exists ! \text{D}`R \cap \text{D}`S$ (1)

$\vdash . (1) . \text{Transp} . \supset \vdash : \text{D}`R \cap \text{D}`S = \Lambda . \supset . \sim \{x (R \dot{\cap} S) y\}$ (2)

$\vdash . (2) . *11·11·3 . \supset \vdash : \text{D}`R \cap \text{D}`S = \Lambda . \supset . (x, y) . \sim \{x (R \dot{\cap} S) y\} .$

[*25·15] $\supset . R \dot{\cap} S = \dot{\Lambda} : \supset \vdash . \text{Prop}$

***33·33.** $\vdash : \text{Q}`R \cap \text{Q}`S = \Lambda . \supset . R \dot{\cap} S = \dot{\Lambda}$ [Proof as in *33·32]

***33·34.** $\vdash : C`R \cap C`S = \Lambda . \supset . R \dot{\cap} S = \dot{\Lambda}$

Dem.

$\vdash . *33·161 . *22·49 . \supset \vdash . \text{D}`R \cap \text{D}`S \subset C`R \cap C`S .$

[*24·13] $\supset \vdash : C`R \cap C`S = \Lambda . \supset . \text{D}`R \cap \text{D}`S = \Lambda .$

[*33·32] $\supset . R \dot{\cap} S = \dot{\Lambda} : \supset \vdash . \text{Prop}$

***33·35.** $\vdash :. \text{D}`R \subset \alpha . \equiv : xRy . \supset_{x,y} . x \in \alpha$

Dem.

$\vdash . *33·13 . \supset \vdash :. \text{D}`R \subset \alpha . \equiv : (\exists y) . xRy . \supset_x . x \in \alpha :$

[*10·23] $\equiv : xRy . \supset_{x,y} . x \in \alpha :. \supset \vdash . \text{Prop}$

***33·351.** $\vdash :. \text{Q}`R \subset \alpha . \equiv : xRy . \supset_{x,y} . y \in \alpha$ [Proof as in *33·35]

***33·352.** $\vdash :. C`R \subset \alpha . \equiv : xRy . \supset_{x,y} . x, y \in \alpha$

Dem.

$\vdash . *33·16 . *22·59 . \supset$

$\vdash :. C`R \subset \alpha . \equiv : \text{D}`R \subset \alpha . \text{Q}`R \subset \alpha :$

[*33·35·351] $\equiv : xRy . \supset_{x,y} . x \in \alpha : xRy . \supset_{x,y} . y \in \alpha :$

[*11·391] $\equiv : xRy . \supset_{x,y} . x, y \in \alpha :. \supset \vdash . \text{Prop}$

The two following propositions (*33·4·41) are very frequently used.

***33·4.** $\vdash . \text{D}`R = \hat{x} \{\exists ! \overleftarrow{R}`x\}$

Dem.

$\vdash . *33·13 . \supset \vdash : x \in \text{D}`R . \equiv . (\exists y) . xRy .$

[*32·181] $\equiv . (\exists y) . y \in \overleftarrow{R}`x .$

[*24·5] $\equiv . \exists ! \overleftarrow{R}`x$ (1)

$\vdash . (1) . *10·11 . *20·33 . \supset \vdash . \text{Prop}$

***33·41.** $\vdash . \text{Q}`R = \hat{y} \{\exists ! \overrightarrow{R}`y\}$ [Similar proof]

***33·42.** $\vdash . C`R = \hat{x} \{\exists ! (\overrightarrow{R}`x \cup \overleftarrow{R}`x)\}$

Dem.

$\vdash . *33·4·41·16 . \supset \vdash . C`R = \hat{x} \{\exists ! \overrightarrow{R}`x\} \cup \hat{x} \{\exists ! \overleftarrow{R}`x\}$

[*22·391] $= \hat{x} \{\exists ! \overrightarrow{R}`x . \vee . \exists ! \overleftarrow{R}`x\}$

[*24·56.*20·15] $= \hat{x} \{\exists ! (\overrightarrow{R}`x \cup \overleftarrow{R}`x)\} . \supset \vdash . \text{Prop}$

***33·43.** $\vdash : \text{E}\,!\,R\text{‘}y \,.\supset .\, y \in \text{Ɑ}\text{‘}R \,.\, R\text{‘}y \in \text{D}\text{‘}R$

Dem.

$\vdash . *30\cdot32 . \supset \vdash : \text{E}\,!\,R\text{‘}y \,.\supset .\, (R\text{‘}y)\, Ry \,.$

$[*33\cdot14] \qquad \supset .\, y \in \text{Ɑ}\text{‘}R \,.\, R\text{‘}y \in \text{D}\text{‘}R : \supset \vdash .\, \text{Prop}$

***33·431.** $\vdash : (y) .\, \text{E}\,!\,R\text{‘}y \,.\supset .\, (\beta) .\, \beta \subset \text{Ɑ}\text{‘}R$

Dem.

$\vdash . *33\cdot43 . \qquad \supset \vdash :. \text{Hp} . \supset : y \in \text{Ɑ}\text{‘}R \,.$

$[\text{Simp}] \qquad \supset : y \in \beta \,.\supset .\, y \in \text{Ɑ}\text{‘}R \qquad (1)$

$\vdash . (1) . *10\cdot11\cdot21 . \supset \vdash : \text{Hp} . \supset .\, \beta \subset \text{Ɑ}\text{‘}R \qquad (2)$

$\vdash . (2) . *10\cdot11\cdot21 . \supset \vdash .\, \text{Prop}$

***33·432.** $\vdash : (y) .\, \text{E}\,!\,R\text{‘}y \,.\supset .\, \text{Ɑ}\text{‘}R = \text{V}$

Dem.

$\vdash . *33\cdot43 . *10\cdot11\cdot27 . \supset \vdash : \text{Hp} . \supset . (y) .\, y \in \text{Ɑ}\text{‘}R \,.$

$[*24\cdot14] \qquad \supset .\, \text{Ɑ}\text{‘}R = \text{V} : \supset \vdash .\, \text{Prop}$

***33·44.** $\vdash : \text{E}\,!\,\breve{R}\text{‘}x \,.\supset .\, x \in \text{D}\text{‘}R \,.\, \breve{R}\text{‘}x \in \text{Ɑ}\text{‘}R$

Dem.

$\vdash . *33\cdot43 \frac{\breve{R}}{R} . \supset \vdash : \text{Hp} . \supset .\, x \in \text{Ɑ}\text{‘}\breve{R} \,.\, \breve{R}\text{‘}x \in \text{D}\text{‘}\breve{R} \,.$

$[*33\cdot2\cdot21] \qquad \supset .\, x \in \text{D}\text{‘}R \,.\, \breve{R}\text{‘}x \in \text{Ɑ}\text{‘}R : \supset \vdash .\, \text{Prop}$

***33·45.** $\vdash :. y \in \text{Ɑ}\text{‘}R \cup \text{Ɑ}\text{‘}S \,.\supset_y .\, R\text{‘}y = S\text{‘}y : \supset .\, R = S$

Note that by our conventions as to denoting expressions, the scope of both $R\text{‘}y$ and $S\text{‘}y$ in the above is "$R\text{‘}y = S\text{‘}y$," and $R\text{‘}y$ is to be first eliminated.

Dem.

$\vdash . *30\cdot11 . \supset \vdash :: R\text{‘}y = S\text{‘}y \,.\equiv :. (\exists b) : xRy \,.\equiv_x .\, x = b : b = S\text{‘}y :.$

$[*30\cdot11] \qquad \equiv :. (\exists b) :. xRy \,.\equiv_x .\, x = b :. (\exists c) : xSy \,.\equiv_x .\, x = c : b = c :.$

$[*13\cdot195] \qquad \equiv :. (\exists b) : xRy \,.\equiv_x .\, x = b : xSy \,.\equiv_x .\, x = b :.$

$[*10\cdot322] \qquad \supset :. xRy \,.\equiv_x .\, xSy \qquad (1)$

$\vdash . (1) . \supset \vdash :: \text{Hp} . \quad \supset :. y \in \text{Ɑ}\text{‘}R \cup \text{Ɑ}\text{‘}S \,.\supset : xRy \,.\equiv .\, xSy :.$

$[*5\cdot32] \qquad \supset :. y \in \text{Ɑ}\text{‘}R \cup \text{Ɑ}\text{‘}S \,.\, xRy \,.\equiv .\, y \in \text{Ɑ}\text{‘}R \cup \text{Ɑ}\text{‘}S \,.\, xSy :.$

$[*33\cdot14 . *4\cdot71] \qquad \supset :. xRy \,.\equiv .\, xSy \qquad (2)$

$\vdash . (2) . *11\cdot11\cdot3 . \supset \vdash :. \text{Hp} . \supset : (x, y) : xRy \,.\equiv .\, xSy :$

$[*21\cdot43] \qquad \supset : R = S :. \supset \vdash .\, \text{Prop}$

***33·46.** $\vdash :. x \in \text{D}\text{‘}R \cup \text{D}\text{‘}S \,.\supset_x .\, \breve{R}\text{‘}x = \breve{S}\text{‘}x : \supset .\, R = S$ [Proof as in *33·45]

***33·47.** $\vdash :. y \in \text{Ɑ}\text{‘}R \cup \text{Ɑ}\text{‘}S \,.\supset_y .\, \overrightarrow{R}\text{‘}y = \overrightarrow{S}\text{‘}y : \supset .\, R = S$

Dem.

$\vdash . *33\cdot41 . \text{Transp} . \supset \vdash : y \sim\in \text{Ɑ}\text{‘}R \cup \text{Ɑ}\text{‘}S \,.\supset .\, \overrightarrow{R}\text{‘}y = \Lambda \,.\, \overrightarrow{S}\text{‘}y = \Lambda \qquad (1)$

$\vdash . (1) . *13\cdot172 . *4\cdot83 . \supset \vdash : \text{Hp} . \supset . (y) . \overrightarrow{R}\text{‘}y = \overrightarrow{S}\text{‘}y .$

[$*30\cdot41$] $\supset . \overrightarrow{R} = \overrightarrow{S} .$

[$*32\cdot14$] $\supset . R = S : \supset \vdash . \text{Prop}$

$*33\cdot48$. $\vdash :. x \in \text{D‘}R \cup \text{D‘}S . \supset_x . \overleftarrow{R}\text{‘}x = \overleftarrow{S}\text{‘}x : \supset . R = S$ [Proof as in $*33\cdot47$]

$*33\cdot5$. $\vdash . C = \overrightarrow{F}$

Dem.

$\vdash . *32\cdot1 . \supset \vdash :. \alpha \overrightarrow{F} R . \equiv . \alpha = \hat{x}(xFR)$

[$*33\cdot103$] $= \hat{x}\{(\exists y) : xRy . \vee . yRx\} .$

[$*33\cdot102$] $\equiv . \alpha CR$ (1)

$\vdash . (1) . *11\cdot11 . *21\cdot43 . \supset \vdash . \text{Prop}$

$*33\cdot51$. $\vdash : x \in C\text{‘}R . \equiv . xFR$ [$*33\cdot132\cdot103$]

F is useful in ordinal arithmetic, where we are concerned with a series generated by a relation P, and "xFP" expresses the fact that x is a member of this series. The above two propositions ($*33\cdot5\cdot51$) will be much used in Part IV, where we deal with the foundations of ordinal arithmetic, but will not often be referred to elsewhere.

$*33\cdot6$. $\vdash : R \in \overleftarrow{\text{D}}\text{‘}\alpha . \equiv . \alpha = \text{D‘}R$

Dem.

$\vdash . *32\cdot181 . \supset \vdash : R \in \overleftarrow{\text{D}}\text{‘}\alpha . \equiv . \alpha \text{D} R .$

[$*33\cdot123$] $\equiv . \alpha = \text{D‘}R : \supset \vdash . \text{Prop}$

$*33\cdot61$. $\vdash : R \in \overleftarrow{\text{Ɑ}}\text{‘}\alpha . \equiv . \alpha = \text{Ɑ‘}R$

$*33\cdot62$. $\vdash : R \in \overleftarrow{C}\text{‘}\alpha . \equiv . \alpha = C\text{‘}R$

*34. THE RELATIVE PRODUCT OF TWO RELATIONS

Summary of *34.

The relative product of two relations R and S is the relation which holds between x and z when there is an intermediate term y such that x has the relation R to y and y has the relation S to z. Thus *e.g.* the relative product of *brother* and *father* is *paternal uncle*; the relative product of *father* and *father* is *paternal grandfather*; and so on. The relative product of R and S is denoted by "$R | S$"; the definition is:

***34·01.** $R | S = \hat{x}\hat{z} \{(\exists y) . xRy . ySz\}$ Df

This definition is only significant when $\text{C\kern-0.4em C}$'R and D'S belong to the same type.

The relative product of R and R is called the square of R; we put

***34·02.** $R^2 = R | R$ Df

***34·03.** $R^3 = R^2 | R$ Df

The most useful propositions in the present number are the following:

***34·2.** $\vdash . \text{Cnv}`(R | S) = \breve{S} | \breve{R}$

I.e. the converse of a relative product is obtained by turning each factor into its converse and reversing the order of the factors.

***34·21.** $\vdash . (P | Q) | R = P | (Q | R)$

I.e. the relative product obeys the associative law.

***34·25.** $\vdash . P | (Q \mathbin{\dot{\cup}} R) = (P | Q) \mathbin{\dot{\cup}} (P | R)$

***34·26.** $\vdash . (P \mathbin{\dot{\cup}} Q) | R = (P | R) \mathbin{\dot{\cup}} (Q | R)$

I.e. the relative product obeys the distributive law with respect to the logical addition of relations. (For logical multiplication instead of logical addition, we only get inclusion instead of identity; cf. *34·23·24.)

***34·34.** $\vdash : R \mathbin{\dot{\subset}} P . S \mathbin{\dot{\subset}} Q . \supset . R | S \mathbin{\dot{\subset}} P | Q$

***34·36.** $\vdash . \text{D}`(P | Q) \subset \text{D}`P . \text{C\kern-0.4em C}`(P | Q) \subset \text{C\kern-0.4em C}`Q$

***34·41.** $\vdash : \text{E} ! P`Q`z . \supset . P`Q`z = (P | Q)`z$

***34·01.** $R | S = \hat{x}\hat{z} \{(\exists y) . xRy . ySz\}$ Df

***34·02.** $R^2 = R | R$ Df

***34·03.** $R^3 = R^2 | R$ Df

***34·1.** $\vdash : x (R | S) z . \equiv . (\exists y) . xRy . ySz$ [*21·3 . (*34·01)]

***34·11.** $\vdash : x(R \mid S)z \,.\equiv.\, \exists ! (\overleftarrow{R}`x \cap \overrightarrow{S}`z)$

Dem.

$\vdash . *34\cdot1 . *32\cdot18\cdot181 . \supset$

$\vdash : x(R \mid S)z \,.\equiv.\, (\exists y) . y \in \overleftarrow{R}`x . y \in \overrightarrow{S}`z .$

$[*22\cdot33] \qquad \equiv . (\exists y) . y \in \overleftarrow{R}`x \cap \overrightarrow{S}`z .$

$[*24\cdot5] \qquad \equiv . \exists ! (\overleftarrow{R}`x \cap \overrightarrow{S}`z) : \supset \vdash . \mathrm{Prop}$

***34·12.** $\vdash . R \mid S = \hat{x}\hat{z}\{\exists ! (\overleftarrow{R}`x \cap \overrightarrow{S}`x)\}$ $[*21\cdot33 . *34\cdot11]$

***34·2.** $\vdash . \mathrm{Cnv}`(R \mid S) = \breve{S} \mid \breve{R}$

Dem.

$\vdash . *31\cdot131 . \supset \vdash : x\{\mathrm{Cnv}`(R \mid S)\}z \,.\equiv.\, z(R \mid S)x .$

$[*34\cdot1] \qquad \equiv . (\exists y) . zRy . ySx .$

$[*31\cdot11] \qquad \equiv . (\exists y) . y\breve{R}z . x\breve{S}y .$

$[*34\cdot1] \qquad \equiv . x(\breve{S} \mid \breve{R})z \qquad (1)$

$\vdash . (1) . *11\cdot11 . *21\cdot43 . \supset \vdash . \mathrm{Prop}$

***34·202.** $\vdash . R \mid S = (\mathrm{Cnv}`\breve{R}) \mid S$

Dem.

$\vdash . *31\cdot131 . \supset \vdash : x(\mathrm{Cnv}`\breve{R})y . ySz \,.\equiv.\, y\breve{R}x . ySz .$

$[*31\cdot11] \qquad \equiv . xRy . ySz \qquad (1)$

$\vdash . (1) . *10\cdot11\cdot281 . *34\cdot1 . \supset \vdash : x\{(\mathrm{Cnv}`\breve{R}) \mid S\}z \,.\equiv.\, x(R \mid S)z \qquad (2)$

$\vdash . (2) . *11\cdot11 . *21\cdot43 . \supset \vdash . \mathrm{Prop}$

***34·203.** $\vdash . R \mid S = R \mid (\mathrm{Cnv}`\breve{S})$ [Similar proof]

***34·21.** $\vdash . (P \mid Q) \mid R = P \mid (Q \mid R)$

Dem.

$\vdash . *34\cdot1 . *10\cdot281 . \supset \vdash :: (\exists z) . x(P \mid Q)z . zRw \,.\equiv :.\, (\exists z) : (\exists y) . xPy . yQz : zRw :.$

$[*11\cdot6] \qquad \equiv :. (\exists y) :. xPy : (\exists z) . yQz . zRw :.$

$[*34\cdot1 . *10\cdot281] \qquad \equiv :. (\exists y) . xPy . y(Q \mid R)w \qquad (1)$

$\vdash . (1) . *11\cdot11 . *34\cdot1 . *21\cdot43 . \supset \vdash . \mathrm{Prop}$

***34·22.** $P \mid Q \mid R = (P \mid Q) \mid R \quad \mathrm{Df}$

This definition serves merely for the avoidance of brackets.

***34·23.** $\vdash . P \mid (Q \dot{\cap} R) \subseteq (P \mid Q) \dot{\cap} (P \mid R)$

Dem.

$\vdash . *34\cdot1 . \supset$

$\vdash :. x\{P \mid (Q \dot{\cap} R)\}y \,.\equiv :\, (\exists z) . xPz . z(Q \dot{\cap} R)y :$

$[*23\cdot33] \qquad \equiv : (\exists z) . xPz . zQy . zRy :$

$[*10\cdot5] \qquad \supset : (\exists z) . xPz . zQy : (\exists z) . xPz . zRy :$

[*34·1] $\supset : x(P \mid Q)y \,.\, x(P \mid R)y :$

[*23·33] $\supset : x\{(P \mid Q) \dot{\cap} (P \mid R)\} y$ (1)

$\vdash . (1) . *11·11 . \supset \vdash .$ Prop

The converse of the above is not true.

***34·24.** $\vdash . (P \dot{\cap} Q) \mid R \,\dot{\subset}\, (P \mid R) \dot{\cap} (Q \mid R)$ [Similar proof]

***34·25.** $\vdash . P \mid (Q \dot{\cup} R) = (P \mid Q) \dot{\cup} (P \mid R)$

Dem.

$\vdash . *23·34 . *10·281 . \supset$

$\vdash :. (\exists z) . xPz . z(Q \dot{\cup} R)y . \equiv : (\exists z) : xPz : zQy . \vee . zRy :$

[*4·4.*10·281] $\equiv : (\exists z) : xPz . zQy . \vee . xPz . zRy :$

[*10·42] $\equiv : (\exists z) . xPz . zQy : \vee : (\exists z) . xPz . zRy :$

[*34·1] $\equiv : x(P \mid Q)y . \vee . x(P \mid R)y :$

[*23·34] $\equiv : x(P \mid Q \dot{\cup} P \mid R)y$ (1)

$\vdash . (1) . *11·11 . *34·1 . \supset \vdash .$ Prop

***34·26.** $\vdash . (P \dot{\cup} Q) \mid R = (P \mid R) \dot{\cup} (Q \mid R)$ [Similar proof]

The above two forms of the distributive law, and the associative law (*34·21), are the only ones of the usual formal laws that hold for the relative product. The commutative law, in particular, does not hold in general.

***34·27.** $\vdash : R = R' . \supset . R \mid P = R' \mid P$

Dem.

$\vdash . *21·43 . \supset \vdash :. \text{Hp} . \supset : (x, y) : xRy . \equiv . xR'y :$

[*11·401] $\supset : (x, y) : xRy . yPz . \equiv_z . xR'y . yPz :$

[*10·281] $\supset : (x) : (\exists y) . xRy . yPz . \equiv_z . (\exists y) . xR'y . yPz :$

[*21·15] $\supset : R \mid P = R' \mid P :. \supset \vdash .$ Prop

***34·28.** $\vdash : R = R' . \supset . P \mid R = P \mid R'$ [Similar proof]

***34·29.** $\vdash : R = R' . \supset . P \mid R \mid Q = P \mid R' \mid Q$

Dem.

$\vdash . *34·27 . \supset \vdash : \text{Hp} . \supset . R \mid Q = R' \mid Q .$

[*34·28] $\supset . P \mid R \mid Q = P \mid R' \mid Q : \supset \vdash .$ Prop

In proving the equality of two relations, say R and S, we usually establish first an asserted proposition of the form

$$xRy . \equiv . xSy$$

or

$$\text{Hp} . \supset : xRy . \equiv . xSy.$$

We then proceed by *11·11 (together with *11·3 in the second case) to

$$(x, y) : xRy . \equiv . xSy \text{ or } \text{Hp} . \supset : (x, y) : xRy . \equiv . xSy,$$

whence the result follows by *21·43. We shall in future omit these steps, and write "$\supset \vdash .$ Prop" after we have established

$$xRy . \equiv . xSy \text{ or } \text{Hp} . \supset : xRy . \equiv . xSy.$$

A similar ellipsis will be made in proving the equality of classes.

***34·3.** $\vdash : \dot{\exists} ! (P \mid Q) . \equiv . \exists ! (\text{ᗡ}‘P \cap \text{D}‘Q)$

Dem.

$\vdash . *25·5 . \supset$

$\vdash :: \dot{\exists} ! (P \mid Q) . \equiv :. (\exists x, y) . x (P \mid Q) y :.$

[*34·1] $\equiv :. (\exists x, y) : (\exists z) . xPz . zQy :.$

[*11·27] $\equiv :. (\exists x, y, z) . xPz . zQy :.$

[*11·24] $\equiv :. (\exists z, x, y) . xPz . zQy :.$

[*11·27] $\equiv :. (\exists z) :. (\exists x, y) . xPz . zQy :.$

[*11·54] $\equiv :. (\exists z) :. (\exists x) . xPz : (\exists y) . zQy :.$

[*33·13·131] $\equiv :. (\exists z) :. z \epsilon \text{ᗡ}‘P . z \epsilon \text{D}‘Q :.$

[*22·33] $\equiv :. (\exists z) :. z \epsilon \text{ᗡ}‘P \cap \text{D}‘Q :.$

[*24·5] $\equiv :. \exists ! (\text{ᗡ}‘P \cap \text{D}‘Q) :: \supset \vdash .$ Prop

***34·301.** $\vdash : \text{ᗡ}‘P \cap \text{D}‘Q = \Lambda . \equiv . P \mid Q = \dot{\Lambda}$ [*34·3 . Transp]

***34·302.** $\vdash : C‘P \cap C‘Q = \Lambda . \supset . P \mid Q = \dot{\Lambda} . Q \mid P = \dot{\Lambda}$

Dem.

$\vdash . *33·16 . \supset \vdash : \text{Hp} . \supset . \text{ᗡ}‘P \cap \text{D}‘Q = \Lambda . \text{ᗡ}‘Q \cap \text{D}‘P = \Lambda .$

[*34·301] $\supset . P \mid Q = \dot{\Lambda} . Q \mid P = \dot{\Lambda} : \supset \vdash .$ Prop

***34·31.** $\vdash : \dot{\exists} ! (P \mid Q) . \supset . \dot{\exists} ! P . \dot{\exists} ! Q$

Dem.

$\vdash . *34·3 . \supset \vdash : \text{Hp} . \supset . \exists ! (\text{ᗡ}‘P \cap \text{D}‘Q) .$

[*24·561] $\supset . \exists ! \text{ᗡ}‘P . \exists ! \text{D}‘Q .$

[*33·24] $\supset . \dot{\exists} ! P . \dot{\exists} ! Q : \supset \vdash .$ Prop

***34·32.** $\vdash :. P = \dot{\Lambda} . \vee . Q = \dot{\Lambda} : \supset . P \mid Q = \dot{\Lambda}$ [*34·31 . Transp . *25·51]

***34·33.** $\vdash : x \epsilon \text{D}‘R . \equiv . x (R \mid \breve{R}) x$

Dem.

$\vdash . *33·13 . \supset \vdash : x \epsilon \text{D}‘R . \equiv . (\exists y) . xRy .$

[*4·24] $\equiv . (\exists y) . xRy . xRy .$

[*31·11] $\equiv . (\exists y) . xRy . y\breve{R}x .$

[*34·1] $\equiv . x (R \mid \breve{R}) x : \supset \vdash .$ Prop

***34·34.** $\vdash : R \subset\!\!\!\cdot\, P . S \subset\!\!\!\cdot\, Q . \supset . R \mid S \subset\!\!\!\cdot\, P \mid Q$

Dem.

$\vdash . *23·1 . \supset \vdash :. \text{Hp} . \supset : xRy . \supset_{x,y} . xPy : ySz . \supset_{y,z} . yQz :$

[*11·2.*10·1·41] $\supset : xRy . \supset . xPy : ySz . \supset . yQz :$

[*3·47] $\supset : xRy . ySz . \supset . xPy . yQz$ (1)

$\vdash . (1) . *10·11·21·28 . \supset$

$\vdash :. \text{Hp} . \supset : (\exists y) . xRy . ySz . \supset . (\exists y) . xPy . yQz :$

[*34·1] $\supset : x (R \mid S) z . \supset . x (P \mid Q) z$ (2)

$\vdash . (2) . *11·11·3 . \supset \vdash .$ Prop

***34·35.** $\vdash : \dot{\exists} ! R . \text{G}'R \subset \text{D}'P . \supset . \dot{\exists} ! R | P$

Dem.

$\vdash . *33·24 . \supset \vdash : \text{Hp} . \supset . \exists ! \text{G}'R$ (1)

$\vdash . *22·621 . \supset \vdash : \text{Hp} . \supset . \text{G}'R = \text{G}'R \cap \text{D}'P$ (2)

$\vdash . (1) . (2) . \supset \vdash : \text{Hp} . \supset . \exists ! \text{G}'R \cap \text{D}'P .$

[*34·3] $\supset . \dot{\exists} ! R | P : \supset \vdash . \text{Prop}$

***34·351.** $\vdash : \dot{\exists} ! R . \text{D}'R \subset \text{G}'P . \supset . \dot{\exists} ! P | R$ [Proof as in *34·35]

***34·36.** $\vdash . \text{D}'(P | Q) \subset \text{D}'P . \text{G}'(P | Q) \subset \text{G}'Q$

Dem.

$\vdash . *33·13 . \supset \vdash :. x \in \text{D}'(P | Q) . \supset : (\exists z) . x (P | Q) z :$

[*34·1] $\supset : (\exists z, y) . xPy . yQz :$

[*11·23] $\supset : (\exists y, z) . xPy . yQz :$

[*11·55.*10·5] $\supset : (\exists y) . xPy :$

[*33·13] $\supset : x \in \text{D}'P$ (1)

Similarly $\vdash :. z \in \text{G}'(P | Q) . \supset : z \in \text{G}'P$ (2)

$\vdash . (1) . (2) . *10·11 . \supset \vdash . \text{Prop}$

The following proposition is a lemma for *95·31.

***34·361.** $\vdash : \dot{\exists} ! R . \text{D}'R \subset \text{G}'P . \text{G}'R \subset \text{D}'Q . \supset . \dot{\exists} ! P | R | Q$

Dem.

$\vdash . *34·35 . \supset \vdash : \text{Hp} . \supset . \dot{\exists} ! R | Q$ (1)

$\vdash . *34·36 . \supset \vdash : \text{Hp} . \supset . \text{D}'(R | Q) \subset \text{G}'P$ (2)

$\vdash . (1) . (2) . *34·351 . \supset \vdash . \text{Prop}$

***34·37.** $\vdash . C'(P | Q) \subset \text{D}'P \cup \text{G}'Q$ [*34·36 . *33·161 . *22·72]

***34·38.** $\vdash . C'(P | Q) \subset C'P \cup C'Q$ [*34·37 . *33·161 . *22·72]

***34·4.** $\vdash : b = P'c . c = Q'z . \supset . b = (P | Q)'z$

Dem.

$\vdash . *30·31 . \supset \vdash : \text{Hp} . \supset . bPc . cQz .$

[*34·1] $\supset . b (P | Q) z$ (1)

$\vdash . *30·31 . \supset \vdash :. \text{Hp} . \supset : yQz . \supset_y . y = c :$

[Fact] $\supset : xPy . yQz . \supset_{x,y} . xPy . y = c .$

[*13·13] $\supset_{x,y} . xPc$ (2)

$\vdash . *30·31 . \supset \vdash :. \text{Hp} . \supset : xPc . \supset_x . x = b$ (3)

$\vdash . (2) . (3) . \supset \vdash :. \text{Hp} . \supset : xPy . yQz . \supset_{x,y} . x = b :$

[*10·23] $\supset : (\exists y) . xPy . yQz . \supset_x . x = b :$

[*34·1] $\supset : x (P | Q) z . \supset_x . x = b$ (4)

$\vdash . (1) . (4) . *30·31 . \supset \vdash . \text{Prop}$

***34·41.** $\vdash : \text{E} ! P'Q'z . \supset . P'Q'z = (P | Q)'z$

Dem.

$\vdash . *30·52 . \supset \vdash : \text{Hp} . \supset . (\exists b, c) . b = P'c . c = Q'z .$

[*30·51.*34·4] $\supset . (\exists b) . b = P'Q'z . b = (P | Q)'z .$

[*14·145] $\supset . P'Q'z = (P | Q)'z : \supset \vdash . \text{Prop}$

The above proposition is no longer true if we change the hypothesis into $E!(P|Q)`z$, since $(P|Q)`z$ may exist when $P`Q`z$ does not. Suppose, *e.g.*, that Q is the relation of child to father, and P the relation of daughter to father. Then $(P|Q)`z =$ the granddaughter of $z$, but $P`Q`z =$ the daughter of the child of z. The first exists whenever z has only one granddaughter, while the second requires further that z should have only one child.

For the same reason we do not have

$$b=(P|Q)`z . \supset . (\exists c) . b=P`c . c=Q`z.$$

This will hold if P, Q are one-many relations (cf. *71), but not in general otherwise.

***34·42.** $\vdash : (z) . R`z = P`Q`z . \supset . R = P|Q$

Dem.

$\vdash . *14{\cdot}21 . \quad \supset \vdash :. \text{Hp} . \supset : (z) . E!R`z : (z) . E!P`Q`z \quad (1)$

$\vdash . (1) . *34{\cdot}41 . \supset \vdash :. \text{Hp} . \supset : (z) . R`z = (P|Q)`z :$

$[*30{\cdot}42.(1)] \quad \supset : R = P|Q :. \supset \vdash . \text{Prop}$

***34·5.** $\vdash : xR^2y . \equiv . (\exists z) . xRz . zRy \quad [*34{\cdot}1 . (*34{\cdot}02)]$

***34·51.** $\vdash : xR^3y . \equiv . (\exists z, w) . xRz . zRw . wRy$

Dem.

$\vdash . *34{\cdot}1 . (*34{\cdot}03) . \supset$

$\vdash :. xR^3y . \equiv : (\exists w) . xR^2w . wRy :$

$[*34{\cdot}5] \quad \equiv : (\exists w) : (\exists z) . xRz . zRw : wRy :$

$[*11{\cdot}55] \quad \equiv : (\exists w, z) . xRz . zRw . wRy :$

$[*11{\cdot}2] \quad \equiv : (\exists z, w) . xRz . zRw . wRy :. \supset \vdash . \text{Prop}$

***34·52.** $\vdash . R^3 = R|R^2 \quad [*34{\cdot}21]$

***34·53.** $\vdash : \dot{\exists}!R^2 . \equiv . \exists!D`R \cap \text{Ꮖ}`R \quad [*34{\cdot}3]$

***34·531.** $\vdash : D`R \cap \text{Ꮖ}`R = \Lambda . \equiv . R^2 = \dot{\Lambda} \quad [*34{\cdot}53 . \text{Transp}]$

***34·54.** $\vdash : xRx . \supset . xR^2x$

Dem.

$\vdash . *4{\cdot}24 . \supset \vdash : xRx . \supset . xRx . xRx .$

$[*10{\cdot}24] \quad \supset . (\exists y) . xRy . yRx .$

$[*34{\cdot}5] \quad \supset . xR^2x : \supset \vdash . \text{Prop}$

***34·55.** $\vdash :. R^2 \in S . \equiv : xRy . yRz . \supset_{x,y,z} . xSz \quad [*34{\cdot}5 . *10{\cdot}23]$

***34·56.** $\vdash . D`R^2 \subset D`R . \text{Ꮖ}`R^2 \subset \text{Ꮖ}`R . C`R^2 \subset C`R \quad [*34{\cdot}36{\cdot}38]$

***34·6.** $\vdash . (R \dot{\cap} S)^2 \in R^2 \dot{\cap} S^2$

Dem.

$\vdash . *34{\cdot}5 . \supset \vdash :. x(R \dot{\cap} S)^2y . \equiv : (\exists z) . x(R \dot{\cap} S)z . z(R \dot{\cap} S)y :$

$[*23{\cdot}33 . *10{\cdot}281] \quad \equiv : (\exists z) . xRz . xSz . zRy . zSy :$

$[*4{\cdot}3 . *10{\cdot}281] \quad \equiv : (\exists z) . xRz . zRy . xSz . zSy :$

$[*10\cdot5] \qquad \supset : (\exists z) . xRz . zRy : (\exists z) . xSz . zSy :$

$[*34\cdot5] \qquad \supset : xR^2y . xS^2y :$

$[*23\cdot33] \qquad \supset : x(R^2 \cap S^2) y \qquad (1)$

$\vdash . (1) . *11\cdot11 . \supset \vdash . \text{Prop}$

***34·62.** $\vdash . (R \cup S)^2 = R^2 \cup R|S \cup S|R \cup S^2$

Dem.

$\vdash . *34\cdot26 . \supset \vdash . (R \cup S)^2 = R|(R \cup S) \cup S|(R \cup S)$

$[*34\cdot25] \qquad = R^2 \cup R|S \cup S|R \cup S^2 . \supset \vdash . \text{Prop}$

The above proposition is a lemma for *160·51, as is also *34·73, which employs the above proposition.

***34·63.** $\vdash . \text{Cnv}\text{‘}(R^2) = (\text{Cnv}\text{‘}R)^2$

Dem.

$\vdash . *31\cdot131 . \supset$

$\vdash :. x\{\text{Cnv}\text{‘}(R^2)\} y . \equiv : yR^2x :$

$[*34\cdot5] \qquad \equiv : (\exists z) . yRz . zRx :$

$[*31\cdot131 . *10\cdot281] \qquad \equiv : (\exists z) . x\breve{R}z . z\breve{R}y :$

$[*31\cdot131 . *34\cdot5] \qquad \equiv : x(\text{Cnv}\text{‘}R)^2 y : \supset \vdash . \text{Prop}$

***34·7.** $\vdash . \text{Cnv}\text{‘}(S|\breve{S}) = S|\breve{S}$

Dem.

$\vdash . *34\cdot2 . \supset \vdash . \text{Cnv}\text{‘}(S|\breve{S}) = (\text{Cnv}\text{‘}\breve{S})|\breve{S}$

$[*34\cdot202] \qquad = S|\breve{S} . \supset \vdash . \text{Prop}$

Thus $S|\breve{S}$ is always a symmetrical relation, *i.e.* one which is equal to its converse.

***34·701.** $\vdash . \text{Cnv}\text{‘}(\breve{S}|S) = \breve{S}|S$ [*34·2·203]

***34·702.** $\vdash . C\text{‘}(S|\breve{S}) = \text{D}\text{‘}S$

Dem.

$\vdash . *34\cdot37 . \supset \vdash . C\text{‘}(S|\breve{S}) \subset \text{D}\text{‘}S \cup \text{C}\text{‘}\breve{S}$

$[*33\cdot21] \qquad \subset \text{D}\text{‘}S \qquad (1)$

$\vdash . *33\cdot13 . \supset \vdash : x \in \text{D}\text{‘}S . \supset . (\exists y) . xSy .$

$[*31\cdot11] \qquad \supset . (\exists y) . xSy . y\breve{S}x .$

$[*34\cdot1] \qquad \supset . x(S|\breve{S})x .$

$[*33\cdot17] \qquad \supset . x \in C\text{‘}(S|\breve{S}) \qquad (2)$

$\vdash . (1) . (2) . *10\cdot11 . \supset \vdash . \text{Prop}$

***34·703.** $\vdash . C\text{‘}(\breve{S}|S) = \text{C}\text{‘}S$ [Similar proof]

***34·73.** $\vdash : C\text{'}P \cap C\text{'}Q = \Lambda . \supset . (P \cup Q)^2 = P^2 \cup Q^2$

Dem.

$\vdash . *34\cdot302 . \supset \vdash : \text{Hp} . \supset . P \mid Q = \dot{\Lambda} . Q \mid P = \dot{\Lambda} .$

$[*25\cdot24] \quad \supset . P^2 \cup Q^2 = P^2 \cup P \mid Q \cup Q \mid P \cup Q^2$

$[*34\cdot62] \quad = (P \cup Q)^2 : \supset \vdash . \text{Prop}$

***34·8.** $\vdash : R = \breve{R} . R^2 \in R . \supset . R = R^2 = R \mid \breve{R}$

Dem.

$\vdash . *34\cdot28 . \quad \supset \vdash : R = \breve{R} . \supset . R^2 = R \mid \breve{R}$ (1)

$\vdash . *34\cdot33 . *33\cdot14 . \quad \supset \vdash : xRy . \supset . x (R \mid \breve{R}) x$ (2)

$\vdash . (1) . (2) . \quad \supset \vdash :. R = \breve{R} . \supset : xRy . \supset . xR^2x$ (3)

$\vdash . (3) . *23\cdot1 . \quad \supset \vdash :. R = \breve{R} . R^2 \in R . \supset : xRy . \supset . xRx :$

$[*4\cdot7] \quad \supset : xRy . \supset . xRx . xRy .$

$[*10\cdot24 . *34\cdot5] \quad \supset . xR^2y$ (4)

$\vdash . (4) . *11\cdot11\cdot3 . \quad \supset \vdash : \text{Hp} . \supset . R \in R^2$ (5)

$\vdash . *3\cdot27 . \quad \supset \vdash : \text{Hp} . \supset . R^2 \in R$ (6)

$\vdash . (5) . (6) . *23\cdot41 . \supset \vdash : \text{Hp} . \supset . R = R^2$ (7)

$\vdash . (1) . (7) . \quad \supset \vdash . \text{Prop}$

The hypothesis of the above proposition is the hypothesis that R is symmetrical ($R = \breve{R}$) and transitive ($R^2 \in R$). These are the formal properties of those relations which can suitably be regarded as expressing equality in some respect.

***34·81.** $\vdash : R = \breve{R} . R^2 \in R . \equiv . R = \breve{R} . R^2 = R \quad [*34\cdot8 . *4\cdot71]$

The following propositions are lemmas for *34·85, which is used in *72·64.

***34·82.** $\vdash :. R = \breve{R} . R^2 \in R . \supset : x \in \text{D}\text{'}R . \equiv . xRx$

Dem.

$\vdash . *34\cdot33 . \supset \vdash : x \in \text{D}\text{'}R . \equiv . x (R \mid \breve{R}) x$ (1)

$\vdash . *34\cdot8 . \quad \supset \vdash :. \text{Hp} . \supset : x (R \mid \breve{R}) x . \equiv . xRx$ (2)

$\vdash . (1) . (2) . \supset \vdash . \text{Prop}$

***34·83.** $\vdash : R = \breve{R} . R^2 \in R . xRy . \supset . \overleftarrow{R}\text{'}x = \overleftarrow{R}\text{'}y$

Dem.

$\vdash . *31\cdot11 . \supset \vdash :. \text{Hp} . \supset : yRx :$

$[*3\cdot2] \quad \supset : xRz . \supset . yRx . xRz .$

$[*34\cdot55 . \text{Hp}] \quad \supset . yRz$ (1)

$\vdash . *3\cdot2 . \quad \supset \vdash :. \text{Hp} . \supset : yRz . \supset . xRy . yRz .$

$[*34\cdot55 . \text{Hp}] \quad \supset . xRz$ (2)

$\vdash . (1) . (2) . \supset \vdash :. \text{Hp} . \supset : xRz . \equiv . yRz :$

$[*10\cdot11\cdot21 . *20\cdot15 . *32\cdot111] \supset : \overleftarrow{R}\text{'}x = \overleftarrow{R}\text{'}y :. \supset \vdash . \text{Prop}$

$*34\cdot84$. $\vdash : R = \breve{R} . R^2 \subseteq R . y \in D'R . \overleftarrow{R}'x = \overleftarrow{R}'y . \supset . xRy$

Dem.

$\vdash . *34\cdot82 . \quad \supset \vdash : \text{Hp} . \supset . yRy$ (1)

$\vdash . *32\cdot181 . *20\cdot31 . \supset \vdash :. \text{Hp} . \supset : xRz . \equiv_z . yRz :$

$[*10\cdot1] \quad \supset : xRy . \equiv . yRy$ (2)

$\vdash . (1) . (2) . \quad \supset \vdash . \text{Prop}$

$*34\cdot841$. $\vdash : R = \breve{R} . R^2 \subseteq R . x \in D'R . \overleftarrow{R}'x = \overleftarrow{R}'y . \supset . xRy$

Dem.

$\vdash . *34\cdot84 \frac{y, x}{x, y} . \supset \vdash : \text{Hp} . \supset . yRx .$

$[*31\cdot11 . \text{Hp}] \quad \supset . xRy : \supset \vdash . \text{Prop}$

$*34\cdot85$. $\vdash :. R = \breve{R} . R^2 \subseteq R . \supset : xRy . \equiv . x \in D'R . \overleftarrow{R}'x = \overleftarrow{R}'y$

$[*34\cdot83\cdot841 . *33\cdot14]$

*35. RELATIONS WITH LIMITED DOMAINS AND CONVERSE DOMAINS

Summary of *35.

In this section, we have to consider the relation derived from a given relation R by limiting either its domain or its converse domain to members of some assigned class. A relation R with its domain limited to members of α is written "$\alpha \rceil R$"; with its converse domain limited to members of β, it is written "$R \upharpoonright \beta$"; with both limitations, it is written "$\alpha \rceil R \upharpoonright \beta$." Thus *e.g.* "brother" and "sister" express the same relation (that of a common parentage), with the domain limited in the first case to males, in the second to females. "The relation of white employers to coloured employees" is a relation limited both as to its domain and as to its converse domain. We put

***35·01.** $\alpha \rceil R = \hat{x}\hat{y}(x \epsilon \alpha . xRy)$ Df

with similar definitions for $R \upharpoonright \alpha$ and $\alpha \rceil R \upharpoonright \beta$.

A particularly important case is the case in which the same limitation is imposed on the domain and on the converse domain, *i.e.* where we have a relation of the form "$\alpha \rceil R \upharpoonright \alpha$." In this case, the limitation to members of α may be more briefly stated as being imposed on the *field*. For this case, it is convenient to adopt "$R \upharpoonright\!\!\!\downharpoonright \alpha$" as an alternative notation. This case will be considered in *36.

It is convenient to consider in the present connection the relation between x and y which is constituted by x being a member of α and y being a member of β. This relation will be denoted by "$\alpha \uparrow \beta$." Thus we put

***35·04.** $\alpha \uparrow \beta = \hat{x}\hat{y}(x \epsilon \alpha . y \epsilon \beta)$ Df

The chief importance of relations with limited *fields* arises in the theory of series. Given a series generated by a relation R, let α be a class consisting of part of this series. Then α is the field of the relation $\alpha \rceil R \upharpoonright \alpha$ or $R \upharpoonright\!\!\!\downharpoonright \alpha$, and it is this relation which is the generating relation of the series of members of α in the same order which they have as parts of the original series. Thus parts of a series, considered not merely as classes but as series, are dealt with by means of serial relations with limited fields.

Relations with limited *domains* are not nearly so much used as relations with limited *converse domains*. Relations with limited converse domains play a great part in arithmetic, especially in establishing the formal laws. What is wanted in such cases is a one-one relation correlating two classes or two series. That is, we want a relation such that not only does R‘y exist whenever $y \epsilon$ Ɑ‘R, but also $\breve{R}$‘x exists whenever $x \epsilon$ D‘R. The kind of relation which is most frequently found to effect such a correlation is some such relation as D or G or C, or some other constant relation for which we always have $\text{E}!\,R\text{‘}y$, with its converse domain so limited that, subject to the limitation, only one value of y gives any given value of $R\text{‘}y$. Thus for example let λ be a class of relations no two of which have the same domain; then $\text{D}\restriction\lambda$ will give a one-one correlation of these relations with their domains: if $R, S\,\epsilon\,\lambda$, we shall have

$$\text{D}\text{‘}R = \text{D}\text{‘}S \,.\supset.\, R = S.$$

We shall also have $\text{D}\text{‘}R = (\text{D}\restriction\lambda)\text{‘}R$ and $\text{D}\text{‘}S = (\text{D}\restriction\lambda)\text{‘}S$. Moreover the converse domain of $\text{D}\restriction\lambda$ is λ, and the domain of $\text{D}\restriction\lambda$ is the class of domains of members of λ. Thus $\text{D}\restriction\lambda$ gives a one-one correlation of λ with the domains of members of λ. It is chiefly in such ways that relations with limited converse domains are useful.

For purposes of reference, a great many propositions are given in the present number, but the propositions that will be used frequently are comparatively few. Among these are the following:

***35·21.** $\vdash .\, \alpha \upharpoonleft R \restriction \beta = (\alpha \upharpoonleft R) \restriction \beta = \alpha \upharpoonleft (R \restriction \beta)$

***35·31.** $\vdash .\, (R \restriction \alpha) \restriction \beta = R \restriction (\alpha \cap \beta)$

***35·354.** $\vdash .\, (R \restriction \alpha) | S = R | \alpha \upharpoonleft S$

I.e. in a relative product it makes no difference whether we limit the converse domain of the first factor, or the domain of the second.

***35·412.** $\vdash .\, R \restriction (\beta \cup \beta') = R \restriction \beta \,\dot{\cup}\, R \restriction \beta'$

***35·452.** $\vdash : \text{G}\text{‘}R \subset \beta \,.\supset.\, R \restriction \beta = R$

***35·48.** $\vdash : \text{G}\text{‘}P \subset \alpha \,.\supset.\, P | (\alpha \upharpoonleft R) = P | R$

***35·52.** $\vdash .\, \text{Cnv}\text{‘}(R \restriction \beta) = \beta \upharpoonleft \breve{R}$

***35·61.** $\vdash .\, \text{D}\text{‘}(\alpha \upharpoonleft R) = \alpha \cap \text{D}\text{‘}R$

***35·64.** $\vdash .\, \text{G}\text{‘}(R \restriction \beta) = \beta \cap \text{G}\text{‘}R$

***35·65.** $\vdash : \beta \subset \text{G}\text{‘}R \,.\supset.\, \text{G}\text{‘}(R \restriction \beta) = \beta$

The hypothesis $\beta \subset \text{G}\text{‘}R$ is fulfilled in the great majority of cases in which we have occasion to use $R \restriction \beta$.

***35·66.** $\vdash : \text{G}\text{‘}R \subset \beta \,.\equiv.\, R \restriction \beta = R$

***35·7.** $\vdash : \phi\,\{(R \restriction \beta)\text{‘}y\} \,.\equiv.\, y\,\epsilon\,\beta \,.\, \phi\,(R\text{‘}y)$

This proposition is used very frequently, owing to the fact that limitation of the converse domain is chiefly applied to such relations as give rise to descriptive functions (*e.g.* D, G, C).

***35·71.** $\vdash :.\, y\,\epsilon\,\beta \,.\supset_y.\, R\text{‘}y = S\text{‘}y : \supset .\, R \restriction \beta = S \restriction \beta$

This proposition is useful for a reason similar to that which makes *35·7 useful.

***35·82.** $\vdash .\, \alpha \uparrow \beta = \alpha \upharpoonleft \dot{\text{V}} \restriction \beta$

Owing to this proposition, the properties of $\alpha \uparrow \beta$ can be deduced from the already proved properties of $\alpha \upharpoonleft R \restriction \beta$, by putting $R = \dot{\text{V}}$.

The relation "$\alpha \uparrow \beta$" is what may be called an "analysable" relation, *i.e.* it holds between x and y when $x \epsilon \alpha$ and $y \epsilon \beta$, *i.e.* when x has a property independent of y, and y has a property independent of x.

***35·85.** $\vdash : \exists ! \beta . \supset . D`(\alpha \uparrow \beta) = \alpha$

***35·86.** $\vdash : \exists ! \alpha . \supset . \text{Ɑ}`(\alpha \uparrow \beta) = \beta$

If either α or β is null, so is $\alpha \uparrow \beta$ (*35·88).

***35·01.** $\alpha \upharpoonleft R = \hat{x}\hat{y}(x \epsilon \alpha . xRy)$ Df

***35·02.** $R \upharpoonright \beta = \hat{x}\hat{y}(xRy . y \epsilon \beta)$ Df

***35·03.** $\alpha \upharpoonleft R \upharpoonright \beta = \hat{x}\hat{y}(x \epsilon \alpha . xRy . y \epsilon \beta)$ Df

***35·04.** $\alpha \uparrow \beta = \hat{x}\hat{y}(x \epsilon \alpha . y \epsilon \beta)$ Df

***35·05.** $R`x \uparrow \beta = (R`x) \uparrow \beta$ Df

The last definition serves merely for the avoidance of brackets.

***35·1.** $\vdash : x(\alpha \upharpoonleft R) y . \equiv . x \epsilon \alpha . xRy$ [*21·3 . (*35·01)]

***35·101.** $\vdash : x(R \upharpoonright \beta) y . \equiv . xRy . y \epsilon \beta$

***35·102.** $\vdash : x(\alpha \upharpoonleft R \upharpoonright \beta) y . \equiv . x \epsilon \alpha . xRy . y \epsilon \beta$

***35·103.** $\vdash : x(\alpha \uparrow \beta) y . \equiv . x \epsilon \alpha . y \epsilon \beta$

***35·11.** $\vdash . \alpha \upharpoonleft R \upharpoonright \beta = (\alpha \upharpoonleft R) \dot{\cap} (R \upharpoonright \beta)$

Dem.

$\vdash . *35·102 . \supset \vdash : x(\alpha \upharpoonleft R \upharpoonright \beta) y . \equiv . x \epsilon \alpha . xRy . y \epsilon \beta .$

[*4·24] $\equiv . x \epsilon \alpha . xRy . xRy . y \epsilon \beta .$

[*35·1·101] $\equiv . x(\alpha \upharpoonleft R) y . x(R \upharpoonright \beta) y .$

[*23·33] $\equiv . x \{(\alpha \upharpoonleft R) \dot{\cap} (R \upharpoonright \beta)\} y : \supset \vdash .$ Prop

***35·12.** $\vdash . (\alpha \upharpoonleft R) \dot{\cap} (S \upharpoonright \beta) = \alpha \upharpoonleft (R \dot{\cap} S) \upharpoonright \beta$

Dem.

$\vdash . *23·33 . \supset \vdash : x \{(\alpha \upharpoonleft R) \dot{\cap} (S \upharpoonright \beta)\} y . \equiv . x(\alpha \upharpoonleft R) y . x(S \upharpoonright \beta) y .$

[*35·1·101] $\equiv . x \epsilon \alpha . xRy . xSy . y \epsilon \beta .$

[*23·33] $\equiv . x \epsilon \alpha . x(R \dot{\cap} S) y . y \epsilon \beta .$

[*35·102] $\equiv . x \{\alpha \upharpoonleft (R \dot{\cap} S) \upharpoonright \beta\} y : \supset \vdash .$ Prop

***35·13.** $\vdash . (\alpha \upharpoonleft R) \dot{\cap} (\beta \upharpoonleft S) = (\alpha \cap \beta) \upharpoonleft (R \dot{\cap} S)$

Dem.

$\vdash . *23·33 . \supset \vdash : x \{(\alpha \upharpoonleft R) \dot{\cap} (\beta \upharpoonleft S)\} y . \equiv . x(\alpha \upharpoonleft R) y . x(\beta \upharpoonleft S) y .$

[*35·1] $\equiv . x \epsilon \alpha . xRy . x \epsilon \beta . xSy .$

[*22·33.*23·33] $\equiv . x \epsilon (\alpha \cap \beta) . x(R \dot{\cap} S) y .$

[*35·1] $\equiv . x \{(\alpha \cap \beta) \upharpoonleft (R \dot{\cap} S)\} y : \supset \vdash .$ Prop

***35·14.** $\vdash . (R \upharpoonright \alpha) \dot{\cap} (S \upharpoonright \beta) = (R \dot{\cap} S) \upharpoonright (\alpha \cap \beta)$ [Similar proof to *35·13]

***35·15.** $\vdash . (\alpha \upharpoonleft R \restriction \beta) \dot{\cap} (\alpha' \upharpoonleft S \restriction \beta') = (\alpha \cap \alpha') \upharpoonleft (R \dot{\cap} S) \restriction (\beta \cap \beta')$

Dem.

$\vdash . *35·11 . \supset$

$\vdash . (\alpha \upharpoonleft R \restriction \beta) \dot{\cap} (\alpha' \upharpoonleft S \restriction \beta') = (\alpha \upharpoonleft R) \dot{\cap} (R \restriction \beta) \dot{\cap} (\alpha' \upharpoonleft S) \dot{\cap} (S \restriction \beta')$

[*35·13·14] $= \{(\alpha \cap \alpha') \upharpoonleft (R \dot{\cap} S)\} \dot{\cap} \{(R \dot{\cap} S) \restriction (\beta \cap \beta')\}$

[*35·11] $= \{(\alpha \cap \alpha') \upharpoonleft (R \dot{\cap} S) \restriction (\beta \cap \beta')\} . \supset \vdash .$ Prop

***35·16.** $\vdash . (\alpha \upharpoonleft R) \dot{\cap} S = \alpha \upharpoonleft (R \dot{\cap} S) = R \dot{\cap} \alpha \upharpoonleft S$ [Similar proof to *35·13]

***35·17.** $\vdash . (R \restriction \beta) \dot{\cap} S = (R \dot{\cap} S) \restriction \beta = R \dot{\cap} S \restriction \beta$ [Similar proof to *35·13]

***35·18.** $\vdash . (\alpha \upharpoonleft R \restriction \beta) \dot{\cap} S = \alpha \upharpoonleft (R \dot{\cap} S) \restriction \beta = R \dot{\cap} \alpha \upharpoonleft S \restriction \beta$
[Similar proof to *35·15]

***35·21.** $\vdash . \alpha \upharpoonleft R \restriction \beta = (\alpha \upharpoonleft R) \restriction \beta = \alpha \upharpoonleft (R \restriction \beta)$

Dem.

$\vdash . *35·102 . \supset \vdash : x (\alpha \upharpoonleft R \restriction \beta) y . \equiv . x \epsilon \alpha . xRy . y \epsilon \beta .$

[*35·1] $\equiv . x (\alpha \upharpoonleft R) y . y \epsilon \beta .$

[*35·101] $\equiv . x \{(\alpha \upharpoonleft R) \restriction \beta\} y$ (1)

$\vdash . *35·102 . \supset \vdash : x (\alpha \upharpoonleft R \restriction \beta) y . \equiv . x \epsilon \alpha . xRy . y \epsilon \beta .$

[*35·101] $\equiv . x \epsilon \alpha . x (R \restriction \beta) y .$

[*35·1] $\equiv . x \{\alpha \upharpoonleft (R \restriction \beta)\} y$ (2)

$\vdash . (1) . (2) . \supset \vdash .$ Prop

***35·22.** $\vdash . (\alpha \upharpoonleft R) | S = \alpha \upharpoonleft (R | S)$

Dem.

$\vdash . *34·1 . \supset \vdash :. x \{(\alpha \upharpoonleft R) | S\} y . \equiv : (\exists z) . x (\alpha \upharpoonleft R) z . zSy :$

[*35·1] $\equiv : (\exists z) . x \epsilon \alpha . xRz . zSy :$

[*10·35] $\equiv : x \epsilon \alpha : (\exists z) . xRz . zSy .$

[*34·1] $\equiv : x \epsilon \alpha . x (R | S) y :$

[*35·1] $\equiv : x \{\alpha \upharpoonleft (R | S)\} y :. \supset \vdash .$ Prop

***35·23.** $\vdash . S | (R \restriction \beta) = (S | R) \restriction \beta$ [Similar proof to *35·22]

***35·24.** $\alpha \upharpoonleft R | S = (\alpha \upharpoonleft R) | S$ Df

***35·25.** $S | R \restriction \beta = (S | R) \restriction \beta$ Df

***35·26.** $\vdash . (\alpha \upharpoonleft R) | (S \restriction \beta) = \alpha \upharpoonleft (R | S) \restriction \beta = \{\alpha \upharpoonleft (R | S)\} \restriction \beta = \alpha \upharpoonleft \{(R | S) \restriction \beta\}$
$= \{(\alpha \upharpoonleft R) | S\} \restriction \beta = \alpha \upharpoonleft \{R | (S \restriction \beta)\}$
$= (\alpha \upharpoonleft R | S) \restriction \beta = \alpha \upharpoonleft (R | S \restriction \beta)$

Dem.

$\vdash . *34·1 . \supset \vdash :. x \{(\alpha \upharpoonleft R) | (S \restriction \beta)\} y . \equiv : (\exists z) . x (\alpha \upharpoonleft R) z . z (S \restriction \beta) y :$

[*35·1·101] $\equiv : (\exists z) . x \epsilon \alpha . xRz . zSy . y \epsilon \beta :$

[*10·35] $\equiv : x \epsilon \alpha . y \epsilon \beta : (\exists z) . xRz . zSy :$

[*34·1] $\equiv : x \epsilon \alpha . x (R | S) y . y \epsilon \beta :$

[*35·102] $\equiv : x \{\alpha \upharpoonleft (R | S) \restriction \beta\} y$ (1)

$\vdash . (1) . *35·21·22·23 . (*35·24·25) . \supset \vdash .$ Prop

***35·27.** $\alpha \upharpoonleft R | S \restriction \beta = (\alpha \upharpoonleft R | S) \restriction \beta$ Df

***35·31.** $\vdash . (R \restriction \alpha) \restriction \beta = R \restriction (\alpha \cap \beta)$

Dem.

$\vdash . *35\cdot101 . \supset \vdash : x \{(R \restriction \alpha) \restriction \beta\} y . \equiv . x (R \restriction \alpha) y . y \epsilon \beta .$

[*35·101] $\equiv . xRy . y \epsilon \alpha . y \epsilon \beta .$

[*22·33] $\equiv . xRy . y \epsilon \alpha \cap \beta .$

[*35·101] $\equiv . x \{R \restriction (\alpha \cap \beta)\} y : \supset \vdash .$ Prop

***35·32.** $\vdash . \alpha \upharpoonleft (\beta \upharpoonleft R) = (\alpha \cap \beta) \upharpoonleft R$ [Proof similar to that of *35·31]

***35·33.** $\vdash . (\alpha \upharpoonleft R \restriction \beta) \restriction \gamma = \{\alpha \upharpoonleft R \restriction (\beta \cap \gamma)\}$ [Proof similar to that of *35·31]

***35·34.** $\vdash . \alpha \upharpoonleft (\beta \upharpoonleft R \restriction \gamma) = \{(\alpha \cap \beta) \upharpoonleft R \restriction \gamma\}$ [Proof similar to that of *35·31]

***35·35.** $\vdash . \alpha \upharpoonleft R = (\alpha \cap \text{D}`R) \upharpoonleft R$

Dem.

$\vdash . *35\cdot1 . \supset \vdash : x (\alpha \upharpoonleft R) y . \equiv . x \epsilon \alpha . xRy .$

[*33·14] $\equiv . x \epsilon \alpha . x \epsilon \text{D}`R . xRy .$

[*22·33.*35·1] $\equiv . x \{(\alpha \cap \text{D}`R) \upharpoonleft R\} y : \supset \vdash .$ Prop

***35·351.** $\vdash . R \restriction \beta = R \restriction (\beta \cap \text{Ɑ}`R)$ [Proof as in *35·35]

***35·352.** $\vdash . \alpha \upharpoonleft R \restriction \beta = (\alpha \cap \text{D}`R) \upharpoonleft R \restriction (\beta \cap \text{Ɑ}`R)$ [Proof as in *35·35]

***35·354.** $\vdash . (R \restriction \alpha) | S = R | \alpha \upharpoonleft S$

Dem.

$\vdash . *34\cdot1 . *35\cdot101 . \supset$

$\vdash : x \{(R \restriction \alpha) | S\} z . \equiv . (\exists y) . xRy . y \epsilon \alpha . ySz .$

[*35·1] $\equiv . (\exists y) . xRy . y (\alpha \upharpoonleft S) z .$

[*34·1] $\equiv . x \{R | (\alpha \upharpoonleft S)\} z : \supset \vdash .$ Prop

***35·41.** $\vdash . (\alpha \cup \alpha') \upharpoonleft R = \alpha \upharpoonleft R \,\dot{\cup}\, \alpha' \upharpoonleft R$ [*35·1 . *22·34]

***35·412.** $\vdash . R \restriction (\beta \cup \beta') = R \restriction \beta \,\dot{\cup}\, R \restriction \beta'$ [*35·101 . *22·34]

***35·413.** $\vdash . (\alpha \cup \alpha') \upharpoonleft R \restriction (\beta \cup \beta') = (\alpha \upharpoonleft R \restriction \beta) \,\dot{\cup}\, (\alpha \upharpoonleft R \restriction \beta')$
$\dot{\cup}\, (\alpha' \upharpoonleft R \restriction \beta) \,\dot{\cup}\, (\alpha' \upharpoonleft R \restriction \beta')$ [*35·102 . *22·34]

***35·42.** $\vdash . \alpha \upharpoonleft (R \,\dot{\cup}\, S) = (\alpha \upharpoonleft R) \,\dot{\cup}\, (\alpha \upharpoonleft S)$ [*35·1 . *23·34]

***35·421.** $\vdash . (R \,\dot{\cup}\, S) \restriction \beta = (R \restriction \beta) \,\dot{\cup}\, (S \restriction \beta)$ [*35·101 . *23·34]

***35·422.** $\vdash . \alpha \upharpoonleft (R \,\dot{\cup}\, S) \restriction \beta = (\alpha \upharpoonleft R \restriction \beta) \,\dot{\cup}\, (\alpha \upharpoonleft S \restriction \beta)$ [*35·102 . *23·34]

***35·43.** $\vdash : \alpha \subset \beta . \supset . \alpha \upharpoonleft R \,\dot{\subset}\, \beta \upharpoonleft R$

Dem.

$\vdash . *35\cdot1 . \supset \vdash :. \alpha \subset \beta . \supset : x (\alpha \upharpoonleft R) y . \equiv . x \epsilon \alpha . xRy .$

[*22·1] $\supset . x \epsilon \beta . xRy .$

[*35·1] $\supset . x (\beta \upharpoonleft R) y :. \supset \vdash .$ Prop

***35·431.** $\vdash : \beta \subset \gamma . \supset . R \restriction \beta \,\dot{\subset}\, R \restriction \gamma$ [Proof similar to that of *35·43]

***35·432.** $\vdash : \alpha \subset \gamma . \beta \subset \delta . \supset . \alpha \upharpoonleft R \restriction \beta \,\dot{\subset}\, \gamma \upharpoonleft R \restriction \delta$

[Proof similar to that of *35·43]

***35·44.** $\vdash . \alpha \upharpoonleft R \Subset R$

Dem.

$\vdash . *35\cdot1 . \supset \vdash : x(\alpha \upharpoonleft R)y . \supset . x \in \alpha . xRy .$

$[*3\cdot27] \qquad \supset . xRy : \supset \vdash . \text{Prop}$

***35·441.** $\vdash . R \upharpoonright \beta \Subset R$ [Proof similar to that of *35·44]

***35·442.** $\vdash . \alpha \upharpoonleft R \upharpoonright \beta \Subset R$ [Proof similar to that of *35·44]

***35·451.** $\vdash : \mathrm{D}‘R \subset \alpha . \supset . \alpha \upharpoonleft R = R$

Dem.

$\vdash . *4\cdot71 . \supset \vdash :. \text{Hp} . \supset : x \in \mathrm{D}‘R . \equiv . x \in \mathrm{D}‘R . x \in \alpha :$

$[*4\cdot36] \qquad \supset : x \in \mathrm{D}‘R . xRy . \equiv . x \in \mathrm{D}‘R . xRy . x \in \alpha$ (1)

$\vdash . *33\cdot14 . *4\cdot71 . \qquad \supset \vdash : xRy . \equiv . x \in \mathrm{D}‘R . xRy$ (2)

$\vdash . (1) . (2) . \supset \vdash :. \text{Hp} . \supset : xRy . \equiv . xRy . x \in \alpha .$

$[*35\cdot1] \qquad \equiv . x(\alpha \upharpoonleft R)y :. \supset \vdash . \text{Prop}$

***35·452.** $\vdash : \text{Ɑ}‘R \subset \beta . \supset . R \upharpoonright \beta = R$ [Similar proof]

***35·453.** $\vdash : \mathrm{D}‘R \subset \alpha . \supset . \alpha \upharpoonleft R \upharpoonright \beta = R \upharpoonright \beta$ [Similar proof]

***35·454.** $\vdash : \text{Ɑ}‘R \subset \beta . \supset . \alpha \upharpoonleft R \upharpoonright \beta = \alpha \upharpoonleft R$ [Similar proof]

***35·46.** $\vdash : R \Subset S . \supset . \alpha \upharpoonleft R \Subset \alpha \upharpoonleft S$

Dem.

$\vdash . *23\cdot1 . \supset \vdash :. \text{Hp} . \supset : xRy . \supset . xSy :$

$[\text{Fact}] \qquad \supset : x \in \alpha . xRy . \supset . x \in \alpha . xSy :$

$[*35\cdot1] \qquad \supset : x(\alpha \upharpoonleft R)y . \supset . x(\alpha \upharpoonleft S)y :. \supset \vdash . \text{Prop}$

***35·461.** $\vdash : R \Subset S . \supset . R \upharpoonright \beta \Subset S \upharpoonright \beta$ [Similar proof]

***35·462.** $\vdash : R \Subset S . \supset . \alpha \upharpoonleft R \upharpoonright \beta \Subset \alpha \upharpoonleft S \upharpoonright \beta$ [Similar proof]

***35·471.** $\vdash : \text{Ɑ}‘P \cap \alpha = \Lambda . \supset . P \,|\, (\alpha \upharpoonleft R) = \dot{\Lambda}$

Dem.

$\vdash . *34\cdot1 . \supset \vdash : x\{P \,|\, (\alpha \upharpoonleft R)\}z . \supset . (\exists y) . xPy . y(\alpha \upharpoonleft R)z .$

$[*35\cdot1] \qquad \supset . (\exists y) . xPy . y \in \alpha . yRz .$

$[*33\cdot14 . *10\cdot5] \qquad \supset . (\exists y) . y \in \text{Ɑ}‘P . y \in \alpha .$

$[*22\cdot33 . *24\cdot5] \qquad \supset . \exists ! \text{Ɑ}‘P \cap \alpha$ (1)

$\vdash . (1) . \text{Transp} . *24\cdot51 . \supset$

$\qquad \vdash : \text{Ɑ}‘P \cap \alpha = \Lambda . \supset . \sim x\{P \,|\, (\alpha \upharpoonleft R)\}z :$

$[*11\cdot11\cdot3] \quad \supset \vdash : \text{Ɑ}‘P \cap \alpha = \Lambda . \supset . (x, z) . \sim x\{P \,|\, (\alpha \upharpoonleft R)\}z .$

$[*25\cdot15] \qquad \supset . P \,|\, (\alpha \upharpoonleft R) = \dot{\Lambda} : \supset \vdash . \text{Prop}$

***35·472.** $\vdash : \mathrm{D}‘P \cap \alpha = \Lambda . \supset . (R \upharpoonright \alpha) \,|\, P = \dot{\Lambda}$

***35·473.** $\vdash : \text{Ɑ}‘P \cap \alpha = \Lambda . \supset . P \,|\, (\alpha \upharpoonleft R \upharpoonright \beta) = \dot{\Lambda}$

***35·474.** $\vdash : \mathrm{D}‘P \cap \beta = \Lambda . \supset . (\alpha \upharpoonleft R \upharpoonright \beta) \,|\, P = \dot{\Lambda}$

***35·48.** $\vdash : \text{Ɑ}`P \subset \alpha . \supset . P \mid (\alpha \upharpoonleft R) = P \mid R$

Dem.

$\vdash . *22·1 . \quad \supset \vdash :. \mathrm{Hp} . \supset : y \in \text{Ɑ}`P . \supset_y . y \in \alpha :$

[*4·71] $\quad \supset : y \in \text{Ɑ}`P . y \in \alpha . \equiv_y . y \in \text{Ɑ}`P :$

[*10·311] $\quad \supset : xPy . y \in \text{Ɑ}`P . y \in \alpha . \equiv_y . xPy . y \in \text{Ɑ}`P$ (1)

$\vdash . *33·14 . *4·71 . \supset \vdash : xPy . y \in \text{Ɑ}`P . \equiv . xPy$ (2)

$\vdash . (1) . (2) . \supset \vdash :. \mathrm{Hp} . \supset : xPy . y \in \alpha . \equiv_y . xPy :$

[*10·311] $\quad \supset : xPy . y \in \alpha . yRz . \equiv_y . xPy . yRz :$

[*35·1] $\quad \supset : xPy . y (\alpha \upharpoonleft R) z . \equiv_y . xPy . yRz :$

[*10·281] $\quad \supset : (\exists y) . xPy . y (\alpha \upharpoonleft R) z . \equiv . (\exists y) . xPy . yRz :$

[*34·1] $\quad \supset : x (P \mid \alpha \upharpoonleft R) z . \equiv . x (P \mid R) z :. \supset \vdash . \mathrm{Prop}$

***35·481.** $\vdash : \mathrm{D}`R \subset \beta . \supset . (P \upharpoonright \beta) \mid R = P \mid R$ [Similar proof]

***35·51.** $\vdash . \mathrm{Cnv}`(\alpha \upharpoonleft R) = \breve{R} \upharpoonright \alpha$

Dem.

$\vdash . *31·131 . \supset \vdash : x \{\mathrm{Cnv}`(\alpha \upharpoonleft R)\} y . \equiv . y (\alpha \upharpoonleft R) x .$

[*35·1] $\quad \equiv . y \in \alpha . yRx .$

[*31·11] $\quad \equiv . x\breve{R}y . y \in \alpha .$

[*35·101] $\quad \equiv . x (\breve{R} \upharpoonright \alpha) y : \supset \vdash . \mathrm{Prop}$

***35·52.** $\vdash . \mathrm{Cnv}`(R \upharpoonright \beta) = \beta \upharpoonleft \breve{R}$ [Proof similar to that of *35·51]

***35·53.** $\vdash . \mathrm{Cnv}`(\alpha \upharpoonleft R \upharpoonright \beta) = \beta \upharpoonleft \breve{R} \upharpoonright \alpha$ [Proof similar to that of *35·51]

***35·61.** $\vdash . \mathrm{D}`(\alpha \upharpoonleft R) = \alpha \cap \mathrm{D}`R$

Dem.

$\vdash . *33·13 . \supset \vdash :. x \in \mathrm{D}`(\alpha \upharpoonleft R) . \equiv : (\exists y) . x (\alpha \upharpoonleft R) y :$

[*35·1] $\quad \equiv : (\exists y) . x \in \alpha . xRy :$

[*10·35] $\quad \equiv : x \in \alpha : (\exists y) . xRy :$

[*33·13] $\quad \equiv : x \in \alpha . x \in \mathrm{D}`R :$

[*22·33] $\quad \equiv : x \in (\alpha \cap \mathrm{D}`R) :. \supset \vdash . \mathrm{Prop}$

***35·62.** $\vdash : \alpha \subset \mathrm{D}`R . \supset . \mathrm{D}`(\alpha \upharpoonleft R) = \alpha$ [*35·61 . *22·621]

***35·63.** $\vdash : \mathrm{D}`R \subset \alpha . \equiv . \alpha \upharpoonleft R = R$

Dem.

$\vdash . *35·61 . \supset \vdash : \alpha \upharpoonleft R = R . \supset . \alpha \cap \mathrm{D}`R = \mathrm{D}`R .$

[*22·621] $\quad \supset . \mathrm{D}`R \subset \alpha$ (1)

$\vdash . (1) . *35·451 . \supset \vdash . \mathrm{Prop}$

***35·64.** $\vdash . \text{Ɑ}`(R \upharpoonright \beta) = \beta \cap \text{Ɑ}`R$ [Proof as in *35·61]

***35·641.** $\vdash : \alpha \cap \mathrm{D}`R = \Lambda . \supset . \alpha \upharpoonleft R = \dot{\Lambda}$ [*35·61 . *33·241]

***35·642.** $\vdash : \alpha \cap \text{Ɑ}`R = \Lambda . \supset . R \upharpoonright \alpha = \dot{\Lambda}$ [*35·64 . *33·241]

***35·643.** $\vdash : \alpha \cap \mathrm{D}`R = \Lambda . \supset . \alpha \upharpoonleft (R \mathbin{\dot{\cup}} S) = \alpha \upharpoonleft S$ [*35·641·42]

∗35·644. $\vdash : \alpha \cap \text{Œ}`R = \Lambda . \supset . (R \cup S) \upharpoonright \alpha = S \upharpoonright \alpha$ [∗35·642·421]

∗35·65. $\vdash : \beta \subset \text{Œ}`R . \supset . \text{Œ}`(R \upharpoonright \beta) = \beta$ [∗35·64 . ∗22·621]

∗35·66. $\vdash : \text{Œ}`R \subset \beta . \equiv . R \upharpoonright \beta = R$ [Proof as in ∗35·63]

∗35·671. $\vdash . \text{D}`(R | S) = \text{D}`(R \upharpoonright \text{D}`S)$

Dem.

$\vdash . ∗33·13 . \supset \vdash :. x \in \text{D}`(R | S) . \equiv : (\exists y) . x (R | S) y :$

[∗34·1] $\equiv : (\exists y, z) . xRz . zSy :$

[∗11·23] $\equiv : (\exists z, y) . xRz . zSy :$

[∗10·35] $\equiv : (\exists z) : xRz : (\exists y) . zSy :$

[∗33·13] $\equiv : (\exists z) . xRz . z \in \text{D}`S :$

[∗35·101] $\equiv : (\exists z) . x (R \upharpoonright \text{D}`S) z :$

[∗33·13] $\equiv : x \in \text{D}`(R \upharpoonright \text{D}`S) :. \supset \vdash . \text{Prop}$

∗35·672. $\vdash . \text{Œ}`(R | S) = \text{Œ}`(\text{Œ}`R \upharpoonleft S)$ [Similar proof]

∗35·68. $\vdash : \alpha \cap \beta = \Lambda . \supset . (\alpha \upharpoonleft R \upharpoonright \beta)^2 = \dot{\Lambda}$

Dem.

$\vdash . ∗35·61·64·21 . \supset \vdash . \text{D}`(\alpha \upharpoonleft R \upharpoonright \beta) \subset \alpha . \text{Œ}`(\alpha \upharpoonleft R \upharpoonright \beta) \subset \beta .$

[∗22·49.∗24·13] $\supset \vdash : \alpha \cap \beta = \Lambda . \supset . \text{D}`(\alpha \upharpoonleft R \upharpoonright \beta) \cap \text{Œ}`(\alpha \upharpoonleft R \upharpoonright \beta) = \Lambda .$

[∗34·531] $\supset . (\alpha \upharpoonleft R \upharpoonright \beta)^2 = \dot{\Lambda} : \supset \vdash . \text{Prop}$

∗35·7. $\vdash : \phi \{(R \upharpoonright \beta)`y\} . \equiv . y \in \beta . \phi (R`y)$

This proposition is very often used in the later parts of the work.

Dem.

$\vdash . ∗14·21 . \quad \supset \vdash : \phi \{(R \upharpoonright \beta)`y\} . \supset . \text{E} ! (R \upharpoonright \beta)`y .$

[∗33·43] $\supset . y \in \text{Œ}`(R \upharpoonright \beta) .$

[∗35·64] $\supset . y \in \beta$ (1)

$\vdash . (1) . ∗4·71 . \supset \vdash : \phi \{(R \upharpoonright \beta)`y\} . \equiv . y \in \beta . \phi \{(R \upharpoonright \beta)`y\}$ (2)

$\vdash . ∗4·73 . ∗35·101 . \supset \vdash :. y \in \beta . \quad \supset : x (R \upharpoonright \beta) y . \equiv_x . xRy :$

[∗14·272] $\supset : \phi \{(R \upharpoonright \beta)`y\} . \equiv . \phi (R`y)$ (3)

$\vdash . (3) . ∗5·32 . \supset \vdash : y \in \beta . \phi \{(R \upharpoonright \beta)`y\} . \equiv . y \in \beta . \phi (R`y)$ (4)

$\vdash . (2) . (4) . \quad \supset \vdash . \text{Prop}$

∗35·71. $\vdash :. y \in \beta . \supset_y . R`y = S`y : \supset . R \upharpoonright \beta = S \upharpoonright \beta$

Dem.

$\vdash . ∗4·7 . \supset \vdash :. \text{Hp} . \supset : y \in \beta . \supset_y . y \in \beta . R`y = S`y :$

[∗35·7] $\supset : y \in \beta . \supset_y . (R \upharpoonright \beta)`y = (S \upharpoonright \beta)`y :$

[∗35·64] $\supset : y \in \text{Œ}`(R \upharpoonright \beta) \cup \text{Œ}`(S \upharpoonright \beta) . \supset_y . (R \upharpoonright \beta)`y = (S \upharpoonright \beta)`y :$

[∗33·45] $\supset : R \upharpoonright \beta = S \upharpoonright \beta :. \supset \vdash . \text{Prop}$

∗35·75. $\vdash . \Lambda \upharpoonleft R = R \upharpoonright \Lambda = \Lambda \upharpoonleft R \upharpoonright \beta = \alpha \upharpoonleft R \upharpoonright \Lambda = \dot{\Lambda}$

Dem.

$\vdash . ∗35·61 . \quad \supset \vdash . \text{D}`(\Lambda \upharpoonleft R) = \Lambda .$

[∗33·241] $\supset \vdash . \Lambda \upharpoonleft R = \dot{\Lambda}$ (1)

$\vdash . *35\cdot64 . \quad \supset \vdash . \text{Ɑ}'(R \upharpoonright \Lambda) = \Lambda .$
$[*33\cdot241] \quad \supset \vdash . R \upharpoonright \Lambda = \dot{\Lambda} \quad (2)$
$\vdash . *35\cdot441\cdot21 . \supset \vdash . \Lambda \upharpoonleft R \upharpoonright \beta \mathrel{\dot{\subset}} \Lambda \upharpoonleft R .$
$[(1) . *25\cdot13] \quad \supset \vdash . \Lambda \upharpoonleft R \upharpoonright \beta = \dot{\Lambda} \quad (3)$
$\vdash . *35\cdot44\cdot21 . \supset \vdash . \alpha \upharpoonleft R \upharpoonright \Lambda \mathrel{\dot{\subset}} R \upharpoonright \Lambda .$
$[(2) . *25\cdot13] \quad \supset \vdash . \alpha \upharpoonleft R \upharpoonright \Lambda = \dot{\Lambda} \quad (4)$
$\vdash . (1) . (2) . (3) . (4) . \supset \vdash . \text{Prop}$

$*35\cdot76$. $\vdash . \mathrm{V} \upharpoonleft R = R \upharpoonright \mathrm{V} = \mathrm{V} \upharpoonleft R \upharpoonright \mathrm{V} = R$

Dem.

$\vdash . *35\cdot1 . \quad \supset \vdash : x(\mathrm{V} \upharpoonleft R) y . \quad \equiv . x \epsilon \mathrm{V} . xRy .$
$[*24\cdot104 . *4\cdot73] \quad \equiv . xRy \quad (1)$
$\vdash . *35\cdot101 . \supset \vdash : x(R \upharpoonright \mathrm{V}) y . \quad \equiv . xRy . y \epsilon \mathrm{V} .$
$[*24\cdot104 . *4\cdot73] \quad \equiv . xRy \quad (2)$
$\vdash . *35\cdot102 . \supset \vdash : x(\mathrm{V} \upharpoonleft R \upharpoonright \mathrm{V}) y . \equiv . x \epsilon \mathrm{V} . xRy . y \epsilon \mathrm{V} .$
$[*24\cdot104 . *4\cdot73] \quad \equiv . xRy \quad (3)$
$\vdash . (1) . (2) . (3) . \supset \vdash . \text{Prop}$

The rest of this number, down to $*35\cdot93$ exclusive, is concerned with $\alpha \uparrow \beta$, except $*35\cdot81\cdot812$.

$*35\cdot81$. $\vdash : x(\alpha \upharpoonleft \dot{\mathrm{V}}) y . \equiv . x \epsilon \alpha \quad [*35\cdot1 . *25\cdot104]$

$*35\cdot812$. $\vdash : x(\dot{\mathrm{V}} \upharpoonright \beta) y . \equiv . y \epsilon \beta \quad [*35\cdot101 . *25\cdot104]$

$*35\cdot82$. $\vdash . \alpha \uparrow \beta = \alpha \upharpoonleft \dot{\mathrm{V}} \upharpoonright \beta$

Dem.

$\vdash . *35\cdot103 . \supset \vdash : x(\alpha \uparrow \beta) y . \equiv . x \epsilon \alpha . y \epsilon \beta .$
$[*25\cdot104] \quad \equiv . x \epsilon \alpha . x \dot{\mathrm{V}} y . y \epsilon \beta .$
$[*35\cdot102] \quad \equiv . x(\alpha \upharpoonleft \dot{\mathrm{V}} \upharpoonright \beta) y : \supset \vdash . \text{Prop}$

$*35\cdot822$. $\vdash . \alpha \upharpoonleft R \upharpoonright \beta = R \mathbin{\dot{\cap}} (\alpha \uparrow \beta)$

Dem.

$\vdash . *35\cdot102 . \supset \vdash : x(\alpha \upharpoonleft R \upharpoonright \beta) y . \equiv . x \epsilon \alpha . xRy . y \epsilon \beta .$
$[*4\cdot3] \quad \equiv . xRy . x \epsilon \alpha . y \epsilon \beta .$
$[*35\cdot103] \quad \equiv . xRy . x(\alpha \uparrow \beta) y .$
$[*23\cdot33] \quad \equiv . x \{R \mathbin{\dot{\cap}} (\alpha \uparrow \beta)\} y : \supset \vdash . \text{Prop}$

$*35\cdot83$. $\vdash : \mathrm{D}'R \subset \alpha . \text{Ɑ}'R \subset \beta . \equiv . R \mathrel{\dot{\subset}} \alpha \uparrow \beta$

Dem.

$\vdash . *33\cdot14 . \quad \supset \vdash :. xRy . \supset : x \epsilon \mathrm{D}'R . y \epsilon \text{Ɑ}'R :$
$[*22\cdot46] \quad \supset : \mathrm{D}'R \subset \alpha . \text{Ɑ}'R \subset \beta . \supset . x \epsilon \alpha . y \epsilon \beta \quad (1)$
$\vdash . (1) . \text{Comm} . \supset \vdash :. \mathrm{D}'R \subset \alpha . \text{Ɑ}'R \subset \beta . \supset : xRy . \supset . x \epsilon \alpha . y \epsilon \beta .$
$[*35\cdot103] \quad \supset . x(\alpha \uparrow \beta) y \quad (2)$
$\vdash . *35\cdot103 . \quad \supset \vdash :. R \mathrel{\dot{\subset}} \alpha \uparrow \beta . \supset : xRy . \supset_{x,y} . x \epsilon \alpha . y \epsilon \beta :$
$[*33\cdot35\cdot351] \quad \supset : \mathrm{D}'R \subset \alpha . \text{Ɑ}'R \subset \beta \quad (3)$
$\vdash . (2) . (3) . \quad \supset \vdash . \text{Prop}$

***35·831.** $\vdash . \dot{-}(\alpha \uparrow \beta) = (-\alpha \uparrow \beta) \cup (\alpha \uparrow -\beta) \cup (-\alpha \uparrow -\beta)$

Dem.

$\vdash . *23·35 . \supset \vdash :: x \{\dot{-}(\alpha \uparrow \beta)\} y . \equiv :. \sim \{x (\alpha \uparrow \beta) y\} :.$

[*35·103] $\equiv :. \sim (x \epsilon \alpha . y \epsilon \beta) :.$

[*4·51] $\equiv :. x \sim \epsilon \alpha . \mathsf{v} . y \sim \epsilon \beta :.$

[*4·42] $\equiv :. x \sim \epsilon \alpha : y \epsilon \beta . \mathsf{v} . y \sim \epsilon \beta :. \mathsf{v} :. x \epsilon \alpha . \mathsf{v} . x \sim \epsilon \alpha : y \sim \epsilon \beta :.$

[*4·4] $\equiv :. x \sim \epsilon \alpha . y \epsilon \beta . \mathsf{v} . x \sim \epsilon \alpha . y \sim \epsilon \beta . \mathsf{v} . x \epsilon \alpha . y \sim \epsilon \beta . \mathsf{v} . x \sim \epsilon \alpha . y \sim \epsilon \beta :.$

[*4·25·31·37]

$\equiv :. x \sim \epsilon \alpha . y \epsilon \beta . \mathsf{v} . x \epsilon \alpha . y \sim \epsilon \beta . \mathsf{v} . x \sim \epsilon \alpha . y \sim \epsilon \beta :.$

[*22·35] $\equiv :. x \epsilon - \alpha . y \epsilon \beta . \mathsf{v} . x \epsilon \alpha . y \epsilon - \beta . \mathsf{v} . x \epsilon - \alpha . y \epsilon - \beta :.$

[*35·103] $\equiv :. x (-\alpha \uparrow \beta) y . \mathsf{v} . x (\alpha \uparrow -\beta) y . \mathsf{v} . x (-\alpha \uparrow -\beta) y :.$

[*23·34] $\equiv :. x \{(-\alpha \uparrow \beta) \cup (\alpha \uparrow -\beta) \cup (-\alpha \uparrow -\beta)\} y :: \supset \vdash .$ Prop

***35·832.** $\vdash . \dot{-}(\alpha \upharpoonleft R \upharpoonright \beta) = (-\alpha \uparrow \beta) \cup (\alpha \uparrow -\beta) \cup (-\alpha \uparrow -\beta) \cup \dot{-} R$

[*35·822·831 . Transp . *23·84]

***35·834.** $\vdash . (\alpha \uparrow \beta) \dot{\cap} (\gamma \uparrow \delta) = (\alpha \cap \gamma) \uparrow (\beta \cap \delta)$

Dem.

$\vdash . *35·103 . \supset$

$\vdash : x \{(\alpha \uparrow \beta) \dot{\cap} (\gamma \uparrow \delta)\} y . \equiv . x \epsilon \alpha . y \epsilon \beta . x \epsilon \gamma . y \epsilon \delta .$

[*22·33.*35·103] $\equiv . x \{(\alpha \cap \gamma) \uparrow (\beta \cap \delta)\} y : \supset \vdash .$ Prop

***35·84.** $\vdash . \text{Cnv}'(\alpha \uparrow \beta) = \beta \uparrow \alpha$ [*35·103 . *31·131]

***35·85.** $\vdash : \exists ! \beta . \supset . \text{D}'(\alpha \uparrow \beta) = \alpha$

Dem.

$\vdash . *35·103 . *10·281 . \supset$

$\vdash :. (\exists y) . x (\alpha \uparrow \beta) y . \equiv : (\exists y) . x \epsilon \alpha . y \epsilon \beta :$

[*10·35] $\equiv : x \epsilon \alpha : (\exists y) . y \epsilon \beta :$

[*24·5] $\equiv : x \epsilon \alpha . \exists ! \beta$ (1)

$\vdash . (1) . *33·13 . *10·35 . \supset \vdash .$ Prop

***35·86.** $\vdash : \exists ! \alpha . \supset . \text{ᗡ}'(\alpha \uparrow \beta) = \beta$ [Similar proof]

***35·87.** $\vdash : \dot{\exists} ! (\alpha \uparrow \beta) . \equiv . \exists ! \alpha . \exists ! \beta$

Dem.

$\vdash . *35·103 . \supset \vdash :. \dot{\exists} ! (\alpha \uparrow \beta) . \equiv : (\exists x, y) . x \epsilon \alpha . y \epsilon \beta :$

[*11·54] $\equiv : (\exists x) . x \epsilon \alpha : (\exists y) . y \epsilon \beta :$

[*24·5] $\equiv : \exists ! \alpha . \exists ! \beta :. \supset \vdash .$ Prop

***35·88.** $\vdash :. \alpha \uparrow \beta = \dot{\Lambda} . \equiv : \alpha = \Lambda . \mathsf{v} . \beta = \Lambda$

[*35·87 . Transp . *24·51 . *25·51]

***35·881.** $\vdash : \text{ᗡ}'R \subset \alpha . \supset . R | (\alpha \uparrow \beta) = \text{D}'R \uparrow \beta$

Dem.

$\vdash . *34·1 . *35·103 . \supset$

$\vdash : x \{R | (\alpha \uparrow \beta)\} y . \equiv . (\exists z) . x R z . z \epsilon \alpha . y \epsilon \beta$ (1)

$\vdash . *33 \cdot 14 . \supset \vdash :. \text{G}`R \subset \alpha . \supset : xRz . \supset . z \epsilon \alpha :$

[*4·73] $\supset : xRz . \equiv . xRz . z \epsilon \alpha$ (2)

$\vdash . (1) . (2) . \supset \vdash :: \text{Hp} . \supset :. x \{R | (\alpha \uparrow \beta)\} y . \equiv : (\exists z) . xRz . y \epsilon \beta :$

[*10·35] $\equiv : (\exists z) . xRz : y \epsilon \beta :$

[*33·13] $\equiv : x \epsilon \text{D}`R . y \epsilon \beta :$

[*35·103] $\equiv : x (\text{D}`R \uparrow \beta) y :: \supset \vdash . \text{Prop}$

***35·882.** $\vdash : \text{D}`R \subset \beta . \supset . (\alpha \uparrow \beta) | R = \alpha \uparrow \text{G}`R$ [Similar proof]

***35·89.** $\vdash : \exists ! \beta . \supset . (\alpha \uparrow \beta) | (\beta \uparrow \gamma) = (\alpha \uparrow \gamma) : \sim \exists ! \beta . \supset . (\alpha \uparrow \beta) | (\beta \uparrow \gamma) = \dot{\Lambda}$

Dem.

$\vdash . *34 \cdot 1 . \supset \vdash :. x \{(\alpha \uparrow \beta) | (\beta \uparrow \gamma)\} z .$

$\equiv : (\exists y) . x (\alpha \uparrow \beta) y . y (\beta \uparrow \gamma) z :$

[*35·103] $\equiv : (\exists y) . x \epsilon \alpha . y \epsilon \beta . y \epsilon \beta . z \epsilon \gamma :$

[*4·24] $\equiv : (\exists y) . x \epsilon \alpha . y \epsilon \beta . z \epsilon \gamma :$

[*10·35] $\equiv : \exists ! \beta : x \epsilon \alpha . z \epsilon \gamma :$

[*35·103] $\equiv : \exists ! \beta : x (\alpha \uparrow \gamma) z$ (1)

$\vdash . (1) . \supset \vdash :: \exists ! \beta . \supset : x \{(\alpha \uparrow \beta) | (\beta \uparrow \gamma)\} z . \equiv . x (\alpha \uparrow \gamma) z :.$

$\sim (\exists ! \beta) . \supset : \sim [x \{(\alpha \uparrow \beta) | (\beta \uparrow \gamma)\} z] :: \supset \vdash . \text{Prop}$

***35·891.** $\vdash :. \exists ! \beta . \vee . \sim \exists ! \alpha : \supset . (\alpha \uparrow \beta) | (\beta \uparrow \alpha) = (\alpha \uparrow \alpha)$

Dem.

$\vdash . *35 \cdot 88 . \supset \vdash : \sim \exists ! \alpha . \supset . \alpha \uparrow \alpha = \dot{\Lambda} . \alpha \uparrow \beta = \dot{\Lambda} .$

[*34·32] $\supset . \alpha \uparrow \alpha = \dot{\Lambda} . (\alpha \uparrow \beta) | (\beta \uparrow \alpha) = \dot{\Lambda} .$

[*21·24] $\supset . (\alpha \uparrow \alpha) = (\alpha \uparrow \beta) | (\beta \uparrow \alpha)$ (1)

$\vdash . (1) . *35 \cdot 89 . \supset \vdash . \text{Prop}$

***35·892.** $\vdash : (\alpha \uparrow \alpha)^2 = (\alpha \uparrow \alpha)$ $\left[*35 \cdot 891 \frac{\alpha}{\beta} \right]$

***35·895.** $\vdash : \alpha \cap \beta = \Lambda . \supset . (\alpha \uparrow \beta)^2 = \dot{\Lambda}$ [*35·68·82]

***35·9.** $\vdash . \text{D}`(\alpha \uparrow \alpha) = \text{G}`(\alpha \uparrow \alpha) = C`(\alpha \uparrow \alpha) = \alpha$

Dem.

$\vdash . *35 \cdot 85 \cdot 86 . \supset \vdash : \exists ! \alpha . \supset . \text{D}`(\alpha \uparrow \alpha) = \alpha . \text{G}`(\alpha \uparrow \alpha) = \alpha$ (1)

$\vdash . *35 \cdot 88 . \supset \vdash : \sim \exists ! \alpha . \supset . \sim \dot{\exists} ! (\alpha \uparrow \alpha) .$

[*33·29] $\supset . \text{D}`(\alpha \uparrow \alpha) = \Lambda . \text{G}`(\alpha \uparrow \alpha) = \Lambda .$

[*24·51] $\supset . \text{D}`(\alpha \uparrow \alpha) = \alpha . \text{G}`(\alpha \uparrow \alpha) = \alpha$ (2)

$\vdash . (1) . (2) . *4 \cdot 83 . \supset \vdash . \text{D}`(\alpha \uparrow \alpha) = \text{G}`(\alpha \uparrow \alpha) = \alpha . \supset \vdash . \text{Prop}$

***35·91.** $\vdash : R \mathrel{\dot{\subset}} \alpha \uparrow \alpha . \equiv . C`R \subset \alpha$

Dem.

$\vdash . *35 \cdot 103 . \supset \vdash :. R \mathrel{\dot{\subset}} \alpha \uparrow \alpha . \equiv : xRy . \supset_{x,y} . x, y \epsilon \alpha :$

[*33·352] $\equiv : C`R \subset \alpha :. \supset \vdash . \text{Prop}$

***35·92.** $\vdash :. (\exists \alpha) . P = \alpha \uparrow \alpha . \supset : R \mathrel{\dot{\subset}} P . \equiv . C`R \subset C`P$ [*35·9·91]

***35·93.** $\vdash : (R) . \phi(\text{D}'R) . \equiv . (\alpha) . \phi\alpha$

Dem.

$\vdash . *33·12 . *14·18 . \supset \vdash : (\alpha) . \phi\alpha . \supset . \phi(\text{D}'R) :$

[*10·11·21] $\supset \vdash : (\alpha) . \phi\alpha . \supset . (R) . \phi(\text{D}'R)$ (1)

$\vdash . *10·1 .$ $\supset \vdash : (R) . \phi(\text{D}'R) . \supset . \phi\{\text{D}'(\alpha \uparrow \alpha)\} .$

[*35·9] $\supset . \phi\alpha :$

[*10·11·21] $\supset \vdash : (R) . \phi(\text{D}'R) . \supset . (\alpha) . \phi\alpha$ (2)

$\vdash . (1) . (2) .$ $\supset \vdash .$ Prop

***35·931.** $\vdash : (R) . \phi(\text{Ɑ}'R) . \equiv . (\alpha) . \phi\alpha$ [Proof as in *35·93]

***35·932.** $\vdash : (R) . \phi(C'R) . \equiv . (\alpha) . \phi\alpha$ [Proof as in *35·93]

***35·94.** $\vdash : (\exists R) . \phi(\text{D}'R) . \equiv . (\exists\alpha) . \phi\alpha$ [*35·93 . Transp]

***35·941.** $\vdash : (\exists R) . \phi(\text{Ɑ}'R) . \equiv . (\exists\alpha) . \phi\alpha$ [*35·931 . Transp]

***35·942.** $\vdash : (\exists R) . \phi(C'R) . \equiv . (\exists\alpha) . \phi\alpha$ [*35·932 . Transp]

$*$36. RELATIONS WITH LIMITED FIELDS

Summary of $*36$.

In this number we are concerned with the special case in which the same limitation is imposed upon the domain and the converse domain of a relation. In this case, the same result is achieved by imposing the limitation on the field. It is convenient to be able to regard $\alpha \rceil P \lceil \alpha$ as a descriptive function of α or of P, which we secure by the notation $P \mathbin{\lceil\!\!\lfloor} \alpha$, whence, as will be explained in $*38$, $P \mathbin{\lceil\!\!\lfloor}\text{‘}\alpha$ and $\mathbin{\lceil\!\!\lfloor} \alpha\text{‘}P$ will both mean $P \mathbin{\lceil\!\!\lfloor} \alpha$. If P is a serial relation, and $\alpha \subset C\text{‘}P$, "$P \mathbin{\lceil\!\!\lfloor} \alpha$" will stand for "the terms of α arranged in the order determined by P," or, as we may call it briefly, "α in the P-order." $P \mathbin{\lceil\!\!\lfloor} \alpha$ is defined as follows:

$*36{\cdot}01$. $P \mathbin{\lceil\!\!\lfloor} \alpha = \alpha \rceil P \lceil \alpha$ Df

We thus have

$*36{\cdot}13$. $\vdash : x (P \mathbin{\lceil\!\!\lfloor} \alpha) y \,.\equiv .\, x, y \in \alpha \,.\, xPy$

Most of the propositions concerning $P \mathbin{\lceil\!\!\lfloor} \alpha$ demand that P should have some at least of the characteristics of a *serial* relation. Hence the propositions concerning $P \mathbin{\lceil\!\!\lfloor} \alpha$ which can be given in the present number are, for the most part, not the most useful propositions concerning $P \mathbin{\lceil\!\!\lfloor} \alpha$. The most useful propositions in the present number are the following:

$*36{\cdot}25$. $\vdash : C\text{‘}P \subset \alpha \,.\equiv .\, P \mathbin{\lceil\!\!\lfloor} \alpha = P$

$*36{\cdot}29$. $\vdash .\, P \mathbin{\lceil\!\!\lfloor} \alpha = P \mathbin{\dot{\cap}} \alpha \uparrow \alpha$

$*36{\cdot}3$. $\vdash .\, P \mathbin{\lceil\!\!\lfloor} \alpha = P \mathbin{\lceil\!\!\lfloor} (\alpha \cap C\text{‘}P)$

$*36{\cdot}33$. $\vdash .\, P \mathbin{\lceil\!\!\lfloor} C\text{‘}P = P$

$*36{\cdot}01$. $P \mathbin{\lceil\!\!\lfloor} \alpha = \alpha \rceil P \lceil \alpha$ Df

$*36{\cdot}11$. $\vdash .\, P \mathbin{\lceil\!\!\lfloor} \alpha = \alpha \rceil P \lceil \alpha$ [($*36{\cdot}01$)]

$*36{\cdot}13$. $\vdash : x (P \mathbin{\lceil\!\!\lfloor} \alpha) y \,.\equiv .\, x, y \in \alpha \,.\, xPy$ [$*36{\cdot}11 \,.\, *35{\cdot}102$]

The following propositions are obtained from those of $*35$ by means of $*36{\cdot}11$, which, as it is used in each case, is not referred to again.

$*36{\cdot}2$. $\vdash .\, P \mathbin{\lceil\!\!\lfloor} \alpha \mathbin{\dot{\cap}} Q \mathbin{\lceil\!\!\lfloor} \beta = (P \mathbin{\dot{\cap}} Q) \mathbin{\lceil\!\!\lfloor} (\alpha \cap \beta)$ [$*35{\cdot}15$]

$*36{\cdot}201$. $\vdash .\, P \mathbin{\lceil\!\!\lfloor} \alpha \mathbin{\dot{\cap}} P \mathbin{\lceil\!\!\lfloor} \beta = P \mathbin{\lceil\!\!\lfloor} (\alpha \cap \beta)$ [$*36{\cdot}2$]

$*36{\cdot}202$. $\vdash .\, P \mathbin{\lceil\!\!\lfloor} \alpha \mathbin{\dot{\cap}} Q \mathbin{\lceil\!\!\lfloor} \alpha = (P \mathbin{\dot{\cap}} Q) \mathbin{\lceil\!\!\lfloor} \alpha$ [$*36{\cdot}2$]

$*36{\cdot}203$. $\vdash .\, P \mathbin{\lceil\!\!\lfloor} \alpha \mathbin{\dot{\cap}} Q = (P \mathbin{\dot{\cap}} Q) \mathbin{\lceil\!\!\lfloor} \alpha$ [$*35{\cdot}18$]

$*36{\cdot}21$. $\vdash .\, (P \mathbin{\lceil\!\!\lfloor} \alpha) \mathbin{\lceil\!\!\lfloor} \beta = P \mathbin{\lceil\!\!\lfloor} (\alpha \cap \beta)$ [$*35{\cdot}33{\cdot}34$]

***36·22.** $\vdash . (P \upharpoonright \alpha) | (Q \upharpoonright \alpha) \mathrel{\dot\subset} (P | Q) \upharpoonright \alpha$

Dem.

$\vdash . *36·13 . *34·1 . \supset \vdash : x \{(P \upharpoonright \alpha) | (Q \upharpoonright \alpha)\} z . \equiv . (\exists y) . x, y, z \in \alpha . xPy . yQz .$

[*10·5] $\supset . (\exists y) . x, z \in \alpha . xPy . yQz$ (1)

$\vdash . (1) . *10·35 . *34·1 . \supset \vdash .$ Prop

***36·23.** $\vdash . (P \mathrel{\dot\cup} Q) \upharpoonright \alpha = P \upharpoonright \alpha \mathrel{\dot\cup} Q \upharpoonright \alpha$ [*35·422]

***36·24.** $\vdash : \alpha \subset \beta . \supset . P \upharpoonright \alpha \mathrel{\dot\subset} P \upharpoonright \beta$ [*35·432]

***36·241.** $\vdash : P \mathrel{\dot\subset} Q . \supset . P \upharpoonright \alpha \mathrel{\dot\subset} Q \upharpoonright \alpha$ [*35·462]

***36·25.** $\vdash : C\text{‘}P \subset \alpha . \equiv . P \upharpoonright \alpha = P$

Dem.

$\vdash . *36·13 . *4·7 . \supset \vdash :. P \upharpoonright \alpha = P . \equiv : xPy . \supset_{x,y} . x, y \in \alpha :$

[*33·352] $\equiv : C\text{‘}P \subset \alpha :. \supset \vdash .$ Prop

***36·26.** $\vdash : C\text{‘}P \cap \alpha = \Lambda . \supset . P | (Q \upharpoonright \alpha) = \dot\Lambda . (Q \upharpoonright \alpha) | P = \dot\Lambda$ [*35·473·474]

***36·27.** $\vdash : P \upharpoonright \Lambda = \dot\Lambda$ [*35·75]

***36·28.** $\vdash . P \upharpoonright V = P$ [*35·76]

***36·29.** $\vdash . P \upharpoonright \alpha = P \mathrel{\dot\cap} \alpha \uparrow \alpha$ [*35·822]

***36·3.** $\vdash . P \upharpoonright \alpha = P \upharpoonright (\alpha \cap C\text{‘}P)$

Dem.

$\vdash . *33·17 . *4·71 . \supset \vdash : xPy . \equiv . x, y \in C\text{‘}P . xPy :$

[Fact] $\supset \vdash : x, y \in \alpha . xPy . \equiv . x, y \in \alpha . x, y \in C\text{‘}P . xPy .$

[*22·33] $\equiv . x, y \in \alpha \cap C\text{‘}P . xPy .$

[*36·13] $\equiv . x \{P \upharpoonright (\alpha \cap C\text{‘}P)\} y$ (1)

$\vdash . (1) . *36·13 . \supset \vdash .$ Prop

***36·31.** $\vdash : \alpha \cap C\text{‘}P = \Lambda . \supset . P \upharpoonright \alpha = \dot\Lambda$ [*36·3·27]

***36·32.** $\vdash : \alpha \cap C\text{‘}P = \beta \cap C\text{‘}P . \supset . P \upharpoonright \alpha = P \upharpoonright \beta$ [*36·3]

***36·33.** $\vdash . P \upharpoonright C\text{‘}P = P$ [*36·25]

***36·34.** $\vdash . \text{Cnv‘}P \upharpoonright \alpha = (\breve{P}) \upharpoonright \alpha$ [*35·53]

***36·35.** $\vdash . (P \upharpoonright \alpha)^2 \mathrel{\dot\subset} (P^2) \upharpoonright \alpha$ [*36·22]

***36·4.** $\vdash :. \alpha \cap D\text{‘}R = \Lambda . \lor . \alpha \cap \text{ᗡ‘}R = \Lambda : \supset . (R \mathrel{\dot\cup} S) \upharpoonright \alpha = S \upharpoonright \alpha$

Dem.

$\vdash . *35·643 . \supset \vdash : \alpha \cap D\text{‘}R = \Lambda . \supset . \alpha \upharpoonleft (R \mathrel{\dot\cup} S) = \alpha \upharpoonleft S .$

[*35·21] $\supset . (R \mathrel{\dot\cup} S) \upharpoonright \alpha = S \upharpoonright \alpha$ (1)

Similarly $\vdash : \alpha \cap \text{ᗡ‘}R = \Lambda . \supset . (R \mathrel{\dot\cup} S) \upharpoonright \alpha = S \upharpoonright \alpha$ (2)

$\vdash . (1) . (2) . \supset \vdash .$ Prop

*37. PLURAL DESCRIPTIVE FUNCTIONS

Summary of *37.

In this number, we introduce what may be regarded as the plural of $R'y$. "$R'y$" was defined to mean "the term which has the relation R to y." We now introduce the notation "$R''\beta$" to mean "the terms which have the relation R to members of β." Thus if β is the class of great men, and R is the relation of wife to husband, $R''\beta$ will mean "wives of great men." If β is the class of fractions of the form $1 - 1/2^n$ for integral values of n, and R is the relation "less than," $R''\beta$ will be the class of fractions each of which is less than some member of this class of fractions, *i.e.* $R''\beta$ will be the class of proper fractions. Generally, $R''\beta$ is the class of those referents which have relata that are members of β.

We require also a notation for the relation of $R''\beta$ to β. This relation we will call R_ϵ. Thus R_ϵ is the relation which holds between two classes α and β when α consists of all terms which have the relation R to some member of β.

A specially important case arises when $R'y$ always exists if $y \in \beta$. In this case, $R''\beta$ is the class of all terms of the form $R'y$ when $y \in \beta$. We will denote the hypothesis that $R'y$ always exists if $y \in \beta$ by the notation $\mathrm{E}\,!!\, R''\beta$, meaning "the R's of β's exist."

The definitions are as follows:

*37·01. $R''\beta = \hat{x}\{(\exists y) . y \in \beta . xRy\}$ Df

*37·02. $R_\epsilon = \hat{\alpha}\hat{\beta}(\alpha = R''\beta)$ Df

*37·03. $\breve{R}_\epsilon = \mathrm{Cnv}'(R_\epsilon)$ Df

This definition serves merely for the avoidance of brackets. Without it, "$\breve{R}_\epsilon$" would be ambiguous as between $(\breve{R})_\epsilon$ and $\mathrm{Cnv}'(R_\epsilon)$, which are not equal. In all cases in which a suffix occurs, we shall adopt the same convention, *i.e.* we shall always put

$$\breve{R}_{\text{suffix}} = \mathrm{Cnv}'(R_{\text{suffix}}).$$

*37·04. $R'''\kappa = R_\epsilon''\kappa$ Df

Thus $R'''\kappa$ consists of all classes which have the relation R_ϵ to some member of κ. $R'''\kappa$ is only significant when κ is a class of classes relatively to members of the converse domain of R; in this case, $R'''\kappa$ is a class of classes relatively to members of the domain of R.

*37·05. $\mathrm{E}\,!!\, R''\beta . = : y \in \beta . \supset_y . \mathrm{E}\,!\, R'y$ Df

Here the symbol "$\mathrm{E}\,!!\, R''\beta$" must be treated as a whole, *i.e.* we must not regard it as making an assertion about $R''\beta$. If $R''\beta = \alpha$, we must not suppose that we shall be able to put "E !! α," which would be nonsense, just as "E ! x" is nonsense even when $x = R'y$ and E ! $R'y$.

The notation $R``\alpha$, introduced in the present number, is extremely useful, and embodies a very important idea. Its use is somewhat different according to the kind of relation concerned. Consider first the kind of relation which leads to a descriptive function, say D. If $\lambda$ is a class of relations, $D``\lambda$ is the class of the domains of these relations. In this case, $D``\lambda$ is a class each of whose members is of the form $D'R$, where $R \epsilon \lambda$. Again, let us denote by "$\times n$" the relation of $m$ to $m \times n$; then if we denote by "$NC$" the class of cardinal numbers, $\times n``NC$ will denote all numbers that result from multiplying a cardinal number by n, *i.e.* all multiples of n. Thus *e.g.* $\times 2``NC$ will be the class of even numbers. If R is a correlation between two classes α and β, *i.e.* a relation such that, if $y \epsilon \beta$, $R'y$ exists and is a member of α, while conversely, if $x \epsilon \alpha$, $\breve{R}'x$ exists and is a member of β, then $\alpha = R``\beta$, and we may regard R as a transformation applied to each member of β and giving rise to a member of α. It is by means of such transformations that two classes are shown to be *similar*, *i.e.* to have the same (cardinal) number of terms.

In the case of serial relations, the utility of the notation $R``\beta$ is somewhat different. Suppose, for example, that $R$ is the relation of less to greater among real numbers. Then if $\beta$ is any class of real numbers, $R``\beta$ will be the segment of real numbers determined by β, *i.e.* the class of real numbers which are less than the limit or maximum of β. In any series, if β is a class contained in the series and R is the generating relation of the series, $R``\beta$ is the segment determined by $\beta$. If $\beta$ has either a limit or a maximum, say $x$, $R``\beta$ will be $\overrightarrow{R}'x$. But if β has neither a limit nor a maximum, $R``\beta$ will be what we may call an "irrational" segment of the series. We shall see at a later stage that the real numbers may be identified with the segments of the series of rationals, *i.e.* if $R$ is the relation of less to greater among rationals, the real numbers will be all classes such as $R``\beta$, for different values of β. The real numbers which correspond to rationals will be those resulting from a β which has a limit or maximum; the irrationals will be those resulting from a β which has no limit or maximum.

The present number may be divided into various sections, as follows: (1) First, we have various elementary properties of the terms defined at the beginning of the number; this section ends with ∗37·29. (2) We have next a set of propositions dealing with relative products, and with such symbols as $P``Q``\gamma$, $P``Q```\kappa$, and so on. The central proposition here is

∗37·33. $\vdash . (P \mid Q)``\gamma = P``Q``\gamma$

By the definition, $Q```\kappa = Q_\epsilon``\kappa$. Thus $P``Q```\kappa = (P \mid Q_\epsilon)``\kappa$. This connects propositions concerning such symbols as $P``Q```\kappa$ with propositions concerning relative products. This second section consists of the propositions from ∗37·3 to ∗37·39. (3) We have next a set of propositions on relations with limited domains and converse domains. The chief of these are

∗37·401. $\vdash . \text{D}`(R \upharpoonright \beta) = R``\beta$

∗37·412. $\vdash . (R \upharpoonright \alpha)``\beta = R``(\alpha \cap \beta)$

∗37·41. $\vdash . \text{D}`(R \upharpoonleft \alpha) = \alpha \cap R``\alpha . \text{Ɑ}`(R \upharpoonleft \alpha) = \alpha \cap \breve{R}``\alpha$

These propositions on relations with limited domains and converse domains, together with certain others naturally connected with them, extend from ∗37·4 to ∗37·52. (4) We next have a number of very important propositions on the consequences of the hypothesis E !! $R``\beta$, *i.e.* the hypothesis that, for any argument which is a member of β, R gives rise to a descriptive function $R`y$. The chief proposition in this section is

∗37·6. $\vdash : \text{E !! } R``\beta . \supset . R``\beta = \hat{x}\{(\exists y) . y \in \beta . x = R`y\}$

Propositions with the hypothesis E !! $R``\beta$ are applied to the cases of $\overrightarrow{R}$ and $\overleftarrow{R}$, in which the hypothesis is verified. This section extends from ∗37·6 to ∗37·791. (5) Finally, we have three propositions on the relative product of $\alpha \uparrow \beta$ with other relations. These propositions are useful in relation-arithmetic (Part IV).

The propositions of the present number which are most used in the sequel, apart from those already mentioned, are the following (omitting such as merely embody definitions):

∗37·15. $\vdash . R``\alpha \subset \text{D}`R$

∗37·16. $\vdash . \breve{R}``\alpha \subset \text{Ɑ}`R$

∗37·2. $\vdash : \alpha \subset \beta . \supset . P``\alpha \subset P``\beta$

∗37·22. $\vdash . P``(\alpha \cup \beta) = P``\alpha \cup P``\beta$

∗37·25. $\vdash . \text{D}`R = R``\text{Ɑ}`R . \text{Ɑ}`R = \breve{R}``\text{D}`R$

∗37·26. $\vdash . R``\beta = R``(\beta \cap \text{Ɑ}`R)$

∗37·265. $\vdash . R``\alpha = R``(\alpha \cap C`R) . \breve{R}``\alpha = \breve{R}``(\alpha \cap C`R)$

∗37·29. $\vdash . R``\Lambda = \Lambda . \breve{R}``\Lambda = \Lambda$

∗37·32. $\vdash . \text{D}`(P \mid Q) = P``\text{D}`Q . \text{Ɑ}`(P \mid Q) = \breve{Q}``\text{Ɑ}`P$

∗37·45. $\vdash :. (y) . \text{E ! } R`y . \supset : \exists ! R``\beta . \equiv . \exists ! \beta$

∗37·46. $\vdash : x \in R``\alpha . \equiv . \exists ! \alpha \cap \overleftarrow{R}`x$

∗37·61. $\vdash :: \text{E !! } R``\beta . \supset :. R``\beta \subset \alpha . \equiv : y \in \beta . \supset_y . R`y \in \alpha$

For example, let R be the relation of father to son, β the class of Etonians, α the class of rich men; then "$R``\beta \subset \alpha$" states "all fathers of Etonians are rich," while "$y \in \beta . \supset_y . R`y \in \alpha$" states "if a boy is an Etonian, his father must be rich." In virtue of the above proposition, these two statements are equivalent.

***37·62.** $\vdash : E ! R`y . y \epsilon \alpha . \supset . R`y \epsilon R``\alpha$

***37·63.** $\vdash :: E !! R``\alpha . \supset :. x \epsilon R``\alpha . \supset_x . \psi x : \equiv : y \epsilon \alpha . \supset_y . \psi (R`y)$

***37·01.** $R``\beta = \hat{x} \{(\exists y) . y \epsilon \beta . xRy\}$ Df

***37·02.** $R_\epsilon = \hat{\alpha}\hat{\beta} (\alpha = R``\beta)$ Df

***37·03.** $\breve{R}_\epsilon = \text{Cnv}`(R_\epsilon)$ Df

***37·04.** $R```\kappa = R_\epsilon``\kappa$ Df

***37·05.** $E !! R``\beta . = : y \epsilon \beta . \supset_y . E ! R`y$ Df

***37·1.** $\vdash : x \epsilon R``\beta . \equiv . (\exists y) . y \epsilon \beta . xRy$ [*20·3 . (*37·01)]

***37·101.** $\vdash : \alpha R_\epsilon \beta . \equiv . \alpha = R``\beta$ [*21·3 . (*37·02)]

***37·102.** $\vdash : \alpha (\breve{R})_\epsilon \beta . \equiv . \alpha = \breve{R}``\beta$ [*37·101]

***37·103.** $\vdash : \alpha \epsilon R```\kappa . \equiv . (\exists \beta) . \beta \epsilon \kappa . \alpha = R``\beta . \equiv . \alpha \epsilon R_\epsilon``\kappa$
[*37·1·101 . (*37·04)]

***37·104.** $\vdash :. E !! R``\beta . \equiv : y \epsilon \beta . \supset_y . E ! R`y$ [*4·2 . (*37·05)]

***37·105.** $\vdash : x \epsilon \breve{R}``\beta . \equiv . (\exists y) . y \epsilon \beta . yRx$ [*37·1 . *31·11]

***37·106.** $\vdash :. E ! R`x . \supset : x \epsilon \breve{R}``\beta . \equiv . R`x \epsilon \beta$

Dem.

$\vdash . \text{*37·105} . \text{*30·4} . \supset \vdash :. \text{Hp} . \supset : x \epsilon \breve{R}``\beta . \equiv . (\exists y) . y \epsilon \beta . y = R`x .$
[*14·205] $\equiv . R`x \epsilon \beta :. \supset \vdash . \text{Prop}$

***37·11.** $\vdash . R_\epsilon`\beta = R``\beta$ [*37·101 . *30·3]

***37·111.** $\vdash . E ! R_\epsilon`\beta$ [*37·11 . *14·21]

***37·12.** $\vdash : (\beta) . R``\beta = Q`\beta . \equiv . R_\epsilon = Q$ [*30·42 . *37·111·11]

***37·13.** $\vdash : P = Q . \supset . P``\beta = Q``\beta$

Dem.

$\vdash . \text{*21·43} . \supset \vdash :. \text{Hp} . \supset : xPy . \equiv_{x,y} . xQy :$
[Fact] $\supset : y \epsilon \beta . xPy . \equiv_{x,y} . y \epsilon \beta . xQy :$
[*10·281] $\supset : (\exists y) . y \epsilon \beta . xPy . \equiv_x . (\exists y) . y \epsilon \beta . xQy :$
[*37·1] $\supset : x \epsilon P``\beta . \equiv_x . x \epsilon Q``\beta :. \supset \vdash . \text{Prop}$

***37·131.** $\vdash : P = Q . \supset . P_\epsilon = Q_\epsilon$

Dem.

$\vdash . \text{*37·13} . \supset \vdash :. \text{Hp} . \supset : \alpha = P``\beta . \equiv_{\alpha,\beta} . \alpha = Q``\beta :$
[*37·101] $\supset : \alpha P_\epsilon \beta . \equiv_{\alpha,\beta} . \alpha Q_\epsilon \beta :. \supset \vdash . \text{Prop}$

***37·14.** $\vdash : P = Q . \equiv . P_\epsilon = Q_\epsilon$

Dem.

$\vdash . *37·101 . *21·15 . \supset$

$\vdash :. P_\epsilon = Q_\epsilon . \equiv : \alpha = P``\beta . \equiv_{\alpha, \beta} . \alpha = Q``\beta :$

[*13·183] $\equiv : (\beta) . P``\beta = Q``\beta :$

[*37·1.*20·15] $\equiv : (\beta, x) : (\exists y) . y \epsilon \beta . xPy . \equiv . (\exists y) . y \epsilon \beta . xQy :$

[*10·1] $\supset : (x) : (\exists y) . y \epsilon \hat{z}(z = w) . xPy . \equiv . (\exists y) . y \epsilon \hat{z}(z = w) . xQy :$

[*20·3] $\supset : (x) : (\exists y) . y = w . xPy . \equiv . (\exists y) . y = w . xQy :$

[*13·195] $\supset : (x) : xPw . \equiv . xQw$ (1)

$\vdash . (1) . *10·11·21 . *11·2 . \supset$

$\vdash :. P_\epsilon = Q_\epsilon . \supset : (x, w) : xPw . \equiv . xQw :$

[*21·43] $\supset : P = Q$ (2)

$\vdash . (2) . *37·131 . \supset \vdash .$ Prop

***37·15.** $\vdash . R``\alpha \subset D`R$

Dem.

$\vdash . *37·1 . \supset \vdash : x \epsilon R``\alpha . \supset . (\exists y) . y \epsilon \alpha . xRy .$

[*10·5] $\supset . (\exists y) . xRy .$

[*33·13] $\supset . x \epsilon D`R : \supset \vdash .$ Prop

***37·16.** $\vdash . \breve{R}``\alpha \subset \text{ᗡ}`R$ [*37·15 $\frac{\breve{R}}{R}$. *33·2]

***37·17.** $\vdash :. R``\beta \subset \alpha . \equiv : y \epsilon \beta . xRy . \supset_{x,y} . x \epsilon \alpha$

Dem.

$\vdash . *37·1 . \supset \vdash :. R``\beta \subset \alpha . \equiv : (\exists y) . y \epsilon \beta . xRy . \supset_x . x \epsilon \alpha :$

[*10·23] $\equiv : y \epsilon \beta . xRy . \supset_{x,y} . x \epsilon \alpha :. \supset \vdash .$ Prop

***37·171.** $\vdash :. \breve{R}``\alpha \subset \beta . \equiv : x \epsilon \alpha . xRy . \supset_{x,y} . y \epsilon \beta$

Dem.

$\vdash . *37·105 . \supset \vdash :. \breve{R}``\alpha \subset \beta . \equiv : (\exists x) . x \epsilon \alpha . xRy . \supset_y . y \epsilon \beta :$

[*10·23] $\equiv : x \epsilon \alpha . xRy . \supset_{x,y} . y \epsilon \beta :. \supset \vdash .$ Prop

***37·18.** $\vdash : y \epsilon \beta . \supset . \overrightarrow{R}`y \subset R``\beta$

Dem.

$\vdash . *32·18 . \supset \vdash :. \text{Hp} . \supset : x \epsilon \overrightarrow{R}`y . \supset . xRy . y \epsilon \beta .$

[*37·1] $\supset . x \epsilon R``\beta :. \supset \vdash .$ Prop

***37·181.** $\vdash : x \epsilon \alpha . \supset . \overleftarrow{R}`x \subset \breve{R}``\alpha$ [Proof as in *37·18]

***37·2.** $\vdash : \alpha \subset \beta . \supset . P``\alpha \subset P``\beta$

Dem.

$\vdash . *22·1 . \supset \vdash :. \text{Hp} . \supset : y \epsilon \alpha . \supset_y . y \epsilon \beta :$

[*10·31] $\supset : y \epsilon \alpha . xPy . \supset_y . y \epsilon \beta . xPy :$

[*10·28] $\supset : (\exists y) . y \epsilon \alpha . xPy . \supset . (\exists y) . y \epsilon \beta . xPy :$

[*37·1] $\supset : x \epsilon P``\alpha . \supset . x \epsilon P``\beta :. \supset \vdash .$ Prop

The above proposition ($*37\cdot2$) is one of the forms of asyllogistic inference due to Leibniz's teacher Jungius. The instance given by Jungius is: "Circulus est figura; ergo qui circulum describit, is figuram describit*." Here the class of circles is our α, the class of figures is our β, and the relation of describing is our P.

$*37\cdot201$. $\vdash : P \mathrel{\dot\subset} Q . \supset . P``\alpha \subset Q``\alpha$ [Similar proof]

$*37\cdot202$. $\vdash : \alpha \subset \beta . P \mathrel{\dot\subset} Q . \supset . P``\alpha \subset Q``\beta$ [$*37\cdot2\cdot201$]

$*37\cdot21$. $\vdash . P``(\alpha \cap \beta) \subset P``\alpha \cap P``\beta$

Dem.

$\vdash . *37\cdot1 . \supset \vdash :. x \epsilon P``(\alpha \cap \beta) . \equiv : (\exists y) . y \epsilon \alpha \cap \beta . xPy :$

[$*22\cdot33$] $\equiv : (\exists y) . y \epsilon \alpha . y \epsilon \beta . xPy :$

[$*10\cdot5$] $\supset : (\exists y) . y \epsilon \alpha . xPy : (\exists y) . y \epsilon \beta . xPy :$

[$*37\cdot1$] $\supset : x \epsilon P``\alpha . x \epsilon P``\beta :$

[$*22\cdot33$] $\supset : x \epsilon P``\alpha \cap P``\beta :. \supset \vdash . \text{Prop}$

$*37\cdot211$. $\vdash . (P \mathbin{\dot\cap} Q)``\alpha \subset P``\alpha \cap Q``\alpha$ [Similar proof]

$*37\cdot212$. $\vdash . (P \mathbin{\dot\cap} Q)``(\alpha \cap \beta) \subset P``\alpha \cap P``\beta \cap Q``\alpha \cap Q``\beta$ [$*37\cdot21\cdot211$]

$*37\cdot22$. $\vdash . P``(\alpha \cup \beta) = P``\alpha \cup P``\beta$

This proposition is very frequently used. The fact that here we have identity, while in $*37\cdot21$ we only have inclusion, is due to the fact that $*10\cdot42$ states an equivalence, while $*10\cdot5$ only states an implication.

Dem.

$\vdash . *37\cdot1 . \supset \vdash :. x \epsilon P``(\alpha \cup \beta) . \equiv : (\exists y) . y \epsilon \alpha \cup \beta . xPy :$

[$*22\cdot34$] $\equiv : (\exists y) : y \epsilon \alpha . \vee . y \epsilon \beta : xPy :$

[$*4\cdot4$] $\equiv : (\exists y) : y \epsilon \alpha . xPy . \vee . y \epsilon \beta . xPy :$

[$*10\cdot42$] $\equiv : (\exists y) . y \epsilon \alpha . xPy : \vee : (\exists y) . y \epsilon \beta . xPy :$

[$*37\cdot1$] $\equiv : x \epsilon P``\alpha . \vee . x \epsilon P``\beta :$

[$*22\cdot34$] $\equiv : x \epsilon P``\alpha \cup P``\beta :. \supset \vdash . \text{Prop}$

$*37\cdot221$. $\vdash . (P \mathbin{\dot\cup} Q)``\alpha = P``\alpha \cup Q``\alpha$ [Similar proof]

$*37\cdot222$. $\vdash . (P \mathbin{\dot\cup} Q)``(\alpha \cup \beta) = P``\alpha \cup P``\beta \cup Q``\alpha \cup Q``\beta$ [$*37\cdot22\cdot221$]

$*37\cdot23$. $\vdash . \text{D}`R_\epsilon = \hat\alpha \{(\exists \beta) . \alpha = R``\beta\}$ [$*37\cdot101 . *33\cdot11$]

$*37\cdot231$. $\vdash . \text{Ɑ}`R_\epsilon = \text{Cls}$

The type of "Cls" here is that type whose members are of the same type as $\text{Ɑ}`R$. In the proof, use is made of the convention that a Greek letter always stands for an expression of the form $\hat z (\phi ! z)$.

Dem.

$\vdash . *37\cdot101 .$ $\supset \vdash : \alpha R_\epsilon \hat z (\phi ! z) . \equiv . \alpha = R``\hat z (\phi ! z) :$

[$*10\cdot11\cdot281$] $\supset \vdash : (\exists \alpha) . \alpha R_\epsilon \hat z (\phi ! z) . \equiv . (\exists \alpha) . \alpha = R``\hat z (\phi ! z) :$

[$*33\cdot131$] $\supset \vdash : \hat z (\phi ! z) \epsilon \text{Ɑ}`R_\epsilon . \equiv . (\exists \alpha) . \alpha = R``\hat z (\phi ! z)$ (1)

* We quote from Couturat, *La Logique de Leibniz*, Chapter III, § 15 (p. 75 n.).

$\vdash . *20\cdot2 . (*37\cdot01) . \supset \vdash : \hat{x}\{(\exists y) . y \in \hat{z}(\phi ! z) . xRy\} = R``\hat{z}(\phi ! z) :$

$[*10\cdot11\cdot24] \quad \supset \vdash : (\phi) : (\exists \alpha) . \alpha = R``\hat{z}(\phi ! z)$ (2)

$\vdash . (1) . (2) . *2\cdot02 . \supset \vdash : \hat{z}(\phi ! z) \in \text{Cls} . \supset . \hat{z}(\phi ! z) \in \text{D}`R_\epsilon$ (3)

$\vdash . *20\cdot41 . *2\cdot02 . \supset \vdash : \hat{z}(\phi ! z) \in \text{D}`R_\epsilon . \supset . \hat{z}(\phi ! z) \in \text{Cls}$ (4)

$\vdash . (3) . (4) . \quad \supset \vdash . \text{Prop}$

As appears in the above proof, it is necessary, when a proposition containing "Cls" is to be proved, to abandon the notation with Greek letters, and revert to the explicit functional notation.

***37·24.** $\vdash : \alpha \in \text{D}`R_\epsilon . \supset . \alpha \subset \text{D}`R$

Dem.

$\vdash . *33\cdot13 . *37\cdot101 . \supset \vdash :: \alpha \in \text{D}`R_\epsilon . \equiv :. (\exists \beta) . \alpha = R``\beta :.$

$[*20\cdot33 . *37\cdot1] \quad \equiv :. (\exists \beta) : x \in \alpha . \equiv_x . (\exists y) . y \in \beta . xRy :.$

$[*11\cdot61] \quad \supset :. x \in \alpha . \supset_x : (\exists \beta, y) . y \in \beta . xRy :$

$[*11\cdot23] \quad \supset_x : (\exists y, \beta) . y \in \beta . xRy :$

$[*11\cdot55] \quad \supset_x : (\exists y) : xRy : (\exists \beta) . y \in \beta :$

$[*10\cdot5] \quad \supset_x : (\exists y) . xRy :$

$[*33\cdot13] \quad \supset_x : x \in \text{D}`R :: \supset \vdash . \text{Prop}$

***37·25.** $\vdash . \text{D}`R = R``\text{ᴅ}`R . \text{ᴅ}`R = \breve{R}``\text{D}`R$

Dem.

$\vdash . *33\cdot13 . \supset \vdash : x \in \text{D}`R . \equiv . (\exists y) . xRy .$

$[*33\cdot14 . *4\cdot71] \quad \equiv . (\exists y) . y \in \text{ᴅ}`R . xRy .$

$[*37\cdot1] \quad \equiv . x \in R``\text{ᴅ}`R$ (1)

$\vdash . *33\cdot131 . \supset \vdash : y \in \text{ᴅ}`R . \equiv . (\exists x) . xRy .$

$[*33\cdot14 . *4\cdot71] \quad \equiv . (\exists x) . x \in \text{D}`R . xRy .$

$[*37\cdot105] \quad \equiv . y \in \breve{R}``\text{D}`R$ (2)

$\vdash . (1) . (2) . \supset \vdash . \text{Prop}$

***37·26.** $\vdash . R``\beta = R``(\beta \cap \text{ᴅ}`R)$

Dem.

$\vdash . *37\cdot1 . \supset \vdash :. x \in R``\beta . \equiv : (\exists y) . y \in \beta . xRy :$

$[*33\cdot14 . *4\cdot71] \quad \equiv : (\exists y) . y \in \beta . y \in \text{ᴅ}`R . xRy :$

$[*22\cdot33] \quad \equiv : (\exists y) . y \in \beta \cap \text{ᴅ}`R . xRy :$

$[*37\cdot1] \quad \equiv : x \in R``(\beta \cap \text{ᴅ}`R) :. \supset \vdash . \text{Prop}$

***37·261.** $\vdash . \breve{R}``\beta = \breve{R}``(\beta \cap \text{D}`R)$ $[*37\cdot26 . *33\cdot21]$

***37·262.** $\vdash : \alpha \cap \text{ᴅ}`R = \beta \cap \text{ᴅ}`R . \supset . R``\alpha = R``\beta$ $[*37\cdot26]$

***37·263.** $\vdash : \alpha \cap \text{D}`R = \beta \cap \text{D}`R . \supset . \breve{R}``\alpha = \breve{R}``\beta$ $[*37\cdot261]$

***37·264.** $\vdash : \exists ! \alpha \cap R``\beta . \equiv . (\exists x, y) . x \in \alpha . y \in \beta . xRy . \equiv . \exists ! \beta \cap \breve{R}``\alpha$

Dem.

$\vdash . *22\cdot33 . *37\cdot1 . \supset \vdash :. \exists ! \alpha \cap R``\beta . \equiv : (\exists x) : x \in \alpha : (\exists y) . y \in \beta . xRy :$ (1)

[*11·55] $\equiv : (\exists x, y) . x \epsilon \alpha . y \epsilon \beta . xRy$ (2)

$\vdash . (1) . *11·6 . \supset \vdash :. \exists ! \alpha \cap R``\beta . \equiv : (\exists y) : y \epsilon \beta : (\exists x) . x \epsilon \alpha . xRy :$

[*37·105] $\equiv : (\exists y) . y \epsilon \beta . y \epsilon \breve{R}``\alpha :$

[*22·33] $\equiv : \exists ! \beta \cap \breve{R}``\alpha$ (3)

$\vdash . (2) . (3) . \supset \vdash .$ Prop

***37·265.** $\vdash . R``\alpha = R``(\alpha \cap C`R) . \breve{R}``\alpha = \breve{R}``(\alpha \cap C`R)$

Dem.

$\vdash . *33·161 . *22·621 . \supset \vdash . \text{Ɑ}`R = C`R \cap \text{Ɑ}`R .$

[*22·481] $\supset \vdash . \alpha \cap \text{Ɑ}`R = \alpha \cap C`R \cap \text{Ɑ}`R .$

[*37·262] $\supset \vdash . R``\alpha = R``(\alpha \cap C`R)$ (1)

$\vdash . (1) . *33·22 . \supset \vdash .$ Prop

***37·27.** $\vdash : \text{Ɑ}`R \subset \beta . \supset . D`R = R``\beta$ [*22·621 . *37·25·26]

***37·271.** $\vdash : D`R \subset \alpha . \supset . \text{Ɑ}`R = \breve{R}``\alpha$ [*22·621 . *37·25·261]

***37·28.** $\vdash . R``V = D`R . \breve{R}``V = \text{Ɑ}`R$ [*37·27·271 . *24·11]

***37·29.** $\vdash . R``\Lambda = \Lambda . \breve{R}``\Lambda = \Lambda$

Dem.

$\vdash . *10·5 . \supset \vdash : (\exists y) . y \epsilon \Lambda . xRy . \supset . (\exists y) . y \epsilon \Lambda$ (1)

$\vdash . (1) .$ Transp $. *24·53 . \supset \vdash . \sim (\exists y) . y \epsilon \Lambda . xRy .$

[*37·1] $\supset \vdash . \sim \exists ! R``\Lambda .$

[*24·51] $\supset \vdash . R``\Lambda = \Lambda$ (2)

$\vdash . (2) \frac{\breve{R}}{R} . \supset \vdash . \breve{R}``\Lambda = \Lambda$ (3)

$\vdash . (2) . (3) . \supset \vdash .$ Prop

***37·3.** $\vdash . \{sg`(P \mid Q)\}`z = P``\overrightarrow{Q}`z$

Dem.

$\vdash . *32·23·13 . \supset$

$\vdash . \{sg`(P \mid Q)\}`z = \hat{x}\{x (P \mid Q) z\}$

[*34·1] $= \hat{x}\{(\exists y) . xPy . yQz\}$

[*32·18] $= \hat{x}\{(\exists y) . xPy . y \epsilon \overrightarrow{Q}`z\}$

[(*37·01)] $= P``\overrightarrow{Q}`z . \supset \vdash .$ Prop

***37·301.** $\vdash . \{gs`(P \mid Q)\}`x = \breve{Q}``\overleftarrow{P}`x$ [Similar proof]

***37·302.** $\vdash : R = P \mid Q . \supset . \overrightarrow{R}`z = P``\overrightarrow{Q}`z . \overleftarrow{R}`x = \breve{Q}``\overleftarrow{P}`x$

[*37·3·301 . *32·23·231·16]

***37·31.** $\vdash . sg`(P \mid Q) = P_\epsilon \mid \overrightarrow{Q}$

Dem.

$\vdash . *37·11·3 . \supset \vdash . (z) . \{sg`(P \mid Q)\}`z = P_\epsilon`\overrightarrow{Q}`z$ (1)

$\vdash . (1) . *34·42 . \supset \vdash .$ Prop

***37·311.** $\vdash . \mathrm{gs}`(P \mid Q) = (\breve{Q})_\epsilon \mid \overleftarrow{P}$ [Similar proof]

***37·32.** $\vdash . \mathrm{D}`(P \mid Q) = P``\mathrm{D}`Q . \text{Ɑ}`(P \mid Q) = \breve{Q}``\text{Ɑ}`P$

Dem.

$\vdash . *33·13 . *34·1 . \supset$

$\vdash :. x \epsilon \mathrm{D}`(P \mid Q) . \equiv : (\exists z) : (\exists y) . xPy . yQz :$

$[*11·23] \quad \equiv : (\exists y) : (\exists z) . xPy . yQz :$

$[*11·55] \quad \equiv : (\exists y) : xPy : (\exists z) . yQz :$

$[*33·13] \quad \equiv : (\exists y) . xPy . y \epsilon \mathrm{D}`Q :$

$[*37·1] \quad \equiv : x \epsilon P``\mathrm{D}`Q$ (1)

$\vdash . (1) . *10·11 . *20·43 . \supset$

$\vdash . \mathrm{D}`(P \mid Q) = P``\mathrm{D}`Q$ (2)

$\vdash . *33·2 . \supset \vdash . \text{Ɑ}`(P \mid Q) = \mathrm{D}`\mathrm{Cnv}`(P \mid Q)$

$[*34·2] \quad = \mathrm{D}`(\breve{Q} \mid \breve{P})$

$[(2)] \quad = \breve{Q}``\mathrm{D}`\breve{P}$

$[*33·2] \quad = \breve{Q}``\text{Ɑ}`P$ (3)

$\vdash . (2) . (3) . \supset \vdash . \mathrm{Prop}$

***37·321.** $\vdash : \text{Ɑ}`P \subset \mathrm{D}`Q . \supset . \mathrm{D}`(P \mid Q) = \mathrm{D}`P$ [*37·32·27]

***37·322.** $\vdash : \mathrm{D}`Q \subset \text{Ɑ}`P . \supset . \text{Ɑ}`(P \mid Q) = \text{Ɑ}`Q$ [*37·32·271]

***37·323.** $\vdash : \text{Ɑ}`P = \mathrm{D}`Q . \supset . \mathrm{D}`(P \mid Q) = \mathrm{D}`P . \text{Ɑ}`(P \mid Q) = \text{Ɑ}`Q$ [*37·321·322]

***37·33.** $\vdash . (P \mid Q)``\gamma = P``Q``\gamma$

Dem.

$\vdash . *37·1 . \supset \vdash :. x \epsilon (P \mid Q)``\gamma . \equiv : (\exists z) . z \epsilon \gamma . x (P \mid Q) z :$

$[*34·1 . *11·55] \quad \equiv : (\exists z, y) . z \epsilon \gamma . xPy . yQz :$

$[*11·23] \quad \equiv : (\exists y, z) . xPy . yQz . z \epsilon \gamma :$

$[*11·55] \quad \equiv : (\exists y) : xPy : (\exists z) . yQz . z \epsilon \gamma :$

$[*37·1] \quad \equiv : (\exists y) . xPy . y \epsilon Q``\gamma :$

$[*37·1] \quad \equiv : x \epsilon P``Q``\gamma :. \supset \vdash . \mathrm{Prop}$

***37·34.** $\vdash . (P \mid Q)_\epsilon = P_\epsilon \mid Q_\epsilon$

Dem.

$\vdash . *37·11 . \supset \vdash . (P \mid Q)_\epsilon`\gamma = (P \mid Q)``\gamma$

$[*37·33] \quad = P``Q``\gamma$

$[*37·11] \quad = P_\epsilon`Q_\epsilon`\gamma$ (1)

$\vdash . (1) . *10·11 . *34·42 . \supset \vdash . \mathrm{Prop}$

***37·341.** $\vdash . \{\mathrm{Cnv}`(P \mid Q)\}_\epsilon = (\breve{Q})_\epsilon \mid (\breve{P})_\epsilon$ [*34·2 . *37·34]

***37·35.** $\vdash : (z) . R`z = P`Q`z . \supset . (\gamma) . R``\gamma = P``Q``\gamma$

Dem.

$\vdash . *34·42 . \supset \vdash : \mathrm{Hp} . \supset . R = P \mid Q .$

$[*37·13] \quad \supset . R``\gamma = (P \mid Q)``\gamma$

$[*37·33] \quad = P``Q``\gamma : \supset \vdash . \mathrm{Prop}$

***37·351.** $\vdash : (\alpha) . R`\alpha = P`Q``\alpha . \supset . (\kappa) . R``\kappa = P``Q```\kappa$

$\left[*37\cdot35 \dfrac{Q_\epsilon}{Q} . *37\cdot11 . (*37\cdot04) \right]$

***37·352.** $\vdash : (\alpha) . R``\alpha = P`Q``\alpha . \supset . (\kappa) . R```\kappa = P``Q```\kappa$

$\left[*37\cdot351 \dfrac{R_\epsilon}{R} . *37\cdot11 . (*37\cdot04) \right]$

***37·353.** $\vdash : (z) . R`S`z = P`Q`z . \supset . (\gamma) . R``S``\gamma = P``Q``\gamma$

Dem.

$\vdash . *14\cdot21 . \supset \vdash : \text{Hp} . \supset . (z) . E ! R`S`z .$

$[*34\cdot41] \quad \supset . (z) . R`S`z = (R | S)`z .$

$[*14\cdot131\cdot144] \quad \supset . (z) . (R | S)`z = P`Q`z .$

$[*37\cdot35] \quad \supset . (\gamma) . (R | S)``\gamma = P``Q``\gamma .$

$[*37\cdot33] \quad \supset . (\gamma) . R``S``\gamma = P``Q``\gamma : \supset \vdash . \text{Prop}$

***37·354.** $\vdash : (\alpha) . R`S`\alpha = P`Q``\alpha . \supset . (\kappa) . R``S``\kappa = P``Q```\kappa \quad \left[*37\cdot353 \dfrac{Q_\epsilon}{Q} \right]$

***37·355.** $\vdash : (z) . R`S`z = P``Q`z . \supset . (\gamma) . R``S``\gamma = P```Q``\gamma \quad \left[*37\cdot353 \dfrac{P_\epsilon}{P} \right]$

***37·36.** $\vdash . D`R^2 = R``D`R . \text{ɑ}`R^2 = \breve{R}``\text{ɑ}`R \quad [*37\cdot32]$

***37·37.** $\vdash . (R^2)_\epsilon = (R_\epsilon)^2 \quad [*37\cdot34]$

***37·371.** $R_\epsilon{}^2 = (R_\epsilon)^2$ Df

This definition serves merely for the avoidance of brackets. Like *37·03, this definition will be extended to all suffixes.

***37·38.** $\vdash . \overrightarrow{R^2}`x = R``\overrightarrow{R}`x \quad [*37\cdot3]$

***37·39.** $\vdash . R^2``\alpha = R``R``\alpha \quad [*37\cdot33]$

***37·4.** $\vdash . \text{ɑ}`(\alpha \upharpoonleft R) = \breve{R}``\alpha$

Dem.

$\vdash . *33\cdot131 . *35\cdot1 . \supset \vdash : y \,\epsilon\, \text{ɑ}`(\alpha \upharpoonleft R) . \equiv . (\exists x) . x \,\epsilon\, \alpha . xRy .$

$[*37\cdot105] \quad \equiv . y \,\epsilon\, \breve{R}``\alpha : \supset \vdash . \text{Prop}$

***37·401.** $\vdash . D`(R \upharpoonright \beta) = R``\beta$ [Similar proof]

***37·402.** $\vdash . D`(\alpha \upharpoonleft R \upharpoonright \beta) = \alpha \cap R``\beta . \text{ɑ}`(\alpha \upharpoonleft R \upharpoonright \beta) = \beta \cap \breve{R}``\alpha$

Dem.

$\vdash . *33\cdot13 . *35\cdot102 . \supset$

$\vdash :. x \,\epsilon\, D`(\alpha \upharpoonleft R \upharpoonright \beta) . \equiv : (\exists y) . x \,\epsilon\, \alpha . xRy . y \,\epsilon\, \beta :$

$[*10\cdot35] \quad \equiv : x \,\epsilon\, \alpha : (\exists y) . xRy . y \,\epsilon\, \beta :$

$[*37\cdot1] \quad \equiv : x \,\epsilon\, \alpha . x \,\epsilon\, R``\beta :$

$[*22\cdot33] \quad \equiv : x \,\epsilon\, \alpha \cap R``\beta \quad (1)$

Similarly

$\vdash : y \,\epsilon\, \text{ɑ}`(\alpha \upharpoonleft R \upharpoonright \beta) . \equiv . y \,\epsilon\, \beta \cap \breve{R}``\alpha \quad (2)$

$\vdash . (1) . (2) . \supset \vdash . \text{Prop}$

***37·41.** $\vdash . \text{D}'(R \upharpoonleft\!\!\!\! \upharpoonright \alpha) = \alpha \cap R``\alpha . \text{Cl}'(R \upharpoonleft\!\!\!\! \upharpoonright \alpha) = \alpha \cap \breve{R}``\alpha$ [*37·402 . *36·11]

***37·411.** $\vdash . (\alpha \upharpoonleft R)``\beta = \text{D}'(\alpha \upharpoonleft R \upharpoonright \beta) = \alpha \cap R``\beta$

Dem.

$\vdash . *37·401 . \supset \vdash . (\alpha \upharpoonleft R)``\beta = \text{D}'(\alpha \upharpoonleft R) \upharpoonright \beta$

$[*35·21] \qquad = \text{D}'(\alpha \upharpoonleft R \upharpoonright \beta)$ (1)

$\vdash . (1) . *37·402 . \supset \vdash . \text{Prop}$

***37·412.** $\vdash . (R \upharpoonright \alpha)``\beta = R``(\alpha \cap \beta)$

Dem.

$\vdash . *37·401 . \supset \vdash . (R \upharpoonright \alpha)``\beta = \text{D}'(R \upharpoonright \alpha) \upharpoonright \beta$

$[*35·31] \qquad = \text{D}'R \upharpoonright (\alpha \cap \beta)$

$[*37·401] \qquad = R``(\alpha \cap \beta) . \supset \vdash . \text{Prop}$

***37·413.** $\vdash . (R \upharpoonleft\!\!\!\! \upharpoonright \alpha)``\beta = \alpha \cap R``(\alpha \cap \beta)$

Dem.

$\vdash . *37·411 . *35·21 . \supset \vdash . (R \upharpoonleft\!\!\!\! \upharpoonright \alpha)``\beta = \alpha \cap (R \upharpoonright \alpha)``\beta$

$[*37·412] \qquad = \alpha \cap R``(\alpha \cap \beta) . \supset \vdash . \text{Prop}$

***37·42.** $\vdash : R``\beta \subset \alpha . \supset . (\alpha \upharpoonleft R)``\beta = R``\beta$ [*37·411 . *22·621]

***37·421.** $\vdash : \beta \subset \alpha . \supset . (R \upharpoonright \alpha)``\beta = R``\beta$ [*37·412 . *22·621]

***37·43.** $\vdash :. \beta \subset \text{Cl}'R . \supset : \exists ! R``\beta . \equiv . \exists ! \beta$

Dem.

$\vdash . *37·401 . *35·65 . \supset \vdash :. \text{Hp} . \supset : R``\beta = \text{D}'(R \upharpoonright \beta) . \beta = \text{Cl}'(R \upharpoonright \beta)$ (1)

$\vdash . (1) . *33·24 . \quad \supset \vdash . \text{Prop}$

***37·431.** $\vdash :. \alpha \subset \text{D}'R . \supset : \exists ! \breve{R}``\alpha . \equiv . \exists ! \alpha$ [Proof as in *37·43]

***37·44.** $\vdash :. \text{Cl}'R = \text{V} . \supset : \exists ! R``\beta . \equiv . \exists ! \beta$ [*37·43 . *24·11]

***37·441.** $\vdash :. \text{D}'R = \text{V} . \supset : \exists ! \breve{R}``\alpha . \equiv . \exists ! \alpha$ [Proof as in *37·44]

***37·45.** $\vdash :. (y) . \text{E} ! R`y . \supset : \exists ! R``\beta . \equiv . \exists ! \beta$ [*33·431 . *37·43]

***37·451.** $\vdash :. (x) . \text{E} ! \breve{R}`x . \supset : \exists ! \breve{R}``\alpha . \equiv . \exists ! \alpha$ [Proof as in *37·45]

***37·46.** $\vdash : x \in R``\alpha . \equiv . \exists ! \alpha \cap \overleftarrow{R}`x$ [*37·1 . *32·181]

***37·461.** $\vdash : x \sim\in R``\alpha . \equiv . \alpha \cap \overleftarrow{R}`x = \Lambda . \equiv . \overleftarrow{R}`x \subset -\alpha$ [*37·46 . *24·311]

***37·462.** $\vdash : x \sim\in \breve{R}``\alpha . \equiv . \alpha \cap \overrightarrow{R}`x = \Lambda . \equiv . \overrightarrow{R}`x \subset -\alpha$ [*37·461 . *32·241]

***37·47.** $\vdash : \exists ! \alpha . \equiv . \exists ! R```\alpha . \equiv . \exists ! \breve{R}```\alpha$

Dem.

$\vdash . *37·45·111 . \supset \vdash : \exists ! \alpha . \equiv . \exists ! R_\epsilon``\alpha .$

$[(*37·04)] \qquad \equiv . \exists ! R```\alpha$ (1)

$\vdash . (1) \frac{\breve{R}}{R} . \qquad \supset \vdash : \exists ! \alpha . \equiv . \exists ! \breve{R}```\alpha$ (2)

$\vdash . (1) . (2) . \qquad \supset \vdash . \text{Prop}$

***37·5.** $\vdash : (\beta) . P``\beta = Q`\beta . \supset . (\kappa) . P```\kappa = Q``\kappa$

Dem.

$\vdash . *37 \cdot 12 . \supset \vdash : \text{Hp} . \supset . P_\epsilon = Q .$

$[*37 \cdot 13] \quad \supset . P_\epsilon``\kappa = Q``\kappa .$

$[(*37 \cdot 04)] \quad \supset . P```\kappa = Q``\kappa : \supset \vdash . \text{Prop}$

***37·501.** $\vdash . \beta \cap \text{Ɑ}`R \subset \breve{R}``R``\beta$

Dem.

$\vdash . *37 \cdot 1 . *10 \cdot 24 . \supset \vdash : y \in \beta . xRy . \supset . x \in R``\beta :$

$[\text{Exp}.*10 \cdot 11 \cdot 21] \quad \supset \vdash :. y \in \beta . \supset : xRy . \supset_x . x \in R``\beta :$

$[*4 \cdot 7] \quad \supset : xRy . \supset_x . xRy . x \in R``\beta :$

$[*10 \cdot 28] \quad \supset : (\exists x) . xRy . \supset . (\exists x) . xRy . x \in R``\beta :$

$[*33 \cdot 131.*37 \cdot 105] \quad \supset : y \in \text{Ɑ}`R . \supset . y \in \breve{R}``R``\beta$ (1)

$\vdash . (1) . \text{Imp} . *22 \cdot 33 . \supset$

$\vdash : y \in \beta \cap \text{Ɑ}`R . \supset . y \in \breve{R}``R``\beta : \supset \vdash . \text{Prop}$

***37·502.** $\vdash . \alpha \cap \text{D}`R \subset R``\breve{R}``\alpha$ [Similar proof]

***37·51.** $\vdash : \beta \subset \text{Ɑ}`R . \equiv . \beta \subset \breve{R}``R``\beta$

Dem.

$\vdash . *37 \cdot 501 . *22 \cdot 621 . \supset \vdash : \beta \subset \text{Ɑ}`R . \supset . \beta \subset \breve{R}``R``\beta$ (1)

$\vdash . *37 \cdot 16 . \quad \supset \vdash : \beta \subset \breve{R}``R``\beta . \supset . \beta \subset \text{Ɑ}`R$ (2)

$\vdash . (1) . (2) . \quad \supset \vdash . \text{Prop}$

***37·52.** $\vdash : \alpha \subset \text{D}`R . \equiv . \alpha \subset R``\breve{R}``\alpha$ [Similar proof]

The following propositions, down to *37·7 exclusive, are concerned with the special properties of $R``\beta$ which result from the hypothesis $\text{E} !! R``\beta$, defined in *37·05. The hypothesis $\text{E} !! R``\beta$ is important, because it has many consequences and is satisfied in many cases with which we wish to deal.

***37·6.** $\vdash : \text{E} !! R``\beta . \supset . R``\beta = \hat{x} \{(\exists y) . y \in \beta . x = R`y\}$

This proposition is very important, and is used constantly.

Dem.

$\vdash . *37 \cdot 104 . \supset \vdash :: \text{Hp} . \supset :. y \in \beta . \supset_y : \text{E} ! R`y :$

$[*30 \cdot 4] \quad \supset_y : x = R`y . \equiv . xRy :.$

$[*5 \cdot 32] \quad \supset :. y \in \beta . x = R`y . \equiv_y . y \in \beta . xRy :.$

$[*10 \cdot 281] \quad \supset :. (\exists y) . y \in \beta . x = R`y . \equiv . (\exists y) . y \in \beta . xRy .$

$[*37 \cdot 1] \quad \equiv . x \in R``\beta$ (1)

$\vdash . (1) . *10 \cdot 11 \cdot 21 . *20 \cdot 33 . \supset \vdash . \text{Prop}$

***37·601.** $\vdash : (x) . \text{E} ! R`x . \supset . R``\text{V} = \hat{x} \{(\exists y) . x = R`y\}$

Dem.

$\vdash . *2 \cdot 02 . *10 \cdot 11 \cdot 27 . \supset \vdash :. \text{Hp} . \supset : x \in \text{V} . \supset_x . \text{E} ! R`x :$

[*37·104] $\supset$: E !! R"V :

[*37·6] $\supset$: R"V $= \hat{x}\{(\exists y) . y \epsilon \mathrm{V} . x = R'y\}$ (1)

$\vdash$. *24·104 . *4·73 . $\supset \vdash : y \epsilon \mathrm{V} . x = R'y . \equiv . x = R'y$:

[*10·11·281] $\supset \vdash : (\exists y) . y \epsilon \mathrm{V} . x = R'y . \equiv . (\exists y) . x = R'y$:

[*20·15] $\supset \vdash : \hat{x}\{(\exists y) . y \epsilon \mathrm{V} . x = R'y\} = \hat{x}\{(\exists y) . x = R'y\}$ (2)

$\vdash$. (1) . (2) . $\supset \vdash$. Prop

***37·61.** $\vdash$ :: E !! R"β . $\supset$:. R"$\beta \subset \alpha . \equiv : y \epsilon \beta . \supset_y . R'y \epsilon \alpha$

Dem.

$\vdash$. *37·17 . $\supset \vdash$ :: R"$\beta \subset \alpha . \equiv :. y \epsilon \beta . xRy . \supset_{x,y} . x \epsilon \alpha$:.

[*11·2·62] $\equiv :. y \epsilon \beta . \supset_y : xRy . \supset_x . x \epsilon \alpha$ (1)

$\vdash$. *37·104 . $\supset \vdash$ ::. Hp . $\supset$ :: $y \epsilon \beta . \supset_y$:. E ! $R'y$:.

[*30·33] $\supset_y$:. $R'y \epsilon \alpha . \equiv : xRy . \supset_x . x \epsilon \alpha$ (2)

$\vdash$. (1) . (2) . $\supset \vdash$ :: Hp . $\supset$:. R"$\beta \subset \alpha . \equiv : y \epsilon \beta . \supset_y . R'y \epsilon \alpha$ :: $\supset \vdash$. Prop

***37·62.** $\vdash$: E ! $R'y . y \epsilon \alpha . \supset . R'y \epsilon R$"$\alpha$

Dem.

$\vdash$. *30·33 . $\supset$

$\vdash$ :: E ! $R'y$. $\supset$:. $R'y \epsilon R$"$\alpha . \equiv : xRy . \supset_x . x \epsilon R$"$\alpha$ (1)

$\vdash$. *3·2 . $\supset \vdash$:. $y \epsilon \alpha . \supset : xRy . \supset . y \epsilon \alpha . xRy$.

[*10·24.*37·1] $\supset . x \epsilon R$"α (2)

$\vdash$. (2) . *10·11·21 . $\supset \vdash$:. $y \epsilon \alpha . \supset : xRy . \supset_x . x \epsilon R$"$\alpha$ (3)

$\vdash$. (1) . (3) . $\supset \vdash$. Prop

The above is the type of inference concerning which Jevons says*: "I remember the late Prof. De Morgan remarking that all Aristotle's logic could not prove that 'Because a horse is an animal, the head of a horse is the head of an animal.'" It must be confessed that this was a merit in Aristotle's logic, since the proposed inference is fallacious without the added premiss "E ! the head of the horse in question." *E.g.* it does not hold for an oyster or a hydra. But with the addition E ! $R'y$, the above proposition gives an important and common type of asyllogistic inference.

***37·63.** $\vdash$ :: E !! R"α . $\supset$:. $x \epsilon R$"$\alpha . \supset_x . \psi x : \equiv : y \epsilon \alpha . \supset_y . \psi(R'y)$

Dem.

$\vdash$. *37·1 . $\supset \vdash$ :: $x \epsilon R$"$\alpha . \supset_x . \psi x : \equiv :. (\exists y) . y \epsilon \alpha . xRy . \supset_x . \psi x$:.

[*10·23] $\equiv :. y \epsilon \alpha . xRy . \supset_{x,y} . \psi x$:.

[*11·2·62] $\equiv :. y \epsilon \alpha . \supset_y : xRy . \supset_x . \psi x$ (1)

$\vdash$. *37·104 . $\supset \vdash$ ::. Hp . $\supset$ :: $y \epsilon \alpha . \supset_y$:. E ! $R'y$:.

[*30·33] $\supset_y$:. $\psi(R'y) . \equiv : xRy . \supset_x . \psi x$ (2)

$\vdash$. (1) . (2) . $\supset \vdash$. Prop

This proposition is very frequently used.

* *Principles of Science*, chap. I. (p. 18 of edition of 1887).

***37·64.** $\vdash :. \mathrm{E} ! ! R``\alpha . \supset : (\exists y) . y \epsilon \alpha . \psi (R`y) . \equiv . (\exists x) . x \epsilon R``\alpha . \psi x$

Dem.

$\vdash . *30·33 . \supset \vdash :: \mathrm{Hp} . \supset :. y \epsilon \alpha . \supset : \psi (R`y) . \equiv . (\exists x) . xRy . \psi x :.$

[*5·32] $\supset :. y \epsilon \alpha . \psi (R`y) . \equiv : y \epsilon \alpha : (\exists x) . xRy . \psi x$ (1)

$\vdash . (1) . *10·11·21·281 . \supset$

$\vdash :: \mathrm{Hp} . \supset :. (\exists y) . y \epsilon \alpha . \psi (R`y) . \equiv : (\exists y) : y \epsilon \alpha : (\exists x) . xRy . \psi x :$

[*11·6] $\equiv : (\exists x) : (\exists y) . y \epsilon \alpha . xRy : \psi x :$

[*37·1] $\equiv : (\exists x) . x \epsilon R``\alpha . \psi x :: \supset \vdash . \mathrm{Prop}$

***37·65.** $\vdash : \mathrm{E} ! ! R``\beta . \alpha \subset R``\beta . \supset . \alpha = R``(\breve{R}``\alpha \cap \beta)$

Dem.

$\vdash . *30·21 . *3·27 . \supset \vdash :: \mathrm{Hp} . \supset :. y \epsilon \beta . \supset_y : zRy . xRy . \supset . z = x$ (1)

$\vdash . *37·1 . \supset \vdash :. \mathrm{Hp} . \supset :$

$x \epsilon R``(\breve{R}``\alpha \cap \beta) . \equiv . (\exists y) . y \epsilon \breve{R}``\alpha \cap \beta . xRy .$

[*37·105.*11·55] $\equiv . (\exists y, z) . z \epsilon \alpha . zRy . y \epsilon \beta . xRy .$

[(1).*4·71] $\equiv . (\exists y, z) . z \epsilon \alpha . zRy . y \epsilon \beta . xRy . z = x .$

[*13·194] $\equiv . (\exists y, z) . z \epsilon \alpha . y \epsilon \beta . xRy . z = x .$

[*13·195] $\equiv . (\exists y) . x \epsilon \alpha . y \epsilon \beta . xRy .$

[*10·35.*37·1] $\equiv . x \epsilon \alpha . x \epsilon R``\beta .$

[*4·71.Hp] $\equiv . x \epsilon \alpha :. \supset \vdash . \mathrm{Prop}$

***37·66.** $\vdash :. \mathrm{E} ! ! R``\beta . \supset : \alpha \subset R``\beta . \equiv . (\exists \gamma) . \gamma \subset \beta . \alpha = R``\gamma$

Dem.

$\vdash . *37·65 . \mathrm{Exp} . *13·195 . *22·43 . \supset$

$\vdash :. \mathrm{Hp} . \supset : \alpha \subset R``\beta . \supset . (\exists \gamma) . \gamma \subset \beta . \alpha = R``\gamma$ (1)

$\vdash . *37·2 . *13·13 . \supset \vdash : \gamma \subset \beta . \alpha = R``\gamma . \supset . \alpha \subset R``\beta :$

[*10·11·23] $\supset \vdash : (\exists \gamma) . \gamma \subset \beta . \alpha = R``\gamma . \supset . \alpha \subset R``\beta$ (2)

$\vdash . (1) . (2) . \supset \vdash . \mathrm{Prop}$

***37·67.** $\vdash :. z \epsilon \gamma . \supset_z . \mathrm{E} ! R`S`z : \supset . R``S``\gamma = \hat{x} \{(\exists z) . z \epsilon \gamma . x = R`S`z\}$

Dem.

$\vdash . *34·41 . \supset \vdash : \mathrm{Hp} . z \epsilon \gamma . \supset_z . R`S`z = (R \mid S)`z$ (1)

$\vdash . (1) . *14·21 . \supset \vdash : \mathrm{Hp} . z \epsilon \gamma . \supset_z . \mathrm{E} ! (R \mid S)`z$ (2)

$\vdash . (2) . *37·6 . \supset \vdash : \mathrm{Hp} . \supset . (R \mid S)``\gamma = \hat{x} \{(\exists z) . z \epsilon \gamma . x = (R \mid S)`\gamma\}$

[(1)] $= \hat{x} \{(\exists z) . z \epsilon \gamma . x = R`S`\gamma\}$ (3)

$\vdash . *37·33 . \supset \vdash . R``S``\gamma = (R \mid S)``\gamma$ (4)

$\vdash . (3) . (4) . \supset \vdash . \mathrm{Prop}$

***37·68.** $\vdash :. z \epsilon \gamma . \supset_z . P`Q`z = R`z : \supset . P``Q``\gamma = R``\gamma$

Dem.

$\vdash . *14·21 . \supset \vdash : \mathrm{Hp} . z \epsilon \gamma . \supset . \mathrm{E} ! P`Q`z . \mathrm{E} ! R`z .$

[*34·41] $\supset . P`Q`z = (P \mid Q)`z . \mathrm{E} ! R`z .$ (1)

[*14·21·131·144.Hp] $\supset . \mathrm{E} ! (P \mid Q)`z . (P \mid Q)`z = R`z$ (2)

$\vdash . *37\cdot33 . \supset \vdash . P``Q``\gamma = (P \mid Q)``\gamma$ (3)

$\vdash . (2) . (3) . *37\cdot6 . \supset$

$\vdash : \text{Hp} . \supset . P``Q``\gamma = \hat{x}\{(\exists z) . z \in \gamma . x = (P \mid Q)`z\}$

$[(2)] \qquad = \hat{x}\{(\exists z) . z \in \gamma . x = R`z\}$

$[*37\cdot6.(1)] \qquad = R``z : \supset \vdash . \text{Prop}$

***37·69.** $\vdash :. y \in \beta . \supset_y . R`y = S`y : \supset . R``\beta = S``\beta$

Dem.

$\vdash . *14\cdot21 . \supset \vdash :: \text{Hp} . \supset :. y \in \beta . \supset . E ! R`y . E ! S`y :.$ (1)

$[*30\cdot4] \qquad \supset :. y \in \beta . \supset : xRy . \equiv . x = R`y .$

$[*14\cdot142] \qquad \equiv . x = S`y .$

$[*30\cdot4.(1)] \qquad \equiv . xSy :.$

$[*5\cdot32] \qquad \supset :. y \in \beta . xRy . \equiv . y \in \beta . xSy$ (2)

$\vdash . (2) . *10\cdot11\cdot21\cdot281 . \supset$

$\vdash :. \text{Hp} . \supset : (\exists y) . y \in \beta . xRy . \equiv . (\exists y) . y \in \beta . xSy :$

$[*37\cdot1] \supset : x \in R``\beta . \equiv . x \in S``\beta :. \supset \vdash . \text{Prop}$

A specially important case of $R``\beta$ is $\overrightarrow{R}``\beta$ or $\overleftarrow{R}``\beta$. This case will be further studied later (in *70); for the present, we shall only give a few preliminary propositions about it. It will be observed that the hypothesis $E !! \overrightarrow{R}``\beta$ or $E !! \overleftarrow{R}``\beta$ is always verified, in virtue of *32·12·121. Hence the following applications of *37·6 ff.:

***37·7.** $\vdash . \overrightarrow{R}``\beta = \hat{\alpha}\{(\exists y) . y \in \beta . \alpha = \overrightarrow{R}`y\}$ [*37·6 . *32·12]

***37·701.** $\vdash . \overleftarrow{R}``\alpha = \hat{\beta}\{(\exists x) . x \in \alpha . \beta = \overleftarrow{R}`x\}$ [*37·6 . *32·121]

***37·702.** $\vdash :. \overrightarrow{R}``\beta \subset \kappa . \equiv : y \in \beta . \supset_y . \overrightarrow{R}`y \in \kappa$ [*37·61]

***37·703.** $\vdash :. \overleftarrow{R}``\beta \subset \kappa . \equiv : x \in \beta . \supset_x . \overleftarrow{R}`x \in \kappa$ [*37·61]

***37·704.** $\vdash : y \in \alpha . \supset . \overrightarrow{R}`y \in \overrightarrow{R}``\alpha$ [*37·62 . *32·12]

***37·705.** $\vdash : x \in \alpha . \supset . \overleftarrow{R}`x \in \overleftarrow{R}``\alpha$ [*37·62 . *32·121]

***37·706.** $\vdash :. \alpha \in \overrightarrow{R}``\beta . \supset_\alpha . \psi\alpha : \equiv : y \in \beta . \supset_y . \psi(\overrightarrow{R}`y)$ [*37·63]

***37·707.** $\vdash :. \beta \in \overleftarrow{R}``\alpha . \supset_\beta . \psi\beta : \equiv : x \in \alpha . \supset_x . \psi(\overleftarrow{R}`x)$ [*37·63]

***37·708.** $\vdash :. (\exists \alpha) . \alpha \in \overrightarrow{R}``\beta . \psi\alpha . \equiv . (\exists y) . y \in \beta . \psi(\overrightarrow{R}`y)$ [*37·64]

***37·709.** $\vdash :. (\exists \alpha) . \alpha \in \overleftarrow{R}``\beta . \psi\alpha . \equiv . (\exists x) . x \in \beta . \psi(\overleftarrow{R}`x)$ [*37·64]

***37·71.** $\vdash : \kappa \subset \overrightarrow{R}``\beta . \supset . \kappa = \overrightarrow{R}``\{(\text{Cnv}`\overrightarrow{R})``\kappa \cap \beta\}$ [*37·65]

***37·711.** $\vdash : \kappa \subset \overleftarrow{R}``\beta . \supset . \kappa = \overleftarrow{R}``\{(\text{Cnv}`\overleftarrow{R})``\kappa \cap \beta\}$ [*37·65]

***37·712.** $\vdash : \kappa \subset \overrightarrow{R}``\beta . \equiv . (\exists \gamma) . \gamma \subset \beta . \kappa = \overrightarrow{R}``\gamma$ [*37·66]

***37·713.** $\vdash : \kappa \subset \overleftarrow{R}``\beta . \equiv . (\exists \gamma) . \gamma \subset \beta . \kappa = \overleftarrow{R}``\gamma$ [*37·66]

***37·72.** $\vdash : R = P \mid Q . \supset . \overrightarrow{R}``\gamma = P_\epsilon``\overrightarrow{Q}``\gamma$

Dem.

$\vdash . *37·11·302 . \supset \vdash : \text{Hp} . \supset . (z) . P_\epsilon`\overrightarrow{Q}`z = \overrightarrow{R}`z .$

[*37·68] $\supset . P_\epsilon``\overrightarrow{Q}``\gamma = \overrightarrow{R}``\gamma .$

[(*37·04)] $\supset . P_\epsilon``\overrightarrow{Q}``\gamma = \overrightarrow{R}``\gamma : \supset \vdash . \text{Prop}$

***37·721.** $\vdash : R = P \mid Q . \supset . \overleftarrow{R}``\gamma = \breve{Q}_\epsilon``\overleftarrow{P}``\gamma$ [Proof as in *37·72]

***37·73.** $\vdash : \exists ! \beta . \equiv . \exists ! \overrightarrow{R}``\beta . \equiv . \exists ! \overleftarrow{R}``\beta$ [*37·45 . *32·12·121]

***37·731.** $\vdash : \beta = \Lambda . \equiv . \overrightarrow{R}``\beta = \Lambda . \equiv . \overleftarrow{R}``\beta = \Lambda$ [*37·73 . Transp]

Observe that the Λ's which occur in this proposition will not be all of the same type. *E.g.* if R relates individuals to individuals, the first Λ will be the class of no individuals, while the second and third will be the class of no classes. Thus the ambiguity which attaches to the type of Λ must be differently determined for different occurrences of Λ in this proposition. In general, when this is the case with our ambiguous symbols, we shall adopt a notation which indicates the fact. But when the ambiguous symbol is Λ, it seems hardly worth while.

***37·74.** $\vdash :. \beta \subset \text{Cl}`R . \equiv : \alpha \in \overrightarrow{R}``\beta . \supset_\alpha . \exists ! \alpha$

Dem.

$\vdash . *37·706 . \supset \vdash :. \alpha \in \overrightarrow{R}``\beta . \supset_\alpha . \exists ! \alpha : \equiv : y \in \beta . \supset_y . \exists ! \overrightarrow{R}`y :$

[*33·31] $\equiv : \beta \subset \text{Cl}`R :. \supset \vdash . \text{Prop}$

***37·75.** $\vdash :. \alpha \subset \text{D}`R . \equiv : \beta \in \overleftarrow{R}``\alpha . \supset_\beta . \exists ! \beta$ [Proof as in *37·74]

***37·76.** $\vdash . \overrightarrow{R}``\beta \subset \text{Cls}$

Dem.

$\vdash . *37·7 . \supset \vdash :. \alpha \in \overrightarrow{R}``\beta . \supset : (\exists y) . y \in \beta . \alpha = \overrightarrow{R}`y :$

[*10·5] $\supset : (\exists y) . \alpha = \overrightarrow{R}`y :$

[*32·13] $\supset : (\exists y) . \alpha = \hat{x}(xRy) :$

[*20·16] $\supset : (\exists \phi) . \alpha = \hat{x}(\phi ! x) :$

[*20·4] $\supset : \alpha \in \text{Cls} :. \supset \vdash . \text{Prop}$

***37·761.** $\vdash . \overleftarrow{R}``\alpha \subset \text{Cls}$ [Proof as in *37·76]

***37·77.** $\vdash : \alpha \in \overrightarrow{R}``\text{Cl}`R . \supset_\alpha . \exists ! \alpha$ [*37·74 . *22·42]

***37·771.** $\vdash : \beta \in \overleftarrow{R}``\text{D}`R . \supset_\beta . \exists ! \beta$ [Proof as in *37·77]

***37·772.** $\vdash . \Lambda \sim\in \overrightarrow{R}``\text{Cl}`R$ [*37·77 . *24·63]

***37·773.** $\vdash . \Lambda \sim\in \overleftarrow{R}``\text{D}`R$ [*37·771 . *24·63]

***37·78.** $\vdash . \text{D}`\overrightarrow{R} = \overrightarrow{R}``\text{V}$ [*37·28]

***37·781.** $\vdash . \text{D}`\overleftarrow{R} = \overleftarrow{R}``\text{V}$ [*37·28]

***37·79.** $\vdash . \overrightarrow{R}``\text{V} = \hat{\alpha}\{(\exists y) . \alpha = \overrightarrow{R}`y\}$ [*37·601 . *32·12]

***37·791.** $\vdash . \overleftarrow{R}``\text{V} = \hat{\beta}\{(\exists x) . \beta = \overleftarrow{R}`x\}$ [*37·601 . *32·121]

***37·8.** $\vdash . (\alpha \uparrow \beta) | S = \alpha \uparrow \breve{S}``\beta$

Dem.

$\vdash . *35·103 . *34·1 . \supset \vdash : x\{(\alpha \uparrow \beta) | S\} z . \equiv . (\exists y) . x \epsilon \alpha . y \epsilon \beta . ySz .$

[*10·35.*37·105] $\equiv . x \epsilon \alpha . z \epsilon \breve{S}``\beta .$

[*35·103] $\equiv . x (\alpha \uparrow \breve{S}``\beta) z : \supset \vdash .$ Prop

***37·81.** $\vdash . R | (\alpha \uparrow \beta) = (R``\alpha) \uparrow \beta$ [Proof as in *37·8]

***37·82.** $\vdash . R | (\alpha \uparrow \beta) | S = (R``\alpha) \uparrow (\breve{S}``\beta)$ [*37·8·81]

*38. RELATIONS AND CLASSES DERIVED FROM A DOUBLE DESCRIPTIVE FUNCTION

Summary of *38.

A double descriptive function is a non-propositional function of two arguments, such as $\alpha \cap \beta$, $\alpha \cup \beta$, $R \dot{\cap} S$, $R \dot{\cup} S$, $R \mid S$, $\alpha \upharpoonleft R$, $R \upharpoonright \alpha$, $R \,\underset{\cdot}{\upharpoonright}\, \alpha$. The propositions of the present number apply to all such functions, assuming the notation to be (as in the above instances) a functional sign placed between the two arguments. In order to deal with all analogous cases at once, we shall in this number adopt the notation

$$x \mathbin{♀} y,$$

where "♀" stands for any such sign as $\cap$, $\cup$, $\dot{\cap}$, $\dot{\cup}$, $\mid$, $\upharpoonleft$, $\upharpoonright$, $\underset{\cdot}{\upharpoonright}$, or any functional sign to be hereafter defined and satisfying the condition

$$(x, y) \,.\, \mathrm{E}\,!\,(x \mathbin{♀} y).$$

The derived relations and classes with which we shall be concerned may be illustrated by taking the case of $\alpha \cap \beta$. The relation of $\alpha \cap \beta$ to β will be written $\alpha \cap$, and the relation of $\alpha \cap \beta$ to α will be written $\cap \beta$. Thus we shall have

$$\vdash \,.\, \alpha \cap \beta = \alpha \cap \text{‘}\beta = \cap \beta \text{‘}\alpha.$$

The utility of this notation is chiefly due to the possibility of such notations as $\alpha \cap \text{“}\kappa$ and $\cap \beta \text{“}\kappa$. For example, take such a phrase as "the foreign members of English Clubs." Then if we put $\alpha =$ foreigners, $\kappa =$ English Clubs, we have

$\alpha \cap \text{“}\kappa =$ the classes of foreign members of the various English Clubs.

Or again, let α be a conic, and κ a pencil of lines; then

$\alpha \cap \text{“}\kappa =$ the various pairs of points in which members of κ meet α.

In this case, since $\alpha \cap \beta = \beta \cap \alpha$, we have $\alpha \cap = \cap \alpha$. But when the function concerned is not commutative, this does not hold. Thus for example we do not have $R \mid = \mid R$.

The notations of this number will be frequently applied hereafter to $R \mid S$. In accordance with what was said above, we write $R \mid$ for the relation of $R \mid S$ to S, and $\mid S$ for the relation of $R \mid S$ to R. Hence we have

$$R \mid \text{‘}S = \mid S \text{‘}R = R \mid S.$$

Hence $\mid S \text{“}\lambda$ will be the class of relations obtained by taking members of λ and relatively multiplying them by S. Thus if λ were the class of relations first cousin, second cousin, etc., and S were the relation of parent to child, $\mid S \text{“}\lambda$ would be the class of relations first cousin once removed, second cousin once removed, etc., taken in the sense which goes from the older to the younger generation.

It is often convenient to be able to exhibit $|S``\lambda$ and kindred expressions as descriptive functions of the first argument instead of the second. For this purpose we put

$$\lambda \underset{,,}{|} S = |S``\lambda$$

with similar notations for other descriptive double functions. We then have, just as in the case of $R|S$,

$$\lambda \underset{,,}{|} `S = \underset{,,}{|} S`\lambda = \lambda \underset{,,}{|} S.$$

This enables us to form the class $\lambda \underset{,,}{|} ``\mu$. This class is chiefly useful because the members of its members (*i.e.* $s`\lambda \underset{,,}{|} ``\mu$, as we shall define it in $*40$) constitute the class of all products $R|S$ that can be formed of a member of λ and a member of μ.

Thus we are led to three general definitions for descriptive double functions, namely (if $x ♀ y$ be any such function)

$x ♀$ is the relation of $x ♀ y$ to y for any y,
$♀ y$,, ,, ,, ,, ,, x ,, x,
$\alpha \underset{,,}{♀} y$ is the class of values of $x ♀ y$ when x is an α.

Since $\alpha \underset{,,}{♀} y$ is again a descriptive double function, the first two of the above definitions can be applied to it. The third definition, for typographical reasons, cannot be applied conveniently, though theoretically it is of course applicable. The relations $x ♀$ and $♀ y$ represent the general idea contained in some of the uses in mathematics of the term "operation," *e.g.* $+1$ is the operation of adding 1.

The uses of the notations introduced in the present number occur chiefly in arithmetic (Parts III and IV). Few propositions can be given at this stage, since most of the important uses of the notation here introduced depend upon the substitution of some special function for the general function "♀" here used. In the present number, the propositions given are all immediate consequences of the definitions.

$*38{\cdot}01$. $x ♀ = \hat{u}\hat{y}(u = x ♀ y)$ Df

$*38{\cdot}02$. $♀ y = \hat{u}\hat{x}(u = x ♀ y)$ Df

$*38{\cdot}03$. $\alpha \underset{,,}{♀} y = ♀ y``\alpha$ Df

$*38{\cdot}1$. $\vdash : u(x ♀) y . \equiv . u = x ♀ y$ [$(*38{\cdot}01)$]

$*38{\cdot}101$. $\vdash : u(♀ y) x . \equiv . u = x ♀ y$ [$(*38{\cdot}02)$]

$*38{\cdot}11$. $\vdash . x ♀ `y = ♀ y`x = x ♀ y$ [$*38{\cdot}1{\cdot}101 . *30{\cdot}3$]

$*38{\cdot}12$. $\vdash . E ! x ♀ `y . E ! ♀ y`x$ [$*38{\cdot}11 . *14{\cdot}21$]

$*38{\cdot}13$. $\vdash : u \epsilon x ♀ ``\alpha . \equiv . (\exists y) . y \epsilon \alpha . u = x ♀ y$ [$*38{\cdot}1 . *37{\cdot}1$]

$*38{\cdot}131$. $\vdash : u \epsilon ♀ y``\alpha . \equiv . (\exists x) . x \epsilon \alpha . u = x ♀ y$ [$*38{\cdot}101 . *37{\cdot}1$]

***38·2.** $\vdash . \alpha \underset{,,}{\text{♀}} y = \text{♀} y \lq\lq \alpha$ [(*38·03)]

***38·21.** $\vdash . \alpha \underset{,,}{\text{♀}} y = \hat{u} \{(\exists x) . x \in \alpha . u = x \,\text{♀}\, y\}$ [*38·2·131]

***38·22.** $\vdash . \alpha \underset{,,}{\text{♀}} \lq y = \underset{,,}{\text{♀}} y \lq \alpha = \alpha \underset{,,}{\text{♀}} y$ [*38·11]

***38·23.** $\vdash . \mathrm{E} ! \, \alpha \underset{,,}{\text{♀}} \lq y . \mathrm{E} ! \underset{,,}{\text{♀}} y \lq \alpha$ [*38·22 . *14·21]

***38·24.** $\vdash : \exists ! \, \alpha \underset{,,}{\text{♀}} y . \equiv . \exists ! \, \alpha$

Dem.

$\vdash . *38\cdot2 . *37\cdot29 . \text{Transp} . \supset \vdash : \exists ! \, \alpha \underset{,,}{\text{♀}} y . \supset . \exists ! \, \alpha$ (1)

$\vdash . *38\cdot21 . \supset \vdash : x \in \alpha . \supset . (x \,\text{♀}\, y) \in \alpha \underset{,,}{\text{♀}} y .$

$[*10\cdot24] \qquad \supset . \exists ! \, \alpha \underset{,,}{\text{♀}} y$ (2)

$\vdash . (1) . (2) . \supset \vdash . \text{Prop}$

***38·3.** $\vdash . \alpha \underset{,,}{\text{♀}} \lq\lq \beta = \hat{\gamma} \{(\exists y) . y \in \beta . \gamma = \alpha \underset{,,}{\text{♀}} y\} = \hat{\gamma} \{(\exists y) . y \in \beta . \gamma = \text{♀} y \lq\lq \alpha\}$

[*38·13·2]

***38·31.** $\vdash . \underset{,,}{\text{♀}} y \lq\lq \kappa = \hat{\gamma} \{(\exists \alpha) . \alpha \in \kappa . \gamma = \alpha \underset{,,}{\text{♀}} y\} = \hat{\gamma} \{(\exists \alpha) . \alpha \in \kappa . \gamma = \text{♀} y \lq\lq \alpha\} = \text{♀} y \lq\lq\lq \kappa$

[*38·131·2 . *37·103]

NOTE TO SECTION D

General Observations on Relations. The notion of "relation" is so general that it is important to realize the different sorts of relations to which the notations defined in the preceding section may be applied. It often happens that a proposition which holds for any relation is only important for relations of certain kinds; hence it is desirable that the reader should have in mind some of the principal kinds of relations. Of the various uses to which different sorts of relations may be put, there are three which are specially important, namely (1) to give rise to descriptive functions, (2) to establish correlations between different classes, (3) to generate series. Let us consider these in succession.

(1) In order that a relation R may give rise to a descriptive function, it must be such that the referent is unique when the relatum is given. Thus, for example, the relations Cnv, $\overrightarrow{R}$, $\overleftarrow{R}$, D, Ɑ, C, R_ϵ, defined above, all give rise to descriptive functions. In general, if R gives rise to a descriptive function, there will be a certain class, namely Ɑ‘R, to which the argument of the function must belong in order that the function may have a value for that argument. For example, taking the sine as an illustration, and writing "sin‘y" instead of "sin y," y must be a number in order that sin‘y may exist. Then $\breve{\text{sin}}$ is the relation of y to x when $x = \text{sin}‘y$. If we put $\alpha =$ numbers between $-\pi/2$ and $\pi/2$, both included, $\text{sin} \upharpoonright \alpha$ will be the relation of x to y when $x = \text{sin}‘y$ and $-\pi/2 \leqslant y \leqslant \pi/2$. The converse of this relation, which is $\alpha \upharpoonleft \breve{\text{sin}}$, will also give rise to a descriptive function; thus $(\alpha \upharpoonleft \breve{\text{sin}})‘x =$ that value of $\text{sin}^{-1} x$ which lies between $-\pi/2$ and $\pi/2$. This illustrates a case which arises very frequently, namely, that a relation R does not, as it stands, give rise to a descriptive function, but does do so when its domain or converse domain is suitably limited. Thus for example the relation "parent" does not give rise to a descriptive function, but does do so when its domain is limited to males or limited to females. The relation "square root," similarly, gives rise to a descriptive function when its domain is limited to positive numbers, or limited to negative numbers. The relation "wife" gives rise to a descriptive function when its converse domain is limited to Christian men, but not when Mohammedans are included. The domain of a relation which gives rise to a descriptive function without limiting its domain or converse domain consists of all possible values of the function; the converse domain consists of all possible arguments to the function. Again, if R gives rise to a descriptive function, $\overleftarrow{R}‘x$ will be the class of those arguments for which the value of the function is x. Thus $\overleftarrow{\text{sin}}‘x$ consists of all numbers

whose sine is x, *i.e.* all values of $\sin^{-1} x$. Again, $\sin``\alpha$ will be the sines of the various members of $\alpha$. If $\alpha$ is a class of numbers, then, by the notation of $*38$, $2 \times ``\alpha$ will be the doubles of those numbers, $3 \times ``\alpha$ the trebles of them, and so on. To take another illustration, let $\alpha$ be a pencil of lines, and let $R`x$ be the intersection of a line $x$ with a given transversal. Then $R``\alpha$ will be the intersections of lines belonging to the pencil with the transversal.

(2) Relations which establish a correlation between two classes are really a particular case of relations giving rise to descriptive functions, namely the case in which the converse relation also gives rise to a descriptive function. In this case, the relation is "one-one," *i.e.* given the referent, the relatum is determinate, and vice versa. A relation which is to be conceived as a correlation will generally be denoted by S or T. In such cases, we are as a rule less interested in the particular terms x and y for which xRy, than in classes of such terms. We generally, in such cases, have some class β contained in the converse domain of our relation S, and we have a class α such that $\alpha = S``\beta$. In this case, the relation $S$ correlates the members of $\alpha$ and the members of $\beta$. We shall have also $\beta = \breve{S}``\alpha$, so that, for such a relation, the correlation is reciprocal. Such relations are fundamental in arithmetic, since they are used in defining what is meant by saying that two classes (or series) have the same cardinal (or ordinal) number of terms.

(3) Relations which give rise to series will in general be denoted by P or Q, and in propositions whose chief importance lies in their application to series we shall also, as a rule, denote a variable relation by P or Q. When P is used, it may be read as "precedes." Then $\breve{P}$ may be read "follows," $\overrightarrow{P}`x$ may be read "predecessors of $x$," $\overleftarrow{P}`x$ may be read "followers of x." $\text{D}`P$ will be all members of the series generated by $P$ except the last (if any), $\text{ᗡ}`P$ will be all members of the series except the first (if any), $C`P$ will be all the members of the series. $P``\alpha$ will consist of all terms preceding some member of $\alpha$. Suppose, for example, that our series is the series of real numbers, and that $\alpha$ is the class of members of an ascending series $x_1, x_2, x_3, \ldots x_\nu, \ldots$. Then $P``\alpha$ will be the segment of the real numbers defined by this series, *i.e.* it will be all the predecessors of the limit of the series. (In the event of the series $x_1, x_2, x_3, \ldots x_\nu, \ldots$ growing without limit, $P``\alpha$ will be the whole series of real numbers.)

It very often happens that a relation has more or less of a serial character, without having all the characteristics necessary for generating series. Take, for example, the relation of son to father. It is obvious that by means of this relation series can be generated which start from any man and end with Adam. But these series are not the field of the relation in question; moreover this relation is not *transitive*, *i.e.* a son of a son of x is not a son of x. If, however, we substitute for "son" the relation "descendant in the direct male line" (which can be defined in terms of "son" by the method explained in *90 and *91), and if we limit the converse domain of this relation to ancestors of x in the direct male line, we obtain a new relation which *is* serial, and has for its field x and all his ancestors in the direct male line. Again, one relation may generate a number of series, as for example the relation "x is east of y." If x and y are points on the earth's surface, and in the eastern hemisphere, this relation generates one series for every parallel of latitude. By confining the field of the relation further to one parallel of latitude, we obtain a relation which generates a series. (The reason for confining x and y to one hemisphere is to insure that the relation shall be transitive, since otherwise we might have x east of y and y east of z, but x west of z.)

A relation may have the characteristics of all the three kinds of relations, provided we include in the third kind all those which lead to series by some such limitations as those just described. For example, the relation $+1$, *i.e.* (in virtue of the notation of *38) the relation of $x+1$ to x, where x is supposed to be a finite cardinal integer, has the characteristics of all three kinds of relations. In the first place, it leads to the descriptive function $(+1)'x$, *i.e.* $x+1$. In the second place, it correlates with any class α of numbers the class obtained by adding 1 to each member of α, *i.e.* $(+1)``\alpha$. This correlation may be used to prove that the number of finite integers is infinite (in one of the two senses of the word "infinite"); for if we take as our class $\alpha$ all the natural numbers including 0, the class $(+1)``\alpha$ consists of all the natural numbers except 0, so that the natural numbers can be correlated with a proper part* of themselves. Again, the relation $+1$ may be used, like that of father to son, to generate a series, namely the usual series of the natural numbers in order of magnitude, in which each has to its immediate predecessor the relation $+1$. Thus this relation partakes of the characteristics of all three kinds of relations.

* *I.e.* a part not the whole. On this definition of infinity, see *124.

SECTION E

PRODUCTS AND SUMS OF CLASSES

Summary of Section E.

In the present section, we make an extension of $\alpha \cap \beta$, $\alpha \cup \beta$, $R \dot{\cap} S$, $R \dot{\cup} S$. Given a class of classes, say κ, the *product* of κ (which is denoted by $p`\kappa$) is the common part of all the members of κ, *i.e.* the class consisting of those terms which belong to every member of κ. The definition is

$$p`\kappa = \hat{x}(\alpha \epsilon \kappa \,.\, \supset_\alpha \,.\, x \epsilon \alpha) \quad \text{Df.}$$

If κ has only two members, α and β say, $p`\kappa = \alpha \cap \beta$. If $\kappa$ has three members, $\alpha, \beta, \gamma$, then $p`\kappa = \alpha \cap \beta \cap \gamma$; and so on. But this process can only be continued to a finite number of terms, whereas the definition of $p`\kappa$ does not require that κ should be finite. This notion is chiefly important in connection with the lower limits of series. For example, let λ be the class of rational numbers whose square is greater than 2, and let "xMy" mean "$x < y$, where x and y are rationals." Then if $x \epsilon \lambda$, $\overrightarrow{M}`x$ will be the class of rationals less than x. Thus $\overrightarrow{M}``\lambda$ will be the class of such classes as $\overrightarrow{M}`x$, where $x \epsilon \lambda$. Thus the product of $\overrightarrow{M}``\lambda$, which we call $p`\overrightarrow{M}``\lambda$, will be the class of rationals which are less than every member of $\lambda$, *i.e.* the class of rationals whose squares are less than 2. Each member of $\overrightarrow{M}``\lambda$ is a segment of the series of rationals, and $p`\overrightarrow{M}``\lambda$ is the lower limit of these segments. It is thus that we prove the existence of lower limits of series of segments.

Similarly the *sum* of a class of classes κ is defined as the class consisting of all terms belonging to *some* member of κ; *i.e.*

$$s`\kappa = \hat{x}\{(\exists \alpha) \,.\, \alpha \epsilon \kappa \,.\, x \epsilon \alpha\} \quad \text{Df,}$$

i.e. x belongs to the sum of κ if x belongs to some κ. This notion plays the same part for upper limits of series of segments as $p`\kappa$ plays for lower limits. It has, however, many more other uses than $p`\kappa$, and is altogether a more important conception. Thus in cardinal arithmetic, if no two members of κ have any term in common, the arithmetical sum of the numbers of members possessed by the various members of κ is the number of members possessed by $s`\kappa$.

The product of a class of relations (λ say) is the relation which holds between x and y when x and y have every relation of the class λ. The definition is

$$\dot{p}`\lambda = \hat{x}\hat{y}(R \epsilon \lambda \,.\, \supset_R \,.\, xRy) \quad \text{Df.}$$

The properties of $\dot{p}`\lambda$ are analogous to those of $p`\kappa$, but its uses are fewer.

The sum of a class of relations (λ say) is the relation which holds between x and y whenever there is a relation of the class λ which holds between x and y. The definition is

$$\dot{s}`\lambda = \hat{x}\hat{y}\{(\exists R) . R \epsilon \lambda . xRy\} \quad \text{Df.}$$

This conception, though less important than $s`\kappa$, is more important than $\dot{p}`\lambda$. The summation of series and ordinal numbers depends upon it, though the connection is less immediate than that of the summation of cardinal numbers with $s`\kappa$.

Instead of defining $p`\kappa$, $s`\kappa$, $\dot{p}`\lambda$, $\dot{s}`\lambda$, it would be formally more correct to define p, s, $\dot{p}$ and $\dot{s}$, which are the relations giving rise to the above descriptive functions. Thus we should have

$$p = \hat{\beta}\hat{\kappa}\{\beta = \hat{x}(\alpha \epsilon \kappa . \supset_\alpha . x \epsilon \alpha)\} \quad \text{Df,}$$

whence we should proceed to

$$\vdash : \beta p \kappa . \equiv . \beta = \hat{x}(\alpha \epsilon \kappa . \supset_\alpha . x \epsilon \alpha),$$

$$\vdash . p`\kappa = \hat{x}(\alpha \epsilon \kappa . \supset_\alpha . x \epsilon \alpha),$$

and $\vdash . \text{E} ! p`\kappa$.

But in cases where the relation, as opposed to the descriptive function, is very seldom required, it is simpler and easier to give the definition of the descriptive function in the first instance. In such cases, the relation is always tacitly assumed to be also defined; *i.e.* when we give a definition of the form

$$R`x = S`x \quad \text{Df,}$$

where S is some previously defined relation, we always assume that this definition is to be regarded as derived from

$$R = \hat{u}\hat{x}(u = S`x) \quad \text{Df.}$$

In addition to products and sums, we deal, in the present section, with certain properties of the relations $R|$ and $|S$, the meanings of which result from the notation introduced in *38. Such relations are very useful in arithmetic. The reason for dealing with them in the present section is that a large proportion of the propositions to be proved involve sums of classes of classes or relations.

*40. PRODUCTS AND SUMS OF CLASSES OF CLASSES

Summary of *40.

In this number, we introduce the two notations (explained above)

$$p'\kappa = \hat{x}(\alpha \epsilon \kappa . \supset_\alpha . x \epsilon \alpha) \quad \text{Df}$$

$$s'\kappa = \hat{x}\{(\exists \alpha) . \alpha \epsilon \kappa . x \epsilon \alpha\} \quad \text{Df}$$

Both these notations will be found increasingly useful as we proceed, but $s'\kappa$ remains more useful than $p'\kappa$ throughout. It is required for the significance of $p'\kappa$ and $s'\kappa$ that κ should be a class of classes.

In the present number, the most useful propositions are the following:

*40·12. $\vdash : \alpha \epsilon \kappa . \supset . p'\kappa \subset \alpha$

I.e. the product of κ is contained in every member of κ.

*40·13. $\vdash : \alpha \epsilon \kappa . \supset . \alpha \subset s'\kappa$

I.e. every member of κ is contained in the sum of κ.

*40·15. $\vdash :. \beta \subset p'\kappa . \equiv : \gamma \epsilon \kappa . \supset_\gamma . \beta \subset \gamma$

I.e. β is contained in the product of κ if β is contained in every member of κ, and vice versa.

*40·151. $\vdash :. s'\kappa \subset \beta . \equiv : \gamma \epsilon \kappa . \supset_\gamma . \gamma \subset \beta$

I.e. the sum of κ is contained in β if every member of κ is contained in β, and vice versa.

*40·2. $\vdash : \kappa = \Lambda . \supset . p'\kappa = V$

I.e. the product of the null-class of classes is the universal class. This may seem paradoxical at first sight, but it is really not so. The fewer members κ has, the larger, speaking generally, $p'\kappa$ becomes. If κ has no members, then κ has no members to which a given term x does not belong, and therefore x belongs to $p'\kappa$.

*40·23. $\vdash : \exists ! \kappa . \supset . p'\kappa \subset s'\kappa$

I.e. unless κ is null, its product is contained in its sum.

*40·38. $\vdash . R``s'\kappa = s'R```\kappa$

This proposition is very often used in arithmetic. What it states is as follows: Given a class of classes κ, take its sum, $s'\kappa$, and then consider all the terms that have the relation R to some member of $s'\kappa$; this gives the class $R``s'\kappa$; next, take each separate member of $\kappa$, say $\alpha$, and form the class $R``\alpha$, consisting of all terms having the relation R to some member of α. The class of all such classes as $R``\alpha$, for various α's which are members of κ, is $R```\kappa$; the sum of this class, by the above proposition, is the same as $R``s'\kappa$.

*40·4. $\vdash :. E !! R``\beta . \supset . s'R``\beta = \hat{x}\{(\exists y) . y \epsilon \beta . x \epsilon R'y\}$

This proposition requires, for significance, that $R'y$ should always be a class. The proposition states that, if $R`y$ always exists when $y \epsilon \beta$, then the sum of all classes which have the relation $R$ to some member of $\beta$ consists of all members of such classes as $R`y$, where $y \epsilon \beta$.

***40·5.** $\vdash . s`\overrightarrow{R}``\beta = R``\beta$

This proposition results from *40·4 by substituting $\overrightarrow{R}$ for R in that proposition.

***40·51.** $\vdash . p`\overrightarrow{R}``\beta = \hat{x}\{y \epsilon \beta . \supset_y . xRy\}$

In virtue of *40·5, $p`\overrightarrow{R}``\beta$ is correlative to $R``\beta$. Thus if R is a serial relation, $p`\overrightarrow{R}``\beta$ consists of terms preceding the whole of $\beta$, and $R``\beta$ consists of terms preceding part of β. If β has a lower limit, it will be the upper limit or maximum of $p`\overrightarrow{R}``\beta$; if $\beta$ has an upper limit, it will be the upper limit of $R``\beta$.

***40·61.** $\vdash : \exists ! \beta . \supset . p`\overrightarrow{R}``\beta \subset R``\beta . p`\overleftarrow{R}``\beta \subset \breve{R}``\beta$

In this proposition the hypothesis is essential, since, if $\beta = \Lambda$, $p`\overrightarrow{R}``\beta = V$ and $R``\beta = \Lambda$.

***40·01.** $p`\kappa = \hat{x}(\alpha \epsilon \kappa . \supset_\alpha . x \epsilon \alpha)$ Df

***40·02.** $s`\kappa = \hat{x}\{(\exists\alpha) . \alpha \epsilon \kappa . x \epsilon \alpha\}$ Df

***40·1.** $\vdash :. x \epsilon p`\kappa . \equiv : \alpha \epsilon \kappa . \supset_\alpha . x \epsilon \alpha$ [*20·3 . (*40·01)]

***40·11.** $\vdash : x \epsilon s`\kappa . \equiv . (\exists\alpha) . \alpha \epsilon \kappa . x \epsilon \alpha$ [*20·3 . (*40·02)]

***40·12.** $\vdash : \alpha \epsilon \kappa . \supset . p`\kappa \subset \alpha$

Dem.

$\vdash .$ *40·1 . *10·1 . $\supset \vdash :. x \epsilon p`\kappa . \supset : \alpha \epsilon \kappa . \supset . x \epsilon \alpha :.$

[Comm] $\supset \vdash :. \alpha \epsilon \kappa . \supset : x \epsilon p`\kappa . \supset . x \epsilon \alpha$ (1)

$\vdash .$ (1) . *10·11·21 . *22·1 . $\supset \vdash .$ Prop

***40·13.** $\vdash : \alpha \epsilon \kappa . \supset . \alpha \subset s`\kappa$

Dem.

$\vdash .$ *40·11 . *10·24 . $\supset \vdash : \alpha \epsilon \kappa . x \epsilon \alpha . \supset . x \epsilon s`\kappa :$

[Exp] $\supset \vdash :. \alpha \epsilon \kappa . \supset : x \epsilon \alpha . \supset . x \epsilon s`\kappa$ (1)

$\vdash .$ (1) . *10·11·21 . *22·1 . $\supset \vdash .$ Prop

***40·14.** $\vdash : \alpha \epsilon \kappa . x \epsilon p`\kappa . \supset . x \epsilon \alpha$ [*40·12 . Imp]

***40·141.** $\vdash : \alpha \epsilon \kappa . x \epsilon \alpha . \supset . x \epsilon s`\kappa$ [*40·11 . *10·24]

***40·15.** $\vdash :. \beta \subset p`\kappa . \equiv : \gamma \epsilon \kappa . \supset_\gamma . \beta \subset \gamma$

Dem.

$\vdash .$ *40·1 . $\supset \vdash :: \beta \subset p`\kappa : \equiv :. x \epsilon \beta . \supset_x : \gamma \epsilon \kappa . \supset_\gamma . x \epsilon \gamma :.$

[*11·62] $\equiv :. (x, \gamma) : x \epsilon \beta . \gamma \epsilon \kappa . \supset . x \epsilon \gamma :.$

[*4·3·84.*11·33] $\equiv :. (x, \gamma) : \gamma \epsilon \kappa . x \epsilon \beta . \supset . x \epsilon \gamma :.$

[*11·2·62] $\equiv :. \gamma \epsilon \kappa . \supset_\gamma : x \epsilon \beta . \supset_x . x \epsilon \gamma :.$

[*22·1] $\equiv :. \gamma \epsilon \kappa . \supset_\gamma . \beta \subset \gamma :: \supset \vdash .$ Prop

***40·151.** $\vdash :. s\text{‘}\kappa \subset \beta . \equiv : \gamma \in \kappa . \supset_\gamma . \gamma \subset \beta$

Dem.

$\vdash . *40\cdot 11 . \supset \vdash :: s\text{‘}\kappa \subset \beta . \equiv :. (\exists \gamma) . \gamma \in \kappa . x \in \gamma . \supset_x . x \in \beta :.$

$[*10\cdot 23] \quad \equiv :. (\gamma, x) :. \gamma \in \kappa . x \in \gamma . \supset . x \in \beta :.$

$[*11\cdot 62] \quad \equiv :. (\gamma) :. \gamma \in \kappa . \supset : (x) : x \in \gamma . \supset . x \in \beta :.$

$[*22\cdot 1] \quad \equiv :. \gamma \in \kappa . \supset_\gamma . \gamma \subset \beta :: \supset \vdash . \text{Prop}$

This proposition is frequently used.

***40·16.** $\vdash : \kappa \subset \lambda . \supset . p\text{‘}\lambda \subset p\text{‘}\kappa$

Dem.

$\vdash . *10\cdot 1 . \supset \vdash :: \text{Hp} . \supset :. \gamma \in \kappa . \supset . \gamma \in \lambda :.$

$[\text{Syll}] \quad \supset :. \gamma \in \lambda . \supset . x \in \gamma : \supset : \gamma \in \kappa . \supset . x \in \gamma \quad (1)$

$\vdash . (1) . *10\cdot 11\cdot 21 . \supset$

$\vdash :: \text{Hp} . \quad \supset :. (\gamma) :. \gamma \in \lambda . \supset . x \in \gamma : \supset : \gamma \in \kappa . \supset . x \in \gamma :.$

$[*10\cdot 27] \quad \supset :. (\gamma) : \gamma \in \lambda . \supset . x \in \gamma : \supset : (\gamma) : \gamma \in \kappa . \supset . x \in \gamma :.$

$[*40\cdot 1] \quad \supset :. x \in p\text{‘}\lambda . \supset . x \in p\text{‘}\kappa \quad (2)$

$\vdash . (2) . *10\cdot 11\cdot 21 . \supset \vdash . \text{Prop}$

***40·161.** $\vdash : \kappa \subset \lambda . \supset . s\text{‘}\kappa \subset s\text{‘}\lambda$

Dem.

$\vdash . *10\cdot 1 . \supset \vdash :. \text{Hp} . \supset : \gamma \in \kappa . \supset . \gamma \in \lambda :$

$[\text{Fact}] \quad \supset : \gamma \in \kappa . x \in \gamma . \supset . \gamma \in \lambda . x \in \gamma :$

$[*10\cdot 11\cdot 28] \quad \supset : (\exists \gamma) . \gamma \in \kappa . x \in \gamma . \supset . (\exists \gamma) . \gamma \in \lambda . x \in \gamma :$

$[*40\cdot 11] \quad \supset : x \in s\text{‘}\kappa . \supset . x \in s\text{‘}\lambda \quad (1)$

$\vdash . (1) . *10\cdot 11\cdot 21 . \supset \vdash . \text{Prop}$

***40·17.** $\vdash . p\text{‘}\kappa \cup p\text{‘}\lambda \subset p\text{‘}(\kappa \cap \lambda)$

Dem.

$\vdash . *22\cdot 34 . \supset \vdash :: x \in p\text{‘}\kappa \cup p\text{‘}\lambda . \equiv :. x \in p\text{‘}\kappa . \vee . x \in p\text{‘}\lambda :.$

$[*40\cdot 1] \quad \equiv :. \gamma \in \kappa . \supset_\gamma . x \in \gamma : \vee : \gamma \in \lambda . \supset_\gamma . x \in \gamma :.$

$[*10\cdot 41] \quad \supset :. (\gamma) :. \gamma \in \kappa . \supset . x \in \gamma : \vee : \gamma \in \lambda . \supset . x \in \gamma :.$

$[*4\cdot 79] \quad \supset :. (\gamma) : \gamma \in \kappa . \gamma \in \lambda . \supset . x \in \gamma :.$

$[*22\cdot 33] \quad \supset :. (\gamma) : \gamma \in \kappa \cap \lambda . \supset . x \in \gamma :.$

$[*40\cdot 1] \quad \supset :. x \in p\text{‘}(\kappa \cap \lambda) \quad (1)$

$\vdash . (1) . *10\cdot 11 . \supset \vdash . \text{Prop}$

***40·171.** $\vdash . s\text{‘}\kappa \cup s\text{‘}\lambda = s\text{‘}(\kappa \cup \lambda)$

Dem.

$\vdash . *22\cdot 34 . \supset \vdash :: x \in s\text{‘}\kappa \cup s\text{‘}\lambda . \equiv :. x \in s\text{‘}\kappa . \vee . x \in s\text{‘}\lambda :.$

$[*40\cdot 11] \quad \equiv :. (\exists \gamma) . \gamma \in \kappa . x \in \gamma : \vee : (\exists \gamma) . \gamma \in \lambda . x \in \gamma :.$

$[*10\cdot 42] \quad \equiv :. (\exists \gamma) : \gamma \in \kappa . x \in \gamma . \vee . \gamma \in \lambda . x \in \gamma :.$

$[*4\cdot 4] \quad \equiv :. (\exists \gamma) :. \gamma \in \kappa . \vee . \gamma \in \lambda : x \in \gamma :.$

[✱22·34] $\equiv :. (\exists \gamma) . \gamma \epsilon \kappa \cup \lambda . x \epsilon \gamma :.$

[✱40·11] $\equiv :. x \epsilon s\text{‘}(\kappa \cup \lambda) :: \supset \vdash . \text{Prop}$

✱40·18. $\vdash . p\text{‘}(\kappa \cup \lambda) = p\text{‘}\kappa \cap p\text{‘}\lambda$

Dem.

$\vdash . ✱40·1 . \supset \vdash :: x \epsilon p\text{‘}(\kappa \cup \lambda) . \equiv :. \gamma \epsilon \kappa \cup \lambda . \supset_\gamma . x \epsilon \gamma :.$

[✱22·34] $\equiv :. (\gamma) :. \gamma \epsilon \kappa . \vee . \gamma \epsilon \lambda : \supset . x \epsilon \gamma :.$

[✱4·77] $\equiv :. (\gamma) :. \gamma \epsilon \kappa . \supset . x \epsilon \gamma : \gamma \epsilon \lambda . \supset . x \epsilon \gamma :.$

[✱10·22·221] $\equiv :. (\gamma) : \gamma \epsilon \kappa . \supset . x \epsilon \gamma :. (\gamma) : \gamma \epsilon \lambda . \supset . x \epsilon \gamma :.$

[✱40·1] $\equiv :. x \epsilon p\text{‘}\kappa . x \epsilon p\text{‘}\lambda :.$

[✱22·33] $\equiv :. x \epsilon p\text{‘}\kappa \cap p\text{‘}\lambda :: \supset \vdash . \text{Prop}$

✱40·181. $\vdash . s\text{‘}(\kappa \cap \lambda) \subset s\text{‘}\kappa \cap s\text{‘}\lambda$

Dem.

$\vdash . ✱40·11 . \supset \vdash :: x \epsilon s\text{‘}(\kappa \cap \lambda) . \equiv :. (\exists \gamma) . \gamma \epsilon \kappa \cap \lambda . x \epsilon \gamma :.$

[✱22·33] $\equiv :. (\exists \gamma) . \gamma \epsilon \kappa . \gamma \epsilon \lambda . x \epsilon \gamma :.$

[✱10·5] $\supset :. (\exists \gamma) . \gamma \epsilon \kappa . x \epsilon \gamma : (\exists \gamma) . \gamma \epsilon \lambda . x \epsilon \gamma :.$

[✱40·11.✱22·33] $\supset :. x \epsilon s\text{‘}\kappa \cap s\text{‘}\lambda :: \supset \vdash . \text{Prop}$

✱40·19. $\vdash :: x \epsilon s\text{‘}\kappa . \equiv :. \gamma \epsilon \kappa . \supset_\gamma . \gamma \subset \beta : \supset_\beta . x \epsilon \beta$

This proposition is the extension of ✱22·6.

Dem.

$\vdash . ✱40·151 . \supset$

$\vdash :: \gamma \epsilon \kappa . \supset_\gamma . \gamma \subset \beta : \supset_\beta . x \epsilon \beta :. \equiv :. s\text{‘}\kappa \subset \beta . \supset_\beta . x \epsilon \beta$ (1)

$\vdash . ✱10·1 . \supset \vdash :. s\text{‘}\kappa \subset \beta . \supset_\beta . x \epsilon \beta : \supset : s\text{‘}\kappa \subset s\text{‘}\kappa . \supset . x \epsilon s\text{‘}\kappa :$

[✱22·42] $\supset : x \epsilon s\text{‘}\kappa$ (2)

$\vdash . ✱22·46 . \supset \vdash :. x \epsilon s\text{‘}\kappa . s\text{‘}\kappa \subset \beta . \supset . x \epsilon \beta :.$

[Exp] $\supset \vdash :. x \epsilon s\text{‘}\kappa . \supset : s\text{‘}\kappa \subset \beta . \supset . x \epsilon \beta :.$

[✱10·11·21] $\supset \vdash :. x \epsilon s\text{‘}\kappa . \supset : s\text{‘}\kappa \subset \beta . \supset_\beta . x \epsilon \beta$ (3)

$\vdash . (2) . (3) . \supset \vdash :. s\text{‘}\kappa \subset \beta . \supset_\beta . x \epsilon \beta : \equiv . x \epsilon s\text{‘}\kappa$ (4)

$\vdash . (1) . (4) . \supset \vdash . \text{Prop}$

✱40·2. $\vdash : \kappa = \Lambda . \supset . p\text{‘}\kappa = \text{V}$

Dem.

$\vdash . ✱24·5·51 . \supset \vdash :. \text{Hp} . \supset : \sim (\exists \alpha) . \alpha \epsilon \kappa :$

[✱10·53] $\supset : (\alpha) : \alpha \epsilon \kappa . \supset . x \epsilon \alpha :$

[✱40·1] $\supset : x \epsilon p\text{‘}\kappa$ (1)

$\vdash . (1) . ✱10·11·21 . \supset \vdash : \text{Hp} . \supset . (x) . x \epsilon p\text{‘}\kappa .$

[✱24·14] $\supset . p\text{‘}\kappa = \text{V} : \supset \vdash . \text{Prop}$

✱40·21. $\vdash : \kappa = \Lambda . \supset . s\text{‘}\kappa = \Lambda$

Dem.

$\vdash . ✱24·51 . \supset \vdash : \text{Hp} . \supset . \sim (\exists \alpha) . \alpha \epsilon \kappa .$

[✱10·5.Transp] $\supset . \sim (\exists \alpha) . \alpha \epsilon \kappa . x \epsilon \alpha .$

$[*40 \cdot 11 . \text{Transp}] \qquad \supset . x \sim \epsilon s'\kappa$ (1)

$\vdash . (1) . *10 \cdot 11 \cdot 21 . \supset \vdash : \text{Hp} . \supset . (x) . x \sim \epsilon s'\kappa .$

$[*24 \cdot 15] \qquad \supset . s'\kappa = \Lambda : \supset \vdash . \text{Prop}$

In the above proposition, the two Λ's are of different types, since κ is of the type next above that of $s'\kappa$. Thus it would be more correct to write

$$\vdash : \kappa = \Lambda \cap \text{Cls} . \supset . s'\kappa = \Lambda \cap \text{V}.$$

But in the case of Λ it is not very important to keep the types distinct.

***40·22.** $\vdash : \Lambda \epsilon \kappa . \supset . p'\kappa = \Lambda$

Dem.

$\vdash . *40 \cdot 12 . \supset \vdash : \text{Hp} . \supset . p'\kappa \subset \Lambda .$

$[*24 \cdot 13] \qquad \supset . p'\kappa = \Lambda : \supset \vdash . \text{Prop}$

In this proposition, the two Λ's are of the same type.

***40·221.** $\vdash : \text{V} \epsilon \kappa . \supset . s'\kappa = \text{V}$

Dem.

$\vdash . *40 \cdot 13 . \supset \vdash : \text{Hp} . \supset . \text{V} \subset s'\kappa .$

$[*24 \cdot 141] \qquad \supset . s'\kappa = \text{V} : \supset \vdash . \text{Prop}$

***40·23.** $\vdash : \exists ! \kappa . \supset . p'\kappa \subset s'\kappa$

Dem.

$\vdash . *40 \cdot 12 \cdot 13 . \supset \vdash : \alpha \epsilon \kappa . \supset . p'\kappa \subset \alpha . \alpha \subset s'\kappa .$

$[*22 \cdot 44] \qquad \supset . p'\kappa \subset s'\kappa :$

$[*10 \cdot 11 \cdot 23] \quad \supset \vdash : (\exists \alpha) . \alpha \epsilon \kappa . \supset . p'\kappa \subset s'\kappa : \supset \vdash . \text{Prop}$

Observe that the hypothesis $\exists ! \kappa$ is essential to this proposition, since when $\kappa = \Lambda$, $p'\kappa = \text{V}$ and $s'\kappa = \Lambda$. Thus

$$\vdash : \exists ! \kappa . \equiv . p'\kappa \subset s'\kappa.$$

***40·24.** $\vdash :. \exists ! \kappa : \gamma \epsilon \kappa . \supset_\gamma . \beta \subset \gamma : \supset . \beta \subset s'\kappa$

Dem.

$\vdash . *40 \cdot 15 . \supset \vdash :. \gamma \epsilon \kappa . \supset_\gamma . \beta \subset \gamma : \supset . \beta \subset p'\kappa$ (1)

$\vdash . *40 \cdot 23 . \supset \vdash : \exists ! \kappa . \supset . p'\kappa \subset s'\kappa$ (2)

$\vdash . (1) . (2) . \supset \vdash : \text{Hp} . \supset . \beta \subset p'\kappa . p'\kappa \subset s'\kappa .$

$[*22 \cdot 44] \qquad \supset . \beta \subset s'\kappa : \supset \vdash . \text{Prop}$

The above proposition is used in the proof of *215·25.

***40·25.** $\vdash : x \epsilon s'\kappa . \equiv . \exists ! \kappa \cap \hat{\alpha}(x \epsilon \alpha)$

Dem.

$\vdash . *22 \cdot 33 . \supset \vdash : \exists ! \kappa \cap \hat{\alpha}(x \epsilon \alpha) . \equiv . (\exists \gamma) . \gamma \epsilon \kappa . \gamma \epsilon \hat{\alpha}(x \epsilon \alpha) .$

$[*20 \cdot 3] \qquad \equiv . (\exists \gamma) . \gamma \epsilon \kappa . x \epsilon \gamma .$

$[*40 \cdot 11] \qquad \equiv . x \epsilon s'\kappa : \supset \vdash . \text{Prop}$

***40·26.** $\vdash : \exists ! s'\kappa . \equiv . (\exists \alpha) . \alpha \epsilon \kappa . \exists ! \alpha$

Dem.

$\vdash . *40 \cdot 11 . \supset \vdash :. \exists ! s'\kappa . \equiv : (\exists x) : (\exists \alpha) . \alpha \epsilon \kappa . x \epsilon \alpha :$

$[*11 \cdot 23 \cdot 55] \qquad \equiv : (\exists \alpha) : \alpha \epsilon \kappa : (\exists x) . x \epsilon \alpha :$

$[*24 \cdot 5] \qquad \equiv : (\exists \alpha) . \alpha \epsilon \kappa . \exists ! \alpha :. \supset \vdash . \text{Prop}$

The following proposition is used in the proof of *216·51.

***40·27.** $\vdash :. \alpha \cap s\text{‘}\kappa = \Lambda . \equiv : \gamma \in \kappa . \supset_\gamma . \alpha \cap \gamma = \Lambda$

Dem.

$\vdash . *24\cdot311 . \supset$

$\vdash :: \alpha \cap s\text{‘}\kappa = \Lambda . \equiv :. s\text{‘}\kappa \subset - \alpha :.$

[*22·1·35] $\equiv :. a \in s\text{‘}\kappa . \supset_x . x \sim\in \alpha :.$

[*40·1] $\equiv :. (\exists\gamma) . \gamma \in \kappa . x \in \gamma . \supset_x . x \sim\in \alpha :.$

[*10·23] $\equiv :. \gamma \in \kappa . x \in \gamma . \supset_{x,\gamma} . x \sim\in \alpha :.$

[*11·2·62] $\equiv :. \gamma \in \kappa . \supset_\gamma : x \in \gamma . \supset_x . x \sim\in \alpha :.$

[*24·39] $\equiv :. \gamma \in \kappa . \supset_\gamma . \alpha \cap \gamma = \Lambda :: \supset \vdash .$ Prop

The following propositions are only significant when R is a relation whose domain consists of classes, for they concern $p\text{‘}R\text{“}\alpha$ or $s\text{‘}R\text{“}\alpha$, and therefore require that $R\text{“}\alpha$ should be a class of classes.

***40·3.** $\vdash . p\text{‘}R\text{“}(\alpha \cup \beta) = p\text{‘}R\text{“}\alpha \cap p\text{‘}R\text{“}\beta$ [*37·22 . *40·18]

***40·31.** $\vdash . s\text{‘}R\text{“}(\alpha \cup \beta) = s\text{‘}R\text{“}\alpha \cup s\text{‘}R\text{“}\beta$ [*37·22 . *40·171]

***40·32.** $\vdash . p\text{‘}R\text{“}\alpha \cup p\text{‘}R\text{“}\beta \subset p\text{‘}R\text{“}(\alpha \cap \beta)$

Dem.

$\vdash . *37\cdot21 . \supset \vdash . R\text{“}(\alpha \cap \beta) \subset R\text{“}\alpha \cap R\text{“}\beta .$

[*40·16] $\supset \vdash . p\text{‘}(R\text{“}\alpha \cap R\text{“}\beta) \subset p\text{‘}R\text{“}(\alpha \cap \beta)$ (1)

$\vdash . *40\cdot17 . \supset \vdash . p\text{‘}R\text{“}\alpha \cup p\text{‘}R\text{“}\beta \subset p\text{‘}(R\text{“}\alpha \cap R\text{“}\beta)$ (2)

$\vdash . (1) . (2) . *22\cdot44 . \supset \vdash .$ Prop

***40·33.** $\vdash . s\text{‘}R\text{“}(\alpha \cap \beta) \subset s\text{‘}R\text{“}\alpha \cap s\text{‘}R\text{“}\beta$ [*37·21 . *40·161 . *40·181]

The following propositions no longer require that the domain of R should be composed of classes.

***40·35.** $\vdash . p\text{‘}R\text{“‘}\kappa = \hat{x}\{\beta \in \kappa . \supset_\beta . x \in R\text{“}\beta\}$

Dem.

$\vdash . *40\cdot1 . \supset \vdash :. x \in p\text{‘}R\text{“‘}\kappa . \equiv : \gamma \in R\text{“‘}\kappa . \supset_\gamma . x \in \gamma :$

[*37·103] $\equiv : (\exists\beta) . \beta \in \kappa . \gamma = R\text{“}\beta . \supset_\gamma . x \in \gamma :$

[*10·23] $\equiv : \beta \in \kappa . \gamma = R\text{“}\beta . \supset_{\beta,\gamma} . x \in \gamma :$

[*13·191] $\equiv : \beta \in \kappa . \supset_\beta . x \in R\text{“}\beta$ (1)

$\vdash . (1) . *10\cdot11 . *20\cdot3 . \supset \vdash .$ Prop

***40·36.** $\vdash . s\text{‘}R\text{“‘}\kappa = \hat{x}\{(\exists\beta) . \beta \in \kappa . x \in R\text{“}\beta\}$ [Similar proof]

***40·37.** $\vdash . R\text{“}p\text{‘}\kappa \subset p\text{‘}R\text{“‘}\kappa$

Dem.

$\vdash . *37\cdot1 . \supset \vdash :: x \in R\text{“}p\text{‘}\kappa . \equiv :. (\exists y) . y \in p\text{‘}\kappa . xRy :.$

[*40·1] $\equiv :. (\exists y) : \beta \in \kappa . \supset_\beta . y \in \beta : xRy :.$

[*10·33] $\equiv :. (\exists y) :. (\beta) : \beta \in \kappa . \supset . y \in \beta : xRy :.$

[*11·26] $\supset :. (\beta) :. (\exists y) : \beta \in \kappa . \supset . y \in \beta : xRy :.$

[*5·31] $\supset :. (\beta) :. (\exists y) : \beta \in \kappa . \supset . y \in \beta . xRy :.$

[*10·37] $\supset :. (\beta) :. \beta \in \kappa . \supset . (\exists y) . y \in \beta . xRy :.$
[*37·1] $\supset :. (\beta) : \beta \in \kappa . \supset . x \in R``\beta :.$
[*40·35] $\supset :. x \in p`R```\kappa :: \supset \vdash .$ Prop

***40·38.** $\vdash . R``s`\kappa = s`R```\kappa$

Dem.

$\vdash . *37·1 . \supset \vdash :: x \in R``s`\kappa . \equiv :. (\exists y) . y \in s`\kappa . xRy :.$
[*40·11] $\equiv :. (\exists y) :. (\exists \alpha) . \alpha \in \kappa . y \in \alpha : xRy :.$
[*11·6] $\equiv :. (\exists \alpha) :. \alpha \in \kappa : (\exists y) . y \in \alpha . xRy :.$
[*37·1] $\equiv :. (\exists \alpha) . \alpha \in \kappa . x \in R``\alpha :.$
[*40·36] $\equiv :. x \in s`R```\kappa :: \supset \vdash .$ Prop

This proposition is frequently used in the proofs of arithmetical propositions.

***40·4.** $\vdash : \text{E}!! R``\beta . \supset . s`R``\beta = \hat{x} \{(\exists y) . y \in \beta . x \in R`y\}$

This proposition is only significant when $\text{D}`R \subset \text{Cls}$.

Dem.

$\vdash . *37·6 . \supset \vdash : \text{Hp} . \supset . R``\beta = \hat{\alpha} \{(\exists y) . y \in \beta . \alpha = R`y\}$ (1)
$\vdash . (1) . *40·11 . \supset$
$\vdash :: \text{Hp} . \supset :. x \in s`R``\beta . \equiv : (\exists \alpha) : (\exists y) . y \in \beta . \alpha = R`y : x \in \alpha :$
[*11·6] $\equiv : (\exists y) : y \in \beta : (\exists \alpha) . \alpha = R`y . x \in \alpha :$
[*14·205] $\equiv : (\exists y) . y \in \beta . x \in R`y :: \supset \vdash .$ Prop

***40·41.** $\vdash : \text{E}!! R``\beta . \supset . p`R``\beta = \hat{x} \{y \in \beta . \supset_y . x \in R`y\}$ [Similar proof]

***40·42.** $\vdash : (x) . R`x = P`x \cup Q`x . \supset . s`R``\alpha = s`(P``\alpha \cup Q``\alpha) = s`P``\alpha \cup s`Q``\alpha$

Dem.

$\vdash . *14·21 . \supset \vdash : \text{Hp} . \supset . (x) . \text{E}! R`x . \text{E}! P`x . \text{E}! Q`x$ (1)
$\vdash . (1) . *40·4 . \supset \vdash : \text{Hp} . \supset . s`R``\alpha = \hat{x} \{(\exists y) . y \in \alpha . x \in R`y\}$
[Hp] $= \hat{x} \{(\exists y) . y \in \alpha . x \in P`y \cup Q`y\}$
[*22·34] $= \hat{x} \{(\exists y) : y \in \alpha : x \in P`y . \vee . x \in Q`y\}$
[*4·4.*10·42] $= \hat{x} \{(\exists y) . y \in \alpha . x \in P`y . \vee . (\exists y) . y \in \alpha . x \in Q`y\}$
[(1).*40·4] $= \hat{x} \{x \in s`P``\alpha . \vee . x \in s`Q``\alpha\}$
[*20·42.*22·34] $= s`P``\alpha \cup s`Q``\alpha$
[*40·171] $= s`(P``\alpha \cup Q``\alpha) : \supset \vdash .$ Prop

This proposition is used in *40·57, where we take $R = C$, $P = \text{D}$, $Q = \text{Cl}$.

***40·43.** $\vdash :: \text{E}!! R``\beta . \supset :. s`R``\beta \subset \alpha . \equiv : y \in \beta . \supset_y . R`y \subset \alpha$

Dem.

$\vdash . *37·63 . \supset \vdash :: \text{Hp} . \supset :. y \in \beta . \supset_y . R`y \subset \alpha : \equiv : \gamma \in R``\beta . \supset_\gamma . \gamma \subset \alpha :$
[*40·151] $\equiv : s`R``\beta \subset \alpha :: \supset \vdash .$ Prop

***40·44.** $\vdash :: \text{E}!! R``\beta . \supset :. \alpha \subset p`R``\beta . \equiv : y \in \beta . \supset_y . \alpha \subset R`y$

Dem.

$\vdash . *37·63 . \supset \vdash :: \text{Hp} . \supset :. y \in \beta . \supset_y . \alpha \subset R`y : \equiv : \gamma \in R``\beta . \supset_\gamma . \alpha \subset \gamma :.$
[*40·15] $\equiv : \alpha \subset p`R``\beta :: \supset \vdash .$ Prop

The following proposition is used in the proof of $*84{\cdot}44$.

$*40{\cdot}45$. $\vdash :. y \epsilon \beta . \supset_y . R`y \subset S`y : \supset . s`R``\beta \subset s`S``\beta$

Dem.

$\vdash . *14{\cdot}21 . \supset \vdash :. \mathrm{Hp} . \supset : \exists ! S``\beta . \exists ! R``\beta :$ (1)

$[*37{\cdot}62 . *40{\cdot}13] \quad \supset : y \epsilon \beta . \supset_y . S`y \subset s`S``\beta :$

$[\mathrm{Hp}] \quad \supset : y \epsilon \beta . \supset_y . R`y \subset s`S``\beta :$

$[*40{\cdot}43 . (1)] \quad \supset : s`R``\beta \subset s`S``\beta :. \supset \vdash . \mathrm{Prop}$

The following proposition is used in the proof of $*94{\cdot}402$.

$*40{\cdot}451$. $\vdash :. y \epsilon \beta . \supset_y . R`y \subset S`y : \supset . p`R``\beta \subset p`S``\beta$

Dem.

$\vdash . *14{\cdot}21 . *37{\cdot}62 . *40{\cdot}12 . \supset \vdash :. \mathrm{Hp} . \supset : y \epsilon \beta . \supset . p`R``\beta \subset R`y .$

$[\mathrm{Hp}] \quad \supset . p`R``\beta \subset S`y .$

$[*40{\cdot}44] \quad \supset : p`R``\beta \subset p`S``\beta :. \supset \vdash . \mathrm{Prop}$

$*40{\cdot}5$. $\vdash . s`\overrightarrow{R}``\beta = R``\beta$

Dem.

$\vdash . *32{\cdot}12 . *40{\cdot}4 . \supset \vdash . s`\overrightarrow{R}``\beta = \hat{x} \{(\exists y) . y \epsilon \beta . x \epsilon \overrightarrow{R}`y\}$

$[*32{\cdot}18] \quad = \hat{x} \{(\exists y) . y \epsilon \beta . xRy\}$

$[(*37{\cdot}01)] \quad = R``\beta . \supset \vdash . \mathrm{Prop}$

$*40{\cdot}51$. $\vdash . p`\overrightarrow{R}``\beta = \hat{x} \{y \epsilon \beta . \supset_y . xRy\}$ $[*32{\cdot}12 . *40{\cdot}41 . *32{\cdot}18]$

$p`\overrightarrow{R}``\beta$ is the class of terms each of which has the relation $R$ to *every* member of $\beta$, just as $R``\beta$ is the class of terms each of which has the relation R to *some* member of β. In the theory of series, $p`\overrightarrow{R}``\beta$ plays an important part, correlative to that played by $R``\beta$ (which is $s`\overrightarrow{R}``\beta$, by $*40{\cdot}5$). If $\beta$ is a class contained in a series whose generating relation is $R$, $p`\overrightarrow{R}``\beta$ will be the predecessors of all members of β, while $R``\beta$ will be the predecessors of some β.

$*40{\cdot}52$. $\vdash . s`\overleftarrow{R}``\beta = \breve{R}``\beta$ [Proof as in $*40{\cdot}5$]

$*40{\cdot}53$. $\vdash . p`\overleftarrow{R}``\beta = \hat{y} \{x \epsilon \beta . \supset_x . xRy\}$ [Proof as in $*40{\cdot}51$]

$*40{\cdot}54$. $\vdash . p`\overrightarrow{R}``\beta = \hat{x} (\beta \subset \overleftarrow{R}`x)$ $[*40{\cdot}51 . *32{\cdot}181]$

$*40{\cdot}55$. $\vdash . p`\overleftarrow{R}``\alpha = \hat{y} (\alpha \subset \overrightarrow{R}`y)$ $[*40{\cdot}53 . *32{\cdot}18]$

From this point onwards to $*40{\cdot}69$, the propositions are inserted on account of their use in the theory of series.

$*40{\cdot}56$. $\vdash . s`C``\lambda = F``\lambda$ $[*33{\cdot}5 . *40{\cdot}5]$

In the above proposition, the conditions of significance require that λ should be a class of relations.

$*40{\cdot}57$. $\vdash . s`C``\lambda = s`(\mathrm{D}``\lambda \cup \mathrm{G}``\lambda) = s`\mathrm{D}``\lambda \cup s`\mathrm{G}``\lambda$ $[*40{\cdot}42 . *33{\cdot}16]$

$*40{\cdot}6.\quad \vdash . p‘\overrightarrow{R}“\Lambda = V . p‘\overleftarrow{R}“\Lambda = V\quad [*37{\cdot}29 . *40{\cdot}2]$

$*40{\cdot}61.\quad \vdash : \exists ! \beta . \supset . p‘\overrightarrow{R}“\beta \subset R“\beta . p‘\overleftarrow{R}“\beta \subset \breve{R}“\beta$

Dem.

$\vdash . *37{\cdot}73 . \supset \vdash : \mathrm{Hp} . \supset . \exists ! \overrightarrow{R}“\beta .$

$[*40{\cdot}23]\quad \supset . p‘\overrightarrow{R}“\beta \subset s‘\overrightarrow{R}“\beta .$

$[*40{\cdot}5]\quad \supset . p‘\overrightarrow{R}“\beta \subset R“\beta \quad (1)$

Similarly $\vdash : \mathrm{Hp} . \supset . p‘\overleftarrow{R}“\beta \subset \breve{R}“\beta \quad (2)$

$\vdash . (1) . (2) . \supset \vdash . \mathrm{Prop}$

$*40{\cdot}62.\quad \vdash : \exists ! \beta . \supset . p‘\overrightarrow{R}“\beta \subset C‘R . p‘\overleftarrow{R}“\beta \subset C‘R$
$[*40{\cdot}61 . *37{\cdot}15{\cdot}16 . *33{\cdot}161]$

The two following propositions ($*40{\cdot}63{\cdot}64$) are used in proving $*40{\cdot}65$, which is used in $*204{\cdot}63$.

$*40{\cdot}63.\quad \vdash : \exists ! \beta - \text{Ɑ}‘R . \supset . p‘\overrightarrow{R}“\beta = \Lambda$

Dem.

$\vdash . *33{\cdot}41 . \mathrm{Transp} . \supset \vdash : x \sim\epsilon\, \text{Ɑ}‘R . \quad \supset . \overrightarrow{R}‘x = \Lambda \quad (1)$

$\vdash . *37{\cdot}704 . \quad \supset \vdash : x \epsilon \beta . \quad \supset . \overrightarrow{R}‘x \,\epsilon\, \overrightarrow{R}“\beta \quad (2)$

$\vdash . (1) . (2) . *22{\cdot}32 . \supset \vdash : x \epsilon \beta - \text{Ɑ}‘R . \supset . \overrightarrow{R}‘x \,\epsilon\, \overrightarrow{R}“\beta . \overrightarrow{R}‘x = \Lambda .$

$[*20{\cdot}57]\quad \supset . \Lambda \,\epsilon\, \overrightarrow{R}“\beta .$

$[*40{\cdot}22]\quad \supset . p‘\overrightarrow{R}“\beta = \Lambda \quad (3)$

$\vdash . (3) . *10{\cdot}11{\cdot}23 . \quad \supset \vdash . \mathrm{Prop}$

$*40{\cdot}64.\quad \vdash : \exists ! \beta - \mathrm{D}‘R . \supset . p‘\overleftarrow{R}“\beta = \Lambda$ [Proof as in $*40{\cdot}63$]

$*40{\cdot}65.\quad \vdash : \exists ! \beta - C‘R . \supset . p‘\overrightarrow{R}“\beta = \Lambda . p‘\overleftarrow{R}“\beta = \Lambda\quad [*40{\cdot}63{\cdot}64 . *33{\cdot}16]$

$*40{\cdot}66.\quad \vdash :. \alpha \subset p‘\overrightarrow{R}“\beta . \equiv : x \epsilon \alpha . y \epsilon \beta . \supset_{x,y} . xRy$

Dem.

$\vdash . *40{\cdot}51 . \supset \vdash :: \alpha \subset p‘\overrightarrow{R}“\beta . \equiv :. \alpha \subset \hat{x}(y \epsilon \beta . \supset_y . xRy) :.$

$[*20{\cdot}3]\quad \equiv :. x \epsilon \alpha . \supset_x : y \epsilon \beta . \supset_y . xRy :.$

$[*11{\cdot}62]\quad \equiv :. (x, y) :. x \epsilon \alpha . y \epsilon \beta . \supset . xRy :: \supset \vdash . \mathrm{Prop}$

$*40{\cdot}67.\quad \vdash :. \beta \subset p‘\overleftarrow{R}“\alpha . \equiv : x \epsilon \alpha . y \epsilon \beta . \supset_{x,y} . xRy : \equiv . \alpha \subset p‘\overrightarrow{R}“\beta$
[Proof as in $*40{\cdot}66$]

$*40{\cdot}68.\quad \vdash . \alpha \cap p‘\overleftarrow{P}“\alpha \subset \breve{P}“p‘\overleftarrow{P}“\alpha$

Dem.

$\vdash . *40{\cdot}53 . \supset \vdash :. x \epsilon \alpha \cap p‘\overleftarrow{P}“\alpha . \supset : x \epsilon \alpha : y \epsilon \alpha . \supset_y . yPx :$

$[*10{\cdot}26]\quad \supset : xPx : y \epsilon \alpha . \supset_y . yPx :$

$[*10{\cdot}24] \qquad \supset : (\exists z) : zPx : y \in \alpha \,.\, \supset_y \,.\, yPz :$

$[*40{\cdot}53 . *37{\cdot}105] \qquad \supset : x \in \breve{P}``p`\overleftarrow{P}``\alpha :. \supset \vdash . \text{Prop}$

This proposition is used in the theory of series (*206·2).

***40·681.** $\vdash . \alpha \cap p`\overrightarrow{P}``\alpha \subset P``p`\overrightarrow{P}``\alpha$ [Proof as in *40·68]

The following proposition is used in *211·56.

***40·682.** $\vdash : \exists ! \alpha \cap p`\overleftarrow{P}``\beta \,.\, \supset . \beta \subset P``\alpha$

Dem.

$\vdash . *40{\cdot}53 . \supset \vdash :. \text{Hp} . \supset : (\exists x) : x \in \alpha : y \in \beta \,.\, \supset_y . yPx :$

$[*5{\cdot}31] \qquad \supset : (\exists x) : y \in \beta \,.\, \supset_y . x \in \alpha \,.\, yPx :$

$[*11{\cdot}61] \qquad \supset : y \in \beta \,.\, \supset_y . (\exists x) . x \in \alpha \,.\, yPx .$

$[*37{\cdot}1] \qquad \supset_y . y \in P``\alpha :. \supset \vdash . \text{Prop}$

***40·69.** $\vdash : \exists ! C`P \cap p`\overleftarrow{P}``\alpha \,.\, \equiv . \dot{\exists} ! P \,.\, \exists ! p`\overleftarrow{P}``\alpha$

Dem.

$\vdash . *33{\cdot}24 . *24{\cdot}561 . \supset \vdash : \exists ! C`P \cap p`\overleftarrow{P}``\alpha \,.\, \supset . \dot{\exists} ! P \,.\, \exists ! p`\overleftarrow{P}``\alpha \quad (1)$

$\vdash . *40{\cdot}62 . \qquad \supset \vdash : \exists ! \alpha \,.\, \exists ! p`\overleftarrow{P}``\alpha \,.\, \supset . \exists ! C`P \cap p`\overleftarrow{P}``\alpha \quad (2)$

$\vdash . *40{\cdot}6 . \qquad \supset \vdash :. \alpha = \Lambda . \supset : C`P \cap p`\overleftarrow{P}``\alpha = C`P :$

$[*33{\cdot}24] \qquad \supset : \dot{\exists} ! P . \supset . \exists ! C`P \cap p`\overleftarrow{P}``\alpha \quad (3)$

$\vdash . (2) . (3) . *4{\cdot}83 . \quad \supset \vdash : \dot{\exists} ! P \,.\, \exists ! p`\overleftarrow{P}``\alpha \,.\, \supset . \exists ! C`P \cap p`\overleftarrow{P}``\alpha \quad (4)$

$\vdash . (1) . (4) . \qquad \supset \vdash . \text{Prop}$

The above propositions concerning $p`\overrightarrow{R}``\beta$ and $p`\overleftarrow{R}``\beta$ of course have analogues for $s`\overrightarrow{R}``\beta$ and $s`\overleftarrow{R}``\beta$. But owing to *40·5, these analogues are more simply stated as properties of $R``\beta$ and $\breve{R}``\beta$. Thus, for example, *37·264 is the analogue of *40·67. The above propositions concerning $p`\overrightarrow{R}``\beta$ and $p`\overleftarrow{R}``\beta$ will be used in the theory of series, but until we reach that stage they will seldom be referred to.

***40·7.** $\vdash . s`\alpha \underset{,,}{\female}``\beta = \hat{z}\{(\exists x, y) . x \in \alpha \,.\, y \in \beta \,.\, z = x \female y\}$

Dem.

$\vdash . *40{\cdot}11 . *38{\cdot}3 . \supset$

$\vdash . s`\alpha \underset{,,}{\female}``\beta = \hat{z}\{(\exists \gamma, y) . y \in \beta \,.\, \gamma = \female y``\alpha \,.\, z \in \gamma\}$

$[*38{\cdot}131] \quad = \hat{z}\{(\exists \gamma, x, y) . y \in \beta \,.\, \gamma = \female y``\alpha \,.\, x \in \alpha \,.\, z = x \female y\}$

$[*13{\cdot}19] \quad = \hat{z}\{(\exists x, y) . x \in \alpha \,.\, y \in \beta \,.\, z = x \female y\} . \supset \vdash . \text{Prop}$

This proposition is of considerable importance, since it gives a compact form for the class of all values of the function $x \female y$ obtained by taking x in the class α and y in the class β. Thus, for example, suppose α is the class of numbers which are multiples of 3, and β is the class of numbers which are multiples of 5, and $x \times y$ represents the arithmetical product of x and y, then $s'\alpha \underset{\text{,,}}{\times}{}''\beta$ will be the class of products of multiples of 3 and multiples of 5, *i.e.* the class of multiples of 15. Again suppose α and β are both classes of relations; then $s'\alpha \underset{\text{,,}}{|}{}''\beta$ will be all relative products $R|S$ obtained by choosing R in the class α and S in the class β.

***40·71.** $\vdash . s'\underset{\text{,,}}{♀}y''\kappa = (s'\kappa)\underset{\text{,,}}{♀}y = ♀y''s'\kappa$

Dem.

$\vdash . *40{\cdot}38 . *38{\cdot}31 . \supset \vdash . s'\underset{\text{,,}}{♀}y''\kappa = ♀y''s'\kappa$

$[*38{\cdot}2] \qquad = (s'\kappa)\underset{\text{,,}}{♀}y . \supset \vdash . \text{Prop}$

The hypothesis $\breve{R}''\alpha \subset \alpha$, which appears in *40·8·81, is one which plays an important part at a later stage. In the theory of induction (Part II, Section E) it characterizes a *hereditary* class, and in the theory of series it characterizes an *upper section* (when combined with $\alpha \subset C'R$).

***40·8.** $\vdash :. \alpha \epsilon \kappa . \supset_\alpha . \breve{R}''\alpha \subset \alpha : \supset . \breve{R}''s'\kappa \subset s'\kappa$

Dem.

$\vdash . *37{\cdot}171 . \supset \vdash :: \text{Hp} . \supset :. \alpha \epsilon \kappa . \supset_\alpha : x \epsilon \alpha . xRy . \supset_{x,y} . y \epsilon \alpha :.$

$[*11{\cdot}62] \qquad \supset :. \alpha \epsilon \kappa . x \epsilon \alpha . xRy . \supset_{\alpha,x,y} . y \epsilon \alpha :.$

$[*40{\cdot}13] \qquad \supset_{\alpha,x,y} . y \epsilon s'\kappa :.$

$[*40{\cdot}11 . *10{\cdot}23] \qquad \supset :. x \epsilon s'\kappa . xRy . \supset_{x,y} . y \epsilon s'\kappa :.$

$[*37{\cdot}171] \qquad \supset :. \breve{R}''s'\kappa \subset s'\kappa :: \supset \vdash . \text{Prop}$

***40·81.** $\vdash :. \alpha \epsilon \kappa . \supset_\alpha . \breve{R}''\alpha \subset \alpha : \supset . \breve{R}''p'\kappa \subset p'\kappa$

Dem.

$\vdash . *37{\cdot}171 . \supset \vdash ::. \text{Hp} . \supset :: \alpha \epsilon \kappa . \supset : x \epsilon \alpha . xRy . \supset . y \epsilon \alpha ::$

$[\text{Exp.Comm}] \qquad \supset :: xRy . \supset :. \alpha \epsilon \kappa . \supset : x \epsilon \alpha . \supset . y \epsilon \alpha :.$

$[*2{\cdot}77] \qquad \supset :. \alpha \epsilon \kappa . \supset . x \epsilon \alpha : \supset : \alpha \epsilon \kappa . \supset . y \epsilon \alpha \qquad (1)$

$\vdash . (1) . *10{\cdot}11{\cdot}21{\cdot}27 . \supset$

$\vdash ::. \text{Hp} . \supset :: xRy . \supset :. \alpha \epsilon \kappa . \supset_\alpha . x \epsilon \alpha : \supset : \alpha \epsilon \kappa . \supset_\alpha . y \epsilon \alpha :.$

$\supset :. x \epsilon p'\kappa . \supset . y \epsilon p'\kappa ::$

$[\text{Imp}] \qquad \supset :: x \epsilon p'\kappa . xRy . \supset . y \epsilon p'\kappa \qquad (2)$

$\vdash . (2) . *37{\cdot}171 . \supset \vdash . \text{Prop}$

$*$41. THE PRODUCT AND SUM OF A CLASS OF RELATIONS

Summary of $*$41.

The propositions to be given in this number, down to $*$41·3 exclusive, are the analogues of those of $*$40, excluding those from $*$40·3 onwards, which have no analogues. Proofs will not be given, in this number, when they are exactly analogous to those of propositions with the same decimal part in $*$40. The smaller importance of $\dot{p}\text{‘}\lambda$ and $\dot{s}\text{‘}\lambda$, as compared with $p\text{‘}\lambda$ and $s\text{‘}\lambda$, is illustrated by the smaller number of propositions in $*$41 as compared with $*$40.

Our definitions are

$*$41·01. $\dot{p}\text{‘}\lambda = \hat{x}\hat{y}\,(R \epsilon \lambda \,.\supset_R .\, xRy)$ Df

$*$41·02. $\dot{s}\text{‘}\lambda = \hat{x}\hat{y}\,\{(\exists R) \,.\, R \epsilon \lambda \,.\, xRy\}$ Df

Of the propositions preceding $*$41·3, which are analogues of propositions in $*$40, the only two that are frequently used are

$*$41·13. $\vdash : R \epsilon \lambda \,.\supset .\, R \mathrel{\dot{\subset}} \dot{s}\text{‘}\lambda$

$*$41·151. $\vdash :.\, \dot{s}\text{‘}\lambda \mathrel{\dot{\subset}} S \,.\equiv :\, R \epsilon \lambda \,.\supset_R .\, R \mathrel{\dot{\subset}} S$

Of the remaining propositions of this number, which have no analogues in $*$40, the most important are $*$41·43·44·45, namely

$$\mathrm{D}\text{‘}\dot{s}\text{‘}\lambda = s\text{‘}\mathrm{D}\text{“}\lambda, \quad \text{Ɑ}\text{‘}\dot{s}\text{‘}\lambda = s\text{‘}\text{Ɑ}\text{“}\lambda, \quad C\text{‘}\dot{s}\text{‘}\lambda = s\text{‘}C\text{“}\lambda.$$

These propositions are constantly required in the theory of selections (Part II, Section D) and in relation-arithmetic. Most of the other propositions of this number are used only once or not at all.

$*$41·01. $\dot{p}\text{‘}\lambda = \hat{x}\hat{y}\,(R \epsilon \lambda \,.\supset_R .\, xRy)$ Df

$*$41·02. $\dot{s}\text{‘}\lambda = \hat{x}\hat{y}\,\{(\exists R) \,.\, R \epsilon \lambda \,.\, xRy\}$ Df

$*$41·1. $\vdash :.\, x\,(\dot{p}\text{‘}\lambda)\,y \,.\equiv :\, R \epsilon \lambda \,.\supset_R .\, xRy$

$*$41·11. $\vdash : x\,(\dot{s}\text{‘}\lambda)\,y \,.\equiv .\, (\exists R) \,.\, R \epsilon \lambda \,.\, xRy$

$*$41·12. $\vdash : R \epsilon \lambda \,.\supset .\, \dot{p}\text{‘}\lambda \mathrel{\dot{\subset}} R$

$*$41·13. $\vdash : R \epsilon \lambda \,.\supset .\, R \mathrel{\dot{\subset}} \dot{s}\text{‘}\lambda$

$*$41·14. $\vdash : R \epsilon \lambda \,.\, x\,(\dot{p}\text{‘}\lambda)\,y \,.\supset .\, xRy$

$*$41·141. $\vdash : R \epsilon \lambda \,.\, xRy \,.\supset .\, x\,(\dot{s}\text{‘}\lambda)\,y$

$*$41·15. $\vdash :.\, S \mathrel{\dot{\subset}} \dot{p}\text{‘}\lambda \,.\equiv :\, R \epsilon \lambda \,.\supset_R .\, S \mathrel{\dot{\subset}} R$

$*$41·151. $\vdash :.\, \dot{s}\text{‘}\lambda \mathrel{\dot{\subset}} S \,.\equiv :\, R \epsilon \lambda \,.\supset_R .\, R \mathrel{\dot{\subset}} S$

$*$41·16. $\vdash : \lambda \subset \mu \,.\supset .\, \dot{p}\text{‘}\mu \mathrel{\dot{\subset}} \dot{p}\text{‘}\lambda$

$*$41·161. $\vdash : \lambda \subset \mu \,.\supset .\, \dot{s}\text{‘}\lambda \mathrel{\dot{\subset}} \dot{s}\text{‘}\mu$

$*$41·17. $\vdash .\, \dot{p}\text{‘}\lambda \mathbin{\dot{\cup}} \dot{p}\text{‘}\mu \mathrel{\dot{\subset}} \dot{p}\text{‘}(\lambda \cap \mu)$

***41·171.** $\vdash . \dot{s}‘\lambda \mathbin{\dot{\cup}} \dot{s}‘\mu = \dot{s}‘(\lambda \cup \mu)$

***41·18.** $\vdash . \dot{p}‘(\lambda \cup \mu) = \dot{p}‘\lambda \mathbin{\dot{\cap}} \dot{p}‘\mu$

***41·181.** $\vdash . \dot{s}‘(\lambda \cap \mu) \mathrel{\dot{\subset}} \dot{s}‘\lambda \mathbin{\dot{\cap}} \dot{s}‘\mu$

***41·19.** $\vdash :: x(\dot{s}‘\lambda)y . \equiv :. R \in \lambda . \supset_R . R \mathrel{\dot{\subset}} S : \supset_S . xSy$

***41·2.** $\vdash : \lambda = \Lambda . \supset . \dot{p}‘\lambda = \dot{V}$

***41·21.** $\vdash : \lambda = \Lambda . \supset . \dot{s}‘\lambda = \dot{\Lambda}$

***41·22.** $\vdash : \dot{\Lambda} \in \lambda . \supset . \dot{p}‘\lambda = \dot{\Lambda}$

***41·221.** $\vdash : \dot{V} \in \lambda . \supset . \dot{s}‘\lambda = \dot{V}$

***41·23.** $\vdash : \exists ! \lambda . \supset . \dot{p}‘\lambda \mathrel{\dot{\subset}} \dot{s}‘\lambda$

***41·24.** $\vdash :. \exists ! \lambda : R \in \lambda . \supset_R . S \mathrel{\dot{\subset}} R : \supset . S \mathrel{\dot{\subset}} \dot{s}‘\lambda$

***41·25.** $\vdash : x(\dot{s}‘\lambda)y . \equiv . \exists ! \lambda \cap \hat{R}(xRy)$

***41·26.** $\vdash : \dot{\exists} ! \dot{s}‘\lambda . \equiv . (\exists R) . R \in \lambda . \dot{\exists} ! R$

***41·27.** $\vdash :. P \mathbin{\dot{\cap}} \dot{s}‘\lambda = \dot{\Lambda} . \equiv : R \in \lambda . \supset_R . P \mathbin{\dot{\cap}} R = \dot{\Lambda}$

***41·3.** $\vdash . \mathrm{Cnv}‘\dot{p}‘\lambda = \dot{p}‘\mathrm{Cnv}“\lambda$

Dem.

$\vdash . *31·131 . \supset$

$\vdash :. y(\mathrm{Cnv}‘\dot{p}‘\lambda)x . \equiv : x(\dot{p}‘\lambda)y :$

$[*41·1] \qquad \equiv : R \in \lambda . \supset_R . xRy :$

$[*31·131] \qquad \equiv : R \in \lambda . \supset_R . y(\mathrm{Cnv}‘R)x :$

$[*37·63 . *31·13] \qquad \equiv : P \in \mathrm{Cnv}“\lambda . \supset_P . yPx :$

$[*41·1] \qquad \equiv : y(\dot{p}‘\mathrm{Cnv}“\lambda)x :. \supset \vdash . \mathrm{Prop}$

***41·31.** $\vdash . \mathrm{Cnv}‘\dot{s}‘\lambda = \dot{s}‘\mathrm{Cnv}“\lambda$ [Proof as in *41·3]

***41·32.** $\vdash . \mathrm{Cnv}“\dot{p}“\kappa = \dot{p}“\mathrm{Cnv}“‘\kappa$ [*41·3 . *37·354]

***41·33.** $\vdash . \mathrm{Cnv}“\dot{s}“\kappa = \dot{s}“\mathrm{Cnv}“‘\kappa$ [*41·31 . *37·354]

***41·34.** $\vdash . \dot{s}‘\alpha \upharpoonleft “\lambda = \alpha \upharpoonleft \dot{s}‘\lambda$

Dem.

$\vdash . *41·11 . *38·13 . *13·195 . \supset \vdash :. x(\dot{s}‘\alpha \upharpoonleft “\lambda)y . \equiv : (\exists P) . P \in \lambda . x(\alpha \upharpoonleft P)y :$

$[*35·1] \qquad \equiv : (\exists P) . P \in \lambda . x \in \alpha . xPy :$

$[*10·35] \qquad \equiv : x \in \alpha : (\exists P) . P \in \lambda . xPy :$

$[*41·11 . *35·1] \qquad \equiv : x(\alpha \upharpoonleft \dot{s}‘\lambda)y :. \supset \vdash . \mathrm{Prop}$

***41·341.** $\vdash . \dot{s}‘\upharpoonright \alpha“\lambda = (\dot{s}‘\lambda) \upharpoonright \alpha$ [Proof as in *41·34]

***41·342.** $\vdash . \dot{s}‘\upharpoonleft\!\upharpoonright \alpha“\lambda = (\dot{s}‘\lambda) \upharpoonleft\!\upharpoonright \alpha$

Dem.

$\vdash . *36·11 . *35·21 . \supset \vdash . \dot{s}‘\upharpoonleft\!\upharpoonright \alpha“\lambda = \dot{s}‘\alpha \upharpoonleft “\upharpoonright \alpha“\lambda$

$[*41·34] \qquad = \alpha \upharpoonleft (\dot{s}‘\upharpoonright \alpha“\lambda)$

$[*41·341] \qquad = \alpha \upharpoonleft (\dot{s}‘\lambda) \upharpoonright \alpha$

$[*36·11] \qquad = (\dot{s}‘\lambda) \upharpoonleft\!\upharpoonright \alpha . \supset \vdash . \mathrm{Prop}$

The following proposition is used in $*85{\cdot}22$.

$*41{\cdot}35$. $\vdash . \dot{s}`M \upharpoonright``\kappa = M \upharpoonright s`\kappa$

Dem.

$$\begin{aligned}
&\vdash . *41{\cdot}11 . *38{\cdot}13 . \supset \vdash : x(\dot{s}`M\upharpoonright``\kappa)y . \equiv . (\exists \alpha) . \alpha \in \kappa . x(M \upharpoonright \alpha) y . \\
&[*35{\cdot}101] \qquad \equiv . (\exists \alpha) . \alpha \in \kappa . y \in \alpha . xMy . \\
&[*40{\cdot}11 . *35{\cdot}101] \qquad \equiv . x(M \upharpoonright s`\kappa) y : \supset \vdash . \text{Prop}
\end{aligned}$$

$*41{\cdot}351$. $\vdash . \dot{s}`\upharpoonleft M``\kappa = (s`\kappa) \upharpoonleft M$ [Proof as in $*41{\cdot}35$]

$*41{\cdot}4$. $\vdash . \text{D}`\dot{p}`\lambda \subset p`\text{D}``\lambda$

Dem.

$$\begin{aligned}
&\vdash . *33{\cdot}13 . \supset \\
&\vdash :: x \in \text{D}`\dot{p}`\lambda . \equiv :. (\exists y) . x(\dot{p}`\lambda) y :. \\
&[*41{\cdot}1] \qquad \equiv :. (\exists y) : R \in \lambda . \supset_R . xRy :. \\
&[*11{\cdot}61] \qquad \supset :. R \in \lambda . \supset_R . (\exists y) . xRy :. \\
&[*33{\cdot}13] \qquad \supset :. R \in \lambda . \supset_R . x \in \text{D}`R :. \\
&[*40{\cdot}41 . *33{\cdot}12] \supset :. x \in p`\text{D}``\lambda :: \supset \vdash . \text{Prop}
\end{aligned}$$

$*41{\cdot}41$. $\vdash . \text{Ɑ}`\dot{p}`\lambda \subset p`\text{Ɑ}``\lambda$ [Proof as in $*41{\cdot}4$]

$*41{\cdot}42$. $\vdash . C`\dot{p}`\lambda \subset p`C``\lambda$

Dem.

$$\begin{aligned}
&\vdash . *33{\cdot}132 . \supset \vdash ::. x \in C`\dot{p}`\lambda . \equiv :: (\exists y) : x(\dot{p}`\lambda) y . \vee . y(\dot{p}`\lambda) x :: \\
&[*41{\cdot}1] \qquad \equiv :: (\exists y) :: R \in \lambda . \supset_R . xRy : \vee : R \in \lambda . \supset_R . yRx :: \\
&[*10{\cdot}41{\cdot}221] \qquad \supset :: (\exists y) :: (R) :. R \in \lambda . \supset . xRy : \vee : R \in \lambda . \supset . yRx :: \\
&[*4{\cdot}78] \qquad \supset :: (\exists y) :: (R) :. R \in \lambda . \supset : xRy . \vee . yRx :: \\
&[*11{\cdot}61] \qquad \supset :: (R) :: R \in \lambda . \supset : (\exists y) : xRy . \vee . yRx : \\
&[*33{\cdot}132] \qquad \supset : x \in C`R :: \\
&[*40{\cdot}41 . *33{\cdot}122] \qquad \supset :: x \in p`C``\lambda ::. \supset \vdash . \text{Prop}
\end{aligned}$$

$*41{\cdot}43$. $\vdash . \text{D}`\dot{s}`\lambda = s`\text{D}``\lambda$

Dem.

$$\begin{aligned}
&\vdash . *33{\cdot}13 . \supset \vdash :. x \in \text{D}`\dot{s}`\lambda . \equiv : (\exists y) . x(\dot{s}`\lambda) y : \\
&[*41{\cdot}11] \qquad \equiv : (\exists y) : (\exists R) . R \in \lambda . xRy : \\
&[*11{\cdot}23{\cdot}55] \qquad \equiv : (\exists R) : R \in \lambda : (\exists y) . xRy : \\
&[*33{\cdot}13] \qquad \equiv : (\exists R) . R \in \lambda . x \in \text{D}`R : \\
&[*40{\cdot}4 . *33{\cdot}12] \qquad \equiv : x \in s`\text{D}``\lambda :. \supset \vdash . \text{Prop}
\end{aligned}$$

$*41{\cdot}44$. $\vdash . \text{Ɑ}`\dot{s}`\lambda = s`\text{Ɑ}``\lambda$ [Proof as in $*41{\cdot}43$]

$*41{\cdot}45$. $\vdash . C`\dot{s}`\lambda = s`C``\lambda$

Dem.

$$\begin{aligned}
&\vdash . *33{\cdot}16 . \supset \vdash . C`\dot{s}`\lambda = \text{D}`\dot{s}`\lambda \cup \text{Ɑ}`\dot{s}`\lambda \\
&[*41{\cdot}43{\cdot}44] \qquad = s`\text{D}``\lambda \cup s`\text{Ɑ}``\lambda \\
&[*40{\cdot}57] \qquad = s`C``\lambda . \supset \vdash . \text{Prop}
\end{aligned}$$

$*41\cdot5.\quad \vdash . \dot{p}`\lambda \,|\, \dot{p}`\mu \mathrel{\dot{\subset}} \dot{p}`s`\lambda \,|_{,,}``\mu$

Dem.

$\vdash . *34\cdot1 . \supset$

$\vdash :: x(\dot{p}`\lambda \,|\, \dot{p}`\mu) z . \equiv :. (\exists y) . x(\dot{p}`\lambda) y . y(\dot{p}`\mu) z :.$

$[*41\cdot1] \quad \equiv :. (\exists y) :. P \epsilon \lambda . \supset_P . xPy : Q \epsilon \mu . \supset_Q . yQz :.$

$[*11\cdot56] \quad \equiv :. (\exists y) :. (P, Q) : P \epsilon \lambda . \supset . xPy : Q \epsilon \mu . \supset . yQz :.$

$[*11\cdot37\cdot39] \quad \supset :. (\exists y) :. (P, Q) : P \epsilon \lambda . Q \epsilon \mu . \supset . xPy . yQz :.$

$[*11\cdot61] \quad \supset :. (P, Q) :. P \epsilon \lambda . Q \epsilon \mu . \supset . (\exists y) . xPy . yQz .$

$[*34\cdot1] \quad \supset . x(P \,|\, Q) z :.$

$[*13\cdot191] \quad \supset :. (P, Q, R) :. P \epsilon \lambda . Q \epsilon \mu . R = P \,|\, Q . \supset . xRz :.$

$[*11\cdot21\cdot35] \quad \supset :. (R) : (\exists P, Q) . P \epsilon \lambda . Q \epsilon \mu . R = P \,|\, Q . \supset . xRz :$

$[*40\cdot7] \quad \supset :. (R) : R \epsilon s`\lambda \,|_{,,}``\mu . \supset . xRz :.$

$[*41\cdot1] \quad \supset :. x(\dot{p}`s`\lambda \,|_{,,}``\mu) z :: \supset \vdash . \mathrm{Prop}$

$*41\cdot51.\quad \vdash . \dot{s}`\lambda \,|\, \dot{s}`\mu = \dot{s}`s`\lambda \,|_{,,}``\mu$

Dem.

$\vdash . *34\cdot1 . \supset$

$\vdash :: x(\dot{s}`\lambda \,|\, \dot{s}`\mu) z . \equiv :. (\exists y) . x(\dot{s}`\lambda) y . y(\dot{s}`\mu) z :.$

$[*41\cdot11] \quad \equiv :. (\exists y) :. (\exists P) . P \epsilon \lambda . xPy : (\exists Q) . Q \epsilon \mu . yQz :.$

$[*11\cdot54] \quad \equiv :. (\exists y) :. (\exists P, Q) : P \epsilon \lambda . xPy . Q \epsilon \mu . yQz :.$

$[*11\cdot24\cdot27] \quad \equiv :. (\exists P, Q) :. (\exists y) . P \epsilon \lambda . xPy . Q \epsilon \mu . yQz :.$

$[*10\cdot35] \quad \equiv :. (\exists P, Q) :. P \epsilon \lambda . Q \epsilon \mu : (\exists y) . xPy . yQz :.$

$[*34\cdot1] \quad \equiv :. (\exists P, Q) : P \epsilon \lambda . Q \epsilon \mu . x(P \,|\, Q) z :.$

$[*13\cdot195] \quad \equiv :. (\exists P, Q, R) . P \epsilon \lambda . Q \epsilon \mu . R = P \,|\, Q . xRz :.$

$[*11\cdot24 . *40\cdot7] \quad \equiv :. (\exists R) . R \epsilon s`\lambda \,|_{,,}``\mu . xRz :.$

$[*41\cdot11] \quad \equiv :. x(\dot{s}`s`\lambda \,|_{,,}``\mu) z :: \supset \vdash . \mathrm{Prop}$

The above proposition, which is used in $*92\cdot31$, states that, if λ and μ are classes of relations, the relative product of the relational sum of λ and the relational sum of μ is the relational sum of all the relative products formed of a member of λ and a member of μ.

The following proposition is used in $*96\cdot111$.

$*41\cdot52.\quad \vdash :. \alpha \upharpoonright \dot{s}`\lambda \mathrel{\dot{\subset}} Q . \equiv : P \epsilon \lambda . \supset_P . \alpha \upharpoonright P \mathrel{\dot{\subset}} Q$

Dem.

$\vdash . *35\cdot1 . *41\cdot11 . \supset$

$\vdash :: \alpha \upharpoonright \dot{s}`\lambda \mathrel{\dot{\subset}} Q . \equiv :. x \epsilon \alpha : (\exists P) . P \epsilon \lambda . xPy : \supset_{x,y} . xQy :.$

$[*10\cdot35\cdot23] \quad \equiv :. x \epsilon \alpha . P \epsilon \lambda . xPy . \supset_{P,x,y} . xQy :.$

$[*35\cdot1] \quad \equiv :. P \epsilon \lambda . x(\alpha \upharpoonright P) y . \supset_{P,x,y} . xQy :.$

$[*11\cdot62] \quad \equiv :. P \epsilon \lambda . \supset_P . \alpha \upharpoonright P \mathrel{\dot{\subset}} Q :: \supset \vdash . \mathrm{Prop}$

The following proposition is used in ∗162·32 and in ∗166·461.

∗41·6. $\vdash :. y \epsilon \beta . \supset_y . P\text{‘}y = Q\text{‘}y \cup R\text{‘}y : \supset . \dot{s}\text{‘}P\text{“}\beta = \dot{s}\text{‘}Q\text{“}\beta \cup \dot{s}\text{‘}R\text{“}\beta$

Dem.

$\vdash . *37\cdot6 . *14\cdot21 . *41\cdot11 . *13\cdot195 . \supset$

$\vdash :: \text{Hp} . \supset :. u (\dot{s}\text{‘}P\text{“}\beta) v . \equiv : (\exists y) . y \epsilon \beta . u (P\text{‘}y) v :$

[Hp] $\equiv : (\exists y) . y \epsilon \beta . u (Q\text{‘}y \cup R\text{‘}y) v :$

[∗23·34.∗10·42] $\equiv : (\exists y) . y \epsilon \beta . u (Q\text{‘}y) v . \vee . (\exists y) . y \epsilon \beta . u (R\text{‘}y) v :$

[∗37·6.∗41·11] $\equiv : u (\dot{s}\text{‘}Q\text{“}\beta) v . \vee . u (\dot{s}\text{‘}R\text{“}\beta) v :: \supset \vdash . \text{Prop}$

$*42$. MISCELLANEOUS PROPOSITIONS

Summary of $*42$.

The present number contains various propositions concerning products and sums of classes. They are concerned chiefly with classes of classes of classes, or with relations of relations of relations. These are required respectively in cardinal and in ordinal arithmetic. Thus $*42\cdot1$ is used in $*112$ and $*113$, which are concerned with cardinal addition and multiplication, while $*42\cdot12\cdot2$ are used in $*160$ and $*162$, which are concerned with ordinal addition. $*42\cdot22$, though not explicitly referred to, is useful in facilitating the comprehension of propositions on series of series of series, or rather on relations between relations between relations, which are required in connection with the associative law of multiplication in relation-arithmetic.

$*42\cdot1$. $\vdash . s`s``\kappa = s`s`\kappa$

Here κ must, for significance, be a class of classes of classes. The proposition states that if we take each member, α, of κ, and form $s`\alpha$, and then form the sum of all the classes so obtained, the result is the same as if we form the sum of the sum of κ. This is the associative law for s, and is (as will appear later) the source of the associative law of addition in cardinal arithmetic. The way in which this proposition comes to be the associative law for s may be seen as follows: Suppose κ consists of two classes, α and β; suppose α in turn consists of the two classes ξ and η, and β of the two classes ξ' and η'. Then $s`\alpha = \xi \cup \eta . s`\beta = \xi' \cup \eta'$. (This will be proved later.) Thus $s``\kappa$ has two members, one of which is $\xi \cup \eta$, while the other is $\xi' \cup \eta'$. Thus

$$s`s``\kappa = (\xi \cup \eta) \cup (\xi' \cup \eta').$$

But $s`\kappa$ has four members, namely $\xi, \eta, \xi', \eta'$. Thus $s`s`\kappa = \xi \cup \eta \cup \xi' \cup \eta'$. Thus our proposition leads to

$$(\xi \cup \eta) \cup (\xi' \cup \eta') = \xi \cup \eta \cup \xi' \cup \eta',$$

which is obviously a case of the associative law.

Our proposition states the associative law generally, including the case where the number of brackets, or of summands in any bracket, is infinite. The proof is as follows.

Dem.

$\vdash . *40\cdot4 . \supset \vdash :: x \in s`s``\kappa . \equiv :. (\exists\alpha) . \alpha \in \kappa . x \in s`\alpha :.$

$[*40\cdot11] \quad \equiv :. (\exists\alpha) : \alpha \in \kappa : (\exists\xi) . \xi \in \alpha . x \in \xi :.$

$[*11\cdot6] \quad \equiv :. (\exists\xi) :. (\exists\alpha) . \alpha \in \kappa . \xi \in \alpha : x \in \xi :.$

$[*40\cdot11] \quad \equiv :. (\exists\xi) . \xi \in s`\kappa . x \in \xi :.$

$[*40\cdot11] \quad \equiv :. x \in s`s`\kappa :: \supset \vdash . \text{Prop}$

***42·11.** $\vdash . p`p``\kappa = p`s`\kappa$

Dem.

$\vdash . *40·41 . \supset \vdash :. x \epsilon p`p``\kappa . \equiv : \beta \epsilon \kappa . \supset_\beta . x \epsilon p`\beta :$

[*40·1.*11·62] $\equiv : \beta \epsilon \kappa . \gamma \epsilon \beta . \supset_{\beta,\gamma} . x \epsilon \gamma :$

[*11·2.*10·23] $\equiv : (\exists\beta) . \beta \epsilon \kappa . \gamma \epsilon \beta . \supset_\gamma . x \epsilon \gamma :$

[*40·11] $\equiv : \gamma \epsilon s`\kappa . \supset_\gamma . x \epsilon \gamma :$

[*40·1] $\equiv : x \epsilon p`s`\kappa :. \supset \vdash$. Prop

This is the associative law for products. Supposing again, for illustration, that κ consists of the two classes α, β, while α consists of the two classes ξ, η and β of the two classes ξ', η', then $p``\kappa$ consists of the two classes $\xi \cap \eta$ and $\xi' \cap \eta'$, so that $p`p``\kappa = (\xi \cap \eta) \cap (\xi' \cap \eta')$, while $p`s`\kappa = \xi \cap \eta \cap \xi' \cap \eta'$. Thus our proposition becomes

$$(\xi \cap \eta) \cap (\xi' \cap \eta') = \xi \cap \eta \cap \xi' \cap \eta'.$$

A descriptive function $R`\kappa$ whose arguments are classes or classes of classes may be said to obey the associative law provided

$$R`R``\kappa = R`s`\kappa.$$

This equation may be interpreted as follows: Given a class α, divide it into any number of subordinate classes, so that no member is left out, though one member may belong to two or more classes. Let the classes into which α is divided make up the class κ, so that κ is a class of classes, and $s`\kappa = \alpha$. Then the above equation asserts that if we first form the R's of the various sub-classes of α, and then the R of the resulting class, the result is the same as if we formed the R of α directly.

In some cases—for example, that of arithmetical addition of cardinals—the above equation holds only when no two members of κ have a common term, *i.e.* when the parts into which α is divided are mutually exclusive.

For a descriptive function whose arguments are relations of relations, we shall find another form for the associative law; this form plays in ordinal arithmetic a part analogous to that played by the above form in cardinal arithmetic.

***42·12.** $\vdash . \dot{s}`\dot{s}``\lambda = \dot{s}`s`\lambda$

Dem.

$\vdash . *41·11 . \supset \vdash : x (\dot{s}`\dot{s}``\lambda) y . \equiv . (\exists\mu) . \mu \epsilon \lambda . x (\dot{s}`\mu) y .$

[*41·11] $\equiv . (\exists\mu, P) . \mu \epsilon \lambda . P \epsilon \mu . xPy .$

[*40·11] $\equiv . (\exists P) . P \epsilon s`\lambda . xPy .$

[*41·11] $\equiv . x (\dot{s}`s`\lambda) y : \supset \vdash$. Prop

***42·13.** $\vdash . \dot{p}`\dot{p}``\lambda = \dot{p}`s`\lambda$

Dem.

$\vdash . *41·1 . \supset \vdash :. x (\dot{p}`\dot{p}``\lambda) y . \equiv : \mu \epsilon \lambda . \supset_\mu . x (\dot{p}`\mu) y :$

[*41·1] $\equiv : \mu \epsilon \lambda . R \epsilon \mu . \supset_{\mu, R} . xRy :$

[*11·2.*10·23] $\equiv : (\exists\mu) . \mu \epsilon \lambda . R \epsilon \mu . \supset_R . xRy :$

[*40·11] $\equiv : R \epsilon s`\lambda . \supset_R . xRy :$

[*41·1] $\equiv : x (\dot{p}`s`\lambda) y :. \supset \vdash .$ Prop

***42·2.** $\vdash . C`\dot{s}`C`P = s`C``C`P = F``C`P = \overrightarrow{F^2}`P$

This proposition assumes that P is a relation between relations. For example, suppose we have a series of series, whose generating relations are ordered by the relation P. Then $C`P$ is the class of these generating relations; $\dot{s}`C`P$ is the relation "one or other of the generating relations which compose $C`P$," and $C`\dot{s}`C`P$ is the class of all the terms occurring in any of the series. $C``C`P$ is the fields of the various series, and $s`C``C`P$ is again all the terms occurring in any of the series. $F``C`P$ is all the terms belonging to fields of series which are members of $C`P$, and $\overrightarrow{F^2}`P$ is all members of fields of members of the field of P; each of these again is all the terms occurring in any of the series. The proof is as follows:

Dem.

$\vdash . *41·45 . \supset \vdash . C`\dot{s}`C`P = s`C``C`P$ (1)

$\vdash . *40·56 . \supset \vdash . s`C``C`P = F``C`P$ (2)

$\vdash . *33·5 . \supset \vdash . F``C`P = F``\overrightarrow{F}`P$

[*37·38] $= \overrightarrow{F^2}`P$ (3)

$\vdash . (1) . (2) . (3) . \supset \vdash .$ Prop

The following propositions apply to a relation of relations of relations. These propositions are useful for proving associative laws in ordinal arithmetic, since these laws deal with series of series of series, and series of series of series are most simply constituted by supposing the generating relations of the constituent series to be ordered by relations which are themselves ordered by a relation P.

***42·21.** $\vdash . s`C```C``C`P = C``s`C``C`P = C``C`\dot{s}`C`P = C``F``C`P = C``\overrightarrow{F^2}`P$

Dem.

$\vdash . *40·38 . \quad \supset \vdash . s`C```C``C`P = C``s`C``C`P$ (1)

$\vdash . (1) . *42·2 . \supset \vdash .$ Prop

***42·22.** $\vdash . s`s`C```C``C`P = s`C``s`C``C`P = s`C``C`\dot{s}`C`P$

$= C`\dot{s}`C`\dot{s}`C`P = s`C``F``C`P$

$= F``F``C`P = F``\overrightarrow{F^2}`P = \overrightarrow{F^3}`P$

[*42·21 . *41·45 . *40·56 . *42·2 . *37·3]

If P, in the above proposition, is a relation which generates a series of series of series, the above gives various forms for the class of ultimate terms of these series. Thus suppose $Q \epsilon C`P$; then Q is a relation between generating relations of series. If now $R \epsilon C'Q$, R is the generating relation of a series which we may regard as composed of individuals. The class of individuals so obtainable may be expressed in any of the above forms, as well as in others which are not given above.

***42·3.** $\vdash . s's''\overrightarrow{R}''\alpha = s'R''\alpha$

Dem.

$\vdash . *42·1 . \supset \vdash . s's''\overrightarrow{R}''\alpha = s's'\overrightarrow{R}''\alpha$

[*40·5] $= s'R''\alpha . \supset \vdash . \text{Prop}$

***42·31.** $\vdash . s's''\overleftarrow{R}''\alpha = s'\breve{R}''\alpha$ [Proof as in *42·3]

∗43. THE RELATIONS OF A RELATIVE PRODUCT TO ITS FACTORS

Summary of ∗43.

The purpose of the present number is to give certain propositions on the relation which holds between P and Q whenever $P = Q \mid R$, or whenever $P = R \mid Q$, or whenever $P = R \mid Q \mid S$, where R and S are fixed. In virtue of the general definitions of ∗38, these relations are respectively $\mid R$, $R \mid$, and $(R \mid) \mid (\mid S)$. Such relations are of great utility both in cardinal and in ordinal arithmetic; they are also much used in the theory of induction (Part II, Section E). In place of the notation $(R \mid) \mid (\mid S)$, which is cumbrous, we adopt the more compact notation $R \| S$. If λ is a class of relations, $R \mid ``\lambda$ will be the class of relations $R \mid P$ where $P \in \lambda$, $\mid R``\lambda$ will be the class of relations $P \mid R$ where $P \in \lambda$, and $(R \| S)``\lambda$ will be the class of relations $R \mid P \mid S$ where $P \in \lambda$. These classes of relations are often required in subsequent work.

In virtue of our definitions, we have

∗43·112. $\vdash . (R \| S)`Q = R \mid Q \mid S$

The propositions most used in the present number (except such as merely embody definitions) are the following:

∗43·302. $\vdash . (P) . P \in \text{Ɑ}`(R \| S)$

∗43·411. $\vdash . \breve{R}```\text{Ɑ}``\lambda = \text{Ɑ}``\mid R``\lambda$

∗43·421. $\vdash . \dot{s}`\mid R``\lambda = (\dot{s}`\lambda) \mid R$

The remaining propositions are used seldom, but their uses, when they are used, are important.

∗43·01. $R \| S = (R \mid) \mid (\mid S)$ Df

At a later stage (in ∗150) we shall introduce a simpler notation for the special case of $R \| \breve{R}$. The following propositions are for the most part immediate consequences of the definitions, and proofs are therefore usually omitted.

∗43·1. $\vdash : P (R \mid) Q . \equiv . P = R \mid Q$

∗43·101. $\vdash : P (\mid R) Q . \equiv . P = Q \mid R$

∗43·102. $\vdash : P (R \| S) Q . \equiv . P = R \mid Q \mid S$

∗43·11. $\vdash . R \mid `Q = R \mid Q$

∗43·111. $\vdash . \mid R`Q = Q \mid R$

∗43·112. $\vdash . (R \| S)`Q = R \mid Q \mid S$

∗43·12. $\vdash . \text{E} ! \, R \mid `Q$

***43·121.** $\vdash . \text{E} ! \,|R`Q$

***43·122.** $\vdash . \text{E} ! (R \| S)`Q$

***43·2.** $\vdash . (R|) | (S|) = (R|S)|$

Dem.

$\vdash . *43·1 . \supset \vdash : L \{(R|)|(S|)\} N . \equiv . (\exists M) . L = R|M . M = S|N .$

$[*13·195 . *34·21] \qquad \equiv . L = R|S|N .$

$[*43·1] \qquad \equiv . L \{(R|S)|\} N : \supset \vdash . \text{Prop}$

***43·201.** $\vdash . (|R)|(|S) = |(S|R)$ [Proof as in *43·2]

***43·202.** $\vdash . (|R)|(S|) = (S|)|(|R) = S \| R$ [Proof as in *43·2]

***43·21.** $\vdash . (P \| Q)|(R|) = (P|R) \| Q$

***43·211.** $\vdash . (R|)|(P \| Q) = (R|P) \| Q$

***43·212.** $\vdash . (P \| Q)|(|R) = P \| (R|Q)$

***43·213.** $\vdash . (|R)|(P \| Q) = P \| (Q|R)$

***43·22.** $\vdash . (P \| Q)|(R \| S) = (P|R) \| (S|Q)$

***43·3.** $\vdash . (P) . P \in \text{Q}`R|$ [*43·12 . *33·43]

***43·301.** $\vdash . (P) . P \in \text{Q}`|R$

***43·302.** $\vdash . (P) . P \in \text{Q}`(R \| S)$

***43·31.** $\vdash . P \upharpoonright \text{Q}`R| = P \upharpoonright C`R| = P$

Dem.

$\vdash . *43·12 . *33·431 . \supset \vdash . \text{Q}`P \subset \text{Q}`R|$ (1)

$[*33·161] \qquad \supset \vdash . \text{Q}`P \subset C`R|$ (2)

$\vdash . (1) . (2) . *35·452 . \supset \vdash . \text{Prop}$

***43·311.** $\vdash . P \upharpoonright \text{Q}`|R = P \upharpoonright C`|R = P$

***43·312.** $\vdash . P \upharpoonright \text{Q}`(R \| S) = P \upharpoonright C`(R \| S) = P$

***43·34.** $\vdash . R|`R = |R`R = R^2$ [*43·11·111]

***43·4.** $\vdash . R``\text{D}`P = \text{D}`R|`P$ [*37·32 . *43·1]

***43·401.** $\vdash . \breve{R}``\text{Q}`P = \text{Q}`|R`P$ [*37·32 . *43·101]

***43·41.** $\vdash . R```\text{D}``\lambda = \text{D}``R|``\lambda$ [*43·4 . *37·355]

***43·411.** $\vdash . \breve{R}```\text{Q}``\lambda = \text{Q}``|R``\lambda$ [*43·401 . *37·355]

***43·42.** $\vdash . \dot{s}`R|``\lambda = R|\dot{s}`\lambda$

Dem.

$\vdash . *41·11 . *37·1 . *43·1 . \supset$

$\vdash :. x (\dot{s}`R|``\lambda) z . \equiv : (\exists T) . T \in \lambda . x (R|T) z :$

$[*34·1] \qquad \equiv : (\exists T) : T \in \lambda : (\exists y) . xRy . yTz :$

$[*11·6] \qquad \equiv : (\exists y) : xRy : (\exists T) . T \in \lambda . yTz :$

$[*41·11 . *34·1] \qquad \equiv : x (R|\dot{s}`\lambda) z :. \supset \vdash . \text{Prop}$

***43·421.** $\vdash . \dot{s}'|R``\lambda = (\dot{s}'\lambda)|R$ [Proof as in *43·42]

***43·43.** $\vdash . \dot{s}'(R \| S)``\lambda = (R \| S)'\dot{s}'\lambda$

Dem.

$\vdash . *37·33 . \supset \vdash . \dot{s}'(R \| S)``\lambda = \dot{s}'R | `` | S``\lambda$

[*43·42] $= R | (\dot{s}' | S``\lambda)$

[*43·421] $= R | \dot{s}'\lambda | S$

[*43·112] $= (R \| S)'\dot{s}'\lambda . \supset \vdash .$ Prop

***43·48.** $\vdash : \text{D}'P \subset \alpha . \supset . Q | 'P = (Q \upharpoonright \alpha) | 'P$ [*35·481]

***43·481.** $\vdash : \text{Cl}'P \subset \beta . \supset . | R'P = | (\beta \upharpoonleft R)'P$ [*35·48]

***43·49.** $\vdash : s'\text{D}``\lambda \subset \alpha . \supset . (Q|) \upharpoonright \lambda = \{(Q \upharpoonright \alpha)|\} \upharpoonright \lambda$

Dem.

$\vdash . *40·43 . \supset \vdash :. \text{Hp} . \supset : P \epsilon \lambda . \supset . \text{D}'P \subset \alpha .$

[*43·48] $\supset . Q|'P = \{(Q \upharpoonright \alpha)|\}'P$ (1)

$\vdash . (1) . *35·71 . \supset \vdash .$ Prop

***43·491.** $\vdash : s'\text{Cl}``\lambda \subset \beta . \supset . (|R) \upharpoonright \lambda = \{|(\beta \upharpoonleft R)\} \upharpoonright \lambda$ [Proof as in *43·49]

***43·5.** $\vdash : \text{D}'P \subset \alpha . \text{Cl}'P \subset \beta . \supset . (Q \| R)'P = \{(Q \upharpoonright \alpha) \| (\beta \upharpoonleft R)\}'P$
[*35·48·481 . *43·112]

***43·51.** $\vdash : s'\text{D}``\lambda \subset \alpha . s'\text{Cl}``\lambda \subset \beta . \supset . (Q \| R) \upharpoonright \lambda = \{(Q \upharpoonright \alpha) \| (\beta \upharpoonleft R)\} \upharpoonright \lambda$

Dem.

$\vdash . *40·43 . \supset \vdash :. \text{Hp} . \supset : P \epsilon \lambda . \supset . \text{D}'P \subset \alpha . \text{Cl}'P \subset \beta .$

[*43·5] $\supset . (Q \| R)'P = \{(Q \upharpoonright \alpha) \| (\beta \upharpoonleft R)\}'P$ (1)

$\vdash . (1) . *35·71 . \supset \vdash .$ Prop

The above proposition is used in the proof of *74·773.

PART II

PROLEGOMENA TO CARDINAL ARITHMETIC

SUMMARY OF PART II, SECTION A

The objects to be studied in this Part are not sharply distinguished from those studied in Part I. The difference is one of degree, the objects in this Part being of somewhat less general importance than those of Part I, and being studied more on account of their bearing on cardinal arithmetic than on their own account. Although cardinal arithmetic is the goal which determines our course in Part II, all the objects studied will be found to be also required in ordinal arithmetic and the theory of series.

Section A of this Part deals with unit classes and couples. A *unit* class is the class of terms identical with a given term, *i.e.* the class whose only member is the given term. (As explained in the Introduction, Chapter III, pp. 76 to 79, the class whose only member is x is not identical with x.) We define 1 as the class of all unit classes. In like manner, we define a (cardinal or ordinal) couple, and then define 2 as the class of all couples.

SECTION A

UNIT CLASSES AND COUPLES

Summary of Section A.

In this section we begin (∗50) by introducing a notation for the *relation* of identity, as opposed to the *function* "$x = y$"; that is, calling the relation of identity I, we put

$$I = \hat{x}\hat{y}(x = y) \quad \text{Df.}$$

The purpose of this definition is chiefly convenience of notation. The definition enables us to speak of $\overrightarrow{I}$, $D'I$, $I \mid R$, $\alpha \rceil I$, $I``\alpha$, etc., which we could not otherwise do.

At the same time we introduce *diversity*, which is defined as the negation of identity, and denoted by the letter J. The properties of I and J result immediately from ∗13, since

$$xIy \, . \equiv . \, x = y.$$

We next introduce a very important notation, due to Peano, for the class whose only member is x. If we took a strictly and purely extensional view of classes, we should naturally suppose this class to be identical with x. But in view of the theory of classes explained in ∗20, it is plain that x can never be identical with a class of which it is a member, even when it is the only member of that class. Peano uses the notation "ιx" for the class whose only member is x; we shall alter this to "$\iota'x$," following our general notation for descriptive functions. Thus we are to have

$$\iota'x = \hat{y}(y = x) = \hat{y}(yIx) = \overrightarrow{I}'x.$$

Hence we take as our definition

$$\iota = \overrightarrow{I} \quad \text{Df,}$$

since this definition gives the desired value of $\iota'x$. The properties of ι are many and important.

It is important to observe that "$\breve{\iota}'\alpha$" means "the only member of α." Thus it exists when, and only when, α has one member and no more, in which case α is of the form $\iota'x$, if x is its only member. Thus "$\breve{\iota}'\alpha$" means the same as "$(\imath x)(x \in \alpha)$," and "$\breve{\iota}'\hat{z}(\phi z)$" means the same as "$(\imath x)(\phi x)$." What we call "$\breve{\iota}'\alpha$" is denoted, in Peano's notation, by "$\imath\alpha$."

Classes of the form $\iota'x$ are called *unit classes*, and the class of all such classes is called 1. This is the cardinal number 1, according to the definition of cardinal numbers which will be given in ∗100. The properties of 1, so far as they do not depend upon other cardinals, or upon the fact that 1 is a cardinal, will be studied in ∗52.

After a number (∗53) containing various propositions involving 1 or ι, we pass to the consideration of cardinal couples (∗54) and ordinal couples (∗55). A cardinal couple is a class $\iota`x \cup \iota`y$, where $x \neq y$. The class of such couples is defined as 2, and will be shown at a later stage (∗101) to be a cardinal number. An ordinal couple, which, unlike a cardinal couple, involves an order as between its members, is defined as a relation $\iota`x \uparrow \iota`y$ (cf. ∗35·04), where we may either add $x \neq y$ or not. The properties of ordinal couples are in part analogous to those of unit classes, in part to those of cardinal couples. In ∗56, we define the ordinal number 2 (which we denote by 2_r, to distinguish it from the cardinal 2) as the class of all ordinal couples $\iota`x \uparrow \iota`y$, where $x \neq y$. It will be shown at a later stage that this is an ordinal number according to our definition of ordinal numbers (∗153 and ∗251).

$*50$. IDENTITY AND DIVERSITY AS RELATIONS

Summary of $*50$.

The purpose of the present number is primarily notational. For notational reasons, we must be able to express identity and diversity as relations, and not merely as propositional functions, *i.e.* we require a notation for $\hat{x}\hat{y}\,(x = y)$ and $\hat{x}\hat{y}\,(x \neq y)$. We therefore put

$$I = \hat{x}\hat{y}\,(x = y) \quad \text{Df},$$

$$J = \dot{-}\, I \quad \text{Df}.$$

In spite of the fact that diversity is merely the negation of identity, the kinds of propositions that employ diversity are quite different from the kinds that employ identity. Identity as a relation is required, to begin with, in the theory of unit classes, which is our reason for treating of it at this stage. It is next required, constantly, in the theory of mathematical induction (Part II, Section E). It is required also in showing that cardinal and ordinal similarity are reflexive. These are its principal uses.

Diversity, on the other hand, is required almost exclusively in the theory of series (Part V), and the first number in that theory will be devoted to diversity. Until that stage, diversity will seldom be referred to, with one important exception, namely in proving the associative law of multiplication in relation-arithmetic ($*174$).

The most important propositions on identity in the present number are the following:

$*50{\cdot}16$. $\vdash . I``\alpha = \alpha$

$*50{\cdot}4$. $\vdash . R \,|\, I = I \,|\, R = R$

$*50{\cdot}5$. $\vdash . \alpha \rceil I = I \upharpoonright \alpha = \alpha \rceil I \upharpoonright \alpha$

$*50{\cdot}51$. $\vdash . \text{Cnv}`(\alpha \rceil I) = \alpha \rceil I$

$*50{\cdot}52$. $\vdash . \text{D}`(\alpha \rceil I) = \text{Ꭰ}`(\alpha \rceil I) = C`(\alpha \rceil I) = \alpha$

$*50{\cdot}62$. $\vdash : \text{Ꭰ}`R \subset \alpha . \supset . R \,|\, (I \upharpoonright \alpha) = R$

$*50{\cdot}63$. $\vdash : \text{D}`R \subset \alpha . \supset . I \upharpoonright \alpha \,|\, R = R$

The most important propositions on diversity in the present number are the following:

$*50{\cdot}23$. $\vdash : R \Subset J . \equiv . \breve{R} \Subset J$

$*50{\cdot}24$. $\vdash : R \Subset J . \equiv . (x) . \sim (xRx)$

$*50{\cdot}43$. $\vdash : R^2 \Subset J . \equiv . R \dot{\cap} \breve{R} = \dot{\Lambda}$

$*50{\cdot}45$. $\vdash : R^2 \Subset J . \supset . R \Subset J$

$*50{\cdot}47$. $\vdash :. R^2 \Subset R . \supset : R \Subset J . \equiv . R^2 \Subset J . \equiv . R \dot{\cap} \breve{R} = \dot{\Lambda}$

It will be observed that all these propositions are concerned with $R \in J$ or $R^2 \in J$, both of which are satisfied if R is a *serial* relation. The hypothesis $R^2 \in J$ or $R \dot{\wedge} \breve{R} = \Lambda$ characterizes an *asymmetrical* relation, *i.e.* one which, if it holds between x and y, cannot hold between y and x.

***50·01.** $I = \hat{x}\hat{y}(x = y)$ Df

***50·02.** $J = \dot{-} I$ Df

Most of the propositions of this number are obvious, and call for no comment.

***50·1.** $\vdash : xIy . \equiv . x = y$ [*21·3 . (*50·01)]

***50·11.** $\vdash : xJy . \equiv . x \neq y$ [*23·35 . *50·1 . (*50·02)]

***50·12.** $\vdash . J = \hat{x}\hat{y}(x \neq y)$ [*50·11 . *21·33]

***50·13.** $\vdash . \dot{\exists} ! I$ [*13·19 . *10·24·281 . *50·1]

***50·14.** $\vdash . I'y = y$ [*30·3 . *50·1 . *10·11]

***50·15.** $\vdash . (y) . E ! I'y$ [*50·14 . *14·21 . *10·11]

***50·16.** $\vdash . I``\alpha = \alpha$

Dem.

$\vdash . *37·1 . \supset \vdash : x \in I``\alpha . \equiv . (\exists y) . y \in \alpha . xIy .$

[*50·1] $\equiv . (\exists y) . y \in \alpha . x = y .$

[*13·195] $\equiv . x \in \alpha : \supset \vdash . \text{Prop}$

***50·17.** $\vdash :. x \in \alpha . \supset_x . R'x = x : \supset . R``\alpha = \alpha$

Dem.

$\vdash . *14·21 . \supset \vdash : \text{Hp} . \supset . E !! R``\alpha$ (1)

$\vdash . *50·14 . \supset \vdash :. \text{Hp} . \supset : x \in \alpha . \supset_x . R'x = I'x :$

[*37·69.(1)] $\supset : R``\alpha = I``\alpha :$

[*50·16] $\supset : R``\alpha = \alpha :. \supset \vdash . \text{Prop}$

***50·2.** $\vdash . I = \breve{I}$

Dem.

$\vdash . *50·1 . \supset \vdash : xIy . \equiv . x = y .$

[*13·16] $\equiv . y = x .$

[*50·1] $\equiv . yIx .$

[*31·11] $\equiv . x\breve{I}y : \supset \vdash . \text{Prop}$

***50·21.** $\vdash . J = \breve{J}$

Dem.

$\vdash . *21·2 . (*50·02) . \supset \vdash . J = \dot{-} I$ (1)

[*50·2.*23·83] $= \dot{-} \breve{I}$

[*31·16] $= \text{Cnv}' \dot{-} I$

[(1).*31·32] $= \breve{J} . \supset \vdash . \text{Prop}$

***50·22.** $\vdash : R \Subset I \,.\equiv.\, \breve{R} \Subset I$ [*31·4 . *50·2]

***50·23.** $\vdash : R \Subset J \,.\equiv.\, \breve{R} \Subset J$ [*31·4 . *50·21]

***50·24.** $\vdash : R \Subset J \,.\equiv.\, (x) \,.\, \sim(xRx)$

Dem.

$\vdash . *50{\cdot}11 . \supset \vdash :. R \Subset J \,.\equiv : xRy \,.\supset_{x,y} .\, x \neq y :$

[Transp] $\equiv : x = y \,.\supset_{x,y} .\, \sim(xRy) :$

[*13·191] $\equiv : (x) \,.\, \sim(xRx) :. \supset \vdash .$ Prop

***50·3.** $\vdash . (x) \,.\, xIx$ [*50·1 . *13·15]

***50·31.** $\vdash . \text{D}`I = \text{V} \,.\, \text{ᗡ}`I = \text{V}$

Dem.

$\vdash . *50{\cdot}3 . *10{\cdot}24 . \supset \vdash :. (x) : (\exists y) \,.\, xIy :. (x) : (\exists y) \,.\, yIx :.$

[*33·13·131] $\supset \vdash : (x) \,.\, x \in \text{D}`I : (x) \,.\, x \in \text{ᗡ}`I :$

[*24·14] $\supset \vdash . \text{D}`I = \text{V} \,.\, \text{ᗡ}`I = \text{V} \,. \supset \vdash .$ Prop

***50·32.** $\vdash . C`I = \text{V}$ [*50·31 . *33·16 . *24·27]

***50·33.** $\vdash : \dot{\exists} ! J \,.\supset.\, \text{D}`J = \text{V} \,.\, \text{ᗡ}`J = \text{V} \,.\, C`J = \text{V}$

Dem.

$\vdash . *13{\cdot}171 . \text{Transp} . \supset \vdash :. y \neq z \,.\supset : x \neq y \,.\vee.\, x \neq z :.$

[*50·11] $\supset \vdash :. yJz \,.\supset : xJy \,.\vee.\, xJz :$

[*33·14] $\supset : x \in \text{D}`J$ (1)

$\vdash . (1) . *11{\cdot}11{\cdot}35 . \supset \vdash : \dot{\exists} ! J \,.\supset.\, x \in \text{D}`J :$

[*10·11·21] $\supset \vdash : \dot{\exists} ! J \,.\supset.\, (x) \,.\, x \in \text{D}`J .$

[*24·14] $\supset .\, \text{D}`J = \text{V}$ (2)

$\vdash . (2) . *50{\cdot}21 . \supset \vdash .$ Prop

In the above proposition (*50·33), the hypothesis $\dot{\exists} ! J$ is equivalent to the hypothesis that more than one object exists of the type in question. This can be proved for all except the lowest type. For the lowest type, we can only prove the existence of at least one object: this is proved in *24·52. For the next type, we can prove the existence of at least two objects, namely Λ and V; these are distinct, by *24·1. For the next type, we can prove the existence of 2^2 objects; for the next, 2^4, etc. But for the class of individuals we cannot prove, from our primitive propositions, that there is more than one object in the universe, and therefore we cannot prove $\dot{\exists} ! J$. We might, of course, have included among our primitive propositions the assumption that more than one individual exists, or some assumption from which this would follow, such as

$$(\exists \phi, x, y) \,.\, \phi ! x \,.\, \sim \phi ! y.$$

But very few of the propositions which we might wish to prove depend upon this assumption, and we have therefore excluded it. It should be observed that many philosophers, being monists, deny this assumption.

***50·34.** $\vdash . \dot{\exists} ! J \upharpoonright \mathrm{Cls}$

Dem.

$\vdash . *20·41 . *22·38 . (*24·01·02) . \supset \vdash . \Lambda, V \epsilon \mathrm{Cls}$.

[*24·1] $\supset \vdash . \Lambda \neq V . \Lambda, V \epsilon \mathrm{Cls}$.

[*36·13.*50·11] $\supset \vdash . \Lambda \{J \upharpoonright \mathrm{Cls}\} V$.

[*10·24] $\supset \vdash$. Prop

***50·35.** $\vdash . \dot{\exists} ! J \upharpoonright \mathrm{Rel}$ [Proof as in *50·34]

***50·4.** $\vdash . R | I = I | R = R$

Dem.

$\vdash . *34·1 . \supset \vdash : x(R | I) z . \equiv . (\exists y) . xRy . yIz$.

[*50·1] $\equiv . (\exists y) . xRy . y = z$.

[*13·195] $\equiv . xRz$ (1)

$\vdash . *34·1 . \supset \vdash : x(I | R) z . \equiv . (\exists y) . xIy . yRz$.

[*50·1] $\equiv . (\exists y) . x = y . yRz$.

[*13·195] $\equiv . xRz$ (2)

$\vdash . (1) . (2) . \supset \vdash$. Prop

***50·41.** $\vdash : R | \breve{P} \subseteq J . \equiv . \breve{R} | P \subseteq J . \equiv . R \dot{\cap} P = \dot{\Lambda}$

Dem.

$\vdash . *34·1 . *50·11 . \supset \vdash :. R | \breve{P} \subseteq J . \equiv : (\exists y) . xRy . y\breve{P}z . \supset_{x,z} . x \neq z :$

[*13·196] $\equiv : (x) : \sim (\exists y) . xRy . y\breve{P}x :$

[*10·252] $\equiv : \sim (\exists x, y) . xRy . y\breve{P}x :$

[*31·11] $\equiv : \sim (\exists x, y) . xRy . xPy :$

[*23·33.*25·51] $\equiv : R \dot{\cap} P = \dot{\Lambda} :$ (1)

[*31·14·24] $\equiv : \breve{R} \dot{\cap} \breve{P} = \dot{\Lambda} :$

$\left[(1) \dfrac{\breve{R}, \breve{P}}{R, P}\right]$ $\equiv : \breve{R} | \mathrm{Cnv}‘\breve{P} \subseteq J :$

[*34·203] $\equiv : \breve{R} | P \subseteq J$ (2)

$\vdash . (1) . (2) . \supset \vdash$. Prop

***50·42.** $\vdash . I^2 = I$

Dem.

$\vdash . *34·5 . \supset \vdash : xI^2z . \equiv . (\exists y) . xIy . yIz$.

[*50·1] $\equiv . (\exists y) . xIy . y = z$.

[*13·195] $\equiv . xIz : \supset \vdash$. Prop

***50·43.** $\vdash : R^2 \subseteq J . \equiv . R \dot{\cap} \breve{R} = \dot{\Lambda}$ $\left[*50·41 \dfrac{\breve{R}}{P}\right]$

This proposition is useful in the theory of series. "$R \dot{\cap} \breve{R} = \dot{\Lambda}$" is the characteristic of an *asymmetrical* relation.

$*50{\cdot}44$. $\vdash : \dot{\exists} ! (R \dot{\frown} I) . \supset . \dot{\exists} ! (R^2 \dot{\frown} I)$

Dem.

$\vdash . *23{\cdot}33 . *50{\cdot}1 . \supset \vdash : \dot{\exists} ! (R \dot{\frown} I) . \equiv . (\exists x, y) . xRy . x = y .$

$[*13{\cdot}195] \quad \equiv . (\exists x) . xRx .$

$[*34{\cdot}54] \quad \supset . (\exists x) . xR^2x .$

$[*13{\cdot}195] \quad \supset . (\exists x, y) . xR^2y . x = y .$

$[*23{\cdot}33 . *50{\cdot}1] \quad \supset . \dot{\exists} ! (R^2 \dot{\frown} I) : \supset \vdash . \text{Prop}$

$*50{\cdot}45$. $\vdash : R^2 \in J . \supset . R \in J \quad [*50{\cdot}44 . \text{Transp} . *25{\cdot}311]$

$*50{\cdot}46$. $\vdash : R \dot{\frown} \breve{R} = \dot{\Lambda} . \supset . R \in J \quad [*50{\cdot}43{\cdot}45]$

$*50{\cdot}47$. $\vdash :. R^2 \in R . \supset : R \in J . \equiv . R^2 \in J . \equiv . R \dot{\frown} \breve{R} = \dot{\Lambda}$

Dem.

$\vdash . *23{\cdot}44 . \supset \vdash :. \text{Hp} . \supset : R \in J . \supset . R^2 \in J \quad (1)$

$\vdash . (1) . *50{\cdot}45{\cdot}43 . \supset \vdash . \text{Prop}$

This proposition is used in the theory of series. If R is a serial relation, we shall have $R^2 \in R$ and $R \in J$.

$*50{\cdot}5$. $\vdash . \alpha \upharpoonleft I = I \upharpoonright \alpha = \alpha \upharpoonleft I \upharpoonright \alpha$

Dem.

$\vdash . *35{\cdot}1 . \supset \vdash : x (\alpha \upharpoonleft I) y . \equiv . x \in \alpha . xIy .$

$[*50{\cdot}1] \quad \equiv . x \in \alpha . x = y .$

$[*13{\cdot}193] \quad \equiv . y \in \alpha . x = y .$

$[*50{\cdot}1] \quad \equiv . xIy . y \in \alpha .$

$[*35{\cdot}101] \quad \equiv . x (I \upharpoonright \alpha) y \quad (1)$

$\vdash . (1) . *23{\cdot}5 . \supset \vdash . \alpha \upharpoonleft I = \alpha \upharpoonleft I \dot{\frown} I \upharpoonright \alpha$

$[*35{\cdot}11] \quad = \alpha \upharpoonleft I \upharpoonright \alpha \quad (2)$

$\vdash . (1) . (2) . \supset \vdash . \text{Prop}$

$*50{\cdot}51$. $\vdash . \text{Cnv}`(\alpha \upharpoonleft I) = \alpha \upharpoonleft I \quad [*35{\cdot}51 . *50{\cdot}2{\cdot}5]$

$*50{\cdot}52$. $\vdash . \text{D}`(\alpha \upharpoonleft I) = \text{ᗡ}`(\alpha \upharpoonleft I) = C`(\alpha \upharpoonleft I) = \alpha$

Dem.

$\vdash . *35{\cdot}61 . \supset \vdash . \text{D}`(\alpha \upharpoonleft I) = \alpha \cap \text{D}`I$

$[*50{\cdot}31] \quad = \alpha \cap \text{V}$

$[*24{\cdot}26] \quad = \alpha \quad (1)$

Similarly $\vdash . \text{ᗡ}`(\alpha \upharpoonleft I) = \alpha \quad (2)$

$\vdash . (1) . (2) . *33{\cdot}18 . \supset \vdash . \text{Prop}$

$*50{\cdot}53$. $\vdash . \alpha \upharpoonleft I \upharpoonright \beta = (\alpha \cap \beta) \upharpoonleft I = I \upharpoonright (\alpha \cap \beta)$

Dem.

$\vdash . *35{\cdot}21 . *50{\cdot}5 . \supset \vdash . \alpha \upharpoonleft I \upharpoonright \beta = \alpha \upharpoonleft (\beta \upharpoonleft I)$

$[*35{\cdot}32] \quad = (\alpha \cap \beta) \upharpoonleft I \quad (1)$

$\vdash . (1) . *50{\cdot}5 . \supset \vdash . \text{Prop}$

***50·54.** $\vdash . (\alpha \upharpoonleft I)^2 = \alpha \upharpoonleft I$

Dem.

$\vdash . *50{\cdot}5 . \supset \vdash . (\alpha \upharpoonleft I)^2 = (\alpha \upharpoonleft I) | (I \restriction \alpha)$

$[*35{\cdot}12] \qquad = \alpha \upharpoonleft I^2 \restriction \alpha$

$[*50{\cdot}42] \qquad = \alpha \upharpoonleft I \restriction \alpha$

$[*50{\cdot}5] \qquad = \alpha \upharpoonleft I . \supset \vdash . \text{Prop}$

***50·55.** $\vdash : \alpha \cap \beta = \Lambda . \equiv . \alpha \uparrow \beta \subset J$

Dem.

$\vdash . *24{\cdot}37 . *50{\cdot}11 . \supset$

$\vdash :. \alpha \cap \beta = \Lambda . \equiv : x \epsilon \alpha . y \epsilon \beta . \supset_{x,y} . xJy :$

$[*35{\cdot}103] \qquad \equiv : \alpha \uparrow \beta \subset J :. \supset \vdash . \text{Prop}$

***50·56.** $\vdash : \exists ! (\alpha \cap \beta) . \equiv . \dot{\exists} ! \{(\alpha \uparrow \beta) \dot{\cap} I\}$

Dem.

$\vdash . *50{\cdot}55 . \text{Transp} . *24{\cdot}54 . \supset$

$\vdash : \exists ! (\alpha \cap \beta) . \qquad \equiv . \sim \{\alpha \uparrow \beta \subset J\} .$

$[*25{\cdot}55] \qquad \equiv . \dot{\exists} (\alpha \uparrow \beta) \dot{-} J .$

$[*23{\cdot}831 . (*50{\cdot}02)] \equiv . \dot{\exists} ! \{(\alpha \uparrow \beta) \dot{\cap} I\} : \supset \vdash . \text{Prop}$

***50·57.** $\vdash . I \dot{\cap} \alpha \upharpoonleft R = I \dot{\cap} R \restriction \alpha = I \dot{\cap} \alpha \upharpoonleft R \restriction \alpha$

Dem.

$\vdash . *35{\cdot}16 . \supset \vdash . I \dot{\cap} \alpha \upharpoonleft R = \alpha \upharpoonleft I \dot{\cap} R$

$[*50{\cdot}5] \qquad = I \restriction \alpha \dot{\cap} R$

$[*35{\cdot}17] \qquad = I \dot{\cap} R \restriction \alpha \qquad (1)$

$[*50{\cdot}5] \qquad = \alpha \upharpoonleft I \restriction \alpha \dot{\cap} R$

$[*35{\cdot}16{\cdot}17{\cdot}21] \qquad = I \dot{\cap} \alpha \upharpoonleft R \restriction \alpha \qquad (2)$

$\vdash . (1) . (2) . \supset \vdash . \text{Prop}$

***50·58.** $\vdash : \alpha \upharpoonleft R \subset J . \equiv . R \restriction \alpha \subset J . \equiv . \alpha \upharpoonleft R \restriction \alpha \subset J$

Dem.

$\vdash . *50{\cdot}57 . \supset \vdash : I \dot{\cap} \alpha \upharpoonleft R = \dot{\Lambda} . \equiv . I \dot{\cap} R \restriction \alpha = \dot{\Lambda} . \equiv . I \dot{\cap} \alpha \upharpoonleft R \restriction \alpha = \dot{\Lambda} \qquad (1)$

$\vdash . (1) . *50{\cdot}41 . \supset \vdash . \text{Prop}$

***50·59.** $\vdash . (I \restriction \alpha) \text{“} \beta = \alpha \cap \beta$

Dem.

$\vdash . *37{\cdot}412 . \supset \vdash . (I \restriction \alpha) \text{“} \beta = I \text{“} (\alpha \cap \beta)$

$[*50{\cdot}16] \qquad = \alpha \cap \beta . \supset \vdash . \text{Prop}$

***50·6.** $\vdash . R | (I \restriction \alpha) = R \restriction \alpha$

Dem.

$\vdash . *35{\cdot}23 . \supset \vdash . R | (I \restriction \alpha) = (R | I) \restriction \alpha$

$[*50{\cdot}4] \qquad = R \restriction \alpha . \supset \vdash . \text{Prop}$

***50·61.** $\vdash . I \restriction \alpha | R = \alpha \upharpoonleft R$

Dem.

$\vdash . *35{\cdot}354 . \supset \vdash . I \restriction \alpha | R = I | (\alpha \upharpoonleft R)$

$[*50{\cdot}4] \qquad = \alpha \upharpoonleft R . \supset \vdash . \text{Prop}$

***50·62.** $\vdash : \text{Ꮷ}`R \subset \alpha . \supset . R \mid (I \upharpoonright \alpha) = R$ [*50·6 . *35·452]

***50·63.** $\vdash : \text{D}`R \subset \alpha . \supset . I \upharpoonright \alpha \mid R = R$ [*50·61 . *35·451]

***50·64.** $\vdash . R \mid (I \upharpoonright \text{Ꮷ}`R) = R \mid (I \upharpoonright C`R) = R$ [*50·62 . *22·42 . *33·161]

***50·65.** $\vdash . I \upharpoonright (\text{D}`R) \mid R = I \upharpoonright (C`R) \mid R = R$ [*50·63 . *22·42 . *33·161]

***50·7.** $\vdash : \text{Ꮷ}`R \subset \alpha . \supset . R \mid `I \upharpoonright \alpha = R$ [*50·62 . *43·11]

***50·71.** $\vdash : \text{D}`R \subset \alpha . \supset . \mid R`I \upharpoonright \alpha = R$ [*50·63 . *43·111]

***50·72.** $\vdash . R \mid `(I \upharpoonright C`R) = \mid R`(I \upharpoonright C`R) = R$ [*50·7·71]

***50·73.** $\vdash . R \mid `I = \mid R`I = R$ [*50·4 . *43·11·111]

***50·74.** $\vdash . R \parallel I = R \mid$

Dem.

$\vdash . *43·112 . \supset \vdash . (R \parallel I)`Q = R \mid Q \mid I$

[*50·4] $= R \mid Q$

[*43·11] $= R \mid `Q$ (1)

$\vdash . (1) . *30·41 . \supset \vdash . \text{Prop}$

***50·75.** $\vdash . I \parallel R = \mid R$ [Proof as in *50·74]

***50·76.** $\vdash : P \mid = R \mid . \equiv . P = R$

Dem.

$\vdash . *34·27 . *30·41 . \supset \vdash : P = R . \supset . P \mid = R \mid$ (1)

$\vdash . *50·73 . *30·36 . \supset \vdash : P \mid = R \mid . \supset . P = R$ (2)

$\vdash . (1) . (2) . \supset \vdash . \text{Prop}$

***50·761.** $\vdash : \mid P = \mid R . \equiv . P = R$ [Proof as in *50·76]

*51. UNIT CLASSES

Summary of *51.

In this number we introduce a new descriptive function $\iota' x$, meaning "the class of terms which are identical with x," which is the same thing as "the class whose only member is x." We are thus to have

$$\iota' x = \hat{y}(y = x).$$

But $\hat{y}(y = x) = \overrightarrow{I}' x$. Hence we secure what we require by the following definition:

*51·01. $\iota = \overrightarrow{I}$ Df

As a matter of notation, it might be thought that $\overrightarrow{I}$ would do as well as ι, and that this definition is superfluous. But we need also the converse of this relation, and "Cnv'$\overrightarrow{I}$" is not a sufficiently convenient symbol.

The propositions of this number are constantly used in what follows. It should be observed that the class whose members are x and y is $\iota' x \cup \iota' y$, the class whose members are x, y, z is $\iota' x \cup \iota' y \cup \iota' z$, the class formed by adding x to α is $\alpha \cup \iota' x$, and the class formed by taking x away from α is $\alpha - \iota' x$. (If x is not a member of α, this is equal to α.)

The distinction between x and $\iota' x$ is one of the merits of Peano's symbolic logic, as well as of Frege's. On the basis of our theory of classes, the necessity for the distinction is of course obvious. But apart from this, the following consideration makes the necessity apparent. Let α be a class; then the class whose only member is α has only one member, namely α, while α may have many members. Hence the class whose only member is α cannot be identical with α*.

The propositions of the present number which are most used are the following:

*51·15. $\vdash : y \epsilon \iota' x . \equiv . y = x$

*51·16. $\vdash . x \epsilon \iota' x$

*51·2. $\vdash : x \epsilon \alpha . \equiv . \iota' x \subset \alpha$

This proposition is useful because it enables us to replace membership of a class ($x \epsilon \alpha$) by inclusion in the class ($\iota' x \subset \alpha$).

*51·211. $\vdash : x \sim \epsilon \alpha . \equiv . \iota' x \cap \alpha = \Lambda$

*51·221. $\vdash : x \epsilon \alpha . \equiv . (\alpha - \iota' x) \cup \iota' x = \alpha$

* This argument is due to Frege. See his article "Kritische Beleuchtung einiger Punkte in E. Schröder's Vorlesungen über die Algebra der Logik," *Archiv für Syst. Phil.*, vol. I. p. 444 (1895).

***51·222.** $\vdash : x \sim \epsilon \alpha . \equiv . \alpha - \iota ` x = \alpha$

***51·23.** $\vdash : \iota ` x = \iota ` y . \equiv . y \epsilon \iota ` x . \equiv . x \epsilon \iota ` y . \equiv . x = y$

***51·4.** $\vdash : \exists ! \alpha . \alpha \subset \iota ` x . \equiv . \alpha = \iota ` x$

I.e. an existent class contained in a unit class must be identical with the unit class. From this proposition it will follow that 0 is the only cardinal which is less than 1.

***51·51.** $\vdash : \alpha = \iota ` x . \equiv . x = \breve{\iota} ` \alpha . \equiv . x \, \breve{\iota} \, \alpha$

For classes, $\breve{\iota} ` \alpha$ has the same uses that $(\imath x)(\phi x)$ has for functions; "$\breve{\iota} ` \alpha$" means "the only member of α." We have

***51·59.** $\vdash : \psi \{ \breve{\iota} ` \hat{z} (\phi z) \} . \equiv . \psi (\imath x)(\phi x)$

***51·01.** $\iota = \overrightarrow{I}$ Df

***51·1.** $\vdash : \alpha \iota x . \equiv . \alpha = \hat{y}(y = x)$

Dem.

$\vdash . *4·2 . (*51·01) . \supset \vdash : \alpha \iota x . \equiv . \alpha \overrightarrow{I} x .$

$[*32·1] \quad \equiv . \alpha = \hat{y}(y I x) .$

$[*50·1] \quad \equiv . \alpha = \hat{y}(y = x) : \supset \vdash . \text{Prop}$

***51·11.** $\vdash . \iota ` x = \hat{y}(y = x)$ [*30·3 . *51·1]

***51·12.** $\vdash . E ! \iota ` x$ [*51·11 . *14·21]

***51·13.** $\vdash : \alpha = \iota ` x . \equiv . \alpha = \hat{y}(y = x)$ [*20·57·2 . *51·11]

***51·131.** $\vdash : \alpha \iota x . \equiv . \alpha = \iota ` x$ [*51·1·13]

***51·14.** $\vdash :. \alpha = \iota ` x . \equiv : y \epsilon \alpha . \equiv_y . y = x$ [*51·13 . *20·33]

***51·141.** $\vdash :. \alpha = \iota ` x . \equiv : \exists ! \alpha : y \epsilon \alpha . \supset_y . y = x : \equiv : x \epsilon \alpha : y \epsilon \alpha . \supset_y . y = x$

[*51·14 . *14·122]

***51·15.** $\vdash : y \epsilon \iota ` x . \equiv . y = x$ [*51·11 . *20·33]

***51·16.** $\vdash . x \epsilon \iota ` x$ [*51·15 . *13·15]

***51·161.** $\vdash . \exists ! \iota ` x$ [*51·16 . *10·24]

***51·17.** $\vdash . \text{Cl} ` \iota = V$

Dem.

$\vdash . *51·1 . *20·2 . \supset \vdash . \{ \hat{y}(y = x) \} \iota x .$

$[*10·24] \quad \supset \vdash . (\exists \alpha) . \alpha \iota x .$

$[*33·131] \quad \supset \vdash . x \epsilon \text{Cl} ` \iota .$

$[*10·11] \quad \supset \vdash . (x) . x \epsilon \text{Cl} ` \iota .$

$[*24·14] \quad \supset \vdash . \text{Cl} ` \iota = V$

The above proposition is used in the theory of selections (*83·71).

***51·2.** $\vdash : x \epsilon \alpha . \equiv . \iota' x \subset \alpha$

Dem.

$\vdash . *13·191 . \supset \vdash :. x \epsilon \alpha . \equiv : y = x . \supset_y . y \epsilon \alpha :$

$[*51·15] \qquad \equiv : y \epsilon \iota' x . \supset_y . y \epsilon \alpha :$

$[*22·1] \qquad \equiv : \iota' x \subset \alpha :. \supset \vdash . \text{Prop}$

The above proposition shows how to replace membership of a class by inclusion in a class; thus for example it gives:

Socrates is a man . ≡ . the class of terms identical with Socrates is included in the class of men.

Before Peano and Frege, the relation of membership (ϵ) was regarded as merely a particular case of the relation of inclusion ($\subset$). For this reason, the traditional formal logic treated such propositions as "Socrates is a man" as instances of the universal affirmative *A*, "All *S* is *P*," which is what we express by "$\alpha \subset \beta$." This involved a confusion of fundamentally different kinds of propositions, which greatly hindered the development and usefulness of symbolic logic. But by means of the above proposition (*51·2), we can always obtain a proposition stating an inclusion (namely "$\iota' x \subset \alpha$") which is equivalent to a given proposition stating membership of a class (namely "$x \epsilon \alpha$").

***51·21.** $\vdash . x \sim\epsilon \alpha - \iota' x$

Dem.

$\vdash . *22·33·35 . \supset \vdash : x \epsilon \alpha - \iota' x . \equiv . x \epsilon \alpha . x \sim\epsilon \iota' x .$

$[*3·27] \qquad \supset . x \sim\epsilon \iota' x \qquad (1)$

$\vdash . (1) . \text{Transp} . *51·16 . \supset \vdash . \text{Prop}$

***51·211.** $\vdash : x \sim\epsilon \alpha . \equiv . \iota' x \cap \alpha = \Lambda$

Dem.

$\vdash . *24·39 . \supset \vdash :. \iota' x \cap \alpha = \Lambda . \equiv : y \epsilon \iota' x . \supset_y . y \sim\epsilon \alpha :$

$[*51·15] \qquad \equiv : y = x . \supset_y . y \sim\epsilon \alpha :$

$[*13·191] \qquad \equiv : x \sim\epsilon \alpha :. \supset \vdash . \text{Prop}$

***51·22.** $\vdash : \alpha \cap \iota' x = \Lambda . \alpha \cup \iota' x = \beta . \equiv . x \epsilon \beta . \alpha = \beta - \iota' x$

Dem.

$\vdash . *24·47 . \supset$

$\vdash : \alpha \cap \iota' x = \Lambda . \alpha \cup \iota' x = \beta . \equiv . \iota' x \subset \beta . \alpha = \beta - \iota' x .$

$[*51·2] \qquad \equiv . x \epsilon \beta . \alpha = \beta - \iota' x : \supset \vdash . \text{Prop}$

***51·221.** $\vdash : x \epsilon \alpha . \equiv . (\alpha - \iota' x) \cup \iota' x = \alpha$

Dem.

$\vdash . *51·2 . \supset \vdash : x \epsilon \alpha . \equiv . \iota' x \subset \alpha .$

$[*22·62] \qquad \equiv . \iota' x \cup \alpha = \alpha .$

$[*22·91] \qquad \equiv . (\alpha - \iota' x) \cup \iota' x = \alpha : \supset \vdash . \text{Prop}$

***51·222.** $\vdash : x \sim \epsilon \alpha . \equiv . \alpha - \iota ' x = \alpha$ [*51·211 . *24·313]

***51·23.** $\vdash : \iota ' x = \iota ' y . \equiv . y \epsilon \iota ' x . \equiv . x \epsilon \iota ' y . \equiv . x = y$

Dem.

$\vdash . *20·31 . *51·15 . \supset$

$\vdash :. \iota ' x = \iota ' y . \equiv : z = x . \equiv_z . z = y :$

[*13·183] $\equiv : x = y :$ (1)

[*51·15] $\equiv : x \epsilon \iota ' y :$ (2)

[(1).*13·16] $\equiv : y \epsilon \iota ' x$ (3)

$\vdash . (1) . (2) . (3) . \supset \vdash .$ Prop

***51·231.** $\vdash : \iota ' x \cap \iota ' y = \Lambda . \equiv . x \neq y$

Dem.

$\vdash . *24·311 . \supset \vdash :. \iota ' x \cap \iota ' y = \Lambda . \equiv : \iota ' x \subset - \iota ' y :$

[*51·15] $\equiv : z = x . \supset_z . z \neq y :$

[*13·191] $\equiv : x \neq y :. \supset \vdash .$ Prop

***51·232.** $\vdash :. z \epsilon (\iota ' x \cup \iota ' y) . \equiv : z = x . \vee . z = y$ [*22·34 . *51·15]

This proposition states that a member of $\iota ' x \cup \iota ' y$ must be either x or y, and vice versa, *i.e.* that $\iota ' x \cup \iota ' y$ is the class whose only members are x and y.

***51·233.** $\vdash :: \alpha = \iota ' x \cup \iota ' y . \supset :. (z) :. z \epsilon \alpha . \equiv : z = x . \vee . z = y$
[*51·232 . *10·11 . *20·18]

***51·234.** $\vdash :: \alpha = \iota ' x \cup \iota ' y . \supset :. z \epsilon \alpha . \supset_z . \phi z : \equiv . \phi x . \phi y$

Dem.

$\vdash . *51·233 . \supset \vdash ::. \text{Hp} . \supset :: z \epsilon \alpha . \supset_z . \phi z : \equiv :. z = x . \vee . z = y : \supset_z . \phi z :.$

[*4·77] $\equiv :. (z) :. z = x . \supset . \phi z : z = y . \supset . \phi z :.$

[*10·22] $\equiv :. z = x . \supset_z . \phi z : z = y . \supset_z . \phi z :.$

[*13·191] $\equiv :. \phi x . \phi y ::. \supset \vdash .$ Prop

***51·235.** $\vdash :: \alpha = \iota ' x \cup \iota ' y . \supset :. (\exists z) . z \epsilon \alpha . \phi z . \equiv : \phi x . \vee . \phi y$

Dem.

$\vdash . *51·233 . \supset$

$\vdash :: \text{Hp} . \supset :. (\exists z) . z \epsilon \alpha . \phi z . \equiv : (\exists z) : z = x . \vee . z = y : \phi z :$

[*4·4] $\equiv : (\exists z) : z = x . \phi z . \vee . z = y . \phi z :$

[*10·42] $\equiv : (\exists z) . z = x . \phi z . \vee . (\exists z) . z = y . \phi z :$

[*13·195] $\equiv : \phi x . \vee . \phi y :: \supset \vdash .$ Prop

***51·236.** $\vdash :. z \epsilon \iota ' x \cup \beta . \equiv : z = x . \vee . z \epsilon \beta$ [*22·34 . *51·15]

***51·237.** $\vdash :: \alpha = \iota ' x \cup \beta . \supset :. (z) :. z \epsilon \alpha . \equiv : z = x . \vee . z \epsilon \beta$
[*51·236 . *10·11 . *20·18]

***51·238.** $\vdash :: \alpha = \iota ' x \cup \beta . \supset :. z \epsilon \alpha . \supset_z . \phi z : \equiv : \phi x : z \epsilon \beta . \supset_z . \phi z$

Dem.

$\vdash . *51·237 . \supset \vdash ::. \text{Hp} . \supset :: z \epsilon \alpha . \supset_z . \phi z : \equiv :. z = x . \vee . z \epsilon \beta : \supset_z . \phi z :.$

[*4·77] $\equiv :. (z) :. z = x . \supset . \phi z : z \epsilon \beta . \supset . \phi z :.$

[*10·22] $\equiv :. z = x . \supset_z . \phi z : z \epsilon \beta . \supset_z . \phi z :.$

[*13·191] $\equiv :. \phi x : z \epsilon \beta . \supset_z . \phi z ::. \supset \vdash .$ Prop

***51·239.** $\vdash :: \alpha = \iota`x \cup \beta . \supset :. (\exists z) . z \epsilon \alpha . \phi z . \equiv : \phi x . \vee . (\exists z) . z \epsilon \beta . \phi z$

Dem.

$\vdash . *51·237 . \supset$

$\vdash :: \text{Hp} . \supset :. (\exists z) . z \epsilon \alpha . \phi z . \equiv : (\exists z) : z = x . \vee . z \epsilon \beta : \phi z :$

[*4·4] $\equiv : (\exists z) : z = x . \phi z . \vee . z \epsilon \beta . \phi z :$

[*10·42] $\equiv : (\exists z) . z = x . \phi z . \vee . (\exists z) . z \epsilon \beta . \phi z :$

[*13·195] $\equiv : \phi x . \vee . (\exists z) . z \epsilon \beta . \phi z :: \supset \vdash . \text{Prop}$

***51·24.** $\vdash :. \iota`y \subset \iota`x \cup \beta . \equiv : y = x . \vee . y \epsilon \beta$

Dem.

$\vdash . *51·236 . \supset$

$\vdash :: \iota`y \subset \iota`x \cup \beta . \equiv :. z \epsilon \iota`y . \supset_z : z = x . \vee . z \epsilon \beta :.$

[*51·15] $\equiv :. z = y . \supset_z : z = x . \vee . z \epsilon \beta :.$

[*13·191] $\equiv :. y = x . \vee . y \epsilon \beta :: \supset \vdash . \text{Prop}$

***51·25.** $\vdash : \alpha \subset \iota`x \cup \beta . x \sim \epsilon \alpha . \supset . \alpha \subset \beta$ [*51·211 . *24·49]

***51·3.** $\vdash : y \epsilon \alpha . y \neq x . \equiv . y \epsilon \alpha - \iota`x$ [*51·15 . *22·33·35]

***51·31.** $\vdash : \exists ! \alpha \cap \iota`x . \equiv . \iota`x \subset \alpha . \equiv . \alpha \cap \iota`x = \iota`x . \equiv . x \epsilon \alpha$

Dem.

$\vdash . *22·33 . *51·15 . \supset \vdash : \exists ! \alpha \cap \iota`x . \equiv . (\exists y) . y \epsilon \alpha . y = x .$

[*13·195] $\equiv . x \epsilon \alpha .$ (1)

[*51·2] $\equiv . \iota`x \subset \alpha .$ (2)

[*22·621] $\equiv . \iota`x = \iota`x \cap \alpha$ (3)

$\vdash . (1) . (2) . (3) . \supset \vdash . \text{Prop}$

***51·34.** $\vdash : x \epsilon \alpha . \equiv . - \alpha \subset - \iota`x$ [*51·2 . *22·81]

***51·35.** $\vdash : x \sim \epsilon \alpha . \equiv . \iota`x \subset - \alpha$ [*51·2 . *22·35]

***51·36.** $\vdash : x \sim \epsilon \alpha . \equiv . \alpha \subset - \iota`x$ [*51·35 . *22·811]

*51·36 is frequently used.

***51·37.** $\vdash . \alpha = \hat{x} (\iota`x \subset \alpha)$ [*51·2 . *20·33]

***51·4.** $\vdash : \exists ! \alpha . \alpha \subset \iota`x . \equiv . \alpha = \iota`x$

Dem.

$\vdash . *24·5 . *51·15 . \supset \vdash :. \exists ! \alpha . \alpha \subset \iota`x . \equiv : (\exists y) . y \epsilon \alpha : y \epsilon \alpha . \supset_y . y = x :$

[*14·122] $\equiv : y \epsilon \alpha . \equiv_y . y = x :$

[*51·11.*20·33] $\equiv : \alpha = \iota`x :. \supset \vdash . \text{Prop}$

***51·401.** $\vdash :. \alpha \subset \iota`x . \equiv : \alpha = \Lambda . \vee . \alpha = \iota`x$

Dem.

$\vdash . *51·4 . *5·6 . \quad \supset \vdash :. \alpha \subset \iota`x . \supset : \alpha = \Lambda . \vee . \alpha = \iota`x$ (1)

$\vdash . *24·12 . *22·42 . \supset \vdash :. \alpha = \Lambda . \vee . \alpha = \iota`x : \supset . \alpha \subset \iota`x$ (2)

$\vdash . (1) . (2) . \supset \vdash . \text{Prop}$

This proposition shows that unit classes are the smallest existent classes.

***51·41.** $\vdash : \iota‘x \cup \iota‘y = \iota‘x \cup \iota‘z . \equiv . y = z$

Dem.

$\vdash . *20·2 . *13·13 . \supset \vdash : y = z . \supset . \iota‘x \cup \iota‘y = \iota‘x \cup \iota‘z$ (1)

$\vdash . *22·58 . \supset \vdash :. \iota‘x \cup \iota‘y = \iota‘x \cup \iota‘z . \supset : \iota‘y \subset \iota‘x \cup \iota‘z . \iota‘z \subset \iota‘x \cup \iota‘y :$

[*51·16·232] $\supset : y = x . \vee . y = z : z = x . \vee . z = y :$

[*13·16.*4·41] $\supset : y = x . z = x . \vee . y = z :$

[*13·172.*2·621] $\supset : y = z$ (2)

$\vdash . (1) . (2) . \supset \vdash .$ Prop

The two following propositions are lemmas for *51·43.

***51·42.** $\vdash :. \iota‘x \cup \iota‘y = \iota‘z \cup \iota‘w . \supset : x = z . y = w . \vee . x = w . y = z$

Dem.

$\vdash . *51·232 . \supset$

$\vdash :: \iota‘x \cup \iota‘y = \iota‘z \cup \iota‘w . \equiv :. a = x . \vee . a = y : \equiv_a : a = z . \vee . a = w :.$

[*10·1] $\supset :. x = x . \vee . x = y : \equiv : x = z . \vee . x = w :.$

[*13·15] $\supset :. x = z . \vee . x = w$ (1)

$\vdash . *20·2 . *13·13 . \supset \vdash : \iota‘x \cup \iota‘y = \iota‘z \cup \iota‘w . x = z . \supset . \iota‘x \cup \iota‘y = \iota‘x \cup \iota‘w .$

[*51·41] $\supset . y = w$ (2)

Similarly $\vdash : \iota‘x \cup \iota‘y = \iota‘z \cup \iota‘w . x = w . \supset . y = z$ (3)

$\vdash . (1) . (2) . (3) . \supset \vdash .$ Prop

***51·421.** $\vdash :. x = z . y = w . \vee . x = w . y = z : \supset . \iota‘x \cup \iota‘y = \iota‘z \cup \iota‘w$ [*51·41]

***51·43.** $\vdash :. \iota‘x \cup \iota‘y = \iota‘z \cup \iota‘w . \equiv : x = z . y = w . \vee . x = w . y = z$ [*51·42·421]

The following propositions are concerned with $\breve{\iota}$, *i.e.* with the relation of the only member of a unit class to that class. If α is a unit class, $\breve{\iota}‘\alpha$ is its only member. $(\imath x)(\phi x)$ and $\breve{\iota}‘\hat{z}(\phi z)$ are equal whenever either exists, and any proposition about the one is equivalent to the same proposition about the other.

***51·51.** $\vdash : \alpha = \iota‘x . \equiv . x = \breve{\iota}‘\alpha . \equiv . x \breve{\iota} \alpha$

Dem.

$\vdash . *51·131 . *31·11 . \supset \vdash : \alpha = \iota‘x . \equiv . x \breve{\iota} \alpha$ (1)

$\vdash . (1) . \supset \vdash : x \breve{\iota} \alpha . y \breve{\iota} \alpha . \supset . \alpha = \iota‘x . \alpha = \iota‘y .$

[*51·23.*20·57·2] $\supset . x = y$ (2)

$\vdash . (2) . \text{Exp} . *10·11 . *4·71 . \supset \vdash :. x \breve{\iota} \alpha . \equiv : x \breve{\iota} \alpha : y \breve{\iota} \alpha . \supset_y . x = y :$

[*30·31] $\equiv : x = \breve{\iota}‘\alpha$ (3)

$\vdash . (1) . (3) . \supset \vdash .$ Prop

***51·511.** $\vdash . \breve{\iota}\text{‘}\iota\text{‘}x = x$ $\quad \left[*51\cdot51 \dfrac{\iota\text{‘}x}{\alpha} . *20\cdot2\right]$

***51·52.** $\vdash : E\,!\,\breve{\iota}\text{‘}\alpha . \equiv . \alpha = \iota\text{‘}\breve{\iota}\text{‘}\alpha$ $\quad \left[*51\cdot51 \dfrac{\breve{\iota}\text{‘}\alpha}{x} . *14\cdot21\cdot18\right]$

***51·53.** $\vdash : E\,!\,\breve{\iota}\text{‘}\alpha . \equiv . \breve{\iota}\text{‘}\alpha \,\epsilon\, \alpha$ $\quad [*51\cdot52\cdot16 . *14\cdot21\cdot18]$

***51·54.** $\vdash : E\,!\,\breve{\iota}\text{‘}\alpha . \equiv . (\exists x) . \alpha = \iota\text{‘}x$ $\quad [*51\cdot51 . *14\cdot204]$

***51·55.** $\vdash : E\,!\,\breve{\iota}\text{‘}\alpha . \equiv . E\,!\,(\imath x)(x \,\epsilon\, \alpha)$

Dem.

$\vdash . *51\cdot54\cdot14 . \supset \vdash :. E\,!\,\breve{\iota}\text{‘}\alpha . \equiv : (\exists x) : y \,\epsilon\, \alpha . \equiv_y . y = x :$

$[*14\cdot11] \qquad \equiv : E\,!\,(\imath x)(x \,\epsilon\, \alpha) :. \supset \vdash . \text{Prop}$

***51·56.** $\vdash : b = \breve{\iota}\text{‘}\hat{y}(\phi y) . \equiv . \hat{y}(\phi y) = \iota\text{‘}b . \equiv . b = (\imath x)(\phi x)$

Dem.

$\vdash . *51\cdot51 . \supset \vdash :. b = \breve{\iota}\text{‘}\hat{y}(\phi y) . \equiv : \hat{y}(\phi y) = \iota\text{‘}b :$ (1)

$[*20\cdot15 . *51\cdot11] \qquad \equiv : \phi y . \equiv_y . y = b :$

$[*14\cdot202] \qquad \equiv : b = (\imath x)(\phi x)$ (2)

$\vdash . (1) . (2) . \supset \vdash . \text{Prop}$

***51·57.** $\vdash : E\,!\,\breve{\iota}\text{‘}\hat{y}(\phi y) . \equiv . \breve{\iota}\text{‘}\hat{y}(\phi y) = (\imath x)(\phi x) . \equiv . E\,!\,(\imath x)(\phi x)$

Dem.

$\vdash . *14\cdot204 . *51\cdot56 . \supset \vdash : E\,!\,\breve{\iota}\text{‘}\hat{y}(\phi y) . \equiv . E\,!\,(\imath x)(\phi x)$ (1)

$\vdash . *14\cdot205 . \supset \vdash : (\imath x)(\phi x) = \breve{\iota}\text{‘}\hat{y}(\phi y) . \equiv . (\exists b) . b = (\imath x)(\phi x) . b = \breve{\iota}\text{‘}\hat{y}(\phi y) .$

$[*51\cdot56 . *4\cdot71] \qquad \equiv . (\exists b) . b = (\imath x)(\phi x) .$

$[*14\cdot204\cdot13] \qquad \equiv . E\,!\,(\imath x)(\phi x)$ (2)

$\vdash . (1) . (2) . \supset \vdash . \text{Prop}$

***51·58.** $\vdash : E\,!\,\breve{\iota}\text{‘}\alpha . \equiv . \breve{\iota}\text{‘}\alpha = (\imath x)(x \,\epsilon\, \alpha)$ $\quad [*51\cdot57 . *20\cdot3 . *14\cdot272]$

***51·59.** $\vdash : \psi\{\breve{\iota}\text{‘}\hat{z}(\phi z)\} . \equiv . \psi(\imath x)(\phi x)$ $\quad [*51\cdot56 . *14\cdot205]$

*52. THE CARDINAL NUMBER 1

Summary of *52.

In this number, we introduce the cardinal number 1, defined as the class of all unit classes. The fact that 1 so defined is a cardinal number is not relevant at present, and cannot of course be proved until "cardinal number" has been defined. For the present, therefore, 1 is to be regarded simply as the class of all unit classes, unit classes being such classes as are of the form $\iota' x$ for some x.

Like Λ and V, 1 is ambiguous as to type: it means "all unit classes of the type in question." The symbol "1 (α)," where α is a type, will mean "all unit classes whose sole members belong to the type α" (cf. *65). Thus *e.g.* "$\xi \in 1$ (Indiv)" will mean "ξ is a class consisting of one individual," if "Indiv" stands for the class of individuals.

The properties of 1 to be proved in the present number are what we may call *logical* as opposed to *arithmetical* properties, *i.e.* they are not concerned with the arithmetical operations (addition, etc.) which can be performed with 1, but with the relations of 1 to unit classes. The arithmetical properties of 1 will be considered later, in Part III.

The propositions of the present number which are most used are the following:

***52·16.** $\vdash :. \alpha \in 1 . \equiv : \exists ! \alpha : x, y \in \alpha . \supset_{x,y} . x = y$

I.e. α is a unit class if, and only if, it is not null, and all its members are identical.

***52·22.** $\vdash . \iota' x \in 1$

***52·4.** $\vdash :. \alpha \in 1 \cup \iota' \Lambda . \equiv : x, y \in \alpha . \supset_{x,y} . x = y$

We shall define 0 as $\iota' \Lambda$. Thus the above proposition states that a class has one member or none when, and only when, all its members are identical.

***52·41.** $\vdash : \exists ! \alpha . \alpha \sim \in 1 . \equiv . (\exists x, y) . x, y \in \alpha . x \neq y$

This proposition is obtainable from *52·4 by transposition, *i.e.* by negating each side of the equivalence.

***52·46.** $\vdash :. \alpha, \beta \in 1 . \supset : \alpha \subset \beta . \equiv . \alpha = \beta . \equiv . \exists ! (\alpha \cap \beta)$

I.e. two unit classes are identical when, and only when, one is contained in the other, and when and only when they have a common part.

***52·01.** $1 = \hat{\alpha} \{(\exists x) . \alpha = \iota' x\}$ Df

***52·1.** $\vdash : \alpha \in 1 . \equiv . (\exists x) . \alpha = \iota' x$ [*20·3 . (*52·01)]

***52·11.** $\vdash :. \alpha \in 1 . \equiv : (\exists x) : y \in \alpha . \equiv_y . y = x$ [*52·1 . *51·14]

***52·12.** $\vdash : \hat{z}(\phi z) \in 1 . \equiv . E ! (\imath x)(\phi x)$

Dem.

$\vdash . *52·11 . \supset \vdash :. \hat{z}(\phi z) \in 1 . \equiv : (\exists x) : y \in \hat{z}(\phi z) . \equiv_y . y = x :$

[*20·3] $\equiv : (\exists x) : \phi y . \equiv_y . y = x :$

[*14·11] $\equiv : E ! (\imath x)(\phi x) :. \supset \vdash .$ Prop

***52·13.** $\vdash . 1 = D'\iota$

Dem.

$\vdash . *51·131 . \supset \vdash : \alpha = \iota' x . \equiv . \alpha \iota x :$

[*10·11·281] $\supset \vdash : (\exists x) . \alpha = \iota' x . \equiv . (\exists x) . \alpha \iota x :$

[*52·1] $\supset \vdash : \alpha \in 1 . \equiv . (\exists x) . \alpha \iota x$

[*33·13] $\equiv . \alpha \in D'\iota : \supset \vdash .$ Prop

***52·14.** $\vdash . 1 = \iota `` V$ [*52·13 . *37·28]

***52·15.** $\vdash : \alpha \in 1 . \equiv . E ! \breve{\iota}' \alpha$ [*51·54 . *52·1]

***52·16.** $\vdash :. \alpha \in 1 . \equiv : \exists ! \alpha : x, y \in \alpha . \supset_{x,y} . x = y$ [*52·15 . *51·55 . *14·203]

***52·17.** $\vdash : \alpha \in 1 . \equiv . \breve{\iota}' \alpha = (\imath x)(x \in \alpha)$ [*51·58 . *52·15]

***52·171.** $\vdash : \alpha \in 1 . \equiv . E ! (\imath x)(x \in \alpha)$ [*51·55 . *52·15]

***52·172.** $\vdash : \alpha \in 1 . \equiv . \iota' \breve{\iota}' \alpha = \alpha$ [*51·52 . *52·15]

***52·173.** $\vdash : \alpha \in 1 . \equiv . \breve{\iota}' \alpha \in \alpha$ [*51·53 . *52·15]

***52·18.** $\vdash :. \alpha \in 1 . \equiv : (\exists x) : x \in \alpha : y \in \alpha . \supset_y . y = x$

Dem.

$\vdash . *51·141 . \supset \vdash :. (\exists x) . \alpha = \iota' x . \equiv : (\exists x) : x \in \alpha : y \in \alpha . \supset_y . y = x$ (1)

$\vdash . (1) . *52·1 . \supset \vdash .$ Prop

***52·181.** $\vdash :. \alpha \sim\in 1 . \equiv : x \in \alpha . \supset_x . (\exists y) . y \in \alpha . y \neq x$ [*52·18 . *10·51]

***52·2.** $\vdash . 1 \subset \mathrm{Cls}$

Dem.

$\vdash . *52·1 . \supset \vdash : \alpha \in 1 . \supset . (\exists x) . \alpha = \iota' x .$

[*51·11] $\supset . (\exists x) . \alpha = \hat{z}(z = x) .$

[*20·54] $\supset . (\exists x, \phi) . \hat{z}(\phi ! z) = \hat{z}(z = x) . \alpha = \hat{z}(\phi ! z) .$

[*10·5] $\supset . (\exists \phi) . \alpha = \hat{z}(\phi ! z) .$

[*20·4] $\supset . \alpha \in \mathrm{Cls} : \supset \vdash .$ Prop

***52·21.** $\vdash . \Lambda \sim\in 1$

Dem.

$\vdash . *52·16 . \supset \vdash : \alpha \in 1 . \supset_\alpha . \exists ! \alpha :$

[*24·63] $\supset \vdash : \Lambda \sim\in 1$

***52·22.** $\vdash . \iota' x \in 1$ [*51·12 . *14·28 . *10·24 . *52·1]

***52·23.** $\vdash . \exists ! 1 . \exists ! -1$

Dem.

$\vdash . *52·22 . *10·24 . \supset \vdash . (\exists x) . \iota 'x \in 1 .$
[*20·54] $\supset \vdash . (\exists x, \alpha) . \alpha = \iota 'x . \alpha \in 1 .$
[*10·5] $\supset \vdash . (\exists \alpha) . \alpha \in 1$ (1)
$\vdash . *52·21 . *22·35 . \supset \vdash . \Lambda \in -1 .$
[*10·24] $\supset \vdash . (\exists \alpha) . \alpha \in -1$ (2)
$\vdash . (1) . (2) . \supset \vdash .$ Prop

***52·24.** $\vdash . 1 \neq \Lambda \cap \text{Cls} . 1 \neq V \cap \text{Cls}$ [*52·23 . *24·54 . *24·17 . Transp]

***52·3.** $\vdash . \iota ``\alpha \subset 1$

Dem.

$\vdash . *52·22 . *2·02 . \supset \vdash : y \in \alpha . \supset . \iota 'y \in 1 :$
[*51·12.*10·11.*37·61] $\supset \vdash . \iota ``\alpha \subset 1$

***52·31.** $\vdash : \kappa \subset 1 . \equiv . (\exists \alpha) . \kappa = \iota ``\alpha$

Dem.

$\vdash . *52·14 . \supset \vdash : \kappa \subset 1 . \equiv . \kappa \subset \iota ``V .$
[*37·66.*51·12] $\equiv . (\exists \alpha) . \alpha \subset V . \kappa = \iota ``\alpha .$
[*24·11] $\equiv . (\exists \alpha) . \kappa = \iota ``\alpha : \supset \vdash .$ Prop

***52·4.** $\vdash :. \alpha \in 1 \cup \iota '\Lambda . \equiv : x, y \in \alpha . \supset_{x,y} . x = y$

Dem.

$\vdash . *52·16 . *24·54 . \supset$
$\vdash :. \alpha \in 1 . \equiv : \alpha \neq \Lambda : x, y \in \alpha . \supset_{x,y} . x = y :.$
[*4·37] $\supset \vdash :: \alpha \in 1 . \vee . \alpha = \Lambda : \equiv :. \alpha = \Lambda :. \vee :. \alpha \neq \Lambda : x, y \in \alpha . \supset_{x,y} . x = y :.$
[*5·63] $\equiv :. \alpha = \Lambda : \vee : x, y \in \alpha . \supset_{x,y} . x = y$ (1)
$\vdash . *24·51 . *10·53 . *11·62 . \supset \vdash :. \alpha = \Lambda . \supset : x, y \in \alpha . \supset_{x,y} . x = y$ (2)
$\vdash . (1) . (2) . *4·72 . \supset \vdash :: \alpha \in 1 . \vee . \alpha = \Lambda : \equiv :. x, y \in \alpha . \supset_{x,y} . x = y$ (3)
$\vdash . (3) . *51·236 . \supset \vdash .$ Prop

This proposition is frequently useful. We shall define the number 0 as $\iota '\Lambda$; thus the above proposition states that a class has one member or none when, and only when, all its members are identical. It will be seen that $x, y \in \alpha . \supset_{x,y} . x = y$ does not imply $\exists ! \alpha$, and therefore allows the possibility of α having no members.

***52·41.** $\vdash : \exists ! \alpha . \alpha \sim\in 1 . \equiv . (\exists x, y) . x, y \in \alpha . x \neq y$

Dem.

$\vdash . *24·54 . \supset \vdash :. \exists ! \alpha . \alpha \sim\in 1 . \equiv : \alpha \neq \Lambda . \alpha \sim\in 1 :$
[*4·56] $\equiv : \sim \{\alpha \in 1 . \vee . \alpha = \Lambda\} :$
[*51·236] $\equiv : \sim (\alpha \in 1 \cup \iota '\Lambda) :$
[*52·4.Transp] $\equiv : \sim \{x, y \in \alpha . \supset_{x,y} . x = y\}$
[*11·52] $\equiv : (\exists x, y) . x, y \in \alpha . x \neq y :. \supset \vdash .$ Prop

∗52·42. $\vdash :. \alpha \epsilon 1 . \supset : \exists ! \alpha \cap \beta . \equiv . \alpha \cap \beta \epsilon 1$

Dem.

$\vdash . ∗51·31 . \quad \supset \vdash :. \exists ! \iota`x \cap \beta . \equiv . \iota`x \cap \beta = \iota`x :.$

$[∗20·53] \quad \supset \vdash :. \alpha = \iota`x . \supset : \exists ! \alpha \cap \beta . \equiv . \alpha \cap \beta = \iota`x :.$

$[∗10·11·28] \quad \supset \vdash :. (\exists x) . \alpha = \iota`x . \supset : (\exists x) : \exists ! \alpha \cap \beta . \equiv . \alpha \cap \beta = \iota`x :$

$[∗10·37] \quad \supset : \exists ! \alpha \cap \beta . \supset . (\exists x) . \alpha \cap \beta = \iota`x \quad (1)$

$\vdash . (1) . ∗52·1 . \supset \vdash :. \alpha \epsilon 1 . \supset : \exists ! \alpha \cap \beta . \supset . \alpha \cap \beta \epsilon 1 \quad (2)$

$\vdash . ∗52·16 . \quad \supset \vdash : \alpha \cap \beta \epsilon 1 . \supset . \exists ! \alpha \cap \beta \quad (3)$

$\vdash . (2) . (3) . \quad \supset \vdash . \mathrm{Prop}$

∗52·43. $\vdash : \alpha \epsilon 1 . \exists ! \alpha \cap \beta . \equiv . \alpha \epsilon 1 . \alpha \cap \beta \epsilon 1 \quad [∗52·42 . ∗5·32]$

∗52·44. $\vdash :. \alpha \epsilon 1 . \supset : \exists ! \alpha \cap \beta . \equiv . \alpha \subset \beta . \equiv . \alpha \cap \beta = \alpha$

Dem.

$\vdash . ∗51·31 . \quad \supset \vdash : \exists ! \iota`x \cap \beta . \equiv . \iota`x \subset \beta :$

$[∗13·13 . \mathrm{Exp}] \quad \supset \vdash :. \alpha = \iota`x . \supset : \exists ! \alpha \cap \beta . \equiv . \alpha \subset \beta :.$

$[∗10·11·23] \quad \supset \vdash :. (\exists x) . \alpha = \iota`x . \supset : \exists ! \alpha \cap \beta . \equiv . \alpha \subset \beta :.$

$[∗52·1] \quad \supset \vdash :. \alpha \epsilon 1 . \supset : \exists ! \alpha \cap \beta . \equiv . \alpha \subset \beta \quad (1)$

$\vdash . (1) . ∗22·621 . \supset \vdash . \mathrm{Prop}$

∗52·45. $\vdash :: \alpha, \beta \epsilon 1 . \supset :. \alpha \subset \beta \cup \gamma . \equiv : \alpha = \beta . \vee . \alpha \subset \gamma$

Dem.

$\vdash . ∗51·236 \dfrac{x, y, \gamma}{z, x, \beta} . \supset$

$\vdash :. x \epsilon \iota`y \cup \gamma . \equiv : x = y . \vee . x \epsilon \gamma :.$

$[∗51·2·23] \quad \supset \vdash :. \iota`x \subset \iota`y \cup \gamma . \equiv : \iota`x = \iota`y . \vee . \iota`x \subset \gamma :.$

$[∗13·21] \quad \supset \vdash :: \alpha = \iota`x . \beta = \iota`y . \supset :. \alpha \subset \beta \cup \gamma . \equiv : \alpha = \beta . \vee . \alpha \subset \gamma ::$

$[∗11·11·35] \quad \supset \vdash :: (\exists x, y) . \alpha = \iota`x . \beta = \iota`y . \supset :. \alpha \subset \beta \cup \gamma . \equiv : \alpha = \beta . \vee . \alpha \subset \gamma \quad (1)$

$\vdash . (1) . ∗52·1 . \supset \vdash . \mathrm{Prop}$

∗52·46. $\vdash :. \alpha, \beta \epsilon 1 . \supset : \alpha \subset \beta . \equiv . \alpha = \beta . \equiv . \exists ! (\alpha \cap \beta)$

Dem.

$\vdash . ∗51·2·23 . \quad \supset \vdash : \iota`x \subset \iota`y . \equiv . \iota`x = \iota`y \quad (1)$

$\vdash . (1) . ∗13·21 . \quad \supset \vdash :. \alpha = \iota`x . \beta = \iota`y . \supset : \alpha \subset \beta . \equiv . \alpha = \beta \quad (2)$

$\vdash . (2) . ∗11·11·35 . ∗52·1 . \supset \vdash :. \alpha, \beta \epsilon 1 . \supset : \alpha \subset \beta . \equiv . \alpha = \beta \quad (3)$

$\vdash . (3) . ∗52·44 . \quad \supset \vdash . \mathrm{Prop}$

∗52·6. $\vdash :. \alpha \epsilon 1 . \supset : x \epsilon \alpha . \equiv . \iota`x = \alpha . \equiv . x = \breve{\iota}`\alpha$

Dem.

$\vdash . ∗51·23 . \quad \supset \vdash : x \epsilon \iota`y . \equiv . \iota`x = \iota`y :$

$[∗13·13 . \mathrm{Exp}] \quad \supset \vdash :. \alpha = \iota`y . \supset : x \epsilon \alpha . \equiv . \iota`x = \alpha :.$

$[∗10·11·23 . ∗52·1] \supset \vdash :. \alpha \epsilon 1 . \quad \supset : x \epsilon \alpha . \equiv . \iota`x = \alpha . \quad (1)$

$[∗51·51] \quad \equiv . x = \breve{\iota}`\alpha \quad (2)$

$\vdash . (1) . (2) . \supset \vdash . \mathrm{Prop}$

***52·601.** $\vdash :: \alpha \in 1 . \supset :. \phi(\breve{\iota}`\alpha) . \equiv : x \in \alpha . \supset_x . \phi x : \equiv : (\exists x) . x \in \alpha . \phi x$

Dem.

$\vdash . *52\cdot15 . \supset \vdash :. \text{Hp} . \supset : E ! \breve{\iota}`\alpha :$ (1)

$[*30\cdot4] \quad \supset : x \breve{\iota} \alpha . \equiv . x = \breve{\iota}`\alpha .$

$[*52\cdot6] \quad \equiv . x \in \alpha$ (2)

$\vdash . (1) . *30\cdot33 . \supset$

$\vdash :: \text{Hp} . \supset :. \phi(\breve{\iota}`\alpha) . \equiv : x \breve{\iota} \alpha . \supset_x . \phi x : \equiv : (\exists x) . x \breve{\iota} \alpha . \phi x$ (3)

$\vdash . (2) . (3) . \supset \vdash . \text{Prop}$

***52·602.** $\vdash :. \hat{z}(\phi z) \in 1 . \supset : \psi(\imath x)(\phi x) . \equiv . \phi x \supset_x \psi x . \equiv . (\exists x) . \phi x . \psi x$ [*52·12 . *14·26]

***52·61.** $\vdash :. \alpha \in 1 . \supset : \breve{\iota}`\alpha \in \beta . \equiv . \alpha \subset \beta . \equiv . \exists ! (\alpha \cap \beta)$ $\left[*52\cdot601 \frac{x \in \beta}{\phi x}\right]$

***52·62.** $\vdash :. \alpha, \beta \in 1 . \supset : \alpha = \beta . \equiv . \breve{\iota}`\alpha = \breve{\iota}`\beta$

Dem.

$\vdash . *52\cdot601 . \supset \vdash :: \text{Hp} . \supset :. \breve{\iota}`\alpha = \breve{\iota}`\beta . \equiv : x \in \alpha . \supset_x . x = \breve{\iota}`\beta :$

$[*52\cdot6] \quad \equiv : x \in \alpha . \supset_x . x \in \beta :$

$[*52\cdot46] \quad \equiv : \alpha = \beta :: \supset \vdash . \text{Prop}$

***52·63.** $\vdash : \alpha, \beta \in 1 . \alpha \neq \beta . \supset . \alpha \cap \beta = \Lambda$ [*52·46 . Transp]

***52·64.** $\vdash : \alpha \in 1 . \supset . \alpha \cap \beta \in 1 \cup \iota`\Lambda$

Dem.

$\vdash . *52\cdot43 . \quad \supset \vdash : \text{Hp} . \exists ! \alpha \cap \beta . \supset . \alpha \cap \beta \in 1 :$

$[*5\cdot6 . *24\cdot54] \supset \vdash :. \text{Hp} . \supset : \alpha \cap \beta = \Lambda . \vee . \alpha \cap \beta \in 1 :$

$[*51\cdot236] \quad \supset : \alpha \cap \beta \in 1 \cup \iota`\Lambda :. \supset \vdash . \text{Prop}$

***52·7.** $\vdash :. \beta - \alpha \in 1 . \alpha \subset \xi . \xi \subset \beta . \supset : \xi = \alpha . \vee . \xi = \beta$

Dem.

$\vdash . *22\cdot41 . \quad \supset \vdash : \text{Hp} . \xi \subset \alpha . \supset . \xi = \alpha$ (1)

$\vdash . *24\cdot55 . \quad \supset \vdash : \sim(\xi \subset \alpha) . \supset . \exists ! \xi - \alpha$ (2)

$\vdash . *22\cdot48 . \quad \supset \vdash : \text{Hp} . \quad \supset . \xi - \alpha \subset \beta - \alpha$ (3)

$\vdash . (2) . (3) . \quad \supset \vdash : \text{Hp} . \sim(\xi \subset \alpha) . \supset . \exists ! \xi - \alpha . \xi - \alpha \subset \beta - \alpha$ (4)

$\vdash . *52\cdot1 . \quad \supset \vdash : \text{Hp} . \supset . (\exists x) . \beta - \alpha = \iota`x$ (5)

$\vdash . (4) . (5) . *51\cdot4 . \supset \vdash : \text{Hp} . \sim(\xi \subset \alpha) . \supset . \xi - \alpha = \beta - \alpha .$

$[*24\cdot411] \quad \supset . \xi = \beta$ (6)

$\vdash . (1) . (6) . \supset \vdash . \text{Prop}$

$*53$. MISCELLANEOUS PROPOSITIONS INVOLVING UNIT CLASSES

Summary of $*53$.

The propositions to be given in this number are mostly such as would have come more naturally at an earlier stage, but could not be given sooner because they involved unit classes. It is to be observed that $\iota`x \cup \iota`y$ is the class consisting of the members x and y, while $\iota`x \uparrow \iota`y$ is the relation which holds only between x and y. If α and β are classes, $\iota`\alpha \cup \iota`\beta$ is a class of classes, its members being α and β. If R and S are relations, $\iota`R \uparrow \iota`S$ is a relation of relations; and so on.

The present number begins by connecting products and sums $p`\kappa$, $s`\kappa$, $\dot{p}`\lambda$, $\dot{s}`\lambda$, in cases where the members of κ or λ are specified, with the products or sums $\alpha \cap \beta$, $\alpha \cup \beta$. $R \dot{\cap} S$, $R \dot{\cup} S$. We have

$*53{\cdot}01$. $\vdash . p`\iota`\alpha = \alpha$

$*53{\cdot}1$. $\vdash . p`(\iota`\alpha \cup \iota`\beta) = \alpha \cap \beta$

$*53{\cdot}14$. $\vdash . p`(\kappa \cup \iota`\alpha) = p`\kappa \cap \alpha$

with similar propositions for s, $\dot{p}$ and $\dot{s}$.

We have next a set of propositions on sums and products of classes of unit classes. The most important of these is

$*53{\cdot}22$. $\vdash . s`\iota``\alpha = \alpha$

We have next a proposition showing that the sum of κ is null when, and only when, κ is either null or has the null-class for its only member, *i.e.*

$*53{\cdot}24$. $\vdash :. s`\kappa = \Lambda . \equiv : \kappa = \Lambda \cap \mathrm{Cls} . \vee . \kappa = \iota`\Lambda$

(Here we write "$\Lambda \cap \mathrm{Cls}$," to show that the "$\Lambda$" in question is of the next type above that of the other two Λ's.)

We have next various propositions on the relations of $\overrightarrow{R}`x$ and $R`x$ and $R``\alpha$ in various cases, first for a general relation R, and then for the particular relation s defined in $*40$. Three of these propositions are very frequently used, namely:

$*53{\cdot}3$. $\vdash : \mathrm{E} ! R`x . \equiv . \overrightarrow{R}`x \in 1$

$*53{\cdot}301$. $\vdash . R``\iota`x = \overrightarrow{R}`x$

$*53{\cdot}31$. $\vdash : \mathrm{E} ! R`x . \supset . R``\iota`x = \iota`R`x = \overrightarrow{R}`x$

The remaining propositions of this number are of less importance, and are seldom referred to.

***53·01.** $\vdash . p`\iota`\alpha = \alpha$

Dem.

$\vdash . *40·1 . \supset \vdash :. x \in p`\iota`\alpha . \equiv : \beta \in \iota`\alpha . \supset_\beta . x \in \beta :$

[*51·15] $\equiv : \beta = \alpha . \supset_\beta . x \in \beta :$

[*13·191] $\equiv : x \in \alpha :. \supset \vdash .$ Prop

***53·02.** $\vdash . s`\iota`\alpha = \alpha$

Dem.

$\vdash . *40·11 . \supset \vdash : x \in s`\iota`\alpha . \equiv . (\exists \beta) . \beta \in \iota`\alpha . x \in \beta .$

[*51·15] $\equiv . (\exists \beta) . \beta = \alpha . x \in \beta .$

[*13·195] $\equiv . x \in \alpha : \supset \vdash .$ Prop

***53·03.** $\vdash . \dot{p}`\iota`R = R$ [Proof as in *53·01]

***53·04.** $\vdash . \dot{s}`\iota`R = R$ [Proof as in *53·02]

***53·1.** $\vdash . p`(\iota`\alpha \cup \iota`\beta) = \alpha \cap \beta$

Dem.

$\vdash . *40·18 . \supset \vdash . p`(\iota`\alpha \cup \iota`\beta) = p`\iota`\alpha \cap p`\iota`\beta$

[*53·01] $= \alpha \cap \beta . \supset \vdash .$ Prop

This proposition can be extended to $\iota`\alpha \cup \iota`\beta \cup \iota`\gamma$, etc. It shows the connection (for finite classes of classes) between the product $p`\kappa$ and the product of the members $\alpha \cap \beta \cap \gamma \cap \ldots$.

***53·11.** $\vdash . s`(\iota`\alpha \cup \iota`\beta) = \alpha \cup \beta$

Dem.

$\vdash . *40·171 . \supset \vdash . s`(\iota`\alpha \cup \iota`\beta) = s`\iota`\alpha \cup s`\iota`\beta$

[*53·02] $= \alpha \cup \beta . \supset \vdash .$ Prop

Similar remarks apply to this proposition as to *53·1.

***53·12.** $\vdash . \dot{p}`(\iota`R \cup \iota`S) = R \dot{\cap} S$ [*41·18 . *53·03]

This proposition shows the connection between the product $\dot{p}`\kappa$ for a class κ consisting of two relations R and S, and the product $R \dot{\cap} S$. The proposition can be extended to the product of any given finite class of relations.

***53·13.** $\vdash . \dot{s}`(\iota`R \cup \iota`S) = R \dot{\cup} S$ [*41·171 . *53·04]

Similar remarks apply to this proposition as to *53·12.

***53·14.** $\vdash . p`(\kappa \cup \iota`\alpha) = p`\kappa \cap \alpha$

Dem.

$\vdash . *40·18 . \supset \vdash . p`(\kappa \cup \iota`\alpha) = p`\kappa \cap p`\iota`\alpha$

[*53·01] $= p`\kappa \cap \alpha$

***53·15.** $\vdash . s`(\kappa \cup \iota`\alpha) = s`\kappa \cup \alpha$ [Proof as in *53·14]

***53·16.** $\vdash . \dot{p}`(\lambda \cup \iota`R) = \dot{p}`\lambda \dot{\cap} R$ [Proof as in *53·14]

***53·17.** $\vdash . \dot{s}`(\lambda \cup \iota`R) = \dot{s}`\lambda \dot{\cup} R$ [Proof as in *53·14]

The above proposition and the next are both used in connection with mathematical induction (*91·55 and *97·46 respectively).

***53·18.** $\vdash . s'(\alpha - \iota'\Lambda) = s'\alpha$

Dem.

$$\begin{array}{lll} \vdash . *51\cdot221 . \supset \vdash : \Lambda \epsilon \alpha . & \supset . (\alpha - \iota'\Lambda) \cup \iota'\Lambda = \alpha . & \\ [*53\cdot15] & \supset . s'(\alpha - \iota'\Lambda) \cup \Lambda = s'\alpha . & \\ [*24\cdot24] & \supset . s'(\alpha - \iota'\Lambda) = s'\alpha & (1) \\ \vdash . *51\cdot222 . \supset \vdash : \Lambda \sim\epsilon \alpha . \supset . & \alpha - \iota'\Lambda = \alpha . & \\ [*30\cdot37] & \supset . s'(\alpha - \iota'\Lambda) = s'\alpha & (2) \\ \vdash . (1) . (2) . \supset \vdash . \text{Prop} & & \end{array}$$

***53·181.** $\vdash . \dot{s}'(\lambda \dot{-} \iota'\dot{\Lambda}) = \dot{s}'\lambda$ [Proof as in *53·18]

***53·2.** $\vdash : \kappa \epsilon 1 . \supset . \breve{\iota}'\kappa = p'\kappa = s'\kappa$

This proposition requires, for significance, that κ should be a class of classes. It is used in *88·47, in the number on the existence of selections and the multiplicative axiom.

Dem.

$$\vdash . *52\cdot601 . \supset \vdash :: \text{Hp} . \supset :. x \epsilon \breve{\iota}'\kappa : \equiv : \alpha \epsilon \kappa . \supset_\alpha . x \epsilon \alpha : \equiv : (\exists \alpha) . \alpha \epsilon \kappa . x \epsilon \alpha \quad (1)$$

$$\vdash . (1) . *40\cdot1\cdot11 . \quad \supset \vdash . \text{Prop}$$

***53·21.** $\vdash : \lambda \epsilon 1 . \supset . \breve{\iota}'\lambda = \dot{p}'\lambda = \dot{s}'\lambda$ [Similar proof]

This proposition requires, for significance, that λ should be a class of relations.

***53·22.** $\vdash . s'\iota''\alpha = \alpha$

Dem.

$$\begin{array}{ll} \vdash . *40\cdot11 . \supset \vdash : x \epsilon s'\iota''\alpha . \equiv . (\exists \gamma) . \gamma \epsilon \iota''\alpha . x \epsilon \gamma . \\ [*37\cdot64 . *51\cdot12] & \equiv . (\exists y) . y \epsilon \alpha . x \epsilon \iota'y . \\ [*51\cdot15] & \equiv . (\exists y) . y \epsilon \alpha . x = y . \\ [*13\cdot195] & \equiv . x \epsilon \alpha : \supset \vdash . \text{Prop} \end{array}$$

***53·221.** $\vdash . \iota''(\iota'x \cup \iota'y) = \iota'\iota'x \cup \iota'\iota'y$

Dem.

$$\begin{array}{ll} \vdash . *37\cdot1 . \supset \vdash :. \alpha \epsilon \iota''(\iota'x \cup \iota'y) . \equiv : (\exists z) . z \epsilon (\iota'x \cup \iota'y) . \alpha \iota z : \\ [*51\cdot131] & \equiv : (\exists z) . z \epsilon (\iota'x \cup \iota'y) . \alpha = \iota'z : \\ [*51\cdot235] & \equiv : \alpha = \iota'x . \vee . \alpha = \iota'y : \\ [*51\cdot232] & \equiv : \alpha \epsilon (\iota'\iota'x \cup \iota'\iota'y) :. \supset \vdash . \text{Prop} \end{array}$$

***53·222.** $\vdash : \kappa = \iota''\alpha . \supset . \alpha = \breve{\iota}''\kappa$

Dem.

$$\begin{array}{ll} \vdash . *13\cdot12 . *20\cdot2 . \supset \vdash : \text{Hp} . \supset . \breve{\iota}''\kappa = \breve{\iota}''\iota''\alpha \\ [*51\cdot511 . *14\cdot21 . *37\cdot67] & = \hat{x} \{(\exists y) . y \epsilon \alpha . x = \breve{\iota}'\iota'y\} \\ [*51\cdot511] & = \hat{x} \{(\exists y) . y \epsilon \alpha . x = y\} \\ [*13\cdot195] & = \alpha : \supset \vdash . \text{Prop} \end{array}$$

***53·23.** $\vdash : \kappa \subset 1 . \supset . s'\kappa = \breve{\iota}``\kappa$

Dem.

$\vdash . *52·31 . \supset \vdash : \text{Hp} . \equiv . (\exists \alpha) . \kappa = \iota``\alpha$ (1)

$\vdash . *53·22 . \supset \vdash : \kappa = \iota``\alpha . \supset . s'\kappa = \alpha$

[*53·222] $= \breve{\iota}``\kappa$ (2)

$\vdash . (1) . (2) . *10·11·23 . \supset \vdash . \text{Prop}$

***53·231.** $\vdash :. x \in \alpha . \supset_x . x = y : \equiv : \alpha = \Lambda . \vee . \alpha = \iota'y$

Dem.

$\vdash . *51·141 . \supset \vdash :. \exists ! \alpha : x \in \alpha . \supset_x . x = y : \equiv : \alpha = \iota'y$ (1)

$\vdash . *10·53 . \supset \vdash :. \sim \exists ! \alpha . \supset : x \in \alpha . \supset_x . x = y :.$

[*4·71] $\supset \vdash :. \sim \exists ! \alpha : x \in \alpha . \supset_x . x = y : \equiv . \sim \exists ! \alpha .$

[*24·51] $\equiv . \alpha = \Lambda$ (2)

$\vdash . (1) . (2) . *4·42·39 . \supset \vdash . \text{Prop}$

***53·24.** $\vdash :. s'\kappa = \Lambda . \equiv : \kappa = \Lambda \cap \text{Cls} . \vee . \kappa = \iota'\Lambda$

Dem.

$\vdash . *24·15 . *40·11 . \supset$

$\vdash :. s'\kappa = \Lambda . \quad \equiv : (x) : \sim \{(\exists \alpha) . \alpha \in \kappa . x \in \alpha\} :$

[*10·51] $\equiv : (x, \alpha) : x \in \alpha . \supset . \alpha \sim \in \kappa :$

[*11·2.*10·23] $\equiv : (\exists x) . x \in \alpha . \supset_\alpha . \alpha \sim \in \kappa :$

[*24·54] $\equiv : \alpha \neq \Lambda . \supset_\alpha . \alpha \sim \in \kappa :$

[Transp] $\equiv : \alpha \in \kappa . \supset_\alpha . \alpha = \Lambda :$

[*53·231] $\equiv : \kappa = \Lambda \cap \text{Cls} . \vee . \kappa = \iota'\Lambda :. \supset \vdash . \text{Prop}$

In the enunciation and the last line of the proof of the above proposition, we write "$\kappa = \Lambda \cap \text{Cls}$" rather than "$\kappa = \Lambda$," because this Λ must be of the type next above that of the Λ in "$\kappa = \iota'\Lambda$."

The following proposition is used in the theory of selections (*83·731).

***53·25.** $\vdash :. s'\kappa \cap s'\lambda = \Lambda . \supset : \kappa \cap \lambda = \Lambda \cap \text{Cls} . \vee . \kappa \cap \lambda = \iota'\Lambda$

Dem.

$\vdash . *40·181 . \supset \vdash :. \text{Hp} . \supset : s'(\kappa \cap \lambda) = \Lambda :$

[*53·24] $\supset : \kappa \cap \lambda = \Lambda \cap \text{Cls} . \vee . \kappa \cap \lambda = \iota'\Lambda :. \supset \vdash . \text{Prop}$

***53·3.** $\vdash : E ! R'x . \equiv . \overrightarrow{R}'x \in 1$

Dem.

$\vdash . *30·2 . \supset \vdash :. E ! R'x . \equiv : (\exists b) : yRx . \equiv_y . y = b :$

[*32·18.*51·15] $\equiv : (\exists b) : y \in \overrightarrow{R}'x . \equiv_y . y \in \iota'b :$

[*20·31] $\equiv : (\exists b) . \overrightarrow{R}'x = \iota'b :$

[*52·1] $\equiv : \overrightarrow{R}'x \in 1 :. \supset \vdash . \text{Prop}$

The above proposition is very frequently used.

***53·301.** $\vdash . R``\iota`x = \overrightarrow{R}`x$

Dem.

$\vdash . *37·1 . *51·15 . \supset \vdash : y \epsilon R``\iota`x . \equiv . (\exists z) . z = x . yRz .$

$[*13·195] \quad \equiv . yRx .$

$[*32·18] \quad \equiv . y \epsilon \overrightarrow{R}`x : \supset \vdash . \text{Prop}$

***53·302.** $\vdash . R``(\iota`x \cup \iota`y) = \overrightarrow{R}`x \cup \overrightarrow{R}`y$ $[*37·22 . *53·301]$

The above proposition is used in the cardinal theory of exponentiation (*116·71).

***53·31.** $\vdash : E ! R`x . \supset . R``\iota`x = \iota`R`x = \overrightarrow{R}`x$

The above proposition is one of which the subsequent use is frequent.

Dem.

$\vdash . *51·11 . *14·18 . \supset \vdash : \text{Hp} . \supset . \iota`R`x = \hat{y}(y = R`x)$

$[*30·4] \quad = \hat{y}(yRx)$

$[*32·13] \quad = \overrightarrow{R}`x \quad (1)$

$\vdash . (1) . *53·301 . \supset \vdash . \text{Prop}$

***53·32.** $\vdash : E ! R`x . E ! R`y . \supset . R``(\iota`x \cup \iota`y) = \iota`R`x \cup \iota`R`y$

Dem.

$\vdash . *37·22 . \supset \vdash . R``(\iota`x \cup \iota`y) = R``\iota`x \cup R``\iota`y \quad (1)$

$\vdash . (1) . *53·31 . \supset \vdash . \text{Prop}$

***53·33.** $\vdash . s``\iota`\kappa = \iota`s`\kappa$ $\left[*53·31 \frac{s}{R}\right]$

***53·34.** $\vdash . s``(\iota`\kappa \cup \iota`\lambda) = \iota`s`\kappa \cup \iota`s`\lambda$ $\left[*53·32 \frac{s}{R}\right]$

***53·35.** $\vdash . s`s``(\iota`\kappa \cup \iota`\lambda) = s`\kappa \cup s`\lambda = s`(\kappa \cup \lambda)$

Dem.

$\vdash . *53·34 . \supset \vdash . s`s``(\iota`\kappa \cup \iota`\lambda) = s`(\iota`s`\kappa \cup \iota`s`\lambda)$

$[*53·11] \quad = s`\kappa \cup s`\lambda$

$[*40·171] \quad = s`(\kappa \cup \lambda) . \supset \vdash . \text{Prop}$

The above proposition may also be proved as follows:

$\vdash . *42·1 . \supset \vdash . s`s``(\iota`\kappa \cup \iota`\lambda) = s`s`(\iota`\kappa \cup \iota`\lambda)$

$[*53·11] \quad = s`(\kappa \cup \lambda)$

$[*40·171] \quad = s`\kappa \cup s`\lambda . \supset \vdash . \text{Prop}$

***53·4.** $\vdash : x = R`y . \equiv . \overrightarrow{R}`y \epsilon 1 . x \epsilon \overrightarrow{R}`y . \equiv . \iota`x = \overrightarrow{R}`y . \equiv . x = \breve{\iota}`\overrightarrow{R}`y$

Dem.

$\vdash . *14·21 . *4·71 . \supset \vdash : x = R`y . \equiv . E ! R`y . x = R`y .$

$[*30·4 . *5·32] \quad \equiv . E ! R`y . xRy .$

$[*53·3 . *32·18] \quad \equiv . \overrightarrow{R}`y \epsilon 1 . x \epsilon \overrightarrow{R}`y . \quad (1)$

$[*52·6 . *5·32] \quad \equiv . \overrightarrow{R}`y \epsilon 1 . \iota`x = \overrightarrow{R}`y .$

[*52·22] $\equiv . \iota`x = \overrightarrow{R}`y .$ (2)

[*51·51] $\equiv . x = \iota`\overrightarrow{\breve{R}}`y$ (3)

$\vdash . (1) . (2) . (3) . \supset \vdash .$ Prop

***53·5.** $\vdash : \exists ! \alpha . \equiv . \alpha \in \text{Cls} - \iota`\Lambda$

Dem.

$\vdash . *20·41 . \supset \vdash : \exists ! \hat{z}(\phi z) . \equiv . \hat{z}(\phi z) \in \text{Cls} . \exists ! \hat{z}(\phi z) .$

[*24·54] $\equiv . \hat{z}(\phi z) \in \text{Cls} . \hat{z}(\phi z) \neq \Lambda .$

[*51·3] $\equiv . \hat{z}(\phi z) \in \text{Cls} - \iota`\Lambda : \supset \vdash .$ Prop

In the above proof, as usually where "Cls" or other type-symbols occur, it is necessary to abandon the notation by Greek letters and revert to the explicit notation.

***53·51.** $\vdash : \dot{\exists} ! R . \equiv . R \in \text{Rel} - \iota`\dot{\Lambda}$ [Proof as in *53·5]

***53·52.** $\vdash : \alpha \in \kappa . \exists ! \alpha . \equiv . \alpha \in \kappa - \iota`\Lambda$

Dem.

$\vdash . *24·54 . \supset \vdash : \alpha \in \kappa . \exists ! \alpha . \equiv . \alpha \in \kappa . \alpha \neq \Lambda .$

[*51·3] $\equiv . \alpha \in \kappa - \iota`\Lambda : \supset \vdash .$ Prop

***53·53.** $\vdash : R \in \lambda . \dot{\exists} ! R . \equiv . R \in \lambda - \iota`\dot{\Lambda}$ [Proof as in *53·52]

The following propositions are inserted because of their connection with the definition of $\alpha \rightarrow \beta$ in *70. $\overrightarrow{R}``\text{Ɑ}`R$ and $\overrightarrow{R}``\text{V}$ are both important classes.

***53·6.** $\vdash : R = \dot{\Lambda} . \exists ! \alpha . \supset . \overrightarrow{R}``\alpha = \iota`\Lambda . \overleftarrow{R}``\alpha = \iota`\Lambda$

Dem.

$\vdash . *33·15·241 . *24·13 . \supset \vdash : \text{Hp} . \supset . \overrightarrow{R}`x = \Lambda$ (1)

$\vdash . (1) . *37·7 . \supset \vdash : \text{Hp} . \supset . \overrightarrow{R}``\alpha = \hat{\beta}\{(\exists x) . x \in \alpha . \beta = \Lambda\}$

[*10·35] $= \hat{\beta}\{\exists ! \alpha . \beta = \Lambda\}$

[*4·73] $= \hat{\beta}(\beta = \Lambda)$

[*51·11] $= \iota`\Lambda$ (2)

Similarly $\vdash : \text{Hp} . \supset . \overleftarrow{R}``\alpha = \iota`\Lambda$ (3)

$\vdash . (2) . (3) . \supset \vdash .$ Prop

***53·601.** $\vdash : \exists ! \alpha . \alpha \cap \text{Ɑ}`R = \Lambda . \supset . \overrightarrow{R}``\alpha = \iota`\Lambda$

Dem.

$\vdash . *33·41 . \supset \vdash : \text{Hp} . x \in \alpha . \supset . \overrightarrow{R}`x = \Lambda$ (1)

$\vdash . (1) . *37·7 . \supset \vdash : \text{Hp} . \supset . \overrightarrow{R}``\alpha = \hat{\beta}\{(\exists x) . x \in \alpha . \beta = \Lambda\}$

[*10·35] $= \hat{\beta}\{\exists ! \alpha . \beta = \Lambda\}$

[*4·73.*51·11] $= \iota`\Lambda : \supset \vdash .$ Prop

***53·602.** $\vdash : \exists ! \alpha . \alpha \cap \text{D}`R = \Lambda . \supset . \overleftarrow{R}``\alpha = \iota`\Lambda$ [Proof as in *53·601]

***53·603.** $\vdash : \exists ! - \text{Ɑ}`R . \supset . \overrightarrow{R}``(- \text{Ɑ}`R) = \iota`\Lambda$ [*24·21 . *53·601]

***53·604.** $\vdash : \exists ! -\mathrm{D}`R . \supset . \overleftarrow{R}``(-\mathrm{D}`R) = \iota`\Lambda$ [*24·21 . *53·602]

***53·61.** $\vdash : \mathrm{G}`R \subset \alpha . \mathrm{G}`R \neq \alpha . \supset . \overrightarrow{R}``\alpha = \overrightarrow{R}``\mathrm{G}`R \cup \iota`\Lambda$

Dem.

$\vdash . *22·92 . \quad \supset \vdash : \mathrm{Hp} . \supset . \alpha = \mathrm{G}`R \cup (\alpha - \mathrm{G}`R)$ (1)

$\vdash . *24·6 . \quad \supset \vdash : \mathrm{Hp} . \supset . \exists ! \alpha - \mathrm{G}`R .$

[*24·21.*53·601] $\supset . \overrightarrow{R}``(\alpha - \mathrm{G}`R) = \iota`\Lambda$ (2)

$\vdash . (1) . *37·22 . \supset \vdash : \mathrm{Hp} . \supset . \overrightarrow{R}``\alpha = \overrightarrow{R}``\mathrm{G}`R \cup \overrightarrow{R}``(\alpha - \mathrm{G}`R)$

[(2)] $= \overrightarrow{R}``\mathrm{G}`R \cup \iota`\Lambda : \supset \vdash . \mathrm{Prop}$

***53·611.** $\vdash : \mathrm{D}`R \subset \alpha . \mathrm{D}`R \neq \alpha . \supset . \overleftarrow{R}``\alpha = \overleftarrow{R}``\mathrm{D}`R \cup \iota`\Lambda$ [Proof as in *53·61]

***53·612.** $\vdash : \mathrm{G}`R \neq \mathrm{V} . \supset . \overrightarrow{R}``\mathrm{V} = \overrightarrow{R}``\mathrm{G}`R \cup \iota`\Lambda$ [*53·61 . *24·11]

***53·613.** $\vdash : \mathrm{D}`R \neq \mathrm{V} . \supset . \overleftarrow{R}``\mathrm{V} = \overleftarrow{R}``\mathrm{D}`R \cup \iota`\Lambda$ [*53·611 . *24·11]

***53·614.** $\vdash . \overrightarrow{R}``\mathrm{G}`R = \overrightarrow{R}``\mathrm{V} - \iota`\Lambda$

Dem.

$\vdash . *53·612 . *22·68 . *24·21 . \supset$

$\vdash : \mathrm{G}`R \neq \mathrm{V} . \supset . \overrightarrow{R}``\mathrm{V} - \iota`\Lambda = \overrightarrow{R}``\mathrm{G}`R - \iota`\Lambda$ (1)

$\vdash . *22·481 . \supset \vdash : \mathrm{G}`R = \mathrm{V} . \supset . \overrightarrow{R}``\mathrm{V} - \iota`\Lambda = \overrightarrow{R}``\mathrm{G}`R - \iota`\Lambda$ (2)

$\vdash . *37·772 . *51·36 . *22·621 . \supset \vdash . \overrightarrow{R}``\mathrm{G}`R - \iota`\Lambda = \overrightarrow{R}``\mathrm{G}`R$ (3)

$\vdash . (1) . (2) . (3) . \supset \vdash . \mathrm{Prop}$

***53·615.** $\vdash . \overleftarrow{R}``\mathrm{D}`R = \overleftarrow{R}``\mathrm{V} - \iota`\Lambda$ [Proof as in *53·614]

The two following propositions are used in *70·12.

***53·62.** $\vdash : \overrightarrow{R}``\mathrm{G}`R \subset \gamma . \equiv . \overrightarrow{R}``\mathrm{V} \subset \gamma \cup \iota`\Lambda$

Dem.

$\vdash . *53·614 . \supset \vdash : \overrightarrow{R}``\mathrm{G}`R \subset \gamma . \equiv . \overrightarrow{R}``\mathrm{V} - \iota`\Lambda \subset \gamma .$

[*24·43] $\equiv . \overrightarrow{R}``\mathrm{V} \subset \gamma \cup \iota`\Lambda : \supset \vdash . \mathrm{Prop}$

***53·621.** $\vdash : \overleftarrow{R}``\mathrm{D}`R \subset \gamma . \equiv . \overleftarrow{R}``\mathrm{V} \subset \gamma \cup \iota`\Lambda$ [Proof as in *53·62]

***53·63.** $\vdash : \mathrm{G}`R \neq \mathrm{V} . \supset . \mathrm{D}`\overrightarrow{R} = \overrightarrow{R}``\mathrm{G}`R \cup \iota`\Lambda$ [*37·78 . *53·612]

***53·631.** $\vdash : \mathrm{D}`R \neq \mathrm{V} . \supset . \mathrm{D}`\overleftarrow{R} = \overleftarrow{R}``\mathrm{D}`R \cup \iota`\Lambda$ [*37·781 . *53·613]

***53·64.** $\vdash : \mathrm{G}`R = \mathrm{V} . \supset . \mathrm{D}`\overrightarrow{R} = \overrightarrow{R}``\mathrm{G}`R$ [*37·78]

***53·641.** $\vdash : \mathrm{D}`R = \mathrm{V} . \supset . \mathrm{D}`\overleftarrow{R} = \overleftarrow{R}``\mathrm{D}`R$ [*37·781]

*54. CARDINAL COUPLES

Summary of *54.

Couples are of two kinds, namely (1) $\iota'x \cup \iota'y$, in which there is no order as between x and y, and (2) $\iota'x \uparrow \iota'y$, in which there is an order. We may distinguish these two kinds of couples as cardinal and ordinal respectively, since (as will be shown hereafter) the class of all couples of the form $\iota'x \cup \iota'y$ (where $x \neq y$) is the cardinal number 2, while the class of all couples of the form $\iota'x \uparrow \iota'y$ (where $x \neq y$) is the ordinal number 2, to which, for the sake of distinction, we assign the symbol "2_r," where the suffix "r" stands for "relational," because the ordinal 2 is a class of relations. In the present and the following numbers, we shall define 2 and 2_r as the classes of cardinal and ordinal couples respectively, leaving it to a later stage to show that 2 and 2_r, so defined, are respectively a cardinal and an ordinal number. An ordinal couple will also be called an *ordered* couple or a *couple with sense*. Thus a couple with sense is a couple of which one comes first and the other second.

We introduce here the cardinal number 0, defined as $\iota'\Lambda$. That 0 so defined is a cardinal number, will be proved at a later stage; for the present, we postpone the proof that 0 so defined has the arithmetical properties of zero.

Cardinal couples are much less important, even in cardinal arithmetic, than ordinal couples, which will be considered in the two following numbers (*55 and *56). It is necessary, however, to prove some of the properties of cardinal couples, and this will be done in the present number. Some properties of cardinal couples which have been already proved are here repeated for convenience of reference. The definitions of 0 and 2 are:

***54·01.** $0 = \iota'\Lambda$ Df

***54·02.** $2 = \hat{\alpha}\{(\exists x, y) . x \neq y . \alpha = \iota'x \cup \iota'y\}$ Df

Most of the propositions of the present number, except those that merely embody the definitions (*54·1·101·102), are used very seldom. The following are among the most important.

***54·26.** $\vdash : \iota'x \cup \iota'y \in 2 . \equiv . x \neq y$

***54·3.** $\vdash . 2 = \hat{\alpha}\{(\exists x) . x \in \alpha . \alpha - \iota'x \in 1\}$

***54·4.** $\vdash :. \beta \subset \iota'x \cup \iota'y . \equiv : \beta = \Lambda . \vee . \beta = \iota'x . \vee . \beta = \iota'y . \vee . \beta = \iota'x \cup \iota'y$

***54·53.** $\vdash : \alpha \in 2 . x, y \in \alpha . x \neq y . \supset . \alpha = \iota'x \cup \iota'y$

***54·56.** $\vdash : \alpha \sim\in 0 \cup 1 \cup 2 . \equiv . (\exists x, y, z) . x, y, z \in \alpha . x \neq y . x \neq z . y \neq z$

***54·01.** $0 = \iota`\Lambda$ Df

***54·02.** $2 = \hat{\alpha}\{(\exists x, y) . x \neq y . \alpha = \iota`x \cup \iota`y\}$ Df

***54·1.** $\vdash . 0 = \iota`\Lambda$ [(*54·01)]

***54·101.** $\vdash : \alpha \in 2 . \equiv . (\exists x, y) . x \neq y . \alpha = \iota`x \cup \iota`y$ [(*54·02)]

***54·102.** $\vdash : \alpha \in 0 . \equiv . \alpha = \Lambda$ [*54·1]

The two following propositions have already occurred in *51, but are here repeated, because they belong to the subject of the present number.

***54·21.** $\vdash : \iota`x \cup \iota`y = \iota`x \cup \iota`z . \equiv . y = z$ [*51·41]

***54·22.** $\vdash :. \iota`x \cup \iota`y = \iota`z \cup \iota`w . \equiv : x = z . y = w . \vee . x = w . y = z$ [*51·43]

***54·25.** $\vdash : \iota`x \cup \iota`y \in 1 . \equiv . x = y$

Dem.

$\vdash . *52·46·1 . *22·58 . \supset \vdash : \iota`x \cup \iota`y \in 1 . \supset . \iota`x \cup \iota`y = \iota`x . \iota`x \cup \iota`y = \iota`y .$

[*20·23] $\supset . \iota`x = \iota`y$ (1)

$\vdash . *22·56 . \supset \vdash : \iota`x = \iota`y . \supset . \iota`x \cup \iota`y = \iota`x .$

[*52·22] $\supset . \iota`x \cup \iota`y \in 1$ (2)

$\vdash . (1) . (2) . \supset \vdash : \iota`x \cup \iota`y \in 1 . \equiv . \iota`x = \iota`y .$

[*51·23] $\equiv . x = y : \supset \vdash .$ Prop

***54·26.** $\vdash : \iota`x \cup \iota`y \in 2 . \equiv . x \neq y$

Dem.

$\vdash . *54·101 . \supset \vdash :: \iota`x \cup \iota`y \in 2 .$

$\equiv :. (\exists z, w) . z \neq w . \iota`x \cup \iota`y = \iota`z \cup \iota`w :.$

[*54·22] $\equiv :. (\exists z, w) : z \neq w : x = z . y = w . \vee . x = w . y = z :.$

[*4·4.*11·41] $\equiv :. (\exists z, w) . z \neq w . x = z . y = w . \vee . (\exists z, w) . z \neq w . x = w . y = z :.$

[*13·22] $\equiv :. x \neq y . \vee . y \neq x :.$

[*13·16] $\equiv :. x \neq y :: \supset \vdash .$ Prop

***54·27.** $\vdash . \iota`x \cup \iota`y \in 1 \cup 2$ [*54·25·26]

***54·271.** $\vdash . 1 \cup 2 = \hat{\alpha}\{(\exists x, y) . \alpha = \iota`x \cup \iota`y\}$

Dem.

$\vdash . *4·42 . \supset$

$\vdash :. \alpha = \iota`x \cup \iota`y . \equiv : x = y . \alpha = \iota`x \cup \iota`y . \vee . x \neq y . \alpha = \iota`x \cup \iota`y$ (1)

$\vdash . (1) . *11·11·341·41 . \supset \vdash :. (\exists x, y) . \alpha = \iota`x \cup \iota`y .$

$\equiv : (\exists x, y) . x = y . \alpha = \iota`x \cup \iota`y . \vee . (\exists x, y) . x \neq y . \alpha = \iota`x \cup \iota`y :$

[*13·195] $\equiv : (\exists x) . \alpha = \iota`x \cup \iota`x . \vee . (\exists x, y) . x \neq y . \alpha = \iota`x \cup \iota`y :$

[*22·56] $\equiv : (\exists x) . \alpha = \iota`x . \vee . (\exists x, y) . x \neq y . \alpha = \iota`x \cup \iota`y :$

[*52·1.*54·101] $\equiv : \alpha \in 1 . \vee . \alpha \in 2 :$

[*22·34] $\equiv : \alpha \in 1 \cup 2 :. \supset \vdash .$ Prop

***54·3.** $\vdash . 2 = \hat{\alpha}\{(\exists x) . x \epsilon \alpha . \alpha - \iota`x \epsilon 1\}$

Dem.

$\vdash . *52·1 . *10·35 . \supset$

$\vdash : (\exists x) . x \epsilon \alpha . \alpha - \iota`x \epsilon 1 . \equiv . (\exists x, y) . x \epsilon \alpha . \alpha - \iota`x = \iota`y .$

$\left[*51·22 \frac{\iota`y, \alpha}{\alpha, \beta}\right]$ $\equiv . (\exists x, y) . \iota`x \cap \iota`y = \Lambda . \iota`x \cup \iota`y = \alpha .$

[*51·231.*54·101] $\equiv . \alpha \epsilon 2 : \supset \vdash . \text{Prop}$

***54·4.** $\vdash :. \beta \subset \iota`x \cup \iota`y . \equiv : \beta = \Lambda . \vee . \beta = \iota`x . \vee . \beta = \iota`y . \vee . \beta = \iota`x \cup \iota`y$

Dem.

$\vdash . *51·2 . \supset \vdash : x, y \epsilon \beta . \supset . \iota`x \cup \iota`y \subset \beta :$

[Fact] $\supset \vdash : \beta \subset \iota`x \cup \iota`y . x, y \epsilon \beta . \supset . \beta \subset \iota`x \cup \iota`y . \iota`x \cup \iota`y \subset \beta .$

[*22·41] $\supset . \beta = \iota`x \cup \iota`y$ (1)

$\vdash . *51·25 . \supset \vdash :. \beta \subset \iota`x \cup \iota`y . y \sim\epsilon \beta . \supset : \beta \subset \iota`x :$

[*51·401] $\supset : \beta = \Lambda . \vee . \beta = \iota`x$ (2)

Similarly $\vdash :. \beta \subset \iota`x \cup \iota`y . x \sim\epsilon \beta . \supset : \beta = \Lambda . \vee . \beta = \iota`y$ (3)

$\vdash . (2) . (3) . *3·48 . \supset$

$\vdash :. \beta \subset \iota`x \cup \iota`y . \sim(x, y \epsilon \beta) . \supset : \beta = \Lambda . \vee . \beta = \iota`x . \vee . \beta = \iota`y$ (4)

$\vdash . (1) . (4) . *3·48 . \supset$

$\vdash :. \beta \subset \iota`x \cup \iota`y . \supset : \beta = \Lambda . \vee . \beta = \iota`x . \vee . \beta = \iota`y . \vee . \beta = \iota`x \cup \iota`y$ (5)

$\vdash . *24·12 . *22·58·42 . \supset$

$\vdash :. \beta = \Lambda . \vee . \beta = \iota`x . \vee . \beta = \iota`y . \vee . \beta = \iota`x \cup \iota`y : \supset . \beta \subset \iota`x \cup \iota`y$ (6)

$\vdash . (5) . (6) . \supset \vdash . \text{Prop}$

This proposition shows that a class contained in a couple is either the null-class or a unit class or the couple itself, whence it will follow that 0 and 1 are the only numbers which are less than 2.

***54·41.** $\vdash :: \alpha \epsilon 2 . \supset :. \beta \subset \alpha . \supset : \beta = \Lambda . \vee . \beta \epsilon 1 . \vee . \beta \epsilon 2$

Dem.

$\vdash . *52·1 . \supset \vdash :. \beta = \iota`x . \vee . \beta = \iota`y : \supset . \beta \epsilon 1$ (1)

$\vdash . *54·26 . \supset \vdash :. x \neq y . \supset : \beta = \iota`x \cup \iota`y . \supset . \beta \epsilon 2$ (2)

$\vdash . (1) . (2) . *54·4 . \supset$

$\vdash :: x \neq y . \supset :. \beta \subset \iota`x \cup \iota`y . \supset : \beta = \Lambda . \vee . \beta \epsilon 1 . \vee . \beta \epsilon 2 ::$

[*13·12] $\supset \vdash :: \alpha = \iota`x \cup \iota`y . x \neq y . \supset :. \beta \subset \alpha . \supset : \beta = \Lambda . \vee . \beta \epsilon 1 . \vee . \beta \epsilon 2 ::$

[11·11·35] $\supset$

$\vdash :. (\exists x, y) . \alpha = \iota`x \cup \iota`y . x \neq y . \supset :. \beta \subset \alpha : \beta = \Lambda . \vee . \beta \epsilon 1 . \vee . \beta \epsilon 2$ (3)

$\vdash . (3) . *54·101 . \supset \vdash . \text{Prop}$

***54·411.** $\vdash :. \alpha \epsilon 2 . \supset : \beta \subset \alpha . \supset . \beta \epsilon 0 \cup 1 \cup 2$ [*54·41·102]

***54·42.** $\vdash :: \alpha \in 2 . \supset :. \beta \subset \alpha . \exists ! \beta . \beta \neq \alpha . \equiv . \beta \in \iota``\alpha$

Dem.

$\vdash . *54·4 . \supset \vdash :: \alpha = \iota`x \cup \iota`y . \supset :.$

$\beta \subset \alpha . \exists ! \beta . \equiv : \beta = \Lambda . \vee . \beta = \iota`x . \vee . \beta = \iota`y . \vee . \beta = \alpha : \exists ! \beta :$

[*24·53·56.*51·161] $\equiv : \beta = \iota`x . \vee . \beta = \iota`y . \vee . \beta = \alpha$ (1)

$\vdash$. *54·25 . Transp . *52·22 . $\supset \vdash : x \neq y . \supset . \iota`x \cup \iota`y \neq \iota`x . \iota`x \cup \iota`y \neq \iota`y :$

[*13·12] $\supset \vdash : \alpha = \iota`x \cup \iota`y . x \neq y . \supset . \alpha \neq \iota`x . \alpha \neq \iota`y$ (2)

$\vdash . (1) . (2) . \supset \vdash :: \alpha = \iota`x \cup \iota`y . x \neq y . \supset :.$

$\beta \subset \alpha . \exists ! \beta . \beta \neq \alpha . \equiv : \beta = \iota`x . \vee . \beta = \iota`y :$

[*51·235] $\equiv : (\exists z) . z \in \alpha . \beta = \iota`z :$

[*37·6] $\equiv : \beta \in \iota``\alpha$ (3)

$\vdash$. (3) . *11·11·35 . *54·101 . $\supset \vdash$. Prop

***54·43.** $\vdash :. \alpha, \beta \in 1 . \supset : \alpha \cap \beta = \Lambda . \equiv . \alpha \cup \beta \in 2$

Dem.

$\vdash . *54·26 . \supset \vdash :. \alpha = \iota`x . \beta = \iota`y . \supset : \alpha \cup \beta \in 2 . \equiv . x \neq y .$

[*51·231] $\equiv . \iota`x \cap \iota`y = \Lambda .$

[*13·12] $\equiv . \alpha \cap \beta = \Lambda$ (1)

$\vdash . (1) . *11·11·35 . \supset$

$\vdash :. (\exists x, y) . \alpha = \iota`x . \beta = \iota`y . \supset : \alpha \cup \beta \in 2 . \equiv . \alpha \cap \beta = \Lambda$ (2)

$\vdash$. (2) . *11·54 . *52·1 . $\supset \vdash$. Prop

From this proposition it will follow, when arithmetical addition has been defined, that $1 + 1 = 2$.

***54·44.** $\vdash :. z, w \in \iota`x \cup \iota`y . \supset_{z,w} . \phi(z, w) : \equiv . \phi(x, x) . \phi(x, y) . \phi(y, x) . \phi(y, y)$

Dem.

$\vdash . *51·234 . *11·62 . \supset \vdash :. z, w \in \iota`x \cup \iota`y . \supset_{z,w} . \phi(z, w) : \equiv :$

$z \in \iota`x \cup \iota`y . \supset_z . \phi(z, x) . \phi(z, y) :$

[*51·234.*10·29] $\equiv : \phi(x, x) . \phi(x, y) . \phi(y, x) . \phi(y, y) :. \supset \vdash$. Prop

***54·441.** $\vdash :: z, w \in \iota`x \cup \iota`y . z \neq w . \supset_{z,w} . \phi(z, w) : \equiv :. x = y : \vee : \phi(x, y) . \phi(y, x)$

Dem.

$\vdash . *5·6 . \supset \vdash :: z, w \in \iota`x \cup \iota`y . z \neq w . \supset_{z,w} . \phi(z, w) : \equiv :.$

$z, w \in \iota`x \cup \iota`y . \supset_{z,w} : z = w . \vee . \phi(z, w) :.$

[*54·44] $\equiv : x = x . \vee . \phi(x, x) : x = y . \vee . \phi(x, y) :$

$y = x . \vee . \phi(y, x) : y = y . \vee . \phi(y, y) :$

[*13·15] $\equiv : x = y . \vee . \phi(x, y) : y = x . \vee . \phi(y, x) :$

[*13·16.*4·41] $\equiv : x = y . \vee . \phi(x, y) . \phi(y, x)$

This proposition is used in *163·42, in the theory of relations of mutually exclusive relations.

***54·442.** $\vdash :: x \neq y . \supset :. z, w \in \iota`x \cup \iota`y . z \neq w . \supset_{z,w} . \phi(z, w) : \equiv . \phi(x, y) . \phi(y, x)$ [*54·441]

***54·443.** $\vdash :: x \neq y : \phi(x,y) . \equiv . \phi(y,x) : \supset :.$
$z, w \in \iota^{\prime}x \cup \iota^{\prime}y . z \neq w . \supset_{z,w} . \phi(z,w) : \equiv . \phi(x,y)$ [*54·442]

***54·45.** $\vdash :. (\exists z, w) . z, w \in \iota^{\prime}x \cup \iota^{\prime}y . \phi(z,w) .$
$\equiv : \phi(x,x) . \vee . \phi(x,y) . \vee . \phi(y,x) . \vee . \phi(y,y)$ [*51·235]

***54·451.** $\vdash :: \sim\phi(x,x) . \sim\phi(y,y) . \supset :. (\exists z, w) . z, w \in \iota^{\prime}x \cup \iota^{\prime}y . \phi(z,w) .$
$\equiv : \phi(x,y) . \vee . \phi(y,x)$ [*54·45]

***54·452.** $\vdash :: \sim\phi(x,x) . \sim\phi(y,y) : \phi(x,y) . \equiv . \phi(y,x) : \supset :$
$(\exists z, w) . z, w \in \iota^{\prime}x \cup \iota^{\prime}y . \phi(z,w) . \equiv . \phi(x,y)$ [*54·451]

***54·46.** $\vdash : (\exists z, w) . z, w \in \iota^{\prime}x \cup \iota^{\prime}y . z \neq w . \equiv . x \neq y$ [*54·452 . *13·15·16]

***54·5.** $\vdash :. \alpha \in 2 . \supset : \alpha \subset \iota^{\prime}z \cup \iota^{\prime}w . \equiv . \alpha = \iota^{\prime}z \cup \iota^{\prime}w$

Dem.

$\vdash . *54 \cdot 4 . \supset$

$\vdash :. \alpha \subset \iota^{\prime}z \cup \iota^{\prime}w . \supset : \alpha = \Lambda . \vee . \alpha = \iota^{\prime}z . \vee . \alpha = \iota^{\prime}w . \vee . \alpha = \iota^{\prime}z \cup \iota^{\prime}w$ (1)

$\vdash . *54 \cdot 3 . *24 \cdot 54 . \quad \supset \vdash : \mathrm{Hp} . \supset . \alpha \neq \Lambda$ (2)

$\vdash . *54 \cdot 26 \frac{z, z}{x, y} . *13 \cdot 15 . \quad \supset \vdash : \mathrm{Hp} . \supset . \alpha \neq \iota^{\prime}z$ (3)

$\vdash . (3) \frac{w}{z} . \quad \supset \vdash : \mathrm{Hp} . \supset . \alpha \neq \iota^{\prime}w$ (4)

$\vdash . (1) . (2) . (3) . (4) . *2 \cdot 53 . \supset \vdash :. \mathrm{Hp} . \supset : \alpha \subset \iota^{\prime}z \cup \iota^{\prime}w . \supset . \alpha = \iota^{\prime}z \cup \iota^{\prime}w$ (5)

$\vdash . *22 \cdot 42 . \quad \supset \vdash : \alpha = \iota^{\prime}z \cup \iota^{\prime}w . \supset . \alpha \subset \iota^{\prime}z \cup \iota^{\prime}w$ (6)

$\vdash . (5) . (6) . \supset \vdash . \mathrm{Prop}$

***54·51.** $\vdash :. \alpha \in 2 . \beta \in 1 \cup 2 . \supset : \alpha \subset \beta . \equiv . \alpha = \beta$

Dem.

$\vdash . *54 \cdot 5 . \supset \vdash :. \alpha \in 2 . \beta = \iota^{\prime}z \cup \iota^{\prime}w . \supset : \alpha \subset \beta . \equiv . \alpha = \beta$ (1)

$\vdash . (1) . *11 \cdot 11 \cdot 35 \cdot 45 . \supset$

$\vdash :. \alpha \in 2 : (\exists z, w) . \beta = \iota^{\prime}z \cup \iota^{\prime}w : \supset : \alpha \subset \beta . \equiv . \alpha = \beta$ (2)

$\vdash . (2) . *54 \cdot 271 . \supset \vdash . \mathrm{Prop}$

***54·52.** $\vdash :. \alpha, \beta \in 2 . \supset : \alpha \subset \beta . \equiv . \alpha = \beta . \equiv . \beta \subset \alpha$ [*54·51]

***54·53.** $\vdash : \alpha \in 2 . x, y \in \alpha . x \neq y . \supset . \alpha = \iota^{\prime}x \cup \iota^{\prime}y$

Dem.

$\vdash . *51 \cdot 2 . \quad \supset \vdash : \mathrm{Hp} . \supset . \iota^{\prime}x \subset \alpha . \iota^{\prime}y \subset \alpha .$

[*22·59] $\quad \supset . \iota^{\prime}x \cup \iota^{\prime}y \subset \alpha$ (1)

$\vdash . *54 \cdot 26 . \quad \supset \vdash : \mathrm{Hp} . \supset . \iota^{\prime}x \cup \iota^{\prime}y \in 2$ (2)

$\vdash . (1) . (2) . *54 \cdot 52 . \supset \vdash . \mathrm{Prop}$

***54·531.** $\vdash :. \alpha \in 2 . \supset : x, y \in \alpha . x \neq y . \equiv . \alpha = \iota^{\prime}x \cup \iota^{\prime}y$

Dem.

$\vdash . *54 \cdot 53 . \mathrm{Exp} . \supset \vdash :. \alpha \in 2 . \supset : x, y \in \alpha . x \neq y . \supset . \alpha = \iota^{\prime}x \cup \iota^{\prime}y$ (1)

$\vdash . *54 \cdot 26 . \quad \supset \vdash :. \alpha \in 2 . \supset : \alpha = \iota^{\prime}x \cup \iota^{\prime}y . \supset . x \neq y$ (2)

$\vdash . *51 \cdot 16 . \quad \supset \vdash : \alpha = \iota^{\prime}x \cup \iota^{\prime}y . \supset . x, y \in \alpha$ (3)

$\vdash . (2) . (3) . \quad \supset \vdash :. \alpha \in 2 . \supset : \alpha = \iota^{\prime}x \cup \iota^{\prime}y . \supset . x, y \in \alpha . x \neq y$ (4)

$\vdash . (1) . (4) . \quad \supset \vdash . \mathrm{Prop}$

***54·54.** $\vdash :. \alpha \epsilon 2 . \equiv : x, y \epsilon \alpha . x \neq y . \supset_{x,y} . \alpha = \iota'x \cup \iota'y : (\exists x, y) . x, y \epsilon \alpha . x \neq y$

Dem.

$\vdash . *54\cdot531 . *11\cdot11\cdot3 . \supset \vdash :. \alpha \epsilon 2 . \supset : x, y \epsilon \alpha . x \neq y . \supset_{x,y} . \alpha = \iota'x \cup \iota'y$ (1)

$\vdash . *51\cdot16 . *54\cdot101 . \supset \vdash : \alpha \epsilon 2 . \supset . (\exists x, y) . x, y \epsilon \alpha . x \neq y$ (2)

$\vdash . *5\cdot3 . *3\cdot27 . \supset \vdash :. x, y \epsilon \alpha . x \neq y . \supset . \alpha = \iota'x \cup \iota'y : \supset :$
$x, y \epsilon \alpha . x \neq y . \supset . x \neq y . \alpha = \iota'x \cup \iota'y :.$

$[*11\cdot11\cdot32\cdot34] \supset \vdash :. x, y \epsilon \alpha . x \neq y . \supset_{x,y} . \alpha = \iota'x \cup \iota'y : \supset :$
$(\exists x, y) . x, y \epsilon \alpha . x \neq y . \supset . (\exists x, y) . x \neq y . \alpha = \iota'x \cup \iota'y$ (3)

$\vdash . (3) . \text{Imp} . *54\cdot101 . \supset \vdash :. x, y \epsilon \alpha . x \neq y . \supset_{x,y} . \alpha = \iota'x \cup \iota'y :$
$(\exists x, y) . x, y \epsilon \alpha . x \neq y : \supset . \alpha \epsilon 2$ (4)

$\vdash . (1) . (2) . (4) . \supset \vdash . \text{Prop}$

In the above proposition, "$x, y \epsilon \alpha . x \neq y . \supset_{x,y} . \alpha = \iota'x \cup \iota'y$" secures that α has not *more* than two members, while "$(\exists x, y) . x, y \epsilon \alpha . x \neq y$" secures that α has not *fewer* than two members.

***54·55.** $\vdash . 0 \cup 1 \cup 2 = \hat{\alpha} \{x, y \epsilon \alpha . x \neq y . \supset_{x,y} . \alpha = \iota'x \cup \iota'y\}$

Dem.

$\vdash . *4\cdot42 . \supset \vdash :: x, y \epsilon \alpha . x \neq y . \supset_{x,y} . \alpha = \iota'x \cup \iota'y : \equiv :.$
$x, y \epsilon \alpha . x \neq y . \supset_{x,y} . \alpha = \iota'x \cup \iota'y : \sim (\exists x, y) . x, y \epsilon \alpha . x \neq y :.$
$\vee :. x, y \epsilon \alpha . x \neq y . \supset_{x,y} . \alpha = \iota'x \cup \iota'y : (\exists x, y) . x, y \epsilon \alpha . x \neq y$ (1)

$\vdash . *11\cdot63 . \supset \vdash :. \sim (\exists x, y) . x, y \epsilon \alpha . x \neq y . \supset : x, y \epsilon \alpha . x \neq y . \supset_{x,y} . \alpha = \iota'x \cup \iota'y :.$

$[*4\cdot71] \supset \vdash :. x, y \epsilon \alpha . x \neq y . \supset_{x,y} . \alpha = \iota'x \cup \iota'y : \sim (\exists x, y) . x, y \epsilon \alpha . x \neq y : \equiv :$
$\sim (\exists x, y) . x, y \epsilon \alpha . x \neq y :$

$[*11\cdot521] \equiv : x, y \epsilon \alpha . \supset_{x,y} . x = y :$

$[*52\cdot4] \equiv : \alpha \epsilon 0 \cup 1$ (2)

$\vdash . (1) . (2) . *54\cdot54 . \supset$

$\vdash :. x, y \epsilon \alpha . x \neq y . \supset_{x,y} . \alpha = \iota'x \cup \iota'y : \equiv : \alpha \epsilon 0 \cup 1 . \vee . \alpha \epsilon 2 :$

$[*22\cdot34] \equiv : \alpha \epsilon 0 \cup 1 \cup 2 :. \supset \vdash . \text{Prop}$

***54·56.** $\vdash : \alpha \sim \epsilon 0 \cup 1 \cup 2 . \equiv . (\exists x, y, z) . x, y, z \epsilon \alpha . x \neq y . x \neq z . y \neq z$

Dem.

$\vdash . *54\cdot55 . *11\cdot52 . \supset$

$\vdash :. \alpha \sim \epsilon 0 \cup 1 \cup 2 . \equiv : (\exists x, y) . x, y \epsilon \alpha . x \neq y . \alpha \neq \iota'x \cup \iota'y :$

$[*51\cdot2 . *22\cdot59] \equiv : (\exists x, y) . \iota'x \cup \iota'y \subset \alpha . x \neq y . \alpha \neq \iota'x \cup \iota'y :$

$[*24\cdot6] \equiv : (\exists x, y) . \iota'x \cup \iota'y \subset \alpha . x \neq y . \exists ! \alpha - (\iota'x \cup \iota'y) :$

$[*51\cdot232 . \text{Transp}] \equiv : (\exists x, y) : \iota'x \cup \iota'y \subset \alpha . x \neq y : (\exists z) . z \epsilon \alpha . z \neq x . z \neq y :$

$[*51\cdot2 . *22\cdot59] \equiv : (\exists x, y, z) . x, y, z \epsilon \alpha . x \neq y . x \neq z . y \neq z :. \supset \vdash . \text{Prop}$

In virtue of this proposition, a class which is neither null nor a unit class nor a couple contains at least three distinct members. Hence it will follow that any cardinal number other than 0 or 1 or 2 is equal to or greater than 3. The above proposition is used in *104·43, which is an existence-theorem of considerable importance in cardinal arithmetic.

***54·6.** $\vdash :. \alpha \cap \beta = \Lambda . x, x' \in \alpha . y, y' \in \beta . \supset :$

$\iota`x \cup \iota`y = \iota`x' \cup \iota`y' . \equiv . x = x' . y = y'$

Dem.

$\vdash . *51\cdot2 . \supset \vdash :. \mathrm{Hp} . \supset : \iota`x \subset \alpha . \iota`x' \subset \alpha . \iota`y \subset \beta . \iota`y' \subset \beta . \alpha \cap \beta = \Lambda :$

[*24·48] $\supset :. \iota`x \cup \iota`y = \iota`x' \cup \iota`y' . \equiv . \iota`x = \iota`x' . \iota`y = \iota`y' .$

[*51·23] $\equiv . x = x' . y = y' :. \supset \vdash . \mathrm{Prop}$

The above proposition is useful in dealing with sets of couples formed of one member of a class α and one member of a class β, where α and β have no members in common. It is used in the theory of cardinal multiplication (*113·148).

*55. ORDINAL COUPLES

Summary of *55.

Ordinal couples, which are now to be considered, are much more important, even in cardinal arithmetic, than cardinal couples. Their properties are in part analogous to those of cardinal couples, but in part also to those of unit classes; for they are the smallest existent relations, just as unit classes are the smallest existent classes. The properties which are analogous to those of unit classes do not demand that the two terms of the couple should be distinct, *i.e.* they hold for $\iota`x \uparrow \iota`x$ as well as for $\iota`x \uparrow \iota`y$ (where $x \neq y$); on the other hand, the properties which are analogous to those of cardinal couples do in general demand that the two terms of the ordinal couple should be distinct.

The notation $\iota`x \uparrow \iota`y$ is cumbrous, and does not readily enable us to exhibit the couple as a descriptive function of x for the argument y, or vice versa. We therefore introduce a new symbol, "$x \downarrow y$," for the couple. In a couple $x \downarrow y$, we shall call x the referent of the couple, and y the relatum. In virtue of the definitions in *38, this gives rise to two relations $x \downarrow$ and $\downarrow y$; hence we obtain the notations $x \downarrow ``\beta$, $\downarrow y``\alpha$, $\alpha \underset{\prime\prime}{\downarrow} y$, $\alpha \underset{\prime\prime}{\downarrow} ``\beta$ and so on, which will be much used in the sequel. It should be observed that $x \downarrow ``\beta$ means the class of ordinal couples in which x is referent and a member of β is relatum, while $\downarrow y``\alpha$ or $\alpha \underset{\prime\prime}{\downarrow} y$ denotes the class of couples having $y$ as relatum and a member of $\alpha$ as referent; $\alpha \underset{\prime\prime}{\downarrow} ``\beta$ denotes all such classes of couples as $\downarrow y``\alpha$, where $y$ is any member of $\beta$; and in virtue of *40·7, $s`\alpha \underset{\prime\prime}{\downarrow} ``\beta$ denotes all ordinal couples of which the referent is a member of α, while the relatum is a member of β. This is a very important class, which will be used to define the product of two cardinal numbers; for it is evident that the number of members of $s`\alpha \underset{\prime\prime}{\downarrow} ``\beta$ is the product of the number of members of α and the number of members of β.

The first few propositions of the present number are immediate consequences of the definition of $x \downarrow y$ and the notations introduced in *38. We then proceed to various elementary properties of the relation $x \downarrow y$, of which the most used are the following:

***55·13.** $\vdash : z\,(x \downarrow y)\,w \,.\equiv .\, z = x \,.\, w = y$

***55·15.** $\vdash .\, \mathrm{D}`(x \downarrow y) = \iota`x \,.\, \mathrm{Ɑ}`(x \downarrow y) = \iota`y \,.\, C`(x \downarrow y) = \iota`x \cup \iota`y$

***55·16.** $\vdash : \mathrm{D}`R = \iota`x \,.\, \mathrm{Ɑ}`R = \iota`y \,.\equiv .\, R = x \downarrow y$

***55·202.** $\vdash : x \downarrow y = z \downarrow w \,.\equiv .\, x = z \,.\, y = w \,.\equiv .\, y \downarrow x = w \downarrow z$

This proposition should be contrasted with *54·22, as giving one reason why ordinal couples are more useful in arithmetic than cardinal couples. In virtue of the above proposition, when two ordinal couples are identical, their referents are identical, and their relata are identical.

We proceed next to various properties of the relations $x \downarrow$ and $\downarrow x$. These relations play a great part in arithmetic. It will be observed that if two terms have the relation $x \downarrow$, the referent is a couple whose relatum is the relatum in the relation $x \downarrow$, *i.e.* when we have $R (x \downarrow) y$, we have $R = x \downarrow y$ (cf. *55·122). Similar remarks apply to the relation $\downarrow x$. The class $\downarrow x``\alpha$, consisting of all couples whose referent is a member of α, while the relatum is x, is important. We have

***55·232.** $\vdash : \exists ! \downarrow x``\alpha \cap \downarrow y``\beta \,.\equiv .\, x = y \,.\, \exists ! \alpha \cap \beta$

This proposition is frequently useful.

We proceed next (*55·3—·51) to give various properties of $x \downarrow y$ which are analogous to the properties of unit classes. Among the more important of these properties are the following:

***55·3.** $\vdash : xRy \,.\equiv .\, x \downarrow y \mathrel{\dot\subset} R \,.\equiv .\, \dot\exists ! (x \downarrow y) \mathrel{\dot\cap} R$

This is the analogue of *51·31.

***55·34.** $\vdash : \dot\exists ! R \,.\, R \mathrel{\dot\subset} x \downarrow y \,.\equiv .\, R = x \downarrow y$

This is the analogue of *51·4.

***55·5.** $\vdash :. R \mathrel{\dot\subset} x \downarrow y \mathrel{\dot\cup} z \downarrow w \,.\equiv :$

$$R = \dot\Lambda \,.\vee .\, R = x \downarrow y \,.\vee .\, R = z \downarrow w \,.\vee .\, R = x \downarrow y \mathrel{\dot\cup} z \downarrow w$$

This is the analogue of *54·4.

We then proceed to such properties of ordinal couples as are not analogous to those of unit classes. For connecting the cardinal number 2 with the ordinal number 2_r, we have the proposition

***55·54.** $\vdash :: x \neq y \,.\supset :.\, C`R = \iota`x \cup \iota`y \,.\, R \mathrel{\dot\cap} \breve{R} = \dot\Lambda \,.\equiv : R = x \downarrow y \,.\vee .\, R = y \downarrow x$

This proposition shows that the only asymmetrical relations which have a given cardinal couple $\iota`x \cup \iota`y$ for their field are the two corresponding ordinal couples $x \downarrow y$ and $y \downarrow x$. We have next a set of propositions on the relative products of couples and other relations, *i.e.* on $R \,|\, (x \downarrow y)$, $(x \downarrow y) \,|\, S$, and $R \,|\, (x \downarrow y) \,|\, S$. These propositions are very useful in arithmetic. The chief of them is

***55·61.** $\vdash : E ! R`z \,.\, E ! S`w \,.\supset .\, (R \parallel \breve{S})`(z \downarrow w) = (R`z) \downarrow (S`w)$

Finally we have four propositions which belong, by their subject, to *43, but could not be given there, because the proofs make use of ordinal couples.

***55·01.** $x \downarrow y = \iota{}'x \uparrow \iota{}'y$ Df

***55·02.** $R'x \downarrow y = R'(x \downarrow y)$ Df

This definition serves merely for the avoidance of brackets.

***55·1.** $\vdash . x \downarrow y = (\iota{}'x) \uparrow (\iota{}'y)$ [(*55·01)]

***55·11.** $\vdash . x \downarrow{}'y = \downarrow y'x = x \downarrow y = \iota{}'x \uparrow \iota{}'y$ [*38·11 . *55·1]

***55·12.** $\vdash . \text{E}!\, x \downarrow{}'y$ [*55·11 . *14·21]

***55·121.** $\vdash . \text{E}!\, \downarrow y'x$

***55·122.** $\vdash : R(x\downarrow)y . \equiv . R = x \downarrow y$ [*55·11]

***55·123.** $\vdash : R(\downarrow y)x . \equiv . R = x \downarrow y$ [*55·11]

***55·13.** $\vdash : z(x \downarrow y)w . \equiv . z = x . w = y$

Dem.

$\vdash . *35·103 . *55·1 . \supset \vdash : z(x\downarrow y)w . \equiv . z \in \iota{}'x . w \in \iota{}'y .$

$[*51·15] \qquad \equiv . z = x . w = y : \supset \vdash . \text{Prop}$

***55·132.** $\vdash . x(x\downarrow y)y$ [*55·13]

***55·134.** $\vdash . \dot{\exists}!\,(x \downarrow y)$ [*55·132]

***55·14.** $\vdash . x \downarrow y = \text{Cnv}'y \downarrow x$ [*55·13 . *31·131]

***55·15.** $\vdash . \text{D}'x\downarrow y = \iota{}'x . \text{Ɑ}'x \downarrow y = \iota{}'y . C'x\downarrow y = \iota{}'x \cup \iota{}'y$
[*35·85·86 . *51·161]

***55·16.** $\vdash : \text{D}'R = \iota{}'x . \text{Ɑ}'R = \iota{}'y . \equiv . R = x \downarrow y$

Dem.

$\vdash . *33·13·131 . *51·15 . \supset$

$\vdash :: \text{D}'R = \iota{}'x . \text{Ɑ}'R = \iota{}'y . \equiv :. (\exists w) . zRw . \equiv_z . z = x : (\exists z) . zRw . \equiv_w . w = y :.$

$[*14·122] \qquad \equiv :. (\exists z, w) . zRw : (\exists w) . zRw . \supset_z . z = x :$

$\qquad\qquad (\exists w, z) . zRw : (\exists z) . zRw . \supset_w . w = y :.$

$[*11·23 . *4·71] \equiv :. (\exists z, w) . zRw : (\exists w) . zRw . \supset_z . z = x : (\exists z) . zRw . \supset_w . w = y :.$

$[*10·23] \qquad \equiv :. (\exists z, w) . zRw : zRw . \supset_{z,w} . z = x : zRw . \supset_{z,w} . w = y :.$

$[*11·391] \qquad \equiv :. (\exists z, w) . zRw : zRw . \supset_{z,w} . z = x . w = y :.$

$[*14·123] \qquad \equiv :. zRw . \equiv_{z,w} . z = x . w = y :.$

$[*55·13] \qquad \equiv :. zRw . \equiv_{z,w} . z(x\downarrow y)w :.$

$[*21·43] \qquad \equiv :. R = x \downarrow y :: \supset \vdash . \text{Prop}$

The above proposition is important, and will be frequently used.

***55·161.** $\vdash . x \downarrow y = \breve{\iota}{}'\hat{R}(\text{D}'R = \iota{}'x . \text{Ɑ}'R = \iota{}'y)$

Dem.

$\vdash . *55·16 . *20·15 . \supset$

$\vdash . \hat{R}(\text{D}'R = \iota{}'x . \text{Ɑ}'R = \iota{}'y) = \hat{R}(R = x \downarrow y)$

$[*51·11] \qquad = \iota{}'(x \downarrow y) \qquad (1)$

$\vdash . (1) . *51·51 . \supset \vdash . \text{Prop}$

***55·17.** $\vdash . x \downarrow y = \breve{\iota}`(\overleftarrow{\text{D}}`\iota`x \cap \overleftarrow{\text{G}}`\iota`y)$ [*55·161 . *33·6·61]

***55·2.** $\vdash : x \downarrow y = x \downarrow z . \equiv . y = z$

Dem.

$\vdash . *30·37 . *55·11·12 . \supset \vdash : y = z . \supset . x \downarrow y = x \downarrow z$ (1)

$\vdash . *30·37 . *33·121 . \supset$

$\vdash : x \downarrow y = x \downarrow z . \supset . \text{G}`x \downarrow y = \text{G}`x \downarrow z .$

[*55·15] $\supset . \iota`y = \iota`z .$

[*51·23] $\supset . y = z$ (2)

$\vdash . (1) . (2) . \supset \vdash .$ Prop

***55·201.** $\vdash : x \downarrow z = y \downarrow z . \equiv . x = y$

***55·202.** $\vdash : x \downarrow y = z \downarrow w . \equiv . x = z . y = w . \equiv . y \downarrow x = w \downarrow z$

Dem.

$\vdash . *55·2·201 . \supset$

$\vdash : x = z . y = w . \supset . x \downarrow y = z \downarrow y . z \downarrow y = z \downarrow w .$

[*13·17] $\supset . x \downarrow y = z \downarrow w$ (1)

$\vdash . *30·37 . *33·12·121 . \supset$

$\vdash : x \downarrow y = z \downarrow w . \supset . \text{D}`x \downarrow y = \text{D}`z \downarrow w . \text{G}`x \downarrow y = \text{G}`z \downarrow w .$

[*55·15] $\supset . \iota`x = \iota`z . \iota`y = \iota`w .$

[*51·23] $\supset . x = z . y = w$ (2)

$\vdash . (1) . (2) . \supset$

$\vdash : x \downarrow y = z \downarrow w . \equiv . x = z . y = w$ (3)

Similarly

$\vdash : y \downarrow x = w \downarrow z . \equiv . x = z . y = w$ (4)

$\vdash . (3) . (4) . \supset \vdash .$ Prop

The above proposition is important.

***55·21.** $\vdash . \text{G}`x \downarrow = \text{V} . \text{G}` \downarrow x = \text{V}$ [*33·432 . *55·12·121]

***55·22.** $\vdash . \text{D}`x \downarrow = \hat{R} \{(\exists y) . R = x \downarrow y\}$ [*55·122]

***55·221.** $\vdash . \text{D}` \downarrow x = \hat{R} \{(\exists y) . R = y \downarrow x\}$ [*55·123]

***55·222.** $\vdash : R \in \text{D}`x \downarrow . \equiv . \text{D}`R = \iota`x . \text{G}`R \in 1$

Dem.

$\vdash . *55·22·16 . \supset \vdash :. R \in \text{D}`x \downarrow . \equiv : (\exists y) . \text{D}`R = \iota`x . \text{G}`R = \iota`y :$

[*10·35] $\equiv : \text{D}`R = \iota`x : (\exists y) . \text{G}`R = \iota`y :$

[*52·1] $\equiv : \text{D}`R = \iota`x . \text{G}`R \in 1 :. \supset \vdash .$ Prop

***55·223.** $\vdash : R \in \text{D}` \downarrow x . \equiv . \text{G}`R = \iota`x . \text{D}`R \in 1$ [Proof as in *55·222]

***55·224.** $\vdash . \text{D}`x \downarrow \cap \text{D}` \downarrow y = \iota`(x \downarrow y)$

Dem.

$\vdash . *55·222·223 . \supset$

$\vdash : R \in \text{D}`x \downarrow \cap \text{D}` \downarrow y . \equiv . \text{D}`R = \iota`x . \text{G}`R \in 1 . \text{G}`R = \iota`y . \text{D}`R \in 1 .$

[*52·22.*4·71] $\equiv . \text{D}'R = \iota'x . \text{Ꭰ}'R = \iota'y .$

[*55·16] $\equiv . R = x \downarrow y .$

[*51·15] $\equiv . R \epsilon \iota'(x \downarrow y) : \supset \vdash . \text{Prop}$

***55·23.** $\vdash . x \downarrow ``\alpha = \hat{R}\{(\exists y) . y \epsilon \alpha . R = x \downarrow y\}$ [*38·13]

***55·231.** $\vdash . \downarrow x``\alpha = \hat{R}\{(\exists y) . y \epsilon \alpha . R = y \downarrow x\}$ [*38·131]

***55·232.** $\vdash : \exists ! \downarrow x``\alpha \cap \downarrow y``\beta . \equiv . x = y . \exists ! \alpha \cap \beta$

Dem.

$\vdash .$ *55·231 . *11·55 . $\supset$

$\vdash :. \exists ! \downarrow x``\alpha \cap \downarrow y``\beta . \equiv : (\exists R) : (\exists z, w) . z \epsilon \alpha . R = z \downarrow x . w \epsilon \beta . R = w \downarrow y :$

[*13·195] $\equiv : (\exists z, w) . z \epsilon \alpha . w \epsilon \beta . z \downarrow x = w \downarrow y :$

[*55·202] $\equiv : (\exists z, w) . z \epsilon \alpha . w \epsilon \beta . x = y . z = w :$

[*13·195] $\equiv : (\exists z) . z \epsilon \alpha \cap \beta . x = y :$

[*10·35] $\equiv : \exists ! \alpha \cap \beta . x = y :. \supset \vdash . \text{Prop}$

***55·233.** $\vdash : x \neq y . \supset . \downarrow x``\alpha \cap \downarrow y``\beta = \Lambda$ [*55·232 . Transp]

The above two propositions are frequently useful in arithmetic.

***55·24.** $\vdash . \dot{s}'x \downarrow ``\alpha = \iota'x \uparrow \alpha$

Dem.

$\vdash .$ *41·11 . $\supset$

$\vdash :. z(\dot{s}'x \downarrow ``\alpha) w . \equiv . (\exists R) . R \epsilon x \downarrow ``\alpha . zRw .$

[*55·23] $\equiv . (\exists R, y) . y \epsilon \alpha . R = x \downarrow y . zRw .$

[*13·195] $\equiv . (\exists y) . y \epsilon \alpha . z(x \downarrow y) w .$

[*55·13] $\equiv . (\exists y) . y \epsilon \alpha . z = x . w = y .$

[*13·195] $\equiv . z = x . w \epsilon \alpha .$

[*51·15.*35·103] $\equiv . z(\iota'x \uparrow \alpha) w :. \supset \vdash . \text{Prop}$

***55·241.** $\vdash . \dot{s}'\downarrow x``\alpha = \alpha \uparrow \iota'x$ [Proof as in *55·24]

***55·25.** $\vdash : \exists ! \alpha . \supset . \text{D}``x \downarrow ``\alpha = \iota'\iota'x$

Dem.

$\vdash .$ *37·67 . *33·12 . *55·12 . $\supset$

$\vdash : \beta \epsilon \text{D}``x \downarrow ``\alpha . \equiv . (\exists y) . y \epsilon \alpha . \beta = \text{D}'x \downarrow y .$

[*55·15] $\equiv . (\exists y) . y \epsilon \alpha . \beta = \iota'x .$

[*10·35] $\equiv . \exists ! \alpha . \beta = \iota'x$ (1)

$\vdash . (1) . \supset \vdash :. \text{Hp} . \supset : \beta \epsilon \text{D}``x \downarrow ``\alpha . \equiv . \beta = \iota'x .$

[*51·15] $\equiv . \beta \epsilon \iota'\iota'x :. \supset \vdash . \text{Prop}$

***55·251.** $\vdash : \exists ! \alpha . \supset . \text{Ꭰ}`` \downarrow x``\alpha = \iota'\iota'x$ [Proof as in *55·25]

This proposition is used in the theory of cardinal multiplication (*113·142).

***55·26.** $\vdash . \text{Ꭰ}``x \downarrow ``\alpha = \iota``\alpha$ [*55·15 . *37·35]

***55·261.** $\vdash . \text{D}`` \downarrow x``\alpha = \iota``\alpha$ [*55·15 . *37·35]

***55·262.** $\vdash : \downarrow x``\alpha = \downarrow y``\beta . \supset . \alpha = \beta$ [*55·261 . *53·22]

***55·27.** $\vdash . C``\downarrow x``\alpha = C``x\downarrow``\alpha = \hat{\beta}\{(\exists y) . y \in \alpha . \beta = \iota`x \cup \iota`y\}$ [*55·15]

***55·28.** $\vdash : \text{Cl}`x \downarrow y = \text{Cl}`x \downarrow z . \equiv . y = z . \equiv . x \downarrow y = x \downarrow z$
[*55·15 . *51·23 . *55·2]

***55·281.** $\vdash : \text{D}`y \downarrow x = \text{D}`z \downarrow x . \equiv . y = z . \equiv . y \downarrow x = z \downarrow x$

***55·282.** $\vdash : C`x \downarrow y = C`x \downarrow z . \equiv . y = z . \equiv . x \downarrow y = x \downarrow z$
[*55·15·2 . *54·21]

***55·283.** $\vdash : C`y \downarrow x = C`z \downarrow x . \equiv . y = z . \equiv . y \downarrow x = z \downarrow x$

***55·29.** $\vdash . \text{Cl} | (x \downarrow) = \iota$ [*55·15 . *34·42]

***55·291.** $\vdash . \text{D} | (\downarrow x) = \iota$ [*55·15 . *34·42]

***55·292.** $\vdash . C | (x \downarrow) = C | (\downarrow x) = \hat{\alpha}\hat{y}(\alpha = \iota`x \cup \iota`y)$ [*55·15 . *34·41]

The following propositions, down to *55·51 inclusive, give properties of ordinal couples which are analogous to the properties of unit classes.

***55·3.** $\vdash : xRy . \equiv . x \downarrow y \subset R . \equiv . \dot{\exists} ! (x \downarrow y) \dot{\wedge} R$ [*13·21·22 . *55·13]

The first half of this proposition is the analogue of *51·2; like that proposition, it gives a means of reducing propositions to the form of inclusions. For the second half, compare *51·31.

***55·31.** $\vdash : x \downarrow y = z \downarrow w . \equiv . z(x \downarrow y)w . \equiv . x(z \downarrow w)y . \equiv . x = z . y = w$

This proposition is the analogue of *51·23.

Dem.

$\vdash . *55·16 . \supset \vdash : x \downarrow y = z \downarrow w . \equiv . \text{D}`x \downarrow y = \iota`z . \text{Cl}`x \downarrow y = \iota`w .$
[*55·15] $\equiv . \iota`x = \iota`z . \iota`y = \iota`w .$
[*51·23] $\equiv . x = z . y = w .$ (1)
[*55·13] $\equiv . x(z \downarrow w)y .$ (2)
[(1).*13·16] $\equiv . z = x . w = y .$
[*55·13] $\equiv . z(x \downarrow y)w$ (3)
$\vdash . (1) . (2) . (3) . \supset \vdash .$ Prop

***55·32.** $\vdash :. x \downarrow y \dot{\wedge} z \downarrow w = \dot{\Lambda} . \equiv : x \neq z . \vee . y \neq w$

Dem.

$\vdash . *55·3 . \supset \vdash : \dot{\exists} ! x \downarrow y \dot{\wedge} z \downarrow w . \equiv . x(z \downarrow w)y .$
[*55·13] $\equiv . x = z . y = w$ (1)
$\vdash . (1) .$ Transp $. \supset \vdash .$ Prop

***55·33.** $\vdash : xRy . \equiv . x \downarrow y \dot{\wedge} R = x \downarrow y$ [*55·3 . *23·621]

***55·34.** $\vdash : \dot{\exists} ! R . R \subset x \downarrow y . \equiv . R = x \downarrow y$

Dem.

$\vdash . *55·13 . \supset \vdash :. \dot{\exists} ! R . R \subset x \downarrow y . \equiv : (\exists z, w) . zRw : zRw . \supset_{z,w} . z = x . w = y :$
[*14·123] $\equiv : zRw . \equiv_{z,w} . z = x . w = y :$
[*55·13] $\equiv : zRw . \equiv_{z,w} . z(x \downarrow y)w :. \supset \vdash .$ Prop

***55·341.** $\vdash :. R \in x \downarrow y . \equiv : R = \dot{\Lambda} . \vee . R = x \downarrow y$

Dem.

$\vdash . *4·42 . \supset \vdash :. R \in x \downarrow y . \equiv : R \in x \downarrow y . R = \dot{\Lambda} . \vee . R \in x \downarrow y . R \neq \dot{\Lambda} :$

[*25·54] $\equiv : R \in x \downarrow y . R = \dot{\Lambda} . \vee . R \in x \downarrow y . \dot{\exists} ! R :$

[*55·34] $\equiv : R \in x \downarrow y . R = \dot{\Lambda} . \vee . R = x \downarrow y :$

[*25·12] $\equiv : R = \dot{\Lambda} . \vee . R = x \downarrow y :. \supset \vdash .$ Prop

***55·35.** $\vdash : R \dot{\cap} x \downarrow y = \dot{\Lambda} . R \dot{\cup} x \downarrow y = S . \equiv . xSy . R = S \dot{-} x \downarrow y$

Dem.

$\vdash . *25·47 . \supset$

$\vdash : R \dot{\cap} x \downarrow y = \dot{\Lambda} . R \dot{\cup} x \downarrow y = S . \equiv . x \downarrow y \in S . R = S \dot{-} x \downarrow y .$

[*55·3] $\equiv . xSy . R = S \dot{-} x \downarrow y : \supset \vdash .$ Prop

***55·36.** $\vdash : xRy . \equiv . (R \dot{-} x \downarrow y) \dot{\cup} x \downarrow y = R$

Dem.

$\vdash . *55·3 . \supset \vdash : xRy . \equiv . x \downarrow y \in R .$

[*23·62] $\equiv . x \downarrow y \dot{\cup} R = R .$

[*23·91] $\equiv . (R \dot{-} x \downarrow y) \dot{\cup} x \downarrow y = R : \supset \vdash .$ Prop

***55·37.** $\vdash : x \epsilon \alpha . y \epsilon \beta . \equiv . x \downarrow y \in \alpha \uparrow \beta$

Dem.

$\vdash . *35·103 . \supset \vdash : x \epsilon \alpha . y \epsilon \beta . \equiv . x (\alpha \uparrow \beta) y .$

[*55·3] $\equiv . x \downarrow y \in \alpha \uparrow \beta : \supset \vdash .$ Prop

The following proposition is the analogue of *51·232.

***55·4.** $\vdash :. a \{x \downarrow y \dot{\cup} z \downarrow w\} b . \equiv : a = x . b = y . \vee . a = z . b = w$ [*55·13 . *23·34]

***55·41.** $\vdash :: R = x \downarrow y \dot{\cup} z \downarrow w . \supset :. aRb . \supset_{a,b} . \phi(a,b) : \equiv . \phi(x,y) . \phi(z,w)$

Dem.

$\vdash . *55·4 . \supset \vdash ::. \text{Hp} . \supset :: aRb . \supset_{a,b} . \phi(a,b) : \equiv :.$
$a = x . b = y . \vee . a = z . b = w : \supset_{a,b} . \phi(a,b) :.$

[*4·77] $\equiv :. (a,b) :. a = x . b = y . \supset . \phi(a,b) : a = z . b = w . \supset . \phi(a,b) :.$

[*11·31] $\equiv :. (a,b) : a = x . b = y . \supset . \phi(a,b) :. (a,b) : a = z . b = w . \supset . \phi(a,b) :.$

[*13·21] $\equiv :. \phi(x,y) . \phi(z,w) ::. \supset \vdash .$ Prop

The above proposition is the analogue of *51·234. The following proposition (*55·42) is the analogue of *51·235.

***55·42.** $\vdash :: R = x \downarrow y \dot{\cup} z \downarrow w . \supset :. (\exists a,b) . aRb . \phi(a,b) . \equiv : \phi(x,y) . \vee . \phi(z,w)$

Dem.

$\vdash . *55·4 . \supset \vdash ::. \text{Hp} . \supset :: (\exists a,b) . aRb . \phi(a,b) . \equiv :.$
$(\exists a,b) :. a = x . b = y . \vee . a = z . b = w : \phi(a,b) :.$

[*4·4] $\equiv :. (\exists a,b) : a = x . b = y . \phi(a,b) : \vee : a = z . b = w . \phi(a,b) :.$

[*11·41] $\equiv :. (\exists a,b) . a = x . b = y . \phi(a,b) . \vee . (\exists a,b) . a = z . b = w . \phi(a,b) :.$

[*13·22] $\equiv :. \phi(x,y) . \vee . \phi(z,w) ::. \supset \vdash .$ Prop

***55·43.** $\vdash : x \downarrow y \cup z \downarrow w = x \downarrow y \cup c \downarrow d . \equiv . z = c . w = d . \equiv . z \downarrow w = c \downarrow d$

This proposition is the analogue of *51·41.

Dem.

$\vdash . *55\cdot202 . \supset \vdash : z = c . w = d . \supset . z \downarrow w = c \downarrow d .$

[*23·551] $\supset . x \downarrow y \cup z \downarrow w = x \downarrow y \cup c \downarrow d$ (1)

$\vdash . *23\cdot58 . \supset \vdash :. x \downarrow y \cup z \downarrow w = x \downarrow y \cup c \downarrow d . \supset :$

$z \downarrow w \in x \downarrow y \cup c \downarrow d . c \downarrow d \in x \downarrow y \cup z \downarrow w :$

[*55·3·13.*23·34] $\supset : z = x . w = y . \vee . z = c . w = d : c = x . d = y . \vee . c = z . d = w :$

[*13·16] $\supset : z = x . w = y . \vee . z = c . w = d : c = x . d = y . \vee . z = c . w = d :$

[*4·41] $\supset : z = x . w = y . c = x . d = y . \vee . z = c . w = d :$

[*13·172] $\supset : z = c . w = d$ (2)

$\vdash . (1) . (2) . \supset \vdash : x \downarrow y \cup z \downarrow w = x \downarrow y \cup c \downarrow d . \equiv . z = c . w = d$ (3)

$\vdash . (3) . *55\cdot202 . \supset \vdash$. Prop

***55·431.** $\vdash :. x \downarrow y \cup z \downarrow w = a \downarrow b \cup c \downarrow d . \supset :$

$x = a . y = b . z = c . w = d . \vee . x = c . y = d . z = a . w = b$

Dem.

$\vdash . *55\cdot4 . \supset \vdash :: \text{Hp} . \equiv :. u = x . v = y . \vee . u = z . v = w :$

$\equiv_{u,v} : u = a . v = b . \vee . u = c . v = d :.$

[*11·1] $\supset :. x = x . y = y . \vee . x = z . y = w :$

$\equiv : x = a . y = b . \vee . x = c . y = d :.$

[*13·15] $\supset :. x = a . y = b . \vee . x = c . y = d$ (1)

$\vdash . *55\cdot43 . \supset \vdash :. x = a . y = b . \supset : x \downarrow y \cup z \downarrow w = a \downarrow b \cup z \downarrow w :$

[*13·171] $\supset : \text{Hp} . \supset . a \downarrow b \cup z \downarrow w = a \downarrow b \cup c \downarrow d .$

[*55·43] $\supset . z = c . w = d$ (2)

$\vdash . (2) . \text{Comm} . *4\cdot7 . \supset \vdash :. \text{Hp} . \supset : x = a . y = b . \supset . x = a . y = b . z = c . w = d$ (3)

Similarly $\vdash :. \text{Hp} . \supset : x = c . y = d . \supset . x = c . y = d . z = a . w = b$ (4)

$\vdash . (1) . (3) . (4) . \supset \vdash$. Prop

***55·44.** $\vdash :. x \downarrow y \cup z \downarrow w = a \downarrow b \cup c \downarrow d .$

$\equiv : x = a . y = b . z = c . w = d . \vee . x = c . y = d . z = a . w = b :$

$\equiv : x \downarrow y = a \downarrow b . z \downarrow w = c \downarrow d . \vee . x \downarrow y = c \downarrow d . z \downarrow w = a \downarrow b$

Dem.

$\vdash . *55\cdot43 . \supset \vdash : x = a . y = b . \supset . x \downarrow y \cup z \downarrow w = a \downarrow b \cup z \downarrow w :$

$z = c . w = d . \supset . a \downarrow b \cup z \downarrow w = a \downarrow b \cup c \downarrow d :$

[*3·47.*13·17] $\supset \vdash : x = a . y = b . z = c . w = d .$

$\supset . x \downarrow y \cup z \downarrow w = a \downarrow b \cup c \downarrow d$ (1)

Similarly $\vdash : x = c . y = d . z = a . w = b .$

$\supset . x \downarrow y \cup z \downarrow w = a \downarrow b \cup c \downarrow d$ (2)

$\vdash . (1) . (2) . *55\cdot431\cdot202 . \supset \vdash$. Prop

The above proposition is the analogue of *51·43

***55·5.** $\vdash :. R \subset x \downarrow y \cup z \downarrow w .$
$\equiv : R = \dot{\Lambda} . \vee . R = x \downarrow y . \vee . R = z \downarrow w . \vee . R = x \downarrow y \cup z \downarrow w$

Dem.

$\vdash . *25·12 . *23·58·42 . \supset$
$\vdash :. R = \dot{\Lambda} . \vee . R = x \downarrow y . \vee . R = z \downarrow w . \vee . R = x \downarrow y \cup z \downarrow w :$
$\supset . R \subset x \downarrow y \cup z \downarrow w$ (1)
$\vdash . *25·49 . \supset \vdash :. R \subset x \downarrow y \cup z \downarrow w . R \dot{\cap} x \downarrow y = \dot{\Lambda} . \supset : R \subset z \downarrow w :$
[*55·341] $\supset : R = \dot{\Lambda} . \vee . R = z \downarrow w$ (2)
$\vdash . *25·43 . \supset \vdash :. R \subset x \downarrow y \cup z \downarrow w . \supset : R \dot{-} x \downarrow y \subset z \downarrow w :$
[*55·341] $\supset : R \dot{-} x \downarrow y = \dot{\Lambda} . \vee . R \dot{-} x \downarrow y = z \downarrow w :$
[*25·24.*23·551] $\supset : (R \dot{-} x \downarrow y) \cup x \downarrow y = x \downarrow y . \vee .$
$(R \dot{-} x \downarrow y) \cup x \downarrow y = x \downarrow y \cup z \downarrow w$ (3)
$\vdash . *55·3·36 . \supset \vdash : \dot{\exists} ! (R \dot{\cap} x \downarrow y) . \supset . (R \dot{-} x \downarrow y) \cup x \downarrow y = R$ (4)
$\vdash . (3) . (4) . \supset \vdash :. R \subset x \downarrow y \cup z \downarrow w . \dot{\exists} ! (R \dot{\cap} x \downarrow y) . \supset :$
$R = x \downarrow y . \vee . R = x \downarrow y \cup z \downarrow w$ (5)
$\vdash . (2) . (5) . \supset \vdash :. R \subset x \downarrow y \cup z \downarrow w . \supset :$
$R = \dot{\Lambda} . \vee . R = x \downarrow y . \vee . R = z \downarrow w . \vee . R = x \downarrow y \cup z \downarrow w$ (6)
$\vdash . (1) . (6) . \supset \vdash .$ Prop

The above proposition is the analogue of *54·4.

***55·51.** $\vdash :. R \subset x \downarrow y \cup S . \supset : xRy . \vee . R \subset S$

Dem.

$\vdash . *55·3 . \supset \vdash : \dot{\exists} ! (R \dot{\cap} x \downarrow y) . \supset . xRy$ (1)
$\vdash . *25·49 . \supset \vdash : \text{Hp} . \sim \dot{\exists} ! (R \dot{\cap} x \downarrow y) . \supset . R \subset S$ (2)
$\vdash . (1) . (2) . \supset \vdash .$ Prop

In the remainder of the present number, we are concerned with properties of ordinal couples which have no analogues for unit classes.

***55·52.** $\vdash . (\iota`x \cup \iota`y) \uparrow (\iota`z \cup \iota`w) = x \downarrow z \cup x \downarrow w \cup y \downarrow z \cup y \downarrow w$ [*35·82·413]

***55·521.** $\vdash : x \neq y . \equiv . x \downarrow y \subset J$ [*55·3 . *50·11]

***55·53.** $\vdash :. x \neq y . \supset : C`R = \iota`x \cup \iota`y . R \subset J . \equiv . \dot{\exists} ! R . R \subset x \downarrow y \cup y \downarrow x$

Dem.

$\vdash . *55·5 . \supset \vdash :. \dot{\exists} ! R . R \subset x \downarrow y \cup y \downarrow x . \equiv :$
$R = x \downarrow y . \vee . R = y \downarrow x . \vee . R = x \downarrow y \cup y \downarrow x$ (1)
$\vdash . *55·15 . \supset \vdash . C`x \downarrow y = \iota`x \cup \iota`y . C`y \downarrow x = \iota`x \cup \iota`y$ (2)
$\vdash . (2) . *33·262 . \supset \vdash . C`(x \downarrow y \cup y \downarrow x) = \iota`x \cup \iota`y$ (3)
$\vdash . *55·521 . \supset \vdash : x \neq y . \supset . x \downarrow y \subset J . y \downarrow x \subset J .$ (4)
[*23·59] $\supset . x \downarrow y \cup y \downarrow x \subset J$ (5)
$\vdash . (1) . (2) . (3) . (4) . (5) . \supset \vdash :.$
$x \neq y . \supset : \dot{\exists} ! R . R \subset x \downarrow y \cup y \downarrow x . \supset . C`R = \iota`x \cup \iota`y . R \subset J$ (6)
$\vdash . *35·91 . \supset \vdash : C`R = \iota`x \cup \iota`y . \supset . R \subset (\iota`x \cup \iota`y) \uparrow (\iota`x \cup \iota`y) .$
[*55·52] $\supset . R \subset x \downarrow x \cup x \downarrow y \cup y \downarrow x \cup y \downarrow y$ (7)

$\vdash . *50\cdot24 . \supset \vdash : R \mathbin{\dot\subset} J . \supset . \sim(xRx) . \sim(yRy) .$

[$*55\cdot3$.Transp] $\supset . R \mathbin{\dot\cap} x \downarrow x = \dot\Lambda . R \mathbin{\dot\cap} y \downarrow y = \dot\Lambda$ (8)

$\vdash . (7) . (8) . *25\cdot49 . \supset \vdash : C\text{‘}R = \iota\text{‘}x \cup \iota\text{‘}y . R \mathbin{\dot\subset} J . \supset . R \mathbin{\dot\subset} x \downarrow y \mathbin{\dot\cup} y \downarrow x$ (9)

$\vdash . *33\cdot24 . *51\cdot161 . \supset \vdash : C\text{‘}R = \iota\text{‘}x \cup \iota\text{‘}y . \supset . \dot\exists ! R$ (10)

$\vdash . (9) . (10) . \supset \vdash : C\text{‘}R = \iota\text{‘}x \cup \iota\text{‘}y . R \mathbin{\dot\subset} J . \supset . \dot\exists ! R . R \mathbin{\dot\subset} x \downarrow y \mathbin{\dot\cup} y \downarrow x$ (11)

$\vdash . (6) . (11) . \supset \vdash .$ Prop

$*55\cdot54$. $\vdash :: x \neq y . \supset :. C\text{‘}R = \iota\text{‘}x \cup \iota\text{‘}y . R \mathbin{\dot\cap} \breve{R} = \dot\Lambda . \equiv : R = x \downarrow y . \vee . R = y \downarrow x$

Dem.

$\vdash . *50\cdot46 . *4\cdot71 . \supset \vdash : R \mathbin{\dot\cap} \breve{R} = \dot\Lambda . \equiv . R \mathbin{\dot\subset} J . R \mathbin{\dot\cap} \breve{R} = \dot\Lambda$ (1)

$\vdash . (1) . *55\cdot53 . \supset \vdash :: x \neq y . \supset :. C\text{‘}R = \iota\text{‘}x \cup \iota\text{‘}y . R \mathbin{\dot\cap} \breve{R} = \dot\Lambda .$

$\equiv : \dot\exists ! R . R \mathbin{\dot\subset} x \downarrow y \mathbin{\dot\cup} y \downarrow x . R \mathbin{\dot\cap} \breve{R} = \dot\Lambda :$

[$*55\cdot5\cdot134$] $\equiv : R = x \downarrow y . \vee . R = y \downarrow x . \vee . R = x \downarrow y \mathbin{\dot\cup} y \downarrow x : R \mathbin{\dot\cap} \breve{R} = \dot\Lambda$ (2)

$\vdash . *55\cdot32 . \supset \vdash :. x \neq y . \supset : x \downarrow y \mathbin{\dot\cap} y \downarrow x = \dot\Lambda :$

[$*55\cdot14$] $\supset : R = x \downarrow y . \supset . R \mathbin{\dot\cap} \breve{R} = \dot\Lambda :$

$R = y \downarrow x . \supset . R \mathbin{\dot\cap} \breve{R} = \dot\Lambda$ (3)

$\vdash . *55\cdot14 . *31\cdot15\cdot33 . \supset \vdash : R = x \downarrow y \mathbin{\dot\cup} y \downarrow x . \supset . R = \breve{R} .$

[$*23\cdot5$] $\supset . R \mathbin{\dot\cap} \breve{R} = R .$

[$*55\cdot134$] $\supset . \dot\exists ! R \mathbin{\dot\cap} \breve{R}$ (4)

$\vdash . (3) . (4) . *4\cdot71 . *5\cdot71 . \supset$

$\vdash :: x \neq y . \supset :. R = x \downarrow y . \vee . R = y \downarrow x . \vee . R = x \downarrow y \mathbin{\dot\cup} y \downarrow x : R \mathbin{\dot\cap} \breve{R} = \dot\Lambda :$

$\equiv : R = x \downarrow y . \vee . R = y \downarrow x$ (5)

$\vdash . (2) . (5) . \supset \vdash .$ Prop

$*55\cdot57$. $\vdash . R \mid (x \downarrow y) = \overrightarrow{R}\text{‘}x \uparrow \iota\text{‘}y$ [$*37\cdot81 . *55\cdot1 . *53\cdot301$]

$*55\cdot571$. $\vdash . (x \downarrow y) \mid S = \iota\text{‘}x \uparrow \overleftarrow{S}\text{‘}y$

$*55\cdot572$. $\vdash . R \mid (x \downarrow y) \mid S = \overrightarrow{R}\text{‘}x \uparrow \overleftarrow{S}\text{‘}y$ [$*55\cdot571 . *37\cdot81$]

$*55\cdot573$. $\vdash . R \mid (x \downarrow y) \mid \breve{S} = \overrightarrow{R}\text{‘}x \uparrow \overrightarrow{S}\text{‘}y$ $\left[*55\cdot572 \frac{\breve{S}}{S}\right]$

$*55\cdot58$. $\vdash : E ! R\text{‘}x . \supset . R \mid (x \downarrow y) = (R\text{‘}x) \downarrow y$ [$*55\cdot57 . *53\cdot31 . *55\cdot1$]

$*55\cdot581$. $\vdash : E ! \breve{S}\text{‘}y . \supset . (x \downarrow y) \mid S = x \downarrow (\breve{S}\text{‘}y)$

$*55\cdot582$. $\vdash : E ! R\text{‘}x . E ! \breve{S}\text{‘}y . \supset . R \mid (x \downarrow y) \mid S = (R\text{‘}x) \downarrow (\breve{S}\text{‘}y)$ [$*55\cdot58\cdot581$]

$*55\cdot583$. $\vdash : E ! R\text{‘}x . E ! S\text{‘}y . \supset . R \mid (x \downarrow y) \mid \breve{S} = (R\text{‘}x) \downarrow (S\text{‘}y)$ $\left[*55\cdot582 \frac{\breve{S}}{S}\right]$

The above propositions are frequently useful in arithmetic. Their use arises as follows. Let α, β, γ, δ be classes of which α is correlated with γ by the relation R, and β with δ by the relation S. Then if $x \in \gamma . y \in \delta$, the couple consisting of the correlate of x and the correlate of y is $(R\text{‘}x) \downarrow (S\text{‘}y)$, *i.e.*, by the above, $R \,|\, (x \downarrow y) \,|\, \breve{S}$, *i.e.* $(R \,\|\, \breve{S})\text{‘}(x \downarrow y)$. Thus the relation $R \,\|\, \breve{S}$ correlates the couples, in α and β, composed of the correlates of terms in γ and δ. The most useful form, in practice, of ∗55·583, is that given below in ∗55·61.

∗55·6. $\vdash . (R \,\|\, \breve{S})\text{‘}(z \downarrow w) = \overrightarrow{R}\text{‘}z \uparrow \overrightarrow{S}\text{‘}w$ [∗55·573 . ∗43·112]

∗55·61. $\vdash : E \,!\, R\text{‘}z \,.\, E \,!\, S\text{‘}w \,.\, \supset . (R \,\|\, \breve{S})\text{‘}(z \downarrow w) = (R\text{‘}z) \downarrow (S\text{‘}w)$
[∗55·583 . ∗43·112]

∗55·62. $\vdash : z \neq w \,.\, S = x \downarrow z \cup y \downarrow w \,.\, \supset . \breve{S}\text{‘}z = x \,.\, \breve{S}\text{‘}w = y$

Dem.

$\vdash . *55 \cdot 13 . \quad \supset \vdash :: \mathrm{Hp} . \supset :. uSz . \equiv : u = x . z = z . \vee . u = y . z = w$ (1)
$\vdash . (1) . *13 \cdot 15 . \supset \vdash :. \mathrm{Hp} . \supset : uSz . \equiv . u = x$ (2)
Similarly $\vdash :. \mathrm{Hp} . \supset : uSw . \equiv . u = y$ (3)
$\vdash . (2) . (3) . *30 \cdot 3 . \supset \vdash . \mathrm{Prop}$

∗55·621. $\vdash : x \neq y \,.\, S = x \downarrow z \cup y \downarrow w \,.\, \supset . \breve{S}\text{‘}x = z \,.\, \breve{S}\text{‘}y = w$
[Proof as in ∗55·62]

The four following propositions belong to ∗43, but are inserted here because the proof uses ∗55·13.

∗55·63. $\vdash : \dot{\exists} \,!\, Q \mathbin{\dot{\cap}} S \,.\, P \,\|\, Q = R \,\|\, S \,.\, \supset . P = R$

Dem.

$\vdash . *43 \cdot 112 . \supset \vdash :: \mathrm{Hp} . \supset :. P \,|\, (y \downarrow z) \,|\, Q = R \,|\, (y \downarrow z) \,|\, S :.$
[∗34·1] $\supset :. (\exists u, v) . xPu . u (y \downarrow z) v . vQw . \equiv_{x,w} .$
$(\exists u, v) . xRu . u (y \downarrow z) v . vSw :.$
[∗55·13.∗13·22] $\supset :. xPy . zQw . \equiv_{x,w} . xRy . zSw :.$
[∗4·73] $\supset :. zQw . zSw . \supset_w : xPy . \equiv_x . xRy$ (1)
$\vdash . (1) . *10 \cdot 11 . *11 \cdot 35 . \supset \vdash :. \mathrm{Hp} . \supset : xPy . \equiv_x . xRy$ (2)
$\vdash . (2) . *10 \cdot 11 \cdot 21 . \supset \vdash . \mathrm{Prop}$

∗55·631. $\vdash : \dot{\exists} \,!\, P \mathbin{\dot{\cap}} R \,.\, P \,\|\, Q = R \,\|\, S \,.\, \supset . Q = S$ [Proof as in ∗55·63]

∗55·632. $\vdash : P \,\|\, Q = R \,\|\, S \,.\, \dot{\exists} \,!\, P \,.\, \dot{\exists} \,!\, Q \,.\, \supset . \dot{\exists} \,!\, P \mathbin{\dot{\cap}} R \,.\, \dot{\exists} \,!\, Q \mathbin{\dot{\cap}} S$

Dem.

$\vdash . *55 \cdot 13 . \quad \supset \vdash : xPy . zQw . \supset . x \{P \,|\, (y \downarrow z) \,|\, Q\} w .$
[∗43·112] $\supset . x \{(P \,\|\, Q)\text{‘}(y \downarrow z)\} w$ (1)
$\vdash . (1) . \supset \vdash :. \mathrm{Hp} . \supset : xPy . zQw . \supset . x \{(R \,\|\, S)\text{‘}(y \downarrow z)\} w .$
[∗43·112] $\supset . x \{R \,|\, (y \downarrow z) \,|\, S\} w .$
[∗34·1] $\supset . (\exists u, v) . xRu . u (y \downarrow z) v . vSw .$
[∗55·13.∗13·22] $\supset . xRy . zSw .$
[∗4·7] $\supset . x (P \mathbin{\dot{\cap}} R) y . z (Q \mathbin{\dot{\cap}} S) w :. \supset \vdash . \mathrm{Prop}$

∗55·64. $\vdash :. \dot{\exists} \,!\, P \,.\, \dot{\exists} \,!\, Q \,.\, \vee . \dot{\exists} \,!\, R \,.\, \dot{\exists} \,!\, S : \supset : P \,\|\, Q = R \,\|\, S \,.\, \equiv . P = R \,.\, Q = S$
[∗55·63·631·632]

*56. THE ORDINAL NUMBER 2_r.

Summary of *56.

In this number, we have to consider the class of those relations which are each constituted by a single couple. In case the two members of this couple are not identical, the class of such relations is (as will be shown later) the ordinal number 2, which, to distinguish it from the cardinal number 2, we denote by "2_r." (Here the suffix is intended to suggest "relational.") The class of all relations consisting of a single couple, without the restriction that the two members of the couple are to be distinct, will be denoted by "$\dot{2}$." This is not an ordinal number. It will be observed that there is no ordinal number 1, because ordinal numbers apply to series, and series must have more than one member if they have any members. This will appear more fully when we come to deal with series.

The properties of $\dot{2}$ are largely analogous to those of 1, while the properties of 2_r are more analogous to those of 2.

Most of the propositions of the present number are seldom referred to in the sequel, but such references as occur are important. The most useful propositions in the present number are the following:

***56·111.** $\vdash : R \epsilon 2_r . \equiv . \mathrm{D}`R, \text{Ɑ}`R \epsilon 1 . \mathrm{D}`R \cap \text{Ɑ}`R = \Lambda$

***56·112.** $\vdash : R \epsilon 2_r . \equiv . \mathrm{D}`R, \text{Ɑ}`R \epsilon 1 . C`R \epsilon 2$

***56·113.** $\vdash . 2_r = \dot{2} \cap \breve{C}``2$

Observe that "$\breve{C}``2$" means "relations whose fields have two terms."

***56·13.** $\vdash . \dot{2} - 2_r = \hat{R}\{(\exists a) . R = a \downarrow a\}$

***56·37.** $\vdash : R \epsilon 2_r . \equiv . C`R \epsilon 2 . R \dot{\cap} \breve{R} = \dot{\Lambda}$

I.e. 2_r is the class of asymmetrical relations whose fields have two terms.

***56·381.** $\vdash : C`R = \iota`x . \equiv . R = x \downarrow x$

***56·39.** $\vdash . \dot{2} - 2_r = \breve{C}``1$

I.e. the relations which are couples whose referent and relatum are identical are the relations whose fields consist of a single term.

***56·01.** $\dot{2} = \hat{R}\{(\exists x, y) . R = x \downarrow y\}$ Df

***56·02.** $2_r = \hat{R}\{(\exists x, y) . x \neq y . R = x \downarrow y\}$ Df

***56·03.** $0_r = \iota`\dot{\Lambda}$ Df

***56·1.** $\vdash : R \epsilon \dot{2} . \equiv . (\exists x, y) . R = x \downarrow y$ [*20·3 . (*56·01)]

***56·101.** $\vdash : R \in \dot{2} . \equiv . \text{D}`R, \text{Ɑ}`R \in 1$

Dem.

$\vdash . *55\cdot16 . *11\cdot11\cdot341 . \supset$

$\vdash :. (\exists x, y) . R = x \downarrow y . \equiv : (\exists x, y) . \text{D}`R = \iota`x . \text{Ɑ}`R = \iota`y :$

$[*11\cdot54] \quad \equiv : (\exists x) . \text{D}`R = \iota`x : (\exists y) . \text{Ɑ}`R = \iota`y :$

$[*52\cdot1] \quad \equiv : \text{D}`R, \text{Ɑ}`R \in 1 \qquad (1)$

$\vdash . (1) . *56\cdot1 . \supset \vdash . \text{Prop}$

***56·102.** $\vdash . \dot{2} = \breve{\text{D}}``1 \cap \breve{\text{Ɑ}}``1$

Dem.

$\vdash . *56\cdot101 . *37\cdot106 . \supset$

$\vdash : R \in \dot{2} . \equiv . R \in \breve{\text{D}}``1 . R \in \breve{\text{Ɑ}}``1 .$

$[*22\cdot33] \quad \equiv . R \in \breve{\text{D}}``1 \cap \breve{\text{Ɑ}}``1 : \supset \vdash . \text{Prop}$

***56·103.** $\vdash : R \in \dot{2} . \supset . \dot{\exists} ! R$

Dem.

$\vdash . *56\cdot101 . \supset \vdash : R \in \dot{2} . \supset . \text{D}`R \in 1 .$

$[*52\cdot16] \quad \supset . \exists ! \text{D}`R .$

$[*33\cdot24] \quad \supset . \dot{\exists} ! R : \supset \vdash . \text{Prop}$

***56·104.** $\vdash : R \in 0_r . \equiv . R = \dot{\Lambda} \quad [(*56\cdot03)]$

***56·11.** $\vdash : R \in 2_r . \equiv . (\exists x, y) . x \neq y . R = x \downarrow y \quad [*20\cdot3 . (*56\cdot02)]$

***56·111.** $\vdash : R \in 2_r . \equiv . \text{D}`R, \text{Ɑ}`R \in 1 . \text{D}`R \cap \text{Ɑ}`R = \Lambda$

Dem.

$\vdash . *51\cdot231 . *55\cdot16 . \supset$

$\vdash : x \neq y . R = x \downarrow y . \equiv . \iota`x \cap \iota`y = \Lambda . \text{D}`R = \iota`x . \text{Ɑ}`R = \iota`y .$

$[*13\cdot193] \quad \equiv . \text{D}`R \cap \text{Ɑ}`R = \Lambda . \text{D}`R = \iota`x . \text{Ɑ}`R = \iota`y \qquad (1)$

$\vdash . (1) . *56\cdot11 . *11\cdot11\cdot341 . \supset$

$\vdash :. R \in 2_r . \equiv : (\exists x, y) . \text{D}`R \cap \text{Ɑ}`R = \Lambda . \text{D}`R = \iota`x . \text{Ɑ}`R = \iota`y :$

$[*11\cdot45] \quad \equiv : \text{D}`R \cap \text{Ɑ}`R = \Lambda : (\exists x, y) . \text{D}`R = \iota`x . \text{Ɑ}`R = \iota`y :$

$[*11\cdot54] \quad \equiv : \text{D}`R \cap \text{Ɑ}`R = \Lambda : (\exists x) . \text{D}`R = \iota`x : (\exists y) . \text{Ɑ}`R = \iota`y :$

$[*52\cdot1] \quad \equiv : \text{D}`R \cap \text{Ɑ}`R = \Lambda . \text{D}`R, \text{Ɑ}`R \in 1 :. \supset \vdash . \text{Prop}$

***56·112.** $\vdash : R \in 2_r . \equiv . \text{D}`R, \text{Ɑ}`R \in 1 . C`R \in 2$

Dem.

$\vdash . *56\cdot111 . *54\cdot43 . \supset$

$\vdash : R \in 2_r . \equiv . \text{D}`R, \text{Ɑ}`R \in 1 . \text{D}`R \cup \text{Ɑ}`R \in 2 .$

$[*33\cdot16] \quad \equiv . \text{D}`R, \text{Ɑ}`R \in 1 . C`R \in 2 : \supset \vdash . \text{Prop}$

***56·113.** $\vdash . 2_r = \dot{2} \cap \breve{C}``2$

Dem.

$\vdash . *56\cdot112\cdot101 . \supset \vdash : R \in 2_r . \equiv . R \in \dot{2} . C`R \in 2 .$

$[*37\cdot106 . *33\cdot122] \quad \equiv . R \in \dot{2} . R \in \breve{C}``2 .$

$[*22\cdot33] \quad \equiv . R \in \dot{2} \cap \breve{C}``2 : \supset \vdash . \text{Prop}$

***56·114.** $\vdash . 2_r = \breve{\mathrm{D}}``1 \cap \breve{\mathrm{Q}}``1 \cap \breve{C}``2$ [*56·113·102]

***56·12.** $\vdash : R \epsilon 2_r . \equiv . R \epsilon \dot{2} . R \mathrel{\mathsf{G}} J$

Dem.

$\vdash . *55·3 . *50·11 . \supset \vdash : x \neq y . \equiv . x \downarrow y \mathrel{\mathsf{G}} J :$

[Fact] $\supset \vdash : R = x \downarrow y . x \neq y . \equiv . R = x \downarrow y . x \downarrow y \mathrel{\mathsf{G}} J .$

[*13·193] $\equiv . R = x \downarrow y . R \mathrel{\mathsf{G}} J$ (1)

$\vdash . (1) . *11·11·341 . \supset$

$\vdash :. (\exists x, y) . R = x \downarrow y . x \neq y . \equiv : (\exists x, y) . R = x \downarrow y . R \mathrel{\mathsf{G}} J :$

[*11·45] $\equiv : (\exists x, y) . R = x \downarrow y : R \mathrel{\mathsf{G}} J :$

[*56·1] $\equiv : R \epsilon \dot{2} . R \mathrel{\mathsf{G}} J$ (2)

$\vdash . (2) . *56·11 . \supset \vdash .$ Prop

***56·121.** $\vdash . 2_r \mathrel{\mathsf{G}} \dot{2}$ [*56·113]

***56·122.** $\vdash : R \epsilon 2_r . \supset . \dot{\exists} ! R$ [*56·121·103]

***56·13.** $\vdash . \dot{2} - 2_r = \hat{R} \{(\exists a) . R = a \downarrow a\}$

Dem.

$\vdash . *56·11 . *11·52 .$ Transp $. \supset$

$\vdash : R \sim \epsilon 2_r . \equiv : R = x \downarrow y . \supset_{x,y} . x = y$ (1)

$\vdash . (1) . *56·1 . \supset$

$\vdash :. R \epsilon \dot{2} - 2_r . \equiv : (\exists a, b) . R = a \downarrow b : R = x \downarrow y . \supset_{x,y} . x = y :$

[*11·45] $\equiv : (\exists a, b) : R = a \downarrow b : R = x \downarrow y . \supset_{x,y} . x = y :$

[*13·193] $\equiv : (\exists a, b) : R = a \downarrow b : a \downarrow b = x \downarrow y . \supset_{x,y} . x = y :$

[*55·202] $\equiv : (\exists a, b) : R = a \downarrow b : a = x . b = y . \supset_{x,y} . x = y :$

[*13·21] $\equiv : (\exists a, b) . R = a \downarrow b . a = b :$

[*13·195] $\equiv : (\exists a) . R = a \downarrow a :. \supset \vdash .$ Prop

$\dot{2} - 2_r$ might be defined as the ordinal number 1, since it is what we shall call a relation number (cf. *153). But we wish our ordinal numbers to be classes of *serial* relations, and such relations have the property of being contained in diversity. Hence if we were to define $\dot{2} - 2_r$ as the ordinal number 1, we should introduce a tiresome exception, from which trivial complications would be introduced into ordinal arithmetic. We have, therefore, not adopted this course.

***56·14.** $\vdash . \mathrm{D}`(x \downarrow) = \dot{2} \cap \overleftarrow{\mathrm{D}}`\iota`x$

Dem.

$\vdash . *33·6 . \supset \vdash : \mathrm{D}`R = \iota`x . \equiv . R \epsilon \overleftarrow{\mathrm{D}}`\iota`x$ (1)

$\vdash . (1) . *56·1 . \supset$

$\vdash :. R \epsilon \dot{2} \cap \overleftarrow{\mathrm{D}}`\iota`x . \equiv : (\exists z, y) . R = z \downarrow y : \mathrm{D}`R = \iota`x :$

[*55·16] $\equiv : (\exists z, y) . \mathrm{D}`R = \iota`z . \mathrm{Q}`R = \iota`y : \mathrm{D}`R = \iota`x :$

[*11·45] $\equiv : (\exists z, y) . \mathrm{D}`R = \iota`z . \mathrm{Q}`R = \iota`y . \mathrm{D}`R = \iota`x :$

[*13·193] $\equiv : (\exists z, y) . \mathrm{D}‘R = \iota‘z . \text{Ɑ}‘R = \iota‘y . \iota‘z = \iota‘x :$

[*51·23] $\equiv : (\exists z, y) . \mathrm{D}‘R = \iota‘z . \text{Ɑ}‘R = \iota‘y . z = x :$

[*13·195] $\equiv : (\exists y) . \mathrm{D}‘R = \iota‘x . \text{Ɑ}‘R = \iota‘y :$

[*55·16] $\equiv : (\exists y) . R = x \downarrow y :$

[*55·22] $\equiv : R \in \mathrm{D}‘(x \downarrow) :. \supset \vdash . \mathrm{Prop}$

***56·141.** $\vdash . \mathrm{D}‘{\downarrow} x = \dot{2} \cap \overleftarrow{\text{Ɑ}}‘\iota‘x$ [Proof as in *56·14]

***56·15.** $\vdash . \mathrm{D}‘(x \downarrow) - \iota‘(x \downarrow x) = 2_r \cap \overleftarrow{\mathrm{D}}‘\iota‘x$

Dem.

$\vdash . *55·22·16 . \supset \vdash :. R \in \{\mathrm{D}‘(x \downarrow)\} - \iota‘(x \downarrow x) .$

$\equiv : (\exists y) . \mathrm{D}‘R = \iota‘x . \text{Ɑ}‘R = \iota‘y : \sim(\mathrm{D}‘R = \iota‘x . \text{Ɑ}‘R = \iota‘x) :$

[*10·35.*4·51.*5·61] $\equiv : (\exists y) . \mathrm{D}‘R = \iota‘x . \text{Ɑ}‘R = \iota‘y . \sim(\text{Ɑ}‘R = \iota‘x) :$

[*13·193] $\equiv : (\exists y) . \mathrm{D}‘R = \iota‘x . \text{Ɑ}‘R = \iota‘y . \sim(\iota‘y = \iota‘x) :$

[*51·23] $\equiv : (\exists y) . \mathrm{D}‘R = \iota‘x . \text{Ɑ}‘R = \iota‘y . x \neq y :$

[*13·195.*51·23] $\equiv : (\exists z, y) . z \neq y . \mathrm{D}‘R = \iota‘z . \text{Ɑ}‘R = \iota‘y . \iota‘z = \iota‘x :$

[*13·193] $\equiv : (\exists z, y) . z \neq y . \mathrm{D}‘R = \iota‘z . \text{Ɑ}‘R = \iota‘y . \mathrm{D}‘R = \iota‘x :$

[*11·45] $\equiv : (\exists z, y) . z \neq y . \mathrm{D}‘R = \iota‘z . \text{Ɑ}‘R = \iota‘y : \mathrm{D}‘R = \iota‘x :$

[*55·16.*33·6] $\equiv : (\exists z, y) . z \neq y . R = z \downarrow y : R \in \overleftarrow{\mathrm{D}}‘\iota‘x :$

[*56·11.*22·33] $\equiv : R \in 2_r \cap \overleftarrow{\mathrm{D}}‘\iota‘x :. \supset \vdash . \mathrm{Prop}$

***56·151.** $\vdash . \mathrm{D}‘({\downarrow} x) - \iota‘(x \downarrow x) = 2_r \cap \overleftarrow{\text{Ɑ}}‘\iota‘x$ [Proof as in *56·15]

***56·16.** $\vdash . x \downarrow y \in \dot{2}$

Dem.

$\vdash . *21·2 . \supset \vdash . x \downarrow y = x \downarrow y .$

[*11·36] $\supset \vdash . (\exists z, w) . x \downarrow y = z \downarrow w .$

[*56·1] $\supset \vdash . x \downarrow y \in \dot{2} . \supset \vdash . \mathrm{Prop}$

***56·17.** $\vdash : x \downarrow y \in 2_r . \equiv . y \downarrow x \in 2_r . \equiv . x \neq y$

Dem.

$\vdash . *56·11 . \supset$

$\vdash :. x \downarrow y \in 2_r . \equiv : (\exists z, w) . z \neq w . x \downarrow y = z \downarrow w :$

[*55·202] $\equiv : (\exists z, w) . z \neq w . x = z . y = w :$

[*13·22] $\equiv : x \neq y$ (1)

Similarly

$\vdash : y \downarrow x \in 2_r . \equiv . x \neq y$ (2)

$\vdash . (1) . (2) . \supset \vdash . \mathrm{Prop}$

***56·18.** $\vdash : x \sim\in \alpha . \equiv . x \downarrow ‘‘\alpha \subset 2_r . \equiv . {\downarrow} x‘‘\alpha \subset 2_r$

Dem.

$\vdash . *13·196 . \supset \vdash :. x \sim\in \alpha . \equiv : y \in \alpha . \supset_y . y \neq x :$

[*56·17] $\equiv : y \in \alpha . \supset_y . x \downarrow y \in 2_r :$

[*37·61.*38·12·11] $\equiv : x \downarrow ‘‘\alpha \subset 2_r$ (1)

Similarly $\vdash : x \sim\in \alpha . \equiv . {\downarrow} x‘‘\alpha \subset 2_r$ (2)

$\vdash . (1) . (2) . \supset \vdash . \mathrm{Prop}$

***56·19.** $\vdash : R \epsilon 2_r . x \epsilon \text{D}`R . \equiv . (\exists y) . x \neq y . R = x \downarrow y . \equiv . R \epsilon x \downarrow `` - \iota`x$

Dem.

$\vdash . *56·11 . *11·45 . \supset \vdash :. R \epsilon 2_r . x \epsilon \text{D}`R . \equiv : (\exists y, z) . y \neq z . R = y \downarrow z . x \epsilon \text{D}`R :$

[*55·15] $\equiv : (\exists y, z) . y \neq z . R = y \downarrow z . x \epsilon \iota`y :$

[*51·23] $\equiv : (\exists y, z) . y \neq z . R = y \downarrow z . x = y :$

[*13·195] $\equiv : (\exists z) . x \neq z . R = x \downarrow z :$ (1)

[*51·15] $\equiv : (\exists z) . z \epsilon - \iota`x . R = x \downarrow z :$

[*38·13] $\equiv : R \epsilon x \downarrow `` - \iota`x$ (2)

$\vdash . (1) . (2) . \supset \vdash .$ Prop

***56·191.** $\vdash : R \epsilon 2_r . x \epsilon \text{C}`R . \equiv . (\exists y) . x \neq y . R = y \downarrow x . \equiv . R \epsilon \downarrow x`` - \iota`x$

[Proof as in *56·19]

***56·2.** $\vdash :. R \epsilon \dot{2} . \equiv : (\exists x, y) : zRw . \equiv_{z,w} . z = x . w = y$ [*55·13 . *56·1]

***56.21.** $\vdash :. R \epsilon \dot{2} . \equiv : \dot{\exists} ! R : xRy . zRw . \supset_{x,y,z,w} . x = z . y = w$ [*56·2 . *14·124]

***56·22.** $\vdash . \dot{\Lambda} \sim \epsilon \dot{2}$ [*56·103 . *25·53]

***56·24.** $\vdash . \exists ! \dot{2} . \exists ! - \dot{2}$ [*56·22·16 . *10·24]

***56·25.** $\vdash . \dot{2} \neq \Lambda \cap \text{Rel} . \dot{2} \neq V \cap \text{Rel}$ [*56·24 . *24·54·17]

***56·26.** $\vdash :. R \epsilon \dot{2} \cup \iota`\dot{\Lambda} . \equiv : xRy . zRw . \supset_{x,y,z,w} . x = z . y = w$

This proposition is the analogue of *52·4.

Dem.

$\vdash . *51·236 . \supset \vdash :: R \epsilon \dot{2} \cup \iota`\dot{\Lambda} .$

$\equiv :. R \epsilon \dot{2} . \vee . R = \dot{\Lambda} :.$

[*25·51] $\equiv :. R \epsilon \dot{2} . \vee . \sim \dot{\exists} ! R :.$

[*56·21] $\equiv :. \dot{\exists} ! R : xRy . zRw . \supset_{x,y,z,w} . x = z . y = w :. \vee :. \sim \dot{\exists} ! R :.$

[*5·62] $\equiv :. xRy . zRw . \supset_{x,y,z,w} . x = z . y = w . \vee . \sim \dot{\exists} ! R$ (1)

$\vdash . *11·36 . \text{Transp} . \supset \vdash :. \sim \dot{\exists} ! R . \supset : \sim (xRy) . \sim (zRw) :$

[*2·21] $\supset : xRy . \supset . x = z : zRw . \supset . y = w :$

[*3·47] $\supset : xRy . zRw . \supset . x = z . y = w$ (2)

$\vdash . (2) . *11·11·3 . \supset \vdash :. \sim \dot{\exists} ! R . \supset : xRy . zRw . \supset_{x,y,z,w} . x = z . y = w$ (3)

$\vdash . (1) . (3) . *4·72 . \supset \vdash .$ Prop

***56·261.** $\vdash :: R \epsilon \dot{2} . \supset :. S \Subset R . \equiv : S = \dot{\Lambda} . \vee . S = R$

Dem.

$\vdash . *55·341 . \supset \vdash :: R = x \downarrow y \supset :. S \Subset R . \equiv : S = \dot{\Lambda} . \vee . S = R$ (1)

$\vdash . (1) . *11·11·35 . *56·1 . \supset \vdash .$ Prop

***56·262.** $\vdash :. R \epsilon \dot{2} . \supset : S \Subset R . \dot{\exists} ! S . \equiv . S = R$

Dem.

$\vdash . *56·22 . \supset \vdash :. R \epsilon \dot{2} . \supset : S = R . \supset . S \neq \dot{\Lambda}$ (1)

$\vdash . (1) . *5·75 . *56·261 . \supset$

$\vdash :. R \epsilon \dot{2} . \supset : S \Subset R . S \neq \dot{\Lambda} . \equiv . S = R$ (2)

$\vdash . (2) . *25·54 . \supset \vdash .$ Prop

***56·27.** $\vdash :. R \in \dot{2} . \supset : \dot{\exists} ! R \dot{\cap} S . \equiv . R \dot{\cap} S \in \dot{2}$

Dem.

$\vdash . *55 \cdot 34 . *23 \cdot 43 . \supset$

$\vdash :. R = x \downarrow y . \supset : \dot{\exists} ! R \dot{\cap} S . \equiv . R \dot{\cap} S = R .$

$[*56 \cdot 16] \qquad \supset . R \dot{\cap} S \in \dot{2}$ (1)

$\vdash . *56 \cdot 103 . \supset \vdash : R \dot{\cap} S \in \dot{2} . \supset . \dot{\exists} ! R \dot{\cap} S$ (2)

$\vdash . (1) . (2) . \supset \vdash :. R = x \downarrow y . \supset : \dot{\exists} ! R \dot{\cap} S . \equiv . R \dot{\cap} S \in \dot{2}$ (3)

$\vdash . (3) . *11 \cdot 11 \cdot 35 . *56 \cdot 1 . \supset \vdash . \text{Prop}$

***56·28.** $\vdash :. R \in \dot{2} . \supset : \dot{\exists} ! R \dot{\cap} S . \equiv . R \mathrel{\dot{\subset}} S . \equiv . R \dot{\cap} S = R$

Dem.

$\vdash . *55 \cdot 3 . \supset \vdash :. R = x \downarrow y . \supset : \dot{\exists} ! R \dot{\cap} S . \equiv . R \mathrel{\dot{\subset}} S .$ (1)

$[*23 \cdot 621] \qquad \equiv . R \dot{\cap} S = R$ (2)

$\vdash . (1) . (2) . *11 \cdot 11 \cdot 35 . *56 \cdot 1 . \supset \vdash . \text{Prop}$

***56·281.** $\vdash :. R \in 2_r . \supset : \dot{\exists} ! R \dot{\cap} S . \equiv . R \mathrel{\dot{\subset}} S . \equiv . R \dot{\cap} S = R . \equiv . R \dot{\cap} S \in 2_r$

Dem.

$\vdash . *56 \cdot 121 . \supset \vdash :. \text{Hp} . \supset : R \in \dot{2} :$

$[*56 \cdot 28] \qquad \supset : \dot{\exists} ! R \dot{\cap} S . \equiv . R \mathrel{\dot{\subset}} S . \equiv . R \dot{\cap} S = R$ (1)

$\vdash . *13 \cdot 13 . \supset \vdash :. \text{Hp} . \supset : R \dot{\cap} S = R . \supset . R \dot{\cap} S \in 2_r :$

$[(1)] \qquad \supset : \dot{\exists} ! R \dot{\cap} S . \supset . R \dot{\cap} S \in 2_r$ (2)

$\vdash . *56 \cdot 122 . \supset \vdash : R \dot{\cap} S \in 2_r . \supset . \dot{\exists} ! R \dot{\cap} S$ (3)

$\vdash . (2) . (3) . \supset \vdash :. \text{Hp} . \supset : \dot{\exists} ! R \dot{\cap} S . \equiv . R \dot{\cap} S \in 2_r$ (4)

$\vdash . (1) . (4) . \supset \vdash . \text{Prop}$

***56·29.** $\vdash :: P, Q \in \dot{2} . \supset :. P \mathrel{\dot{\subset}} Q \dot{\cup} R . \equiv : P = Q . \vee . P \mathrel{\dot{\subset}} R$

Dem.

$\vdash . *55 \cdot 51 . \supset$

$\vdash :. x \downarrow y \mathrel{\dot{\subset}} z \downarrow w \dot{\cup} R . \supset : x (z \downarrow w) y . \vee . x \downarrow y \mathrel{\dot{\subset}} R :$

$[*55 \cdot 31] \qquad \supset : x \downarrow y = z \downarrow w . \vee . x \downarrow y \mathrel{\dot{\subset}} R$ (1)

$\vdash . (1) . *13 \cdot 12 . \supset$

$\vdash ::. P = x \downarrow y . \supset :: Q = z \downarrow w . \supset :. P \mathrel{\dot{\subset}} Q \dot{\cup} R . \supset : P = Q . \vee . P \mathrel{\dot{\subset}} R$ (2)

$\vdash . (2) . *11 \cdot 11 \cdot 35 . *56 \cdot 1 . \supset$

$\vdash ::. P \in \dot{2} . \supset :: Q = z \downarrow w . \supset :. P \mathrel{\dot{\subset}} Q \dot{\cup} R . \supset : P = Q . \vee . P \mathrel{\dot{\subset}} R$ (3)

$\vdash . (3) . *11 \cdot 11 \cdot 3 \cdot 35 . *56 \cdot 1 . \supset$

$\vdash ::. P \in \dot{2} . \supset :: Q \in \dot{2} . \supset :. P \mathrel{\dot{\subset}} Q \dot{\cup} R . \supset : P = Q . \vee . P \mathrel{\dot{\subset}} R$ (4)

$\vdash . *23 \cdot 58 \cdot 61 . \supset \vdash :. P = Q . \vee . P \mathrel{\dot{\subset}} R : \supset . P \mathrel{\dot{\subset}} Q \dot{\cup} R$ (5)

$\vdash . (4) . \text{Imp} . (5) . \supset \vdash . \text{Prop}$

***56·3.** $\vdash :. P, Q \in \dot{2} . \supset : P \mathrel{\dot{\subset}} Q . \equiv . P = Q . \equiv . \dot{\exists} ! P \dot{\cap} Q$

Dem.

$\vdash . *55 \cdot 3 \cdot 31 . \supset$

$\vdash : x \downarrow y \mathrel{\dot{\subset}} z \downarrow w . \equiv . x \downarrow y = z \downarrow w . \equiv . \dot{\exists} ! (x \downarrow y) \dot{\cap} (z \downarrow w)$ (1)

$\vdash . (1) . *13 \cdot 12 . \supset$

$\vdash :. P = x \downarrow y . Q = z \downarrow w . \supset : P \mathrel{\dot{\subset}} Q . \equiv . P = Q . \equiv . \dot{\exists} ! P \dot{\wedge} Q$ (2)

$\vdash . (2) . *11 \cdot 11 \cdot 35 . *56 \cdot 1 . \supset \vdash .$ Prop

The steps from (2) to the conclusion are analogous to those from (2) of $*56 \cdot 29$ to the conclusion of $*56 \cdot 29$. Analogous steps in succeeding proofs will be merely indicated as above.

$*56 \cdot 31$. $\vdash :. P, Q \epsilon \dot{2} . \supset : P \neq Q . \equiv . P \dot{\wedge} Q = \dot{\Lambda}$ [$*56 \cdot 3$. Transp]

$*56 \cdot 32$. $\vdash : P \epsilon \dot{2} . \supset . P \dot{\wedge} Q \epsilon \dot{2} \cup \iota ` \dot{\Lambda}$

Dem.

$\vdash . *56 \cdot 27 . \supset \vdash :. \text{Hp} . \supset : \dot{\exists} ! P \dot{\wedge} Q . \supset . P \dot{\wedge} Q \epsilon \dot{2} :$

[$*2 \cdot 54 . *25 \cdot 54$] $\supset : P \dot{\wedge} Q = \dot{\Lambda} . \vee . P \dot{\wedge} Q \epsilon \dot{2} :$

[$*51 \cdot 236$] $\supset : P \dot{\wedge} Q \epsilon \dot{2} \cup \iota ` \dot{\Lambda} :. \supset \vdash .$ Prop

$*56 \cdot 33$. $\vdash :: P, Q \epsilon \dot{2} . \supset :. R \mathrel{\dot{\subset}} P \mathrel{\dot{\cup}} Q . \equiv : R = \dot{\Lambda} . \vee . R = P . \vee . R = Q . \vee . R = P \mathrel{\dot{\cup}} Q$

Dem.

$\vdash . *55 \cdot 5 . *13 \cdot 12 . \supset \vdash :: P = x \downarrow y . Q = z \downarrow w . \supset :.$

$R \mathrel{\dot{\subset}} P \mathrel{\dot{\cup}} Q . \equiv : R = \dot{\Lambda} . \vee . R = P . \vee . R = Q . \vee . R = P \mathrel{\dot{\cup}} Q$ (1)

$\vdash . (1) . *11 \cdot 11 \cdot 35 . *56 \cdot 1 . \supset \vdash .$ Prop

$*56 \cdot 34$. $\vdash :: P, Q \epsilon \dot{2} . P \neq Q . \supset :. R \mathrel{\dot{\subset}} P \mathrel{\dot{\cup}} Q . \dot{\exists} ! R . R \neq P \mathrel{\dot{\cup}} Q . \equiv : R = P . \vee . R = Q$

Dem.

$\vdash . *56 \cdot 33 \cdot 103 . *5 \cdot 75 . *25 \cdot 54 . \supset$

$\vdash :: P, Q \epsilon \dot{2} . \supset :. R \mathrel{\dot{\subset}} P \mathrel{\dot{\cup}} Q . \dot{\exists} ! R . \equiv : R = P . \vee . R = Q . \vee . R = P \mathrel{\dot{\cup}} Q$ (1)

$\vdash . *23 \cdot 62 . \supset \vdash : P = P \mathrel{\dot{\cup}} Q . \equiv . Q \mathrel{\dot{\subset}} P :$

[$*56 \cdot 3$] $\supset \vdash :. P, Q \epsilon \dot{2} . \supset : P = P \mathrel{\dot{\cup}} Q . \equiv . P = Q :$

[Transp] $\supset : P \neq Q . \supset . P \neq P \mathrel{\dot{\cup}} Q :.$

[$*13 \cdot 181$] $\supset \vdash :. P, Q \epsilon \dot{2} . P \neq Q . \supset : R = P . \supset . R \neq P \mathrel{\dot{\cup}} Q$ (2)

$\vdash . (2) \dfrac{Q, P}{P, Q} . \supset \vdash :. P, Q \epsilon \dot{2} . P \neq Q . \supset : R = Q . \supset . R \neq P \mathrel{\dot{\cup}} Q$ (3)

$\vdash . (2) . (3) . \supset \vdash :: P, Q \epsilon \dot{2} . P \neq Q . \supset :. R = P . \vee . R = Q : \supset . R \neq P \mathrel{\dot{\cup}} Q$ (4)

$\vdash . (1) . (4) . *5 \cdot 75 . \supset \vdash .$ Prop

$*56 \cdot 35$. $\vdash : C ` R \epsilon 2 . R \dot{\wedge} \breve{R} = \dot{\Lambda} . \supset . R \epsilon 2_r$

Dem.

$\vdash . *55 \cdot 54 . \supset$

$\vdash :. x \neq y . C ` R = \iota ` x \cup \iota ` y . R \dot{\wedge} \breve{R} = \dot{\Lambda} . \supset : R = x \downarrow y . \vee . R = y \downarrow x :$

[$*56 \cdot 17$] $\supset : R \epsilon 2_r$ (1)

$\vdash . (1) . *11 \cdot 11 \cdot 35 . *54 \cdot 101 . \supset \vdash .$ Prop

***56·36.** $\vdash : R \epsilon 2_r . \supset . C'R \epsilon 2 . R \dot{\wedge} \breve{R} = \dot{\Lambda}$

Dem.

$\vdash . *55·54 . \supset$

$\vdash : x \neq y . R = x \downarrow y . \supset . x \neq y . C'R = \iota'x \cup \iota'y . R \dot{\wedge} \breve{R} = \dot{\Lambda}$ (1)

$\vdash . (1) . *11·11·34 . *56·11 . \supset$

$\vdash :. R \epsilon 2_r . \supset : (\exists x, y) . x \neq y . C'R = \iota'x \cup \iota'y . R \dot{\wedge} \breve{R} = \dot{\Lambda} :$

$[*54·101 . *11·45] \supset : C'R \epsilon 2 . R \dot{\wedge} \breve{R} = \dot{\Lambda} :. \supset \vdash .$ Prop

The following proposition, in addition to being used in *56·38, is used in the elementary theory of series (*204·463).

***56·37.** $\vdash : R \epsilon 2_r . \equiv . C'R \epsilon 2 . R \dot{\wedge} \breve{R} = \dot{\Lambda}$ [*56·35·36]

***56·38.** $\vdash . 2_r = \breve{C}``2 \cap \hat{R}(R \dot{\wedge} \breve{R} = \dot{\Lambda})$

Dem.

$\vdash . *37·106 . *33·122 . \supset \vdash : C'R \epsilon 2 . \equiv . R \epsilon \breve{C}``2$ (1)

$\vdash . *20·3 . \supset \vdash : R \dot{\wedge} \breve{R} = \dot{\Lambda} . \equiv . R \epsilon \hat{R}(R \dot{\wedge} \breve{R} = \dot{\Lambda})$ (2)

$\vdash . (1) . (2) . *56·37 . \supset \vdash : R \epsilon 2_r . \equiv . R \epsilon \breve{C}``2 . R \epsilon \hat{R}(R \dot{\wedge} \breve{R} = \dot{\Lambda}) .$

$[*22·33] \equiv . R \epsilon \breve{C}``2 \cap \hat{R}(R \dot{\wedge} \breve{R} = \dot{\Lambda}) : \supset \vdash .$ Prop

This proposition is important as establishing the connection between the cardinal and ordinal 2. It shows that the ordinal 2 consists of those asymmetrical relations whose fields have (cardinal) 2 terms. It is used in the theory of well-ordered series (*250·44).

The following proposition, in addition to being used in *56·39, is used in relation-arithmetic (*165·38) and in the theory of series (*205·4).

***56·381.** $\vdash : C'R = \iota'x . \equiv . R = x \downarrow x$

Dem.

$\vdash . *33·24·161 . *51·161 . \supset \vdash : C'R = \iota'x . \supset . \exists ! D'R . D'R \subset \iota'x .$

$[*51·4] \supset . D'R = \iota'x$ (1)

Similarly $\vdash : C'R = \iota'x . \supset . \text{Ꙇ}'R = \iota'x$ (2)

$\vdash . (1) . (2) . *55·16 . \supset \vdash : C'R = \iota'x . \supset . R = x \downarrow x$ (3)

$\vdash . *55·15 . \supset \vdash : R = x \downarrow x . \supset . C'R = \iota'x$ (4)

$\vdash . (3) . (4) . \supset \vdash .$ Prop

***56·39.** $\vdash . \dot{2} - 2_r = \breve{C}``1$

Dem.

$\vdash . *56·381 . \supset \vdash : C'R \epsilon 1 . \equiv . (\exists x) . R = x \downarrow x .$

$[*56·13] \equiv . R \epsilon \dot{2} - 2_r$ (1)

$\vdash . (1) . *37·106 . \supset \vdash .$ Prop

This proposition establishes the connection between $\dot{2} - 2_r$ and 1, showing that $\dot{2} - 2_r$ is the class of those relations whose fields consist of a single term. It is used in the discussion of 0_r and 2_r and $\dot{2} - 2_r$ as relation-numbers (*153·301).

***56·4.** $\vdash :. \mu \subset \dot{2} . \supset : x \downarrow y \in \mu . \equiv . x (\dot{s}`\mu) y$

Dem.

$\vdash . *41 \cdot 11 . \supset \vdash :. \text{Hp} . \supset : x (\dot{s}`\mu) y . \equiv . (\exists R) . R \in \dot{2} . R \in \mu . xRy .$

[*56·1] $\equiv . (\exists z, w) . z \downarrow w \in \mu . x (z \downarrow w) y .$

[*55·13] $\equiv . (\exists z, w) . z \downarrow w \in \mu . z = x . w = y .$

[*13·22] $\equiv . x \downarrow y \in \mu :. \supset \vdash . \text{Prop}$

This proposition is the analogue of *53·23. It is used in the number on exponentiation in relation-arithmetic (*176·19).

APPENDIX A

✱8. THE THEORY OF DEDUCTION FOR PROPOSITIONS CONTAINING APPARENT VARIABLES*

ALL propositions, of whatever order, are derived from a matrix composed of elementary propositions combined by means of the stroke. Given such a matrix, any constituent may be left constant or turned into an apparent variable; the latter may be done in two ways, by taking "all values" or "some values." Thus, if p and q are elementary propositions, giving rise to $p \mid q$, we may replace p by ϕx or q by ψy or both, where ϕx, ψy are propositional functions whose values are elementary propositions. We thus arrive, to begin with, at four new propositions:

$$(x) \,.\, (\phi x \mid q), \quad (\exists x) \,.\, (\phi x \mid q), \quad (y) \,.\, (p \mid \psi y), \quad (\exists y) \,.\, (p \mid \psi y).$$

By means of definitions, we can separate out the constant and the variable part in these expressions; we put

✱8·01. $\{(x) \,.\, \phi x\} \mid q \,.\!=\!.\, (\exists x) \,.\, (\phi x \mid q)$ Df

✱8·011. $\{(\exists x) \,.\, \phi x\} \mid q \,.\!=\!.\, (x) \,.\, (\phi x \mid q)$ Df

✱8·012. $p \mid \{(y) \,.\, \psi y\} \,.\!=\!.\, (\exists y) \,.\, (p \mid \psi y)$ Df

✱8·013. $p \mid \{(\exists y) \,.\, \psi y\} \,.\!=\!.\, (y) \,.\, (p \mid \psi y)$ Df

These definitions define the meaning of the stroke when it occurs between two propositions of which one is elementary while the other is of the first order.

When the stroke occurs between two propositions which are both of the first order, we shall adopt the rule that the above definitions are to be applied first to the one on the left, treating the one on the right as if it were elementary, and are then to be applied to the one on the right. Thus

$$\{(x) \,.\, \phi x\} \mid \{(y) \,.\, \psi y\} \,.\!=\!:\, (\exists x) : \phi x \mid \{(y) \,.\, \psi y\} :$$
$$=: (\exists x) : (\exists y) \,.\, (\phi x \mid \psi y).$$

The same rule can be applied to n propositions; they are to be eliminated from left to right. If a proposition occurs more than once, its occurrences must be eliminated successively as if they were different propositions. These rules are only required for the sake of definiteness, as different orders of elimination give equivalent results. This is only true because we are dealing with various functions each containing one variable, and no variable occurs on both sides of the stroke; it would not be true if we were dealing with functions of several variables. We have *e.g.*

$$(\exists x) : (y) \,.\, (\phi x \mid \psi y) : \equiv : (y) : (\exists x) \,.\, (\phi x \mid \psi y).$$

* This chapter is to replace ✱9 of the text.

But we do not have in general

$$(\exists x) : (y) . \chi(x, y) : \equiv : (y) : (\exists x) . \chi(x, y) ;$$

here the right-hand side is more likely to be true than the left-hand side. For the present, however, we are not concerned with variable functions of two variables.

It should be observed that this possibility of changing the order of the variables is a merit of the stroke. We have

$$(\exists x) : (y) . \phi x | \psi y : \equiv : (y) : (\exists x) . \phi x | \psi y : \equiv : (\exists x) . \sim \phi x . \vee . (y) . \sim \psi y.$$

That is, these equivalent propositions are true when, and only when, either ϕ is sometimes false or ψ is always false. But if we take *e.g.*

$$\phi x \vee \psi y . \sim \phi x \vee \sim \psi y$$

we shall not get the same result. For

$$(\exists x) : (y) . \phi x \vee \psi y . \sim \phi x \vee \sim \psi y : \supset : (y) . \psi y . \vee . (y) . \sim \psi y,$$

whereas $(y) : (\exists x) . \phi x \vee \psi y . \sim \phi x \vee \sim \psi y$ does not imply this.

Written in stroke notation, after some reduction, the above matrix is

$$\{\phi x | (\psi y | \psi y)\} | \{\psi y | (\phi x | \phi x)\}.$$

Here both x and y occur on both sides of the principal matrix. Thus in order to be able to change the order of "$(\exists x)$" and "(y)," it is sufficient (though not *always* necessary) that the matrix should contain some part of the form $\phi x | \psi y$, and that x and y should not occur in any other part of the matrix. (This part may of course be the whole matrix.) We assume the legitimacy of this interchange by a primitive proposition, and in practice arrange to have all the $\exists$-prefixes as far to the right as possible, because this facilitates proofs.

Our primitive propositions are the following:

***8·1.** $\vdash . (\exists x, y) . \phi a | (\phi x | \phi y)$ Pp

On applying the definitions, this is seen to be

$$\vdash : \phi a . \supset . (\exists x) . \phi x.$$

***8·11.** $\vdash . (\exists x) . \phi x | (\phi a | \phi b)$ Pp

On applying the definitions, this becomes

$$\vdash : (x) . \phi x . \supset . \phi a . \phi b.$$

We have $\qquad \phi a | (\phi a | \phi b) . \vee . \phi b | (\phi a | \phi b)$

and by *8·1 $\qquad \vdash : \phi a | (\phi a | \phi b) . \supset . (\exists x) . \phi x | (\phi a | \phi b) :$

$$\phi b | (\phi a | \phi b) . \supset . (\exists x) . \phi x | (\phi a | \phi b),$$

but we cannot deduce $(\exists x) . \phi x | (\phi a | \phi b)$ without *8·11 or an equivalent.

***8·12.** From "$(x) . \phi x$" and "$(x) . \phi x \supset \psi x$" we can infer "$(x) . \psi x$," even when ϕ and ψ are not elementary. Pp

***8·13.** If all occurrences of x are separated from all occurrences of y by a certain stroke, we can change the order of x and y in the prefix, *i.e.* we can replace "$(y) : (\exists x) . \phi x | \psi y$" by "$(\exists x) : (y) . \phi x | \psi y$" and *vice versa*. Pp

The above primitive propositions are to be assumed, not only for one or two variables, but for any number. Thus *e.g.* *8·1 allows us to assert

$\vdash : \phi(a_1, a_2, \ldots a_n) . \supset . (\exists x_1, x_2, \ldots x_n) . \phi(x_1, x_2, \ldots x_n).$

***8·2.** $\vdash : (x) . \phi x . \supset . \phi a$ $\left[*8{\cdot}11 \frac{a}{b} \right]$

In what follows, the method of proof is invariably the same. We first apply the definitions until the whole asserted proposition is brought into the form of a matrix with a prefix. If necessary, we apply *8·13 to change the order of the variables in the prefix. When the proposition to be proved has been brought into this form, we deduce it by means of *8·1·11, using *8·12 in the deduction if necessary. It will be observed that *8·1 is $\vdash : \phi a . \supset . (\exists x) . \phi x$. Hence, by *8·12, whenever we know ϕa, we can assert $(\exists x) . \phi x$; *8·1 is often used in this way.

***8·21.** $\vdash :. (x) . \phi x \supset \psi x . \supset : (\exists x) . \phi x . \supset . (\exists x) . \psi x$

Dem.

Applying the definitions, and using *8·13, the proposition to be proved becomes

$(y, y') : (\exists x, z, w, z', w') . \{\phi x \mid (\psi x \mid \psi x)\} \mid [\{\phi y \mid (\psi z \mid \psi w)\} \mid \{\phi y' \mid (\psi z' \mid \psi w')\}].$

Putting $z = w = z' = w' = x$, the above becomes

$(y, y') : (\exists x) . \{\phi x \mid (\psi x \mid \psi x)\} \mid [\{\phi y \mid (\psi x \mid \psi x)\} \mid \{\phi y' \mid (\psi x \mid \psi x)\}].$

By *8·1, the proposition to be proved is true if this is true. But this is true by *8·11, putting y, y' for a, b and $\phi y \mid (\psi x \mid \psi x)$ for ϕa. Hence the proposition is true.

***8·22.** $\vdash : \phi a \vee \phi b . \supset . (\exists x) . \phi x$

Dem.

$\vdash . *8{\cdot}11 . \supset \vdash . (\exists z) . (\sim \phi z) \mid (\sim \phi a \mid \sim \phi b)$ (1)

Transp. $\supset \vdash : (\sim \phi z) \mid (\sim \phi a \mid \sim \phi b) . \supset . (\phi a \vee \phi b) \mid (\phi z \mid \phi z)$ (2)

$\vdash . (1) . (2) . *8{\cdot}21 . \supset \vdash . (\exists z) . (\phi a \vee \phi b) \mid (\phi z \mid \phi z)$ (3)

$\vdash . (3) . *8{\cdot}1{\cdot}21 . \quad \supset \vdash . (\exists z, w) . (\phi a \vee \phi b) \mid (\phi z \mid \phi w) .$

[(*8·012·013)] $\quad \supset \vdash : \phi a \vee \phi b . \supset . (\exists x) . \phi x : \supset \vdash .$ Prop

These propositions, as well as all the others in *8, apply to any number of variables, since the primitive propositions do so.

***8·23.** $\vdash : (\exists x) . \phi x \vee \phi c . \supset . (\exists x) . \phi x$

Dem.

Applying the definitions, this proposition is

$(x) : (\exists y, z) . (\phi x \vee \phi c) \mid (\phi y \mid \phi z),$

i.e. $(x) : \phi x \vee \phi c . \supset . (\exists x) . \phi x,$

which follows from *8·22.

The following propositions are concerned with forms of the syllogism.

***8·24.** $\vdash :: p \supset q . \supset :. q . \supset . (\exists x) . \phi x : \supset : p . \supset . (\exists x) . \phi x$

Dem.

Applying the definitions, we obtain a matrix

$(p \supset q) \mid [\{(q \mid (\phi x \mid \phi y)) \mid (p \mid (\phi z \mid \phi w) \mid p \mid (\phi u \mid \phi v))\} \mid$
{the same with accented letters}]

with a prefix

$$(x, y, x', y') : (\exists z, w, u, v, z', w', u', v').$$

By *8·1, this will be true if it is true for chosen values of $z, w, u, v, z', w', u', v'$. Put $z = u = x . w = v = y . z' = u' = x' . w' = v' = y'$. Then what has to be proved becomes

$p \supset q . \supset :. q . \supset . \phi x . \phi y : \supset : p . \supset . \phi x . \phi y :. q . \supset . \phi x' . \phi y' : \supset : p . \supset . \phi x' . \phi y'$,

which is true by Syll. Hence the proposition follows.

***8·241.** $\vdash :: (x) . \phi x . \supset . p : \supset :. p \supset q . \supset : (x) . \phi x . \supset . q$

Putting $f(y, z) . = . \{p \mid (q \mid q)\} \mid [\{\phi y \mid (q \mid q)\} \mid \{\phi z \mid (q \mid q)\}]$,

the matrix of the proposition to be proved is

$$\{\phi x \mid (p \mid p)\} \mid \{f(y, z) \mid f(y', z')\}$$

and the prefix is $(x) : (\exists y, z, y', z')$. Putting $y = z = y' = z' = x$, the matrix reduces to $\phi x \supset p . \supset : p \supset q . \supset . \phi x \supset q$, which is true by Syll. Hence the proposition is true by *8·1.

***8·25.** $\vdash :: p . \supset . (\exists x) . \phi x : \supset :. (\exists x) . \phi x . \supset . (\exists x) . \psi x : \supset : p . \supset . (\exists x) . \psi x$

Dem.

Put $f(x, y, z, u, v, m, n) . = . \{\phi x \mid (\psi y \mid \psi z)\} \mid [\{p \mid (\psi u \mid \psi v)\} \mid \{p \mid (\psi m \mid \psi n)\}]$. Then the proposition to be proved, on applying the definitions, is found to have a matrix

$$\{p \mid (\phi a \mid \phi b)\} \mid \{f(x, y, z, u, v, m, n) \mid f(x', y', z', u', v', m', n')\}$$

with the prefix

$$(a, b, y, z, y', z') : (\exists x, u, v, m, n, x', u', v', m', n').$$

Put $x = a . x' = b . u = v = y . m = n = z . u' = v' = y' . m' = n' = z'$.

Then the matrix reduces to

$$p . \supset . \phi a . \phi b : \supset :. \phi a . \supset . \psi y . \psi z : \supset : p . \supset . \psi y . \psi z :.$$
$$\phi b . \supset . \psi y' . \psi z' : \supset : p . \supset . \psi y' . \psi z',$$

which is true by Syll. Hence our proposition results by repeated applications of *8·1·13.

Analogous proofs apply to other forms of the syllogism.

***8·26.** $\vdash : \phi a \vee \phi b \vee \phi c . \supset . (\exists x) . \phi x \vee \phi c$

Dem.

$\vdash : \phi a \vee \phi b \vee \phi c . \supset . (\phi a \vee \phi c) \vee (\phi b \vee \phi c)$ (1)

$\vdash . *8{\cdot}22 . \supset \vdash : (\phi a \vee \phi c) \vee (\phi b \vee \phi c) . \supset . (\exists x) . \phi x \vee \phi c$ (2)

$\vdash . (1) . (2) . *8{\cdot}24 . \supset \vdash . \text{Prop}$

***8·261.** $\vdash : \phi a \lor \phi b \lor \phi c . \supset . (\exists x) . \phi x$
[*8·25·26·23]

It is obvious that we can prove in like manner

$$\phi a \lor \phi b \lor \phi c \lor \phi d . \supset . (\exists x) . \phi x$$

and so on.

***8·27.** $\vdash :: q . \supset . (\exists x) . \phi x : \supset :. p \supset q . \supset : p . \supset . (\exists x) . \phi x$

Dem.

Put $f(x, y, u, v) . = . \{p \mid (\phi x \mid \phi y)\} \mid \{p \mid (\phi u \mid \phi v)\}$.

Then the matrix is

$$\{q \mid (\phi a \mid \phi b)\} \mid [\{(p \supset q) \mid f(x, y, u, v)\} \mid \{(p \supset q) \mid f(x', y', u', v')\}]$$

and the prefix is $(a, b) : (\exists x, y, u, v, x', y', u', v')$.

Putting $x = u = x' = u' = a . y = v = y' = v' = b$, the matrix becomes

$$q . \supset . \phi a . \phi b : \supset :. p \supset q . \supset : p . \supset . \phi a . \phi b,$$

which is true. Hence the proposition.

***8·271.** $\vdash :: q . \supset . (\exists x, y) . \phi(x, y) : \supset :. p \supset q . \supset : p . \supset . (\exists x, y) . \phi(x, y)$
[Proof as in *8·27]

It is obvious that we can prove similarly the analogous proposition with $\phi(x_1, x_2, \ldots x_n)$ in place of $\phi(x, y)$.

***8·272.** $\vdash ::. p . \supset : q . \supset . (\exists x) . \phi x :. \supset :: r \supset p . \supset :. r . \supset : q . \supset . (\exists x) . \phi x$

Dem.

$q . \supset . (\exists x) . \phi x$ is $(\exists x, y) . q \mid (\phi x \mid \phi y)$. Hence the proposition results from *8·271 by the substitution of p for q, r for p, and $q \mid (\phi x \mid \phi y)$ for $\phi(x, y)$.

***8·28.** $\vdash :: p . \supset . (\exists x) . \phi x : \supset :. q . \supset . (\exists x) . \phi x : \supset : p \lor q . \supset . (\exists x) . \phi x$

Dem.

Put $f(x, y, z, w) . = . \{(p \lor q) \mid (\phi x \mid \phi y)\} \mid \{(p \lor q) \mid (\phi z \mid \phi w)\}$.

Then the matrix is

$$\{p \mid (\phi a \mid \phi b)\} \mid [\{(q \mid (\phi c \mid \phi d)) \mid f(x, y, z, w)\} \mid \{(q \mid (\phi c' \mid \phi d')) \mid f(x', y', z', w')\}]$$

and the prefix is

$$(a, b, c, d, c', d') : (\exists x, y, z, w, x', y', z', w').$$

The matrix is

$$p . \supset . \phi a . \phi b : \supset :. q . \supset . \phi c . \phi d : \supset . f(x, y, z, w) :.$$
$$q . \supset . \phi c' . \phi d' . \supset . f(x', y', z', w'),$$

while $f(x, y, z, w) . \equiv : p \lor q . \supset . \phi x . \phi y . \phi z . \phi w$.

Call the matrix $F(x, y, z, w, x', y', z', w')$.

Then $\vdash : p . \supset . F(a, b, a, b, a, b, a, b)$,

$\vdash : \sim p . \supset . F(c, d, c, d, c', d', c', d')$.

Hence $\vdash : F(a, b, a, b, a, b, a, b) . \lor . F(c, d, c, d, c', d', c', d')$.

Hence, by the extension of ∗8·261 to eight variables,

$$\vdash . (\exists x, y, z, w, x', y', z', w') . F(x, y, z, w, x', y', z', w'),$$

which was to be proved.

∗8·29. $\vdash :. (x) . \phi x \supset \psi x . \supset : (x) . \phi x . \supset . (x) . \psi x$

Dem.

Applying the definitions, our proposition is found to have a matrix

$$(\phi x \supset \psi x) \mid [\{\phi y \mid (\psi u \mid \psi v)\} \mid \{\phi y' \mid (\psi u' \mid \psi v')\}]$$

with a prefix (after using ∗8·13)

$$(u, v, u', v') : (\exists x, y, y').$$

The matrix is equivalent to

$$\phi x \supset \psi x . \supset : \phi y . \supset . \psi u . \psi v : \phi y' . \supset . \psi u' . \psi v'.$$

Calling this $M(x, y, y')$, we have to prove

$$(\exists x, y, y') . M(x, y, y').$$

If $\psi u . \psi v . \psi u' . \psi v'$, $M(x, y, y')$ is always true. (1)

If $\sim \psi u$, put $x = y = y' = u$. Then if ϕu is true, $\phi u \supset \psi u$ is false and $M(u, u, u)$ is true. But if ϕu is false, $\phi u . \supset . \psi u . \psi v$ and $\phi u . \supset . \psi u' . \psi v'$ are true, so that $M(u, u, u)$ is true. Hence

$$\sim \psi u . \supset . M(u, u, u) . \supset . (\exists x, y, y') . M(x, y, y'). \quad (2)$$

Similarly if $\sim \psi v \lor \sim \psi u' \lor \sim \psi v'$. (3)

(1), (2), and (3) exhaust possible cases. Hence the result by ∗8·28.

We are now in a position to prove that all the propositions of ∗1—∗5 remain true when one or more of the propositions $p, q, r, \ldots$ are first-order propositions instead of being elementary propositions. For this purpose, we take, not the one primitive proposition which Nicod has shown to be sufficient, but the two which he has shown to be equivalent to it, namely:

$$p \supset p \text{ and } p \supset q . \supset . s \mid q \supset p \mid s.$$

We show that these are true when one, or two, or three, of the propositions p, q, s are first-order propositions. From this, the rest follows. The first of these primitive propositions, $p \supset p$, gives rise to two cases, according as we substitute $(x) . \phi x$ or $(\exists x) . \phi x$ for p; the second primitive proposition gives rise to 26 cases. These have to be considered one by one.

∗8·3. $\vdash : (x) . \phi x . \supset . (x) . \phi x$

Applying the definitions, this is $(\exists x) : (y, z) . \phi x \mid (\phi y \mid \phi z)$, which follows from ∗8·11 by ∗8·13.

∗8·31. $\vdash : (\exists x) . \phi x . \supset . (\exists x) . \phi x$

Applying the definitions, this is $(x) : (\exists y, z) . \phi x \mid (\phi y \mid \phi z)$. This is ∗8·1.

This completes the proof of $p \supset p$.

***8·32.** $\vdash :. (x) . \phi x . \supset . q : \supset : s \mid q . \supset . \{(x) . \phi x\} \mid s$

Putting $p . = . (x) . \phi x$, the proposition to be proved is

$$(p \mid \sim q) \mid \sim \{(s \mid q) \mid \sim (p \mid s)\}.$$

By the definitions,

$$p \mid \sim q . = . (\exists a) . \phi a \mid (q \mid q), \qquad (1)$$

$$p \mid s . = . (\exists x) . \phi x \mid s,$$

$$\sim (p \mid s) . = . (x, y) . (\phi x \mid s) \mid (\phi y \mid s),$$

$$(s \mid q) \mid \sim (p \mid s) . = . (\exists x, y) . (s \mid q) \mid \{(\phi x \mid s) \mid (\phi y \mid s)\}.$$

Put $$f(x, y) . = . (s \mid q) \mid \{(\phi x \mid s) \mid (\phi y \mid s)\}.$$

Then $$\sim \{(s \mid q) \mid \sim (p \mid s)\} . = . (x, y, x', y') . f(x, y) \mid f(x', y'). \qquad (2)$$

By (1) and (2), the proposition to be proved is

$$(a) : (\exists x, y, x', y') . \{\phi a \mid (q \mid q)\} \mid \{f(x, y) \mid f(x', y')\}.$$

Putting $x = y = x' = y' = a$, the matrix of this proposition reduces to

$$\phi a \supset q . \supset . s \mid q \supset \phi a \mid s,$$

which is our primitive proposition with ϕa substituted for p, and is therefore true. Hence the proposition follows by *8·1.

In what follows, the reduction of the proposition to be proved to a matrix and prefix, by means of the definitions, proceeds always by the same method, and the steps will usually be omitted.

***8·321.** $\vdash :. (\exists x) . \phi x . \supset . q : \supset : s \mid q . \supset . \{(\exists x) . \phi x\} \mid s$

We obtain the same matrix as in *8·32, but the opposite prefix, *i.e.* the prefix is

$$(x, y, x', y') : (\exists a).$$

The matrix is equivalent to

$$\phi a \supset q . \supset : q \supset \sim s . \supset . \phi x \supset \sim s . \phi y \supset \sim s . \phi x' \supset \sim s . \phi y' \supset \sim s.$$

Calling this fa, we have to prove $(\exists a) . fa$, for any x, y, x', y'. We have

$$\phi a . \sim q . \supset . fa.$$

Also $\phi a . q . \supset :. fa . \equiv : \sim s . \supset . \phi x \supset \sim s . \phi y \supset \sim s . \phi x' \supset \sim s . \phi y' \supset \sim s :.$
$\supset :. fa.$

Hence $$\phi a . \supset . fa.$$

Hence by *8·1·24 $$\phi x . \supset . (\exists a) . fa,$$

and similarly for ϕy, $\phi x'$, $\phi y'$. Hence by *8·261

$$\phi x \vee \phi y \vee \phi x' \vee \phi y' . \supset . (\exists a) . fa.$$

Also $$\sim \phi x . \sim \phi y . \sim \phi x' . \sim \phi y' . \supset . fa .$$

[*8·1·24] $$\supset . (\exists a) . fa.$$

Hence by *8·28

$$\phi x \vee \phi y \vee \phi x' \vee \phi y' \vee \sim \phi x . \sim \phi y . \sim \phi x' . \sim \phi y' : \supset . (\exists a) . fa.$$

Hence, by *8·12, $(\exists a) . fa$, which was to be proved.

***8·322.** $\vdash :. p . \supset . (x) . \psi x : \supset : s \mid \{(x) . \psi x\} . \supset . p \mid s$

Dem.

Put $$fy . = . (s \mid \psi y) \mid \{(p \mid s) \mid (p \mid s)\}.$$

Then the proposition to be proved is

$$(y, y') : (\exists b, c) . \{p \mid (\psi b \mid \psi c)\} \mid (fy \mid fy').$$

The matrix here is equivalent to

$$p . \supset . \psi b . \psi c : \supset : s \mid \psi y . \supset . p \mid s : s \mid \psi y' . \supset . p \mid s.$$

Putting $b = y . c = y'$, this follows at once from the primitive proposition, which gives

$$p \supset \psi y . \supset : s \mid \psi y . \supset . p \mid s,$$
$$p \supset \psi y' . \supset : s \mid \psi y' . \supset . p \mid s.$$

Hence the proposition.

***8·323.** $\vdash :. p . \supset . (\exists x) . \psi x : \supset : s \mid \{(\exists x) . \psi x\} . \supset . p \mid s$

We have the same matrix as in *8·322, but the opposite prefix, *i.e.*

$$(b, c) : (\exists y, y').$$

Putting $y = b . y' = c$, the matrix is satisfied, as in *8·322.

***8·324.** $\vdash :. p \supset q . \supset : \{(x) . \chi x\} \mid q . \supset . p \mid \{(x) . \chi x\}$

Dem.

Put $f(x, y, z) . = . (\chi x \mid q) \mid \{(p \mid \chi y) \mid (p \mid \chi z)\}$. Then the matrix is

$$\{p \mid (q \mid q)\} \mid \{f(x, y, z) \mid f(x', y', z')\}$$

and the prefix is $(x, x') : (\exists y, z, y', z')$. Putting

$$y = z = x . y' = z' = x',$$

the matrix is equivalent to

$$p \supset q . \supset : \chi x \mid q . \supset . p \mid \chi x : \chi x' \mid q . \supset . p \mid \chi x',$$

which follows from our primitive proposition by Comp.

***8·325.** $\vdash :. p \supset q . \supset : \{(\exists x) . \chi x\} \mid q . \supset . p \mid \{(\exists x) . \chi x\}$

Dem.

The matrix is the same as in *8·324, but the prefix is the opposite, *i.e.*

$$(y, z, y', z') : (\exists x, x').$$

Calling the matrix $M(x, x')$, we have, if $\theta w . \equiv_w . \sim \chi w$,

$$M(x, x') . \equiv :: p \supset q . \supset :. q \supset \theta x . \supset : p . \supset . \theta y . \theta z :. q \supset \theta x' . \supset : p . \supset . \theta y' . \theta z'.$$

Hence $$\theta y . \theta z . \theta y' . \theta z' . \supset . M(x, x') . \supset . (\exists x, x') . M(x, x') \quad (1)$$

But $\sim \theta x . \sim \theta x' . \supset . M(x, x')$. Hence

$$\sim \theta x . \supset . M(x, x) . \supset . (\exists x, x') . M(x, x') \quad (2)$$

Similarly with $\theta y, \theta x', \theta y'$. Hence the result follows as in *8·321.

This ends the cases in which only one of p, q, r in

$$p \supset q . \supset : s \mid q . \supset . p \mid s$$

is of the first order instead of being elementary. We have now to deal with the cases in which two, but not three, are of the first order.

$*8{\cdot}33$. $\vdash :. (x) . \phi x . \supset . (x) . \psi x : \supset : s \mid \{(x) . \psi x\} . \supset . \{(x) . \phi x\} \mid s$

Putting $f(x, y, z) . = . (s \mid \psi x) \mid \{(\phi y \mid s) \mid (\phi z \mid s)\}$, the matrix is

$$\{\phi a \mid (\psi b \mid \psi c)\} \mid \{f(x, y, z) \mid f(x', y', z')\}$$

and the prefix is $(a, x, x') : (\exists b, c, y, z, y', z')$. The matrix is satisfied by

$$b = x' . c = x' . y = z = y' = z' = a,$$

in which case it is equivalent to

$$\phi a . \supset . \psi x . \psi x' : \supset :. \psi x \supset \sim s . \supset . \phi a \supset \sim s : \psi x' \supset \sim s . \supset . \phi a \supset \sim s.$$

Hence Prop.

We have the same matrix in the three following propositions, only with different prefixes.

$*8{\cdot}331$. $\vdash :. (x) . \phi x . \supset . (\exists x) . \psi x : \supset : s \mid \{(\exists x) . \psi x\} . \supset . \{(x) . \phi x\} \mid s$

Here the prefix to the matrix is $(a, b, c) : (\exists x, y, z, x', y', z')$. The matrix is satisfied by $x = b . x' = c . y = z = y' = z' = a$. Hence Prop.

$*8{\cdot}332$. $\vdash :. (\exists x) . \phi x . \supset . (x) . \psi x : \supset : s \mid \{(x) . \psi x\} . \supset . \{(\exists x) . \phi x\} \mid s$

The prefix here is $(x, y, z, x', y', z') : (\exists a, b, c)$. Writing r for $\sim s$, matrix becomes

$$\phi a . \supset . \psi b . \psi c : \supset :. \psi x \supset r . \supset . \phi y \vee \phi z \supset r : \psi x' \supset r . \supset . \phi y' \vee \phi z' \supset r.$$

(Here only a, b, c can be chosen arbitrarily.) This is true if $\phi y, \phi z, \phi y', \phi z'$ are all false. Suppose ϕy is true. Put $a = y$. Then if ψb or ψc is false, $\phi a . \supset . \psi b . \psi c$ is false, and the matrix is true. Therefore if ψx is false, put $b = c = x$; if $\psi x'$ is false, put $b = c = x'$. If ψx and $\psi x'$ are both true, putting $a = y . b = c = x$, the matrix becomes equivalent to

$$r . \supset . \phi y \vee \phi z \supset r : r . \supset . \phi y' \vee \phi z' \supset r,$$

which is true. Hence if ϕy is true, the matrix can be made true. Similarly for z, y', z'. This exhausts possible cases. Hence Prop, by $*8{\cdot}28$.

$*8{\cdot}333$. $\vdash :. (\exists x) . \phi x . \supset . (\exists x) . \psi x : \supset : s \mid \{(\exists x) . \psi x\} . \supset . \{(\exists x) . \phi x\} \mid s$

Dem.

The matrix is as before, and the prefix (after using $*8{\cdot}13$) is

$$(b, c, y, z, y', z') : (\exists a, x, x').$$

Call the matrix $M(a, x, x')$. Then

$\vdash : \psi b . \supset . M(a, b, b) . \supset . (\exists a, x, x') . M(a, x, x')$ (1)

$\vdash : \psi c . \supset . M(a, c, c) . \supset . (\exists a, x, x') . M(a, x, x')$ (2)

$\vdash : \sim \psi b . \sim \psi c . \phi y . \supset . M(y, b, c) . \supset . (\exists a, x, x') . M(a, x, x')$ (3)

$(1) . (2) . (3) . \supset \vdash : \phi y . \supset . (\exists a, x, x') . M(a, x, x')$ [using $*8{\cdot}28$] (4)

Similarly for $\phi y', \phi z, \phi z'$. Hence by $*8{\cdot}28$

$\vdash : \phi y \vee \phi y' \vee \phi z \vee \phi z' . \supset . (\exists a, x, x') . M(a, x, x')$ (5)

But $\vdash :. \sim \phi y . \sim \phi y' . \sim \phi z . \sim \phi z' . \supset : \phi y \vee \phi z \supset r . \phi y' \vee \phi z' \supset r :$

$\supset : M(a, x, x')$

[$*8{\cdot}1$] $\supset : (\exists a, x, x') . M(a, x, x')$ (6)

$\vdash . (5) . (6) . *8{\cdot}28 . \supset \vdash . (\exists a, x, x') . M(a, x, x') . \supset \vdash .$ Prop

This ends the cases in which p and q but not s contain apparent variables. We take next the four cases in which p and s, but not q, contain apparent variables.

***8·34.** $\vdash :. (x) . \phi x . \supset . q : \supset : \{(x) . \chi x\} \mid q . \supset . \{(x) . \phi x\} \mid \{(x) . \chi x\}$

Putting $f(x, y, z, u, v) . = . (\chi x \mid q) \mid \{(\phi y \mid \chi z) \mid (\phi u \mid \chi v)\}$, the matrix is

$$(\phi a \mid \sim q) \mid \{f(x, y, z, u, v) \mid f(x', y', z', u', v')\}.$$

(This is also the matrix of the three following propositions.)

The prefix is $(a, x, x') : (\exists y, z, u, v, y', z', u', v')$.

The matrix is equivalent to

$$\phi a \supset q . \supset . f(x, y, z, u, v) . f(x', y', z', u', v')$$

and $\quad f(x, y, z, u, v) . \equiv : \chi x \mid q . \supset . \phi y \mid \chi z . \phi u \mid \chi v :$

$$\equiv : q \supset \sim \chi x . \supset . \phi y \supset \sim \chi z . \phi u \supset \sim \chi v.$$

Putting $y = u = y' = u' = a . z = v = x . z' = v' = x'$, the matrix is satisfied. Hence Prop.

***8·341.** $\vdash :. (x) . \phi x . \supset . q : \supset : \{(\exists x) . \chi x\} \mid q . \supset . \{(x) . \phi x\} \mid \{(\exists x) . \chi x\}$

Matrix as in *8·34. Prefix $(a, z, v, z', v') : (\exists x, y, u, x', y', u')$.

Matrix is equivalent to

$$\phi a \supset q . \supset :. q \supset \sim \chi x . \supset . \phi y \supset \sim \chi z . \phi u \supset \sim \chi v :$$
$$q \supset \sim \chi x' . \supset . \phi y' \supset \sim \chi z' . \phi u' \supset \sim \chi v'.$$

If ϕa is false, this becomes true by putting $y = u = y' = u' = a$. If ϕa is true the matrix is true if q is false. Suppose q true. Then the matrix is equivalent to

$$\sim \chi x . \supset . \phi y \supset \sim \chi z . \phi u \supset \sim \chi v : \sim \chi x' . \supset . \phi y' \supset \sim \chi z' . \phi u' \supset \sim \chi v'.$$

This is true if $\chi z, \chi v, \chi z', \chi v'$ are false. If one of them, say χz, is true, put $x = x' = z$, and the matrix is true. This exhausts possible cases. Hence Prop, by *8·28.

***8·342.** $\vdash :. (\exists x) . \phi x . \supset . q : \supset : \{(x) . \chi x\} \mid q . \supset . \{(\exists x) . \phi x\} \mid \{(x) . \chi x\}$

Matrix as before. Prefix (after using *8·13) $(x, y, u, x', y', u') : (\exists a, z, v, z', v')$.

Call the matrix $M(a, z, v, z', v')$. Then

$$\vdash : \sim \chi x . \quad \supset . M(a, x, x, x, x) \qquad (1)$$
$$\vdash : \sim \chi x' . \quad \supset . M(a, x', x', x', x') \qquad (2)$$
$$\vdash : q . \chi x . \chi x' . \supset . \sim (q \supset \sim \chi x) . \sim (q \supset \sim \chi x') .$$
$$\supset . M(a, z, v, z', v') \qquad (3)$$
$$\vdash : \sim q . \phi y . \quad \supset . \sim (\phi y \supset q) .$$
$$\supset . M(y, z, v, z', v') \qquad (4)$$

Similarly if $\sim q . \phi u$ or $\sim q . \phi y'$ or $\sim q . \phi u'$. Hence by *8·1·28

$$\vdash : \sim q . \phi y \vee \phi u \vee \phi y' \vee \phi u' . \supset . (\exists a, z, v, z', v') . M(a, z, v, z', v') \qquad (5)$$
$$\vdash : \sim \phi y . \sim \phi u . \sim \phi y' . \sim \phi u' . \supset . \phi y \supset \sim \chi z . \phi u \supset \sim \chi v . \phi y' \supset \sim \chi z' . \phi u' \supset \sim \chi v'$$
$$\supset . M(a, z, v, z', v') \qquad (6)$$
$$(5) . (6) . \supset \vdash : \sim q . \quad \supset . (\exists a, z, v, z', v') . M(a, z, v, z', v') \qquad (7)$$

$\vdash . (1) . (2) . (3) . (7) . \supset \vdash .$ Prop

***8·343.** $\vdash :. (\exists x) . \phi x . \supset . q : \supset : \{(\exists x) . \chi x\} \mid q . \supset . \{(\exists x) . \phi x\} \mid \{(\exists x) . \chi x\}$

Prefix to matrix is $(y, z, u, v, y', z', u', v') : (\exists a, x, x')$.

Call the matrix $\quad f(a, x, x')$.

It is true if $\quad \sim\chi z . \sim\chi v . \sim\chi z' . \sim\chi v' \quad$ (1)

Also $\quad \chi z . q . \supset . f(a, z, z) . \supset . (\exists a, x, x') . f(a, x, x') \quad$ (2)

Similarly if we have $\quad \chi v . q$ or $\chi z' . q$ or $\chi v' . q \quad$ (3)

From (1) . (2) . (3), by *8·28, $q . \supset . (\exists a, x, x') . f(a, x, x') \quad$ (4)

Now $\phi a . \sim q . \supset . f(a, x, x')$. Hence

$$\phi y . \sim q . \supset . f(y, x, x') . \supset . (\exists a, x, x') . f(a, x, x')$$

Similarly for $\phi z . \sim q$, $\phi y' . \sim q$, $\phi z' . \sim q$. Hence

$$\phi y \lor \phi z \lor \phi y' \lor \phi z' . \sim q . \supset . (\exists a, x, x') . f(a, x, x') \quad (5)$$

But $\quad \sim\phi y . \sim\phi z . \sim\phi y' . \sim\phi z' . \supset . f(a, x, x') \quad$ (6)

By (5) and (6), $\quad \sim q . \supset . (\exists a, x, x') . f(a, x, x') \quad$ (7)

$\vdash . (4) . (7) .$ *8·28 $. \supset \vdash .$ Prop

In the next four propositions, q and r are replaced by propositions containing apparent variables, while p remains elementary.

***8·35.** $\vdash :. p . \supset . (x) . \psi x : \supset : \{(x) . \chi x\} \mid \{(x) . \psi x\} . \supset . p \mid \{(x) . \chi x\}$

Putting $q . = . (x) . \psi x$, $s . = . (x) . \psi x$, the proposition is

$$(p \mid \sim q) \mid \sim \{(s \mid q) \mid \sim (p \mid s)\}.$$

We have by the definitions

$$\sim q . = . (\exists b, c) . \psi b \mid \psi c,$$
$$p \mid \sim q . = . (b, c) . p \mid (\psi b \mid \psi c),$$
$$s \mid q . = . (\exists x, y) . \chi y \mid \psi x,$$
$$p \mid s . = . (\exists z) . p \mid \chi z,$$
$$\sim (p \mid s) . = . (z, w) . (p \mid \chi z) \mid (p \mid \chi w),$$
$$(s \mid q) \mid \sim (p \mid s) . = : (x, y) : (\exists z, w) . (\chi y \mid \psi x) \mid \{(p \mid \chi z) \mid (p \mid \chi w)\}.$$

Put $\quad f(x, y, z, w) . = . (\chi y \mid \psi x) \mid \{(p \mid \chi z) \mid (p \mid \chi w)\}$.

Then $\quad \sim \{(s \mid q) \mid \sim (p \mid s)\} . = : (\exists x, y, x', y') : (z, w, z', w') . f(x, y, z, w) \mid f(x', y', z', w')$,

$(p \mid \sim q) \mid \sim \{(s \mid q) \mid \sim (p \mid s)\} . = : (x, y, x', y') : (\exists b, c, z, w, z', w') .$

$$\{p \mid (\psi b \mid \psi c)\} \mid \{f(x, y, z, w) \mid f(x', y', z', w')\}.$$

Writing $\theta \hat{x}$ for $\sim \chi \hat{x}$, the matrix is equivalent to

$p . \supset . \psi b . \psi c : \supset :. \psi x \supset \theta y . \supset : p . \supset . \theta z . \theta w :. \psi x' \supset \theta y' . \supset : p . \supset . \theta z' . \theta w'$.

This is satisfied by putting $b = x . c = x' . z = w = y . z' = w' = y'$. Hence Prop. The same matrix appears in the next three propositions; only the prefix changes.

***8·351.** $\vdash :. p . \supset . (x) . \psi x : \supset : \{(\exists x) . \chi x\} \mid \{(x) . \psi x\} . \supset . p \mid \{(\exists x) . \chi x\}$

Same matrix as in *8·35, but prefix $(x, z, w, x', z', w') : (\exists b, c, y, y')$.

Matrix is true if $\quad \theta z . \theta w . \theta z' . \theta w'$.

Assume $\sim \theta z$, and put $y = y' = z . b = x . c = x'$.

We now have $\psi x \supset \theta y . \equiv . \sim\psi x$ and $p . \supset . \theta z . \theta w : \equiv . \sim p$. Hence matrix is equivalent to

$$p . \supset . \psi x . \psi x' : \supset :. \sim\psi x . \supset . \sim p :. \sim\psi x' . \supset : p . \supset . \theta z' . \theta w',$$

which is true. Similarly if $\sim\theta w \vee \sim\theta z' \vee \sim\theta w'$. Hence Prop, by *8·1·28.

***8·352**. $\vdash :. p . \supset . (\exists x) . \psi x : \supset : \{(x) . \chi x\} \mid \{(\exists x) . \psi x\} . \supset . p \mid \{(x) . \chi x\}$

Same matrix, but prefix $(b, c, y, y') : (\exists x, z, w, x', z', w')$.

Satisfied by $x = b . x' = c . z = w = y . z' = w' = y'$. Hence Prop.

***8·353**. $\vdash :. p . \supset . (\exists x) . \psi x : \supset : \{(\exists x) . \chi x\} \mid \{(\exists x) . \psi x\} . \supset . p \mid \{(\exists x) . \chi x\}$

Same matrix, with prefix $(b, c, z, w, z', w') : (\exists x, y, x', y')$.

If ψb is true and θz false, matrix is satisfied by $x = x' = b . y = y' = z$, because these values make $\psi x \supset \theta y$ and $\psi x' \supset \theta y'$ false. Similarly if ψb is true and θw or $\theta z'$ or $\theta w'$ is false, and if ψc is true and θz, θw, $\theta z'$ or $\theta w'$ is false. It remains to consider $\sim\psi b . \sim\psi c : \vee : \theta z . \theta w . \theta z' . \theta w'$.

The second alternative makes the matrix true, because it gives

$$p . \supset . \theta z . \theta w : p . \supset . \theta z' . \theta w'.$$

The first alternative gives

$$p . \supset . \psi b . \psi c : \supset : \sim p :$$
$$\supset : p . \supset . \theta z . \theta w : p . \supset . \theta z' . \theta w',$$

so that again the matrix is true. Hence Prop.

This finishes the cases in which one or two of the three constituents of $p \supset q . \supset . s \mid q \supset p \mid s$ remain elementary. It remains to consider the eight cases in which none remains elementary. These all have the same matrix.

***8·36**. $\vdash :. (x) . \phi x . \supset . (x) . \psi x : \supset : \{(x) . \chi x\} \mid \{(x) . \psi x\} . \supset . \{(x) . \phi x\} \mid \{(x) . \chi x\}$

Putting $p . = . (x) . \phi x$, $q . = . (x) . \psi x$, $s . = . (x) . \chi x$, we have

$$\sim q . = . (\exists b, c) . \psi b \mid \psi c,$$
$$p \mid \sim q . = : (\exists a) : (b, c) . \phi a \mid (\psi b \mid \psi c),$$
$$s \mid q . = . (\exists x, y) . \chi y \mid \psi x,$$
$$p \mid s . = . (\exists z, w) . \phi z \mid \chi w,$$
$$\sim(p \mid s) . = . (z, w, u, v) . (\phi z \mid \chi w) \mid (\phi u \mid \chi v),$$
$$(s \mid q) \mid \sim(p \mid s) . = : (x, y) : (\exists z, w, u, v) . (\chi y \mid \psi x) \mid \{(\phi z \mid \chi w) \mid (\phi u \mid \chi v)\}.$$

Put $f(x, y, z, w, u, v) . = . (\chi y \mid \psi x) \mid \{(\phi z \mid \chi w) \mid (\phi u \mid \chi v)\}$. Then

$$\sim\{(s \mid q) \mid \sim(p \mid s)\} . = : (\exists x, y, x', y') : (z, w, u, v, z', w', u', v') .$$
$$f(x, y, z, w, u, v) \mid f(x', y', z', w', u', v'),$$
$$(p \mid \sim q) \mid \sim\{(s \mid q) \mid \sim(p \mid s)\} . = : (a, x, y, x', y') : (\exists b, c, z, w, u, v, z', w', u', v') .$$
$$\{\phi a \mid (\psi b \mid \psi c)\} \mid \{f(x, y, z, w, u, v) \mid f(x', y', z', w', u', v')\}.$$

Writing $\theta\hat{x}$ for $\sim\chi\hat{x}$, the matrix is equivalent to

$$\phi a . \supset . \psi b . \psi c : \supset :. \psi x \supset \theta y . \supset . \phi z \supset \theta w . \phi u \supset \theta v :$$
$$\psi x' \supset \theta y' . \supset . \phi z' \supset \theta w' . \phi u' \supset \theta v'.$$

This is satisfied by $b = x . c = x' . z = u = z' = u' = a . w = v = y . w' = v' = y'$. Hence Prop.

∗8·361. $\vdash :. (x) . \phi x . \supset . (x) . \psi x : \supset : \{(\exists x) . \chi x\} \mid \{(x) . \psi x\} . \supset . \{(x) . \phi x\} \mid \{(\exists x) . \chi x\}$

Same matrix, but "all" and "some" are interchanged in arguments to χ, *i.e.* in y, w, v, y', w', v'. The $\exists$-variables are therefore $b, c, y, y', z, z', u, u'$.

If $\sim\phi a$, put $z = u = z' = u' = a$, and matrix is satisfied.

If ϕa is true, matrix is true if $\sim\psi b \vee \sim\psi c$, *i.e.* if $\sim\psi x \vee \sim\psi x'$, since b, c are arbitrary. Assume $\psi x . \psi x'$. Then matrix reduces to

$$\theta y . \supset . \phi z \supset \theta w . \phi u \supset \theta v : \theta y' . \supset . \phi z' \supset \theta w' . \phi u' \supset \theta v'.$$

If $\theta w, \theta v, \theta w', \theta v'$ are all true, this is true.

If $\sim\theta w$, put $y = y' = w$, and matrix is satisfied.

Similarly if $\sim\theta v$, $\sim\theta w'$ or $\sim\theta v'$. Hence Prop.

∗8·362. $\vdash :. (x) . \phi x . \supset . (\exists x) . \psi x : \supset : \{(x) . \chi x\} \mid \{(\exists x) . \psi x\} . \supset . \{(x) . \phi x\} \mid \{(x) . \chi x\}$

Matrix as in ∗8·36. Prefix results from ∗8·36 by interchanging "all" and "some" among ψ-arguments, *i.e.* b, c, x, x'. Hence Prop results from same substitutions as in ∗8·36.

∗8·363. $\vdash :. (x) . \phi x . \supset . (\exists x) . \psi x : \supset : \{(\exists x) . \chi x\} \mid \{(\exists x) . \psi x\} .$
$$\supset . \{(x) . \phi x\} \mid \{(\exists x) . \chi x\}$$

Results from interchanging "all" and "some," in ∗8·361, in the ψ-arguments, viz. b, c, x, x'. The $\exists$-variables are therefore $x, x', y, y', z, z', u, u'$, and the proof proceeds exactly as in ∗8·361, interchanging x, x' and b, c.

∗8·364. $\vdash :. (\exists x) . \phi x . \supset . (x) . \psi x : \supset : \{(x) . \chi x\} \mid \{(x) . \psi x\} . \supset . \{(\exists x) . \phi x\} \mid \{(x) . \chi x\}$

The proposition is what results from ∗8·36 by interchanging "all" and "some" in the ϕ-arguments, viz. a, z, u, z', u'. Hence the $\exists$-arguments are a, b, c, w, v, w', v'. If θy is true, put $w = v = w' = v' = y$, and the matrix is satisfied. If $\theta y'$ is true, put $w = v = w' = v' = y'$, and the matrix is satisfied. Assume $\sim\theta y . \sim\theta y'$. The matrix is true if $\psi x \supset \theta y$ and $\psi x' \supset \theta y'$ are false, *i.e.*, since $\theta y, \theta y'$ are false, if ψx and $\psi x'$ are true. If ψx is false, put $b = c = x$ and $a = y$; then $\phi a . \supset . \psi b . \psi c$ is false, and the matrix is true. If $\psi x'$ is false, similarly. Hence Prop.

∗8·365. $\vdash :. (\exists x) . \phi x . \supset . (x) . \psi x : \supset : \{(\exists x) . \chi x\} \mid \{(x) . \psi x\} .$
$$\supset . \{(\exists x) . \phi x\} \mid \{(\exists x) . \chi x\}$$

Prop is what results from ∗8·364 by interchanging "all" and "some" in the χ-arguments, viz. y, w, v, y', w', v'. Hence the $\exists$-arguments are a, b, c, y, y'. Matrix is true if $\theta w . \theta v . \theta w' . \theta v'$. Assume $\sim\theta w$, and put $y = y' = w$. Matrix is true if $\psi x \supset \theta y$ and $\psi x' \supset \theta y'$ are false, *i.e.*, in the present case, if ψx and $\psi x'$ are true. Suppose one of them false, and put $b = x . c = x'$. Then $\psi b . \psi c$ is false. Therefore $\phi a . \supset . \psi b . \psi c$ is false if ϕa is true; therefore the matrix is true if ϕa is true. Therefore if ϕz is true, the matrix is true for $a = z$. Similarly if ϕu, $\phi z'$ or $\phi u'$ is true. But if all are false, matrix is also true. Hence matrix is true when we have $\sim\theta w$ and $\sim\psi x \vee \sim\psi x'$. Similarly for $\sim\theta v$, $\sim\theta w'$ or $\sim\theta v'$ with $\sim\psi x \vee \sim\psi x'$. We saw that matrix can be satisfied for $\sim\theta w$, $\sim\theta v$, $\sim\theta w'$ or $\sim\theta v'$ with $\psi x . \psi x'$. Hence it can be satisfied for $\sim\theta w \vee \sim\theta v \vee \sim\theta w' \vee \sim\theta v'$. And we saw that it is true for $\theta w . \theta v . \theta w' . \theta v'$. This completes the cases. Hence Prop.

***8·366.** $\vdash :. (\exists x) . \phi x . \supset . (\exists x) . \psi x : \supset : \{(x) . \chi x\} | \{(\exists x) . \psi x\} .$
$\supset . \{(\exists x) . \phi x\} | \{(x) . \chi x\}$

Prop is what results from *8·364 by interchanging "all" and "some" among ψ-arguments, viz. b, c, x, x'. Hence $\exists$-arguments are a, x, x', w, v, w', v'. The proof proceeds as in *8·364, interchanging b, c and x, x'.

***8·367.** $\vdash :. (\exists x) . \phi x . \supset . (\exists x) . \psi x : \supset : \{(\exists x) . \chi x\} | \{(\exists x) . \psi x\} .$
$\supset . \{(\exists x) . \phi x\} | \{(\exists x) . \chi x\}$

Prop is what results from *8·365 by interchanging "all" and "some" among ψ-arguments, viz. b, c, x, x'. Hence the $\exists$-arguments are a, x, x', y, y'. The proof proceeds as in *8·365, interchanging b, c and x, x'.

This completes the 26 cases of $p \supset q . \supset . s|q \supset p|s$. Hence in all the propositions of *1—*5 we can substitute propositions containing one variable. The proofs for propositions containing 2 or 3 or 4 or ... variables are step-by-step the same. Hence the propositions of *1—*5 hold of all first-order propositions.

The extension to second-order propositions, and thence to third-order propositions, and so on, is made by exactly analogous steps. Hence all stroke-functions which can be demonstrated for elementary propositions can be demonstrated for propositions of any order.

It remains to prove $\sim\{(x) . \phi x\} . \equiv . (\exists x) . \sim\phi x$ and similar propositions.

***8·4.** $\vdash : \sim\{(x) . \phi x\} . \equiv . (\exists x) . \sim\phi x$

Dem.

$\vdash . *8·1 . \quad \supset \vdash : \phi x | \phi x . \supset . (\exists y) . \phi x | \phi y \quad (1)$

$\vdash . (1) . *8·21 . \quad \supset \vdash : (\exists x) . \phi x | \phi x . \supset . (\exists x, y) . \phi x | \phi y :$

$[(*8·01·012)] \quad \supset \vdash : (\exists x) . \sim\phi x . \supset . \sim\{(x) . \phi x\} \quad (2)$

We have $\quad \vdash : p|q . \equiv . p|p \vee q|q \quad (3)$

$\vdash . (3) . \quad \supset \vdash : \phi x | \phi y . \equiv . \phi x | \phi x \vee \phi y | \phi y \quad (4)$

$\vdash . (4) . *8·22·24 . \supset \vdash : \phi x | \phi y . \supset . (\exists x) . \phi x | \phi x \quad (5)$

$[(*8·011)] \quad \vdash :. (\exists x, y) . f(x, y) . \supset . p : \equiv : (x, y) . f(x, y) \supset p \quad (6)$

$\vdash . (5) . (6) . \quad \supset \vdash : (\exists x, y) : \phi x | \phi y . \supset . (\exists x) . \phi x | \phi x :$

$[(*8·01·012)] \quad \supset \vdash : \sim\{(x) . \phi x\} . \supset . (\exists x) . \sim\phi x \quad (7)$

$\vdash . (2) . (7) . \quad \supset \vdash . \text{Prop}$

***8·41.** $\vdash : \sim\{(\exists x) . \phi x\} . \equiv . (x) . \sim\phi x$

[Similar proof]

***8·42.** $\vdash :. p . \supset . (\exists x) . \phi x : \equiv : (\exists x) . p \supset \phi x$

Dem.

$\vdash :. p . \supset . (\exists x) . \phi x : \equiv : p | \{\sim(\exists x) . \phi x\} :$

$[*8·41] \quad \equiv : p | \{(x) . \sim\phi x\} :$

$[(*8·011)] \quad \equiv : (\exists x) . p | \sim\phi x :$

$[*8·21] \quad \equiv : (\exists x) . p \supset \phi x :. \supset \vdash . \text{Prop}$

***8·43.** $\vdash :. p . \supset . (x) . \phi x : \equiv : (x) . p \supset \phi x$

[Similar proof]

Other propositions of this type may be taken for granted.

***8·44.** $\vdash :. (x) . \phi x . \supset : (x) . \psi x . \supset . (x) . \phi x . \psi x$

Dem.

$\vdash :. \phi z . \supset : \psi z . \supset . \phi z . \psi z$ (1)

$\vdash . (1) . *8\cdot1 . \supset \vdash ::. (\exists x) ::. (\exists y) :: (z) :. \phi x . \supset : \psi y . \supset . \phi z . \psi z$ (2)

$\vdash . (2) . *8\cdot42\cdot43 . \supset \vdash .$ Prop

***8·5.** If $F(p, q, r, \ldots)$ is a stroke-function of elementary propositions, and $p, q, r, \ldots$ are replaced by first-order propositions $p_1, q_1, r_1, \ldots$, we shall have

$$p \equiv p_1 . q \equiv q_1 . r \equiv r_1, \ldots \supset : F(p, q, r, \ldots) . \equiv . F(p_1, q_1, r_1, \ldots).$$

This follows from

$$p_1 . = . (x) . \phi x : \supset : p \equiv p_1 . \supset . p_1 | q \equiv p | q . q | p_1 \equiv q | p,$$

$$p_1 . = . (\exists x) . \phi x : \supset : p \equiv p_1 . \supset . p_1 | q \equiv p | q . q | p_1 \equiv q | p,$$

both of which are very easily proved.

APPENDIX C

TRUTH-FUNCTIONS AND OTHERS

In the Introduction to the present edition we have assumed that a function can only enter into a proposition through its values. We have in fact assumed that a matrix $f\,!\,(\phi\,!\,\hat{z})$ always arises through some stroke-function

$$F(p, q, r, \ldots)$$

by substituting $\phi\,!\,a$, $\phi\,!\,b$, $\phi\,!\,c$, ... for some or all of $p, q, r, \ldots$, and that all other functions of functions are derivable from such matrices by generalization —*i.e.* by replacing some or all of $a, b, c, \ldots$ by variables, and taking "all values" or "some value."

The uses which we have made of this assumption can be validated by definition, even if the assumption is not universally true. That is to say, we can decide that mathematics is to confine itself to functions of functions which obey the above assumption. This amounts to saying that mathematics is essentially extensional rather than intensional. We might, on this ground, abstain from the inquiry whether our assumption is universally true or not. The inquiry, however, is important on its own account, and we shall, in what follows, suggest certain considerations without arriving at a dogmatic conclusion.

There is a prior question, which is simpler, and that is the question whether all functions of propositions are truth-functions. Or, more precisely, can all propositions which do not contain apparent variables be built up from atomic propositions by means of the stroke? If this were the case, we should have, if $f\hat{p}$ is any function of propositions,

$$p \equiv q \,.\supset .\, fp \equiv fq.$$

Consequently, according to the definition $*13{\cdot}01$,

$$p \equiv q \,.\supset .\, p = q.$$

There will thus be only two propositions, one true and one false. This was Frege's point of view, but it is one which cannot easily be accepted. Frege maintained that every proposition is a proper name, either for the true or for the false. On grounds not connected with our present question, we cannot regard propositions as names; but that does not decide the question whether equivalent propositions are identical. It is this latter question that concerns us. That is to say, we have to consider whether, or in what sense, there are functions fp which are true for some true values of p and false for other true values of p.

Two obvious *primâ facie* instances are "A believes p" and "p is about A." We may take these instances as crucial. If A believes p and p is true, it does not follow that A believes every other true proposition q; nor, if A believes p, and p is false, does it follow that A believes every other false proposition q. Again, the proposition "A is mortal" is about A; but the proposition "B is mortal," which is equally true, is not about A. Thus the function "p is about A" is not a truth-function of p. This instance is important, because the notation "ϕx" is used to denote a proposition about x, and thus the conception involved *seems* to be presupposed in the whole procedure of propositional functions.

We must, to begin with, distinguish between a proposition as a fact and a proposition as a vehicle of truth or falsehood. The following series of black marks: "Socrates is mortal," is a fact of geography. The noise which I should make if I were to say "Socrates is mortal" would be a fact of acoustics. The mental occurrence when I entertain the belief "Socrates is mortal" is a fact of psychology. None of these introduces the notion of truth or falsehood, which is, for logic, the essential characteristic of propositions. We shall return in a moment to the consideration of propositions as facts.

When we say that truth or falsehood is, for logic, the essential characteristic of propositions, we must not be misunderstood. It does not matter, for mathematical logic, what constitutes truth or falsehood; all that matters is that they divide propositions into two classes according to certain rules. Let us take a set of marks

$$x_1, x_2, \ldots x_{2n-1}, x_{2n}.$$

Let us put, as unexplained assertions,

$$T(x_{2m+1}) \quad (m < n),$$
$$F(x_{2m}) \quad (m \leqslant n).$$

Let us further introduce the symbol $x_r \mid x_s$, and assume

$$T(x_r \mid x_s) \text{ if } F(x_r) \text{ or } F(x_s);$$
$$F(x_r \mid x_s) \text{ if } T(x_r) \text{ and } T(x_s).$$

Assume further that, if p, q, s are any one of the x's or any combination of them by means of the stroke, the above rules are to apply to $p \mid q$, etc., and further we are to have:

$$T\{p \mid (p \mid p)\},$$
$$T\{p \supset q \,.\, \supset .\, s \mid q \supset p \mid s\},$$

where "$p \supset q$" means "$p \mid (q \mid q)$." Further: given $T\{p \mid (q \mid r)\}$ and $T(p)$, we are to have $T(r)$.

Taking the above as mere conventional rules, all the logic of molecular propositions follows, replacing "$\vdash . p$" by "$T(p)$."

Thus from the formal point of view it is irrelevant what constitutes truth or falsehood: all that matters is that propositions are divided into two classes according to certain rules. It does not matter what propositions are, so long as we are content to regard our primitive propositions as defining hypotheses, not as truths. (From a philosophical point of view, this formal procedure may be shown to presuppose the non-formal interpretation of our primitive propositions; but that does not matter for our present purpose.)

Throughout the logic of molecular propositions, we do not want to know anything about propositions except whether they are true or false. Further, we are concerned only with those combinations of propositions which are true in virtue of the rules, whether their constituent propositions are true or false. That is—to take the simplest illustration—we assert $p\,|(p\,|\,p)$, but we never assert any proposition p that has not some suitable molecular structure, although we believe that half of such propositions are true. Our assertions depend always upon structure, never upon the mere fact that some proposition is true.

A new situation arises, however, when we replace p by $\phi\,!\,x$. For example, we have $\vdash . p\,|\,(p\,|\,p)$
and we infer $\vdash . \phi\,!\,x\,|\,(\phi\,!\,x\,|\,\phi\,!\,x)$.
We cannot *explain* the notation $\phi\,!\,x$ without introducing characteristics of propositions other than their truth or falsehood. Take for example the primitive proposition (∗8·11)

$$\vdash . (\exists x) . \phi\,!\,x\,|\,(\phi\,!\,a\,|\,\phi\,!\,b).$$

The truth of this proposition depends upon the *form* of the constituent propositions $\phi\,!\,x$, $\phi\,!\,a$, $\phi\,!\,b$, not simply upon their truth or falsehood. It cannot be replaced by

$$\text{“}\vdash . (\exists p) . p\,|\,(q\,|\,r),\text{”}$$

which is true but does not have the desired consequences. We are therefore compelled to consider what is meant by saying that a proposition is of the form $\phi\,!\,a$ (where a is some constant). This brings us back to "A occurs in p," which we gave above as an example of a function which is not a truth-function. And this, we shall find, brings us back to the proposition as fact, in opposition to the proposition as true or false.

Let us revert to our two instances: "A believes p" and "p is about A." We shall avoid certain psychological difficulties if we take, to begin with, "A asserts p" instead of "A believes p." Suppose "p" is "Socrates is Greek." A word is a class of similar noises. Thus a person who asserts "Socrates is Greek" is a person who makes, in rapid succession, three noises, of which the first is a member of the class "Socrates," the second a member of the class "is," and the third a member of the class "Greek." This series of events is part of the series of events which constitutes the person. If A is the series of events constituting the person, α is the class of noises "Socrates," β the class "is," and γ the class "Greek," then "A asserts that Socrates is Greek" is (omitting the rapidity of the succession)

$$(\exists x, y, z) . x \epsilon \alpha . y \epsilon \beta . z \epsilon \gamma . x \downarrow y \cup x \downarrow z \cup y \downarrow z \subset A.$$

It is obvious that this is not a function of p as p occurs in a truth-function.

If we now take up "*A* believes *p*," we find the matter rather more complicated, owing to doubt as to what constitutes belief. Some people maintain that a proposition must be expressed in words before we can believe it; if that were so, there would not, from our point of view, be any vital difference between believing and asserting. But if we adopt a less unorthodox standpoint, we shall say that when a man believes "Socrates is Greek" he has simultaneously two thoughts, one of which "means" Socrates while the other "means" Greek, and these two thoughts are related in the way we call "predication." It is not necessary for our purposes to define "meaning," beyond noticing that two different thoughts may "have the same meaning." The relation "having the same meaning" is symmetrical and transitive; moreover, if two thoughts "have the same meaning," either can replace the other in any belief without altering its truth-value. Thus we have one class of thoughts, called "Socrates," which all "have the same meaning"; call this class α. We have another class of thoughts, called "Greek," which all "have the same meaning"; call this class β. Call the relation of predication between two thoughts P. (This is the relation which holds between our thought of the subject and our thought of the predicate when we believe that the subject has the predicate. It is wholly different from the relation which holds between the subject and the predicate when our belief is true.) Then "*A* believes that Socrates is Greek" is

$$(\exists x, y) . x \epsilon \alpha . y \epsilon \beta . xPy . x, y \epsilon C`A.$$

Here, again, the proposition as it occurs in truth-functions has disappeared.

It is not necessary to lay any stress upon the above analysis of belief, which may be completely mistaken. All that is intended is to show that "*A* believes *p*" may very well not be a function of *p*, in the sense in which *p* occurs in truth-functions.

We have now to consider "*p* is about *A*," *e.g.* "'Socrates is Greek' is about Socrates." Here we have to distinguish (1) the fact, (2) the belief, (3) the verbal proposition. The fact and the belief, however, do not raise separate problems, since it is fairly clear that Socrates is a constituent of the fact in the same sense in which the thought of Socrates is a constituent of the belief. And the verbal proposition raises no difficulty, since each instance of the verbal proposition is a series containing a part which is an instance of "Socrates." That is to say, "Socrates" (the word) is a class of series of noises, say λ; and "Socrates is Greek" is another class of series, say μ; and the fact that "Socrates" occurs in "Socrates is Greek" is

$$P \epsilon \mu . \supset . (\exists Q) . Q \epsilon \lambda . Q \Subset P.$$

Thus we are left with the question: What do we mean by saying that Socrates is a constituent of the fact that Socrates is Greek? This raises the whole problem of analysis. But we do not need an ultimate answer; we only need an answer sufficient to throw light on the question whether there are functions of propositions which are not truth-functions.

There are those who deny the legitimacy of analysis. Without admitting that they are in the right, we can frame a theory which they need not reject. Let us assume that facts are capable of various kinds of resemblances and differences. Two facts may have particular-resemblance; then we shall say that they are about the same particular. Again they may have predicate-resemblance, or dyadic-relation-resemblance, or etc. We shall say that a fact is about only one particular if any two facts which have particular-resemblance to the given fact have particular-resemblance to each other. Given such a fact, we may define its one particular as the class of all facts having particular-resemblance to the given fact. In that case, to say that Socrates is a constituent of the fact that Socrates is Greek (assuming conventionally that Socrates is a particular) is to say that the fact is a member of the class of facts which is Socrates. In the case of a belief about Socrates, which is itself a fact composed of thoughts, we shall say that a belief is about Socrates if it is one of the class of facts constituting a certain idea which "means" Socrates in whatever sense we may give to "meaning." Here an "idea" is taken to be a class of psychical facts, say all the beliefs which "refer to" Socrates.

We can define predicates by a similar procedure. Take a fact which is only capable of two kinds of resemblance such as we are considering, namely particular-resemblance and predicate-resemblance; such a fact will be a subject-predicate fact. The predicate involved in it is the class of facts to which it has predicate-resemblance.

We shall assume also various kinds of difference: particular-difference, predicate-difference, etc. These are not necessarily incompatible with the corresponding kind of resemblance; *e.g.* $R(x, x)$ and $R(x, y)$ have both particular-resemblance in respect of x and particular-difference in respect of y. This enables us to define what is meant by saying that a particular occurs twice in a fact, as x occurs twice in $R(x, x)$. First: $R(x, x)$ is a dyadic-relation-fact because it is capable of dyadic-relation-resemblance to other facts; second: any two facts having particular-resemblance to $R(x, x)$ have particular-resemblance to each other. This is what we mean by saying that $R(x, x)$ is a dyadic-relation-fact in which x occurs twice, not a subject-predicate fact. Take next a triadic-relation-fact $R(x, x, z)$. This is, by definition, a triadic-relation-fact because it is capable of triadic-relation-resemblance. The facts having particular-resemblance to $R(x, x, z)$ can be divided into two groups (not three) such that any two members of one group have particular-resemblance to each other. This shows that there is repetition, but not whether it is x or z that is repeated. The facts of the one group are $R(x, x, c)$ for varying c; the facts of the other are $R(a, b, z)$ for varying a and b. Each fact of the group $R(x, x, c)$ belongs to only two groups constituted by particular-resemblance, whereas the facts of the group $R(a, b, z)$, except when it happens that $a = b$, belong to three groups constituted by particular-resemblance. This defines what is meant by saying that x occurs twice and z once in the fact $R(x, x, z)$. It is obvious that we can deal with tetradic etc. relations in the same way.

According to the above, when we say that Socrates is a constituent of the fact that Socrates is Greek, we mean that this fact is a member of the class of facts which is Socrates.

When we use the notation "$\phi ! x$" to denote a proposition in which "x" occurs, it is a fact that "x" occurs in "$\phi ! x$," but we do not need to assert the fact; the fact does its work without having to be asserted. It is also a fact that, if "x" occurs in a proposition p, and p asserts a fact, then x is a constituent of that fact. This is not a law of logic, but a law of language. It might be false in some languages. For instance, in former days, when a crime was committed in India, the indictment stated that it was committed "in the manor of East Greenwich." These words did not denote any constituents of the fact. But a logical language avoids fictions of this kind.

The notation for functions is an illustration of Wittgenstein's principle, that a logical symbol must, in certain formal respects, resemble what it symbolizes. All the facts of which x is a constituent, according to the above, constitute a certain class defined by particular-resemblance. The various symbols ϕx, ψx, χx, ... also all resemble each other in a certain respect, namely that their right-hand halves are very similar (not *exactly* similar, because no two x's are exactly alike). The symbols $R(x, x)$, $R(x, x, z)$, etc. are appropriate to their meanings for similar reasons. The symbols are *used* before their suitability can be explained. To explain *why* "ϕx" is a suitable symbol for a proposition about x is, as we have seen, a complicated matter. But to use the symbol is not a complicated matter. Our symbolism, as a set of facts, resembles, in certain logical respects, the facts which it is to symbolize. This makes it a good symbolism. But in using it we do not presuppose the explanation of why it is good, which belongs to a later stage. And so the notation "ϕx" can be used without first explaining what we mean by "a proposition about x."

We are now in a position to deal with the difference between propositions considered factually and propositions as vehicles of truth and falsehood. When we say "'Socrates' occurs in the proposition 'Socrates is Greek,'" we are taking the proposition factually. Taken in this way, it is a class of series, and 'Socrates' is another class of series. Our statement is only true when we take the proposition and the name as classes. The particular 'Socrates' that occurs at the beginning of our sentence does not occur in the proposition 'Socrates is Greek'; what is true is that another particular closely resembling it occurs in the proposition. It is therefore absolutely essential to all such statements to take words and propositions as classes of similar occurrences, not as single occurrences. But when we assert a proposition, the single occurrence is all that is relevant. When I assert "Socrates is Greek," the particular occurrences of the words have meaning, and the assertion is made by the particular occurrence of that sentence. And to say of that sentence "'Socrates' occurs in it" is simply false, if I mean the 'Socrates' that I have just written down, since it was a different 'Socrates' that occurred in it. Thus we conclude:

A proposition as the vehicle of truth or falsehood is a particular occurrence, while a proposition considered factually is a class of similar occurrences. It is the proposition considered factually that occurs in such statements as "A believes p" and "p is about A."

Of course it is possible to make statements about the particular fact "Socrates is Greek." We may say how many centimetres long it is; we may say it is black; and so on. But these are not the statements that a philosopher or logician is tempted to make.

When an assertion occurs, it is made by means of a particular fact, which is an instance of the proposition asserted. But this particular fact is, so to speak, "transparent"; nothing is said about it, but by means of it something is said about something else. It is this "transparent" quality which belongs to propositions as they occur in truth-functions. This belongs to p when p is asserted, but not when we say "p is true." Thus suppose we say: "All that Xenophon said about Socrates is true." Put

$$X(p) . = . \text{Xenophon asserted } p,$$
$$S(p) . = . p \text{ is about Socrates.}$$

Then our statement is

$$X(p) . S(p) . \supset_p . p \text{ is true.}$$

Here the occurrence of p is not "transparent." But if we say

$$x \epsilon \alpha . \supset_x . \phi ! x$$

we are asserting $\phi ! x$ for a whole class of values of x, and yet "$\phi ! x$" still has a "transparent" occurrence. The essential difference is that in the former case we speak *about* the symbol or belief, whereas in the latter we merely use it to speak about something else. This is the point which distinguishes the occurrences of propositions in mathematical logic from their occurrences in non-truth-functions.

Let us endeavour to give greater definiteness to this point. Take the statement "Socrates had all the predicates that Xenophon said he had." Let the series of events which was Xenophon be called X. Then if Xenophon attributed the predicate α to Socrates, we might appear to have (writing $x \downarrow y \downarrow z \downarrow w$ for the series x, y, z, w)

$$\text{Socrates} \downarrow \text{had} \downarrow \text{predicate} \downarrow \alpha \Subset X.$$

Thus our assertion would be

$$\text{Socrates} \downarrow \text{had} \downarrow \text{predicate} \downarrow \alpha \Subset X . \supset_\alpha . \text{Socrates had predicate } \alpha.$$

Here, however, there is an ambiguity. On the left, "Socrates," "had," "predicate," and "α" occur as noises; on the right they occur as symbols. This ambiguity amounts to a fallacy. For, in fact, what I write on paper is not the noise that Xenophon made, but a symbol for that noise. Thus I am using one symbol "Socrates" in two senses: (*a*) to mean the noise that Xenophon made on a certain occasion, (*b*) to mean a certain man. We must say:

If Xenophon made a series of noises which mean what is meant by "Socrates had the predicate α," then what this means is true.

For example: If Xenophon said "Socrates was wise," then what is meant by "Socrates was wise" is true.

But this does not assert that Socrates was wise. When I actually assert that Socrates was wise, I say something which cannot be said by talking about the words I use in saying it; and when I assert that Socrates was wise, although an instance of the proposition occurs, yet I do not say anything whatever about the proposition—in particular I do not say that it is true. This is an inference, not logical, but linguistic.

If the above considerations in any way approximate to the truth, we see that there is an absolute gulf between the assertion of a proposition and an assertion about the proposition. The p that occurs when we assert p and the p that occurs in "A asserts p" are by no means identical. The occurrence of propositions as asserted is simpler than their occurrence as something spoken about. In the assertion of a proposition, and in the assertion of any molecular function of a proposition, the proposition does not occur, if we mean by the proposition the p that occurs in such propositions as "A asserts p" or "p is about A." When these latter are analysed, they are found not to conflict with the view that propositions, in the sense in which they occur when they are asserted, only occur in truth-functions.

When p is asserted, p does not really occur, but the constituents of p occur, or an instance of p occurs. The same is true when a molecular proposition containing p is asserted. Thus we cannot infer $p = q$, because here p and q occur in a sense in which they do not occur when molecular propositions containing them are asserted.

Similar considerations apply to propositional functions. Suppose there are two predicates α and β which are always found together; we may still say that they are two, on the ground that $\alpha(x)$ and $\beta(x)$ are facts which do not have predicate-resemblance. But the propositional function $\alpha(\hat{x})$ is solely to be used in building up matrices by means of the stroke. The predicate α is a class of facts, whereas the propositional function $\alpha(\hat{x})$ is merely a symbolic convenience in speaking about certain propositions. Thus we may have $\alpha(\hat{x}) = \beta(\hat{x})$ without having $\alpha = \beta$. In this way we escape the *primâ facie* paradoxes of the theory that propositions only occur in truth-functions and propositional functions only occur through their values. The paradoxes rest on the confusion between factual and assertive propositions.

LIST OF DEFINITIONS

1·01. $p \supset q$

2·33. $p \vee q \vee r$

3·01. $p \,.\, q$

3·02. $p \supset q \supset r$

4·01. $p \equiv q$

4·02. $p \equiv q \equiv r$

4·34. $p \,.\, q \,.\, r$

9·01. $\sim\{(x) \,.\, \phi x\}$

9·011. $\sim(x) \,.\, \phi x$

9·02. $\sim\{(\exists x) \,.\, \phi x\}$

9·021. $\sim(\exists x) \,.\, \phi x$

9·03. $(x) \,.\, \phi x \,.\vee.\, p$

9·04. $p \,.\vee.\, (x) \,.\, \phi x$

9·05. $(\exists x) \,.\, \phi x \,.\vee.\, p$

9·06. $p \,.\vee.\, (\exists x) \,.\, \phi x$

9·07. $(x) \,.\, \phi x \,.\vee.\, (\exists y) \,.\, \psi y$

9·08. $(\exists y) \,.\, \psi y \,.\vee.\, (x) \,.\, \phi x$

10·01. $(\exists x) \,.\, \phi x$

10·02. $\phi x \supset_x \psi x$

10·03. $\phi x \equiv_x \psi x$

11·01. $(x, y) \,.\, \phi(x, y)$

11·02. $(x, y, z) \,.\, \phi(x, y, z)$

11·03. $(\exists x, y) \,.\, \phi(x, y)$

11·04. $(\exists x, y, z) \,.\, \phi(x, y, z)$

11·05. $\phi(x, y) \,.\supset_{x, y}.\, \psi(x, y)$

11·06. $\phi(x, y) \,.\equiv_{x, y}.\, \psi(x, y)$

13·01. $x = y$

13·02. $x \neq y$

13·03. $x = y = z$

14·01. $[(\iota x)(\phi x)] \,.\, \psi(\iota x)(\phi x)$

14·02. $E\,!\,(\iota x)(\phi x)$

14·03. $[(\iota x)(\phi x), (\iota x)(\psi x)] \,.\, f\{(\iota x)(\phi x), (\iota x)(\psi x)\}$

14·04. $[(\iota x)(\psi x)] \,.\, f\{(\iota x)(\phi x), (\iota x)(\psi x)\}$

20·01. $f\{\hat{z}(\psi z)\}$

20·02. $x \in (\phi\,!\,\hat{z})$

20·03. Cls

20·04. $x, y \in \alpha$

20·05. $x, y, z \in \alpha$

20·06. $x \sim\in \alpha$

20·07. $(\alpha) \,.\, f\alpha$

20·071. $(\exists \alpha) \,.\, f\alpha$

20·072. $[(\iota\alpha)(\phi\alpha)] \,.\, f(\iota\alpha)(\phi\alpha)$

20·08. $f\{\hat{\alpha}(\psi\alpha)\}$

20·081. $\alpha \in \psi\,!\,\hat{\alpha}$

21·01. $f\{\hat{x}\hat{y}\,\psi(x, y)\}$

21·02. $a\,\{\phi\,!\,(\hat{x}, \hat{y})\}\,b$

21·03. Rel

21·07. $(R) \,.\, fR$

21·071. $(\exists R) \,.\, fR$

21·072. $[(\iota R)(\phi R)] \,.\, f(\iota R)(\phi R)$

21·08. $f\{\hat{R}\hat{S}\,\psi(R, S)\}$

21·081. $P\,\{\phi\,!\,(\hat{R}, \hat{S})\}\,Q$

21·082. $f\{\hat{R}(\psi R)\}$

21·083. $R \in \phi\,!\,\hat{R}$

22·01. $\alpha \subset \beta$

22·02. $\alpha \cap \beta$

22·03. $\alpha \cup \beta$

22·04. $-\alpha$

22·05. $\alpha - \beta$

22·53. $\alpha \cap \beta \cap \gamma$

22·71. $\alpha \cup \beta \cup \gamma$

23·01. $R \mathrel{\dot{\subset}} S$

23·02. $R \mathbin{\dot{\cap}} S$

23·03. $R \mathbin{\dot{\cup}} S$

23·04. $\dot{-} R$

23·05. $R \mathbin{\dot{-}} S$

23·53. $R \mathbin{\dot{\cap}} S \mathbin{\dot{\cap}} T$

23·71. $R \mathbin{\dot{\cup}} S \mathbin{\dot{\cup}} T$

24·01. V

24·02. Λ

24·03. $\exists ! \alpha$

25·01. $\dot{V}$

25·02. $\dot{\Lambda}$

25·03. $\dot{\exists} ! R$

30·01. $R'y$

30·02. $R'S'y$

31·01. Cnv

31·02. $\breve{P}$

32·01. $\overrightarrow{R}$

32·02. $\overleftarrow{R}$

32·03. sg

32·04. gs

33·01. D

33·02. Ɑ

33·03. C

33·04. F

34·01. $R \mid S$

34·02. R^2

34·03. R^3

35·01. $\alpha \upharpoonleft R$

35·02. $R \upharpoonright \beta$

35·03. $\alpha \upharpoonleft R \upharpoonright \beta$

35·04. $\alpha \uparrow \beta$

35·05. $R'x \uparrow \beta$

35·24 $\alpha \upharpoonleft R \mid S$

35·25. $S \mid R \upharpoonright \beta$

36·01. $P \sqsubset \alpha$

37·01. $R''\beta$

37·02. R_ϵ

37·03. $\breve{R}_\epsilon$

37·04. $R'''\kappa$

37·05. $E!! R''\beta$

38·01. x ♀

38·02. ♀ y

38·03. α ♀,, y

40·01. $p'\kappa$

40·02. $s'\kappa$

41·01. $\dot{p}'\lambda$

41·02. $\dot{s}'\lambda$

43·01. $R \parallel S$

50·01. I

50·02. J

51·01. ι

52·01. 1

54·01. 0

54·02. 2

55·01. $x \downarrow y$

55·02. $R'x \downarrow y$

56·01. $\dot{2}$

56·02. 2_r

56·03. 0_r

Printed in the United States
By Bookmasters